Brain Norepinephrine: Neurobiology and Therapeutics

Drugs that manipulate central nervous system (CNS) norepinephrine have been used therapeutically for over 50 years. Even today, drugs are being developed that target noradrenergic neurons to deliver therapeutic effects, and new disease indications are being identified for existing and newer noradrenergic drugs. Recent discoveries about norepinephrine's contribution to health, disease, and therapy make this synthesis of evidence, practice, and research very timely. A team of experts provides the reader with a thorough understanding of the anatomy, physiology, molecular biology, pharmacology, and therapeutics of norepinephrine in the brain, including an extensive review of the role of norepinephrine in brain diseases. The book is divided into four sections: the basic biology of norepinephrine; the role that norepinephrine plays in behavior; evidence of pathology of norepinephrine in CNS diseases, and the pharmacology and therapeutics of noradrenergic drugs in the treatment of psychiatric and neurological disorders.

Gregory A. Ordway is Chair of the Department of Pharmacology at the James H. Quillen College of Medicine at East Tennessee State University.

Michael A. Schwartz is Clinical Professor of Psychiatry at the University of Hawai'i at Manoa.

Alan Frazer is Chair of the Department of Pharmacology at the University of Texas Health Sciences Center at San Antonio.

Brain Norepinephrine

Neurobiology and Therapeutics

Edited by

Gregory A. Ordway, Ph.D.
Department of Pharmacology, East Tennessee State University

Michael A. Schwartz, M.D.
Department of Psychiatry, University of Hawai'i at Manoa

Alan Frazer, Ph.D.
Department of Pharmacology, University of Texas Health Science Center, San Antonio

CAMBRIDGE
UNIVERSITY PRESS

CAMBRIDGE UNIVERSITY PRESS
Cambridge, New York, Melbourne, Madrid, Cape Town, Singapore, São Paulo

Cambridge University Press
The Edinburgh Building, Cambridge CB2 8RU, UK

Published in the United States of America by Cambridge University Press, New York

www.cambridge.org
Information on this title: www.cambridge.org/9780521834919

First published 2007

Printed in the United Kingdom at the University Press, Cambridge

A catalog record for this publication is available from the British Library

ISBN-13 978-0-521-83491-9 hardback

Contents

Contributors

Subramaniam Apparsundaram, Ph.D.
Department of Anatomy and Neurobiology
University of Kentucky Chandler Medical Center
Rm MN 225
Lexington KY 40536
USA

Amy F. T. Arnsten, Ph.D.
Yale University School of Medicine
Section of Neurobiology
333 Cedar Street
New Haven, CT 06510
USA

Gary S. Aston-Jones, Ph.D.
Department of Psychiatry
University of Pennsylvania
VAMC, Bldg. 21, Room 520, Mail Code 151
University Woodland Aves
Philadelphia, PA 19104
USA

Inna Belfer, M.D., Ph.D.
Pain and Neurosensory Mechanisms Branch
National Institute of Dental and Craniofacial Research
Bethesda
USA

Susan L. Broom, Ph.D.
Department of Pharmacology
Boston University School of Medicine
Room L-613
80 E. Concord Street
Boston, MA 02118
USA

David B. Bylund, Ph.D.
Department of Pharmacology
University of Nebraska Medical Center
986260 Nebraska Medical Center
Omaha, NE 68198-6260
USA

Edwin Clayton
Department of Psychiatry
University of Pennsylvania
VAMC, Bldg. 21, Room 520, Mail Code 151
University Woodland Aves
Philadelphia, PA 19104
USA

Cristinel Coconcea, M.D.
Department of Psychiatry
University Hospital of Cleveland and Case Western Reserve University School of Medicine
Hanna Pavillion
11110 Euclid Avenue
Cleveland OH 44106-5000
USA

Jonathan Cohen
Department of Psychology
Princeton University

Pedro L. Delgado, M.D
Chair, Department of Psychiatry
University of Texas Health Science Center
Dept Psychiatry
7703 Floyd Curl Drive
San Antonio TX 78229-3900
USA

Scott Doran
Department of Psychiatry
University of Pennsylvania
VAMC, Bldg. 21, Room 520, Mail Code 151
University Woodland Aves
Philadelphia, PA 19104
USA

Ronald S. Duman, Ph.D.
Department of Psychiatry
Yale University School of Medicine
34 Park Street
New Haven, CT 06508
USA

Francesco E. Fornai, M.D., Ph.D.
University of Pisa
Department of Human Morphology and Applied Biology
Via ROMA 55
Pisa 55100
Italy

Alan Frazer, Ph.D.
Professor and Chairman
Department of Pharmacology
University of Texas Health Science Center San Antonio
7703 Floyd Curl Drive MSC 7764
San Antonio, TX 78229-3900
USA

David Goldman, M.D.
12420 Parklawn Drive
Suite 451, MSC 8110
Rockville, MD 20852
USA

Monica Gonzalez
Department of Psychiatry
University of Pennsylvania
VAMC, Bldg. 21, Room 520, Mail Code 151
University Woodland Aves
Philadelphia, PA 19104
USA

Katherine A. Halmi, M.D.
Professor of Psychiatry
Weill Medical College of Cornell University
21 Bloomingdale Road
White Plains, NY 10605
USA

E. Higgins, M.D.
Medical University of South Carolina
USA

Michiyo Iba
Department of Psychiatry
University of Pennsylvania
VAMC, Bldg. 21, Room 520, Mail Code 151
University Woodland Aves
Philadelphia, PA 19104
USA

Rick C. S. Lin, Ph.D.
Department of Anatomy
University of Mississippi Medical Center
2500 North State Street
Jackson, MS 39216
USA

B. K. Marchant
University of Utah
USA

David A. Morilak, Ph.D.
Department of Pharmacology
University of Texas Health Science Center
7703 Floyd Curl Drive
San Antonio, TX 78229-3900
USA

J. Craig Nelson, M.D.
Professor of Psychiatry
Director, Geriatric Psychiatry
University of California San Francisco
401 Parnassus Ave, PO Box 0984-F
San Francisco, CA 94143
USA

Samuel S. Newton, Ph.D.
Department of Psychiatry
Yale University School of Medicine
34 Park Steet
New Haven, CT 06508
USA

Gregory A. Ordway, Ph.D.
Professor and Chairman
Department of Pharmacology
James H. Quillen College of Medicine
East Tennessee State University
P.O. Box 70577
Johnson City, TN 37614
USA

Antti Pertovaara, M.D., Ph.D.
Department of Physiology
Institute of Biomedicine
University of Turku
Turku
Finland

Paul M. Plotsky, Ph.D.
Department of Psychiatry; Behavioral Science
Emory University School of Medicine
1639 Pierce Dr, WMRB 4105
Atlanta, GA 30322
USA

Janusz Rajkowski
Department of Psychiatry
University of Pennsylvania
VAMC, Bldg. 21, Room 520, Mail Code 151
University Woodland Aves
Philadelphia, PA 19104
USA

Fred W. Reimherr, M.D.
University of Utah
USA

Elliott Richelson, M.D.
Mayo Clinic of Jacksonville
4500 San Pablo Road
Jacksonville, FL 32224
USA

Benno Roozendaal, Ph.D.
Center for the Neurobiology of Learning and Memory
Bonney Research Laboratories, Rm #218
University of California,
Irvine, CA 92697-3800
USA

Michael Alan Schwartz, M.D.
Clinical Professor of Psychiatry
University of Hawai'i at Manoa
1106 Blackacre Trail
Austin, Texas 78746
USA

Kimberly L. Simpson, Ph.D.
Department of Anatomy
University of Mississippi Medical Center
2500 North State Street
Jackson, MS 39216
USA

Craig A. Stockmeier, Ph.D.
Department of Psychiatry G127
University of Mississippi Medical Center
2500 North State Street
Jackson, MS 39216-4505
USA

Leonie M. Welberg, Ph.D.
Department of Psychiatry; Behavioral Science
Emory University School of Medicine
1639 Pierce Drive, WMRB 4105
Atlanta, GA 30322
USA

Bryan K. Yamamoto, Ph.D.
Department of Pharmacology
Boston University School of Medicine
Room L-613
80 E. Concord Street
Boston, MA 02118
USA

Sun Young Yum
Weill Medical College of Cornell University
21 Bloomingdale Road
White Plains, NY 10605
USA

Acknowledgements and Disclosures

Authors of this book were asked to disclose affiliations of financial involvement with any organization or entity having a financial interest in or financial conflict with the subject matter or materials discussed in their chapters. For purposes of disclosure, financial involvement is defined as having occurred in the last 3 years and being an amount greater than or equal to $10,000 per year, or having the potential to generate that amount per year in the future. The editors of the book have also disclosed any relevant affiliations or financial involvements.

Amy Arnsten and Yale University have license agreements with Shire Pharmaceuticals for the development of guanfacine, and with Marinus Pharmaceuticals for the development of chelerythrine for the treatment of prefrontal cortical disorders. These companies have also supported research in the Arnsten lab aimed at understanding the mechanism of action of these compounds.

Pedro Delgado has had funding for his research from The National Institutes of Health (R01 MH 48977 and R01 MH54648). He has had a financial interest or other relationships with the manufacturers of some commercial products as specified below:

Eli Lilly & Co	Grant Support/Advisory Board/Speakers bureau
GlaxoSmithKline	Grant Support/Advisory Board/Speakers bureau
Organon, Inc	Grant Support/Advisory Board/Speakers bureau
Wyeth-Ayerst	Advisory Board/Speakers bureau
Forest Laboratories	Grant Support/Advisory Board
Pierre Fabre	Honorarium
Bristol Meyers Squibb	Advisory Board
Paramount Biosciences	Consultant
Concept Therapeutics	Grant Support
Pfizer Inc	Stock

Alan Frazer has had a financial interest or other relationship with the manufacturers of some commercial products as specified below:

Cyberonics Inc	Speakers bureau/Mechanism of Action Advisory Board Recipient of research grant
Forest labs/Lundbeck	Advisory Board member/Recipient of research grant
Lilly Labs	Consultant/Recipient of research grant
Servier	Recipient of research grant
Wyeth Labs	Consultant

Edward S. Higgins has been a consultant and member of the speakers bureau for Wyeth.

Craig Nelson has had a financial interest or other relationship with the manufacturers of some commercial products as specified below:

Abbott, Astra Zeneca, Bristol Myers Squibb, Eli Lilly, Forest, GlaxoSmith Kline, Janssen, Pfizer, Wyeth	Lecture honoraria
Abbott, Biovail, Bristol Myers Squibb Corcept, Eli Lilly, GlaxoSmithkline, Orexigen, Organon, Pfizer, Sepracor, Shire	Consultation/Advisory Board
Eli Lilly	Research support

Gregory A. Ordway has funding for his research from The National Institutes of Health (MH46692, MH63187, MH02031, RR17701) and from Aventis.

Antii Pertovaara has had the following involvement with OrionPharma Inc, Helsinki, Finland: Inventor in one received US patent (no 5 541 211) and one pending US patent (no 2000 2011 5704), both of which are owned by OrionPharma Inc. Additionally, he is an inventor of one European patent (EP1198477) the owner of which is JuvantiaPharma Inc, Turku, Finland. He owns OrionPharma stocks worth USD 16,000. OrionPharma is an industrial partner in a research grant which he received from the Academy of Finland (grant no 105709; USD 180,000 for the period of 2004–2007).

Paul M. Plotsky has had a financial interest or other relationship with the manufacturers of some commercial products as specified below:

Pharmacia Upjohn, Inc (now part of Pfizer)	Research contract entitled "Effects of Chronic Reboxetine Treatment on the Central Noradrenergic System: A Functional and Biochemical Study" (2000–2001).
Janssen, Johnson & Johnson Pharmaceutical Research & Development, LLC	Research contract entitled "Evaluation of Topiramate in a Rodent Neonatal Maternal Separation Model" (2002–2003).

H. Lundbeck A/S	Research contract entitled "Evaluation of Compounds in a Rodent Neonatal Maternal Separation Model" (2004–2006).
Cyberonics, Inc	Research contract entitled "Evaluation of VNS in a Rat Neonatal Maternal Separation Model" (2006–2007).

Fred W. Reimherr has had a financial interest or other relationship with the manufacturers of some commercial products as specified below:

Eli Lilly	Consultant and multiple research grants Contracts through the University of Utah
Shire	Consultant and independent research grant
McNeil (Johnson & Johnson)	Consultant, speaker fees and independent research grant
Celgene	Significant stock ownership

Elliott Richelson has had a financial interest or other relationship with the manufacturers of some commercial products as specified below:

Eli Lilly	Advisory board for depression
Pfizer Inc	Grant funding and Speaker bureau
Janssen Pharmaceuticals	Grant funding

Michael A. Schwartz has had a financial interest or other relationship with the manufacturers of some commercial products as specified below:

Abbott	Grant Support
Astra Zeneca	Grant Support/Speakers bureau
Bristol Meyers Squibb	Grant Support/Speakers bureau
Cephalon	Speakers bureau
Eli Lilly	Grant Support/Advisory Board/Speakers bureau
GlaxoSmithKline	Grant Support/Speakers bureau
Jannsen	Grant Support
King	Speakers bureau
Shire	Speakers bureau
Takeda	Speakers bureau
Wyeth	Grant Support/Speakers bureau

The following authors had no relevant financial interests to disclose.

Subbu Apparsundaram
Gary S. Aston-Jones
Inna Belfer
Susan L. Broom
David B. Bylund
Edwin Clayton

Cristinel Coconcea
Jonathan Cohen
Ronald S. Duman
Katherine A. Halmi
Francesco Fornai
Monica Gonzalez
David Goldman
Michiyo Iba
C. S. Lin
Barrie K. Marchant
Samuel S. Newton
Janusz Rajkowski
Benno Roozendaal
David A. Morilak
Kimberly L. Simpson
Craig A. Stockmeier
Leonie A. M. Welberg
Bryan K. Yamamoto
Sun Young Yum

Scott Doran did not provide disclosures.

Introduction: revision of an old transmitter

Gregory A. Ordway, Ph.D.

The discovery of norepinephrine dates back to the late 1940s when the Swedish scientist, Ulf Svante von Euler first demonstrated that neurons of the sympathetic nervous system use norepinephrine, rather than epinephrine, as a neurotransmitter. Shortly thereafter in 1947, Peter Wilhelm Joseph Holtz demonstrated that norepinephrine occurred in the brain. Today, we know it is one of three major catecholamine (dopamine, norepinephrine, epinephrine) neurotransmitters found in the central nervous system (CNS). Over 50 years of subsequent research has led to an enormous accumulation of information regarding norepinephrine and its role in physiological and behavioral processes. In addition, drugs that directly manipulate brain norepinephrine have been used therapeutically for over 50 years, and even today, drugs are being developed that target noradrenergic neurons to deliver therapeutic effects. In fact, new disease indications continue to be identified for existing and newer noradrenergic drugs.

Given the revered tenure of this relatively old neurotransmitter and the recent advances and subsequent theories about its contribution to health and disease in the CNS, the authors of this book decided the time was right to bring together historical and recent information about norepinephrine in one book. The intention of this volume is to provide the reader with a thorough understanding of the anatomy, physiology, molecular biology, pharmacology, and therapeutics of norepinephrine in the CNS, including an extensive review of the role of norepinephrine in diseases of the CNS.

The book is divided into four parts, each of which can be read individually for focused information, or collectively to obtain a thorough understanding of the neurotransmitter. The first part of the book, *The neurobiology of norepinephrine*, is a comprehensive review of the anatomy and neurochemistry of norepinephrine neurons in the brain, including discussion of neurotransmitter systems directly affected by noradrenergic input. In Chapter 1, Kimberly Simpson and Rick Lin

Brain Norepinephrine: Neurobiology and Therapeutics, ed. Gregory A. Ordway, Michael A. Schwartz and Alan Frazer. Published by Cambridge University Press.

describe in detail the neuroanatomy of the locus coeruleus, the major source of norepinephrine in the brain. Simpson and Lin discuss neuropeptide transmitters that are colocalized with norepinephrine in the locus coeruleus, and provide detailed neuroanatomical descriptions of the afferent and efferent pathways of this nucleus. In Chapter 2, Craig Stockmeier and I provide a review of neurotransmitter systems that provide modulatory input to central noradrenergic neurons. Our focus in this chapter was on transmitter systems that provide input to the locus coeruleus, the specific receptors for those transmitters that are expressed by noradrenergic neurons, and the effects of drugs that activate or inhibit those receptors on locus coeruleus activity. We also describe the reciprocal connections between the locus coeruleus and other brainstem monoamine cell groups such as the dopaminergic ventral tegmental area and the serotonergic raphe nuclei. In Chapters 3 and 4, David Bylund, Ron Duman, and Sam Newton provide a comprehensive review of receptors for norepinephrine as well as their intracellular signaling systems, including the role of norepinephrine in regulating nuclear events such as gene transcription. Bylund also reviews several polymorphisms known to exist in genes encoding noradrenergic receptors, along with their biological consequence. Duman and Newton give due consideration to the effects of norepinephrine, induced as a consequence of norepinephrine uptake inhibition secondary to antidepressant treatment, on gene expression through noradrenergic receptor intracellular signaling pathways. In Chapter 5, Subbu Apparsundaram reviews the norepinephrine transporter, the protein that is largely responsible for the temporal control of noradrenergic transmission through the synapse. This protein is a target of many psychotherapeutic compounds. Considerable recent evidence demonstrates that this protein is not a static entity on the norepinephrine neuron as once believed. Rather, the norepinephrine transporter is regulated in concert with changes in noradrenergic transmission.

The second part of this book, *Norepinephrine and behavior*, is a compilation of reviews of roles that norepinephrine plays in a key output of the brain, i.e. behavior. In Chapter 6, Gary Aston-Jones and coauthors discuss the role of norepinephrine and the locus coeruleus in arousal state maintenance and sleep–wake regulation, and they consider how the locus coeruleus participates in sleep alterations that accompany psychiatric disorders. In Chapter 7, Aston-Jones and coauthors describe different modes of activity that the noradrenergic locus coeruleus exhibits and their relationship to behavioral states. These include a hypoactive mode where locus coeruleus activity is low and relatively unresponsive to external stimuli and behaviorally manifests as drowsiness or inattentiveness. A phasic mode is characterized by phasic activation of the locus coeruleus to discriminated stimuli, associated with high-level performance. Finally, a tonic mode of activity occurs where locus coeruleus neurons do not respond phasically to external (task) stimuli, but are tonically active. This latter mode is associated with poor performance. Together, the behaviors associated with different modes of activity illustrate that the relationship between locus coeruleus activity and performance in a task is biphasic. Aston-Jones discusses the clinical implications of locus coeruleus activity modes with regard to the biological basis of several disorders of the CNS.

In Chapter 8, Benno Roozendahl reviews the animal and human literature that demonstrates the critical role that norepinephrine plays in the consolidation of long-term memory formation. Roozendahl describes the effects of pharmacological manipulation of the central noradrenergic system on memory and the various areas, particularly the amygdala and hippocampus, of the brain that play a role in mediating norepinephrine's effects on memory. In Chapter 9, David Morilak describes how stress modifies central noradrenergic activity, and how pharmacological agents modulate the effects of stress on the locus coeruleus. Morilak introduces the idea that dysregulation of this stress-sensitive system may be a biological substrate for the vulnerability of certain individuals to stress-related psychopathologies.

Part III of this book, *The biology of norepinephrine in CNS pathology*, is focused on evidence that pathology of the central noradrenergic system is a component of clinical CNS pathologies. In Chapter 10, Leonie Welberg and Paul Plotsky consider animal models that have been used to investigate the neurobiology of psychiatric and neurological disorders and the mechanisms of action of neuropsychiatric pharmacotherapy. Welberg and Plotsky describe animal studies that illustrate that environmental stresses encountered during the prenatal period through adulthood and genetic diversity can alter stress-responsive neural circuitry, particularly the noradrenergic system, and ultimately contribute to psychopathology. In Chapter 11, I consider how research studying brain tissue collected postmortem has contributed to an understanding of the role of norepinephrine in the neuropathology of depression and schizophrenia. I discuss evidence of neurochemical pathology of the noradrenergic locus coeruleus in depression that supports the concept that depression is associated with an elevation in stress-sensitive excitatory input to the noradrenergic locus coeruleus that can be dampened by antidepressant drug administration. In Chapter 12, Pedro Delgado examines human studies (in live patients) that have investigated the role of norepinephrine in the pathophysiology of major depression and mechanisms underlying antidepressant drug action. Delgado reviews studies utilizing peripheral markers and neuroendocrine challenges, and more recent studies that employed catecholamine (norepinephrine and dopamine) depletion to investigate the relationship between mood and brain catecholamine concentrations. This avenue of research has confirmed a very strong association between brain catecholamines and mood.

In Chapter 13, Antti Pertovaara reviews literature that demonstrates modulatory actions of norepinephrine on pain processes in the CNS under normal physiological and pathophysiological conditions. Pertovaara details the pharmacological basis for the actions of noradrenergic agonists and antagonists on pain. In Chapter 14, Amy Arnsten reviews how norepinephrine influences cortical and subcortical functions that contribute to cognition. Arnsten provides evidence suggesting that norepinephrine acts as a chemical switch that strengthens prefrontal cortical control of behavior under non-stressful alert conditions, but impairs prefrontal function while enhancing posterior cortical and subcortical activity during stress. Arnsten discusses numerous neurological and psychiatric disorders according to whether norepinephrine or noradrenergic activity is low and insufficient, or high and disruptive to normal cognitive functioning. Francesco Fornai reviews the neuropathological findings involving brain noradrenergic neurons in specific neurological disorders in Chapter 15. Several neurological disorders, as well as normal aging, are characterized by degeneration or loss of noradrenergic neurons. Fornai considers studies utilizing experimental animal models that have investigated the neurobehavioral sequelae resulting from noradrenergic neuron loss. In addition, Fornai reviews research investigating the neuropathology of noradrenergic neurons and noradrenergic pharmacotherapy in Parkinson's disease, epilepsy, multiple systems atrophy, dementia, and several other neurological disorders.

In Chapter 16, Inna Belfer and David Goldman provide a thorough description of norepinephrine-related genes, norepinephrine gene variation and the effects of gene variation on the expression and function of norepinephrine genes. The identification of gene variants is a rapidly expanding field and in the time between completion of this chapter and press, several new gene variants have been identified and it is likely that many more will be discovered. Nevertheless, Belfer and Goldman provide a review of the impact of variation of genes encoding proteins that are found in noradrenergic neurons and proteins (e.g. receptors) that receive information, via norepinephrine, from noradrenergic neurons. Detailed tables listing a host of polymorphisms of noradrenergic genes are included in this chapter. The association of polymorphisms of noradrenergic genes to diseases of the brain and periphery are discussed, since, as noted by Belfer and Goldman, "The brain, the body and the genome are not conveniently compartmentalized by disease."

The last part of the book, *Psychopharmacology of norepinephrine*, is devoted to a review of the pharmacology and therapeutics of psychoactive drugs that act on norepinephrine-binding proteins (receptors, transporters, metabolic and catabolic enzymes). This section focuses on the use of noradrenergic drugs in the treatment of depression, anxiety, attention-deficit, eating disorders, and substance abuse. In Chapter 17, Elliot Richelson provides a history of the discovery of antidepressant drugs and how the elucidation of their pharmacological actions historically shaped (and continues to shape) scientific opinion of the biology of affective disease. Richelson reviews the pharmacological actions of drugs that bind the norepinephrine transporter, including antidepressant and antipsychotic drugs, and also discusses the non-transporter binding effects of antidepressant drugs.

In Chapter 18, Craig Nelson addresses the role of antidepressants that have inhibitory actions at the norepinephrine transporter in the treatment of depression and anxiety disorders. Nelson provides strong evidence for the efficacy of selective norepinephrine uptake inhibitors in the treatment of depression, and also reviews literature suggesting that norepinephrine transporter (uptake) inhibitors share a similar pattern of effects on the core symptoms of depression with selective serotonin uptake inhibitors. Nelson discusses a growing literature that implicates synergistic effects of norepinephrine and serotonin uptake inhibition in the treatment of depression. The lesser role of norepinephrine uptake inhibitors and mixed serotonin/norepinephrine uptake inhibitors in the treatment of panic, obsessive compulsive, posttraumatic stress, and generalized anxiety disorders is also discussed.

The recent success of atomoxetine (a selective norepinephrine transporter blocker) for the treatment of attention-deficit/hyperactivity disorder has reawakened researchers to the importance and role of norepinephrine in attention disorders. To address this, Fred Reimherr and his coauthors in Chapter 19 discuss the use of drugs that modulate noradrenergic transmission in the treatment of attention-deficit/hyperactivity disorder. Reimherr and colleagues provide a detailed review of clinical studies that examine the efficacy of noradrenergic drugs in the treatment of attention disorders. They carefully consider the use of stimulants, tricyclic antidepressants, monoamine oxidase inhibitors, α_2 adrenergic receptor agonists, newer mixed action norepinephrine/serotonin uptake inhibitors, and atomoxetine. Overall, the clinical studies reviewed demonstrate efficacy of medications with norepinephrine actions and these studies provide evidence implicating norepinephrine in the pathophysiology of attention-deficit/hyperactivity disorder.

In Chapter 20, Katherine Halmi and Sun Young Yum address the use of drugs that modulate the central noradrenergic system in the treatment of eating disorders, including anorexia nervosa and bulimia. Halmi and Yum review basic research demonstrating that norepinephrine influences feeding behavior through several complex mechanisms, and examines clinical studies that have investigated various biological and molecular markers of norepinephrine or noradrenergic activity in anorexia and bulimic patients. In addition, Halmi and Yum examine the use of noradrenergic active drugs, in particular tricyclic antidepressants, in the treatment of anorexia and bulimia. The authors conclude that despite robust effects of norepinephrine on feeding behavior, drugs that affect norepinephrine neurotransmission have modest effects in treating bulimia nervosa and essentially no effect in treating anorexia nervosa, although it is worth mentioning here that highly selective norepinephrine uptake inhibitors have not been tested in blinded trials to date.

In Chapter 21, Susan Broom and Bryan Yamamoto provide an overview of the role of norepinephrine in the development and maintenance of substance abuse and summarize the therapeutic potential of noradrenergic drugs to alleviate adverse consequences of substance abuse. Although dopamine is the catecholamine commonly referred to as mediating reward effects of abused drugs, Broom and Yamamoto review evidence that norepinephrine contributes to the dopaminergic system in producing changes in reward pathway activity induced by psychostimulants. The authors discuss evidence that norepinephrine released in select brain regions contributes to withdrawal symptoms and stress-induced drug relapse, and review evidence of the efficacy of drugs that modulate norepinephrine in decreasing adverse consequences of withdrawal.

The chapters in this book reveal a wealth of information about the basic neurobiology of norepinephrine, its role in normal and abnormal behaviors and neurological disease, and its usefulness as a pharmacological target to alleviate symptoms of CNS diseases. The relatively recent development of highly selective drugs to inhibit the norepinephrine transporter further expands the possibility for medical discoveries of the utility of noradrenergic transmission modulation for the treatment of disease. The past 50 years of research elucidating multiple roles of norepinephrine in brain function have simultaneously revealed countless mysteries about norepinephrine's molecular and behavioral actions and these are likely to continue to stimulate research on this transmitter for many years in the future.

Part I

The neurobiology of norepinephrine

1

Neuroanatomical and chemical organization of the locus coeruleus

Kimberly L. Simpson, Ph.D. and Rick C. S. Lin, Ph.D.
University of Mississippi Medical Center, Jackson, USA

Early history of the locus coeruleus

The first description of the norepinephrine (NE)-containing locus coeruleus (LC) dates back to 1809 in a report by Reil.[1] A depression in the anterior floor of the human fourth ventricle was noted that corresponded with a concentration of blue-black substance just below the tissue surface. The pigmented area remained nameless until 1812 when the Wenzel brothers introduced the term "locus coeruleus."[2] Today this designation is widely utilized. However, in early investigations, several other names were applied to this melanin-containing group of cells. Arnold, in 1838 and 1851, referred to the LC as the "substantia ferruginea."[3,4] Jacobsohn, in 1909, coined the name "nucleus pigmentosus pontis."[5] In recognition of a cerebellar subcomponent of the LC, Meynert (1872) and Jacobsohn (1909) assigned the classifications "substantia ferruginea superior" and "nucleus pigmentosus tegmento-cerebellaris," respectively.[5,6] Despite its many names, most studies during this period placed the LC within the dorsolateral portion of the rostral mesencephalic and caudal mesencephalic tegmentum of man and higher primates. Localization of the LC in lower species, such as rodents, was more challenging initially because pigmentation was found to be lacking in areas that were readily observed in higher mammals. Consequently, a considerable amount of uncertainty surrounded the actual position of the LC cell group within these animals. The LC, for example, was once considered to be a caudal extension of the nucleus laterodorsalis tegmenti, a region now acknowledged as a major pontine cholinergic center.[7]

An interesting historical perspective, covering the above information in detail and providing a comparative synopsis of the cytoarchitectonic characteristics of the LC, appears in a review written by Russell in 1955.[8] Within the body of this work, Russell presents a thorough timeline of LC-related findings and offers evidence

Brain Norepinephrine: Neurobiology and Therapeutics, ed. Gregory A. Ordway, Michael A. Schwartz and Alan Frazer. Published by Cambridge University Press.

for the potential involvement of the LC in trigeminal somatosensation, visceral sensibility, and respiration.

Later progress in the field stemmed from the ability of researchers to further elucidate the chemical composition of the LC. As a result of this advancement, it became possible to identify the LC in several mammalian species, including humans, without visualization of melanin. Monoamine oxidase, aromatic monoamine, ascorbic acid, copper, and acetylcholinesterase were among the first substances to be detected.[9–13] A particularly notable event, however, relating to the demonstration of the LC was the development of the Falck–Hillarp method.[14] This technique, which involved exposing freeze-dried tissue sections to formaldehyde vapor, was found to impart fluorescence to catecholamine-containing cells through the formation of tetrahydroisoquinoline derivatives, enabling subgroups of neurons to be differentiated based upon their color. Cells in the substantia nigra and LC were found to exhibit an intense green fluorescence, while profiles of raphe nuclei appeared yellow. Dopaminergic and noradrenergic neurons were, therefore, classified under the A group designation (A1–A12), while serotonergic neurons were categorized in the B series (B1–B9). In 1964, Dahlstrom and Fuxe[15] more specifically labeled the LC as A6 and noted an additional collection of multipolar cells that extended ventrolaterally in an arch toward A5.

Species comparisons

In most mammals including humans, the LC is generally located within the periventricular gray of the isthmus, medial to the mesencephalic nucleus of the trigeminal nerve. It is composed of a cluster of cells that assemble within the lateral boundaries of the central gray, sometimes extending either dorsorostrally along the roof of the fourth ventricle (IV) or ventrally toward the core of the brainstem. Although studies of the LC have been performed on animals such as the rabbit, chimpanzee, baboon, dog, sea lion, horse, and bat, many investigations have focused on the human, primate, cat, and rodent.[8,16] The following section will provide a brief account of the distribution of the LC nucleus in species that have been more routinely studied and will address architectural characteristics that tend to distinguish one group of animals from another. It is worthwhile to note that many of the features that define the LC were initially assessed in Falck–Hillarp-, Golgi-, or Nissl-stained material. However, a number of findings have been immunohistochemically confirmed using antibodies directed against tyrosine hydroxylase (TH), the rate-limiting enzyme in catecholamine synthesis, or dopamine-β-hydroxylase (DBH), the final enzyme in the biosynthesis of NE.[17–19]

A large majority of studies have utilized rats to determine the basic organization of the LC.[19,20] Reports indicate that, on the basis of cytoarchitecture, this nucleus can be segregated into as many as four contiguous subregions.[21] Among these neuronal subsets are cell populations that appear in the cerebellum (A4) and the mesencephalic periaqueductal gray (PAG) at a level just behind the decussation of the brachium conjunctivum (BC). However, the most often recognized of these groups is the LC proper and the ventral subdivision. The LC proper extends approximately 1 mm in the rostrocaudal dimension from the genu of the facial nerve to the midportion of the dorsal tegmental nucleus; the ventral subdivision measures just under 500 μm. The former is characterized by densely packed aggregates of somata, which, as a general rule, express NE and localize within rather restricted nuclear subfields.[22] Within the caudal pole, these dorsally distributed cells are separated from the more loosely arranged cells of the ventral division by the medial vestibular nucleus. At rostral levels, the ventral A6 population merges with the principal LC division and becomes less distinct. Together with the NE-containing cells of A7, the ventral portion of the LC is considered by Olson and Fuxe[23] to constitute the subcoeruleus (sub-LC). Grzanna and Molliver, however, point out that the sub-LC lies outside of the central gray and should, therefore, not be included as part of the LC group.[21] They describe the sub-LC as a group of neurons that is situated approximately 1 mm rostral to the cells of the ventral LC subdivision and extends ventrally from the LC proper along the medial aspect of the trigeminal motor nucleus into the region of the rostral superior olivary complex. Despite these minor interpretational differences, regional estimates of cell number in Nissl-stained paraffin sections indicate that one LC nucleus contains roughly 1643 neurons, the ventral and dorsal subdivisions of which contribute approximately 210 and 1430 neurons, respectively.[24] Although results obtained from DBH-labeled material suggest that LC cell populations are slightly smaller (about 1439 neurons, total), these discrepancies can be explained by methodological limitations associated with antibody penetration and the use of thicker tissue sections.

Studies in nonrodent animal models support the idea that the basic structural organization of the LC is conserved across species. However, these same investigations have also revealed that important distinctions exist in the nuclear construct of different experimental models. The cat LC is particularly notable in this regard, due to the fact that it varies from rat LC on several counts.[25,26] One readily apparent feature relates to the diffuse arrangement of cells in the cat LC. Both neurons and glial elements tend to spread out in an irregular fashion over large areas of the periventricular gray and to distribute within deep aspects of the pontine tegmentum. In its entirety, the nucleus extends a distance of 4 mm from the trochlear nucleus to the trigeminal motor nucleus and includes within its boundaries dorsolateral areas surrounding the BC and neuronal territories previously identified as the Kölliker–Fuse nucleus and the magnocellular component of the marginal nucleus of the BC.[25] Unlike the rat LC, which has been described as a collection of four subregions, the cat LC is composed predominantly of two divisions. The first subgroup, the dorsal LC (LCD), bears resemblance to its rodent counterpart in that it is located within the periventricular gray medial to the mesencephalic root of the fifth nerve. The second subpopulation, LCα, consists of cells that gather in the tegmentum ventral to the gray and occupy the area along the lower half of the trigeminal root.[27] Although not technically considered a part of the LC, the sub-LC is represented within the area lateral to the LC and ventral to the BC. More often than not, cell intermingling within transitional zones prevents the formation of clear partitions between subregions. A final distinguishing characteristic deserving mention relates to the prevalence of non-NE-containing cells within the cat LC. This large population of cells, identified in Falck–Hillarp stained material by its lack of fluorescence,[25] is remarkable in that it has been mapped to all areas of the LC and shown to unilaterally coexist with approximately 9150 catecholaminergic neurons.[28]

Locus coeruleus studies, which have been conducted on pygmy marmosets and on rhesus, macaque, and squirrel monkeys have revealed that primates share more features in common with the rat (and human) than with the cat.[29–32] The characteristics that most directly demonstrate this relationship are the dense population of the nucleus along its most central portion[29] and the expression of catecholamine by virtually all LC cells.[32] Furthermore, LC neurons in the primate (labeled M6 by Garver and Sladek[32]) form an elongated band, which, at caudal levels, merges with cells in the overlying M4 cell group.[30–32] This is a point worth emphasizing for three reasons. First, a region analogous to M4, designated A4 by Dahlstrom and Fuxe,[15] has been described in rats. Second, this neuronal subset is thought to represent a dorsal extension of LC into the cerebellar roof of the fourth ventricle. Third, a feline homolog to M4 has not been detected. An additional discrepancy between primates and cats relates to the dorsolateral distribution of LC cells. More specifically, cells of the primate A6 cell subgroup appear to be limited in their localization around the superior cerebellar peduncle. When present, these cells reside along the superior medial[31] and lateral[29] margins of the BC. An exception to this finding concerns the cells of the sub-LC. These neurons have been found to populate areas along the inferior tip of the BC, and correspond closely to cells located within the parabrachialis (medialis and lateralis) nuclei.[30] Unlike models discussed previously, primate sub-LC neurons have been shown to begin caudally, surrounding the medial and dorsomedial aspects of the trigeminal motor nucleus, and transition rostrally into a dorsal and ventral subdivision.[30,32] Although an overall quantitative estimation of sub-LC population size remains to be provided, reports indicate that *Macaca mulatta* subjects exhibit approximately 7316 cells in the LC proper.[33]

As stated earlier, pigmentation is a prominent property of the LC in man and higher primates. Gorillas, baboons, chimpanzees, macaques, and manguebye monkeys have all been found to exhibit cells with melanin granules. Interestingly, pigmentation in both humans and primates develops progressively towards the onset of puberty. Human infants, examined at 6 days, 2.5 to 3 months, 16 months, and 2 years of age, display no evidence of melanin. However, by the ages of 8 and 9 years some pigmentation becomes apparent.[8] Despite the fact that pigmentation had long been recognized as a hallmark of the human LC, it was quite some time after the initial detection of the nucleus that either pigmentation or immunohistochemistry was utilized to map the organizational pattern of central catecholamine neurons.[34–36] Saper and Petito[35] were among those using melanin as an intracellular marker to describe the A6 cell group. In brain material that they acquired from adults 68 to 80 years old, the LC was found to consist of a compact accumulation of cells that extended from the level of the rostral end of the trigeminal complex to the inferior colliculus. Subsequent studies employing immunohistochemical techniques further revealed that the dimensions of the LC vary in accordance with cell orientation along the rostrocaudal axis. At midrostral intervals, cells assume a more linear arrangement within the horizontal plane.[36] At midcaudal levels, however, the nuclear boundary shifts, and cells conform to a cylindrical cluster.[37] Additional findings have indicated that with continued caudal progression, TH-immunoreactive cells stream away in a ventrolateral direction and enter the region of the sub-LC. Kemper *et al.*[36] have classified neurons that localize within the caudal pons region of the sub-LC subnucleus ventralis as belonging to the A5 cell group, while those in the midpons have been designated A7. Dopamine-β-hydroxylase-positive neurons of the subnucleus dorsalis were, however, identified as part of A6. Included in this account of the human brainstem, Kemper notes that, in comparison to other species, the LC is less densely packed and often appears more segregated from the sub-LC.[36] A majority of DBH-positive LC neurons situate within the rostral half of the pons, while most sub-LC neurons localize caudally.

Several reports have provided quantitative estimates of cell number within human LC, and at least three have presented detailed three-dimensional reconstructions that illustrate the spatial distribution of these cells within the brainstem. Based on plots of pigmented neurons from five brains of subjects ranging from ages 60 to 104 years old, German *et al.* indicate that the human LC is "tube-like" in shape and extends rostrocaudally for approximately 16 mm.[38] The nucleus was further found to demonstrate bilateral symmetry, a span width of 2.5 to 3 mm, and to contain a total of 45 562 to 18 940 cells (youngest to oldest). These calculations are in agreement with other studies that have additionally utilized TH and DBH immunocytochemistry to evaluate the LC cell population.[37,39–41] In these investigations approximately 55 000 to 65 000 cells were reported within the LC proper, and 6 000 to 7 000 cells were identified within the sub-LC. The close correspondence between numbers of pigmented and immunoreactive cells led several

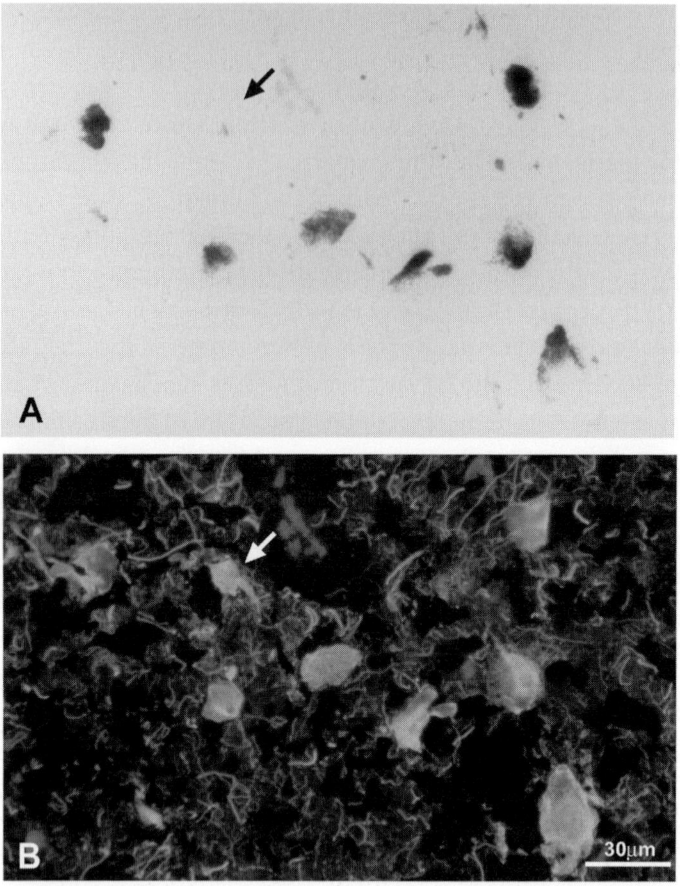

Figure 1.1 Photomicrographs demonstrate the relationship between melanin accumulation and TH-immunoreactivity within human LC cells. The solid arrows in A and B point to a cell that is labeled for TH using an anti mouse IgG conjugated to Cy-3 (B), but is devoid of pigmentation (A).

investigators[37,40] to conclude that counts of pigmented cells accurately reflect the total number of neurons in the human LC complex. However, Chan-Palay and Asan[41] found pigmented cells that did not stain for TH, and TH-immunoreactive cells that did not exhibit pigment. Many of the latter were large multipolar neurons that resided within the rostral portion of the LC. Using immunofluorescent methods, our laboratory has also observed human DBH- and TH-containing LC neurons that lack the appearance of pigment (Figure 1.1). This, plus the fact that melanin tends to accumulate unevenly within the cell body, raises concerns regarding the validity of counting pigmented cells in the LC as the only means of assessing total cell numbers within catecholaminergic nuclei of the human brain.

Ontogeny

There have been many studies over the past decades to address the ontogenetic development of the LC. Using [^{3}H]thymidine autoradiography, Lauder and Bloom found that LC neurogenesis peaks at embryonic day (E)12 in the rat.[42] Shortly thereafter, on fetal day 14, NE fluorescence first appears,[43,44] with the observance[45] of TH around E12.5. By E19, LC synaptogenesis is evident.[46] Early events in LC innervation of the cortex have further been examined with histofluorescence techniques.[47] Labeled fibers were found to penetrate the outer superficial layer on E16, and to span the entire cortical mantle in a rostral-to-caudal direction over the next few days. Later studies showed that, by the first postnatal week, NE fibers distribute throughout the cortex and projection patterns strongly resemble those seen in the adult.[48,49]

More recently, molecular genetic approaches have been employed to reveal gene expression and mechanisms of differentiation in LC neurons during early development. Bone morphogenic proteins (BMPs), in particular BMP5, as well as the transcription factors, Phox2a and 2b, have been demonstrated exclusively within neurons that are involved in the noradrenergic synthesis pathway, i.e., neurons that express either TH- or DBH-immunoreactivity.[50–54] Although specifics regarding the early events in the formation of the LC are beyond the scope of this chapter, it is important to emphasize that this field of research has advanced significantly in the recent past.

Previous studies have also focused on the development of human LC and the emergence of axonal projections from this nucleus. Following the use of either fluorescence histochemistry or TH immunohistochemical techniques, it has been reported that LC neurons are not evident at 5 weeks of gestation, but become identifiable by 12 weeks.[55,56] At this fetal age, axon terminals with granular (dense-cored) vesicles can be detected within the LC, suggesting the presence of NE-containing fibers.[55] In later weeks (17 to 21), TH-immunoreactive fibers course within the vicinity of the neostriatum.[56] These studies, thereby, indicate that, like their rodent counterparts, human LC neurons are also established prenatally and in a position to exert important trophic influences upon target areas during early developmental periods.

Morphological and ultrastructural characteristics

Two types of histological approaches have been applied to characterize the morphology of LC neurons. Swanson[24] employed Golgi impregnation techniques, while Grzanna and Molliver[21] utilized DBH immunohistochemical staining methods. Together, these studies have shown that LC neurons are a heterogeneous population of cells. The two major cell classes that have been identified are small fusiform and medium-sized multipolar neurons. Fusiform cells (20 to 25 μm in diameter) are found mainly in the dorsal portion of LC and often have long, thin dendrites that extend beyond the limits of the nucleus. Multipolar cells (25 to 30 μm) are mainly observed in the ventral portion of the LC as well as the sub-LC. They tend to have long dendrites that radiate in all directions and extend considerable distances. Secondary and tertiary branches have even been observed to reach as far as the cellular components of the mesencephalic nucleus of V and pontine central gray. A few large multipolar cells (35 to 45 μm in diameter) have also been noted in the most rostral portion of the LC. Most neurons in the frontal plane are aligned diagonally in a dorsolateral to ventromedial orientation, and when viewed in the horizontal plane, appear with their long axes directed anteroposteriorly.[24] Both dendrites and somata have been reported to demonstrate evidence of spines that consist of a thin stalk of variable length and a small bulb at the end. Very thin, locally ramifying axon-like plexuses (collaterals) have also been observed, and these likely play a role in the suppression of LC cell activity.[57]

Potassium permanganate and glyoxylic acid fixatives have been utilized to reveal ultrastructural features of LC neurons.[58,59] In accord with findings at the light microscopic level, electron microscopic (EM) preparations have indicated the presence of both medium- and small-sized cells within the LC. The cytoplasm of medium-sized cells was found to contain a prominent aggregation of endoplasmic reticulum and a well-developed Golgi apparatus. A reasonable number of small and large granular vesicles were also noted. Small, oval-shaped cells were characterized by highly indented nuclei, smaller amounts of endoplasmic reticulum, a less-developed Golgi apparatus, and a lack of small and large granular vesicles. Both dendro-somatic and dendro-dendritic contacts were encountered, the latter displaying small clusters of pleomorphic vesicles near the zone of apposition. With the introduction of the aldehyde-osmium fixation technique, several other features of LC neurons were confirmed.[60,61] In parasagittal section, cell bodies were further described as elongate, possessing disc-shaped dendritic fields, which parallel the anterior-posterior axis of the brainstem. The authors suggested that, in addition to dendro-dendritic synapses, recurrent axon collaterals of LC neurons probably participate in the intrinsic regulation of the nucleus.[61] Given the expression of connexin 32 along portions of the plasmalemma between apposed TH-positive dendrites in 1- to 3-week-old rats,[62] it is likely that gap junctions also exert a significant influence over local neuron-to-neuron interactions. It is important to point out that LC dendrites often extend beyond the limits of the LC proper and invade two peri-coerulear zones (peri-LC), i.e., the rostromedial region and the caudal juxtaependymal region.[63] Interestingly, 94% of the labeled profiles within the juxtaependymal area have been shown to be dendrites, which appear as postsynaptic elements in contact with other structures. The functional significance of these peri-LC dendritic arrays will be addressed later in this chapter. Presynaptic profiles in the LC neuropil, on the other hand, have been classified into several subtypes.[60] The most frequently encountered synaptic ending has been shown to contain small, round, densely packed vesicles (41% of sample). However, those with large, round (20%), large, flattened (23%), or small, flattened (23%) vesicles have also been observed. Terminals with dense-cored vesicles were present, but this type of ending represents a much smaller proportion of the population (on the order of 2%).

The morphological characteristics and topography of the human LC have been described in several reports. It is apparent from these studies that, like the rat, human LC neurons also constitute a heterogeneous population.[36,37,39,41,64] Based on both DBH and TH immunostaining, as many as four neuronal subtypes have been observed. These cell groups include large multipolar, small multipolar, large, elliptical, bipolar and small, ovoid, bipolar neurons.[41] The authors indicate that all neurons possess extensive dendritic arbors with three to five major branches, and that large multipolar neurons preferentially localize within the rostral LC. Comparisons between pigment-containing and immunolabeled cells further revealed that small neurons are usually heavily pigmented, while larger cells contain less pigment, and in some cases are completely devoid of pigment. A topographic arrangement of human LC neurons is supported by the findings of Marcyniuk *et al.*, who describe a selective loss of cells in the LC of patients with Alzheimer's disease,[64] and the findings of Chan-Palay and Asan,[41] who indicate greater cell loss within the rostral LC of aged individuals (78 to 86 years old). It is additionally noted in the latter report that small cells are more vulnerable than large cells. In addition, depressed subjects without dementia display an even more profound cell loss.

Transmitter identity and receptor distribution

Numerous studies show that in addition to NE, the LC is richly invested with and influenced by a variety of transmitter agents. In the following section, we provide a brief summary of the various neurochemicals and receptor subtypes that have been identified within the LC to date.

Classical transmitters

Pharmacological studies indicate four major receptor subtypes for NE. These are the α_1-, α_2-, β_1-, and β_2-adrenoceptors. Experiments have shown that the LC contains high concentrations of α_2-, low levels of α_1-, and the absence of β-receptors.[65,66] Investigations have further revealed that alterations in the expression of these receptor subtypes often correspond to the brains of individuals with mental illness. In particular, significantly higher densities of the α_2-adrenoceptor have been found in the frontal cortex, hippocampus, temporal cortex, hypothalamus, and LC in depressed suicides.[67,68] Aghajanian[57] demonstrated that the action of the α_2-adrenoceptor is to inhibit the firing of LC neurons. This autoinhibitory role is thought to suppress the release of NE from LC terminals. It has been hypothesized that NE imbalances produced by dysregulation of the α_2-receptor contribute to the etiology of mood abnormalities such as depression.

Acetylcholinesterase (AChE) histochemical techniques have been employed to examine the presence of acetylcholine (ACh) within the LC. These studies have revealed that AChE-containing neurons are evenly distributed throughout the LC and scattered in the sub-LC.[69] However, in experiments that utilized antibodies directed against choline acetyltransferase (ChAT), neither cell group was found to be labeled.[69] It has been suggested by the authors that these staining differences may be attributable to cholinoceptive activity. In association with these results, nicotinic cholinergic receptors (nAChRs) have been found in high concentration within LC. These ligand-gated channels appear to be assembled primarily from the $\alpha_{3,6,7}$, and β_3 subunits.[70,71] Using single-cell reverse transcription-polymerase chain reaction techniques, Lena *et al.* have further demonstrated that distinct subpopulations of LC neurons can be discriminated on the basis of nAChR subunit expression.[72] Small cells (type A) displayed high levels of α_3 and β_4 mRNA, whereas large cells (type B) preferentially exhibited evidence of the α_6 and β_3 subtypes.

Although only a small number of γ-amino butyric acid (GABA)-immunoreactive cells have been detected in LC, GABAergic terminals are highly prevalent within and around the LC complex.[73] Reports indicate that some of these terminals coexpress enkephalin (ENK).[74] At present, the action of GABA upon the LC is believed to be mediated by $GABA_A$ receptors. By increasing the conductance of Cl^-, GABA has been shown to exert a tonic inhibition on LC firing during sleep.[75,76] This finding corresponds to the fact that GABA release is enhanced in the LC during slow-wave and paradoxical sleep.[77] Further studies[78] have shown that $GABA_A$ receptors within the LC are composed of a unique combination of subunits, $\alpha_3\beta_3\gamma_1$. More recent evidence, however, has revealed the presence of mRNA for the ϵ and θ subunits. The ϵ subunit appears to be preferentially expressed in rat,[79] while the θ subunit is a trait shared by rats, primates, and humans.[79,80] A second, less prominent response has been shown to derive from the activation of $GABA_B$ receptors. This subtype promotes an increase in the conductivity of K^+ and inhibits the spontaneous discharge of LC neurons.[75]

Neuropeptides

Galanin (Gal) is a 29-amino-acid peptide that was originally isolated from porcine intestine tissue.[81] Since its discovery, a number of reports have identified the mouse, rat, primate, and human LC as a Gal-enriched site.[82–86] These studies have revealed marked differences in the cellular localization of Gal between species. Both Gal-immunoreactive neurons and fibers have been detected in mice, rats, and humans, while only the presence of Gal-positive fibers have been demonstrated in primates and felines.[83,85] Gal has been the focus of intense investigation, in part because a majority (70% to 80%) of TH-positive LC neurons coexpress this putative peptide transmitter.[82] The high degree of coexistence between Gal and DBH is illustrated in Figure 1.2 by the large number of neurons that stain positive for both substances following immunohistochemical processing. The small proportion of cells that lack Gal-immunoreactivity are usually found within restricted regions of the ventral LC. Interestingly, it is this population of cells (i.e., those lacking Gal mRNA) that is most susceptible to insults related to Alzheimer's disease.[86,87] Studies that have examined the effect of bath-applied peptide indicate that Gal tends to inhibit the firing of LC neurons. Using intracellular recording techniques, Pieribone *et al.* have shown that Gal increases conductance through K^+ channels and hyperpolarizes LC neurons.[88] At present, three different Gal receptors have been identified, GalR1, GalR2, and GalR3. The LC is known to contain all three receptor subtypes; however, GalR1 appears to be the most abundantly expressed.[89–91]

In general, cells that contain neuropeptide Y (NPY) constitute a small fraction of the LC neuronal population in rodents, and these cells appear to be lacking in sub-LC.[92,93] However, it should be noted that numerical values vary, depending upon whether or not colchicine treatment is introduced into the experimental protocol. Moore and Gustafson[94] have shown that in normal vs. colchicine treated animals, 0% and 50% of LC neurons stain positive for NPY, respectively. Discrepancies can also be found in reports in the human LC literature. These reports indicate both the absence[95] and presence[86] of NPY-expressing cells. Utilizing antibodies directed against the C-terminal flanking peptide of NPY (C-PON), the latter authors further describe a correlation between numbers of labeled neurons and senile dementia of the Alzheimer's type. Tissue collected from affected individuals exhibited higher numbers of C-PON-positive cells and large, tortuous fibers. Axon processes that were evaluated in controls, normal rodents,[92] and in other human studies were by comparison found to have smaller varicosities and to be present in moderate densities.[95] This work, in a manner reminiscent of findings related to Gal, points to the preservation of neuropeptide-containing cells and implicates the neuropeptides in mechanisms that may protect neurons during pathological conditions.

The presence of ENK has been investigated in rat and cat LC.[96–100] Although essentially similar in design, these studies have offered mixed findings. Leger[98] and Charnay[97] reported the existence of ENK-immunoreactive cell bodies in colchicine-treated cats, while experiments performed in rats by Drolet[99] revealed only the presence of ENK-positive terminals. In a more recent examination, Van Bockstaele and

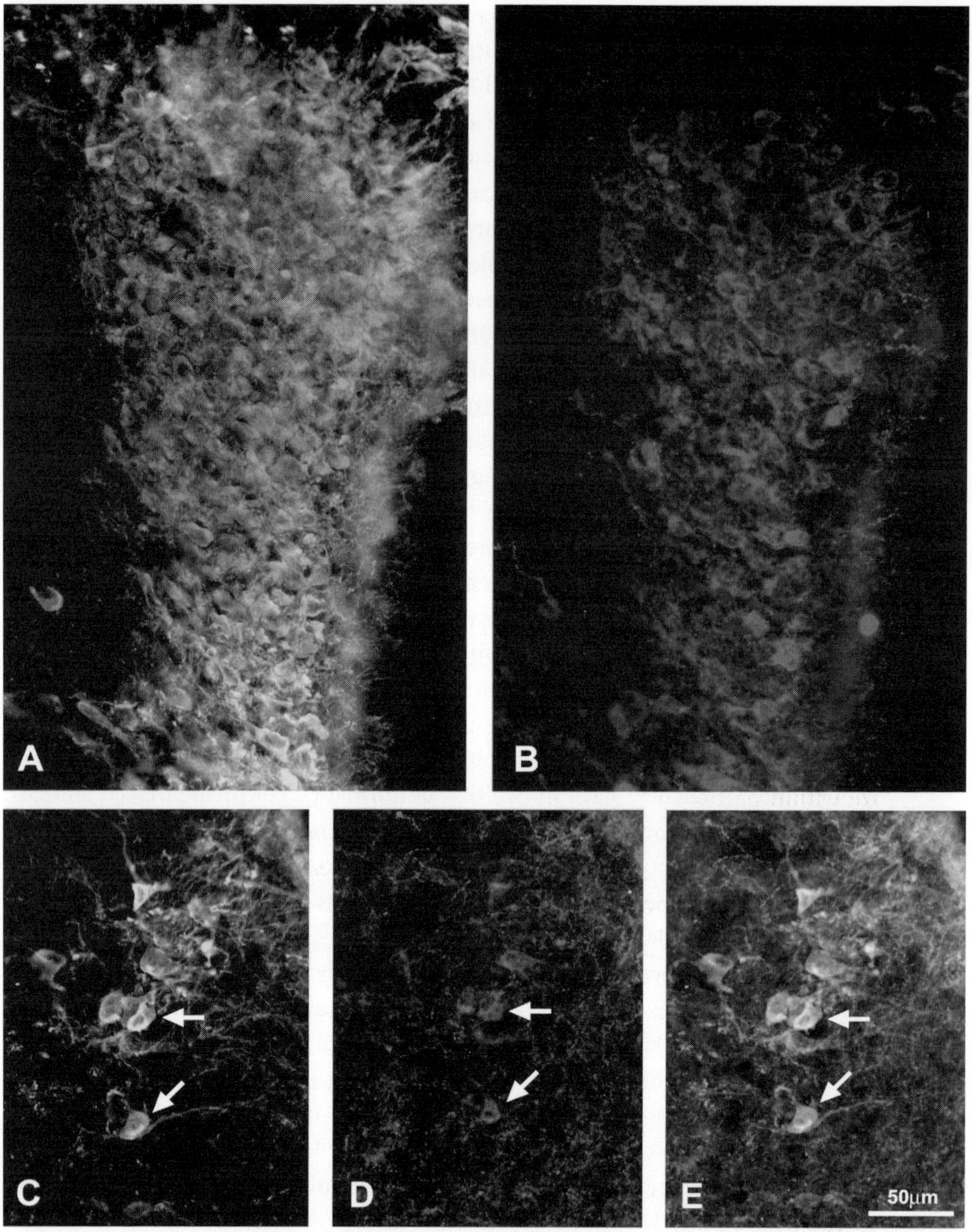

Figure 1.2 Photomicrographs show the overlapping distribution of DBH (A) and Gal (B) within cells of rat LC. The preponderance of Gal expression in the dorsal aspect of the nucleus is revealed by antirabbit IgG conjugated to Cy-3 (B), while the nuclear boundaries of the LC are defined by Cy-2 conjugated to antimouse IgG (A). Note that transmitter co-localization is depicted in C, D, and E (image overlay) by solid arrows pointing to representative double-labeled neurons.

Chan[74] confirm the ENK innervation of LC and reinforce the notion that LC lacks enkephalinergic perikarya. These authors further suggest that internalized peptide bound to the receptor may lead to a false positive signal, and that detection may depend on the extent to which antiserum can recognize degradation products of the peptide. Nonetheless, ENK-immunoreactive neurons have been identified within neighboring structures, such as the parabrachial, sphenoid, and Barrington's nuclei, as well as in the central gray and in the sub-LC area. Additional evidence has also been presented, which supports the coexistence of glutamate (Glu) or GABA in individual terminals within the LC that contain ENK.[74,100] These axonal endings have been shown to target small- or medium-sized LC dendrites.[74] Physiologically, LC neurons respond to the endogenous opioids, methionin^5- or leucine^5-enkephalin, with a potent decrease in spontaneous activity.[101,102] Reports indicate that both the μ- and κ-, but not δ-, opioid receptor subtypes are expressed within the LC and have further demonstrated that μ-receptors localize extrasynaptically on the dendrites of noradrenergic neurons.[103,104]

A number of other neuropeptides have been demonstrated in the LC.[69] These substances, which include somatostatin (SS),[105,106] substance P (SP),[98,107] neurotensin (NT),[108] vasoactive intestinal peptide (VIP),[109,110] vasopressin (VP), as well as neurophysin (NP),[111–113] tend to be found in small subpopulations of neurons. Due to previous requirements of colchicine treatment for visualization, it is generally accepted that these agents are only available in low concentrations. In some instances neurochemically identified cells have been reported to localize within particular regions of the LC. Somatostatin-containing cells have been detected in ventral and medial territories of the LC proper, while SP-positive cells have been found both within the core as well as outside the nuclear boundaries in ventromedial positions. However, in the human LC, only SP-expressing fibers have been observed.[95] Using double immunofluorescent techniques, Chen *et al.*[114] have indicated that the neurokinin-1 receptor (NK1) for SP is present within the LC. Moderate densities of NT cells and fibers have been shown in rodents, although in humans only a few NT-immunoreactive fibers were noted.[95] Young *et al.*[115] suggest that NT influences the LC by exerting an inhibitory action on neuronal firing. Vasoactive intestinal peptide cells have been mostly found in the dorsomedial sector of the LC, with a few cells residing in the sub-LC. Fibers containing VIP appear more evenly distributed within the LC complex. Also noted to be more broadly spread were VP and NP cells in LC and sub-LC. Neurons that express VP have been observed in larger quantities than NP cells, but both subpopulations possess cell bodies that are medium in size. Fiber labeling for these substances in LC has also been observed, albeit in moderate to low densities.

In 1983, Cummings *et al.*[116] reported the existence of corticotropin-releasing factor (CRF) within a group of loosely packed cells in the ventral LC of rodents.

However, later studies have suggested that CRF-positive cells are not situated in the LC, but rather are located in Barrington's nucleus, the nucleus positioned ventral and medial to the LC.[69,117] Accordingly, dense fiber arrays of CRF-immunoreactive fibers have been observed in the peri-LC area. Although two classes of CRF receptors (CRF-R1 and CFR-R2) have been identified and cloned, mRNAs for either subtype have not been detected within the LC.[118] One possibility is that these receptors are expressed presynaptically. Work is currently underway to determine the precise location of CRF-Rs in the LC.[119]

Connections: sources of afferent input to LC

Two basic neuroanatomical techniques have been applied to identify the origin of cells that innervate the LC. These methods have stipulated the placement of retrograde tracers into the LC proper or the injection of anterograde tracers into suspected source nuclei. From these types of studies, various views have emerged regarding the number of structures that project to the LC. Some reports indicate that a very limited number of nuclei innervate the LC, while others propose that the LC is regulated by a cast of different brain structures. The discrepancy between findings is likely to be explained by differences in the sensitivity of the tracers employed and difficulty in determining the precise spread of tracer after application. A major consideration is whether injections are confined to the core boundaries of the LC or whether injections encroach upon the pericoerulear territory. In the next section, a description and summary of retrograde studies will be provided. Outcomes of anterograde studies will follow with a discussion regarding an overall consensus.

Retrograde studies

The retrograde tracer horseradish peroxidase (HRP) was initially used to demonstrate brain regions that project to the LC.[120,121] These early studies suggested that a variety of cell groups that subserve very different functions innervate the LC. Rostrally located structures that were identified included the insular cortex, central nucleus of the amygdala (CeA), preoptic area, bed nucleus of the stria terminalis (BNST), and the hypothalamus. Caudally situated nuclei, such as the raphe, vestibular, solitary tract (NTS), A1, A2, A5, as well as the marginal zones of the spinal cord, were also reported. With more restricted injections of fluoro-gold and wheat germ agglutinin conjugated-HRP (WGA-HRP), Aston-Jones *et al.*[122] found that only two nuclei – the nucleus paragigantocellularis (PGi) and prepositus hypoglossi (PrH) – provide the major afferent input to the LC. The dorsal cap of the paraventricular nucleus and the spinal cord intermediate gray were found to supply minor inputs. Inconsistent labeling was noted in several places. The ventral tegmental area (VTA), raphe, spinal cord, and amygdala were among the sites that were listed.

The authors suggested that variable results might be attributable to tracer spread into the peri-LC subregion, an area poised to receive differing sets of afferents. Continued study with a yet more sensitive tracer, the cholera-toxin B subunit (CTb), confirmed that a host of nuclei project to the LC.[123] In addition to PGi and PrH, the preoptic area, the posterior hypothalamus, the Kölliker–Fuse nucleus, and the mesencephalic reticular formation were found to contribute substantially to the afferent innervation of the LC. A more limited input was found to derive from limbic cortex, the dorsal and median raphe, the lateral parabrachial nucleus, CeA, NTS, vestibular nuclei, the dorsal aspect of the PAG, the lateral portion of BNST, as well as superficial/intermediate layers of the spinal cord. This latter study implicates at least ten different brain regions in the afferent regulation of the LC. The authors of this report indicate that tracer placement outside the nuclear core region produces some variations in labeling, but that overall, patterns are quite similar to those obtained from injections restricted to the LC proper. Needless to say, retrograde tracing studies have elucidated a number of brain regions that may target the LC to influence the activity of noradrenergic neurons.

Anterograde studies

Axon processes have been mapped in an attempt to verify projection pathways of the LC. When WGA-HRP (also an anterograde tracer) was injected into PGi or PrH, dense networks of labeled fibers were observed in the core region of LC.[122] In contrast, placement of tracer into the CeA, the dorsal horn of the thoracic spinal cord, VTA, or rostral NTS tended to reveal fibers in the dorsolateral peri-LC area. No obvious terminal labeling was detected within nuclear boundaries. Investigations with the anterograde tracers, *Phaseolus vulgaris* leucoagglutinin (PHA-L) or CTb, showed that after injections into the preoptic area, Kölliker–Fuse nucleus, and ventrolateral PAG, a large number of fibers distribute more or less evenly between the LC core and peri-LC.[123] However, tracer deposition into the posterior hypothalamus yielded smaller numbers of labeled fibers (small to moderate labeling) within core regions as opposed to peri-LC territory. Additional studies with PHA-L have revealed a minor, nonbiased projection from rat medial prefrontal cortex,[124] as well as a projection from the cat and primate spinal trigeminal nucleus to the LC and peri-LC.[125] The latter are supported by unpublished findings from our laboratory, which indicate a direct input from the caudalis subdivision in the rat to the same targets. Based on available data, it stands to reason that the afferent projection to the LC can be subcategorized into at least three fiber populations. The first appears to be comprised of axons that preferentially target core regions of the LC, and is exemplified by inputs that originate in PGi and PrH. The second can be distinguished by the selective ordering of labeled fibers within the peri-LC neuropil. This group is represented by afferents that arise in the NTS, BNST, CeA, spinal cord, and posterior hypothalamus. The last consists of projections that appear to lack recognition for boundaries between the LC core and peri-LC. Afferents from the preoptic area, Kölliker–Fuse nucleus, ventrolateral PAG, prefrontal cortex, and spinal trigeminal nucleus meet the criteria established for this last subset. Taken together, retrograde and anterograde studies have provided strong evidence in favor of the view that the LC complex is innervated by several brain regions. The fact that these areas are functionally diverse has suggested that the LC is not merely a relay for the transfer of sensory, motor, or autonomic information, but a site for polymodal integration. It follows then, that the differential innervation of the LC may provide a mechanism or substrate by which to code, assimilate, and possibly prioritize converging streams of information.

Ultrastructural examinations

Recently, several reports have begun to reveal the synaptic organization of LC. These studies, mostly performed by Van Bockstaele *et al.*,[126] have placed a strong emphasis on the examination of dendritic contacts in the peri-LC region. In these investigations, a combination of anterograde tracing and TH-immunohistochemical techniques has been used to simultaneously identify both the afferent inputs to the LC, as well as the dendritic processes of LC cells. Projections from CeA, NTS, and PGi are among those that have been evaluated. When biotinylated dextran amine (BDA) or PHA-L was injected into CeA, 20% of the labeled terminals that were detected in the dorsolateral peri-LC area were found to make symmetrical contacts with TH-immunoreactive dendrites.[126] It was further indicated by the authors, that labeled endings often formed associations with unlabeled axon terminals having asymmetrical characteristics. This suggests that CeA afferents may also indirectly modulate the excitatory output of other nuclei that innervate the LC. Similar injections of the NTS in combination with gold-silver TH immunostaining revealed labeled terminals in the dorsolateral and ventromedial sectors of the peri-LC that formed heterogeneous synaptic contacts characteristic of both excitatory- and inhibitory-type transmitters. Approximately 19% of anterogradely labeled terminals formed synaptic specializations with catecholamine-containing dendrites.[127] These findings have provided the first indication that barosensitive NTS neurons may directly modulate LC activity. Additional studies, which have examined inputs to LC from PGi, have demonstrated that projections target LC somata, but preferentially innervate TH-immunoreactive dendrites.[128] Like axonal endings from NTS, these terminals appear in apposition to dendrites with symmetrical as well as asymmetrical postsynaptic membrane specializations. Such findings support the notion that NTS and PGi are composed of mixed populations of neurons, and that these cell groups express a variety of transmitter substances and/or neuropeptides[129] (see Chemoarchitecture: neurochemically identified pathways).

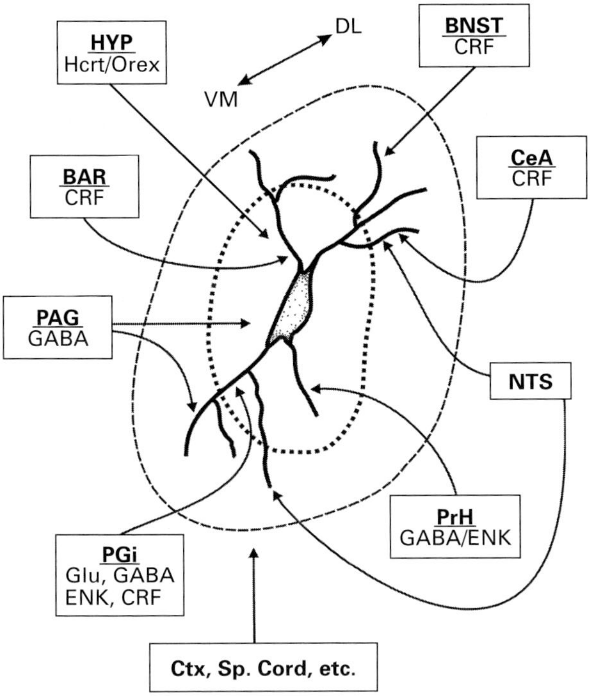

Figure 1.3 Schematic diagram summarizing the overall topography and neurochemical identity of afferent inputs to core and peri-LC regions of rodent LC. The area enclosed within the dotted line denotes the core region of LC, while the area between the dotted and dashed lines designates the peri-LC territory. Note that a majority of LC dendrites extend into the peri-LC region and that they tend to receive innervation from BNST, CeA, and NTS. In contrast, PrH, Barrington's nucleus (BAR), and hypocretinergic/orexinergic (Hcrt/Orex) inputs from hypothalamus (HYP) appear to preferentially target the core region. The precise distribution of afferents arising from brain regions such as the cerebral cortex and the spinal cord remain to be examined. Ctx, cortex; DL, dorsolateral; VM, ventromedial.

Taken together, EM investigations suggest that extranuclear and intranuclear dendrites are important termination sites for LC afferents. A schematic diagram summarizing patterns of afferent input to LC and peri-LC is shown in Figure 1.3.

Two critical questions have emerged that have direct bearing on ultrastructural findings. The first relates to whether most LC neurons have dendrites that extend into the peri-LC region. The second relates to whether regions such as CeA and NTS can directly impact LC activity via axo-dendritic interactions. Aston-Jones *et al.* have used anatomical and electrophysiological approaches to examine each of these issues.[130–132] In these investigations, LC neurons were intracellularly filled with biocytin to reveal the distribution of individual dendritic processes. Of the 20 cells that were sampled, all but one cell had at least one dendrite that extended into the peri-LC region. Using an electrical stimulation paradigm, these researchers further compared the responses of LC neurons to activation of PrH or NTS. Stimulation of PrH was found to evoke a potent, uniform inhibition of LC discharge, whereas no consistent effect was observed following activation of NTS. Fifteen cells were found to be unaffected, three cells were synaptically activated, and four were inhibited. Such findings call into question the functional significance of peri-LC dendrites and their afferent innervation. Interestingly, recent studies profiling subtle responses of cortical neurons have begun to reveal that incoming information may be more effectively transmitted to the cell body if received upon distal portions of the dendrite.[133] Magee and Cook, utilizing CA1 hippocampal pyramidal neuron slice preparations, found that dendritic excitatory postsynaptic potential amplitude increases with distance from the soma.[134] By placing an intracellular recording electrode at varying locations along the extent of an individual apical dendrite, they showed that responses to a high osmolar external solution were preferentially enhanced when applied beyond the proximal segment of the dendrite. Larkhum and Zhu[135] further reported that layer 5 pyramidal neurons have a low-threshold zone in the distal apical dendrite that initiates predominantly Ca^{2+}-dependent regenerative potentials. These potentials appear to be generated within a spatially restricted zone, approximately 550 to 900 μm from the soma, and can propagate without decrement along the apical dendrite. It is, therefore, likely that the distal dendrites serve as a site for the amplification and integration of synaptic inputs. The existence of such mechanisms in LC dendrites and their potential influence over LC operations are currently unknown; however, the data summarized above provide a compelling case for similar studies of dendritic processes within the peri-LC.

The LC efferent projection

Histochemistry and anterograde labeling

The output circuits of the LC have been the focus of several investigations. Since the earliest of studies, there has been general agreement that the LC fiber system is prominent and broadly distributed. Included among the methods utilized to arrive at this consensus are retrograde and anterograde tracing techniques as well as various histochemical and immunocytochemical staining procedures. As these approaches have been refined and improved over time, differing views regarding patterns of efferent organization have evolved. The more widely held notion maintains that the LC projection is diffuse and nonspecific, enabling this nucleus to impart a global impact. The second ongoing perspective favors the concept that the LC selectively innervates its efferent targets and provides for specific downstream influences. The following section will examine evidence related to both schools of thought.

The diverse, noradrenergic projections of the LC were initially revealed in studies that employed histofluorescence techniques. Later immunohistochemical studies confirmed that these monoaminergic fibers extend throughout the central nervous system (CNS). Although some regional differences have been noted between rats and humans, axon processes were found to be distributed throughout several levels of the neuraxis, from the rostrally located olfactory bulbs to the caudally situated spinal cord.[23,136,137] Connectional relationships were further examined in studies that utilized autoradiographic anterograde tracing strategies.[138,139] Data revealed a large group of fibers that originated in the LC and ascended as a well-defined dorsal tegmental bundle. The bundle was found to turn ventrally and traverse the prerubral field of Forel, before joining the dorsal portion of the medial forebrain bundle. Soon after merging, several distinct groups of fibers were observed to branch off from the pathway and proceed in the direction of various forebrain targets. In addition to these findings, radioisotope injections also revealed a smaller, much less obvious group of fibers constituting the ventral tegmental bundle. This descending fiber tract entered the medulla and continued caudally. Fibers either terminated around brainstem nuclei or passed into the ventral portion of the spinal cord. When considered in light of earlier evidence, these results also indicate that the LC-NE system is represented throughout the CNS by a diffuse array of fibers. However, it should be noted that there are major drawbacks associated with the use of such types of neuroanatomical approaches. The first relates to the fact that histochemical and immunocytochemical techniques cannot impart information regarding projections from individual LC neurons or specific subgroups. Second, radioisotope injections have a tendency to produce a significant amount of spread. Therefore, in addition to obtaining transport from the LC, there is a potential for adjacent nuclei to contribute to fiber labeling. With the development of more sensitive anterograde tracers, such as PHA-L, it is now possible to place restricted injections within the nuclear boundaries of the LC. A combined strategy, which utilizes PHA-L in conjunction with DBH immunohistochemistry, would permit axon termination patterns to be more precisely assessed after neurochemical identification. Using this methodology, Fritschy and Grzanna[140] found that the LC preferentially projects to the dorsal horn and intermediate zone of the spinal cord, while the A5 and A7 cell groups selectively target the ventral horn and intermediolateral cell column. From their study it is apparent that specificity is embedded within the LC efferent projection. The pattern of innervation, which was observed, led to the conclusion that output from the LC may have a greater impact on the processing of sensory inputs, particularly nociceptive afferents, than on motoneurons.

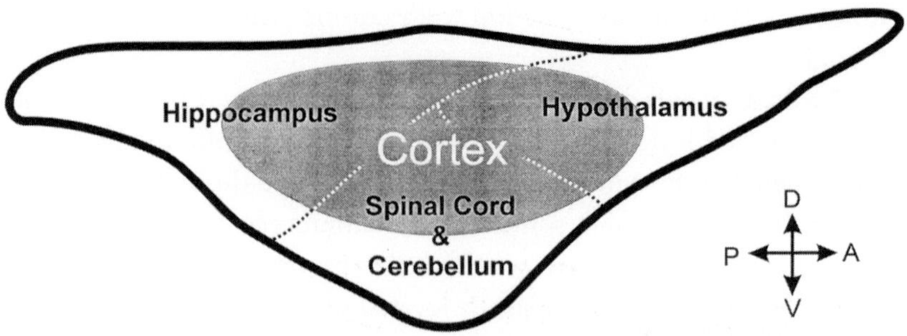

Figure 1.4 Schematic diagram illustrating the intranuclear topography of LC efferents with respect to different target sites. The LC complex may be subdivided into at least four major zones according to the distribution of LC projection neurons along the anterior-posterior and dorso ventral axes. As depicted by the zone of shading, cortical projection neurons tend to be found more dorsally in centralized portions of the nucleus, overlapping with cells that project to hippocampus, hypothalamus and spinal cord, and cerebellum. A, anterior; D, dorsal; P, posterior; V, ventral.

Retrograde labeling

Experiments with retrograde tracers have also proven very valuable for assessing the organization of the LC efferent projection. Besides indicating that the LC has multiple targets, early HRP studies revealed that differences in LC regional labeling vary in accord with the location of tracer injection.[141] Loughlin *et al.*[142,143] conducted an in-depth investigation and demonstrated quite convincingly that cells within the nucleus can be topographically and morphologically subdivided based upon their connectional relationship with different efferent target structures (Figure 1.4). Using HRP and DBH immunohistochemical techniques, the authors reported that the ventral and dorsal hippocampus receive projections solely from the dorsal segment of the LC, while projections to the spinal cord originate from the ventral-posterior portion of the nucleus. It was further noted that large, rostrally located neurons had a propensity for labeling after hypothalamic tracer injections, while fusiform cells were identified in association with regions of the cortex and hippocampus. Large, multipolar cells in the ventral LC were found to innervate the spinal cord and cerebellum.

Given the broad projections of the LC, several studies were undertaken to examine the extent to which individual neurons provide axon collaterals to different brain regions. In these investigations, relative proportions of double labeling were assessed as a means of determining the propensity for neurons to innervate more than one structure. Using HRP and [^{3}H]N-acetyl-WGA, Steindler[144] performed paired injections of cerebellum and visual cortex, or cerebellum and hippocampus. Out of the 41 neurons that were found to project to cerebellum, 61% demonstrated

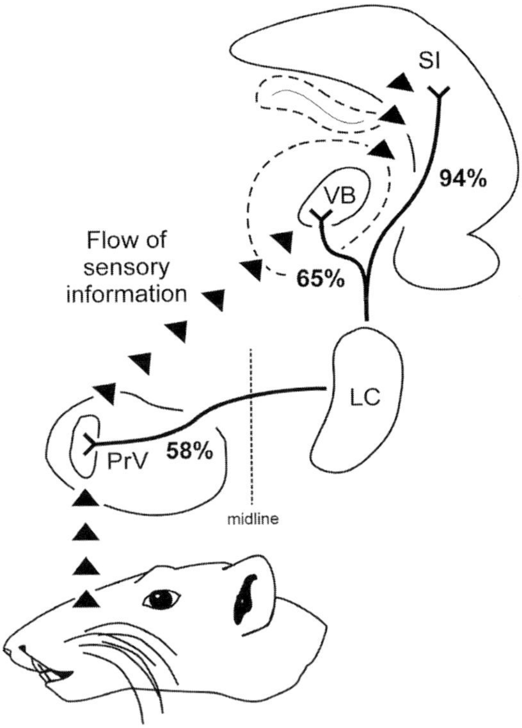

Figure 1.5 Schematic diagram showing the weighted distribution of efferents from one LC nucleus to relay sites along the trigeminal system. Solid arrow heads indicate the flow of ascending sensory information from the left whisker pad and principal trigeminal nucleus (PrV) to the contralateral (right) thalamus (VB) and cerebral cortex (SI). Numerical values were derived by averaging percentages of retrograde labeling across cases with similar injections. Note that the major outflow from the right LC nucleus follows the crossed trajectory of the vibrissae pathway.

evidence of axonal branching to the forebrain. These findings indicated to the authors that a substantial number of LC neurons have a tendency to collateralize. Employing a similar approach, but with the fluorescent tracers primuline (Pr) and diamidinophenylindole dihydrochloride (DAPI), Nagai *et al.*[145] reported that not all LC neurons demonstrate a similar degree of collateralization. When paired injections were placed into the cerebellum and spinal cord, only a few double-labeled neurons were detected. However, dual injections of cerebellum and frontal cortex, or frontal cortex and the olfactory bulb, revealed a larger contingent of double-labeled neurons. Approximately 10% to 30% of the labeled population exhibited both tracers. These data thereby suggest that LC cells may be further segregated on the basis of their capacity to collateralize, or on the basis of the particular combination of efferent structures to which they provide axon collaterals.

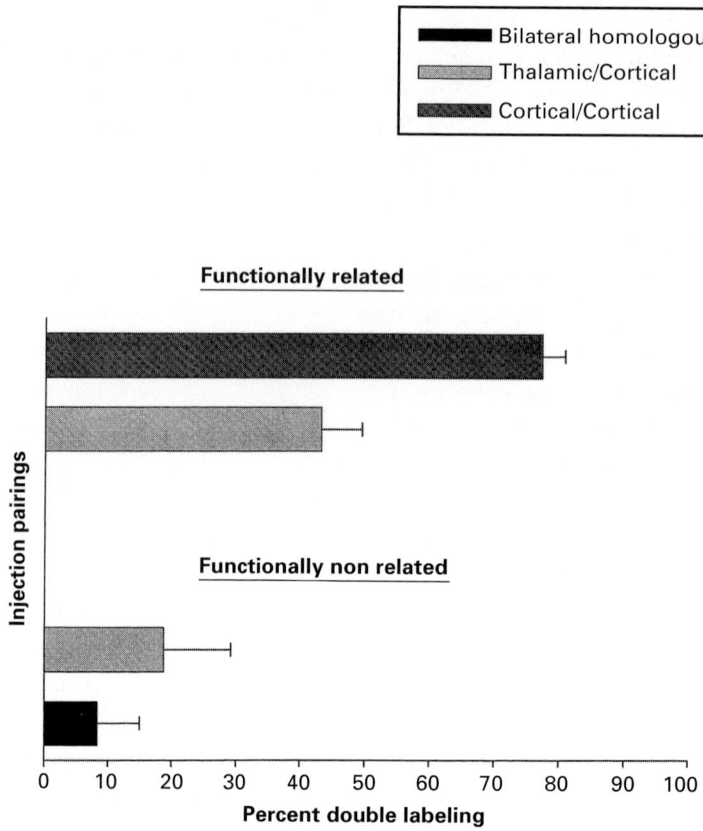

Figure 1.6 Collateral innervation of functionally related vs. functionally nonrelated forebrain structures by individual LC neurons. Higher percentages of double-labeled LC neurons were observed when retrograde tracers were injected into sites that process similar types of information (cortical/cortical pairings – SI/SII; thalamus/cortex pairings – VB/SI; VB/SII), than when injection pairings included sites not linked by modality (thalamus/cortex pairings – dLGN/SI) or not connected within the same functional circuit (bilateral homologous pairings –VB/VB; PrV/PrV). Each horizontal bar and corresponding horizontal line represents the mean percentage of double labeling and standard deviation for cases within similar treatment groups. Differences between functionally affiliated and functionally nonrelated pairings were determined to be statistically significant.

In order to extend the concept that the LC may coordinate actions through a patterned network of efferent connections, Simpson *et al.*[146] examined labeling characteristics of LC neurons that project to functionally related targets. The modality specific, rat ascending trigeminal somatosensory pathway was selected as a model system for studying the construct of LC efferent circuits due to the fact that relay areas along this path have been anatomically mapped and physiologically well defined. In these studies, the intranuclear distribution and collateralization of individual LC neurons was assessed following paired injections of different fluorescent retrograde tracers. Tracing materials were placed into the principal trigeminal nucleus (PrV), ventrobasal thalamic nucleus (VB), the barrel fields of primary somatosensory cortex (BFC or SI), and/or secondary somatosensory cortex (SII) in such a way as to simulate the flow of incoming information from the periphery. It was observed that the major output from one LC nucleus follows the crossed trajectory of the ascending "whisker" pathway. The PrV was found to receive a higher proportion of innervation from neurons located in the LC contralateral to tracer injection, while VB and SI were found to be preferentially supplied by the ipsilateral nucleus. In fact, the projection to the cortex was almost exclusively lateralized, with 94% of coeruleo-cortical neurons originating in the ipsilateral nucleus (Figure 1.5). When double labeling was assessed, it became apparent that LC neurons tend to provide axon collaterals to structures engaged in similar sensory functions. Paired injections of VB and SI, or SI and SII on the same side, produced double labeling in approximately 40% to 80% of the total retrograde-labeled population. In contrast, injection pairings of structures that were not functionally affiliated, i.e. the dorsal lateral geniculate nucleus (dLGN) in combination with SI, or VB in combination with primary visual cortex, resulted in lower percentages (~10% to 20%) of double labeling (Figure 1.6). Collectively, these data indicate that the projections of LC are not randomly arranged, but directed towards target-specific interactions. An organized framework of efferents would permit LC-mediated shifts in noradrenergic tone to selectively modulate activity within specified terminal fields or along channels transmitting particular types of information.

Chemoarchitecture: neurochemically identified pathways

By further exploring the transmitter content of LC afferents and efferents, several laboratories have been able to provide a working picture of the pathways regulating LC function. Advances in tract tracing, neurochemical staining reagents, and imaging technology have been mentioned previously, but it is worthwhile to emphasize that improvements in methodology have afforded researchers the opportunity to dissect LC circuits and examine several attributes of a particular projection. In the following section, we review neurochemically identified connections and discuss their points of origin/termination in relation to LC.

Excitatory inputs

The LC has been reported to receive excitatory amino acid input from two different sources. The PGi in the rostral ventrolateral medulla is one area that has been shown to provide glutamate-containing afferents to LC. Ennis and Aston-Jones[147] showed that single-pulse electrical stimulation of the PGi synaptically activated a large majority of LC neurons (70% to 80%). This excitatory action was further found to be mediated by the kainate receptor subtype.[148] Findings indicated that the *N*-methyl-D-aspartate (NMDA) receptor antagonist, AP7, did not block the LC response to PGi stimulation, whereas the kainate receptor antagonists, kynurenic acid and γ-D-glutamylglycine, completely blocked the excitatory effect. Work performed by Drolet *et al.*[99] further demonstrated that projections from the PGi utilize the neurotransmitter glutamate.

Afferents arising from the medial prefrontal cortex have been suggested as a second source of glutamatergic input. As reported by Jodo and Aston-Jones,[149] electrical stimulation of this area elicits activation of LC neurons. These authors concluded that both NMDA and non-NMDA receptors mediate these excitatory actions, since the NMDA antagonist, AP5, and non-NMDA antagonist, 6-cyano-2, 3-dihydroxy-7-niroquiinoxaline (CNQX), significantly suppressed the evoked response.

Ultrastructural studies have been conducted to reveal the precise location of glutamate receptors. These studies have indicated that the NMDA-R1 subunit is selectively distributed in astrocytic processes and within presynaptic axon terminals.[150] Expression was rarely seen in catecholamine-containing somata or dendrites. In another study, the same authors demonstrated that the non-NMDA receptor subunits GluR5/6/7 are mainly situated at synaptic and extrasynaptic portions of the plasmalemma in LC dendrites and somata.[151] The synaptic organization of specific glutamatergic inputs in relation to these receptor subtypes may, perhaps, be elucidated in future investigations.

Inhibitory inputs

Physiological and pharmacological studies have suggested that neurons in the PrH provide inhibitory projections to the LC. In 1989, Ennis and Aston-Jones[131] reported that electrical stimulation of PrH suppressed the spontaneous activity of 42/47 LC neurons (~80%). Because the $GABA_A$ antagonist, bicuculline, blocked this effect, they concluded that outflow from PrH is conveyed via $GABA_A$ receptors. Pieribone *et al.*[152] later found that approximately 60% of LC-projecting PrH neurons express either GABA or glutamic acid decarboxylase immunoreactivity. However, it should be noted that GABA projections have also been reported to arise from other brain areas. The preoptic nucleus, posterior hypothalamus, ventrolateral PAG, and the lateral PGi (PGCl) are among the additional sources that have been identified.[153] Interestingly, Ennis and Aston-Jones[147] found that 20% of LC neurons are inhibited by electrical stimulation of the PGi. A few reports have also suggested the presence of a small group of GABAergic neurons within the LC proper that may utilize axon collaterals to exert a local inhibitory influence on LC activity.[74,154]

In addition to GABA, immunohistochemical and electrophysiological evidence suggests that the inhibitory transmitter, glycine, is contained within projections to the LC. Luppi *et al.*[155] found that glycine-positive terminals form a dense, varicose plexus within the LC. When glycine was iontophoretically applied, the spontaneous discharge of LC neurons was inhibited. This action was blocked by the selective glycine receptor antagonist, strychnine, but not by bicuculline. Although there is ample evidence to support a role for glycine in the afferent regulation of the LC, the origin of these projections remains unknown at the present time.

Serotonergic inputs

The origin of the serotonergic projection to the LC, first demonstrated by Pickel[156] in 1977, has been investigated using retrograde tracer injections and 5-HT immunohistochemical techniques. Following the placement of WGA-apoHRP gold into the LC proper, Maeda *et al.*[157] reported the presence of double-labeled cells in the nucleus raphe magnus. Although dual labeling was observed after similar injections of the peri-LC, serotonergic projection neurons were detected in the dorsal and median raphe nuclei. Investigations by Pieribone *et al.*,[158] however, suggested the involvement of the rostromedial PGi, PrH, and the ventromedial rostral border of the LC. Application of fluoro-gold to the core region of the LC, revealed serotonergic cells in PGi that belonged to the lateral division of the B3 cell group.[132] Smaller numbers of cells were detected in the other two locations. Taken together, these studies suggest that the LC complex receives parallel innervation from different populations of serotonergic neurons. The fact that these inputs can be discriminated on the basis of target selectivity, i.e. LC core vs. peri-LC region, suggests that these afferents may have a differential impact on LC firing.

Dopaminergic, noradrenergic, and adrenergic inputs

Reports indicate that the LC is the recipient of several catecholaminergic projections. The A13 cell group of the hypothalamus and the more posteriorly situated A11 cell group are among the dopaminergic nuclei known to target the LC.[157,159] Other dopamine-containing neurons, such as the A8 (retrorubral), A9 (substantia nigra), and A10 (ventral tegmental area) cell groups have not been implicated in the innervation of the LC. In addition to the recurrent collaterals of LC neurons, extrinsic noradrenergic inputs have been shown to arise from the A2 and A5 cell groups in rats and cats.[157,159] Adrenergic inputs, on the other hand, have been shown to derive predominantly from PGi (C1) and PrH (C3).[88,157] In these experiments retrograde tracing techniques were combined with phenylethanolamine-*N*-methyltransferase (PNMT) immunohistochemistry. Quantitative analysis revealed that of the two projections, the projection from PGi was more substantial. Approximately 21% of the retrograde-labeled cells in PGi exhibited PNMT immunoreactivity, while PNMT expression was only noted in 4% of LC-projecting cells in PrH.

Neuropeptide inputs

Interest in the hypocretins developed shortly after preprohypocretin (hcrt) mRNA was detected in the tuberal region of the hypothalamus.[160,161] This 130-residue protein was found to be the precursor of two neuroactive peptides, hcrt1 (residues 28 to 66) and hcrt2 (residues 69 to 96). In a study by Sakurai *et al.*,[162] these peptides were identified as orexin A and B, respectively, and shown to activate two distinct G-protein-coupled receptors, OX1 and OX2. Peyron *et al.*[163] later used antisera directed against hcrt to further characterize hcrt-expressing neurons and gain insight into neuron groups that may be influenced by hcrt. Concentrations of hcrt-positive neurons were found in the perifornical nucleus and in the dorsal and lateral hypothalamic areas. Fiber analysis revealed the presence of dense immunoreactive fiber arrays in LC, and indicated the potential for a direct hypocretinergic projection from the hypothalamus. Synaptic contacts between hypocretin-positive terminals and LC cells were demonstrated in rats and monkeys by the studies of Horvath *et al.*[164] More specifically, TH-immunopositive cell bodies and proximal dendrites with asymmetric membrane specializations were found in apposition to hypocretin-containing terminals. When the action of hcrt2 on the LC was examined, all cells showed excitatory responses. Hypocretin-2 uniformly increased the frequency of action potentials in these cells, presumably by reducing the slow component of the afterhyperpolarization and eliciting a modest depolarization. Since this effect was noted even in the presence of tetrodotoxin (TTX), it was concluded that the response was mediated by receptors on LC neurons. Using *in situ* hybridization, Bourgin *et al.*[165] confirmed the existence of hcrt-receptors (hcrtr) in the LC. Although hcrtr2 mRNA was not detected in the LC, hcrtr1 mRNA was highly expressed and, furthermore, implicated in the actions of hcrt1. For example, local administration of hcrt1 in the LC suppressed rapid eye movement (REM) sleep in a dose-dependent manner and increased wakefulness at the expense of deep slow-wave sleep. Iontophoretic application of hcrt1 was also shown to enhance the firing rate of LC neurons in vivo. Interestingly, a number of these findings corresponded with the results of Hagan *et al.*[166] Extracellular recordings of the spontaneous activity of rat LC neurons revealed an increase of firing rate in response to application of orexin A over the range of 30 nM to 3 μM. The maximum increase in firing rate occurred at 1 μM. When filled with neurobiotin, neurons that responded to orexin A were shown to have the characteristic morphology of principal neurons of the LC. Double labeling of these cells with orexin A antisera revealed that extensive portions of the distal dendrites extend into regions of orexin A immunoreactivity. These dendrites appeared in close apposition to immunolabeled fibers that possessed large (up to 2.5 μm in diameter) varicosities. Administration of orexin A at the onset of the normal sleep period produced a significant increase in the proportion of arousal during the second and third hour after dosing. During the same period there was a significant decrease in the proportion of paradoxical sleep, no change in light slow-wave sleep, and a nonsignificant decrease in deep slow-wave sleep. Orexin A caused a significant increase in locomotor activity at all doses and throughout a 90 minute observation period. These findings suggested to the authors that orexin A plays an important role in orchestrating the electrophysiological and behavioral aspects of the sleep–wake cycle. In line with this view, knockout experiments in mice have demonstrated that the absence of hypocretin causes alterations of sleep architecture, particularly on the amount of REM sleep during the dark period.[167] In terms of pathological evidence, seven of nine patients with narcolepsy had undetectable hcrt1 in cerebrospinal fluid.[168]

Corticotropin-releasing factor is considered to be a mediator of LC activation during stress, and has been implicated in signaling events that elicit scanning attention. Studies have shown that CRF increases tonic LC activity,[169] and blunts the brief activation of LC neurons to phasically presented sensory stimuli (e.g. sciatic nerve stimulation, auditory stimulation[170,171]). Corticotropin-releasing factor fibers are more densely distributed in the rostral and lateral peri-LC region, than in the LC proper.[172] Reports indicate that the CRF innervation of the nuclear region is light to moderate,[173] while the periventricular region medial to the LC contains a comparatively larger number of fibers. Further examination at the ultrastructural level has revealed that CRF-containing terminals synaptically contact LC dendrites.[174,175] Quantitative analysis indicates that, 20% to 25% of CRF-positive terminals in the ventromedial peri-LC region form contacts with TH-immunoreactive dendrites. Asymmetric synapses tend to be more prevalent than symmetric synapses.[173] In addition to the CeA,[175] BNST,[176] and Barrington's nucleus,[177] PGi and cells in the dorsal cap of the paraventricular hypothalamic nucleus[172] have been shown to be potential sources of CRF to the LC. Corticotropin-releasing factor-containing cell bodies are located in these areas and have been found to demonstrate retrograde labeling after tracer injections of WGA-Au-HRP or fluorescent latex beads. Valentino *et al.*,[173,177] found that one-third of the retrogradely labeled neurons in Barrington's nucleus stain positive for corticotropin-releasing hormone[173,177] (CRH; interchangeable with CRF), while in other studies, it was shown that 8% of retrogradely labeled PGi neurons contain CRF.[172] Placement of retrograde tracer into the rostrolateral peri-LC has furthermore provided evidence of some neurons in BNST that express CRF.[178] By utilizing a combined anterograde tract-tracing and immunoelectron microscopic approach, CRF-positive afferents from BNST and CeA have been confirmed to target noradrenergic neurons of the LC. In both cases, BDA-labeled axon terminals were found to form symmetric specializations (primarily inhibitory) with TH-labeled dendrites and dendrites that lacked TH-immunoreactivity.[176] However, a large number of synapses were not clearly differentiated. For example, following injections of BNST, 19% of BDA-labeled terminals were found in direct apposition to dendrites that contained gold-silver labeling for TH. Of these terminals, 19% formed asymmetric synapses, 38% formed symmetric synapses, and 41% did not exhibit specialization.

Enkephalin

The rostral medulla has been strongly implicated in the ENK innervation of the LC. Drolet *et al.*[99], reported that following placement of retrograde tracers into the LC, 56% to 57% of labeled projection neurons in the PGi and PrH stained positive for ENK. Further studies at the light microscopic level have shown that, in comparison to CRF fibers, a greater density of enkephalinergic fibers appear within the LC nuclear core. However, like CRF processes, ENK axons tend to be more concentrated in peri-LC dendritic zones, particularly at the rostral level of LC.[117] Ultrastructural evidence supporting a role for the opioid regulation of LC indicates that 62% of opioid-containing afferents target noradrenergic dendrites in the dorsolateral peri-LC.[173,179] Although both asymmetric- and symmetric-type synapses have been observed, asymmetric are more common. Interestingly, ENK has been found to colocalize with other transmitter substances. In their review, Valentino and Van Bockstaele[117] cite evidence of CRF and ENK double labeling both within terminals of the dorsolateral LC, and somata of the PGi, and paraventricular hypothalamic nucleus. However, the coexpression of ENK and GABA, and ENK and Glu has been more thoroughly studied. Quantitative analysis has revealed that 38% of ENK terminals in the LC contain GABA.[74] Of these profiles, 12% formed asymmetrical synapses, 23% formed symmetrical synapses, and 65% did not display evidence of synaptic specialization. In another study, this laboratory reported that 28% of ENK terminals in the LC contain Glu.[100] Some of these synapses clearly demonstrated asymmetric specializations. It is tempting to speculate on the sources of these inputs. Evidence gathered from physiological and retrograde tracing experiments would seem to suggest that GABA-containing ENK-positive terminals most likely originate from neurons in PrH, and Glu-containing ENK-positive terminals most likely originate from neurons in PGi. However, such interpretations await confirmation from combined anterograde tracing and immunohistochemical studies, especially in light of the fact that CeA and NTS are enriched in opioid- and Glu-containing neurons.[126,127] Given the inhibitory effects of opioids on LC activity, it seems rather curious for ENK to be colocalized with an excitatory transmitter such as Glu, simply because the actions of one cotransmitter would likely negate the opposing effect of the other. Van Bockstaele *et al.*[99] raise an interesting point in that ENK release may occur under certain conditions or stimulus frequencies. Providing this to be the case, one might expect the response of an LC neuron to differentially reflect the influence of ENK or Glu, depending upon which transmitter was liberated in greater abundance at the synapse.

Collectively, the above findings suggest that: (1) noradrenergic dendrites are an important site for the integration of information; and (2) several neurochemicals and afferent pathways influence the activity of the LC. According to recent accounts, this regulation may be regarded as controlled and selective. For example, Van Bockstaele *et al.*[173] indicate that afferent input to the LC is topographically organized. Afferents from PGi, PrH, and Barrington's nucleus preferentially innervate the LC core, while neurons originating in the CeA and BNST principally target the rostrolateral peri-LC.[177] Studies have also shown that neurons in the PAG and NTS tend to terminate more frequently in the ventromedial peri-LC. In fact, injections of BDA into the ventrolateral PAG primarily labeled processes in the rostroventromedial peri-LC area, 20% of which formed synapses (predominantly symmetric) with TH-containing dendrites.[173] In considering the functional affiliations of afferent source nuclei, these authors have postulated that each input may serve to selectively influence LC activity in response to a particular type of stimuli. Information concerning emotional or external stressors appears to be conveyed to the rostrolateral peri-LC via limbic projections, while inputs transmitting events related to autonomic and nociceptive processes seem to impinge upon the ventromedial aspect of the peri-LC. The monosynaptic pathway from PAG to LC places the LC in a position to mediate analgesic effects produced by stimulation of the ventrolateral PAG, while the circuitry linking the NTS to the LC may underlie responses of the noradrenergic system to physiological challenges. Superimposed on the topography of these anatomical connections, Van Bockstaele and colleagues[173] have found a neurochemical organization that may further encode streams of incoming information and permit LC inputs to selectively engage their targets. For example, in quantitative studies[173] that examined CRF-immunoreactivity in BDA-labeled axon terminals from BNST and CeA, a higher percentage of CeA terminals (35%) were found to express CRF than BNST terminals (13%). Interestingly, these differences in neurochemical content correspond well with the distinct functional roles that have been proposed for CeA and BNST by Walker and Davis.[180] These investigators found that CeA preferentially mediates the expression of conditioned fear, whereas BNST mediates behavioral responses to unconditioned fear. Output from these nuclei would be expected to impact on LC activity, but the qualitative nature of the impact would be expected to differ, in part, because of the differential representation of CRF within these projections.[173,176] However, we might expect CRF-containing systems to differentially impact LC firing based on the fact that CRF-positive terminals exhibit a range of synaptic profiles. The large number of unspecialized terminals might also suggest that CRF-containing inputs (and other peptides as well) may play more of a modulatory role in the regulation of LC, rather than providing a strict excitatory or inhibitory influence. In support of this view, Page and Abercrombie[181] have reported that direct infusion of CRF into the LC increases the incidence of LC neuron burst firing from 14% to 33%. This finding broadens the range of effects reported previously by Valentino *et al.*[177] Nonetheless, it is interesting that one transmitter may be utilized in several afferent systems, but that its overall expression, timing of release, source of release, and context of release may enable it to elicit a different qualitative impact. It is tempting to further speculate about other peptides (such as Gal) that are present in LC afferent projections (Figure 1.7) and afferent source nuclei.[82] Galaninergic inputs from NTS, BNST, and CeA could conceivably regulate LC activity in parallel with CRF systems.

Efferent connections

The neurochemical content of specific subpopulations of LC projection neurons has, likewise, been examined in an effort to determine the potential contribution of different transmitter substances to the efferent regulation of several LC targets. These studies have not only revealed that neuropeptides are differentially represented in the efferent projections of the LC, but that the LC projection system can be organized according to the chemical composition of its connections. Using a combination of retrograde tracing and immunohistochemical techniques, Holets[182] reported that a very low proportion (2% to 4%) of spinal cord-, cerebral cortex-, and hypothalamus-projecting LC cells stain for NPY. However, when Gal expression was analyzed, hypothalamic areas were found to preferentially receive higher proportions of galaninergic input from the LC. Only 3% of Gal-containing cells projected to the spinal cord and cerebral cortex, while 14% innervated the hypothalamus. Lechner *et al.*[183] extended these findings by undertaking an investigation of the peptide identity of brainstem neurons that project to functionally distinct subregions of the thalamus. Injections of colloidal gold-WGA-HRP were placed into the medial thalamus (limbic associations), lateral thalamus (somatosensory relay area), or zona incerta (sensorimotor integration region) to elucidate projection neurons that were further screened for the presence of Gal, NPY, SP, NT, ENK, dynorphin (DYN), calcitonin gene-related peptide (CGRP), cholecystokinin (CCK), or bombesin (BB). Data showed that LC neurons containing NT and NPY preferentially projected to the medial thalamus, while SP-positive and Gal-expressing LC cells tended to innervate the zona incerta. Similarly, CGRP-containing neurons were found to significantly contribute to the LC-lateral thalamus pathway. It can be deduced from these results that the neurochemical content of groups of LC projection neurons varies in accordance with the functional affiliation of the efferent target. Evidence from our laboratory supports this notion, and further demonstrates similar proportions of peptide expression in LC neurons that project to targets with similar functions. For example, following transport from retrograde tracer injections placed into PrV, VB, or SI, approximately 50% of labeled projection neurons in LC exhibited Gal-immunoreactivity.[83] These findings complement

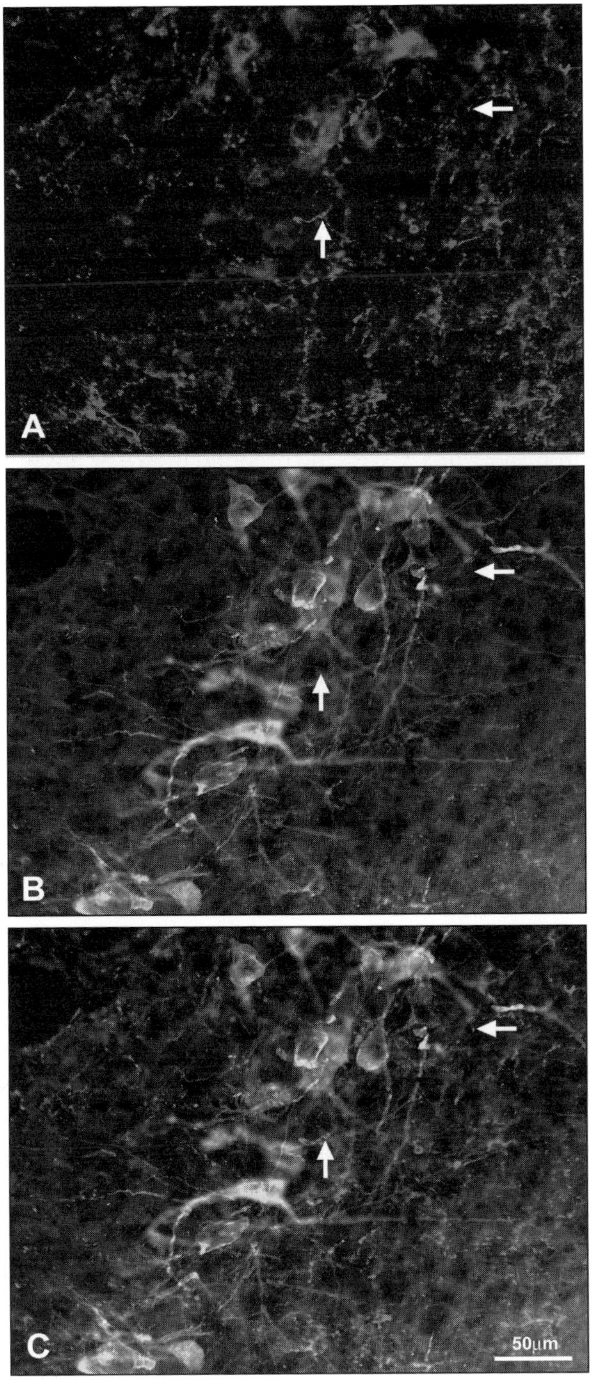

Figure 1.7 Photomicrographs demonstrate Gal-immunoreactive fibers (arrows) within the rostral sector of rat LC. The fluorescent markers Cy-3 and Cy-2 are used to depict the distribution of galaninergic axons in relation to LC cell bodies in A and B, respectively. The merged image in C further illustrates that Gal-positive processes lack in the expression of DBH.

those of Lechner,[183] who found 30%, 57%, and 76% of retrogradely labeled neurons in the LC contained Gal after injections of medial thalamus, lateral thalamus, and zona incerta, respectively. In both studies, injections of VB/lateral thalamus revealed similar proportions of Gal expression in LC-somatosensory thalamus projection neurons. Furthermore, and most importantly, the fact that only injections of the somatosensory brainstem, thalamus, and cortex yielded similar proportions of Gal expression among retrogradely labeled LC neurons suggests LC projection neurons recognize functional distinctions between nuclei at different levels of the central neuraxis. These arrays of neurochemically patterned projections may be one way in which the LC coordinates circuit function within multiple terminal fields.

Taken together with information presented above (LC efferent projection: retrograde labeling), evidence points to a topographical and neurochemical organization of LC efferent projections. Given the preponderance of NE within the LC, such results introduce the possibility of NE–peptide cotransmission and modulatory interactions at specific presynaptic or postsynaptic targets. In fact, recent reports indicate, that Gal, for example, inhibits the NE-induced accumulation of cyclic adenosine monophosphate (cAMP) in the rat cerebral cortex[184] and inhibits the release of NE in hypothalamic slices.[185] The precise functional significance of corelease from LC terminals is a topic of great interest that awaits clarification in future studies.

Overview

The purpose of this chapter was to review and summarize early findings and recent progress regarding the anatomical and neurochemical characteristics of projection systems that impact on LC function. It is apparent from the extensive projections of the LC and the multiple sources of its afferent input that the LC is positioned to exert global effects on neuronal processing and behavior. However, closer inspection of the wiring of this complex network reveals an architecture that is likely to permit the LC to more selectively engage its efferent targets. The organization of this circuit is marked by: (1) chemical coding within the afferent and efferent projections of the LC; (2) differences in the extent of collateralization of individual LC neurons; (3) differences in the intranuclear distribution of LC projection neurons with respect to given targets; and (4) regional differences in the innervation density of LC by particular source nuclei. When considered together with the variable modes of LC discharge and the potential for LC cell burst firing to preferentially elicit the release of neuropeptides,[186–188] this information strongly suggests directed utilization of particular transmitters in combinations that promote site-specific interactions in LC terminal fields. Furthermore, previous physiological studies have demonstrated the capacity of NE to augment the response of a neuron to excitatory and inhibitory synaptic input (increase of signal to noise ratio),[189–191] as well as alter the feature extraction properties of individual sensory neurons. Current findings are in keeping with the selective actions of NE, but also support a mechanism whereby synchronous activation of the LC (as has been indicated in the literature) may translate into peptidergic influences that may vary the impact of NE across its sphere of influence. It is tempting to additionally speculate that, under certain conditions, the organized connections of the LC may be utilized to limit outflow from this brainstem center and permit only the selective activation of specific subsets of LC efferents. Either way, the structural relationships that have been discussed in the present chapter surely play an important role in enabling the LC to participate in a number of critical brain processes, namely those involved in attention, cognition, learning, memory, stress, and mood regulation.

REFERENCES

1. Reil, J. C. Untersuchungen uber den Bau des grossen Gehirns im Menschen. *Arch. Physiol.*, **9** (1809) (Halle), 136–524.
2. Wenzel, J. and Wenzel, C. *De Penitiori Structura Cerebri Hominis et Brutorum.* (Tübingen, 1812).
3. Arnold, F. A. *Bemerkungen uber den Bau den Hirnes und Ruckenmarkes.* (Zurich, 1838).
4. Arnold, F. A. *Handbuch d Anat d Menschen.* Freiburg 1851;2(2):699.
5. Jacobsohn, L. Meddulla oblongata, pons und pedunculus cerebri. In *Uber die Kerne des Menschlichen Hirnstamms Anhang zu den Abhandlungen der Kgl.* (Preuss, Akad d. Wiss. Phys-Mathem Kasse CP, 1909).
6. Meynert, T. The brain of mammals. In *Stricker's Manual of Histology.* (New York: Wm Wood Co., 1872).
7. Crosby, E. C. and Woodburne R. T. The mammalian midbrain and isthmus regions. Part 1. The nuclear pattern. General summary. *J. Comp. Neurol.*, **78** (1943), 505–520.
8. Russell, G. V. The nucleus locus coeruleus (dorsalislateralis tegmenti). *Tex. Rep. Biol. Med.*, **13** (1955), 939–988.
9. Shimizu, N., Morikawa, N. and Okada, K. Histochemical demonstration of monoamine oxidase of the brain of rodents. *Z. Zellforsch.*, **49** (1959), 389–400.
10. Maeda, T., Abe, T. and Shimizu, N. Histochemical demonstration of aromatic monoamine in the locus coeruleus of the mammalian brain. *Nature*, **188** (1960), 326–327.
11. Shimizu, N., Matsunami, T. and Onishi, S. Histochemical demonstration of ascorbic acid in the locus coeruleus of the mammalian brain. *Nature*, **186** (1960), 479–480.
12. Yoshinaga, V. T. and Shimizu, N. Histochemische untersuchung des farbbaren zinks und kupfers im gehirn des menschen. *Acta Histochem.*, **30** (1968), 90–95.
13. Lewis, P. R. and Schon, F. E. G. The localization of acetylcholinesterase in the locus coeruleus of the normal rat and after 6-hydroxydopamine treatment. J. Anat., **120** (1975), 373–385.

14. Falck, B., Hillarp, N. A., Thieme, G. and Torp, A. Fluorescence of catechol amines and related compounds condensed with formaldehyde. *J. Histochem. Cytochem.*, **10** (1962), 348–354.
15. Dahlstrom, A. and Fuxe, K. Evidence for the existence of monoamine-containing neurons in the central nervous system. *Acta Physiol. Scand.*, **62** (1964), 3–55.
16. Blessing, W. W., Chalmers, J. P. and Howe, P. R. C. Distribution of catecholamine-containing cell bodies in the rabbit central nervous system. *J. Comp. Neurol.*, **179** (1978), 407–424.
17. Pickel, V. M., Joh, T. H. and Reis, D. J. Ultrastructural localization of tyrosine hydroxylase in noradrenergic neurons of brain. *Proc. Natl. Acad. Sci. U. S. A.*, **72** (1975), 659–663.
18. Hökfelt, T., Johansson, O., Fuxe, K., Goldstein, M. and Park, D. Immunohistochemical studies on the localization and distribution of monoamine neuron systems in the rat brain. I. Tyrosine hydroxylase in the mes- and diencephalon. *Med. Biol.*, **54** (1976), 427–453.
19. Hartman, B. K., Zide, D. and Udenfriend, S. The use of dopamine-β-hydroxylase as a marker for the central noradrenergic nervous system in rat brain. *Proc. Natl. Acad. Sci. U. S. A.*, **69** (1972), 2722–2726.
20. Descarries, L. and Saucier, G. Disappearance of the locus coeruleus in the rat after intraventricular 6-hydroxydopamine. *Brain Res.*, **37** (1972), 310–316.
21. Grzanna, R. and Molliver, M. E. The locus coeruleus in the rat: an immunohistochemical delineation. *Neuroscience*, **5** (1980), 21–40.
22. Swanson, L. W. and Hartman, E. K. The central adrenergic system. An immunofluorescence study of the location of cell bodies and their efferent connections in the rat utilizing dopamine-β-hydroxylase as a marker. *J. Comp. Neurol.*, **163** (1975), 467–506.
23. Olson, L. and Fuxe, K. Further mapping out of central noradrenaline neuron systems: projections of the subcoeruleus area. *Brain Res.*, **43** (1972), 289–295.
24. Swanson, L. W. The locus coeruleus: a cytoarchitectonic, golgi and immunohistochemical study in the albino rat. *Brain Res.*, **110** (1976), 39–56.
25. Jones, B. E. and Moore, R. Y. Catecholamine-containing neurons of the nucleus locus coeruleus in the cat. *J. Comp. Neurol.*, **157** (1974), 43–52.
26. Leger, L. and Hernandez-Nicaise, M.-L. The cat locus coeruleus: light and electron microscopic study of the neuronal somata. *Anat. Embryol.*, **159** (1980), 181–198.
27. Meesen, H. and Olszewski, J. *A Cytoarchitectonic Atlas of the Rhombencephalon of the Rabbit.* (Basel: Karger, 1949).
28. Wiklund, L., Leger, L. and Persson, M. Monoamine cell distribution in the cat brain stem. A fluorescence histochemical study with quantification of indolaminergic and locus coeruleus cell groups. *J. Comp. Neurol.*, **203** (1981), 613–647.
29. Jacobowitz, D. M. and MacLean, P. D. A brainstem atlas of catecholaminergic neurons and serotonergic perikarya in a pygmy primate (*Cebuella pygmaea*). *J. Comp. Neurol.*, **177** (1978), 397–416.
30. Schofield, S. P. M. and Everitt, B. J. The organization of catecholamine-containing neurons in the brain of the rhesus monkey (*Macaca mulatta*). *J. Anat.*, **132** (1981), 391–418.
31. Felten, D. L., Laties, A. M. and Carpenter, M. B. Monoamine-containing cell bodies in the squirrel monkey brain. *Am. J. Anat.*, **139** (1974), 153–166.
32. Garver, D. L. and Sladek, J. R. Monoamine distribution in primate brain. I. Catecholamine-containing perikarya in the brain stem of macaca speciosa. *J. Comp. Neurol.*, **159** (1975), 289–304.

33. Hwang, B.-H., Chiba, T., Black, A. C. and Williams, T. H. Quantification of catecholamine-containing cell groups in the brainstem of the rhesus monkey. *Soc. Neurosci. Abstr.*, **1** (1975), 563.
34. Braak, H. On the pars cerebellaris loci coerulei within the cerebellum of man. *Cell Tissue Res.*, **160** (1975), 279–282.
35. Saper, C. B. and Petito, C. K. Correspondence of melanin-pigmented neurons in human brain with A1-A14 catecholamine cell groups. *Brain*, **105** (1982), 87–101.
36. Kemper, C. M., O'Connor, D. T. and Westlund, K. N. Immunocytochemical localization of dopamine-β-hydroxylase in neurons of the human brain stem. *Neuroscience*, **23** (1987), 981–989.
37. Baker, K. G., Tork, I., Hornung, J. P. and Halasz, P. The human locus coeruleus complex: an immunohistochemical and three dimensional reconstruction study. *Exp. Brain Res.*, **77** (1989), 257–270.
38. German, D. C., Walker, B. S., Manaye, K. *et al.* The human locus coeruleus: computer reconstruction of cellular distribution. *J. Neurosci.*, **8** (1988), 1776–1788.
39. Pearson, J., Goldstein, M., Markey, K. and Brandeis, L. Human brainstem catecholamine neuronal anatomy as indicated by immunocytochemistry with antibodies to tyrosine hydroxylase. *Neuroscience*, **8** (1983), 3–32.
40. Iversen, L. L., Rossor, M. N., Reynolds, G. P. *et al.* Loss of pigmented dopamine-β-hydroxylase positive cells from locus coeruleus in senile dementia of Alzheimer's type. *Neurosci. Lett.*, **39** (1983), 95–100.
41. Chan-Palay, V. and Asan, E. Quantitation of catecholamine neurons in the locus coeruleus in human brains of normal young and older adults and in depression. *J. Comp. Neurol.*, **287** (1989), 357–372.
42. Lauder, J. M. and Bloom, F. E. Ontogeny of monoamine neurons in the locus coeruleus, raphe nuclei, and substantia nigra of the rat, I. Cell differentiation. *J. Comp. Neurol.*, **155** (1974), 469–482.
43. Maeda, T. and Dresse, A. Recherches sur le developpement du locus coruleus. Etude des catecholamines au microscope de fluorescence. *Acta Neurol. Belg.*, **69** (1969), 5–10.
44. Olson, L., Nystrom, B. and Seiger, A. Monoamine fluorescence histochemistry of human postmortem brain. *Brain Res.*, **63** (1973), 231–247.
45. Specht, L. A., Pickel, V. M., Joh, T. H. and Reis, D. J. Light-microscopic immunocytochemical localization of tyrosine hydroxylase in prenatal rat brain. II. Late ontogeny. *J. Comp. Neurol.*, **199** (1981), 255–276.
46. Lauder, J. M. and Bloom, F. E. Ontogeny of monoamine neurons in the locus coeruleus, raphe nuclei and substantia nigra of the rat, II. Synaptogenesis. *J. Comp. Neurol.*, **163** (1975), 251–264.
47. Schlumpf, M., Shoemaker, W. J. and Bloom, F. E. Innervation of embryonic rat cerebral cortex by catecholamine-containing fibers. *J. Comp. Neurol.*, **192** (1980), 361–376.
48. Lidov, H. G. W., Molliver, M. E. and Zecevic, N. R. Characterization of the monoaminergic innervation of immature rat neocortex: a histofluorescence analysis. *J. Comp. Neurol.*, **181** (1978), 663–680.
49. Levitt, P. and Moore, R. Y. Noradrenaline neuron innervation of the neocortex in the rat. *Brain Res.*, **139** (1978), 219–231.

50. Morin, X., Cremer, H., Hirsch, M.-R. *et al.* Defects in sensory and autonomic ganglia and absence of locus coeruleus in mice deficient for the homeobox gene *Phox2a*. *Neuron*, **18** (1997), 411–423.
51. Lo, L. C., Tiveron, M. C. and Anderson, D. J. MASH1 activates expression of the paired homeodomain transcription factor Phox2a, and couples pan-neuronal and subtype-specific components of autonomic neuronal identity. *Development*, **125** (1998), 609–620.
52. Hirsch, M.-R., Tiveron, M.-C., Guillemot, F., Brunet, J.-F. and Goridis, C. Control of noradrenergic differentiation and Phox2a expression by MASH1 in the central and peripheral nervous system. *Development*, **125** (1998), 599–608.
53. Pattyn, A., Goridis, C. and Brunet, J.-F. Specification of the central noradrenergic phenotype by the homeobox gene *Phox2b*. *Mol. Cell. Neurosci.*, **15** (2000), 235–243.
54. Vogel-Hopker, A. and Rohrer, H. The specification of noradrenergic locus coeruleus (LC) neurons depends on bone morphogenetic proteins (BMPs). *Development*, **129** (2002), 983–991.
55. Choi, B. H., Antanitus, D. S. and Lapham, L. W. Fluorescence histochemical and ultrastructural studies of locus coeruleus of human fetal brain. *J. Neuropathol. Exp. Neurol.*, **34** (1975), 507–516.
56. Pickel, V. M., Specht, L. A., Sumal, K. K. *et al.* Immunocytochemical localization of tyrosine hydroxylase in the human fetal nervous system. *J. Comp. Neurol.*, **194** (1980), 465–474.
57. Aghajanian, G. K., Cedarbaum, J. M. and Wang, R. Y. Evidence for norepinephrine-mediated collateral inhibition of locus coeruleus neurons. *Brain Res.*, **136** (1977), 570–577.
58. Hökfelt, T. On the ultrastructural localization of noradrenaline in the central nervous system of the rat. *Z. Zellforsch.*, **79** (1967), 110–117.
59. Shimizu, N., Katoh, Y., Hida, T. and Satoh, K. The fine structural organization of the locus coeruleus in the rat with reference to noradrenaline contents. *Exp. Brain Res.*, **37** (1979), 139–148.
60. Groves, P. M. and Wilson, C. J. Fine structure of rat locus coeruleus. *J. Comp. Neurol.*, **193** (1980), 841–852.
61. Groves, P. M. and Wilson, C. J. Monoaminergic presynaptic axons and dendrites in rat locus coeruleus seen in reconstructions of serial sections. *J. Comp. Neurol.*, **198** (1980), 853–862.
62. Alvarez-Maubecin, V., Garcia-Hernandez, F., Williams, J. T. and Van Bockstaele, E. J. Functional coupling between neurons and glia. *J. Neurosci.*, **20** (2000), 4091–4098.
63. Shipley, M. T., Fu, L. F., Ennis, M., Liu, W.-L. and Aston-Jones, G. Dendrites of locus coeruleus neurons extend preferentially into two pericoerulear zones. *J. Comp. Neurol.*, **365** (1996), 56–68.
64. Marcyniuk, B., Mann, D. M. A. and Yates, P. O. The topography of cell loss from locus coeruleus in Alzheimer's disease. *J. Neurol. Sci.*, **76** (1986), 335–345.
65. Young, W. S. and Kuhar, M. J. Noradrenergic $\alpha1$ and $\alpha2$ receptors: light microscopic autoradiographic localization. *Proc. Natl. Acad. Sci. U. S. A.*, **77** (1980), 1696–1700.
66. Jones, C. R. and Palacious, J. M. Autoradiography of adrenoceptors in rat and human brain: α-adrenoceptor and idazoxan binding sites. *Prog. Brain Res.*, **88** (1991), 271–291.
67. Ordway, G. A., Widdowson, P. S., Smith, K. S. and Halaris, A. Agonist binding to $\alpha2$-adrenoceptors is elevated in the locus coeruleus from victims of suicide. *J. Neurochem.*, **63** (1994), 617–624.

68. Harro, J. and Oreland, L. Depression as a spreading adjustment disorder of monoaminergic neurons: a case for primary implication of the locus coeruleus. *Brain Res. Rev.*, **38** (2001), 79–128.
69. Sutin, E. L. and Jacobowitz, D. M. Neurochemicals in the dorsal pontine tegmentum. *Prog. Brain Res.*, **88** (1991), 3–14.
70. Le Novere, N., Zoli, M. and Changeux, J.-P. Neuronal nicotinic receptor $\alpha 6$ subunit mRNA is selectively concentrated in catecholaminergic nuclei of the rat brain. *Eur. J. Neurosci.*, **8** (1996), 2428–2439.
71. Bitner, R. S. and Nikkel, A. L. Alpha-7 nicotinic receptor expression by two distinct cell types in the dorsal raphe nucleus and locus coeruleus of rat. *Brain Res.*, **938** (2002), 45–54.
72. Lena, C., de Kerchove d'Exaerde, A., Cordero-Erausquin, M. *et al.* Diversity and distribution of nicotinic acetylcholine receptors in the locus coeruleus neurons. *Proc. Natl. Acad. Sci. U. S. A.*, **96** (1999), 12126–12131.
73. Jones, B. E. Noradrenergic locus coeruleus neurons: their distant connections and their relationship to neighboring (including cholinergic and GABAergic) neurons of the central gray and reticular formation. *Prog. Brain Res.*, **88** (1991), 15–30.
74. Van Bockstaele, E. J. and Chan, J. Electron microscopic evidence for coexistence of leucine^5-enkephalin and γ-aminobutyric acid in a subpopulation of axon terminals in the rat locus coeruleus region. *Brain Res.*, **746** (1997), 171–182.
75. Shefner, S. A. and Osmanovic, S. S. $GABA_A$ and $GABA_B$ receptors and the ionic mechanisms mediating their effects on locus coeruleus neurons. *Prog. Brain Res.*, **88** (1991), 187–195.
76. Gervasoni, D., Darracq, L., Fort, P. *et al.* Electrophysiological evidence that noradrenergic neurons of the rat locus coeruleus are tonically inhibited by GABA during sleep. *Eur. J. Neurosci.*, **10** (1998), 964–970.
77. Nitz, D. and Siegel, J. M. GABA release in the locus coeruleus as a function of sleep/wake state. *Neuroscience*, **78** (1997), 795–801.
78. Luque, J. M., Malherbe, P. and Richards, J. G. Localization of $GABA_A$ receptor subunit mRNAs in the rat locus coeruleus. *Mol. Brain Res.*, **24** (1994), 219–226.
79. Sinkkonen, S. T., Hanna, M. C., Kirkness, E. F. and Korpi, E. R. $GABA_A$ receptor ϵ and θ subunits display unusual structural variation between species and are enriched in the rat locus coeruleus. *J. Neurosci.*, **20** (2000), 3588–3595.
80. Bonnert, T. P., McKernan, R. M., Farrar, S. *et al.* θ, a novel γ-aminobutyric acid type A receptor subunit. *Proc. Natl. Acad. Sci. U. S. A.*, **96** (1999), 9891–9896.
81. Tatemoto, K., Rokaeus, A., Jornvall, J., McDonald, T. J. and Mutt, V. Galanin – a novel biologically active peptide from porcine intestine. *FEBS Lett.*, **164** (1983), 124–128.
82. Melander, T., Hökfelt, T. and Rokaeus, A. Distribution of galanin-like immunoreactivity in the rat central nervous system. *J. Comp. Neurol.*, **248** (1986), 475–517.
83. Simpson, K. L., Waterhouse, B. D. and Lin, R. C. S. Origin, distribution, and morphology of galaninergic fibers in the rodent trigeminal system. *J. Comp. Neurol.*, **411** (1999), 524–534.
84. Perez, S. E., Wynick, D., Steiner, R. A. and Mufson, E. J. Distribution of galaninergic immunoreactivity in the brain of the mouse. *J. Comp. Neurol.*, **434** (2001), 158–185.
85. Kordower, J. H., Le, H. K. and Mufson, E. J. Galanin immunoreactivity in the primate central nervous system. *J. Comp. Neurol.*, **319** (1992), 479–500.

86. Chan-Palay, V., Jentsch, B., Lang, W., Hochli, M. and Asan, E. Distribution of C-terminal flanking peptide of NPY and galanin and coexistence with catecholamine in the locus coeruleus of normal human, Alzheimer's dementia and Parkinson's disease brains. *Dementia*, **1** (1990), 18–31.
87. Miller, M. A., Kolb, P. E., Leverenz, J. B., Peskind, E. R. and Raskind, M. A. Preservation of noradrenergic neurons in the locus coeruleus that coexpress galanin mRNA in Alzheimer's disease. *J. Neurochem.*, **73** (1999), 2028–2036.
88. Pieribone, V. A. and Aston-Jones, G. Adrenergic innervation of the rat nucleus locus coeruleus arises predominantly from the C1 adrenergic cell group in the rostral medulla. *Neuroscience*, **41** (1991), 525–542.
89. Parker, E. M., Izzarelli, D. G., Nowak, H. P. *et al.* Cloning and characterization of the rat GALR1 galanin receptor Rin14B insulinoma cells. *Mol. Brain Res.*, **34** (1995), 179–189.
90. O'Donnell, D., Ahmad, S., Wahlestedt, C. and Walker, P. Expression of the novel galanin receptor subtype GalR2 in the adult rat CNS: distinct distribution from GalR1. *J. Comp. Neurol.*, **409** (1999), 469–481.
91. Mennicken, F., Hoffert, C., Pelletier, M., Ahmad, S. and O'Donnell, D. Restricted distribution of galanin receptor 3 (GalR3) mRNA in the adult rat central nervous system. *J. Chem. Neuroanat.*, **24** (2002), 257–268.
92. Everitt, B. J., Hökfelt, T., Terenius, L. *et al.* Differential co-existence of neuropeptide Y (NPY)-like immunoreactivity with catecholamines in the central nervous system of the rat. *Neuroscience*, **11** (1984), 443–462.
93. Chronwall, B. M., DiMaggio, D. A., Massari, V. J. *et al.* The anatomy of neuropeptide-Y containing neurons in rat brain. *Neuroscience*, **15** (1985), 1159–1181.
94. Moore, R. Y. and Gustafson, E. L. The distribution of dopamine-β-hydroxylase, neuropeptide Y, and galanin in locus coeruleus neurons. *J. Chem. Neuroanat.*, **2** (1989), 95–106.
95. Pammer, C., Gorcs, T. and Palkovits, M. Peptidergic innervation of the locus coeruleus cells in the human brain. *Brain Res.*, **515** (1990), 247–255.
96. Pickel, V. M., Joh, T. H., Reis, D. J., Leeman, S. E. and Miller, R. J. Electron microscopic localization of substance P and enkephalin in axon terminals related to dendrites of catecholaminergic neurons. *Brain Res.*, **160** (1979), 387–400.
97. Charnay, Y., Leger, L., Dray, F. *et al.* Evidence for the presence of enkephalin in catecholaminergic neurons of cat locus coeruleus. *Neurosci. Lett.*, (1982), **30** 147–151.
98. Leger, L., Charnay, Y., Chayvialle, J. A. *et al.* Localization of substance P- and enkephalin-like immunoreactivity in relation to catecholamine-containing cell bodies in the cat dorsolateral pontine tegmentum: an immunofluorescence study. *Neuroscience*, **8** (1983), 525–546.
99. Drolet, G., Van Bockstaele, E. J. and Aston-Jones, G. Robust enkephalin innervation of the locus coeruleus from the rostral medulla. *J. Neurosci.*, **12** (1992), 3162–3147.
100. Van Bockstaele, E. J., Saunders, A., Commons, K. G., Liu, X.-B. and Peoples, J. Evidence for coexistence of enkephalin and glutamate in axon terminals and cellular sites for functional interactions of their receptors in the rat locus coeruleus. *J. Comp. Neurol.*, **417** (2000), 103–114.
101. Bird, S. J. and Kuhar, M. J. Iontophoretic application of opiates to the locus coeruleus. *Brain Res.*, **122** (1977), 523–533.

102. Williams, J. T., Egan, T. M. and North, R. A. Enkephalin opens potassium channels on mammalian central neurons. *Nature*, **299** (1982), 74–77.
103. Tempel, A. and Zukin, R. S. Neuroanatomical patterns of the μ, δ, and κ opioid receptors of rat brain as determined by quantitative in vitro autoradiography. *Proc. Natl. Acad. Sci. U. S. A.*, **84** (1987), 4308–4312.
104. Van Bockstaele, E. J., Colago, E. E. O., Cheng, P. *et al.* Ultrastructural evidence for prominent distribution of the μ-opioid receptor at extrasynaptic sites on noradrenergic dendrites in the rat nucleus locus coeruleus. *J. Neurosci.*, **16** (1996), 5037–5048.
105. Finley, J. C. W., Maderdrut, J. L., Roger, L. J. and Petrusz, P. The immunocytochemical localization of somatostatin-containing neurons in the rat central nervous system. *Neuroscience*, **6** (1981), 2173–2192.
106. Johansson, O., Hökfelt, T. and Elde, R. P. Immunohistochemical distribution of somatostatin-like immunoreactivity in the central nervous system of the adult rat. *Neuroscience*, **13** (1984), 265–339.
107. Ljungdahl, A., Hökfelt, T., Nilsson, G. and Goldstein, M. Distribution of substance P-like immunoreactivity in the central nervous system of the rat – II. Light microscopic localization in relation to catecholamine-containing neurons. *Neuroscience*, **3** (1978), 945–976.
108. Jennes, L., Stumpf, W. E. and Kalivas, P. W. Neurotensin: topographical distribution in rat brain by immunohistochemistry. *J. Comp. Neurol.*, **210** (1982), 211–224.
109. Eiden, L. E., Nilaver, G. and Palkovits, M. Distribution of vasoactive intestinal polypeptide (VIP) in the rat brain stem nuclei. *Brain Res.*, **231** (1982), 472–477.
110. Martin, J.-L., Dietl, M. M., Hof, P. R., Palacios, J. M. and Magistretti, P. J. Autoradiographic mapping of [mono[^{125}I]iodo-tyr^{10}, MetO^{17}]vasoactive intestinal peptide binding sites in the rat brain. *Neuroscience*, **23** (1987), 539–565.
111. Caffe, A. R. and van Leeuwen, F. W. Vasopressin-immunoreactive cells in the dorsomedial hypothalamic region, medial amygdaloid nucleus and locus coeruleus of the rat. *Cell Tissue Res.*, **233** (1983), 23–33.
112. Caffe, A. R., van Leeuwen, F. W., Buijs, R. M., deVries, G. J. and Geffard, M. Coexistence of vasopressin, neurophysin and noradrenaline immunoreactivity in medium-sized cells of the locus coeruleus and subcoeruleus in the rat. *Brain Res.*, **338** (1985), 160–164.
113. deVries, G. J., Buijs, R. M., van Leeuwen, F. W., Caffe, A. R. and Swaab, D. F. The vasopressinergic innervation of the brain in normal and castrated rats. *J. Comp. Neurol.*, **233** (1985), 236–254.
114. Chen, L.-W., Wei, L.-C., Liu, H.-L. and Rao, Z.-R. Noradrenergic neurons expressing substance P receptor (NK1) in the locus coeruleus complex: a double immunofluorescence study in the rat. *Brain Res.*, **873** (2000), 155–159.
115. Young, W. S., Uhl, G. R. and Kuhar, M. J. Iontophoresis of neurotensin in the area of the locus coeruleus. *Brain Res.*, **150** (1978), 431–435.
116. Cummings, S., Elde, R., Ells, J. and Lindall, A. Corticotropin-releasing factor immunoreactivity is widely distributed within the central nervous system of the rat: an immunohistochemical study. *J. Neurosci.*, **3** (1983), 1355–1368.
117. Valentino, R. J. and Van Bockstaele, E. J. Opposing regulation of the locus coeruleus by corticotropin-releasing factor and opioids. *Psychopharmacology*, **158** (2001), 331–342.

118. Chalmers, D. T., Lovenberg, T. W. and De Souza, E. B. Localization of novel corticotrophin-releasing factor receptor (CRF_2) mRNA expression to specific subcortical nuclei in rat brain: comparison with CRF_1 receptor mRNA expression. *J. Neurosci.*, **15** (1995), 6340–6350.
119. Fox, K., Wolff, I., Curtis, A. *et al.* Multiple lines of evidence for the existence of corticotrophin-releasing factor (CRF) receptors on locus coeruleus (LC) neurons. *Soc. Neurosci. Abstr.*, **28** (2002), Program No. 637.9.
120. Cedarbaum, J. M. and Aghajanian, G. K. Afferent projections to the rat locus coeruleus as determined by a retrograde tracing technique. *J. Comp. Neurol.*, **178** (1978), 1–16.
121. Clavier, R. M. Afferent projections to the self-stimulation regions of the dorsal pons, including the locus coeruleus, in the rat as demonstrated by the horseradish peroxidase technique. *Brain Res. Bull.*, **4** (1979), 497–504.
122. Aston-Jones, G., Ennis, M., Pieribone, V. A., Nickell, W. T. and Shipley, M. T. The brain nucleus locus coeruleus: restricted afferent control of a broad efferent network. *Science*, **234** (1986), 734–737.
123. Luppi, P.-H., Aston-Jones, G., Akaoka, H., Chouvet, G. and Jouvet, M. Afferent projections to the rat locus coeruleus demonstrated by retrograde and anterograde tracing with cholera-toxin B subunit and phaseolus vulgaris leucoagglutinin. *Neuroscience*, **65** (1995), 119–160.
124. Sesack, S. R., Deutch, A. Y., Roth, R. H. and Bunney, B. S. Topographical organization of the efferent projections of the medial prefrontal cortex in the rat: an anterograde tract-tracing study with phaseolus vulgaris leucoagglutinin. *J. Comp. Neurol.*, **290** (1989), 213–242.
125. Craig, A. D. Spinal and trigeminal lamina I input to the locus coeruleus anterogradely labeled with phaseolus vulgaris leucoagglutinin (PHA-L) in the cat and the monkey. *Brain Res.*, **584** (1992), 325–328.
126. Van Bockstaele, E. J., Chan, J. and Pickel, V. M. Input from central nucleus of the amygdala efferents to pericoerulear dendrites, some of which contain tyrosine hydroxylase immunoreactivity. *J. Neurosci. Res.*, **45** (1996), 289–302.
127. Van Bockstaele, E. J., Peoples, J. and Telegan, P. Efferent projections of the nucleus of the solitary tract to peri-locus coeruleus dendrites in rat brain: evidence for a monosynaptic pathway. *J. Comp. Neurol.*, **412** (1999), 410–428.
128. Van Bockstaele, E. J., Colago, E. E. and Aicher, S. Light and electron microscopic evidence for topographic and monosynaptic projections from neurons in the ventral medulla to noradrenergic dendrites in the rat locus coeruleus. *Brain Res.*, **784** (1998), 123–138.
129. Sakanaka, M., Shibasaki, T. and Lederis, K. Corticotropin releasing factor-like immunoreactivity in the rat brain as revealed by a modified cobalt-glucose oxidase-diaminobenzidine method. *J. Comp. Neurol.*, **260** (1987), 256–298.
130. Ennis, M. and Aston-Jones, G. GABA-mediated inhibition of locus coeruleus from the dorsomedial rostral medulla. *J. Neurosci.*, **9** (1989), 2973–2981.
131. Ennis, M. and Aston-Jones, G. Potent inhibitory input to locus coeruleus from the nucleus prepositus hypoglossi. *Brain Res. Bull.*, **22** (1989), 793–803.
132. Aston-Jones, G., Shipley, M. T., Chouvet, G. *et al.* Afferent regulation of locus coeruleus neurons: anatomy, physiology, and pharmacology. *Prog. Brain Res.*, **88** (1991), 47–75.
133. Spruston, N. Distant synapses raise their voices. *Nat. Neurosci.*, **3** (2000), 849–851.

134. Magee, J. C. and Cook, E. P. Somatic EPSP amplitude is independent of synapse location in hippocampal pyramidal neurons. *Nat. Neurosci.*, **3** (2000), 895–903.
135. Larkum, M. E. and Zhu, J. J. Signaling of Layer 1 and whisker-evoked Ca^{2+} and Na^{+} action potentials in distal and terminal dendrites of rat neocortical pyramidal neurons in vitro and in vivo. *J. Neurosci.*, **22** (2002), 6991–7005.
136. Lindvall, O. and Bjorklund, A. The organization of the ascending catecholamine neuron systems in the rat brain as revealed by the glyoxylic acid fluorescence method. *Acta Physiol. Scand. Suppl.*, **412** (1974), 1–48.
137. Moore, R. Y. and Card, J. P. Noradrenaline-containing neuron systems. In ed. *Handbook of Chemical Neuroanatomy*, vol. 2, A. Bjorklund and T. Hökfelt. (Amsterdam: Elsevier, 1984), pp. 123–156.
138. Pickel, V. M., Segal, M. and Bloom, F. E. A radiographic study of the efferent pathways of the nucleus locus coeruleus. *J. Comp. Neurol.*, **155** (1974), 15–42.
139. Jones, B. E. and Moore, R. Y. Ascending projections of the locus coeruleus in the rat. II. Autoradiographic study. *Brain Res.*, **127** (1977), 23–53.
140. Fritschy, J.-M. and Grzanna, R. Demonstration of two separate descending noradrenergic pathways to the rat spinal cord: evidence for an intragriseal trajectory of locus coeruleus axons in the superficial layers of the dorsal horn. *J. Comp. Neurol.*, **291** (1990), 553–582.
141. Mason, S. T. and Fibiger, H. S. Regional topography within noradrenergic locus coeruleus as revealed by retrograde transport of horseradish peroxidase. *J. Comp. Neurol.*, **187** (1979), 703–724.
142. Loughlin, S. E., Foote, S. L. and Bloom, F. E. Efferent projections of nucleus locus coeruleus: topographic organization of cells of origin demonstrated by three-dimensional reconstruction. *Neuroscience*, **18** (1986), 291–306.
143. Loughlin, S. E., Foote, S. L. and Grzanna, R. Efferent projections of nucleus locus coeruleus: morphologic subpopulations have different efferent targets. *Neuroscience*, **18**: 2 (1986), 307–319.
144. Steindler, D. A. Locus coeruleus neurons have axons that branch to the forebrain and cerebellum. *Brain Res.*, **223** (1981), 367–373.
145. Nagai, T., Satoh, K., Imamoto, K. and Maeda, T. Divergent projections of catecholamine neurons of the locus coeruleus as revealed by fluorescent retrograde double labeling technique. *Neurosci Lett.*, **23** (1981), 117–123.
146. Simpson, K. L., Altman, D. A., Wang, L. *et al.* Lateralization and functional organization of the locus coeruleus projection to the trigeminal somatosensory pathway in the rat. *J. Comp. Neurol.*, **385** (1997), 135–147.
147. Ennis, M. and Aston-Jones, G. A potent excitatory input to the nucleus locus coeruleus from the ventrolateral medulla. *Neurosci. Lett.*, **71** (1986), 299–305.
148. Ennis, M. and Aston-Jones, G. Activation of locus coeruleus from nucleus paragigantocellularis: a new excitatory amino acid pathway in brain. *J. Neurosci.*, **8** (1988), 3644–3657.
149. Jodo, E. and Aston-Jones, G. Activation of locus coeruleus by prefrontal cortex is mediated by excitatory amino acid inputs. *Brain Res.*, **768** (1997), 327–332.

150. Van Bockstaele, E. J. and Colago, E. E. Selective distribution of the NMDA-R1 glutamate receptor in astrocytes and presynaptic axon terminals in the nucleus locus coeruleus of the rat brain: an immunoelectron microscopic study. *J. Comp. Neurol.*, **369** (1996), 483–496.
151. Van Bockstaele, E. J. and Colago, E. E. Ultrastructural localization of the kainate selective glutamate receptor in noradrenergic perikarya and dendrites of the nucleus locus coeruleus in the rat brain. *Brain Res.*, **732** (1996), 223–231.
152. Pieribone, V. A., Shipley, M. T., Ennis, M. and Aston-Jones, G. Anatomic evidence for GABA-ergic afferents to the rat locus coeruleus in the dorsal medial medulla: an immunocytochemical and retrograde transport study. *Soc. Neurosci. Abstr.*, **16** (1990), 300.
153. Peyron, C., Luppi, P.-H., Rampon, C. and Jouvet, M. Localization of the GABA-ergic neurons projecting to the rat dorsal raphe and locus coeruleus. *Soc. Neurosci. Abstr.*, **21** (1995), Program No. 153.12.
154. Ijima, K. and Ohtomo, K. Immunocytochemical study using a GABA antiserum for the demonstration of inhibitory neurons in the rat locus coeruleus. *Am. J. Anat.*, **188** (1988), 43–52.
155. Luppi, P.-H., Charlety, P. J., Fort, P. *et al.* Anatomical and electrophysiological evidence for a glycinergic inhibitory innervation of the rat locus coeruleus. *Neurosci. Lett.*, **128** (1991), 33–36.
156. Pickel, V. M., Joh, T. H. and Reis, D. J. A serotonergic innervation of noradrenergic neurons in nucleus locus coeruleus: demonstration by immunocytochemical localization of the transmitter specific enzymes tyrosine and tryptophan hydroxylase. *Brain Res.*, **131** (1977), 197–214.
157. Maeda, T., Kojima, Y., Arai, R. *et al.* Monoaminergic interaction in the central nervous system: a morphological analysis in the locus coeruleus of the rat. *Comp. Biochem. Physiol.*, **98** (1991), 193–202.
158. Pieribone, V. A., Van Bockstaele, E. J., Shipley, M. T. and Aston-Jones, G. Serotonergic innervation of rat locus coeruleus derives from non-raphe brain areas. *Soc. Neurosci. Abstr.*, **15** (1989), Program No. 168.17.
159. Sakai, K. Physiological properties and afferent connections of the locus coeruleus and adjacent tegmental neurons involved in the generation of paradoxical sleep in the cat. *Prog. Brain Res.*, **88** (1991), 31–45.
160. Gautvik, K. M., de Lecea, L., Gautvik, V. T. *et al.* Overview of the most prevalent hypothalamus-specific mRNAs, as identified by directional tag PCR subtraction. *Proc. Natl. Acad. Sci. U. S. A.*, **93** (1996), 8733–8738.
161. de Lecea, L., Kilduff, T. S., Peyron, C. *et al.* The hypocretins: hypothalamus-specific peptides with neuroexcitatory activity. *Proc. Natl. Acad. Sci. U. S. A.*, **95** (1998), 322–327.
162. Sakurai, T., Amemiya, A., Ishii, M. *et al.* Orexins and orexin receptors: a family of hypothalamic neuropeptides and G-protein coupled receptors that regulate feeding behavior. *Cell*, **92** (1998), 573–585.
163. Peyron, C., Tighe, D. K., van den Pol, A. N. *et al.* Neurons containing hypocretin (orexin) project to multiple neuronal systems. *J. Neurosci.*, **18** (1998), 9996–10 015.

164. Horvath, T. L., Peyron, C., Diano, S. *et al.* Hypocretin (orexin) activation and synaptic innervation of the locus coeruleus noradrenergic system. *J. Comp. Neurol.*, **415** (1999), 145–159.
165. Bourgin, P., Huitron-Resendiz, S., Spier, A. D. *et al.* Hypocretin-1 modulates rapid eye movement sleep through activation of locus coeruleus neurons. *J. Neurosci.*, **20** (2000), 7760–7765.
166. Hagan, J. J., Leslie, R. A., Patel, S. *et al.* Orexin A activates locus coeruleus cell firing and increases arousal in the cat. *Proc. Natl. Acad. Sci. U. S. A.*, **96** (1999), 10 911–10 916.
167. Chemelli, R. M., Willie, J., Sinton, C. M. *et al.* Narcolepsy in orexin knockout mice: molecular genetics of sleep regulation. *Cell*, **98** (1999), 437–451.
168. Nishino, S., Ripley, B., Overeem, S., Lammers, G. J. and Mignot, E. Hypocretin (orexin) deficiency in human narcolepsy. *Lancet*, **355** (2000), 39–40.
169. Curtis, A. L., Florin-Lechner, S. M., Pavcovich, L. A. and Valentino, R. J. Activation of the locus coeruleus noradrenergic system by intracoerulear microinfusion of corticotrophin-releasing factor: effects on discharge rate, cortical norepinephrine levels, and corticoencephalographic activity. *J. Pharmacol. Exp. Ther.*, **281** (1997), 163–172.
170. Valentino, R. J. and Foote, S. L. Corticotropin-releasing factor disrupts sensory responses of brain noradrenergic neurons. *Neuroendocrinology*, **45** (1987), 28–36.
171. Valentino, R. J. and Foote, S. L. Corticotropin-releasing factor increases tonic but not sensory-evoked activity of noradrenergic locus coeruleus neurons in unanesthetized rats. *J. Neurosci.*, **8** (1988), 1016–1025.
172. Valentino, R. J., Page, M., Van Bockstaele, E. J. and Aston-Jones, G. Corticotropin-releasing factor innervation of the locus coeruleus region: distribution of fibers and sources of input. *Neuroscience*, **48** (1992), 689–705.
173. Van Bockstaele, E. J., Bajic, D., Proudfit, H. and Valentino, R. J. Topographic architecture of stress-related pathways targeting the noradrenergic locus coeruleus. *Physiol. Behav.*, **73** (2001), 273–283.
174. Van Bockstaele, E. J., Colago, E. E. and Valentino, R. J. Corticotropin-releasing factor-containing axon terminals synapse onto catecholamine dendrites and may presynaptically modulate other afferents in the rostral pole of the nucleus locus coeruleus in the rat brain. *J. Comp. Neurol.*, **364** (1996), 523–534.
175. Van Bockstaele, E. J., Colago, E. E. O. and Valentino, R. J. Amygdaloid corticotropin-releasing factor targets locus coeruleus dendrites: substrate for the coordination of emotional and cognitive limbs of the stress response. *J. Neuroendocrinol.*, **10** (1998), 743–757.
176. Van Bockstaele, E. J., Peoples, J. and Valentino, R. J. Anatomic basis for differential regulation of the rostrolateral peri-locus coeruleus region by limbic afferents. *Biol. Psychiatry*, **46** (1999), 1352–1363.
177. Valentino, R. J., Chen, S., Zhu, Y. and Aston-Jones, G. Evidence of divergent projections to the brain noradrenergic system and the spinal parasympathetic system from Barrington's nucleus. *Brain Res.*, **732** (1996), 1–15.
178. Lechner, S. M. and Valentino, R. J. Glucocorticoid receptor-immunoreactivity in corticotrophin-releasing factor afferents to the locus coeruleus. *Brain Res.*, **816** (1999), 17–28.

179. Van Bockstaele, E. J., Branchereau, P. and Pickel, V. M. Morphologically heterogeneous met-enkephalin terminals form synapses with tyrosine hydroxylase-containing dendrites in the rat nucleus locus coeruleus. *J. Comp. Neurol.*, **363** (1995), 423–38.
180. Walker, D. L. and Davis, M. Double dissociation between the involvement of the bed nucleus of the stria terminalis and the central nucleus of the amygdala in startle increases produced by conditioned versus unconditioned fear. *J. Neurosci.*, **17** (1997), 9375–9383.
181. Page, M. E. and Abercrombie, E. D. Discrete local application of corticotropin-releasing factor increases locus coeruleus discharge and extracellular norepinephrine in rat hippocampus. *Synapse*, **33** (1999), 304–313.
182. Holets, V. R., Hökfelt, T., Rokaeus, A., Terenius, L. and Goldstein, M. Locus coeruleus neurons in the rat containing neuropeptide Y, tyrosine hydroxylase or galanin and their efferent projections to the spinal cord, cerebral cortex and hypothalamus. *Neuroscience*, **24** (1988), 893–906.
183. Lechner, J., Leah, J. D. and Zimmerman, M. Brainstem peptidergic neurons projecting to the medial and lateral thalamus and zona incerta in the rat. *Brain Res.*, **603** (1993), 47–56.
184. Nishibori, M., Oishi, R., Itoh, Y. and Saeki, K. Galanin inhibits noradrenaline-induced accumulation of cyclic AMP in the rat cerebral cortex. *J. Neurochem.*, **51** (1988), 1953–1955.
185. Tsuda, K., Yokoo, H. and Goldstein, M. Neuropeptide Y and galanin in norepinephrine release in hypothalamic slices. *Hypertension*, **14** (1989), 81–86.
186. Lundberg, J. M. and Hökfelt, T. Coexistence of peptides and classical neurotransmitters. *Trends Neurosci.*, **6** (1983), 325–333.
187. Grenhoff, J., Nisell, M., Ferre, S., Aston-Jones, G. and Svensson, T. H. Noradrenergic modulation of midbrain dopamine cell firing elicited by stimulation of the locus coeruleus in the rat. *J. Neural Transm.*, **93** (1993), 11–25.
188. Florin-Lechner, S. M., Druhan, J. P., Aston-Jones, G. and Valentino, R. J. Enhanced norepinephrine release in prefrontal cortex with burst stimulation of the locus coeruleus. *Brain Res.*, **742** (1996), 89–97.
189. Foote, S. L., Freedman, R. and Oliver, A. P. Effects of putative neurotransmitters on neuronal activity in monkey auditory cortex. *Brain Res.*, **86** (1975), 229–242.
190. Waterhouse, B. D. and Woodward, D. J. Interaction of norepinephrine with cerebrocortical activity evoked by stimulation of somatosensory afferent pathways in the rat. *Exp. Neurol.*, **67** (1980), 11–34.
191. Waterhouse, B. D., Sessler, F. M., Cheng, J.-T. *et al.* New evidence for a gating action of norepinephrine in central neuronal circuits of mammalian brain. *Brain Res. Bull.*, **21** (1988), 425–432.

2

Interactions of norepinephrine with other neurotransmitter systems: anatomical basis and pharmacology

Craig A. Stockmeier, Ph.D.
University of Mississippi Medical Center, Jackson, USA

Gregory A. Ordway, Ph.D.
East Tennessee State University, Johnson City, USA

Introduction

Norepinephrine-containing neurons clustered within the locus coeruleus (LC) provide most of the norepinephrine present within the central nervous system. These cells have tonic pacemaker activity and this activity is regulated by a variety of neurotransmitter inputs. The focus of this review is primarily on classical, non-neuropeptide, neurotransmitter input to the LC and the reciprocal projections of noradrenergic neurons to those classical neurotransmitter systems. Input to the LC from serotonin-, dopamine-, γ-aminobutyric acid (GABA)-, glutamate-, and acetylcholine-containing neurons are described. In addition, input from the neuropeptide, substance P, receives attention because of the interest in this neuropeptide in psychiatric disease. Special attention is given to reciprocal projections from the LC to the monoamine neurotransmitters dopamine and serotonin. See Chapter 1 for a detailed description of the anatomy of the LC.

Noradrenergic circuitry: input to the LC

Early tract-tracing studies suggested that the LC received widespread input from many sites in the brain. A combination of techniques, however, including discrete injections of a more sensitive tract-tracing compound, anterograde labeling studies, and single-pulse stimulation studies forced a reconsideration of brain areas with direct input to the LC.[1] The major afferents to the LC are rostral medullary in

Brain Norepinephrine: Neurobiology and Therapeutics, ed. Gregory A. Ordway, Michael A. Schwartz and Alan Frazer. Published by Cambridge University Press.

origin, with cell bodies located in the nucleus paragigantocellularis (PGi, using excitatory amino acid neurotransmitters) and nucleus prepositus hypoglossi (PrH, using GABA).[1] However, dense projections from many brain regions terminate in the pericoerulear area, an area heavily invested with dendrites from LC neurons, and in the PGi and PrH. Thus, neurons from the dorsal raphe nucleus, ventral tegmental area, the central nucleus of the amygdala, and frontal cortex send significant axonal projections to the pericoerulear area, a region replete with dendrites of LC neurons and other neurons that target the LC directly.

Serotonin

Anatomical and physiological studies demonstrate innervation of the noradrenergic LC by serotonergic neurons. Anatomically, the cell body region and the dendritic field of the LC receive a dense innervation by serotonin fibers. The most dense innervation by serotonergic neurons appears to be outside the immediate cell body region of the LC, in its dendritic field.[2,3] In an extensive study by Luppi and coworkers,[3] the injection of an anterograde tracer into the median or dorsal raphe nucleus labeled few fibers in the core of the LC, but many fibers in the area immediately surrounding the LC. Serotonergic innervation to the LC originates from other sources as well, including the raphe pontis, magnus and others.[4]

Using serotonin transporters as a marker of serotonergic innervation, Klimek and coworkers demonstrated that serotonergic innervation of the human LC is evenly distributed along the rostrocaudal axis of the LC, and as such is not proportional to the number of noradrenergic neurons along the LC axis.[5] Hence, a proportional relationship between serotonergic terminals and LC target cells does not exist. Serotonin fibers do make direct contact with dendrites of LC neurons.[6] However, it appears that these direct contacts are a small percentage of the total number of serotonergic contacts made in the region of the LC and that other cells, including possibly glial cells, are innervated by serotonergic fibers in this region.[6] Such findings indicate a complex functional relationship between serotonergic and noradrenergic neurons.

Pharmacological and gene expression studies demonstrate that LC neurons receive functional serotonergic innervation. Messenger RNAs encoding 5-HT_{1A}, 5-HT_{1C}, and 5-HT_{2C} serotonin receptors have been observed in rat LC neurons.[7] Immunoreactivity for 5-HT_{1A} has been reported on cell soma as well as on processes of primate LC neurons.[8] A high density of [^{3}H]MDL100907 binding to 5HT_{2A} receptors has been described in the human LC.[9] Pharmacologically, the effects of serotonin drugs on LC activity are complex, and depend on whether drugs used to manipulate the serotonin system are administered directly into the LC or are administered systemically. Systemic administration of 5-HT_1 and 5-HT_2 receptor agonists, or antagonists, modulates LC activity. These effects, in part, appear to be mediated indirectly rather than by actions at serotonin receptors within the LC.[10–12] The microiontophoretic application of serotonin into the LC fails to modify the spontaneous activity of noradrenergic neurons, while a lesion of serotonin neurons is associated with an increase in the spontaneous firing of LC neurons.[12] Electrical or chemical stimulation of the dorsal raphe nucleus enhances, while a lesion of the dorsal raphe nucleus diminishes, the release of serotonin in the LC.[13] Neurotoxic destruction of serotonin terminals results in an increase in LC firing,[13] and increases in tyrosine hydroxylase mRNA and tyrosine hydroxylase activity in the LC.[14] Serotonin affects LC activity, to some degree, by attenuating glutamatergic activation of the LC,[15] and serotonin release in the LC is partially regulated by *N*-methyl-D-aspartate (NMDA) receptor activation.[16] Serotonergic input to the LC is partially modified by stress, given that acute tail pinch and noise-induced stress increase the release of serotonin within the LC.[16] Together, these findings demonstrate the existence of a complex and functional input of serotonin on noradrenergic activity in the brain. Generally, serotonin elicits inhibitory effects on LC activity.

Dopamine

Several lines of evidence demonstrate dopaminergic innervation of the LC.[17] Dopamine-immunoreactive nerve fibers are located in the LC[18] and these originate in part from the hypothalamus.[19] In addition, dopamine-containing fibers projecting to the LC originate in the ventral tegmental area, inasmuch as a lesion of the ventral tegmental area with 6-hydroxydopamine results in a halving of dopamine levels in the LC.[20] Despite evidence from these lesion experiments in the ventral tegmental area, Aston-Jones *et al.*[2] reported that only a few weakly labeled cells were identified in the ventral tegmental area after targeted injections of retrograde tracer into the LC. Further, the injection of an anterograde tracer into the ventral tegmental area yielded significant labeling of cells in the parabrachial area located adjacent to the LC.

Locus coeruleus neurons express D_3 dopamine receptor mRNA and moderate levels of D_2/D_3 dopamine receptor binding.[21–23] The functional significance of dopamine in the LC is revealed in a study showing that activation of the LC is induced by selective stimulation of the ventral tegmental area.[24] Selective stimulation of cell bodies in the ventral tegmental area with kainic acid results in an increase in the norepinephrine metabolite 3-methoxy-4-hydroxyphenylglycol (MHPG) in the prefrontal cortex and hippocampus. Transecting the LC-derived dorsal noradrenergic bundle blocked the effects of kainic acid on MHPG in the prefrontal cortex. In light of the preferential projection of dopamine fibers from the ventral tegmental area to parabrachial areas adjacent to the LC, the effects of ventral tegmental area stimulation with kainic acid on MHPG in the prefrontal cortex are likely mediated by areas immediately adjacent to the LC.[25]

γ-Aminobutyric acid

There is significant innervation of the LC by fibers and terminals containing the inhibitory neurotransmitter GABA. Cell bodies originating from GABA-containing fibers are localized to the PrH,[1] and norepinephrine neurons within the LC are innervated by GABA axon terminals.[26,27] Gamma-aminobutyric acid input to the LC from other brain areas has also been described (see Chapter 1). Locus coeruleus neurons express $GABA_A$ receptors, which have been subdivided based on affinities for specific benzodiazepines as BZ_1 and BZ_2. In the human LC, both BZ_1 and BZ_2 $GABA_A$ receptors exist.[28] Locus coeruleus neurons have been shown to express a number of the $GABA_A$ receptor subunits, including α_2, α_3, ϵ, and θ subunits.[29,30] They also express $GABA_B$ receptors[31] and respond to $GABA_B$ agonists.[32]

Gamma-aminobutyric acid has a tonic inhibitory effect on the LC and regulates levels of norepinephrine in the cerebral cortex. Iontophoretic application of GABA to the LC reduces LC firing.[33] Focal electrical stimulation of the GABAergic PrH inhibits the discharge of neurons in the LC, an effect that is blocked by local infusion of a $GABA_A$ receptor antagonist.[34] The infusion into the LC of muscimol, a $GABA_A$ receptor agonist, or bicuculline, a $GABA_A$ receptor antagonist, decreases or increases, respectively, the levels of norepinephrine in the medial prefrontal cortex.[35] Interestingly, serotonergic activity modulates GABAergic input to the LC.[33]

Gamma-aminobutyric acid clearly plays a physiologically significant role in regulating LC neuronal activity. During rapid eye movement (REM) sleep, GABA mediates inhibition of the LC.[36] Administration of the $GABA_A$ receptor facilitator, diazepam, does not affect spontaneous activity of the LC, but reduces excitatory responses of LC neurons to behavioral stimuli.[37] Withdrawal following chronic treatment with benzodiazepines results in elevations of LC activity that are closely associated with the withdrawal syndrome.[38] Thus, GABA appears to exert a powerful inhibitory influence on the LC and thereby, on norepinephrine levels in LC projection areas. Drugs affecting GABAergic transmission impart at least some of their central effects through indirect actions on the central noradrenergic system.

Glutamate

The excitatory amino acid neurotransmitter, glutamate, has robust effects on the activity of noradrenergic neurons in the LC. Ultrastructural studies provide evidence that glutamate may locally modulate noradrenergic neurons in the LC through kainate-type receptors located at postsynaptic and extrasynaptic cellular sites, and at glutamatergic NMDA-R1 receptor sites selectively distributed within astrocytic processes and presynaptic axon terminals within the LC.[39,40] Terminals containing glutamate-immunoreactivity form a synaptic contact with about one-third of LC dendrites immunohistochemically labeled with an antibody for tyrosine hydroxylase.[27] Although little NMDA-R1-immunoreactivity was found in the LC,[40] LC neurons express a unique constellation of NMDA-R1 and -R2 subunit mRNAs.[41]

The source of excitatory amino acid input to the LC has been identified as originating primarily in the brain stem, and secondarily in the cerebral cortex. Most neurons within the LC are activated by the electrical stimulation of the PGi.[42,43] The PGi-mediated excitation of LC neurons is thought to be mediated via a kainate-type receptor since the response is blocked by the excitatory amino acid receptor antagonists kynurenic acid, D-glutamyl glycine or 6-cyano-2,3-dihydroxy-7-niroguiinoxaline (CNQX), but not by the NMDA antagonists AP5 or AP7, or a quisqualate receptor antagonist.[1] In an elegant series of experiments where afferents from the LC to the forebrain were lesioned, another source of excitatory input to the LC was identified in the medial prefrontal cortex.[44,45] Although the medial prefrontal cortex does not project directly to the core of the LC, it projects fibers to brain stem regions containing LC dendrites. Electrical stimulation of the medial prefrontal cortex activates the majority of neurons in the LC, via NMDA and non-NMDA receptors. Inactivation of the medial prefrontal cortex by local injection of lidocaine suppresses LC neuron firing and reveals that the medial prefrontal cortex exerts a resting tonic excitatory influence on LC neuron activity.[45]

The LC also responds to metabotropic glutamate receptor activation. Group III presynaptic mGluR activation provides a feedback mechanism by which excitatory synaptic transmission in the LC can be negatively modulated during high-frequency synaptic activity.[46] Overall, glutamate plays a major role regulating LC activity, and is one of several transmitter inputs to the LC that is activated as a result of stressful stimuli.[47,48]

Acetylcholine

Cholinergic fibers have been localized to the area immediately adjacent to the LC[1] and acetylcholine can modulate LC activity through both muscarinic and nicotinic receptors. Infusion of the muscarinic cholinergic agonist, oxotremorine, into the LC results in a significant increase in extracellular norepinephrine in the medial prefrontal cortex.[35] Kawahara *et al.* note that neurons of the LC appear to be phasically regulated by the activation of muscarinic receptors since the muscarinic receptor antagonist, atropine, was without significant effect.[35] M_1, M_2 and M_3 muscarinic-binding sites have been described in the rat LC.[49] Acetylcholine appears to excite LC neurons by acting at non-M1 muscarinic receptors.[50]

Noradrenergic neurons in the LC express the α_7 nicotinic acetylcholine receptor subunit,[51] which can form a pentameric ligand-gated ion channel permeable to Ca^{2+}. Vincler and Eisenach[52] confirmed and extended the findings of Bitner and Nikkel[51] in reporting the distribution of nicotinic acetylcholine receptor subunit proteins within the LC. Multiple subunits were expressed within the LC on neuronal soma. Evidence that the effects of acetylcholine are mediated by nicotinic receptors within the LC comes from studies where epibatidine, an agonist ligand at the α_4-subunit nicotinic receptor, produces a large increase in c-Fos expression in the LC.[53]

Intravenous administration of nicotine causes a dose-dependent, short-lasting increase in firing rate of LC neurons[54] and nicotine stimulates release of norepinephrine.[55] Chronic administration of nicotine stimulates norepinephrine turnover in the rat brain, an effect that is blocked by the antagonist, mecamylamine.[56] Interestingly, local injection of a nicotinic cholinergic receptor agonist or antagonist into the LC does not significantly alter extracellular norepinephrine in the prefrontal cortex. Given the role of norepinephrine in vigilance and attention, it seems likely that at least some of the stimulant properties of nicotine may be mediated by actions on norepinephrine neurons.

Substance P

A relatively dense network of substance P-immunoreactive fibers innervates the human LC and surrounding regions.[57] Many of these fibers originate in the nucleus of the solitary tract.[57–59] There is a high density of binding of radiolabeled substance P to neurokinin-1 (NK_1) receptors in the rat [60] and human LC (as observed in our own lab). Of neurokinin receptors,[61–63] the NK_1 receptor preferentially binds substance P, the likely endogenous activator of this receptor subtype. There is evidence from dual-labeling immunofluorescence and *in situ* hybridization studies that LC neurons express substance P receptors.[64,65] There is also evidence of NK_3 receptors in the LC,[66] and this receptor appears to play a permissive role to NK_1 receptor activation.[67] Substance P potently stimulates the firing of the LC neurons.[68] Considerable evidence suggests that substance P plays a role in the central response to stress,[69–71] as do other stress-associated activators of the LC (e.g. glutamate, corticotropin-releasing factor). For example, substance P antagonists (in particular, selective NK_1 receptor antagonists) when administered intracerebroventricular, attenuate restraint-stress-induced biochemical indices of LC activation.[70]

Recent studies suggest that substance P antagonists may have antidepressant properties,[72,73] although there have been questions regarding their efficacy.[74] The human LC contains the highest levels of substance P receptors (NK_1 receptors) in the human brain.[65] Interestingly, substance P is present within a majority of serotonin-containing cell bodies in the human midbrain dorsal raphe nucleus.[75,76] The chronic administration of an antagonist at NK_1 receptors increases the firing of serotonin neurons in the midbrain.[77] Oddly, the effect of NK_1 receptor antagonists on serotonin neurons appears to be mediated by α_2 adrenergic receptors located on norepinephrine cell bodies in the LC.[78] Clearly, modulation of noradrenergic activity is at least one important action of drugs that activate or antagonize substance P neurotransmission, and this modulation likely plays some role in the therapeutic effect of drugs that antagonize substance P receptors.

Noradrenergic circuitry: reciprocal output from the LC

A variety of techniques, including histofluorescence, immunohistochemistry, retrograde tract tracing, and anterograde autoradiographic tracing reveal that noradrenergic fibers from the LC project widely and diffusely throughout the central nervous system.[79–84] The ascending projections of the LC include the dorsal noradrenergic bundle, the central gray dorsal longitudinal fasciculus, and the ventral tegmental medial forebrain bundle. Other projections from the LC innervate the cerebellar cortex and ventrolateral spinal cord. The LC sends reciprocal fibers to regions that project to the LC and pericoerulear area, including the dorsal raphe nucleus, ventral tegmental area, basal forebrain, and cerebral cortex.

The dorsal raphe, which innervates the LC, receives one of the richest noradrenergic innervations in the brain.[85–88] Evidence that some of this noradrenergic innervation originates in the LC comes from results of studies in which injections of tritiated leucine were made into the LC and subcoeruleus nucleus and fibers distributing to the dorsal raphe nucleus were labeled.[82] Electron microscopy reveals a large number of norepinephrine terminals directly innervating serotonin neurons in the dorsal raphe.[87] Numerous studies demonstrate that noradrenergic neurons regulate the activity of the serotonergic raphe nuclei. Administration of α-adrenergic receptor antagonists locally into the dorsal raphe nucleus suppresses the firing of serotonin neurons.[89] Both α_1- and α_2-adrenergic receptors play a role in mediating noradrenergic regulation of serotonin release in the raphe as well as in raphe projection areas. Generally, α_1-adrenergic receptors stimulate serotonin release, while α_2-adrenergic receptor activation inhibits release.[90] Norepinephrine is believed to regulate the pacemaker firing of dorsal raphe nuclei neurons, probably via α_{1B}-adrenergic receptors in the dorsal raphe nuclei.[91] The direct functional interactions between the noradrenergic LC and serotonergic raphe nuclei at the brain stem level are complemented by overlapping terminal distributions of these two systems and illustrates the close alignment of these systems in modulating behavior.

Like raphe nuclei, there is reciprocal innervation of the ventral tegmental area by the LC. Fibers originating in the LC travel rostrally into and through the ventral tegmental area and the compact part of the substantia nigra.[82] Noradrenergic innervation to the ventral tegmental area from the LC is excitatory and mediated largely by α_1-adrenergic receptors.[92] Electrical stimulation of the LC elicits an excitation, followed by an inhibition, of electrical activity in dopamine neurons in the midbrain. The excitation induced in the ventral tegmental area by stimulation of the LC is attenuated by pretreatment with prazosin, an antagonist at α_1-adrenergic receptors. Hence, norepinephrine appears to play a direct role via the ventral tegmental area in regulating dopamine release in the nucleus accumbens and limbic cortical regions, thereby indirectly influencing reward, emotional and cognitive states.

It is worth noting that norepinephrine may exert actions on dopaminergic output from the ventral tegmental area indirectly by its action on serotonin. Serotonin axon terminals are present in the ventral tegmental area,[93] and microinjection of serotonin in this area results in an increase in dopamine release in the ipsilateral nucleus accumbens.[94] Thus, LC neurons projecting to the limbic midbrain directly regulate dopamine neuron activity, and are likely to indirectly regulate dopamine neuron activity by control of serotonin neuron firing in the midbrain dorsal raphe nuclei.

Summary and conclusions

Norepinephrine-containing neurons clustered within the LC provide most of the norepinephrine present within the central nervous system. Limbic regions in the brain are densely innervated by the LC and norepinephrine from the nucleus plays a critical role in regulating mood, attention, sleep and stress.

The major afferents to the LC are excitatory from cell bodies located in the PGi and inhibitory from cell bodies in the PrH. Dense projections from many brain regions, however, terminate in the pericoerulear region, an area heavily invested with dendrites from LC neurons, and in the PGi and PrH. For example, neurons from the dorsal raphe nucleus, ventral tegmental area, the central nucleus of the amygdala, and frontal cortex send significant axonal projections to the pericoerulear area, a region replete with dendrites of LC neurons and other neurons that target the LC directly. In turn, the LC has significant projections to the dorsal raphe nucleus, ventral tegmental nucleus, amygdala, frontal cortex, as well as many other brain regions. The LC and norepinephrine play a significant role through direct projections, as well as indirect projections via other midbrain monoaminergic systems, in modulating areas of the brain that regulate mood, attention, sleep, and stress.

Given the reciprocal relationships between the LC and brain stem monoaminergic nuclei and limbic brain regions (Figure 2.1), dysfunction of the LC would likely impact a variety of behaviors and may lead or contribute to many psychiatric and neurological disorders. There is growing evidence of a role of LC-norepinephrine dysfunction in a variety of central nervous system disorders, particularly clinical depression and anxiety. There are certainly robust effects from drugs used to treat

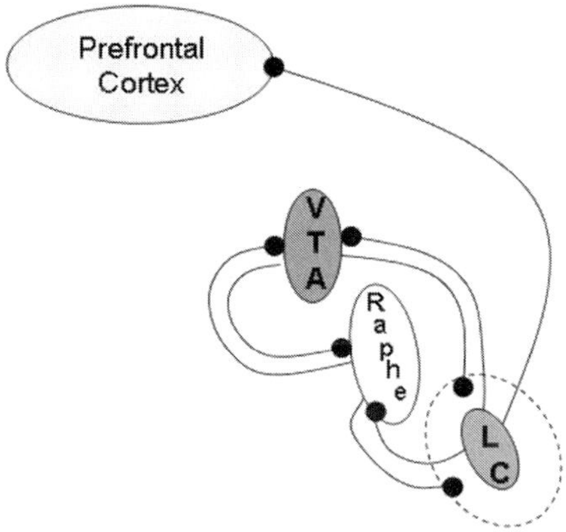

Figure 2.1 The reciprocal relationships between the LC, other brain stem monoaminergic nuclei, and the limbic cortex. LC, locus coeruleus; Raphe, dorsal raphe nucleus; VTA, ventral tegmental area.

depressive and anxiety disorders on LC activity. These actions strongly imply that some of the therapeutic benefits of these drugs are a result of modulation of LC activity through actions at receptors directly on noradrenergic LC neurons, such as substance P, GABA, and dopamine receptors. In addition, actions of drugs on other neurotransmitter systems, e.g. serotonin, dopamine, glutamate, GABA, etc., also produce robust effects on noradrenergic activity in the brain because of innervation of the LC by these systems. These indirect noradrenergic effects may also contribute secondary benefits to the treatment of central nervous system disorders.

ACKNOWLEDGEMENTS

The authors acknowledge the support of The National Institute of Mental Health (MH63187, MH67996, MH46692, MH/AG02031 and RR17701), The American Foundation for Suicide Prevention, and The National Alliance for Research on Schizophrenia and Depression.

REFERENCES

1. Aston-Jones, G., Shipley, M. T., Chouvet, G. *et al.* Afferent regulation of locus coeruleus neurons: anatomy, physiology and pharmacology. *Prog. Brain Res.*, **88** (1991), 47–75.
2. Aston-Jones, G., Ennis, M., Pieribone, V. A., Nickell, W. T. and Shipley, M. T. The brain nucleus locus coeruleus: restricted afferent control of a broad efferent network. *Science*, **234** (1986), 734–737.

3. Luppi, P. H., Aston-Jones, G., Akaoka, H., Chouvet, G. and Jouvet, M. Afferent projections to the rat locus coeruleus demonstrated by retrograde and anterograde tracing with cholera-toxin B subunit and *Phaseolus vulgaris* leucoagglutinin. *Neuroscience*, **65** (1995), 119–160.
4. Morgane, P. J. and Jacobs, M. S. Raphe projections to the locus coeruleus in the rat. *Brain Res. Bull.*, **4**: 4 (1979), 519–534.
5. Klimek, V., Roberson, G., Stockmeier, C. and Ordway, G. A. Serotonin transporter and MAO-B levels in monoamine nuclei of the human brainstem are normal in major depression. *J. Psychiatr. Res.*, **37** (2003), 387–397.
6. Van Bockstaele, E. J. Multiple substrates for serotonergic modulation of rat locus coeruleus neurons and relationships with kainate receptors. *Brain Res. Bull.*, **51**: 5 (2000), 433–442.
7. Pompeiano, M., Palacios, J. M. and Mengod, G. Distribution of the serotonin 5-HT2 receptor family mRNAs: comparison between 5-HT2A and 5-HT2C receptors. *Mol. Brain Res.*, **23** (1994), 163–178.
8. Azmitia, E. C., Gannon, P. J., Kheck, N. M. and Whitaker-Azmitia, P. M. Cellular localization of the 5-HT1A receptor in primate brain neurons and glial cells. *Neuropsychopharmacology*, **14** (1996), 35–46.
9. Lopez-Gimenez, J. F., Mengod, G., Palacios, J. M. and Vilaro, M. T. Human striosomes are enriched in 5-HT2A receptors: autoradiographical visualization with [3H]MDL100,907,[125I](+/-)DOI and [3H]ketanserin. *Eur. J. Neurosci.*, **11**: 10 (1999), 3761–3765.
10. Chiang, C. and Aston-Jones, G. A 5-hydroxytryptamine2 agonist augments gamma-aminobutyric acid and excitatory amino acid inputs to noradrenergic locus coeruleus neurons. *Neuroscience*, **54** (1993), 409–420.
11. Gorea, E., Davenne, D., Lanfumey, L., Chastanet, M. and Adrien, J. Regulation of noradrenergic cerulean neuronal firing mediated by 5-HT2 receptors: involvement of the prepositus hypoglossal nucleus. *Neuropharmacology*, **30** (1991), 1309–1318.
12. Haddjeri, N., de Montigny, C. and Blier, P. Modulation of the firing activity of noradrenergic neurones in the rat locus coeruleus by the 5-hydroxtryptamine system. *Br. J. Pharmacol.*, **120** (1997), 865–875.
13. Kaehler, S. T., Singewald, N. and Philippu, A. Dependence of serotonin release in the locus coeruleus on dorsal raphe neuronal activity. *Naunyn-Schmiedeberg's Arch. Pharmacol.*, **359** (1999), 386–393.
14. Sturtz, F., Rollet, D., Faucon Biguet, F., Mallet, J. and Buda, M. Long term alteration in tyrosine hydroxylase mRNA levels in rat locus coeruleus after intraventricular injection of 5,6-dihydroxytryptamine. *Mol. Brain Res.*, **22** (1994), 107–112.
15. Aston-Jones, G., Akaoka, H., Charlety, P. and Chouvet, G. Serotonin selectively attenuates glutamate-evoked activation of noradrenergic locus coeruleus neurons. *J. Neurosci.*, **11** (1991), 760–769.
16. Singewald, N., Kaehler, S. T., Hemeida, R. and Philippu, A. Influence of excitatory amino acids on basal and sensory stimuli-induced release of 5-HT in the locus coeruleus. *Br. J. Pharmacol.*, **123** (1998), 746–752.
17. Oades, R. D. and Halliday, G. M. Ventral tegmental (A10) system: neurobiology. 1. Anatomy and connectivity. *Brain Res.*, **434** (1987), 117–165.

18. Kitahama, K., Nagatsu, I., Geffard, M. and Maeda, T. Distribution of dopamine-immunoreactive fibers in the rat brainstem. *J. Chem. Neuroanat.*, **18** (2000), 1–9.
19. Maeda, T., Kojima, Y., Arai, R. *et al.* Monoaminergic interaction in the central nervous system: a morphological analysis in the locus coeruleus of the rat. *Comp. Biochem. Physiol. C*, **98** (1991), 193–202.
20. McRae-Degueurce, A. and Milon, H. Serotonin and dopamine afferents to the rat locus coeruleus: a biochemical study after lesioning of the ventral mesencephalic tegmental-A10 region and the raphe dorsalis. *Brain Res.*, **263** (1993), 344–347.
21. Suzuki, M., Hurd, Y. L., Sokoloff, P., Schwartz, J. C. and Sedvall, G. D3 dopamine receptor mRNA is widely expressed in the human brain. *Brain Res.*, **779**: 1–2 (1998), 58–74.
22. Herroelen, L., De Backer, J. P., Wilczak, N. *et al.* Autoradiographic distribution of D3-type dopamine receptors in human brain using 7-[3H]hydroxyl-N,N-di-n-propyl-2-aminotetralin. *Brain Res.*, **648** (1994), 222–228.
23. Yokoyama, C., Okamura, H., Nakajima, T., Taguchi, J. and Ibata, Y. Autoradiographic distribution of [3H]YM-09151–2, a high-affinity and selective antagonist ligand for the dopamine D2 receptor group, in the rat brain and spinal cord. *J. Comp. Neurol.*, **344** (1994), 121–136.
24. Deutch, A. Y., Goldstein, M. and Roth, R. H. Activation of the locus coeruleus induced by selective stimulation of the ventral tegmental area. *Brain Res.*, **363** (1986), 307–314.
25. Chieng, L. C. B. and Bekkers, J. M. GABA-B, opioid and alpha2 receptor inhibition of calcium channels in acutely-dissociated locus coeruleus neurons. *Br. J. Pharmacol.*, **127**: 7 (1999), 1533–1538.
26. Van Bockstaele, E. J. Morphological substrates underlying opioid, epinephrine and gamma-aminobutyric acid inhibitory actions in the rat locus coeruleus. *Brain Res. Bull.*, **47** (1998), 1–15.
27. Somogyi, J. and Llewellyn-Smith, I. J. Patterns of colocalization of GABA, glutamate and glycine immunoreactivities in terminals that synapse on dendrites of noradrenergic neurons in rat locus coeruleus. *Eur. J. Neurosci.*, **14** (2001), 219–228.
28. Zezula, J., Cortes, R., Probst, A. and Palacios, J. M. Benzodiazepine receptor sites in the human brain: autoradiographic mapping. *Neuroscience*, **25**: 3 (1988), 771–795.
29. Rodriguez-Pallares, J., Caruncho, H. J., Lopez-Real, A. *et al.* Rat brain cholinergic, dopaminergic, noradrenergic and serotonergic neurons express GABA-A receptors derived from the alpha3 subunit. *Recept. Channels*, **7**: 6 (2001), 471–478.
30. Moragues, N., Ciofi, P., Tramu, G. and Garret, M. Localisation of GABA(A) receptor epsilon-subunit in cholinergic and aminergic neurones and evidence for co-distribution with the theta-subunit in rat brain. *Neuroscience*, **111**: 3 (2002), 657–669.
31. Lu, X. Y., Ghasemzadeh, M. B. and Kalivas, P. W. Regional distribution and cellular localization of gamma-aminobutyric acid subtype 1 receptor mRNA in the rat brain. *J. Comp. Neurol.*, **407**: 2 (1999), 166–182.
32. Olpe, H. R., Steinmann, M. W., Hall, R. G., Brugger, F. and Pozza, M. F. GABA-A and GABA-B receptors in locus coeruleus: effects of blockers. *Eur. J. Pharmacol.*, **149**: 1,2 (1988), 183–185.
33. Szabo, S. T. and Blier, P. Serotonin (1A) receptor ligands act on norepinephrine neuron firing through excitatory amino acid and GABA(A) receptors: a microiontophoretic study in the rat locus coeruleus. *Synapse*, **42**: 4 (2001), 203–212.

34. Ennis, M. and Aston-Jones, G. GABA-mediated inhibition of locus coeruleus from the dorsomedial rostral medulla. *J. Neurosci.*, **9** (1989), 2973–2981.
35. Kawahara, Y., Kawahara, H. and Westerink, B. H. Tonic regulation of the activity of noradrenergic neurons in the locus coeruleus of the conscious rat studied by dual-probe microdialysis. *Brain Res.*, **823** (1999), 42–48.
36. Gervasoni, D., Darracq, L., Fort, P. *et al.* Electrophysiological evidence that noradrenergic neurons of the rat locus coeruleus are tonically inhibited by GABA during sleep. *Eur. J. Neurosci.*, **10**: 3 (1998), 964–970.
37. Rasmussen, K. and Jacobs, B. L. Single unit activity of locus coeruleus neurons in the freely moving cat. II. Conditioning and pharmacologic studies. *Brain Res.*, **371**: 2 (1986), 335–344.
38. Perez, M. F., Nasif, F. J., Marchesini, G. R., Maglio, L. E. and Ramirez, O. A. Hippocampus and locus coeruleus activity on rats chronically treated with diazepam. *Pharmacol. Biochem. Behav.*, **69**: 3–4 (2001), 431–438.
39. Van Bockstaele, E. J. and Colago, E. E. Ultrastructural localization of the kainate selective glutamate receptor in noradrenergic perikarya and dendrites of the nucleus locus coeruleus in the rat brain. *Brain Res.*, **732** (1996), 223–231.
40. Van Bockstaele, E. J. and Colago, E. E. Selective distribution of the NMDA-R1 glutamate receptor in astrocytes and presynaptic axon terminals in the nucleus locus coeruleus of the rat brain: an immunoelectron microscopic study. *J. Comp. Neurol.*, **369** (1996), 483–496.
41. Luque, J. M., Malherbe, P. and Richards, J. G. Localization of NMDA receptor subunit mRNAs in the rat locus coeruleus. *Brain Res. Mol. Brain Res.*, **29**: 2 (1995), 224–232.
42. Ennis, M. and Aston-Jones, G. A potent excitatory input to the nucleus locus coeruleus from the ventrolateral medulla. *Neurosci. Lett.*, **71** (1986), 299–305.
43. Ennis, M. and Aston-Jones, G. Activation of locus coeruleus from nucleus paragigantocellularis: a new excitatory amino acid pathway in brain. *J. Neurosci.*, **8** (1988), 3644–3657.
44. Jodo, E. and Aston-Jones, G. Activation of locus coeruleus by prefrontal cortex is mediated by excitatory amino acid inputs. *Brain Res.*, **768** (1997), 327–332.
45. Jodo, E., Chiang, C. and Aston-Jones, G. Potent excitatory influence of prefrontal cortex activity on noradrenergic locus coeruleus neurons. *Neuroscience*, **83** (1998), 63–79.
46. Dube, G. R. and Marshall, K. C. Activity-dependent activation of presynaptic metabotropic glutamate receptors in locus coeruleus. *J. Neurophysiol.*, **83**: 3 (2000), 1141–1149.
47. Singewald, N., Zhou, G. Y. and Schneider, C. Release of excitatory and inhibitory amino acids from the locus coeruleus of conscious rats by cardiovascular stimuli and various forms of acute stress. *Brain Res.*, **704** (1995), 42–50.
48. Timmerman, W., Cisci, G., Nap, A., de Vries, J. B. and Westerink, B. H. Effects of handling on extracellular levels of glutamate and other amino acids in various areas of the brain measured by microdialysis. *Brain Res.*, **833** (1999), 150–160.
49. Baghdoyan, H. A. Location and quantification of muscarinic receptor subtypes in rat pons: implications for REM sleep generation. *Am. J. Physiol.*, **273**: 3 Pt 2 (1997), R896–R904.
50. Egan, T. M. and North, R. A. Acetylcholine acts on m2-muscarinic receptors to excite rat locus coeruleus neurones. *Br. J. Pharmacol.*, **85**: 4 (1985), 733–735.

51. Bitner, R. S. and Nikkel, A. L. Alpha-7 nicotinic receptor expression by two distinct cell types in the dorsal raphe nucleus and locus coeruleus of rat. *Brain Res.*, **938** (2002), 45–54.
52. Vincler, M. A. and Eisenach, J. C. Immunocytochemical localization of the alpha3, alpha4, alpha5, alpha7, beta2, beta3 and beta4 nicotinic acetylcholine receptor subunits in the locus coeruleus of the rat. *Brain Res.*, **974** (2003), 25–36.
53. Cucchiaro, G. and Commons, K. G. Alpha 4 nicotinic acetylcholine receptor subunit links cholinergic to brainstem monoaminergic neurotransmission. *Synapse*, **49** (2003), 195–205.
54. Svensson, T. H. and Engberg, G. Effect of nicotine on single cell activity in the noradrenergic nucleus locus coeruleus. *Acta Physiol. Scand. Suppl.*, **479** (1980), 31–34.
55. Gallardo, K. A. and Leslie, F. M. Nicotine-stimulated release of [3H] norepinephrine from fetal rat locus coeruleus cells in culture. *J. Neurochem.*, **70** (1998), 663–670.
56. Morgan, W. W. and Pfeil, K. A. Mecamylamine blockade of nicotine enhanced noradrenaline turnover in rat brain. *Life Sci.*, **24** (1979), 417–420.
57. Fodor, M., Gorc, T. J. and Palkovits, M. Immunohistochemical study on the distribution of neuropeptides within the pontine tegmentum–particularly the parabrachial nuclei and the locus coeruleus of the human brain [published erratum appears in *Neuroscience*, **48**: 3 (1992), 753. *Neuroscience*, **46** (1992), 891–908.
58. Mantyh, P. W. and Hunt, S. P. Neuropeptides are present in projection neurons at all levels in visceral and taste pathways: from periphery to sensory cortex. *Brain Res.*, **299** (1984), 297–312.
59. Riche, D., De Pommery, J. and Menetrey, D. Neuropeptides and catecholamines in efferent projections of the nuclei of the solitary tract in the rat. *J. Comp. Neurol.*, **293** (1990), 399–424.
60. Dam, T. V., Martinelli, B. and Quirion, R. Autoradiographic distribution of brain neurokinin-1/substance P receptors using a highly selective ligand [3H]-[Sar9,Met(O2)11]-substance P. *Brain Res.*, **531** (1990), 333–337.
61. Dam, T. V., Escher, E. and Quirion, R. Evidence for the existence of three classes of neurokinin receptors in brain. Differential ontogeny of neurokinin-1, neurokinin-2 and neurokinin-3 binding sites in rat cerebral cortex. *Brain Res.*, **453** (1988), 372–376.
62. Maggio, J. E. Tachykinins. *Annu. Rev. Neurosci.*, **11** (1988), 13–28.
63. Regoli, D., Drapeau, G., Dion, S. and D'Orleans-Juste, P. Pharmacological receptors for substance P and neurokinins. *Life Sci.*, **40** (1987), 109–117.
64. Chen, L. W., Wei, L. C., Liu, H. L. and Rao, Z. R. Noradrenergic neurons expressing substance P receptor (NK1) in the locus coeruleus complex: a double immunofluorescence study in the rat. *Brain Res.*, **873**: 1 (2000), 155–159.
65. Caberlotto, L., Hurd, Y. L., Murdock, P. *et al.* Neurokinin 1 receptor and relative abundance of the short and long isoforms in the human brain. *Eur. J. Neurosci.*, **17** (2003), 1736–1746.
66. Jung, M., Michaud, J. C., Steinberg, R. *et al.* Electrophysiological, behavioural and biochemical evidence for activation of brain noradrenergic systems following neurokinin NK3 receptor stimulation. *Neuroscience*, **74** (1996), 403–414.
67. Bert, L., Rodier, D., Bougault, I. *et al.* Permissive role of neurokinin NK(3) receptors in NK(1) receptor-mediated activation of the locus coeruleus revealed by SR 142801. *Synapse*, **43**: 1 (2002), 62–69.

68. Guyenet, P. G. and Aghajanian, G. K. Excitation of neurons in the nucleus locus coeruleus by substance P and related peptides. *Brain Res.*, **136** (1977), 178–184.
69. Bannon, M. J., Elliott, P. J., Alpert, J. E. *et al.* Role of endogenous substance P in stress induced activation of mesocortical dopamine neurons. *Nature*, **306** (1983), 791–792.
70. Hahn, M. K. and Bannon, M. J. Stress-induced C-fos expression in the rat locus coeruleus is dependent on neurokinin 1 receptor activation. *Neuroscience*, **94** (1999), 1183–1188.
71. Takayama, H., Ota, Z. and Ogawa, N. Effect of immobilization stress on neuropeptides and their receptors in rat central nervous system. *Regul. Pept.*, **15** (1986), 239–248.
72. Kramer, M. S., Cutler, N., Feighner, J. *et al.* Distinct mechanism for antidepressant activity by blockade of central substance P receptors. *Science*, **281**: 5383 (1998), 1640–1645.
73. Kramer, M. S., Winokur, A., Kelsey, J. *et al.* Demonstration of the efficacy and safety of a novel substance P (NK1) receptor antagonist in major depression. *Neuropsychopharmacology*, **29**: 2 (2004), 385–392.
74. Enserink, M. Can the placebo be the cure? *Science*, **284** (1999), 238–240.
75. Baker, K. G., Halliday, G. M., Hornung, J. P. *et al.* Distribution, morphology and number of monoamine-synthesizing and substance P-containing neurons in the human dorsal raphe nucleus. *Neuroscience*, **42** (1991), 757–775.
76. Sergeyev, V., Hökfelt, T. and Hurd, Y. Serotonin and substance P co-exist in dorsal raphe neurons of the human brain. *NeuroReport*, **10** (1999), 3967–3970.
77. Haddjeri, N. and Blier, P. Sustained blockade of neurokinin-1 receptors enhances serotonin neurotransmission. *Biol. Psychiatry*, **50** (2001), 191–199.
78. Haddjeri, N. and Blier, P. Effect of neurokinin-I receptor antagonists on the function of 5-HT and noradrenaline neurons. *NeuroReport*, **11** (2000), 1323–1327.
79. Ungerstedt, U. Stereotaxic mapping of the monoamine pathways in the rat brain. *Acta Physiol. Scand. Suppl.*, **367** (1971), 1–48.
80. Jones, B. E. and Moore, R. Y. Ascending projections of the locus coeruleus in the rat. II. Autoradiographic study. *Brain Res.*, **127** (1977), 25–53.
81. Foote, S. L., Bloom, F. E. and Aston-Jones, G. Nucleus locus ceruleus: new evidence of anatomical and physiological specificity. *Physiol. Rev.*, **63** (1983), 844–914.
82. Jones, B. E. and Yang, T. Z. The efferent projections from the reticular formation and the locus coeruleus studied by anterograde and retrograde axonal transport in the rat. *J. Comp. Neurol.*, **242** (1985), 56–92.
83. Loughlin, S. E., Foote, S. I. and Bloom, F. E. Efferent projections of nucleus locus coeruleus: topographic organization of cells of origin demonstrated by three-dimensional reconstruction. *Neuroscience*, **18** (1986), 291–306.
84. Loughlin, S. E., Foote, S. L. and Grzanna, R. Efferent projections of the nucleus locus coeruleus: morphologic subpopulations have different efferent targets. *Neuroscience*, **18** (1986), 307–319.
85. Saavedra, J. M., Grobecker, H. and Zivin, J. Catecholamines in the raphe nuclei of a rat. *Brain Res.*, **114** (1976), 339–345.
86. Levitt, P. and Moore, R. Y. Origin and organization of brain stem catecholamine innervation in the rat. *J. Comp. Neurol.*, **186** (1979), 505–528.

87. Baraban, J. M. and Aghajanian, G. K. Noradrenergic innervation of serotonergic neurons in the dorsal raphe: demonstration by electron microscopic autoradiography. *Brain Res.*, **204** (1981), 1–11.
88. Ordway, G. A., Stockmeier, C. A., Cason, G. W. and Klimek, V. Pharmacology and distribution of norepinephrine transporters in the human locus coeruleus and raphe nuclei. *J. Neurosci.*, **17** (1997), 1710–1719.
89. Baraban, J. M. and Aghajanian, G. K. Suppression of firing activity of 5-HT neurons in the dorsal raphe by alpha-adrenoceptor antagonists. *Neuropharmacology*, **19** (1980), 355–363.
90. Pudovkina, O. L., Cremers, T. I. and Westerink, B. H. Regulation of the release of serotonin in the dorsal raphe nucleus by alpha1 and alpha2 adrenoceptors. *Synapse*, **50**: 1 (2003), 77–82.
91. Aghajanian, G. K., Sprouse, J. S. and Rasmussen, K. Physiology of the midbrain 5HT system. In *Psychopharmacology: The Third Generation of Progress*, ed. H. Y. Meltzer. New York: (Raven Press, 1987), pp. 141–148.
92. Grenhoff, J. and Svensson, T. H. Prazosin modulates the firing pattern of dopamine neurons in rat ventral tegmental area. *Eur. J. Pharmacol.*, **233** (1993), 79–84.
93. Herve, D., Pickel, V. M., Joh, T. H. and Beaudet, A. Serotonin axon terminals in the ventral tegmental area of the rat: fine structure and synaptic input to dopaminergic neurons. *Brain Res.*, **435** (1987), 71–83.
94. Guan, X. M. and McBride, W. J. Serotonin microinfusion into the ventral tegmental area increases accumbens dopamine release. *Brain Res. Bull.*, **23** (1989), 541–547.

3

Receptors for norepinephrine and signal transduction pathways

David B. Bylund, Ph.D.
University of Nebraska Medical Center, Omaha, USA

Introduction

Adrenergic receptors mediate the central and peripheral actions of the neurohormones norepinephrine and epinephrine. Both of these catecholamine messengers play important roles in the regulation of diverse physiological systems, and thus adrenergic receptors are widely distributed throughout the body. Stimulation of adrenergic receptors by catecholamines released from the sympathetic autonomic nervous system results in a variety of effects, such as increased heart rate, regulation of vascular tone, and bronchodilatation. In the central nervous system, adrenergic receptors are involved in many functions, including memory, learning, alertness, and the response to stress.

Although adrenergic receptors were originally divided into two major types, α and β, the current classification scheme[1] is based on three major types, α_1, α_2 and β. Each of these three receptor types is further divided into three subtypes: α_{1A}, α_{1B}, α_{1D}; α_{2A}, α_{2B}, α_{2C}; β_1, β_2, β_3. All of these adrenergic receptors are seven transmembrane receptors, each consisting of a single polypeptide chain with seven hydrophobic regions that are thought to form α helical structures and span or transverse the membrane.

Because the mechanism of action of adrenergic receptors includes the activation of guanine nucleotide regulatory binding proteins (G proteins), these receptors are also called G protein-coupled receptors. The binding of norepinephrine (or another agonist) to an adrenergic receptor induces (or stabilizes) a conformational change that allows the receptor to interact with and activate a G protein. The activated receptor facilitates the exchange of guanine diphosphate (GDP) for guanine triphosphate (GTP), leading to the dissociation of the α and $\beta\gamma$ subunits of the G protein, which in turn stimulates or inhibits the activity of various effectors. Importantly, each of the three types of adrenergic receptors couples to a distinct

Brain Norepinephrine: Neurobiology and Therapeutics, ed. Gregory A. Ordway, Michael A. Schwartz and Alan Frazer. Published by Cambridge University Press.

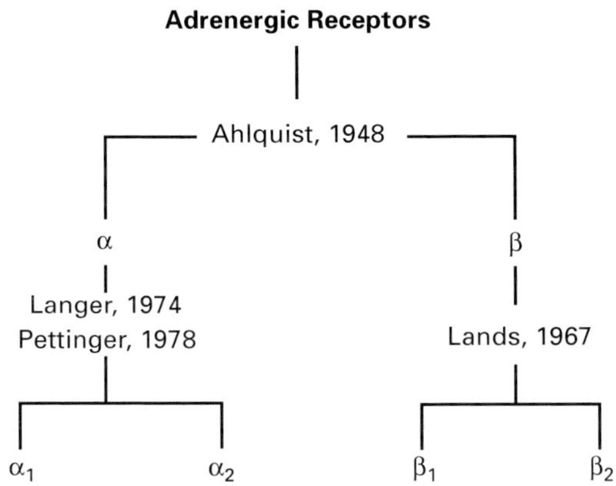

Figure 3.1 Historical nomenclature of adrenergic receptors.

class of G proteins: α_1 to G_q; α_2 to $G_{i/o}$ and β to G_s. In addition to G proteins, adrenergic receptors interact with other signaling proteins and pathways, such as those involving tyrosine kinases.

Adrenergic receptor subtypes

Historical development

Adrenergic receptors were originally divided into two major types, α and β, based on their pharmacological characteristics (i.e., rank order potency of agonists)[2] as shown in Figure 3.1. Subsequently, the β-adrenergic receptors were subdivided into β_1- and β_2-subtypes.[3] The β_1-adrenergic receptor, the dominant receptor in heart and adipose tissue, is equally sensitive to epinephrine and norepinephrine; whereas the β_2-adrenergic receptor, responsible for relaxation of vascular, uterine, and airway smooth muscle, is less sensitive to norepinephrine as compared to epinephrine. Subsequently, it became apparent that not all of the β-adrenergic receptor-mediated responses could be classified as either β_1 or β_2, suggesting the existence of at least one additional β-subtype[4] as shown in Figure 3.2. The β_3-receptor is insensitive to the commonly used β-adrenergic receptor antagonists and was previously referred to as the "atypical" β-adrenergic receptor.[5] A β_4-receptor has been postulated. However, definitive evidence of its existence is lacking, and it is now thought to be a "state" of the β_1-adrenergic receptor.[6]

The α-adrenergic receptors were first subdivided into postsynaptic (α_1) and presynaptic (α_2) subtypes.[7] After realization that not all α-receptors with α_2 pharmacological characteristics were presynaptic (or prejunctional), the pharmacological definition was used.[8] Three genetic and four pharmacological α_2-adrenergic receptor subtypes have been defined (Figure 3.2). The evidence for α_2-receptor

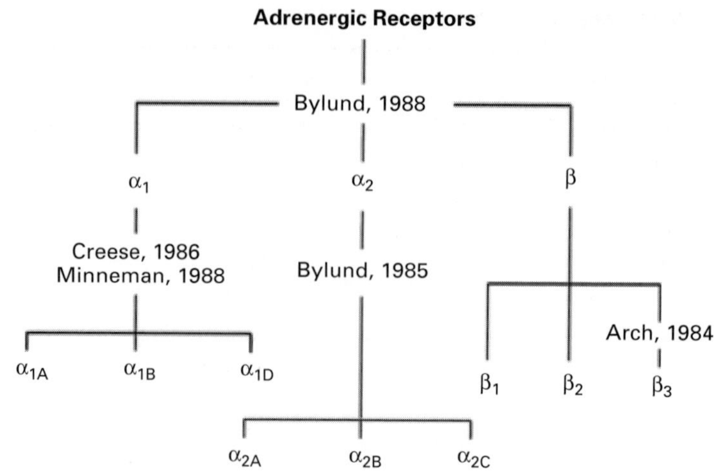

Figure 3.2 Nomenclature of adrenergic receptor subtypes.

subtypes initially came from binding and functional studies in various tissues and cell lines.[9] The α_{2A}- and α_{2B}-subtypes were initially defined based on their differential affinity for adrenergic agents such as prazosin and oxymetazoline.[10] These subtypes were subsequently cloned from human, rat, mouse, and other species. A third subtype, α_{2C}, was identified originally in an opossum kidney cell line[11] and has also been cloned from several species. A fourth pharmacological subtype, the alpha-2D, has been identified in the rat and cow.[12,13] This pharmacological subtype is a species ortholog of the human α_{2A}-subtype, and thus is not considered to be a separate genetic subtype.

Three genetic and four pharmacological α_1-adrenergic receptor subtypes have also been defined.[14] The evidence for α_1-receptor subtypes initially came from binding and functional studies in various tissues and cell lines. The α_{1A}- and α_{1B}-subtypes were initially defined based on their differential affinity for adrenergic agents such as WB4101 and phentolamine,[15] and their differential sensitivities to the site-directed alkylating agent chloroethylclonidine.[16] The α_{1B}-subtype was subsequently cloned from the hamster and the α_{1A} was cloned from bovine brain, although it was originally called the α_{1C}-adrenergic receptor. A third subtype, α_{1D}-adrenergic receptor was subsequently cloned from the rat cerebral cortex, although this clone was originally called the α_{1A}-adrenergic receptor by some workers.[14] A fourth pharmacological subtype, the α_{1L}, has been identified in vascular tissues from several species,[17] but may represent a conformational state of the α_{1A}-receptor.[18] The current classification scheme includes α_{1A}, the α_{1B}, and the α_{1D}, but there is no α_{1C} (Figure 3.2).

Table 3.1 Alpha$_1$-adrenergic receptor agents.

Receptor	Agonists	Antagonists
Alpha$_1$	Metaraminol (Aramine)	Doxazosin (Cardura), Prazosin
	Methoxamine (Vasoxyl)	(Minipress),
	Ozymetazoline (Afrin)	Tamsulosin (Flomax),
	Phenylephrine (Neosynephrine)	Terazosin (Hytrin),
Alpha$_{1A}$	A6160	A131701
	Cirazoline	KMD 3213
	SDZ NVI 086	5-Methylurapidil
		Niguldipine,
		RS 17053
		SNAP 5089
Alpha$_{1B}$		
Alpha$_{1D}$		BMY 7378

The issue of how different the affinities of a drug for different subtypes need to be in order for that drug to be considered selective is clearly a subjective one. The rule of thumb of the author is that a drug needs to be at least ten-fold selective for one subtype over the other subtypes to be considered a useful subtype-selective agent. This definition is based partly on experience with radioligand binding assay, in which at least a ten-fold selectivity is needed in order to clearly discern the two affinities in competition experiments.[19] This is consistent with clinical experience, which also suggests that generally at least a ten-fold selectivity is needed in order for a drug to be a useful subtype-selective agent.

Characteristics of α_1-adrenergic receptors

Pharmacological characteristics of α_1-adrenergic receptors

In addition to norepinephrine and epinephrine, α_1-receptors are activated by a variety of nonselective and subtype-selective agonists (Table 3.1). Agents such as phenylephrine (Neosynephrine), methoxamine (Vasoxyl), and metaraminol (Aramine) are relatively selective for α_1-receptors and have little or no activity at α_2- and β-receptors, except at high concentrations. By contrast, they have similar affinities for the three α_1-subtypes and are thus considered to be nonsubtype-selective agonists. Similarly, antagonists including prazosin (Minipress), terazosin (Hytrin), doxazosin (Cardura) and tamsulosin (Flomax) are relatively selective for α_1-receptors and block α_2- and β-receptors only at high concentrations. These antagonists all have similar affinities for the three α_1-subtypes, although

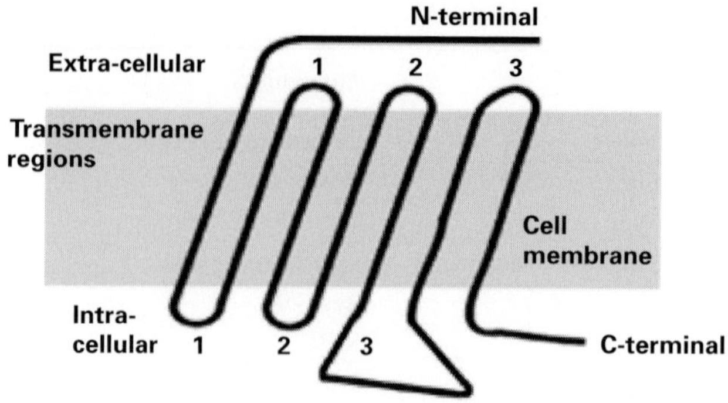

Figure 3.3 Two-dimensional representation of an adrenergic receptor.

tamsulosin is considered to be slightly selective for the α_{1A}- and α_{1D}-subtypes as compared to the α_{1B}-subtype. Several other antagonists such as phentolamine (Regitine) and phenoxybenzamine (Dibenzyline) block both α_1- and α_2-adrenergic receptors with similar affinities.

Several agents have been developed that are at least 10- or 20-fold selective for the α_{1A}-subtype as compared to the α_{1B}- and α_{1D}-subtypes.[20] However, agents that are clearly selective for either the α_{1B}- or α_{1D}-subtypes have not yet been developed, with the exception of BMY 7378, which is an α_{1D}-selective antagonist. α_{1A}-selective antagonists include 5-methylurapidil, niguldipine, SNAP 5089, KMD 3213, RS 17053, and A131701.[20] Potential α_{1A}-selective agonists include cirazoline, SDZ NVI 086, and A6160.

Molecular characteristics of α_1-adrenergic receptors

The α_1-adrenergic receptors are single polypeptide chains of 446 to 572 amino acid residues that span the membrane seven times, with the amino terminal being extracellular and the carboxy terminal intracellular. Thus, there are three intracellular loops and three extracellular loops (Figure 3.3). Some of the molecular characteristics of the α_1-adrenergic receptors in comparison to the α_2- and β-receptor types are compiled in Table 3.2. In contrast to the α_2-receptors, but similar to the β-receptors, the α_1-receptors have a long carboxy-terminal tail (137 to179 amino acid residues) and a short third intracellular loop (68 to 73 amino acid residues). The amino terminals of the α_{1A}- and α_{1B}-subtypes have three (α_{1A}) or four (α_{1B}) consensus sites for N-linked glycosylation. The carboxy-terminal tails of all three subtypes are potentially palmitolyated, thus anchoring the tail to the membrane and forming a small fourth intracellular loop. The carboxy-terminal tails also have multiple sites of phosphorylation, which are thought to be important in the desensitization, recycling, and down-regulation of the receptor.

Table 3.2 Molecular characteristics of human adrenergic receptors.

Subtype	Chromosome localization	Accession number	Number of amino acids	Number of N-linked glycosylation sites	Number of amino acids in		Number of introns	Number of splice variants
					3_{rd} loop	C-terminal tail		
α_{1A}	8p21–p11.2	P35348	446	3	68	137	1	4
α_{1B}	5q33	P35368	519	4	71	179	1	0
α_{1D}	20p13	P25100	572	0	73	167	1	0
α_{2A}	20q23–25	P08913	450	2	157	20	0	0
α_{2B}	2	P18089	450	0	179	21	0	0
α_{2C}	4p16.	P18825	462	2	148	21	0	0
β_1	10q24–26	P08588	477	1	80	97	0	0
β_2	5q31–32	P07550	413	2	54	84	0	0
β_3	8p11–12	p13945	408	2	67	61	1	2

The human α_1-adrenergic receptor genes consist of two exons and a single large intron of at least 20 kilobases in the region corresponding to the sixth transmembrane domain. No splice variants are known for the α_{1B}- and α_{1D}-subtypes. By contrast, at least ten splice variants of human α_{1A}-subtype have been reported, but only four produce full-length receptors. These three variants differ in the length and sequence of the carboxy-terminal domains. Ligand binding and signal transduction does not appear to be altered by splicing of these full-length receptors. Splice variant 1 predominates in most human tissues and cell lines studied.[20]

A comprehensive listing of the nucleotide sequences available in GenBank has been compiled for all three adrenergic receptor types, which includes the presence of a 5′ flanking region upstream of the transcription start site, and the nucleotides defined as introns, coding regions, or 3′ and/or 5′ untranslated but transcribed (UTR) regions.[21]

Regulation of α_1-adrenergic receptors

When adrenergic receptors are stimulated by an agonist, their subsequent responsiveness is decreased or blunted. This biological phenomenon is called desensitization or tachyphylaxis. Several different cellular processes, with different time-courses, seem to be involved. These include modulation of receptor function (minutes), receptor internalization (sequestration and endocytosis; minutes to hours), recycling to the plasma membrane (hours), degradation (hours to days), and regulation of mRNA levels (hours to days).

Two major types of desensitization have been identified. With homologous desensitization, reduced responsiveness is observed to the agonist that originally stimulated the receptor. In heterologous desensitization, a decreased responsiveness is observed with an agent unrelated to the initial stimulus. It is generally accepted that homologous desensitization involves receptor phosphorylation by a G protein-coupled receptor kinase of the agonist-occupied receptor, resulting in the binding of β-arrestin.[22,23]

The α_{1B}-adrenergic receptor has been the most studied of the α_1-subtypes with respect to desensitization.[22] Agonist stimulation results in the phosphorylation of specific sites located in the carboxyl tail of the receptor (serines 404, 408 and 410) that is associated with desensitization and internalization. Direct activation of protein kinase $_C$ leads to α_{1B}-receptor desensitization and internalization associated with phosphorylation by protein kinase C (serines 394 and 400). Activation of other G_q-coupled receptors as well as G_i-coupled receptors can also induce α_{1B}-receptor desensitization and involves phosphorylation by protein kinase C.[22,24] It appears that the α_{1A}- and the α_{1D}-subtypes also undergo desensitization that involves phosphorylation, although evidence is less complete.[25,26]

Table 3.3 Alpha$_2$-adrenergic receptor agents.

Receptor	Agonists	Antagonists
α_2	Brimonidine (Alphagan)	Atipamizole
	Clonidine (Catapres)	Idazoxan
	Dexmedetomidine (Primadex)	RX821002
	Guanabenz (Wytensin)	Yohimibine (Yohimex)
	Guanfacine (Tenex)	
α_{2A}	Ozymetazoline (Afrin)	BRL44408
α_{2B}		ARC-239
		Prazosin
α_{2C}		Rauwolscine

Transcriptional regulation of α_1-adrenergic receptors has been shown to occur under a variety of conditions including hypoxia, ischemic reperfusion, agonist stimulation, cyclic adenosine monophosphate (cyclic AMP) levels, and growth factors.[24] The α_{1B} gene is up-regulated under hypoxic conditions in vascular smooth muscle cells, whereas the α_{1D}-receptor is selectively down-regulated (both mRNA and protein) by platelet-derived growth factor-BB in this same cell type. Stimulation of neonatal rat cardiomyocytes with norepinephrine leads to an increase in α_{1A} message levels, whereas α_{1B} and α_{1D} levels are reduced. Several other agents, including phorbol esters, endothelin-1 and prostaglandin F_2, also cause hypertrophy as well as aortic banding in the whole animal, creating hypertrophy through pressure overload. These agents also cause similar effects on messengers for α_1-receptor subtypes.[27]

Characteristics of α_2-adrenergic receptors

Pharmacological characteristics of α_2-adrenergic receptors

In addition to norepinephrine and epinephrine, α_2-receptors are activated by several nonselective and subtype selective agonists (Table 3.3). Agents such as clonidine (Catapres), guanfacine (Tenex), guanabenz (Wytensin), and brimonidine (Alphagan) are relatively selective for α_2-receptors and have lower affinity at α_1- and β-receptors. By contrast, they have similar affinities for the three α_2-subtypes and are thus considered to be nonsubtype-selective agonists. Similarly, antagonists including yohimibine (Yohimex), idazoxan and RX821002 are relatively selective for α_2-receptors and block α_1- and β-receptors only at higher concentrations. These antagonists have similar affinities for the three α_2-subtypes. Antagonist that are, at least, somewhat selective for one of the α_2-subtypes include BRL44408 for the α_{2A}, prazosin and ARC-239 for the α_{2B} (note, however, that these two agents have much higher affinities for α_1-receptors), and rauwolscine for the α_{2C}-subtype.[28] Ozymetazoline is a partial agonist that has a higher affinity for the α_{2A}-subtype as compared to the α_{2B}- and α_{2C}-subtypes.

Molecular characteristics of α_2-adrenergic receptors

The α_2-adrenergic receptors are single polypeptide chains of 450 to 462 amino acid residues that span the membrane seven times, with the amino terminal being extracellular and the carboxy terminal intracellular (Figure 3.3). Some of the molecular characteristics of the α_2-adrenergic receptors in comparison to the α_1- and β-receptor types are compiled in Table 3.2. In contrast to the α_1- and β-receptors, the α_2-receptors tend to have long third intracellular loops (148 to 179 amino acid residues) and a short carboxy-terminal tail (20 to 21 amino acid residues). The amino terminals of the α_{2A}- and α_{2C}-subtypes have two consensus sites for N-linked glycosylation, and the carboxy-terminal tails of all three subtypes are potentially palmitolyated, thus anchoring the tail to the membrane and forming a fourth intracellular loop. The third intracellular loops have multiple sites of phosphorylation that are thought to be important in the desensitization, recycling, and down-regulation of the receptor. The α_2-adrenergic receptor genes do not contain introns, and thus there are no splice variants.

Regulation of α_2-adrenergic receptors

Most α_2-adrenergic receptors also undergo desensitization and down-regulation (a decrease in receptor density) following stimulation by epinephrine, norepinephrine, or other α_2-agonists, as described above for the α_1-receptors (see Regulation of α_1-adrenergic receptors). There are, however, subtype, tissue, and species differences in the extent of desensitization.[29] Desensitization after short-term agonist exposure appears to involve G protein-coupled receptor kinase-dependent phosphorylation of sites in the third intracellular loop of the human α_{2A}- and α_{2B}-receptors. In contrast, the human α_{2C}-receptor does not desensitize, although the opossum α_{2C}-receptor does desensitize.[30,31] Agonist-induced down-regulation in response to long-term (approximately 24 hour) epinephrine or norepinephrine exposure occurs for all three subtypes in cell lines normally expressing the receptor as well as in cell lines transfected with the receptors. The time course of down-regulation varies with subtype, species, and cell type. In cell lines, it appears that an increase in the rate of receptor degradation is the mechanism for down-regulation of α_{2A}- and α_{2B}-receptors.[32] A common polymorphism has been identified in the third intracellular loop of the α_{2B} (a deletion of three glutamate residues), resulting in a loss of short-term agonist-induced desensitization.[33]

Receptor regulation appears to be important in the etiology and treatment of clinical depression. A characteristic of most antidepressant drugs is that they increase the

Table 3.4 Beta-adrenergic receptor agents.

Receptor	Agonists	Antagonists
β	Isoproterenol (Isuprel)	Carvedilol (Coreg) Nadolol (Corgard) Pindolol (Visken) Propranolol (Inderal) Timolol (Blocadren)
β_1	Prenalterol	Acebutolol (Sectral) CGP 207712 Esmolol (Brevibloc) Metoprolol (Lopressor),
β_2	Albuterol (Ventolin), Metaproterenol (Alupent), Ritodrine (Yutpor), Salmeterol (Serevent), Terbutaline (Brethine)	ICI 118551
β_3	BRL37344	

synaptic concentration of norepinephrine (and other monoamines) by inhibiting metabolism (monoamine oxidase inhibitors), blocking the high affinity reuptake (classical tricyclic antidepressants, selective norepinephrine reuptake inhibitors), or blocking the presynaptic α_2-adrenergic inhibition of neurotransmitter release (α_2-antagonists). Thus, the mechanisms of down-regulation produced by these drugs are thought to be similar to those of agonist-induced down-regulation. Alpha$_2$-adrenergic receptors in the rat brain are desensitized and down-regulated following chronic, but not acute, administration of antidepressants.[34–36] Recent evidence from knockout mice suggests that the effects of the antidepressant imipramine are mediated by the α_{2A}-receptor subtype.[37]

Characteristics of β-adrenergic receptors

Pharmacological characteristics of β-adrenergic receptors

In addition to norepinephrine and epinephrine, β-adrenergic receptors are activated by a variety of nonselective and subtype-selective agonists (Table 3.4). Isoproterenol (Isuprel) is the prototypic nonsubtype-selective β-agonist which has no activity at α_1- and α_2-receptors, except at high concentrations. Epinephrine is 10- to 100-fold more potent at the β_2-receptor as compared to the β_1-subtype; whereas norepinephrine is more potent than epinephrine at the β_3-subtype. Many β_2-selective agonists have been developed for the treatment of asthma. Due to their subtype selectivity, they have a lower incidence of side effects mediated by the β_1-receptor. These β_2-selective agonists include metaproterenol (Alupent), terbutaline

(Brethine), albuterol (Ventolin), salmeterol (Serevent) and ritodrine (Yutpor). A relatively selective agonist for the β_3-receptor is BRL37344.[4]

Propranolol (Inderal) is the prototypic nonsubtype-selective β-antagonist which has equal affinities at the β_1- and β_2-subtypes and no activity at α_1- and α_2-receptors, except at high concentrations. Other nonselective β-adrenergic antagonists include nadolol (Corgard), timolol (Blocadren), pindolol (Visken), which is actually a weak partial agonist) and carvedilol (Coreg), which is also an α_1-antagonist. Propranolol, as well as other classical β-antagonists, has low affinity for the β_3-subtype, which has previously been referred to as the atypical β-receptor.[28] Several β_1-selective antagonists have been developed such as metoprolol (Lopressor), esmolol (Brevibloc), acebutolol (Sectral) and CGP 207712. A useful β_2-selective antagonist is ICI 118551.[28]

Molecular characteristics of β-adrenergic receptors

The β-adrenergic receptors are single polypeptide chains of 408 to 477 amino acid residues that span the membrane seven times, with the amino terminal being extracellular and the carboxy terminal intracellular. Some of the molecular characteristics of the β-adrenergic receptors in comparison to the α_1- and α_2-receptor types are compiled in Table 3.2. In contrast to the α_2-receptors, but similar to the α_1-receptors, the β-receptors tend to have longer carboxy-terminal tails (61 to 97 amino acid residues) and shorter third intracellular loops (54 to 80 amino acids). The amino terminal of the β-receptors has one or two consensus sites for N-linked glycosylation. The carboxy-terminal tails of all three subtypes are potentially palmitolyated, thus anchoring the tail to the membrane and forming a small fourth intracellular loop. The carboxy-terminal tails also have multiple sites of phosphorylation, which are thought to be important in the desensitization, recycling, and down-regulation of the receptor.

The β_1- and β_2-adrenergic receptor genes do not contain introns, and thus they have no splice variants. By contrast, the β_3-receptor has one intron, resulting in two splice variants. However, no functional differences have been found between the splice variants.[5]

Regulation of β-adrenergic receptors

The processes involved in desensitization have been extensively investigated for the β_2-adrenergic receptor, which is the prototypic G protein-coupled receptor model with cells in culture. Initial uncoupling of the β_2-receptor from the G protein after agonist binding is mediated by phosphorylation of specific residues in the carboxyl tail of the receptor. The phosphorylated β_2-receptor serves as a substrate for the binding of β-arrestin, which not only uncouples the receptor from the signal transduction process but also serves as an adapter protein that mediates the binding of additional signaling proteins and entry into the internalization pathway.[38,39] The mechanisms of β_2-adrenergic receptor down-regulation appear to involve both an increase in the rate of degradation of the receptor as well as a decrease in the levels of β-receptor mRNA.[40]

A common result of chronic, but not acute, antidepressant drug administration to rats is down-regulation of β-adrenergic receptors and/or desensitization of β-adrenergic-stimulated adenylate cyclase. This has been observed with most classes of antidepressant drugs as well as with electroconvulsive shock.[41,42] This down-regulation is not immediate, but occurs only after days to weeks of drug treatment, consistent with the therapeutic time course. Down-regulation is more likely to be observed with the β_1-, as compared to the β_2-, adrenergic receptor.[43]

Signal transduction pathways

α_1-adrenergic receptors

All three of the α_1-adrenergic receptor subtypes activate the $G_{q/11}$ family of G proteins leading to the dissociation of the α and β subunits, and the subsequent stimulation of the enzyme phospholipase C (Table 3.5). This enzyme hydrolyzes phosphatidylinositol in the membrane, producing inositol 1, 4, 5-trisphosphate (IP_3) and diacylglycerol. These molecules act as second messengers mediating intracellular Ca^{2+} release via the IP_3 receptor and activating protein kinase C. Other signaling pathways that have also been shown to be activated by α_1-receptors include Ca^{2+} influx via voltage-dependent and -independent Ca^{2+} channels; arachidonic acid release; and activation of phospholipase A_2, phospholipase D activation and mitogen-activated protein kinase (MAPK).[20,44,45] In general, the α_{1A}-subtype appears to couple most efficiently to intracellular signaling cascades whereas the α_{1D} couples least efficiently.[46] All three subtypes activate $G_{q/11}$, but only the α_{1A}- and α_{1B}-subtypes couple to G_{14}, and only α_{1B} couples to G_{16}.[47] Alpha$_{1B}$-receptors (but not the other two) can also couple to G_o.

In addition to regulating Ca^{2+} movements and smooth muscle contraction, the α_1-adrenergic receptors also help regulate cell proliferation through the three major subfamilies of the MAPK family (extracellular regulated kinases or ERK, c-Jun N-terminal kinases or JNK, and the p38 kinases). The activation of the various MAPK members appears to be highly dependent on the particular α_1-receptor subtype(s) present and also on the particular cell line or tissue being studied.[20,47]

α_2-adrenergic receptors

All three of the α_2-adrenergic receptor subtypes activate the $G_{i/o}$ family of G proteins, leading to the dissociation of the α and β subunits, and a subsequent alteration (classically inhibition) in the activity of the enzyme adenylate cyclase. This

Table 3.5 Functional characteristics of adrenergic receptor subtypes.

Subtype	Physiological roles	Rank order of potency[a]	G protein family	Effector	2nd Messengers[b]
α_{1A}	Vascular contraction Attenuate hypertrophy	E = NE >> ISO	$G_{q/11}$	Activation of phospholipase C, MAP kinase	Ca^{2+}, DAG, IP_3
α_{1B}	Vascular contraction Induce hypertrophy	E = NE >> ISO	$G_{q/11}$ (G_o)	Activation of phospholipase C, MAP kinase	Ca^{2+}, DAG, IP_3
α_{1D}	Vascular contraction	E = NE >> ISO	$G_{q/11}$	Activation of phospholipase C, MAP kinase	Ca^{2+}, DAG, IP_3
α_{2A}	Hypotension, sedation, analgesia, hypothermia,	E = NE >> ISO	$G_{i/o}$	Inhibition of adenylate cyclase. Inhibits NE release at high stimulation frequencies Inhibits Ca^{2+} channels Activation of K^+ channels	Cyclic AMP, Ca^{2+}
α_{2B}	Vascular contraction (developmental processes)	E = NE >> ISO	$G_{i/o}$	Inhibits adenylate cyclase	Cyclic AMP
α_{2C}	Cognitive processes (hypothermia)	E = NE >> ISO	$G_{i/o}$	Inhibits adenylate cyclase. Inhibits NE release at low stimulation frequencies	Cyclic AMP
β_1	Increase in rate and force of cardiac contraction	ISO > E = NE	G_s	Activation of adenylate cyclase	Cyclic AMP
β_2	Relaxation of smooth muscle; glycogenolysis	ISO > E > NE	G_s	Activation of adenylate cyclase	Cyclic AMP
β_3	Thermogenesis, lipolysis	ISO = NE > E	G_s	Activation of adenylate cyclase	Cyclic AMP

[a] E, epinephrine; NE, norepinephrine; ISO, isoproterenol

[b] DAG, diacylglycerol; IP_3, inositol 1,4,5-trisphosphate; MAP mitogen-activated protein

process leads to changes in the concentration of the second messenger cyclic AMP (Table 3.5). In addition, the stimulation of α_2-receptors can affect several other effector systems including the activation of K^+ channels, inhibition or activation of Ca^{2+} channels, and activation of phospholipase A_2, phospholipase C, and Na^+/H^+ exchange.[48]

The regulation of adenylate cyclase by receptors that activate the G_i family of G proteins is complex due to the existence of nine differentially regulated isozymes (or isoforms), each the product of a different gene.[49] The α_i subunit inhibits the activity of isoforms 5 and 6 by acting at a site symmetrical to the G_s binding site (which stimulates enzyme activity). In addition, the released $\beta\gamma$ subunits are strong modulators of adenylate cyclase activity and can either be stimulatory (at isoforms 2, 4, and 7) or inhibitory (at isoforms 1 and 8). In fact, the $\beta\gamma$ subunit is among the most potent of all negative regulators of isoforms 1 and 8, and it can markedly inhibit the enzyme activity stimulated by forskolin, G_s, and Ca^{2+}-calmodulin. This is particularly relevant to the functioning of the central nervous system, because both the G_i proteins and cyclase isoforms 1 and 8 are highly expressed in the brain.[49] Released $\beta\gamma$ subunits stimulate the cyclase activity of isoforms 2, 4, and 7, but only when adenylate cyclase has already been activated by G_s. This synergistic activation of cyclase by receptors coupled to G_i and by receptors coupled to G_s has been termed "coincidence signaling" and may play a prominent role in the integration of the various signals received by a given neuron.

Alpha$_2$-adrenergic activation of G_i protein-gated K^+ channels results in membrane hyperpolarization, which in turn causes a decrease in the firing rate of excitable cells in the central nervous system. This effect is Ca^{2+} dependent in many, but not all, systems.[48] Decreased Ca^{2+} conductance mediates α_2-adrenergic inhibition of neurotransmitter release, one of the most prominent physiological actions of α_2-receptors. This effect involves direct inhibition of N-type voltage-gated Ca^{2+} channels, and can be independent of cyclic AMP and protein phosphorylation.[48]

β-adrenergic receptors

All three of the β-adrenergic receptor subtypes activate the G_s family of G proteins leading to the dissociation of the α and β_γ subunits, and a subsequent activation of the enzyme adenylate cyclase, which then leads to increases in the concentration of the second messenger cyclic AMP (Table 3.5). In addition to G proteins, β-adrenergic receptors interact with many other signaling proteins. For example, the cytoplasmic tail of the β_2-adrenergic receptor binds to the phosphoprotein EBP50 (ezrinradixin-moesin-binding phosphoprotein-50) through a PDZ domain and to the Na^+/H^+ exchanger regulatory factor (NHERF) also through a PDZ domain,[50] whereas the β_1-adrenergic receptor binds to CNrasGEF (cyclic AMP-dependent Ras exchange factor), which leads to Ras activation.[51] The identification and functional significance of these interactions is currently an area of intense study.[50] Following binding of agonist, the β_2-receptor is phosphorylated by β-adrenergic receptor kinase (βARK), and the phosphorylated receptor binds β-arrestin.[38,39] The β-arrestin serves as an adapter or scaffolding protein, which binds a variety of other proteins such as clathrin (important in the internalization of the receptor) and various signaling proteins including Src family tyrosine kinases and the ERK and JNK MAP kinases.[39,50]

Physiological roles

α_1-adrenergic receptors

In the periphery, a primary function of α_1-adrenergic receptors appears to be in maintaining resting vascular tone (Table 3.5). This is due primarily to the presence of postjunctional α_1-receptors in a wide array of vascular beds in the systemic arterial circulation. By contrast, in the venous system both α_1- and α_2-receptors contribute to vasoconstriction.[44] The question of which α_1-subtype is involved in vasoconstriction is not easy to determine because most vascular smooth muscle tissues express mixtures of α_1-subtypes. Thus, in most cases, responses to α_1-agonists are probably due to activation of more than one subtype. For example, in the rat the α_{1A}- and α_{1D}-subtypes appear to regulate the larger vessels, whereas the α_{1B}-subtype controls the small resistance vessels.[44] Recent studies with knockout mice also indicate that all subtypes play a role in blood pressure response to α_1-agonists through their vasocontractile responses. The extent of the contribution of each of the α_1-subtypes to the contraction of vascular smooth muscle involved has not yet been completely resolved by these studies.[52] In addition, α_1-receptors may also be involved in the regulation of vascular smooth muscle growth as α_{1B}-receptors appear to induce hypertrophy of arterial smooth muscle cells, whereas the α_{1A}-receptors attenuate this growth response.[44]

Selective α_{1A}-adrenergic receptor antagonists appear to have significant therapeutic advantages over nonsubtype-selective α_1-antagonists in the treatment of benign prostatic hypertrophy. Although the α_{1A}-subtype is the predominant α_1-subtype in human prostatic smooth muscle and does mediate contraction in this tissue, functional studies have not yet shown that this is the primary subtype mediating the beneficial response.[44]

Although the predominant postjunctional adrenergic receptor type in the heart is the β-adrenergic receptor, the α_1-receptors play an important subsidiary role, particularly in increasing the force of contraction at low rates of beating. This ionotropic effect appears to be mediated by both the α_{1A}- and α_{1B}-subtypes.[53,54] The α_{1A}-subtype also appears to be the subtype primarily responsible for the events leading to cardiac hypertrophy caused by adrenergic agents.[54]

Representative estimates for the density of adrenergic receptor types and subtypes are given in Table 3.6. For the three types of adrenergic receptors, the densities are

Table 3.6 Density of adrenergic receptors in the brain.

Type/subtype	Human			Rat		
	Tissue	fmol/mg protein	Reference		fmol/mg protein	Reference
α_1	Cortex	147	(79)	Cortex	210	(83)
α_1	Caudate	20	(80)	Cortex	132	(84)
α_{1A}	Cortex	59	(79)	Cortex	46	(84)
α_{1B}	Cortex	88	(79)	Cortex	86	(84)
α_2	Cortex	197	(81)	Cortex	90	(85)
α_2	Caudate	110	(80)	Striatum	159	(86)
α_{2A}	Cortex	197	(81)	Cortex	90	(85)
α_{2A}	Caudate	57	(80)	Striatum	95	(86)
α_{2C}	Caudate	53	(80)	Striatum	64	(86)
β	Cortex	25	(82)	Cortex	300	(87)
β_1	Cortex	17	(82)	Cortex	250	(88)
β_2	Cortex	6	(82)	Cortex	60	(88)

in the same general range (up to several hundred fmol/mg of particulate protein). For the α_1-receptors, the α_{1B}-subtypes predominate in the cortex in both human and rat brains.

The role of the α_1-receptors in central nervous system function is poorly understood. These receptors appear to be postsynaptic, stimulatory receptors and effect many brain functions, at least in part through nonneuronal mechanisms because they are also present in glial cells.[52] Recent data suggest that the α_1-receptors are involved in locomotion and the control of motor activity, as well as in cognitive functions and behaviors such as learning, memory, and fear. Interestingly, the α_{1B}-subtype may enhance apoptotic neurodegeneration, which suggests that α_{1B}-antagonists may have therapeutic potential in the treatment of human neurodegenerative disorders.[52]

Several α_1-adrenergic receptor polymorphisms have been noted, including a C to T polymorphism in the α_{1A}-subtype. No linkage, however, has been found between this polymorphism and attention-deficit/hyperactivity disorder, Alzheimer's disease, or the reward-dependence personality trait.

Alpha$_2$-adrenergic receptors

Subtype specific functions of α_2-adrenergic receptors

Alpha$_2$-adrenergic receptors, which are found throughout the peripheral and central nervous systems, mediate a variety of effects including hypotension, sedation, and antinociception, as well as a variety of behavioral effects (Table 3.5). In addition,

α_{2A}-receptors on platelets mediate aggregation, and stimulation of α_2-receptor in the pancreas inhibits insulin release.

Activation of α_{2A}-receptors in the rostral ventrolateral medulla decreases sympathetic outflow, causing a reduction in arterial blood pressure and heart rate.[55] This mechanism is the means by which α_2-agonists such as clonidine lower blood pressure in hypertensive patients. In addition to this centrally mediated response, there is a transient hypertensive response following rapid intravenous injection, caused mainly by α_{2B}-adrenergic receptor-mediated vasoconstriction of vascular smooth muscle.[56,57] The α_2-subtype mediating these effects has been determined through the use of mice lacking each of the subtypes (knockout animals). The hypotensive response to administration of α_2-agonists was abolished in α_{2A}-knockout animals, demonstrating that the α_{2A}-subtype plays a principal role in this response. The bradycardic response to agonist was also blunted.[58] The hypertensive response was abolished in α_{2B} knockout mice, and the hypotensive effect was immediate and accentuated. The bradycardic response in α_{2B} knockout mice was normal, and α_{2C} knockout mice showed no differences from wild-type strains in their hypertensive, hypotensive, and bradycardic effects.[56,57]

Due to the sedative effects mediated by α_{2A}-adrenergic receptors, α_2-agonists have been used as anesthetic agents in veterinary practice for the past 25 years, and are now being used in humans as adjuncts to anesthesia.[59] The sedative effects of α_2-agonists were abolished in mice lacking the α_{2A}-subtype as determined by rotarod, loss of righting reflex, and spontaneous locomotor activity tests,[60] indicating that the α_{2A}-adrenergic receptor mediates the sedative effects of agonist administration.[56,57] In contrast, both α_{2B} and α_{2C} knockout mice show dose-dependent sedative effects that are indistinguishable from wild-type mice. α_2-agonists appear to induce sedation by activating presynaptic autoreceptors in the locus coeruleus, reducing its spontaneous rate of firing. In addition, α_2-agonists can reduce the requirement for other anesthetic agents by as much as 90%.[59] This anesthetic-sparing effect also appears to be mediated by the α_{2A}-subtype.

Another therapeutic effect mediated by the α_{2A}-adrenergic receptor is analgesia. α_2-adrenergic agonists have analgesic properties when given parenterally, epidurally, or intrathecally. Descending noradrenergic antinociceptive systems originating in the brain stem contribute to pain control by suppressing the spinal centripetal transmission of nociceptive impulses.[59] Once again the α_{2A}-subtype appears to mediate this effect, based on studies in genetically engineered mice using the ramped hot plate, hot water immersion, and tail-flick latency tests.[56,57]

Similar to the hypertensive, sedative, and analgesic effects, the hypothermic effect also appears to be mediated primarily by the α_{2A}-subtype with a small contribution from the α_{2C}-subtype. The α_{2A}-adrenergic receptor also mediates the antiepileptogenic actions of norepinephrine in the kindling model of epileptogenesis.[56]

The α_{2A}-adrenergic receptor also appears to be the subtype mediating the effects of antidepressant drugs in the mouse.[37] A genetic knockout of the α_{2A}-receptor makes mice less active in the forced-swim test and insensitive to the antidepressant effects of the tricyclic drug imipramine. Thus, the α_{2A}-adrenergic receptor may play a protective role in some forms of human depression and anxiety and the antidepressant effects of imipramine and other antidepressants may be mediated by the α_{2A}-adrenergic receptor.

Presynaptic inhibition of norepinephrine release is a classic α_2 adrenergic receptor function. α_2-agonists can inhibit release of norepinephrine from nerve terminals and antagonists can enhance release. These negative feedback effects are mediated by terminal and by somadendritic α_2-receptors.[61] There is, however, some debate as to whether this extends to endogenously released norepinephrine. Some authors question if, under normal physiological conditions, there is local regulation of neuronal transmitter release by autoreceptors as a result of these receptors being activated by the perineuronal concentration of previously released transmitter.[62] Initial data from knockout animals indicated that the α_{2A}-subtype was the most important in mediating presynaptic α_2-adrenergic receptor inhibition of neurotransmitter release, although a role for at least one other subtype seemed probable. Subsequent studies on the sympathetic nerves in the hearts of mice lacking one or both (double knockout) of the α_{2A}- and the α_{2C}-subtypes indicated that the α_{2A}-receptor inhibits transmitter release at high stimulation frequencies, whereas the α_{2C}-subtype regulates release at lower levels. The regulation at both high and low frequencies appears to be physiologically important.[56,57]

In comparison to the α_{2A}-subtype, relatively less has been discovered about the functions of the α_{2B}- and α_{2C}-subtypes through knockout experiments. As noted above, the α_{2B}-subtype appears to have a dominant role in eliciting the vasoconstrictor response to α_2-adrenergic agonists. The α_{2B}-adrenergic receptor has also been implicated in salt-induced hypertension and may be important in developmental processes, although the role it plays is currently unknown.

The α_{2C}-subtype does not appear to play a significant role in cardiovascular regulation or the other classical effects of α_2-adrenergic receptors, with the exceptions of the hypothermic and presynaptic effects noted above. Studies with genetically modified mice suggest that the α_{2C}-subtype may play a role in stress-dependent depression, in startle responses and prepulse inhibition (which may relate to schizophrenia, attention-deficit-hyperactivity disorder and posttraumatic stress disorder), in modulating motor behavior, and perhaps in memory processes.[56,57]

The α_{2C}-receptor has a higher density in the striatum compared to other brain regions, but its physiologic role is unknown. Recent studies with knockout mice suggest that the α_{2C}-subtype regulates striatal GABA release and that the endogenous catecholamine may be dopamine rather than norepinephrine.[63]

The α_{2C}-subtype may also play a role in modulation of schizophrenia symptoms by antipsychotic drugs. Although the antipsychotic agents display rather equivalent efficacy against the positive symptoms of schizophrenia, they differ with respect to improvement of the negative symptoms such as cognitive dysfunction. These differences are potentially related to the relative affinities of the antipsychotics for the dopamine D_2-receptor and for receptors for other neurotransmitter systems. Interestingly, the affinity of clozapine, one of the so-called atypical antipsychotics, is about 12 times more potent at the α_{2C}-receptor than at the dopamine D_2-receptor, suggesting that α_{2C}-adrenergic receptor blockade may contribute to the improvement of cognitive function produced by clozapine.[64]

Representative estimates for the density of α_2-adrenergic receptors and the subtypes are given in Table 3.6. Except for the caudate (or striatum in the rat), the predominate subtype is the α_{2A} (or α_{2D} in the rat) in nearly all brain areas.[65]

Alpha$_2$-adrenergic receptor polymorphisms

Polymorphisms have been identified in each of the α_2-adrenergic receptor subtypes. For the α_{2A}-subtype, there are two known polymorphisms in the promoter region (identified by Msp-I and Dra-I). No evidence was found for linkage of the α_{2A} gene with attention-deficit-hyperactivity disorder. In addition, there is an infrequent polymorphism (lysine for asparagine at amino acid residue 252) that results in enhanced G_i coupling and a six residue (255 to 260) deletion.[66]

A common polymorphism has been identified in the third intracellular loop of the α_{2B}-receptor, which consists of a deletion of three glutamate residues (301 to 303) and is a risk factor for acute coronary events but not for hypertension. This deletion results in a loss of short-term agonist-induced desensitization.[66]

For the α_{2C}-subtype, a common polymorphism has been identified in the third intracellular loop, which consists of a deletion of four amino acid residues (322 to 325), and results in an impaired coupling to several effectors.[66] No evidence for linkage of a dinucleotide repeat polymorphism located approximately 6 kb from the gene was found with attention-deficit-hyperactivity disorder.[67]

β-adrenergic receptors

β_1- and β_2-adrenergic receptors

β-adrenergic receptors modulate a wide range of physiological responses in the intact animal, including cardiac contraction, vascular and smooth muscle tone, and carbohydrate and lipid metabolism (Table 3.5). The β_1-subtype is considered to be the "cardiac" β-receptor because these receptors predominate in the heart, and in vivo stimulation of these receptors with agonists increases both force and rate of cardiac contraction. β_2-adrenergic receptors are also expressed in the heart and, in some species, influence heart rate and contractility.[68] There is considerable heterogeneity in the tissue distribution and function of β receptor subtypes in the heart. In the adult human left ventricle, the ratio of β_1- to β_2-receptors is 80:20; whereas in the atria the ratio decreases to 70:30.[69] The β_2-adrenergic receptor is the predominant subtype mediating peripheral vascular, as well as bronchial and uterine smooth muscle, relaxation. Thus, β_2-agonists are useful in the treatment of asthma and premature labor. There are also differences in β-adrenergic signaling during development, and cardiac β_2-adrenergic receptors may play a more substantial role in mediating contractile changes in the noninnervated fetal and neonatal heart.[68]

In the central nervous system, β_1 adrenergic receptors are generally associated with forebrain structures such as the cerebral cortex, striatum, and hippocampus. The β_2-subtype predominates in the cerebellum, as determined by radioligand binding, *in situ* hybridization, and immunohistochemistry.[70] There is some evidence that indicates that a significant fraction of the β adrenergic receptors may be located on glial cells.

A decrease in norepinephrine-stimulated cyclic AMP accumulation in the cortex is observed following chronic (but not acute) treatment with a variety of typical and atypical antidepressant agents, as well as for monamine oxidase inhibitors and electroconvulsive therapy. Selective serotonin reuptake inhibitors such as zimelidine and clomipramine as well as norepinephrine reuptake inhibitors reduce norepinephrine-stimulated adenylate cyclase activity. Most, but not all, antidepressant treatments also down-regulate β-adrenergic receptor densities in many brain regions. In spite of intense research, the role played by β-adrenergic receptors in depression and in the mechanism of antidepressant drugs is not well understood.[71]

Gene knockout approaches have been utilized to disrupt expression of all three β adrenergic receptors, individually as well as in combinations. The role of each of the subtypes regulating cardiovascular physiology and metabolism has been determined in these mice.[68] Neither the β_1- nor the β_2-subtype is required for normal cardiac development. Mice lacking the β_1, the β_2, or both subtypes have normal resting heart rate, blood pressure, and cardiac output, suggesting that although these β-adrenergic receptors are responsible for regulating changes in heart rate, blood pressure, and contractility during activity or stress, they are not required for maintaining normal function.[68] The β_1-adrenergic receptor is primarily responsible for sympathetically mediated changes in heart rate during stress or exercise.

Representative estimates for the density of β adrenergic receptors and the subtypes are given in Table 3.6. The cortex is mostly β_1, whereas the cerebellum has a higher density of β_2 as compared to other regions.

β_1- and β_2-adrenergic receptor polymorphisms

The gene encoding the human β_1 adrenergic receptor is quite polymorphic, with 18 single nucleotide polymorphisms (SNPs), 17 within the coding exon for the receptor, and 7 leading to amino acid substitutions. It has been proposed that there may be 11 different genotypes.[72] The two that are of most interest are glycine for serine at amino acid residue 49 and arginine for glycine at residue 389. The latter polymorphism variant shows a gain of function phenotype in that it has an enhanced coupling to G_s and an increased isoproterenol-stimulated cyclic AMP response.[72] The other consistently reported alteration is an enhanced susceptibility to agonist-promoted down-regulation of the glycine-49 genotype. The current data suggest an enhanced function of those two alleles, but the magnitude of the enhancement does not appear to be sufficient to allow consistent detection.[72]

A total of 13 polymorphisms in the β_2 adrenergic receptor gene and its transcriptional regulator upstream peptide have been identified.[73] Three closely linked polymorphisms, two coding the region at amino acid positions 16 and 27, and one in the upstream peptide, are common (i.e., allele frequency > 0.15) in the general white population. The glycine-16 receptor exhibits enhanced down-regulation in vitro after agonist exposure. In contrast, arginine-16 receptors are more resistant to down-regulation.[73] Although initial studies suggested a relationship between the glycine-16 polymorphism and increased risk of severe asthma and increased airway responsiveness, subsequent associations with clinical asthma, atopy, and airway responsiveness have been inconsistent.[74] Liggett has proposed a dynamic model of receptor kinetics to explain the polymorphism data. Glycine-16/glycine-16 homozygous individuals are already down-regulated as a result of exposure to endogenous cathecholamines. Thus, the tachyphylaxis caused by recurring exogenous exposure to a β-adrenergic agonist would be more apparent in the arginine-16/arginine-16 patients because their receptors had not yet been down-regulated. In this model, the initial response to albuterol in β-agonist-naïve patients would be depressed in those who are glycine-16/glycine-16 homozygous, because their receptors had been endogenously down-regulated to a greater extent than the receptors of patients who are arginine-16/arginine-16.[73]

β_3-adrenergic receptors

β_3-adrenergic receptor agonists are very effective thermogenic antiobesity and antidiabetic (insulin-sensitizing) agents in rodents. Their main sites of action are white and brown adipose tissue, and muscle. Although adult humans have little brown adipose tissue, β_3-receptors are expressed in human white, as well as brown, adipose tissue and in skeletal muscle. They play a role in the regulation of energy balance and glucose homeostasis.[75] Early evidence for the β_3-adrenergic receptor stemmed from the discovery that β_1- and β_2-receptor antagonists lacked potency in various gut preparations and as antagonists of β-agonist-driven lipolysis. Novel β-agonists were then discovered, which were more potent as stimulants of lipolysis than of atrial contraction. These compounds were found to stimulate metabolic rate and to have antiobesity and antidiabetic activity in rodents.[75] Although the β_3 is the major β-subtype controlling lipolysis, it also plays a role in regulating vascular tone and cardiac contractility.[68]

Targeted disruption of the mouse β_3-adrenergic receptor creates conditions that predispose the animal to the development of obesity.[76] Somewhat surprisingly, however, none of the mice lacking any one of the three β receptors, nor mice lacking both the β_1- and β_2-subtypes (double knockout), are cold-sensitive or become overtly obese. By contrast, triple knockout mice (β_1, β_2 and β_3 all deleted) exhibit obesity and cold-intolerance.[77] Thus, β-adrenergic signaling through at least one of the subtypes is essential for the resistance to obesity and cold.

A tryptophan-64 to arginine polymorphism has been identified in the β_3-adrenergic receptor. The allele frequence is approximately 30% in the Japanese population, higher in Pima Indians, and lower in Whites.[78] Type 2 diabetic patients with this mutation showed a significantly younger onset-age of diabetes and an increased tendency to obesity, hyperinsulinemia, and hypertension.[78]

Conclusions/summary

Although the existence of norepinephrine and the adrenergic receptors have been known for many years, our understanding of these fascinating receptors is still far from complete. The role of the α_1-receptor in the brain is particularly poorly understood, and the investigations in this area are likely to be very productive. Of special interest are the three subtypes for each of the three types of adrenergic receptors. We suspect that differential regulation of these is an important reason for their existence, and modulation of their regulation provides a prime target for therapeutic intervention. In addition, a fruitful area of study in the next few years will be the delineation of the "protein partners" for each of the nine subtypes and a better understanding of the "signaling complexes" that result from these interactions.

REFERENCES

1. Bylund, D. B. Subtypes of α_2-adrenoceptors: Pharmacological and molecular biological evidence converge. *Trends Pharmacol. Sci.*, **9** (1988), 356–361.
2. Ahlquist, R. P. A study of adrenotropic receptors. *Am. J. Physiol.*, **153** (1948), 586–600.

3. Lands, A. M., Arnold, A., McAuliff, J. P., Luduena, F. P. and Brown, T. G. Differentiation of receptor systems activated by sympathomimetic amines. *Nature*, **214** (1967), 597–598.
4. Arch, J. R. S., Ainsworth, M. A., Cawthorne, M. A. *et al.* Atypical β-adrenoceptors on brown adipocytes as a target for anti-obesity drugs. *Nature*, **309** (1984), 163–165.
5. Strosberg, A. D. Structure and function of the beta-3 adrenergic receptor. *Annu. Rev. Pharmacol. Toxicol.*, **37** (1997), 421–450.
6. Granneman, J. G. The putative beta-4 adrenergic receptor is a novel state of the beta-1 adrenergic receptor. *Am. J. Physiol. Endocrinol. Metab.*, **280** (2001), E199–E202.
7. Langer, S. Z. Presynaptic regulation of catecholamine release. *Biochem. Pharmacol.*, **23** (1974), 1793–1800.
8. Berthelsen, S. and Pettinger, W. A. A functional basis for classification of alpha-adrenergic receptors. *Life Sci.*, **21** (1977), 595–606.
9. Bylund, D. B. Heterogeneity of alpha-2 adrenergic receptors. *Pharmacol. Biochem. Behav.*, **22** (1985), 835–843.
10. Bylund, D. B., Ray-Prenger, C. and Murphy, T. J. Alpha-2A and alpha-2B adrenergic receptor subtypes: antagonist binding in tissues and cell lines containing only one subtype. *J. Pharmacol. Exp. Ther.*, **245** (1988), 600–607.
11. Murphy, T. J. and Bylund, D. B. Characterization of alpha-2 adrenergic receptors in the OK cell, an opossum kidney cell line. *J. Pharmacol. Exp. Ther.*, **244** (1988), 571–578.
12. Michel, A. D., Loury, D. N. and Whiting, R. L. Differences between α_{2A} adrenoceptor in rat submaxillary gland and the α_{2A} and α_{2B} adrenoceptor subtypes. *Br. J. Pharmacol.*, **98** (1989), 890–897.
13. Simonneaux, V., Ebadi, M. and Bylund, D. B. Identification and characterization of α_{2D}-adrenergic receptors in bovine pineal gland. *Mol. Pharmacol.*, **40** (1991), 235–241.
14. Hieble, J. P., Bylund, D. B., Clarke, D. E. *et al.* International union of pharmacology. X. Recommendation for nomenclature of α_1-adrenoceptors: consensus update. *Pharmacol. Rev.*, **47** (1995), 267–270.
15. Morrow, A. L. and Creese, I. Characterization of alpha-1 adrenergic receptor subtypes in rat brain: a reevaluation of [^{3}H]WB 4101 and [^{3}H]prazosin binding. *Mol. Pharmacol.*, **29** (1986), 321–330.
16. Minneman, K. P., Han, C. and Abel, P. W. Comparison of alpha-1 adrenergic receptor subtypes distinguished by chloroethylclonidine and WB4101. *Mol. Pharmacol.*, **33** (1988), 509–514.
17. Muramatsu, I., Murata, S., Isaka, M. *et al.* Alpha-1 adrenoceptor subtypes and two receptor systems in vascular tissues. *Life Sci.*, **62** (1998), 1461–1465.
18. Ford, A. P. D. W., Daniels, D. V., Chang, D. J. *et al.* Pharmacological pleiotropism of the human recombinant α_{1A}-adrenoceptor: implications for α_1-adrenoceptor classification. *Br. J. Pharmacol.*, **121** (1997), 1127–1135.
19. De Lean, A., Hancock, A. A. and Lefkowitz, R. J. Validation and statistical analysis of a computer modeling method for quantitative analysis of radioligand binding data for mixtures of pharmacological receptor subtypes. *Mol. Pharmacol.*, **21** (1982), 5–16.
20. Zhong, H. Y. and Minneman, K. P. Alpha-1adrenoceptor subtypes. *Eur. J. Pharmacol.*, **375** (1999), 261–276.

21. Scofield, M. A., Deupree, J. D. and Bylund, D. B. Adrenergic receptor genes: cDNA and genomic library construction. *Mol. Biotechnol.*, **21** (2002), 171–197.
22. García-Sáinz, J. A., Vázquez-Prado, J. and del Carmen, M. L. Alpha 1-adrenoceptors: function and phosphorylation. *Eur. J. Pharmacol.*, **389** (2000), 1–12.
23. Rockman, H. A., Koch, W. J. and Lefkowitz, R. J. Seven transmembrane-spanning receptors and heart function. *Nature*, **415** (2002), 206–212.
24. Michelotti, G. A., Price, D. T. and Schwinn, D. A. α_1-Adrenergic receptor regulation: basic science and clinical implications. *Pharmacol. Ther.*, **88** (2000), 281–309.
25. García-Sáinz, J. A., Vázquez-Prado, J. and Medina, L. D. α_1-Adrenoceptors: function and phosphorylation. *Eur. J. Pharmacol.*, **389** (2000), 1–12.
26. García-Sáinz, J. A., Vázquez-Cuevas, F. G. and Romero-Avila, M. T. Phosphorylation and desensitization of alpha-1D adrenergic receptors. *Biochem. J.*, **353** (2001), 603–610.
27. Rokosh, D. G., Stewart, A. F. R., Chang, K. C. *et al.* α_1-Adrenergic receptor subtype mRNAs are differentially regulated by α_1-adrenergic and other hypertrophic stimuli in cardiac myocytes in culture and in vivo-repression of α_{1B} and α_{1D} but induction of α_{1C}. *J. Biol. Chem.*, **271** (1996), 5839–5843.
28. Bylund, D. B., Eikenberg, D. C., Hieble, J. P. International Union of Pharmacology nomenclature of adrenoceptors. *Pharmacol. Rev.*, **46** (1994), 121–136.
29. Saunders, C. and Limbird, L. E. Localization and trafficking of α_2-adrenergic receptor subtypes in cells and tissues. *Pharmacol. Ther.*, **84** (1999), 193–205.
30. Eason, M. G. and Liggett, S. B. Subtype-selective desensitization of alpha-2 adrenergic receptors. Different mechanisms control short and long-term agonist-promoted desensitization of alpha 2C10, alpha 2C4, and alpha 2C2. *J. Biol. Chem.*, **267** (1992), 25 473–25 479.
31. Jones, S. B., Leone, S. L. and Bylund, D. B. Desensitization of the alpha-2 adrenergic receptor in H. T29 and opossum kidney cell lines. *J. Pharmacol. Exp. Ther.*, **254** (1990), 294–300.
32. Heck, D. A. and Bylund, D. B. Mechanism of downregulation of alpha-2 adrenergic receptor subtypes. *J. Pharmacol. Exp. Ther.*, **282** (1997), 1219–1227.
33. Small, K. M., Brown, K. M., Forbes, S. L. and Liggett, S. B. Polymorphic deletion of three intracellular acidic residues of the alpha-2B adrenergic receptor decreases GRK-mediated phosphorylation and desensitization. *J. Biol. Chem.*, **276** (2001), 4917–4922.
34. Invernizzi, R. W., Parini, S., Sacchetti, G. *et al.* Chronic treatment with reboxetine by osmotic pumps facilitates its effect on extracellular noradrenaline and may desensitize α_2-adrenoceptors in the prefrontal cortex. *Br. J. Pharmacol.*, **132** (2001), 183–188.
35. Nomura, S., Duman, R. S. and Enna, S. J. In vivo or in vitro exposure to imipramine reduces alpha 2 adrenoceptor-mediated inhibition of cyclic AMP production in rat brain cerebral cortical slices. *Brain Res.*, **410** (1987), 195–198.
36. Smith, C. B., Garcia-Sevilla, J. A. and Hollingsworth, P. J. Alpha 2-adrenoreceptors in rat brain are decreased after long-term tricyclic antidepressant drug treatment. *Brain Res.*, **210** (1981), 413–418.
37. Schramm, N. L., McDonald, M. P. and Limbird, L. E. The α_{2A}-adrenergic receptor plays a protective role in mouse behavioral models of depression and anxiety. *J. Neurosci.*, **21** (2001), 4875–4882.

38. Ferguson, S. S. G. Evolving concepts in G protein-coupled receptor endocytosis: the role in receptor desensitization and signaling. *Pharmacol. Rev.*, **53** (2001), 1–24.
39. Luttrell, L. M. and Lefkowitz, R. J. The role of β-arrestins in the termination and transduction of G-protein-coupled receptor signals. *J. Cell Sci.*, **115** (2002), 455–465.
40. Bouvier, M., Collins, S., O'Dowd, B. F. *et al.* Two distinct pathways for cAMP-mediated down-regulation of the beta-2 adrenergic receptor: phosphorylation of the receptor and regulation of its mRNA level. *J. Biol. Chem.*, **264** (1989), 16 786–16 792.
41. Bylund, D. B. Regulation of central adrenergic receptors. *Adv. Exp. Med. Biol.*, **116** (1979), 133–162.
42. Caldecott-Hazard, S., Morgan, D. G., DeLeon-Jones, F., Overstreet, D. H. and Janowsky, D. Clinical and biochemical aspects of depressive disorders: II. Transmitter/receptor theories. *Synapse*, **9** (1991), 251–301.
43. Ordway, G. A., Gambarana, C., Tejani-Butt, S. M. *et al.* Preferential reduction of binding of 125I-iodopindolol to beta-1 adrenoceptors in the amygdala of rat after antidepressant treatments. *J. Pharmacol. Exp. Ther.*, **257** (1991), 681–690.
44. Guimaraes, S. and Moura, D. Vascular adrenoceptors: an update. *Pharmacol. Rev.*, **53** (2001), 319–356.
45. Piascik, M. T. and Perez, D. M. Alpha-1 adrenergic receptors: new insights and directions. *J. Pharmacol. Exp. Ther.*, **298** (2001), 403–410.
46. Theroux, T. L., Esbenshade, T. A., Peavy, R. D. and Minneman, K. P. Coupling efficiencies of human alpha 1-adrenergic receptor subtypes: titration of receptor density and responsiveness with inducible and repressible expression vectors. *Mol. Pharmacol.*, **50** (1996), 1376–1387.
47. Wu, D., Katz, A., Lee, C. H. and Simon, M. I. Activation of phospholipase C by alpha 1-adrenergic receptors is mediated by the alpha subunits of Gq family. *J. Biol. Chem.*, **267** (1992), 25 798–25 802.
48. Aantaa, R., Marjamäki, A. and Scheinin, M. Molecular pharmacology of α_2-adrenoceptor subtypes. *Ann. Med.*, **27** (1995), 439–449.
49. Sunahara, R. K. and Taussig, R. Isoforms of mammalian adenylyl cyclase: multiplicities of signaling. *Mol Interv.*, **2** (2002), 168–184.
50. Milligan, G. and White, J. H. Protein–protein interactions at G-protein-coupled receptors. *Trends Pharmacol. Sci.*, **22** (2001), 513–518.
51. Pak, Y., Pham, N. and Rotin, D. Direct binding of the β_1 adrenergic receptor to the cyclic AMP-dependent guanine nucleotide exchange factor CNrasGEF leads to Ras activation. *Mol. Cell. Biol.*, **22** (2002), 7942–7952.
52. Tanoue, A., Koshimizu, T. A. and Tsujimoto, G. Transgenic studies of alpha-1 adrenergic receptor subtype function. *Life Sci.*, **71** (2002), 2207–2215.
53. Calzada, B. C. and de Artinano, A. A. Alpha-adrenoceptor subtypes. *Pharmacol. Res.*, **44** (2001), 195–208.
54. Varma, D. R. and Deng, X. F. Cardiovascular α_1-adrenoceptor subtypes: functions and signaling. *Can. J. Physiol. Pharmacol.*, **78** (2000), 267–292.
55. Guyenet, P. G. Is the hypotensive effect of clonidine and related drugs due to imidazoline binding sites? *Am. J. Physiol. Regul. Integr. Comp. Physiol.*, **273** (1997), R1580–R1584.

56. Hein, L. Transgenic models of α_2-adrenergic receptor subtype function. *Rev. Physiol. Biochem. Pharmacol.*, **142** (2001), 161–185.
57. Kable, J. W., Murrin, L. C. and Bylund, D. B. In vivo gene modification elucidates subtype-specific functions of 2-adrenergic receptors. *J. Pharmacol. Exp. Ther.*, **293** (2000), 1–7.
58. Macmillan, L. B., Hein, L., Smith, M. S., Piascik, M. T. and Limbird, L. E. Central hypotensive effects of the α_{2a}-adrenergic receptor subtype. *Science*, **273** (1996), 801–803.
59. Khan, Z. P., Ferguson, C. N. and Jones, R. M. Alpha-2 and imidazoline receptor agonists. Their pharmacology and therapeutic role. *Anaesthesia*, **54** (1999), 146–165.
60. Hunter, J. C., Fontana, D. J., Hedley, L. R. *et al.* Assessment of the role of α_2-adrenoceptor subtypes in the antinociceptive, sedative and hypothermic action of dexmedetomidine in transgenic mice. *Br. J. Pharmacol.*, **122** (1997), 1339–1344.
61. Starke, K. Presynaptic autoreceptors in the third decade: focus on alpha-2 adrenoceptors. *J. Neurochem.*, **78** (2001), 685–693.
62. Kalsner, S. Autoreceptors do not regulate routinely neurotransmitter release: focus on adrenergic systems. *J. Neurochem.*, **78** (2001), 676–684.
63. Zhang, W. and Ordway, G. A. The alpha-2C adrenoreceptors modulates GABA release in mouse striatum. *Brain Res. Mol. Brain Res.*, **112** (2003), 24–32.
64. Kalkman, H. O. and Loetscher, E. Alpha-2C adrenoceptor blockade by clozapine and other antipsychotic drugs. *Eur. J. Pharmacol.*, **462** (2003), 33–40.
65. Ordway, G. A., Jaconetta, S. M. and Halaris, A. E. Characterization of subtypes of alpha-2 adrenoceptors in the human brain. *J. Pharmacol. Exp. Ther.*, **264** (1993), 967–976.
66. Small, K. M. and Liggett, S. B. Identification and functional characterization of α_2-adrenoceptor polymorphisms. *Trends Pharmacol. Sci.*, **22** (2001), 471–477.
67. Barr, C. L., Wigg, K., Zai, G. *et al.* Attention-deficit hyperactivity disorder and the adrenergic receptors alpha-1C and alpha-2C. *Mol. Psychiatry*, **6** (2001), 334–337.
68. Bernstein, D. Cardiovascular and metabolic alterations in mice lacking β_1- and β_2-adrenergic receptors. *Trends Cardiovasc. Med.*, **12** (2002), 287–294.
69. Brodde, O. E., Bruck, H., Leineweber, K. and Seyfarth, T. Presence, distribution and physiological function of adrenergic and muscarinic receptor subtypes in the human heart. *Basic Res. Cardiol.*, **96** (2001), 528–538.
70. Nicholas, A. P., Hökfelt, T. and Pieribone, V. A. The distribution and significance of CNS adrenoceptors examined with in situ hybridization. *Trends Pharmacol. Sci.*, **17** (1996), 245–255.
71. Leonard, B. E. Stress, norepinephrine and depression. *J. Psychiatry Neurosci. Suppl.*, **26** (2001), S11–S16.
72. Michel, M. C. and Insel, P. A. Receptor gene polymorphisms: lessons on functional relevance from the beta-1 adrenoceptor. *Br. J. Pharmacol.*, **138** (2003), 279–282.
73. Liggett, S. B. Beta-2 adrenergic receptor pharmacogenetics. *Am. J. Respir. Crit. Care. Med.*, **161** (2000), S197–S201.
74. Palmer, L. J., Silverman, E. S., Weiss, S. T. and Drazen, J. M. Pharmacogenetics of asthma. *Am. J. Respir. Crit. Care. Med.*, **165** (2002), 861–866.
75. Arch, J. R. S. β_3-Adrenoceptor agonists: potential, pitfalls and progress. *Eur. J. Pharmacol.*, **440** (2002), 99–107.

76. Revelli, J. P., Preitner, F., Samec, S. *et al.* Targeted gene disruption reveals a leptin-independent role for the mouse beta-3 adrenoceptor in the regulation of body composition. *J. Clin. Invest.*, **100** (1997), 1098–1106.
77. Jimenez, M., Leger, B., Canola, K. *et al.* Beta-1/beta-2/beta-3 adrenoceptor knockout mice are obese and cold-sensitive but have normal lipolytic responses to fasting. *FEBS Lett.*, **530** (2002), 37–40.
78. Inukai, T., Tayama, K., Inukai, Y. *et al.* Clinical features of a polymorphism of the beta-3 adrenergic receptor gene in patients with type 2 diabetes mellitus – a study using a pin-point sequencing method. *Exp. Clin. Endocrinol. Diabetes*, **109** (2001), 386–388.
79. Gross, G., Hanft, G. and Mehdorn, D. Vascular adrenoceptors: an update. *Pharmacol. Rev.*, **53** (2001), 319–356.
80. Pereash, A. C. and Bylund, D. B. Alpha-2 adrenoreceptor subtypes indicated by [^{3}H]yohimbine binding in human brain. *Life Sci.*, **38** (1986), 2129–2137.
81. Callado, L. F. and Meana, J. J. Low-affinity conditions for agonists increase the binding of the antagonist [^{3}H]R. X821002 to the alpha$_{2B\text{-}C}$-adrenoceptor subtypes in human brain and rat kidney. *Eur. J. Pharmacol.*, **332** (1997), 109–112.
82. Sastre, M., Guimón, J., García-Sevilla, J. A. Relationships between beta- and alpha$_2$ -adrenoceptors and G coupling proteins in the human brain: effects of age and suicide. *Brain Res.*, **898** (2001), 242–255.
83. Engel, G. and Hoyer, D. [125I]BE 2254, a new high affinity radioligand for [alpha]1-adrenoceptors. *Eur. J. Pharmacol.*, **73** (1981), 221–224.
84. Salles, J. and Badia, A. Selective enrichment with alpha 1A- and alpha 1B-adrenoceptor subtypes in rat brain cortical membranes. *Eur. J. Pharmacol.*, **266** (1994), 301–308.
85. Ribas, C., Miralles, A., Busquets, X. and García-Sevilla, J. A. Brain alpha$_2$-adrenoceptors in monoamine-depleted rats: increased receptor density, G coupling proteins, receptor turnover and receptor mRNA. *Br. J. Pharmacol.*, **132** (2001), 1467–1476.
86. Lu, L. and Ordway, G. A. Reduced expression of alpha$_{2C}$-adrenoceptors in rat striatum following antisense oligodeoxynucleotide infusion. *Mol. Brain Res.*, **47** (1997), 267–274.
87. Bylund, D. B. and Snyder, S. H. Beta adrenergic receptor binding in membrane preparations from mammalian brain. *Mol. Pharmacol.*, **12** (1976), 568–580.
88. Minneman, K. P., Hegstrand, L. R. and Molinoff, P. B. Simultaneous determination of beta-1 and beta-2 adrenergic receptors in tissues containing both receptor subtypes. *Mol. Pharmacol.*, **16** (1979), 34–46.

4

Regulation of gene transcription in the central nervous system by norepinephrine

Ronald S. Duman, Ph.D. and Samuel S. Newton, Ph.D.
Yale University School of Medicine, New Haven, USA

Introduction

During the past few years we have witnessed tremendous advances in sequencing the human genome, as well as the coding sequences of several other species. The next major hurdle is to characterize and understand the mechanisms that control the transcriptional expression and posttranslational processing of these genes. Gene expression in the brain is regulated by many neurotransmitters, hormones, and psychotropic drugs, each of which can induce discrete and unique patterns of gene expression in different brain regions and even within a single neuron. These unique patterns of expression are involved in shaping the function of the brain and its ability to adapt and generate long-term and informed responses to subsequent stimuli.

Norepinephrine (NE) is a well-known modulator of neuronal activity, and has been implicated in many aspects of brain function as well as the etiology and treatment of psychiatric illness, including depression, posttraumatic stress disorder, fear, stress, and anxiety. In addition to modulating neuronal firing and other short-term cellular processes, NE also produces long-term modulating effects via regulation of neuronal gene expression. This chapter will provide an overview of the fundamental aspects of gene transcription, including general concepts of DNA-binding elements and regulation of transcription factor function. Then we will discuss how NE, via regulation of diverse receptor-coupled signaling pathways, controls gene transcription with particular emphasis on two families of transcription factors: the cyclic adenosine monophosphate (cAMP) response element binding protein (CREB) and the c-Fos/activator protein-1 (AP-1) systems. Finally, the role of NE and regulation of gene transcription in the actions of antidepressant drugs, stress, and learning and memory will be examined.

Brain Norepinephrine: Neurobiology and Therapeutics, ed. Gregory A. Ordway, Michael A. Schwartz and Alan Frazer. Published by Cambridge University Press.

Fundamental concepts of gene transcription

The basic mechanisms underlying the regulation of gene transcription will be reviewed in this section. For a more complete description please see reviews by Armstrong, Nestler, and Nestler and Hyman. [1–3]

Initiation of gene transcription

Nuclear DNA is made of four nucleotides: adenine (A), thymine (T), guanine (G) and cytosine (C). Single strands of these nucleotides pair (A with T and G with C) to form double helix strands. Most DNA in the nucleus is found in an inactive state where it is tightly coiled around nucleosomes, the major components of chromosomes. For initiation of gene transcription to occur, the nucleosomes, which are comprised primarily of histone proteins, must open and expose the DNA. This process requires enzymes that modify the histones, transcription factors, and cofactors. Together, this process allows portions of the nucleosomes to be opened so that the DNA is exposed. This allows binding of RNA polymerase, the primary factor mediating gene transcription and the formation of RNA from which all proteins are synthesized. Only about 4% of the chromosomal DNA contains the sequence for the ~40 000 genes that encode messenger RNA.

Gene transcription occurs in a relaxed region of DNA that often requires an activator protein to further release the DNA/protein complex that forms a nucleosome. For example, histone acetylases modify core DNA-binding proteins that form nucleosomes and are critical for remodeling nucleosomes. This displacement allows for a complex of proteins called transcription factors to bind to the DNA at a core promoter site and to recruit RNA polymerase. This is the first step or initiation phase of gene transcription, which is then followed by elongation and termination.

Another class of histone-modifying enzymes, histone deacetylases, is involved in returning DNA to the coiled/bound nucleosome state. Interestingly, histone deacetylases are inhibited by valproic acid, one of the major therapeutic drugs used for the treatment of bipolar disorder.[4] Inhibition of histone deacetylase leads to increased expression of genes, such as β-*catenin*, thought to contribute to the therapeutic action of valproic acid and other mood-stabilizing drugs. However, many other genes are also activated, and this nonspecific increase in gene expression most likely underlies the in utero teratogenic effect of this drug, which is restricted from use in pregnant women.

The process of initiating gene transcription is a critical first step in the synthesis of RNA and is very tightly regulated. The DNA promoter elements that determine the binding site of transcription factors are referred to as the *cis*-regulatory elements, and the transcription factors are the *trans*-regulatory elements. The *cis* elements that determine where in the DNA molecule initiation will occur are referred to as core

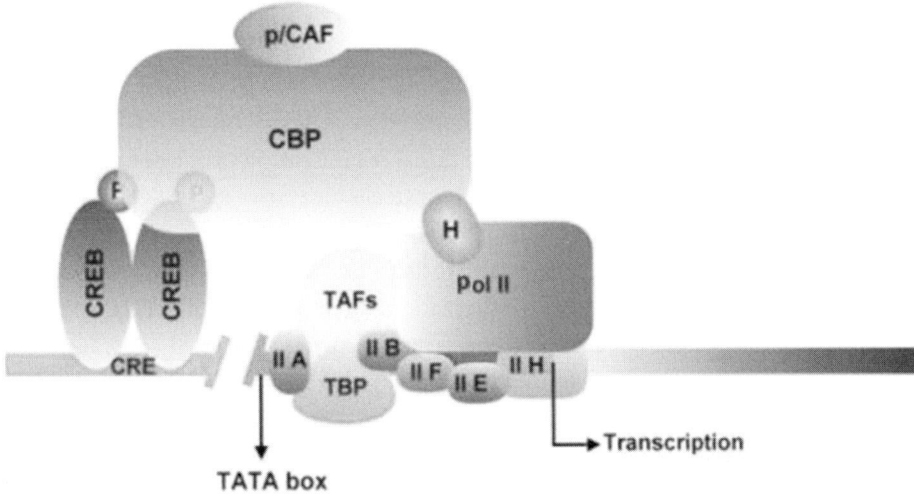

Figure 4.1 Model depicting *cis*- and *trans*-acting DNA elements involved in regulation of gene transcription. This model shows binding of RNA polymerase II (pol II) to a transcription initiation site that is determined by the presence of a TATA box. Several additional binding proteins also contribute to the complex that binds to the TATA box. Also shown in this figure is an upstream promoter site, a cAMP response element (CRE) and the CRE-binding protein (CREB), the transcription factor that recognizes this element. Promoter elements such as the CRE can be located several hundred base pairs upstream from the TATA box and are required for significant activation of pol II and gene transcription. CREB binds to the CRE but does not become fully active until it is phosphorylated. This allows for binding with the CREB-binding protein (CBP), which in turn couples to and activates the pol II transcription complex. IIA to IIH, transcription factors IIA to IIH; pCAF, p300KREB-binding protein associated factor; P, phosphorylation; TBP, TATA-box-bindng protein; TAFs, TBP associated factors.

elements, and these control basal transcriptional activity. Additional *cis*-regulatory elements, which generally lie further upstream from the initiator *cis* elements, help to recruit activator and/or repressor proteins to the transcriptional complex. These additional *cis* elements and their respective transcription factors control higher rates of RNA synthesis and gene expression.

Transcriptional initiation requires the association of RNA polymerase to the core promoter, which often contains a region rich in T and A nucleotides, and is referred to as the TATA box. One form, referred to as RNA polymerase II (pol II) is utilized for the transcription of mRNA (Figure 4.1). Other isoforms of RNA polymerases are utilized for the transcription of very large RNA molecules (I) or for small nuclear RNA (snRNA for I and III). Another protein, referred to as the TATA-binding protein, is responsible for binding of pol II, as well as other transcription factors and cofactors and is thereby critical to the initiation of basal transcription.

However, many genes do not contain a TATA box but have a poorly conserved *cis* domain referred to as an initiator.

Once mRNA is synthesized, there is generally a large degree of posttranscriptional processing. This will not be discussed in detail in this chapter, but it is important to understand that posttranscriptional modifications are important for further regulation and fine tuning of specific patterns and types of genes expressed. For example, there can be multiple splice variants formed from a single gene product, and each splice variant can have unique functional features. Ribonucleic acid editing, where a single nucleotide can be altered and result in the switch of a single amino acid, is another example of posttranscriptional processing that can lead to the formation of unique signaling molecules. One final example is stabilization of RNA via modification of the 3′ ends, which provides a mechanism for increasing or decreasing the half-life of RNA in the cell.

Transcription factors

Although initiation of gene transcription occurs at a basal rate when the TATA-binding protein recruits pol II and other factors to the initiation site, further activation or repression requires interactions with additional DNA elements and their corresponding transcription factors.[2,3] These DNA elements are referred to as promoter and enhancer elements, and are defined by their proximity to the TATA box (i.e., promoters are generally located closer to the TATA box than enhancers). However, functionally there is little difference between promoters and enhancers, particularly since distal enhancer sites are capable of gaining access to the TATA box when long DNA strands form loops that bring the sites into close proximity with the TATA box.

Transcription factors contain several different functional domains, including a DNA-binding domain, transcription activation domains (i.e., sites for binding with other factors), and phosphorylation domains for regulation by protein kinases (see example of CREB, Figure 4.2). Many transcription factors require binding with other factors for complete activation. This can include homo- or heterodimers of two of the same or related transcription factors that may act in concert to enhance transcription, or, conversely, that may result in repression of activity compared to that which occurs in response to a homodimer. The ability to form dimers with different compositions provides a mechanism for fine-tuning gene transcription, particularly when combined with differential expression of transcription factors as will be discussed below for those factors that are regulated by altered gene expression.

In this chapter we will discuss two major families of transcription factors that are regulated by NE and that mediate the transcriptional activity of this catecholamine. These are the CREB-like transcription factors and the AP-1 family. The general

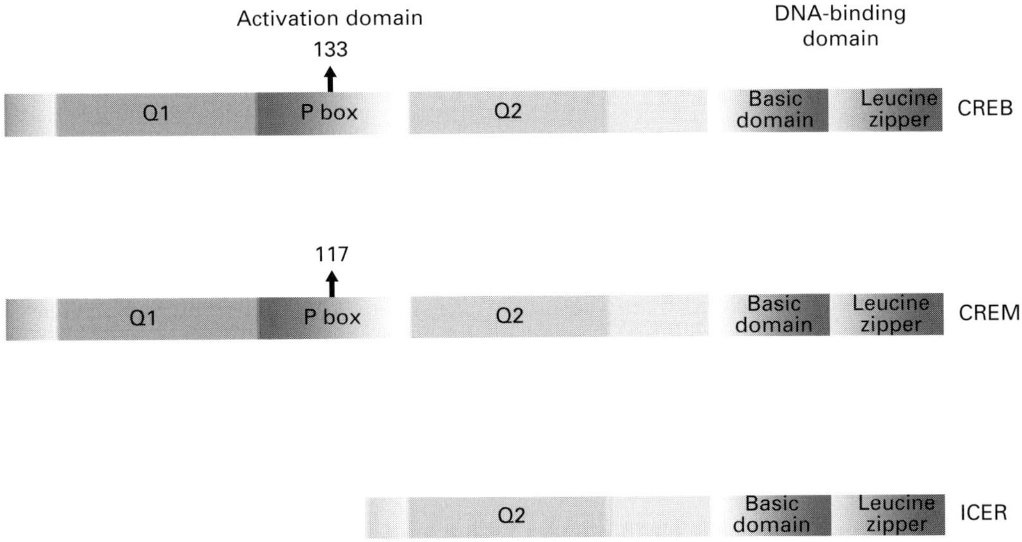

Figure 4.2 Schematic structure of the CREB and cAMP response element modulator (CREM) transcription factors. These models depict the primary functional domains of CREB and CREM. This includes the activation domain, also referred to as the P box or kinase-inducible domain, which contains a phosphorylation site for transcriptional activation. The amino terminus of the proteins contains the basic domain that binds to DNA and the leucine zipper domain that is responsible for dimerization of CREB and related proteins. Also shown at the bottom is the structure of ICER, or the inducible cAMP early repressor. ICER is generated by use of an alternative start site in the *CREM* gene that is activated by cAMP/Ca^{2+} signaling and a CRE site. This results in the synthesis of a truncated form of the transcription factor that lacks the kinase-inducible domain. ICER is able to dimerize with CREB, CREM and related proteins but cannot be activated by phosphorylation. ICER thereby serves as a negative feedback transcription factor that shuts down CRE-mediated gene expression.

mechanisms for regulating the transcriptional activity of the members of these two families are different. Most CREB family members are regulated by phosphorylation of specific amino acid residues in the regulatory domain, while members of the AP-1 family are regulated primarily by induction of the total amount of the transcription factor protein. Another major class of transcription factors is the steroid hormone receptor superfamily. Steroid hormone receptors are soluble cytoplasmic proteins that translocate to the nucleus upon binding to a specific class of steroid. Family members include receptor/transcription factors for adrenal glucocorticoids, thyroid hormones, sex steroids (estrogen, progesterone, and testosterone), retinoic acid, and vitamin D. Although NE-coupled signal transduction pathways do not directly bind to and regulate this class of steroid-transcription factors, there are examples of catecholamine signaling molecules (e.g., protein kinase A) that interact with steroid receptors, indicating that indirect regulation can occur in certain cellular systems.

Noradrenergic receptors couple to multiple signal transduction cascades and transcription factors

Norepinephrine couples to multiple subtypes of G protein-coupled receptors, which in turn activate a diverse array of signal transduction cascades (Figure 4.3). (See Chapter 3 by Bylund in this volume.) The major signal transduction pathways activated by NE include β-adrenergic receptor (βAR) stimulation of cAMP and α_1-adrenergic receptor (α_1AR) activation of Ca^{2+} and protein kinase C (PKC). In addition, NE activation of G protein-coupled receptors can also lead to stimulation of the extracellular-regulated protein kinase (ERK) and the mitogen-associated protein (MAP) kinase cascade, which is one of the major signaling pathways of neurotrophic factors. The influence of these signal transduction pathways on gene expression has been extensively studied in peripheral systems and in cultured cells. Although there is much less known about the influence of these pathways on gene transcription in the brain, the actions of NE on signal transduction and gene expression are likely to be similar. It is more difficult to know what other pathways are also regulated in neurons and glia that, in combination with NE-stimulated signals, create a unique signature that defines the response in one set of cells relative to that in another cell. These unique, cell-specific signatures are being studied using a number of technological approaches, from single-cell real-time polymerase chain reaction and genechip analysis to double and triple labeling of proteins in cells. In this section we will discuss the influence of NE signaling on CREB and AP-1 and the NE-coupled signaling cascades that influence these transcription factors.

Noradrenergic receptor regulation of the cAMP–CREB cascade

Cyclic AMP response element-binding protein is one of the major mediators of cAMP-regulated gene transcription and thereby of βAR stimulation of the cAMP pathway. CREB and related family members bind to the cAMP response element (CRE) generally located in the promoter regions of genes. The consensus CRE is an 8-base pair, palindromic sequence, TGACGTCA (i.e., the reverse sequence, ACTGCAGT, represents the complementary DNA strand). Cyclic AMP response element sequences that have one or two substitutions also bind CREB, although with lower affinity. There are many regulatory genes that contain CRE, including neuropeptides, neurotrophic factors, enzymes that are rate-limiting for neurotransmitter synthesis, neurotransmitter receptors, and transcription factors themselves.

Most of the members of the CREB family of related transcription factors activate gene transcription, although there are a few examples of inhibitors. Transcription

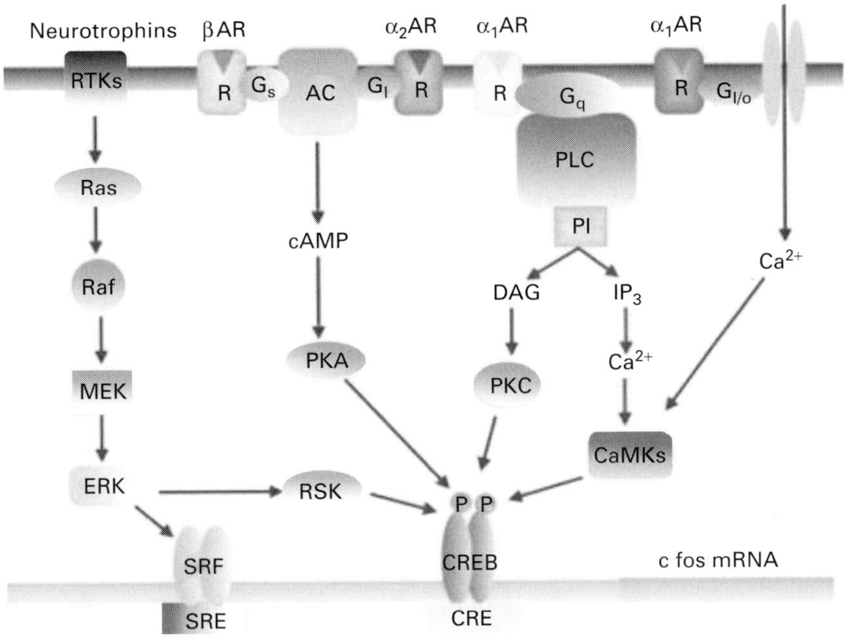

Figure 4.3 Diagram of NE receptor-coupled signal transduction and regulation of gene expression. Norepinephrine acts as a ligand for several different subtypes of G protein-coupled receptors, including β-adrenergic receptor (βAR), α_1AR and α_2AR. These receptors are capable of coupling to several different signaling transduction pathways via interactions with specific G protein subtypes. The βAR and α_2AR stimulate or inhibit adenylate cyclase (AC) and cAMP formation via the stimulatory and inhibitory G proteins, G_s and G_i. The α_1AR couples to phospholipase C (PLC) and the phosphatidylinositol (PI) pathway via G_q, but can also couple to Ca^{2+} channels via $G_{i/o}$. Activation of PLC leads to the formation of diacylglycerol (DAG) and inositol 1, 4, 5-trisphosphate (IP_3), which act as second messengers. These second messengers stimulate their respective protein kinases. This includes cAMP-dependent protein kinase (PKA), protein kinase C (PKC), and Ca^{2+}/calmodulin-dependent protein kinase (CaMK). All three of these pathways are capable of regulating gene expression via phosphorylation of specific transcription factors. Shown here is a model of the *c-fos* gene and a few of the promoter elements that control the expression of this immediate early gene transcription factor. The *c-fos* gene contains several CRE promoter sites that bind CREB, and CREB can be phosphorylated by PKA, CaMK, and PKC to stimulate *c-fos* gene expression. Also shown is the mitogen-activated protein (MAP) kinase cascade that is activated by neurotrophic factors. This occurs via activation of receptors that contain tyrosine kinase activity (RTK), and activation of a cascade of kinases, including Raf, mitogen-activated protein kinase (MEK), and extracellular-regulated protein kinase (ERK). This pathway can also be activated by G-protein-coupled receptors. Moreover, one of the downstream targets of ERK is another CREB kinase, ribosomal S6 kinase (RSK). ERK can also stimulate *c-fos* gene expression via activation of a serum response element (SRE) and the serum response factor (SRF).

factors, which like CREB lead to activation of transcription, are referred to as activating transcription factors (ATFs). Another closely related group are referred to as CRE modulators or CREMs. Many of the latter also activate transcription although there are some CREMs that inhibit CREB-mediated gene expression. The best example of a CREM inhibitor is the inducible cAMP early repressor (ICER). The expression of ICER is itself induced by activation of a CRE in an alternative promoter of the *CREM* gene. Inducible cAMP early repressor is a truncated form of CREB that lacks the kinase-inducible domain and is capable of forming a dimer with, and repressing, other CREB-like proteins. The induction of ICER by stimuli that increase CRE-mediated gene expression is thought to act as a negative feedback to control activated levels of gene transcription.

All of the CREB-like proteins bind to DNA as homo- or heterodimers, and the makeup of the dimer can influence the level of transcriptional activity. For example, ICER dimerizes, but inhibits, gene transcription, because it lacks a kinase-inducible domain. Dimerization occurs via the leucine zipper domain, which is composed of an α-helix in which every seven amino acids is a leucine. The aligned leucine residues of two transcription factors interact hydrophobically to form what is referred to as a leucine zipper that stabilizes the dimer. This domain is located adjacent to a highly basic-DNA-binding domain, leading to the name basic-leucine zipper, or bZIP, which is commonly used to describe many transcription factors that utilize these DNA-binding and dimerization properties.

The transcriptional activity of CREB and related transcription factors is induced by phosphorylation of serine 133. Phosphorylation of this site is necessary for interaction with the CREB-binding protein (CBP) that in turn interacts with the basal transcription complex at the TATA or initiator of a gene and thereby results in transcriptional activation (Figure 4.1). Beta-adrenergic receptor activation of the cAMP production results in stimulation of cAMP-dependent protein kinase (PKA) that is then translocated to the nucleus. In the absence of serine 133 phosphorylation, CREB can dimerize and bind to CRE sites, but it is only upon phosphorylation that transcriptional activation occurs.

Activation of α_1-adrenergic receptors also leads to up-regulation of CREB-mediated gene transcription

Phosphorylation of CREB at serine 133 can also occur via activation of Ca^{2+}/calmodulin-dependent protein kinases (CaMK) and PKC. Therefore α_1AR stimulation can also result in activation of CREB via regulation of these kinases. It is also worth mentioning yet another pathway by which PKC can influence gene expression. Nuclear factor κB (NFκB) and its regulation of gene expression is also controlled by PKC, as well as by immunologic signals. Phosphorylation of a NFκB inhibitory binding protein (IκB) results in the release of NFκB and thereby allows it to be translocated to the nucleus. In the nonphosphorylated state, IκB binds to NFκB and keeps it localized to the cytoplasm thereby masking its nuclear translocation signal. Upon PKC-mediated phosphorylation, IκB and NFκB dissociate, and the nuclear signal leads to translocation, DNA binding, and transcriptional activity of NFκB.

The α_1AR can also influence CREB via activation of the MAP kinase cascade. This can occur via interaction of internalized receptor with soluble tyrosine kinases, or via regulation of small GTP-binding proteins such as Ras or Rap. These pathways then lead to regulation of Raf and the MAP kinase pathway. The downstream kinase in this pathway that regulates CREB is ribosomal S6 kinase, referred to as RSK[5]. Ribosomal S6 kinase is a CREB kinase, and the activity of this kinase has been demonstrated to be critical for neurotrophic factor-mediated survival of neurons.[6]

Noradrenergic receptor regulation of AP-1 family transcription factors

Activator protein-1 transcription factors represent another family of proteins that regulate gene expression. The primary mechanism for regulation of these proteins is via induction of gene expression. The AP-1 proteins bind to the AP-1 promoter element, which is a sequence of seven nucleotides and is similar to the consensus CRE sequence, TGACTCA. Although this sequence only differs from the CRE sequence by one nucleotide, it is sufficient to confer a relatively high degree of selective binding of the AP-1 transcription factors over the CREB-like factors. There are many genes in the brain that contain AP-1 elements, including those for neuropeptides, neurotransmitter synthetic enzymes, receptors, and neurotrophic factors.

The AP-1 transcription factors also belong to the superfamily of basic-leucine zipper DNA-binding proteins. These factors thereby form homo- or heterodimers via the leucine zipper domain and bind to DNA via the adjacent basic-amino-acid-rich domain. There are two major families of AP-1 factors, referred to as Fos and Jun. Members of the Fos family include c-Fos, Fos-related antigen-1 and -2 (FRA-1 and FRA-2), and FosB. The Jun family includes c-Jun, JunB, and JunD. In most cases a functional dimer is made up of one Fos and one Jun family member. Jun homodimers are not uncommon, but the DNA binding affinity of these complexes is lower than for the Fos/Jun heterodimers.

Cellular levels of most Fos and Jun transcription factors are low under basal, unstimulated conditions, but formation of AP-1 complexes is stimulated by activation of signal transduction pathways that increase the expression levels of Fos and Jun. One exception is JunD that is constitutively expressed. Because most AP-1 transcription factors are rapidly induced they are also referred to as immediate early genes (IEGs). The *c-fos* gene is the prototype IEG, and it can be induced rapidly (within minutes) of the presentation of an extracellular stimulus. The rapid activation of c-Fos is due to the presence of multiple response elements in the promoter. These include three CREs. These sites confer rapid induction of *c-fos* gene expression in response to activation of cAMP and/or Ca^{2+}-dependent protein kinases and CREB as described above. In addition, induction of *c-fos* gene expression also occurs via additional signaling cascades, including the MAP kinase–ERK signaling pathway. Activation of ERK can lead to phosphorylation of CREB via activation of RSK. In addition, ERK signaling also regulates another response element located on the *c-fos* gene, the serum response element (SRE). This element is activated when ERK translocates to the nucleus and phosphorylates Elk-1, another transcription factor, which in turn forms a complex with the serum response factor (SRF) that binds to the SRE (Figure 4.3).

Although the AP-1 transcription factors are regulated at the level of gene expression, these factors can also be influenced by phosphorylation. For example, there are several serine residues in the carboxy-terminus of c-Fos that are substrates for PKA, CaMK, and PKC. Many of these protein kinases can also lead to increased *c-fos* gene expression, and sustained activation of these pathways will then result in phosphorylation of carboxy-terminus serine sites. However, in this case phosphorylation is thought to be part of a negative feedback mechanism to turn off AP-1-mediated gene expression.

Repeated stimulation of the *c-fos* gene eventually leads to a refractory state where further transcription is limited. This is likely due to counter-regulatory mechanisms including the induction of inhibitory transcription factors such as ICER that block CRE-mediated gene expression. However, under conditions of repeated and long-term activation of the *c-fos* gene, the accumulation of alternative splice variants known as ΔFosB is observed. Unlike Fos, which degrades very rapidly (i.e., half-life of minutes to hours), ΔFosB is relatively stable and has a very long half-life of days to weeks.[2] This different pattern of expression is thought to contribute to the long-term adaptations to repeated stimulation. Such differences in the temporal stability of AP-1 transcription factors, combined with different patterns of stimulation, provide mechanisms for discrete regulation of gene expression and cellular function.

Role of noradrenergic receptor regulation of transcription factors and gene expression in the actions of antidepressant treatment

Many drugs used for the treatment of major psychiatric illnesses have direct or indirect effects on noradrenergic receptors and their signal transduction pathways. Antidepressants, one of the most widely prescribed classes of drugs for the treatment of depression, anxiety, and other mood disorders, block the reuptake or metabolism of NE, and/or Serotonin (5-HT), and thereby increase synaptic levels of these monoamines. There are several different classes of antidepressants that influence levels of NE, including the NE-selective reuptake inhibitors desipramine, reboxetine, nisoxetine, and atomoxetine, as well as the nonselective reuptake inhibitors, imipramine, venlefaxine and duloxetine that block both NE and 5-HT transporters. Psychostimulants such as cocaine block dopamine reuptake but also act at the NE and 5-HT transporters to increase levels of monoamines. The monoamine oxidase inhibitors represent another class of antidepressants that increase levels of NE and 5-HT nonselectively. In addition to these indirectly acting agents, there are a few drugs that act as agonists or antagonists for noradrenergic receptor subtypes. These include propranolol, a βAR antagonist used for social anxiety and stage fright; clonidine, an α_2AR agonist, used for opiate withdrawal and hypertension; and yohimbine, an α_2AR antagonist, used primarily for challenge studies to increase NE transmission.

Antidepressants increase synaptic levels of NE within a relatively short time – a few hours to a few days depending on how quickly therapeutic blood levels of the drugs are attained. However, treatment for several weeks or even months is required for a therapeutic response. This has lead to the hypothesis that adaptation to elevated NE and 5-HT is necessary for a therapeutic response.[7–11] Adaptations at the levels of signal transduction and gene expression have been areas of special interest. Regulation of transcription factors and gene expression have been particularly attractive as mechanisms underlying the long-term adaptive changes, also referred to as neuronal plasticity, that could account for the therapeutic actions of chronic antidepressant treatment. In this section, the influence of NE-selective reuptake inhibitors on transcription factors and gene expression will be discussed. This work has helped to elucidate the molecular mechanisms underlying the actions of antidepressants, and has provided information for the development of novel therapeutic targets. In addition, these drugs are the most useful pharmacological tools for studying the influence of the NE neurotransmitter system on gene expression, as there are few good noradrenergic receptor agonists that have access to the brain after systemic administration.

Antidepressant treatment up-regulates the cAMP–CREB cascade

The cAMP signal transduction pathway is one of the second messenger systems that are regulated by chronic administration of NE-selective reuptake inhibitors (Figure 4.4). This is thought to occur via sustained activation of βARs that stimulate cAMP formation and PKA. There are also 5-HT receptors (5-$HT_{4,6,7}$) that also couple to the cAMP–CREB cascade. The coupling of the stimulatory G protein, Gs, with adenylate cyclase and particulate levels of PKA are both increased in limbic brain structures.[8,9] A role for increased gene expression in response to activation of the cAMP–PKA pathway was first suggested by studies demonstrating that levels of PKA were elevated in nuclear fractions of brain.[12] Increased PKA levels in the

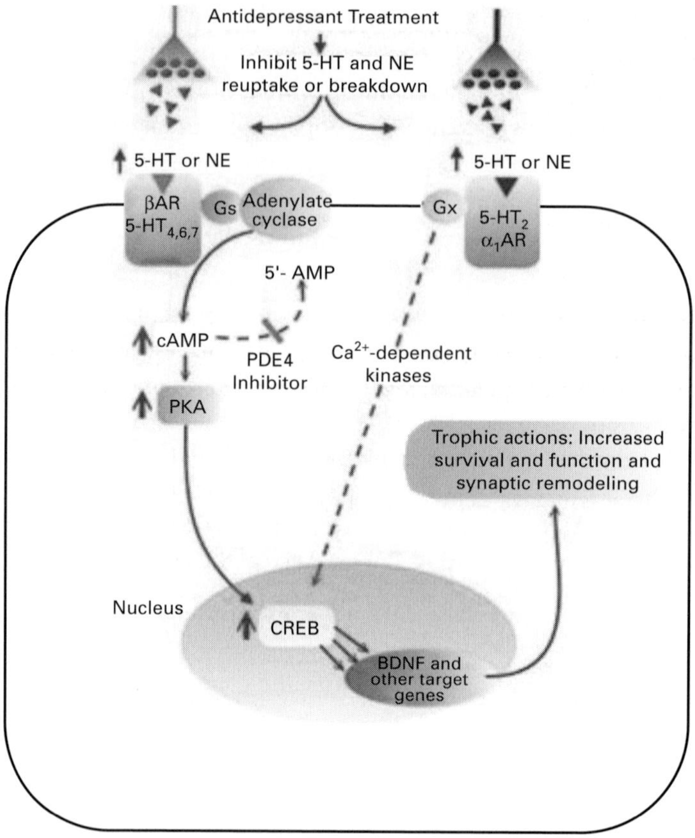

Figure 4.4 Model depicting antidepressant regulation of the cAMP–CREB cascade and gene targets. Antidepressants block the reuptake or metabolism of NE, as well as serotonin (5-HT), and increase synaptic levels of these monoamines. Chronic antidepressant administration results in adaptations of the receptor-coupled signal transduction pathways and regulation of gene expression. One of the pathways that is regulated by antidepressant treatment is the cAMP–CREB cascade. Chronic antidepressant treatment leads to up-regulation of cAMP-dependent protein kinase (PKA), increased levels of this kinase in nuclear fractions, and increased function and expression of the cAMP response element-binding protein (CREB). Inhibitors of phosphodiesterase type IV (PDE4) increase levels of cAMP and are known to have antidepressant efficacy in behavioral models and in clinical trials. CREB can also be regulated by Ca^{2+}-stimulated protein kinases, as well as ribosomal S6 kinase (see Figure 4.3) and thereby acts as a common transcription factor target for multiple signal transduction pathways. Chronic antidepressant treatment and the cAMP–CREB cascade also increase the expression of specific gene targets in limbic brain structures, most notably brain-derived neurotrophic factor (BDNF) and its receptor TrkB. Up-regulation of the cAMP–CREB cascade and increased expression of BDNF/TrkB are thought to produce antidepressant effects, in part, by blocking or reversing the atrophy and decreased neurogenesis resulting from stress.

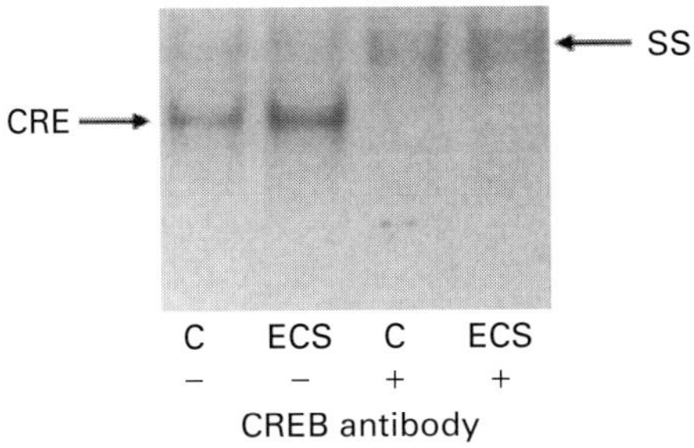

Figure 4.5 Diagram demonstrating the DNA gel mobility shift assay. The binding of transcription factors to promoter sequences can be assessed using a gel mobility shift assay. A synthetic oligonucleotide containing the promoter element of interest, in this case a CRE, is prepared and annealed with the corresponding DNA strand to make double-stranded DNA (ds-DNA). The ds-DNA is labeled with radioactive nucleotides and is then incubated with nuclear-enriched cell extracts. Binding of a transcription factor complex retards the mobility of the radiolabeled ds-DNA when analyzed on a nondenaturing gel, while in the absence of binding the ds-DNA band runs near the gel front. The presence of a particular transcription factor can be confirmed by addition of an antibody to the incubation. This will either disrupt or supershift the complex depending on the binding site of the antibody. In the case shown here the CREB antibody causes supershift to a higher molecular weight complex. Also shown is the effect of chronic electroconvulsive seizure administration, which increases levels of CRE binding, consistent with the up-regulation of CREB expression by antidepressant treatment. C, control; ECS, electroconvulsive shock; SS, supershift.

nucleus could result in phosphorylation of CREB and other transcription factors, and altered regulation of gene expression. The possibility that CREB is regulated by sustained elevation of NE as a result of antidepressant treatment has been substantiated by several studies. The first report demonstrated that chronic administration of desipramine, as well as other antidepressants, increases the expression of CREB mRNA and protein in the hippocampus.[13,14] In addition, this study demonstrates that antidepressant treatment increases the total amount of CREB bound to DNA containing a CRE.

The assay used for the latter investigation is called a gel shift assay and takes advantage of the ability of a transcription factor to bind to a specific DNA sequence that defines a promoter element, such as the CRE. This approach has been extremely useful for studying DNA–transcription factor interactions and is worth taking a moment to briefly describe (Figure 4.5). A piece of double-stranded DNA containing a CRE or other promoter can be synthesized and then radiolabeled so that it can be monitored by autoradiography. The labeled DNA is then incubated with a cell extract, usually a nuclear-enriched fraction. The sample is then subjected to gel electrophoresis under nondenaturing conditions so the DNA–protein complex remains intact. When the radiolabeled-DNA is bound to a transcription factor, it forms a large complex, and its mobility on the gel is retarded or shifted. The specificity of this complex can be confirmed in several ways: first by demonstrating that unlabeled DNA containing the CRE site competes with the radiolabeled-DNA–protein complex; second, by addition of an antibody for CREB, which will either disrupt the DNA–protein complex, further retard the mobility of the complex, or cause a "supershift." Disruption or supershift of the DNA–protein complex will depend on what transcription factor domain is recognized by the antibody: the DNA-binding site or some other portion of the protein. The gel shift assay has been used to characterize many different DNA promoter elements and transcription factors, and the supershift assay has been critical for investigating the interaction of transcription factors and binding at a particular DNA promoter site.

These studies demonstrate that NE-selective reuptake inhibitors increase the expression and binding of CREB, but the transcriptional activity of CREB was not directly determined. This is a critical point because the activity of CREB is controlled primarily by its phosphorylation at serine 133, not just by DNA binding. This question has been addressed by three different approaches. First, using an antibody that distinguishes phospho-CREB from CREB, it was determined that desipramine treatment increases levels of phospho-CREB in several limbic structures, including the amygdala and hippocampus.[15] This demonstrates that desipramine treatment increases the transcriptional active form of CREB. Second, desipramine treatment was demonstrated to increase CRE-mediated gene expression in a line of transgenic mice that expressed a CRE-regulated reporter gene, β-galactosidase.[15] These mice can be used to visualize CRE-mediated gene expression in vivo in response to various types of pharmacological or behavioral stimuli. A similar approach has been used to generate lines of transgenic mice for studying Fos/Jun–AP-1-mediated gene expression. A third approach to demonstrate increased CREB transcriptional activity is to identify target genes that contain CRE sites and that are regulated. There are several such genes that have now been identified, including brain-derived neurotrophic factor (BDNF) and its receptor TrkB (Figure 4.4). Chronic administration of desipramine increases the expression of BDNF and TrkB mRNA in the hippocampus and this effect is blocked in CREB-null mutant mice.[13,16,17]

The *BDNF* gene contains four exon-specific promoters each of which can promote the transcription of a fifth exon that contains the sequence for the protein-coding region of this neurotrophic factor.[18,19] The first four exon-specific promoters control the expression of exon V and allow for different types and patterns of BDNF expression in response to diverse cellular stimuli. The exon III-specific promoter contains a Ca^{2+}/CRE site, referred to as a CaRE that is responsible for the induction of BDNF expression in response to activation of the cAMP–PKA cascade, as well as in response to neuronal depolarization and stimulation of

Ca^{2+}/calmodulin-dependent protein kinase.[19] Recent studies confirm that administration of desipramine increases activation of the exon III promoter, providing direct evidence that sustained elevation of NE leads to up-regulation of the cAMP–CREB-regulated gene promoter.[20] Information has emerged regarding mutations and/or variants of the *BDNF* gene. One variant, which results in a substitution of methionine for valine at postion 66 (val66met), results in altered processing of BDNF and is associated with decreased episodic memory.[21] Studies are currently underway to determine if this variant is linked with psychiatric illness.

CREB produces antidepressant effects in behavioral models of depression

The role of CREB in the actions of antidepressant treatments has been directly tested in two behavioral models of depression, the learned helplessness paradigm and forced-swim test.[22] These behavioral despair models are responsive to antidepressant treatments, including desipramine. In the learned helplessness paradigm, animals that have been exposed to an inescapable stress become "helpless" when they are subsequently tested in an escapable situation. If the animals are administered an antidepressant, they regain the ability to escape. Viral expression of CREB in the hippocampus after exposure to inescapable stress produced an antidepressant-like effect, providing direct evidence that CREB is sufficient to produce an antidepressant response in this paradigm. In the forced-swim test, animals are placed in a pool of water and after swimming and climbing to escape for a few minutes become immobile and float on the surface. Antidepressant administration increases the amount of time that an animal will swim or climb, and viral expression of CREB in the hippocampus produces a similar antidepressant effect.[22]

Cyclic AMP response element-binding protein is expressed throughout the brain and is involved in synaptic plasticity and other cellular functions in many brain regions. It is not surprising that the antidepressant effects of CREB are specific to the brain region in which it is expressed. Transgenic or viral expression of CREB in the nucleus accumbens does not produce an antidepressant effect; in fact, when expressed in this brain region, CREB produces a depressive-like phenotype in the learned helplessness paradigm,[23] and in the forced-swim test.[22,24] Brain-derived neurotrophic factor is not expressed in the nucleus accumbens, probably because of specific silencer elements that restrict its regional expression. However, there are other gene targets of CREB in this region, most notably *prodynorphin*. Increased expression of CREB in nucleus accumbens increases the expression of *prodynorphin* while expression of a dominant negative phosphorylation mutant of CREB (mCREB) decreases the expression of *prodynorphin*.[23,25,26] *Dynorphin* expression produces behavioral aversion in animals and humans that could contribute to the depressive-like phenotype in the CREB-over-expressing animals.[23,24] This possibility is supported by a study demonstrating that local administration of a dynorphin antagonist produces an antidepressant effect in the learned helplessness paradigm.[23]

CREB increases neurogenesis in adult hippocampus

In addition to these behavioral studies, the role of CREB in a recently described cellular response to antidepressant treatment has been examined. Several laboratories have reported that chronic antidepressant treatment increases neurogenesis in the adult hippocampus of rats.[27–29] One of these studies has also demonstrated that chronic administration of an NE-selective reuptake inhibitor, reboxetine, increases adult neurogenesis.[27] Up-regulation of adult neurogenesis in hippocampus is particularly interesting in light of studies demonstrating cellular atrophy and cell loss in experimental animals and in depressed patients.[9,30] Preclinical studies have demonstrated that acute stress decreases neurogenesis in the adult hippocampus, and that repeated stress causes atrophy of CA3 pyramidal neurons. Clinical brain-imaging studies report that the volume of hippocampus is reduced in patients with depression or posttraumatic stress disorder, and that antidepressant treatment partially reverses this atrophy. Based on these findings, it is possible that up-regulation of adult neurogenesis could counteract the atrophy and cell loss resulting from exposure to stress and from depression.[9,27,30] This possibility is supported by a study demonstrating that blockade of antidepressant induction of neurogenesis results in blockade of the actions of antidepressants in behavioral models of depression.[31]

The possibility that CREB underlies the induction of adult neurogenesis in response to antidepressant treatment has been examined. The first evidence that CREB may play a role in adult neurogenesis was provided by studies of phospho-CREB localization in the hippocampus (Figure 4.6). Although CREB is found throughout the granule cell layer, phospho-CREB immunoreactivity is found only in the subgranular zone, a thin layer of cells at the border between the granule cell layer and the hilus (Figure 4.6).[32] It is the subgranular zone where the neural progenitor cells are located and where newborn cells are found in the adult hippocampus. Double-labeling studies demonstrate that phospho-CREB immunoreactivity is colocalized in cells that are also positive for bromodeoxyuridine (BrdU), a thymidine analog that is incorporated into the DNA and is used as a marker of dividing cells. Phospho-CREB is also colocalized with markers of immature neurons in the subgranular zone, such as polysialic acid neural cell adhesion molecular (PSA-NCAM), indicating that this transcription factor is involved in the differentiation and/or maturation of the newborn hippocampal neurons (Figure 4.6).[32]

Two approaches have been used to study the influence of the cAMP–CREB cascade on different phases of adult neurogenesis, including proliferation, survival, and maturation of newborn neurons. First, the influence of rolipram, a drug that

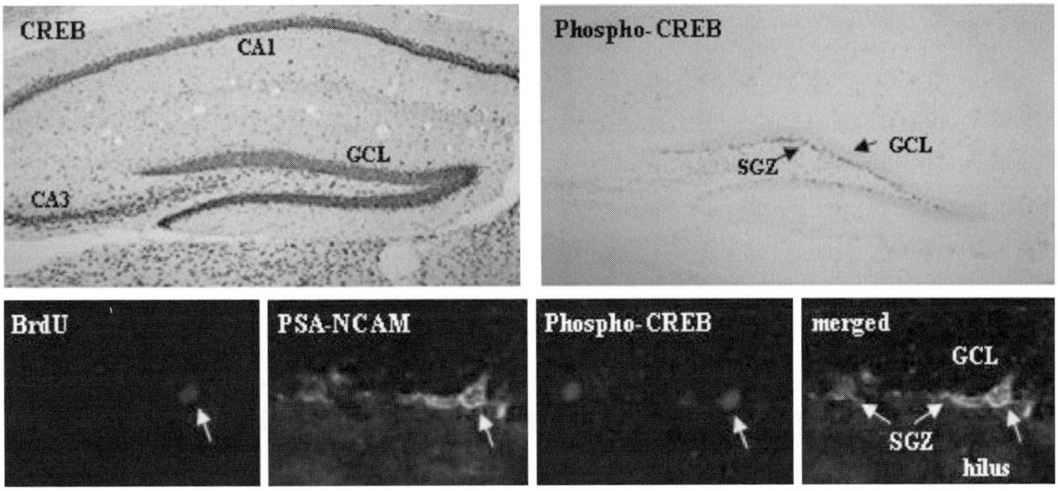

Figure 4.6 Localization of phospho-CREB in newborn neurons in adult hippocampus. Immunohistochemistry for CREB or phospho-CREB in the hippocampus is shown in the upper two panels using a peroxidase staining procedure. CREB-immunoreactivity is found throughout the hippocampus, including the granule cell layer (GCL) and the CA1 and CA3 pyramidal cell layers. In contrast, under basal conditions (untreated) phospho-CREB-immunoreactivity has a very localized pattern, and is found only in cells that are in or adjacent to the subgranular zone (SGZ). The SGZ is the proliferative zone where neural progenitor cells are located in the adult hippocampus. The lower series of panels shows the results of triple immunohistochemistry for bromodeoxy-uridine (BrdU), polysialic acid neural cell adhesion molecule (PSA-NCAM), and phospho-CREB, conducted using an immunofluorescence procedure. PSA-NCAM is a marker for immature neurons. The merged image demonstrates that newborn neurons expressing BrdU/PSA-NCAM also express phospho-CREB. Quantitation of these triple-labeled cells demonstrates that most of the newborn neurons express phospho-CREB as they mature and differentiate. This accounts for the high degree of phospho-CREB staining in the SGZ shown in the upper right panel.

activates the cAMP–CREB cascade by blocking phosphodiesterases type IV (PDE4) and the degradation of cAMP, was examined. Chronic administration of rolipram increased the proliferation of newborn granule cells in the hippocampus, similar to the effect of antidepressant treatment.[32] Second, a dominant negative mutant of CREB was expressed in the hippocampus using the tetracycline-inducible system. Over-expression of mutant CREB in the hippocampus decreased the rate of proliferation in the granule cell layer of adult mice.[32] Chronic rolipram or antidepressant treatment also increases the survival of newborn neurons in the hippocampus.[32] Finally, a recent study demonstrates that chronic rolipram increases the maturation of newborn granule cells.[33] In this study, the dendritic arborization of newborn neurons in the granule cell layer was determined as a measure of cellular maturation. Chronic rolipram treatment increased the number of branch points and the length of dendrites, indicating that activation of the cAMP–CREB cascade increases the rate of neuronal maturation. Taken together, these studies demonstrate that the cAMP–CREB cascade, which is activated by NE and NE-selective reuptake inhibitors, increases adult neurogenesis at several different stages of proliferation and maturation.

Antidepressants and norepinephrine regulation of *c-fos* gene expression and AP-1 DNA binding

The IEG transcription factors, including c-Fos and related AP-1 proteins, have also been identified as targets of NE and NE-selective reuptake inhibitor antidepressants. The regulation of c-Fos- and AP-1-mediated gene transcription, when combined with regulation of CREB, provides a means for fine-tuning the gene expression profile resulting from activation of NE-stimulated neurotransmission. As discussed for CREB, multiple molecular approaches have been developed to study the cellular and behavioral actions of c-Fos and related family members, including null mutants, inducible transgenics, and viral expression vectors. These studies demonstrate that AP-1 transcription factors influence the function of many neuronal systems and the behavioral manifestations of those pathways. However, the influence of these systems in the context of NE neurotransmission has not been examined.

The induction of c-Fos and other IEGs has also proven very useful for mapping regions of the brain that are regulated by NE or other stimuli. The rapid and robust induction of these transcription factors by a variety of receptor-coupled signal transduction pathways makes them useful for brain mapping studies. Moreover, basal levels (i.e., unstimulated conditions) of c-Fos/IEG expression are very low or nonexistent, and this contributes to a very high signal-to-noise ratio. However, there are several limitations, including lack of expression in certain cell types and nonspecific induction, although this has not restricted investigators from using IEGs for brain mapping. A brief overview of the effects of NE and desipramine on c-Fos and IEG expression will be presented below. However, it should be noted that further studies are needed to determine if these AP-1 transcription factors underlie the actions of NE-stimulated gene expression and function. This must be done using approaches similar to those described for CREB, particularly inducible transgenic and knockout mice combined to determine the role of c-Fos/AP-1 transcription factors in NE-stimulated gene expression.

Regulation of c-Fos and IEG expression by desipramine

There have been several reports on the regulation of c-Fos and other IEGs by desipramine administration, in both acute and repeated situations. These studies were undertaken to characterize the transcription factors that are regulated by antidepressant treatment and to map brain regions that are activated by these drugs. Acute administration of desipramine is reported to increase levels of c-Fos in several forebrain regions, most notably the amygdala and hippocampus.[34,35] Chronic desipramine treatment also increases c-Fos expression in several brain regions, although the pattern is different than observed after acute antidepressant administration.[36] In the latter study, chronic desipramine increased levels of c-Fos in the amygdala and the bed nucleus of the stria terminalis, but not in the hippocampus. Direct application of NE into the amygdala is also reported to increase the expression of c-Fos in this structure.[37] Induction of c-Fos in the amygdala is blocked by α_1AR and β_1AR antagonist treatment, demonstrating a direct role for these NE receptor subtypes.

Another pharmacological approach that has been used to study NE regulation of c-Fos is by blockade of α_2ARs, presumably those located on presynaptic terminals. Administration of the α_2AR antagonist yohimbine produces a robust induction of c-Fos in cerebral cortex.[38] This study also demonstrates that yohimbine increases the expression of several other IEGs in cerebral cortex, including nur77, tis-7, zif-268, and tis-21. Induction of c-Fos and these other IEGs was blocked by administration of a β_1AR antagonist, providing further evidence that NE neurotransmission and the β_1AR subtype mediate the action of yohimbine.

Regulation of c-Fos in these limbic brain structures could have long-term effects on the behaviors regulated by these structures. In particular, the amygdala plays a key role in emotion, fear, and anxiety and is considered one of the key structures involved in the pathophysiology and treatment of depression. Norepinephrine neurotransmission in the amygdala is also known to play a critical role in regulation of behavioral models of fear and anxiety. Future studies will be required to further test the role of c-Fos-regulated gene expression in the behavioral actions of NE in this and other limbic structures.

Regulation of c-Fos by stress/NE

Expression of c-Fos and other IEGs has also been utilized to map brain regions that are activated by stress. Moreover, many of the actions of stress on c-Fos are mediated by noradrenergic receptors, supporting previous reports that NE neurotransmission is activated under stress conditions.[38] Restraint stress produces rapid and robust induction of c-Fos expression in many forebrain regions, including the frontal cortex.[38,39] In addition, exposure of animals to the forced-swim test increases c-Fos in several limbic structures, including septum, amygdala, and the paraventricular nucleus of the hypothalamus.[36] The induction of c-Fos by stress is blocked by administration of α_1AR and β_1AR antagonists, providing evidence that the effects of stress are mediated by NE-stimulated receptors.[38] In addition, chronic desipramine treatment attenuates the effects of stress in several brain regions, presumably by causing down-regulation of NE receptors such as the βAR.[36,39]

Although the induction of c-Fos by both antidepressants and stress appears counterintuitive because stress is often used as a model of depression, there are several points to consider. First, both antidepressants and stress also increase NE and 5-HT under certain conditions, yet the therapeutic action of antidepressants is linked to the elevation of these monoamines. This suggests that elevations in c-Fos may be a natural response to stressful stimuli that are designed to contribute to coping mechanisms. Second, these studies are conducted in normal animals and it is possible that there are defects in the signal transduction and gene expression pathways that dampen c-Fos expression in depressed patients and thereby compromise the ability of these individuals to respond to stressful conditions. Further studies will be required to elucidate the role of c-Fos in the response to stress, as well as antidepressant treatment.

Role of NE–CREB signaling in learning and memory

The cAMP–CREB cascade plays a major role in both cellular and behavioral models of learning and memory.[40] In addition, neurotransmitter systems, including NE, that activate CREB via cAMP, Ca^{2+}, or other signal transduction pathways influence learning and memory.[40–42] The influence of NE in the amygdala and hippocampus on learning and memory has been demonstrated using different behavioral models.[43,44] These studies demonstrate increased levels of NE in the amygdala during inhibitory avoidance training, and that levels of NE are highly correlated with conserved memory consolidation. In addition, administration of a βAR antagonist blocks memory consolidation.[41] Moreover, administration of a cAMP analog into the amygdala can mimic the effects of NE, indicating that the cAMP–CREB cascade is involved in memory consolidation in this brain structure.

The role of NE in mediating the influence of emotional arousal and stress on learning and memory is particularly interesting with regard to identifying the molecular mechanisms that underlie neuropsychiatric disorders such as posttraumatic stress disorder and depression.[41,42,44] Studies demonstrating that administration of a βAR antagonist blocks emotional or stress-related memories have led to the suggestion that βAR antagonists could be useful for the treatment of posttraumatic stress disorder. One caveat is that the βAR antagonist must be administered within a short time frame after stress/trauma, during the time period when consolidation of the memory is taking place.[41] Nevertheless, this provides a potentially powerful therapeutic strategy for posttraumatic stress disorder, for which there are currently no effective treatments. Identification of the cAMP–CREB cascade in reconsolidation of emotional/stress memories may also provide additional mechanisms for therapeutic intervention. Recent studies demonstrate that recall and reconsolidation of emotional memories require some of the same neurotransmitter receptor-coupled intracellular signals that underlie the initial consolidation process. It may be possible to disrupt the reconsolidation of emotional/stress memories after recall of these episodes by blockade of NE-stimulated cAMP and/or Ca^{2+}-stimulated signaling.[45] A detailed discussion of the role of NE in learning appears in Chapter 8.

Conclusions

The studies and reports discussed in this chapter demonstrate some of the molecular mechanisms that underlie the actions of NE in the brain. However, studies of the NE system and how it controls neuronal gene expression have not received as much attention as the 5-HT and dopamine monoamine systems. The role of the cAMP–CREB cascade is the only pathway and transcription factor that has been examined in a functional manner, and this has been done primarily in the context of drugs that indirectly increase NE neurotransmission (i.e., NE reuptake blockers). Additional studies with directly acting agonist and antagonists are needed to further characterize the NE receptor subtypes that control gene expression by CREB and related transcription factors. Moreover, studies of c-Fos and other AP-1 transcription factors have been used primarily for mapping purposes, and the relevant downstream targets of these transcription factors and the functional consequences of these genes must be determined. This will be a good start, but it should be kept in mind that these are only a small number of the estimated 1900 transcription factors that are encoded by the human genome, most of which are expressed in the brain. Although the task seems daunting, characterization of neuronal transcription factors and target genes is also likely to yield major breakthroughs in understanding how the brain functions, as well as defining the pathophysiology and treatment of neuropsychiatric illnesses.

ACKNOWLEDGEMENTS

This work is supported by USPHS grants MH45481 and 2 PO1 MH25642, a Veterans Administration National Center Grant for PTSD, and by the Connecticut Mental Health Center.

REFERENCES

1. Armstrong, R. and Montminy, M. R. Transsynaptic control of gene expression. *Annu. Rev. Neurosci.*, **16** (1992), 17–29.
2. Nestler, E., Hyman, S. E. and Malenka, R. C. Signaling to the nucleus. In *Molecular Neuropharmacology*, ed. E. Nestler, S. E. Hyman and R. C. Malenka. (New York: McGraw-Hill, 2001), pp. 115–137.

3. Nestler, E. and Hyman, S. E. Regulation of gene expression. In *Neuropsychopharmacology: The Fifth Generation of Progress*, ed. K. L. Davis, D. Charney, J. T. Coyle and C. Nemeroff. (Philadelphia: Lippincott Williams & Wilkins, 2002), pp. 217–228.
4. Phiel, C., Zhang, F., Huang, E. Y. *et al.* Histone deacetylase is a direct target of valproic acid, a potent anticonvulsant, mood stabilizer, and teratogen. *J. Biol. Chem.*, **276** (2001), 36 734–36 741.
5. Russell, D. and Duman, R. S. Neurotrophic factors and intracellular signal transduction pathways. In *Neuropsychopharmacology: The Fifth Generation of Progress*, ed. K. L. Davis, D. Charney, J. T. Coyle and C. Nemeroff. (Philadelphia: Lippincott Williams & Wilkins, 2002), pp. 207–215.
6. Riccio, A., Ahn, S., Davenport, C. M., Blendy, J. A. and Ginty, D. D. Mediation by a CREB family transcription factor of NGF-dependent survival of sympathetic neurons. *Science*, **286** (1999), 2358–2361.
7. Sulser, F. New perspectives on the molecular pharmacology of affective disorders. *Eur. Arch. Psychiatry Neurol. Sci.*, **238** (1989), 231–239.
8. Duman, R., Heninger, G. R. and Nestler, E. J. A molecular and cellular theory of depression. *Arch. Gen. Psychiatry*, **54** (1997), 597–606.
9. Duman, R., Malberg, J., Nakagawa, S. and D'Sa, C. Neuronal plasticity and survival in mood disorders. *Biol. Psychiatry*, **48** (2000), 732–739.
10. Manji, H., Drevets, W. C. and Charney, D. S. The cellular neurobiology of depression. *Nat. Med.*, **7** (2001), 541–547.
11. Nestler, E. J., Barrot, M., DiLeone, R. J. *et al.* Neurobiology of depression. *Neuron*, **4** (2002), 13–25.
12. Nestler, E., Terwilliger, R. Z. and Duman, R. S. Chronic antidepressant administration alters the subcellular distribution of cAMP-dependent protein kinase in rat frontal cortex. *J. Neurochem.*, **53** (1989), 1644–1647.
13. Nibuya, M., Nestler, E. J. and Duman, R. S. Chronic antidepressant administration increases the expression of cAMP response element binding protein (CREB) in rat hippocampus. *J. Neurosci.*, **16** (1996), 2365–2372.
14. Frechilla, D., Otano, A. and Del Rio, J. Effect of chronic antidepressant treatment on transcription factor binding activity in rat hippocampus and frontal cortex. *Prog. Neuropsychopharmacol. Biol. Psychiatry*, **22** (1998), 787–802.
15. Thome, J., Sakai, N., Shin, K. H. *et al.* cAMP response element-mediated gene transcription is upregulated by chronic antidepressant treatment. *J. Neurosci.*, **20** (2000), 4030–4036.
16. Nibuya, M., Morinobu, S. and Duman, R. S. Regulation of BDNF and trkB mRNA in rat brain by chronic electroconvulsive seizure and antidepressant drug treatments. *J. Neurosci.*, **15** (1995), 7539–7547.
17. Conti, A., Cryan, J. F., Dalvi, A., Lucki, L. and Blendy, J. A. CREB is essential for the upregulation of BDNF transcription, but not the behavioral or endocrine responses to antidepressant drugs. *J. Neurosci.*, **22** (2002), 3262–3268.
18. Timmusk, T., Palm, K., Metsis, M. *et al.* Multiple promoters direct tissue-specific expression of the rat BDNF gene. *Neuron*, **10** (1993), 475–489.

19. Tao, X., Finkbeiner, S., Arnold, D. B., Shaywitz, A. J. and Greenberg, M. E. Ca^{2+} influx regulates BDNF transcription by a CREB family transcription factor-dependent mechanism. *Neuron*, **20** (1998), 709–726.
20. Dias, B., Banerjee, S. B., Duman, R. S. and Vaidya, V. A. Differential regulation of brain derived neurotrophic factor transcripts by antidepressant treatments in the adult rat brain. *Neuropharmacology*, **45** (2003), 553–63.
21. Egan, M., Kojima, M., Callicott, J. H. *et al.* The BDNF val66met polymorphism affects activity-dependent secretion of BDNF and human memory and hippocampal function. *Cell*, **112** (2003), 257–269.
22. Chen, A.-H., Shirayama, Y., Shin, K.-H., Neve, R. L. and Duman, R. S. Expression of the cAMP response element binding protein (CREB) in hippocampus produces antidepressant effect. *Biol. Psychiatry*, **49** (2001), 753–762.
23. Newton, S. S., Thome, J., Wallace, T. L. *et al.* Inhibition of cAMP response element-binding protein or dynorphin in the nucleus accumbens produces an antidepressant-like effect. *J. Neurosci.*, **24** (2002), 10 883–10 890.
24. Pliakas, A., Carlson, R. R., Neve, R. L. *et al.* Altered responsiveness to cocaine and increased immobility in the forced swim test associated with elevated CREB expression in the nucleus accumbens. *J. Neurosci.*, **21** (2001), 7397–7403.
25. Carlezon, W. J., Thome, J., Olson, V. G. *et al.* Regulation of cocaine reward by CREB. *Science*, **282** (1998), 2272–2275.
26. Sakai, N., Thome, J., Chen, J. *et al.* Inducible and brain region-specific CREB transgenic mice. *Mol. Pharmacol.*, **61**: 6 (2002), 1453–1464.
27. Malberg, J., Eisch, A. J., Nestler, E. J. and Duman, R. S. Chronic antidepressant treatment increases neurogenesis in adult hippocampus. *J. Neurosci.*, **20** (2000), 9104–9110.
28. Manev, H., Uz, T., Smalheiser, N. R. and Manev, R. Antidepressants alter cell proliferation in the adult brain in vivo and in neural cultures in vitro. *Eur. J. Pharmacol.*, **411** (2000), 67–70.
29. Czeh, B., Michaelis, T., Watanabe, T. *et al.* Stress-induced changes in cerebral metabolites, hippocampal volume, and cell proliferation are prevented by antidepressant treatment with tianeptine. *Proc. Natl. Acad. Sci. U. S. A.*, **98** (2001), 12 796–12 801
30. Duman, R., Malberg, J. and Nakagawa, S. Regulation of adult neurogenesis by psychotropic drugs and stress. *J. Pharmacol. Exp. Ther.*, **299** (2001), 401–407.
31. Santarelli, L., Saxe, M., Gross, C. *et al.* Requirement of hippocampal neurogenesis for the behavioral effects of antidepressants. *Science*, **301**: 5634 (2003), 805–809.
32. Nakagawa, S., Kim, J.-E., Lee, R. *et al.* Regulation of neurogenesis in adult mouse hippocampus by cAMP and cAMP response element-binding protein. *J. Neurosci.*, **22** (2002), 9868–9876.
33. Fujioka, T., Fujioka, A., Chen, J. and Duman, R. S. Activation of cyclic AMP-CREB cascade facilitates the differentiation of newborn neurons in adult mouse hippocampus. *Soc. Neurosci. Abstr.*, **28**, (2002), Program No. 102.14.
34. Beck, C. Acute treatment with antidepressant drugs selectively increases the expression of c-fos in the rat brain. *J. Psychiatry Neurosci.*, **20** (1995), 25–32.
35. Dahmen, N., Fehr, C., Reuss, S. and Hiemke, C. Stimulation of immediate early gene expression by desipramine in rat brain. *Biol. Psychiatry*, **42** (1997), 317–323.

36. Duncan, G., Knapp, D. J., Johnson, K. B. and Breese, G. R. Functional classification of antidepressants based on antagonism of swim stress-induced fos-like immunoreactivity. *J. Pharmacol. Exp. Ther.*, **277** (1996), 1076–1080.
37. Stone, E., Zhang, Y., Hiller, J. M., Simon, E. J. and Hillman, D. E. Activation of fos in mouse amygdala by local infusion of norepinephrine or atipamezole. *Brain Res.*, **778** (1997), 1–5.
38. Bing, G., Filer, D., Miller, J. C. and Stone, E. A. Noradrenergic activation of immediate early genes in rat cerebral cortex. *Brain Res. Mol. Brain Res.*, **11** (1992), 43–46.
39. Morinobu, S., Nibuya, M. and Duman, R. S. Chronic antidepressant treatment down-regulates the induction of c-fos mRNA in response to acute stress in rat frontal cortex. *Neuropsychopharmacology*, **12** (1995), 221–228.
40. Silva, A., Kogan, J. H., Frankland, P. W. and Kida, S. CREB and memory. *Annu. Rev. Neurosci.*, **21** (1998), 127–148.
41. Ferry, B., Roozendaal, B., and McGaugh, J. L. Role of norepinephrine in mediating stress hormone regulation of long-term memory storage: a critical involvement of the amygdala. *Biol. Psychiatry*, **46** (1999), 1140–1152.
42. McGaugh, J. M. C. and Power, A. E. Amygdala modulation of memory consolidation: interaction with other brain systems. *Neurobiol. Learn. Mem.*, **78** (2002), 539–552.
43. Bevilaqua, L., Ardenghi, P., Schroder, N. *et al.* Drugs acting upon the cyclic adenosine monophosphate/protein kinase A signalling pathway modulate memory consolidation when given late after training into rat hippocampus but not amygdala. *Behav. Pharmacol.*, **8** (1997), 331–338.
44. Roozendaal, B., Quirarte, G. L., and McGaugh, J. L. Glucocorticoids interact with the basolateral amygdala B-adrenoceptor-cAMP/PKA system in influencing memory consolidation. *Eur. J. Neurosci.*, **15** (2002), 553–560.
45. Przybyslawski, J. R. P. and Sara, S. J. Attenuation of emotional and nonemotional memories after their reactivation: role of beta adrenergic receptors. *J. Neurosci.*, **19** (1999), 6623–6628.

5

The norepinephrine transporter and regulation of synaptic transmission

Subbu Apparsundaram, Ph.D.
University of Kentucky Medical Center, Lexington, USA

Introduction

Norepinephrine (NE) signaling involves a coordinated function of a variety of biosynthetic enzymes, ion channels, and pre- and postsynaptic receptors. The action of NE after release is terminated via a reuptake process into a presynaptic site, a process referred to as "neuronal reuptake" or "uptake-1".[1] Neuronal reuptake involves the transfer of NE across the neuronal membrane. Julius Axelrod discovered the existence of the neuronal reuptake process, and provided evidence documenting this process as the primary mechanism for the efficient clearance of NE from the extracellular space.[2–5] It is now well established that neuronal reuptake is mediated by the neuronal plasma membrane resident transporter protein, the norepinephrine transporter (NET).[6–8]

Nonneuronal cells take up NE via uptake-2, an alternative mechanism for the inactivation of NE. This process is mediated by distinct transporter molecules, including OCT-1, OCT-2, and EMT.[9,10] Nonneuronal uptake primarily clears circulating catecholamines, rather than neuronal NE.[10] Uptake-2 may contribute to the clearance of synaptic NE under conditions of diminished NET function.[10] The NET is of prime interest because it is a molecular target for many antidepressants, including the tricyclics, and commonly abused substances, including cocaine and amphetamines. [6,11,12]

The recent identification of a coding mutation in the *NET* gene resulting in compromised NET function, [13] the existence of genetic variants,[14–17] and the emergence of newer NET-specific inhibitors of clinical significance (reboxetine and atomoxetine)[18,19] have stimulated new interest in understanding the function and regulation of NETs in physiological and pathophysiological states.

Brain Norepinephrine: Neurobiology and Therapeutics, ed. Gregory A. Ordway, Michael A. Schwartz and Alan Frazer. Published by Cambridge University Press.

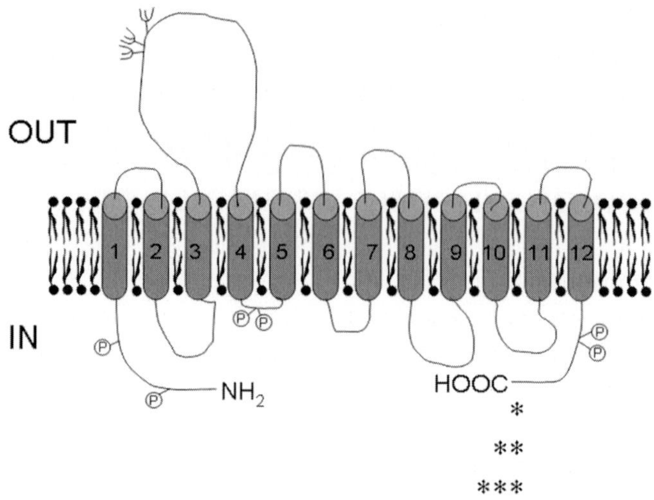

Figure 5.1 Putative transmembrane topology of the norepinephrine transporter. Note the 12 proposed transmembrane domains (TMDs), the large extracellular loop between TMD3 and TMD4 containing three N-glycosylation sites, cytoplasmic NH_2 and COOH termini, and potential phosphorylation sites in internal domains of the transporter. The asterisks below the COOH terminus indicate the identification of three alternative splices yielding NET variants with differing COOH terminus.

The norepinephrine transporter and the importance of norepinephrine uptake

Molecular structure of norepinephrine transporters

The human NET (hNET) was first identified by Pacholczyk *et al.*[6] Subsequently, based on the homology between hNET and the homologous γ-aminobutyric acid (GABA) transporter,[20] transporters for other monoamine neurotransmitters, including dopamine (DA) and serotonin (5-HT), were identified.[7,8,21–23] These transporters are members of the Na^+ and Cl^--coupled cotransporter gene family that also includes transporters for glycine, taurine, creatinine, and betaine [7,8,21–23]. These carrier molecules are predicted to contain 12 hydrophobic transmembrane spanning domains (TMDs) with cytoplasmic amino and carboxy termini, and a large putative extracellular loop positioned between TMD3 and TMD4 (Figure 5.1). This topology is very distinct from that of G-protein-coupled receptors. Experimental evidence derived thus far supports the topological model predicted for NET and other members of the Na^+ and Cl^--coupled cotransporter gene family.[7,8,21,22]

The hNET (GenBank accession # M65105) contains 617 amino acids, with three N-glycosylation sites within the large extracellular loop.[6] The N-glycosylation sites influence the activity and structural stability of the carrier molecule. The hNET also contains protein kinase C (PKC), protein kinase A (PKA), and Ca^{2+}/calmodulin-dependent protein kinase (CaMK) consensus sites for serine/threonine phosphorylation within the amino and carboxy termini and within its internal domains.[6] These canonical phosphorylation sites may be potential sites involved for presynaptic modulation of NET activity. To date, NETs from multiple species, including medaka fish, quail (AF230787), chicken (AF237673), mouse (U76306), rat (Y13223), cow (X79015), sheep (AF246894), and nonhuman primate (AF286026) have been identified.[24–31] These orthologs have structural aspects that are similar to hNET. Amino acid conservation and divergence of these transporters with other monoamine transporters may help to define residues and domains likely to be involved in NE recognition, translocation, and drug sensitivity of these transporters.

Localization of the norepinephrine transporter

Although the diffuse distribution of noradrenergic pathways in the brain and the periphery has presented difficulties for studying the function and distribution of NETs, a variety of techniques, including functional assays, microdialysis, voltammetry, metabolite measurements, radiotracer studies, and immunolocalization studies have shed light on the distribution of NETs.[10] Studies using NET-selective ligand nisoxetine,[32–36] and NET-specific antibodies[37–39] have revealed the distribution of NET in the mammalian brain that is consistent with the known distribution of *NET* gene expression.[40]

Although NET distribution in humans has been documented via uptake studies in vivo[41–46] and autoradiography studies in postmortem tissues,[32,33,36,47–50] most of the current knowledge of the distribution of NETs is derived from immunolocalization studies in rodents. In general, the distribution of NETs is consistent with the distribution of noradrenergic pathways. In the brain, NET is expressed abundantly in the locus coeruleus, which sends projections to the entire cortex, thalamus, and cerebellum.[37–40] Expression of NET is also seen in the medullary A1 and A2 cell groups, which provide projections to the hypothalamus, basal forebrain, and preoptic area.[37–40] In the periphery, NET is expressed in sympathetic neurons of the autonomic nervous system, adrenal medulla, lung, and the placenta.[37,51,52]

In rat noradrenergic neurons, NET-immunoreactivity is localized to noradrenergic somata, dendrites, and axons.[37,40] In axons, NETs exhibit a punctate localization along the fibers. In terminal regions, NET expression is abundant in the hippocampus and cerebellum.[37] Norepinephrine transporter-immunoreactivity is enriched in varicosities along axons, indicating a subcellular compartmentalization that may implicate specific sites involving NET function.[37] Ultramicroscopic studies reveal extrasynaptic localization of NETs reflecting a high spatial organization of these carriers for the efficient clearance of released NE.[37] Norepinephrine transporters are also present in the intracellular compartment of axon terminals and somata,[37] suggesting the potential existence of NETs in multiple compartments in neurons, and potential for redistribution in response to external stimuli. Although NET expression has been shown in cultured astroctyes,[53,54] there is no evidence for NET expression in glial cells in the rodent brain.[37] Norepinephrine transporter-immunoreactivity was absent in epinephrine, dopamine, and 5-HT storing neurons,[37] suggesting restricted localization of NET in noradrenergic neurons. Thus, the pattern of NET expression documented so far is highly consistent with a role for NET in the clearance of NE after release as originally proposed.[1]

Function of the norepinephrine transporter

The reuptake of NE by active transport into presynaptic neurons via NET is the predominant mechanism for terminating the actions of NE. The reuptake of NE into presynaptic sites may have several important implications on noradrenergic function: (1) the concentration of NE at the synapse is reduced more rapidly than simple diffusion, thereby providing for temporal control of NE signaling; (2) the effect of released NE is spatially restricted to a smaller area, permitting spatially restricted signaling; and (3) NE can be recycled for another round of release once it is transported back across the presynaptic membrane and into synaptic vesicles.

Most knowledge regarding NET function was drawn from initial studies performed in peripheral tissues.[10] Later studies revealed NET-mediated NE clearance from the extracellular space in the brain. These studies documented a functional role of transporters in modulating noradrenergic neurotransmission.[1,55] The removal of NE from the extracellular space also limits the activation of auto- and hetero-adrenoceptors expressed on different neurons, and controls NE-mediated volume transmission, a mode of signaling provided by NE that diffuses into nonsynaptic sites.[56,57] Thus, NETs maintain low extracellular NE concentrations, regulating both synaptic and nonsynaptic NE signaling.

The NET transports NE with an estimated Km of 100 to 1000 nm for NE.[6,10,11] Besides NE, NETs also transport other catecholamine neurotransmitters, including DA and epinephrine (EPI).[6,10,11] Indeed, NETs have a higher affinity for DA than NE, but lower affinity for EPI.[6,10] Thus, the affinity rank order of uptake-1 (DA>NE>EPI) is distinct from that of uptake-2 (EPI>NE>DA).[6,10,55] Although immunolocalization studies reveal a lack of NET expression in dopaminergic nerve terminals,[37] NET contributes to the clearance of DA in certain brain regions. Such heterologous uptake may be important for the clearance of DA in regions like the prefrontal cortex, where there are lower levels of DA transporter available for the clearance of DA.[58–60] Moreover, the localization of NETs in DA-rich regions in the ventral tegmental area (VTA) may further support a neuromodulatory role for NETs in regulating brain dopamine function.[39] Norepinephrine transporters also transport other monoamine neurotransmitters, including 5-HT, though with much lower affinity than NE.[61]

It is estimated that about 90% of the released NE is normally removed by the neuronal recapture mechanism.[62] Subsequent to transport into the presynaptic site, NE can be either metabolized by monoamine oxidase located in the mitochondria or transported into the synaptic vesicles via the vesicular monoamine transporter for reutilization. Thus, NET serves as a gateway for NE metabolism and recycling. Since the vesicular monoamine transporter has a much higher affinity for NE than monoamine oxidase, more than 70% of recaptured NE is sequestered into storage vesicles rather than deaminated.

Although >90% of NE is recaptured by presynaptic neurons, the exact extent of NE recapture by NET may depend on the density of innervation and the distance between the release sites and target cells in the effector tissue.[63,64] The recapture mechanism may be more efficient in tissues and regions with dense innervation than in tissues in which there is only sparse innervation.[62–64] Such differences may account for the regional variations in NE uptake, NE content, NE turnover, and NE signaling in the brain.[65–71] Moreover, the presence of NET in the cell bodies and terminal sites suggests potential involvement of NET in multiple functions.[37–39] It has been suggested that the transporters present on the terminal may contribute to the clearance of NE and termination of synaptic signaling, whereas transporters present on the cell body may clear NE that could otherwise stimulate autoreceptors to inhibit neuronal firing.[17,72] Thus, the level of transporter activity at the terminal and cell body may orchestrate the fine-tuning of NE signaling.

The importance of NET in controlling NE homeostasis is further supported by profound changes in NE homeostasis, and synaptic plasticity in NET-deficient mice (NET(−/−)). In NET(−/−) mice, extracellular NE levels are reduced in various brain regions between 55% and 70%. The NET(−/−) mice are smaller than their wild-type littermates and exhibit a reduced body temperature.[73] Despite an elevation in the tyrosine hydroxylase (TH, the rate-limiting enzyme in NE biosyntheis) in NET(−/−) mice, there is a reduction in the total content of NE. This finding suggests that NET plays a key role in maintaining NE levels at the synapse. Investigation of NE kinetics reveals a 60% reduction in release and a six fold reduction in NE clearance, resulting in an overall two fold elevation in extracellular NE levels. These observations in NET(−/−) mice provide genetic evidence for the role of NET in regulating noradrenergic neurotransmission in physiological states.[74] The availability of NET(−/−) mice with expected changes in NE homeostasis coupled with the recent studies documenting changes in NET expression during development[75] may facilitate the investigation of functional roles of the NET in synaptic plasticity in physiological, pathophysiological and developmental states.[73,76–79]

Norepinephrine transporter as a drug target

Besides considering the role of NETs in limiting the concentrations of NE, DA, and EPI, the existence of NETs in noradrenergic neurons has many implications for understanding drug actions. Neuronal uptake serves as a gateway for drugs, including metaraminol, guanthedine, and bretylium, which depend on a selective accumulation in adrenergic nerve terminals for producing their pharmacological actions, and clinical effects. Norepinephrine transporters are also important targets for tricyclic antidepressants (e.g., desipramine) and newer selective NE reuptake inhibitors (SNRIs), including reboxetine, and atomoxetine.[80,81] Recent studies reveal that reboxetine is effective in the treatment of clinical depression in subjects who were unresponsive to other antidepressants,[82] whereas atomoxetine has been recently approved for the treatment of attention-deficit/hyperactivity disorder (ADHD).[83,84]

Tricyclics and SNRIs block NE reuptake. The resultant increase in the temporal and spatial spreading, and potentiation of noradrenergic signaling are thought to be responsible for the psychotherapeutic effects of these agents.[22] The selectivity of NET–antidepressant interaction is further supported by the observation that NET(−/−) mice behaved in the same manner as mice who were administered tricyclic antidepressants.[73] Since NETs contribute to the clearance of DA in certain brain regions, blockade of NETs also augments extracellular DA levels,[85] thus providing for additional mechanisms for the clinical efficacy of SNRIs. Such cross-reaction between antidepressants and multiple neurotransmitter systems may provide an understanding of the complexity of pharmacotherapy in psychiatric diseases and synaptic plasticity.

Norepinephrines transporter are also targets for psychoactive agents, including cocaine and amphetamine.[11,86,87] Cocaine effectively blocks the reuptake of NE, whereas amphetamines, acting via NETs, not only block reuptake but also induce transporter-mediated nonvesicular release (Figure 5.2).[11,86,87] Both cocaine and amphetamine also interact with the DA transporter and 5-HT transporter. Therefore administration of these agents would increase the extracellular concentrations of NE, DA, and 5-HT,[23,87] and the composite effect of these neurotransmitters may dictate both the psychostimulant action and the drug abuse potential of cocaine and amphetamine.

Mechanism of norepinephrine uptake

Although several of the biochemical properties of the NE transport process were evident before the molecular cloning of NETs, the elucidation of NET structure has provided some insights into the mechanism of NE transport. The primary driving force for the transport of NE from the extracellular space to the presynaptic site is the inwardly directed Na^+ gradient. The norepinephrine transporter-mediated

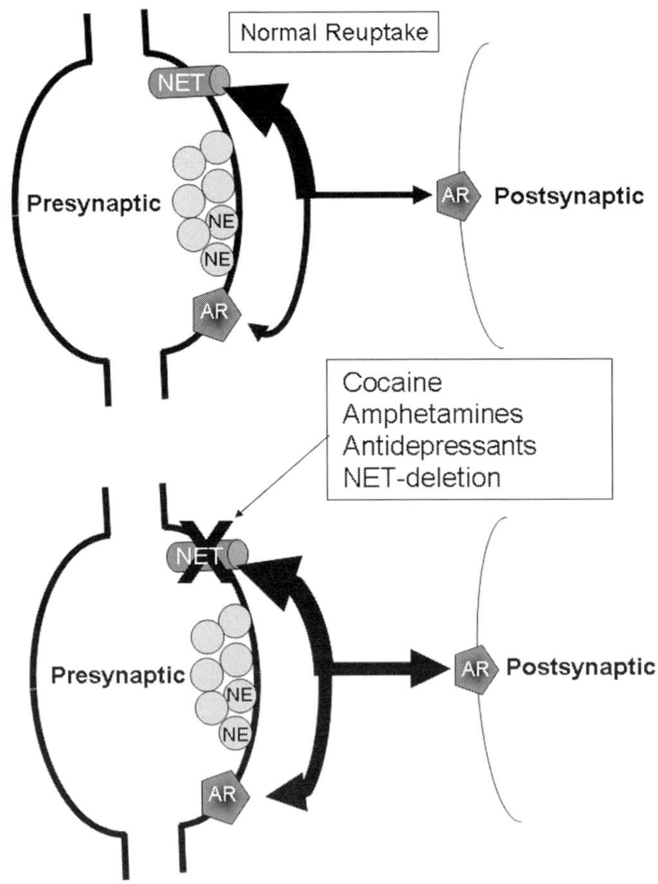

Figure 5.2 Function of the norepinephrine transporter (NET) and action of drugs that block norepinephrine reuptake (NE) in modulating NE actions at adrenergic receptors (AR).

NE reuptake process is also dependent on the availability of Cl^- ions.[6,11] The transport of NE does not depend on adenosine triphosphate (ATP) directly, but, the process of maintaining the ionic gradient across cell membranes that regulates NET activity depends on Na^+/K^+-ATPase. Removal of either Na^+ or Cl^- from the extracellular milieu eliminates NET-mediated NE uptake. Transport of NE by NET is electrogenic, with a net gain of a positive charge.[11] The NET can also function in the reverse direction: in other words, NE can be transported out of the neuron when the Na^+ gradient is reversed or during inward transport of amines like tyramine or amphetamine. This reversal of NET function produces Ca^{2+}-independent NE release.[88]

Recent investigations on how transporters operate have led to the proposal that transporters may have at least two modes of operation: an alternating-access and a

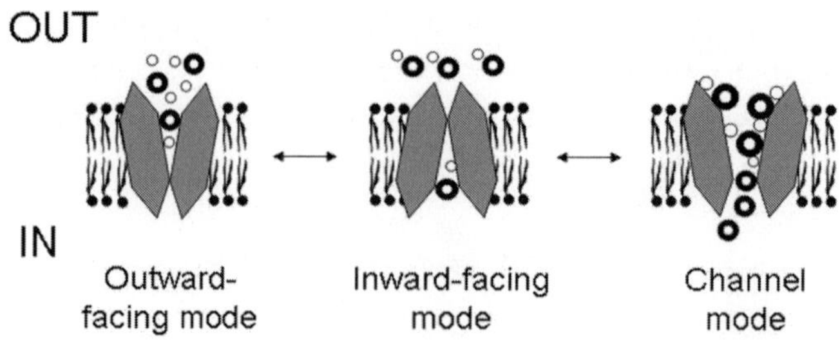

Figure 5.3 Schematic representation of the norepinephrine transporter in the alternating-access and channel mode. In the alternating-access mode, the permeation pathway of the transporter is exposed alternatively to the extracellular or intracellular milieu, and the transport of monoamines (black rings) and ions (clear rings) is stoichiometrically coupled. The channel mode is a low-probability event that allows permeation of ions and substrates down their electrochemical gradient.

channel mode.[89–93] In the alternating-access mode, the carrier is in the outward-facing or inward-facing confirmation. In the outward-facing confirmation, the putative binding pockets for Na^+, Cl^-, and the neurotransmitter are accessible to the extracellular environment. Once the ions and substrate bind, the transporter molecule undergoes a conformational change such that cotransported ions and substrates can gain access to the cytoplasm. As the ions and substrates are released into the cytoplasm, the molecule reverts to the outward-facing state. This reversal of transporter may be the step that contributes to the release of monoamines by amphetamine-like drugs. In contrast, in the channel mode, the protein functions as an ion channel, allowing bulk movement of cotransported ions and substrates.[94] Like ion channels, monoamine transporters have been shown to exhibit conductance and voltage-dependence.[94] Due to their ability to transport many more molecules in a single channel event, transporters in channel mode may profoundly contribute to clearance of neurotransmitters despite low probability (Figure 5.3).

More recent studies in heterologous model systems also provide evidence that monoamine transporters, NET, DA transporter, and 5-HT transporter, may exist as single units and as multimeric complexes.[95–98] Biochemical, and functional studies in cell line models have revealed the potential existence of an oligomeric structure for NETs.[97] Whether or not NETs expressed in native tissues exist as functional oligomers, and exert the properties of such oligomeric structures on drug sensitivities, remains to be explored.

The identification of NET, DA transporter, and 5-HT transporter proteins has also facilitated the investigation of functional residues and domains within the monoamine transporter proteins. Within the Na^+/Cl^--coupled cotransporter gene family, the TMD1, TMD2, and TMDs 4–8 regions show the highest degree of amino acid sequence identity. These highly conserved regions may contain residues important for transport activity, and Na^+- and Cl^--dependence. The Cl^--binding site may be located in TMD1, whereas the Na^+-binding site may be located in TMD2.[96] Aspartate (Asp75) and arginine (Arg 81) in TMD1 of hNET have been identified as being critically involved in ligand binding and substrate transport. Glutamate (E113), glycine residues (G117 and G123) in TMD2, and tyrosine (Y271) in TMD5 are important determinants of NET function.[99–101] Mutagenesis studies also have identified a discrete region between TMD6 and TMD8 as an important determinant of cocaine and tricyclic antidepressants.[102–105] The carboxy-terminal region, from TMD9 through the carboxy-terminal tail, may contribute to high affinity for substrates.[103,104] The amino terminus of the NETs may encompass structural determinants that may contain localization signal for NETs,[106] whereas the carboxy terminus may have residues that contribute to transporter trafficking, stability, and function.[107] Thus, the role of certain sites and domains important for NET function is beginning to emerge. Clearly, additional work is needed to understand the residues and domains important to NET function, ionic dependence, substrate selectivity, substrate translocation, and inhibitor sensitivity.

Regulation of norepinephrine uptake

Given that NETs control the spatial and temporal aspects of NE actions after release, modulation of NET activity by presynaptic mechanisms would be expected to influence the characteristics of noradrenergic neurotransmission. Soon after the identification of uptake-1, several investigators have documented evidence supporting both acute (short-term) and chronic (long-term) regulation of NET activity. Subsequent cloning of hNET cDNA and the identification of NET protein has yielded important tools for probing NET regulation at the gene and protein levels. Employing these tools to delineate the molecular mechanisms underlying regulation of NETs would facilitate the development of strategies for the control of NET function.

Evidence has steadily accumulated documenting presynaptic regulatory mechanisms controlling NET activity and expression.[72,87,102] Neuronal activity, second messengers, neurotransmitters, and hormones have been shown to alter NET activity. Norepinephrine transporter function and expression is also altered in response to substrate and inhibitors. In general, acute regulation of NETs was found to involve rapid changes in the posttranslational modifications of NETs via either silencing or activation of plasma membrane resident transporters. Regulation of NET activity may also involve endocytosis or recruitment of active NETs. Such modulatory mechanisms may involve changes in the NET phosphorylation state, trafficking, and protein–protein interactions (Figure 5.4). Alternately, chronic regulation may

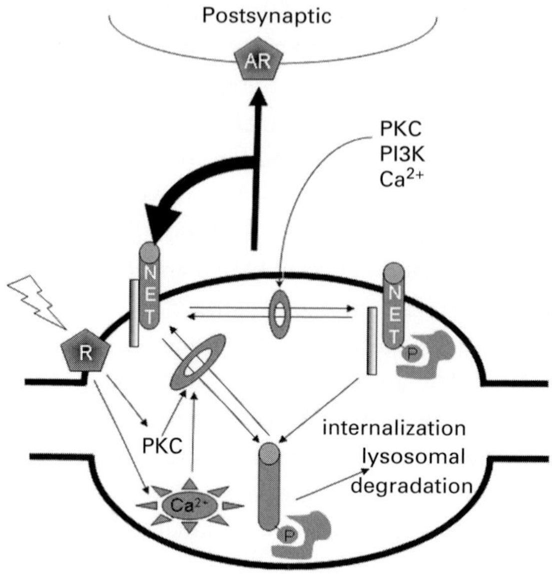

Figure 5.4 Regulation of norepinephrine transporter (NET) via transporter phosphorylation, trafficking and protein interactions. Norepinephrine transporter can exist as dephosphorylated and phosphorylated forms in a complex with cellular proteins. Transporter substrates, presynaptic receptors (R), and second messengers, including protein kinase C (PKC) and cellular Ca^{2+}, can alter the steady-state equilibrium among multiple forms, thereby altering transporter function and regulation. PI3K, phosphatidylinositol-3-kinase.

involve transcriptional changes resulting in either up- or down-regulation of NET density. These complex acute and chronic regulatory events could influence the role of NET in dictating synaptic and nonsynaptic NE signaling and delayed drug efficacy.

Acute regulation of neuronal norepinephrine reuptake

The identification of hNET revealed the presence of serine/threonine PKC, PKA, PKG, and CaMK consensus phosphorylation sites in NETs,[6] and stimulated research in understanding the molecular and biochemical basis underlying acute regulation of NET activity. Activation of PKC and inhibition of serine/threonine protein phosphatases have been shown to modulate transporter phosphorylation state and activity in native tissues, primary cell cultures, noradrenergic neuroblastoma, and heterologous expression cell lines.[108–110] These studies also reveal that NETs and other monoamine transporters, similar to G-protein-coupled receptors, can be sequestered/internalized to intracellular compartments, rendering the transporters unavailable for the extracellular substrates. Apart from changes in NET trafficking,

Table 5.1 Substrates and inhibitors of the norepinephrine transporter.

SUBSTRATES
Catecholamines
Dopamine (Intropin)
Norepinephrine (Levophed)
Epinephrine (Adrenalin)
Other amines
Tyramine
Metaraminol (Aramine)
Meta-iodobenzylguanidine (Iobenguane I-131)
Guanethidine (Ismelin)
Bretylium (Bretylate)
Psychostimulants
Amphetamine (Adderall)
Cocaine
Antidepressants
Desipramine (Norpramin)
Protriptyline (Vivactil)
Atomoxetine (Strattera)
Nortriptyline (Aventyl)
Lofepramine (Emdalen)
Reboxetine (Edronax)
Maprotiline (Ludiomil)
Nomifensine (Merital)
Amoxapine (Asendin)
Doxepin (Adapin)
Mianserin (Lantanon)
Viloxazine (Vivalan)
Mirtazapine (Remeron)
NEUROTOXINS
6-hydroxydopamine
ALKYLATING AGENTS
Phenoxybenzamine (Dibenzyline)
Tranylcypramine (Parnate)
Deprenyl
Xylamine

NET function can also be regulated independent of changes in the surface expression of NET. Thus, changes in posttranslational modification of NETs may involve silencing or activation of plasma membrane resident transporters or endocytosis/recruitment of active NETs.

Modulators of PKC, PKA, PKB, intracellular free Ca^{2+}, phosphatidylinositol-3-kinase, and tyrosine kinase have been shown to alter NET activity. [108,109,111–123] Although most of these studies utilize cell lines or peripheral tissues, these reports present evidence for potential interaction between signal transduction pathways and NETs.

Presynaptic auto- and hetero receptors that modulate neurotransmitter release also have been shown to modulate NET activity, and surface expression. Such regulation may play an important role in fine coordination between neurotransmitter release and the clearance process. Peptide hormones and neurotransmitters, including angiotensin II and III,[124–134] bradykinin,[135,136] endothelin-1,[137] insulin,[120,138–142] natriuretic peptides,[132,136,143–147] and melatonin[146,147] also rapidly modulate NET activity. Other small-molecule neurotransmitters, including acetylcholine,[108,148,149] GABA,[150] opiates,[151,152] ATP,[152–157] histamine,[158,159] nitric oxide,[115,119] gonadal hormones,[160–166] and alcohol[167–170] also alter NET activity. These findings suggest that NETs may be a potential target molecule mediating neurotransmitter – neurotransmitter and neurotransmitter – hormone interactions.

Behavioral stimuli, including stress,[171–177] feeding,[178] and exercise[179] have also been shown to alter NE clearance. These stimuli may alter NET activity via changes in chemical mediators, including changes in neurotransmitter, hormone, and second messenger signaling. Furthermore, these investigations corroborate the physiological relevance of monoamine transporter regulation involving presynaptic proteins.

The clearance of NE is rapidly regulated by electrical activity invading the presynaptic nerve terminal and by extracellular levels of NE. Depending on the experimental condition, stimulation paradigm, and tissue, neuronal depolarization enhances or attenuates NET activity.[63,180–185] Electrical stimulation-evoked changes in NET activity may occur as a result of changes in the intracellular free Ca^{2+}, and other second messenger pathways. Additionally, studies investigating the electrophysiological characteristics of NETs reveal that changes in membrane depolarization may influence NETs' activity.[94] Since catecholamine depletion alters NE uptake, changes evoked by electrical activity in NET activity may be due to changes in the synaptic levels of NE.[186,187] Studies on the DA transporter and 5-HT transporter have shown that substrates might influence transporter surface expression and phosphorylation,[188,189] and such regulation might involve changes in the accessibility of plasma membrane resident transporters to protein kinases, phosphatases, and other regulatory proteins.

Chronic regulation of neuronal norepinephrine reuptake

Besides acute changes in NET surface expression, the density and expression of endogenous NETs are also up-regulated or down-regulated in response to a variety of stimuli. Sustained changes in extracellular NE concentrations, as produced by prolonged exposure to psychoactive agents including antidepressants, amphetamines, and cocaine, may influence NET density.[190–198] Whereas some studies have shown changes in NET protein density and mRNA expression, other studies have reported no change.[199,200] Such inconsistencies might have arisen from variations in the treatment regimen, including drug profile, dosage, route of drug delivery/duration of drug treatment, and model systems employed in the study. Considering the large body of preclinical and clinical evidence indicating marked alterations in the activity of cellular kinases and NET expression following antidepressant administration,[200–202] exploration of relationships between ligand occupancy and changes in density of NETs may be warranted for evaluating the mechanism of action of psychopharmacological agents. Understanding these cellular and molecular events might also lead to an unraveling of the mystery behind the delayed clinical efficacy of psychotherapeutic agents, pathophysiology of mental illness, drug abuse, and synaptic plasticity.

Prolonged suppression of NE synthesis,[203] depletion of tissue NE content,[186] activation of neuromodulatory presynaptic α_2-adrenoceptors,[204] and β-adrenoceptor receptors[205] alter NET expression. Depletion of 5-HT enhances, and chronic lithium administration reduces NET activity.[206–208] Changes in insulin also reduce NET mRNA, and desipramine binding in the locus coeruleus.[209,210] Hypophysectomy, adrenalectomy, corticosterone treatment, and castration, procedures that alter gonadal hormonal levels, also alter NET activity and nisoxetine binding in brain regions.[162,211–214] Chronic administration of a variety of agents, including ketamine,[215] pentylenetetrazol,[216] and pentobarbital[217] affects NET density and expression.[215–217] Long-term intake of excess salt reduces NET expression in sympathetic ganglia innervating heart and blood vessels.[218] Although these studies document regulation of NETs, the role of changes in NET regulation in pharmacological effects of these modulators remains to be investigated.

Norepinephrine transporter-associated proteins

Biochemical and pharmacological studies reveal that the NET exists in complex with presynaptic regulatory proteins. Formation of such complexes might regulate transporter activity, and surface expression of NETs. Recently, Bauman *et al.*[110] reported the PKC regulated association of the catalytic subunit of protein phosphatase PP2A (PP2Ac) with NETs. Norepinephrine transporters also form complexes with the PDZ domain-containing protein PICK1, a protein involved in PKC translocation, resulting in enhanced activity and surface expression of these transporters.[219] Proteomic studies reveal the associations of NET with several isoforms of 14–3–3,[220] a family of signaling molecules that are abundantly expressed in the brain and are known to regulate the receptor function. Recent studies also reveal that a presynaptic SNARE protein, syntaxin 1A, which is involved in synaptic vesicle exocytosis, is found to exist in complexes with NETs.[221] The interaction of syntaxin 1A and NET has been shown to alter intrinsic activity and trafficking of NETs.[221] Additionally, modulators of neurotransmitter release, including cytoplasmic Ca^{2+} and PKC, are found to regulate the association of syntaxin1A with NETs. This finding suggests another mechanism for cross talk among modulators of neurotransmitter release and clearance. Together, these findings raise questions about the functional organization of NETs and the microenvironment in neuronal compartments. Moreover, these reports also suggest that drug treatments that influence the function of NET-associated proteins may also alter NET function.

Norepinephrine transporter gene, polymorphisms and splice variants

The human NET (*SLC6A2*) gene is located on chromosome 16q12.2 and consists of 16 exons.[17,222,223] Although all species examined to date appear to possess a single *NET* gene, coding variants arising from alternative mRNA splicing have been reported.[15,16,224] These alternatively spliced NET mRNAs, differing only in the exons encoding distal residues within the carboxy terminus, result in the expression of NETs that differ in carboxy-terminal regions.[15,225]

Amino acids in the hNET carboxy-terminus region may directly influence the catalytic function of hNET.[107] These residues also may contribute to efficient maturation of hNET protein and trafficking of NETs to the cell surface.[107] In an attempt to gain insight into the function of these splicing variants, Burton *et al.*[226] expressed a bovine NET splice variant in a cell line and found that this variant was retained intracellularly, leading to its lack of function. Kitayama and colleagues[16] also described a rat NET carboxy-terminus splice variant that lacked activity in transfected cells. Coexpression of these variants reveal a dominant negative influence of splice variants on NETs suggesting a possible complex formation between alternative splicing variants, possibly forming structural and functional oligomers.[16,225] In view of the recent studies indicating a role of transporter cytoplasmic domains in protein trafficking and targeting, NET variants that differ in cytoplasmic carboxy terminus might dictate differential trafficking and targeting itineraries. Therefore, alternative splicing could add one more dimension in the genetic variability associated with NE dysfunction and drug effects. To fully understand the implications of alternative splicing variants in noradrenergic function and dysfunction, it is essential to establish the tissue distribution, function, regulation, and pharmacological aspects of these splicing variants.

Apart from alternative splicing variants, recent studies also document the existence of several polymorphisms in the coding and noncoding regions of the *NET* gene.[14,17,227–231] A single mutation resulting in the substitution of Ala457Pro in TMD9 of NET provided the first demonstration of a coding mutation in the monoamine transporter gene that results in alterations in transporter activity.[13] This mutation has been linked to orthostatic intolerance, a disease associated with increase in standing heart rate of at least 30 beats/min and no hypotension. The proband exhibits increased NE spill over, decreased NE clearance, and diminished response to tyramine, a compound that enters noradrenergic neurons via NETs and increases extracellular NE concentrations. In vitro studies in cell line models reveal that Ala457Pro lacks transport activity, due to diminished expression of the mutant protein on the cell surface and affinity for NE.[13,230,232,233] Also, coexpression of a NET mutant (Ala457Pro) identified in proband and wild-type hNET results in a dominant negative effect of the nonfunctional NET mutant.[13,233] Recent studies also report additional variants, including Ala369Pro that have defective trafficking and function.[230] The impact of these findings on NE homeostasis, transporter regulation, the pathophysiology of mental disorders, and the efficacy of psychotherapeutic agents warrants evaluation in preclinical and clinical studies.

Although other coding and noncoding polymorphisms have been identified,[14,17,227–229] genetic association studies reveal no relationship between NET variants and brain disorders and personality traits.[14,17,228,229,234–237] Apart from these coding polymorphisms, several noncoding polymorphisms have been identified.[17] These noncoding polymorphisms may influence NET expression levels, and thereby the clearance of NE, and thus may contribute to genetic variability in noradrenergic function and heterogeneity of response to therapeutic agents.

Clearance of norepinephrine and aging

Normal aging is associated with alterations in noradrenergic signaling in the central and peripheral nervous system. Apart from changes in NE synthesis, NE release, and receptor sensitivity, human and rodent studies also suggest a reduction in NET function. In humans, NET density in the locus coeruleus has been shown to decline with increase in age.[32] Such reduction in NET density may occur due to a loss of noradrenergic projections, rather than the selective loss of NET sites.[32,238,239] Age-related decline in NET function has been reported in the rat and mouse brain regions, including locus coeruleus, hypothalamus, brainstem, and cerebellum.[240–242]

In the periphery, the plasma clearance of NE decreases in humans in an age-related manner, suggesting an age-related reduction in uptake-1.[243] There is also a reduction in NET density in human heart tissue.[244] In rodents, studies also reveal a decline in the function and expression of NETs in multiple peripheral tissues, including rat heart,[245–248] rat vas deferens,[249] and rat trachea.[249,250]

To date, mechanisms underlying alterations in NET function in aging have not been elucidated. One might speculate that there are age-related changes in NET transcription. However, recent studies on the DA transporter reveal a decline in the surface expression of DA transporter with no change in the total DA transporter density, suggesting potential changes in DA transporter trafficking and distribution.[251] In view of these reports, evaluating the mechanism that governs age-related changes in NET activity might shed light on the impairment of noradrenergic signaling in aging.

Norepinephrine transport and disease

Several lines of evidence also support a role for compromised NET function in disease. As revealed in human postmortem studies, NET-binding sites in brain noradrenergic pathways are reduced in depression and suicide.[36,47–50] There is also a reduction in NET expression in Alzheimer's disease.[252] The function of the noradrenergic system is also compromised in ADHD, Parkinson's disease,[253] and schizophrenia.[254–256]

Extensive literature suggesting impaired NET function in humans has been noted in cardiovascular disorders, including hypertension, cardiac failure, myocardial ischemia, pheochromocytoma, porphyria, cardiomyopathy, and prolonged stress.[257–266]

In animal experimental models, decreases in uptake sites and function are observed following cerebral ischemia, hypoxia, myocardial ischemia, cardiac failure, hypertension, hypothyroidism, and acute liver failure.[267–280] Moreover, acidosis and hypothermia also alter NET activity.[281,282] The activity of the NET is also compromised in a diabetes model in rats,[248,272,283] whereas an increase in NET activity has been noted in diabetic cardiomyopathy.[284] Norepinephrine transporter expression is also altered in seizure models.[285] Certain pathophysiological conditions, including hypoxia and glucose deprivation,[286] rather than altering forward transport of NE, induce NET-mediated outward transport of NE.[287] These findings are of great interest in the study of the consequences of ischemia as mechanisms that control reverse transport of NE, and findings may be useful in minimizing risk associated with elevated nonvesicular NE release in myocardial or cerebral ischemia.

Although changes in NET function and expression have been documented in disease states in human and animal models, whether the observed changes in NET function are the cause or the effect remains unclear. The above cited changes in NET function in disease states may involve both transcriptional and posttranslational changes. Elucidation of these mechanisms may allow the evaluation of cause or effect.

Conclusions

During the past four decades, there has been tremendous growth in our understanding of the role of NET in the modulation of noradrenergic function. The availability of molecular and biochemical tools has facilitated the determination of the impact of *NET* gene mutations on NET expression, localization, and posttranslational changes in NET function. We have gained enormous knowledge regarding the molecular basis of NET activity, and its regulation. Norepinephrine transporters and other monoamine transporters are not only a constitutive determinant of synaptic transmission, but also a dynamically regulated component of synaptic plasticity. The ability of multiple presynaptic proteins to regulate NET activity, NET expression, and NET function offers new opportunities to enhance the efficacy of current psychotherapeutic agents, to develop novel therapeutic agents, and to advance our understanding of the diseases that are associated with disrupted noradrenergic signaling. Understanding both short-term and long-term regulatory pathways involved in transporter phosphorylation, trafficking, targeting, and gene expression, as well as organization of the transporter microenvironment, may permit the identification of molecular factors governing the predisposition to and the pathogenesis of several psychiatric, neurological, and neurodegenerative disorders.

ACKNOWLEDGEMENTS

This work was supported by NIH P20RR15592, NIH K12DA14040 and the National Alliance for Research on Schizophrenia and Depression (NARSAD) awards.

REFERENCES

1. Axelrod, J. Noradrenaline: fate and control of its biosynthesis. *Science*, **173** (1971), 598–606.
2. Axelrod, J., Weil-Malherbe, H. and Tomchick, R. The physiological disposition of H3-epinephrine and its metabolite metanephrine. *J. Pharmacol. Exp. Ther.*, **127** (1959), 251–256.
3. Whitby, L. G., Hertting, G. and Axelrod, J. Effect of cocaine on the disposition of noradrenaline labelled with tritium. *Nature*, **187** (1960), 604–605.

4. Hertting, G., Axelrod, J., Kopin, I. J. and Whitby, L. G. Lack of uptake of catecholamines after chronic denervation of sympathetic nerves. *Nature*, **189** (1961), 66.
5. Whitby, L. G., Axelrod, J. and Weil-Malherbe, H. The fate of H3-norepinephrine in animals. *J. Pharmacol. Exp. Ther.*, **132** (1961), 193–201.
6. Pacholczyk, T., Blakely, R. D. and Amara, S. G. Expression cloning of a cocaine- and antidepressant-sensitive human noradrenaline transporter. *Nature*, **350** (1991), 350–354.
7. Amara, S. G. and Kuhar, M. J. Neurotransmitter transporters: recent progress. *Annu. Rev. Neurosci.*, **16** (1993), 73–93.
8. Brownstein, M. J. and Hoffman, B. J. Neurotransmitter transporters. *Recent Prog. Horm. Res.*, **49** (1994), 27–42.
9. Iversen, L. L. *The Uptake and Storage of Noradrenaline in Sympathetic Nerves* (Cambridge, UK: University Press, 1967).
10. Eisenhofer, G. The role of neuronal and extraneuronal plasma membrane transporters in the inactivation of peripheral catecholamines. *Pharmacol. Ther.*, **91** (2001), 35–62.
11. Trendelenburg, U. The TiPS lecture: functional aspects of the neuronal uptake of noradrenaline. *Trends Pharmacol. Sci.*, **12** (1991), 334–337.
12. Iversen, L. Neurotransmitter transporters: fruitful targets for CNS drug discovery. *Mol. Psychiatry*, **5** (2000), 357–362.
13. Shannon, J. R., Flattem, N. L., Jordan, J. *et al.* Orthostatic intolerance and tachycardia associated with norepinephrine-transporter deficiency. *New Engl. J. Med.*, **342** (2000), 541–549.
14. Stober, G., Nothen, M. M., Porzgen, P. *et al.* Systematic search for variation in the human norepinephrine transporter gene: identification of five naturally occurring missense mutations and study of association with major psychiatric disorders. *Am. J. Med. Genet.*, **67** (1996), 523–532.
15. Porzgen, P., Bonisch, H., Hammermann, R. and Bruss, M. The human noradrenaline transporter gene contains multiple polyadenylation sites and two alternatively spliced C-terminal exons. *Biochim. Biophys. Acta*, **1398** (1998), 365–370.
16. Kitayama, S., Ikeda, T., Mitsuhata, C. *et al.* Dominant negative isoform of rat norepinephrine transporter produced by alternative RNA splicing. *J. Biol. Chem.*, **274** (1999), 10 731–10 736.
17. Hahn, M. K. and Blakely, R. D. Monoamine transporter gene structure and polymorphisms in relation to psychiatric and other complex disorders. *Pharmacogenomics J.*, **2** (2002), 217–235.
18. Frazer, A. Norepinephrine involvement in antidepressant action. *J. Clin. Psychiatry*, **61**:Suppl 10 (2000), 25–30.
19. Witcher, J. W., Long, A., Smith, B. *et al.* Atomoxetine pharmacokinetics in children and adolescents with attention deficit hyperactivity disorder. *J. Child Adolesc Psychopharmacol*, **13** (2003), 53–63.
20. Guastella, J., Nelson, N., Nelson, H. *et al.* Cloning and expression of a rat brain GABA transporter. *Science*, **249** (1990), 1303–1306.
21. Shafqat, S., Velaz-Faircloth, M., Guadano-Ferraz, A. and Fremeau, R. T., Jr. Molecular characterization of neurotransmitter transporters. *Mol. Endocrinol.*, **7** (1993), 1517–1529.

22. Nelson, N. The family of Na^+/Cl^- neurotransmitter transporters. *J. Neurochem.*, **71** (1998), 1785–1803.
23. Masson, J., Sagne, C., Hamon, M. and El Mestikawy, S. Neurotransmitter transporters in the central nervous system. *Pharmacol. Rev.*, **51** (1999), 439–464.
24. Roubert, C., Sagne, C., Kapsimali, M. *et al.* A Na(+)/Cl(−)-dependent transporter for catecholamines, identified as a norepinephrine transporter, is expressed in the brain of the teleost fish medaka (*Oryzias latipes*). *Mol. Pharmacol.*, **60** (2001), 462–473.
25. Miller, G. M., Yatin, S. M., De La Garza, R., 2nd, Goulet, M. and Madras, B. K. Cloning of dopamine, norepinephrine and serotonin transporters from monkey brain: relevance to cocaine sensitivity. *Brain Res. Mol. Brain Res.*, **87** (2001), 124–143.
26. Ren, Z. G., Porzgen, P., Zhang, J. M. *et al.* Autocrine regulation of norepinephrine transporter expression. *Mol. Cell. Neurosci.*, **17** (2001), 539–550.
27. Brodski, C., Schnurch, H. and Dechant, G. Neurotrophin-3 promotes the cholinergic differentiation of sympathetic neurons. *Proc. Natl. Acad. Sci. U. S. A.*, **97** (2000), 9683–9688.
28. Fritz, J. D., Jayanthi, L. D., Thoreson, M. A. and Blakely, R. D. Cloning and chromosomal mapping of the murine norepinephrine transporter. *J. Neurochem.*, **70** (1998), 2241–2251.
29. Bruss, M., Porzgen, P., Bryan-Lluka, L. J. and Bonisch, H. The rat norepinephrine transporter: molecular cloning from P. C12 cells and functional expression. *Brain Res. Mol. Brain Res.*, **52** (1997), 257–262.
30. Lingen, B., Bruss, M. and Bonisch, H. Cloning and expression of the bovine sodium- and chloride-dependent noradrenaline transporter. *FEBS Lett.*, **342** (1994), 235–238.
31. Padbury, J. F., Tseng, Y. T., McGonnigal, B. *et al.* Placental biogenic amine transporters: cloning and expression. *Brain Res. Mol. Brain Res.*, **45** (1997), 163–168.
32. Tejani-Butt, S. M. and Ordway, G. A. Effect of age on [3H]nisoxetine binding to uptake sites for norepinephrine in the locus coeruleus of humans. *Brain Res.*, **583** (1992), 312–315.
33. Ordway, G. A., Stockmeier, C. A., Cason, G. W. and Klimek, V. Pharmacology and distribution of norepinephrine transporters in the human locus coeruleus and raphe nuclei. *J. Neurosci.*, **17** (1997), 1710–1719.
34. Charnay, Y., Leger, L., Vallet, P. G. *et al.* [3H]Nisoxetine binding sites in the cat brain: an autoradiographic study. *Neuroscience*, **69** (1995), 259–270.
35. Gehlert, D. R., Schober, D. A. and Gackenheimer, S. L. Comparison of (R)-[3H]tomoxetine and (R/S)-[3H]nisoxetine binding in rat brain. *J. Neurochem.*, **64** (1995), 2792–2800.
36. Klimek, V., Stockmeier, C., Overholser, J. *et al.* Reduced levels of norepinephrine transporters in the locus coeruleus in major depression. *J. Neurosci.*, **17** (1997), 8451–8458.
37. Schroeter, S., Apparsundaram, S., Wiley, R. G. *et al.* Immunolocalization of the cocaine- and antidepressant-sensitive l-norepinephrine transporter. *J. Comp. Neurol.*, **420** (2000), 211–232.
38. Miner, L. H., Schroeter, S., Blakely, R. D. and Sesack, S. R. Ultrastructural localization of the norepinephrine transporter in superficial and deep layers of the rat prelimbic prefrontal cortex and its spatial relationship to probable dopamine terminals. *J. Comp. Neurol.*, **466** (2003), 478–494.

39. Liprando, L. A., Miner, L. H., Blakely, R. D., Lewis, D. A. and Sesack, S. R. Ultrastructural interactions between terminals expressing the norepinephrine transporter and dopamine neurons in the rat and monkey ventral tegmental area. *Synapse*, **52** (2004), 233–244.
40. Lorang, D., Amara, S. G. and Simerly, R. B. Cell-type-specific expression of catecholamine transporters in the rat brain. *J. Neurosci.*, **14** (1994), 4903–4914.
41. Esler, M., Leonard, P., Jackman, G., Bobik, A. and Skews, H. Study of noradrenaline uptake and spillover to plasma in normal subjects and patients with essential hypertension. *Prog. Biochem. Pharmacol.*, **17** (1980), 75–83.
42. Christensen, N. J., Galbo, H., Gjerris, A. *et al.* Whole body and regional clearances of noradrenaline and adrenaline in man. *Acta Physiol. Scand. Suppl.*, **527** (1984), 17–20.
43. Chang, P. C., van Brummelen, P., Vermeij, P. and van der Krogt, J. A. Neuronal and extraneuronal uptake of norepinephrine in the human forearm. *J. Hypertens. Suppl.*, **3** (1985), S141–S143.
44. Goldstein, D. S., Brush, J. E., Jr., Eisenhofer, G., Stull, R. and Esler, M. In vivo measurement of neuronal uptake of norepinephrine in the human heart. *Circulation*, **78** (1988), 41–48.
45. Eisenhofer, G., Esler, M. D., Cox, H. S. *et al.* Differences in the neuronal removal of circulating epinephrine and norepinephrine. *J. Clin. Endocrinol. Metab.*, **70** (1990), 1710–1720.
46. Esler, M., Ferrier, C., Lambert, G. *et al.* Biochemical evidence of sympathetic hyperactivity in human hypertension. *Hypertension*, **17** (1991), III29–III35.
47. Stanley, M., Virgilio, J. and Gershon, S. Tritiated imipramine binding sites are decreased in the frontal cortex of suicides. *Science*, **216** (1982), 1337–1339.
48. Crow, T. J., Cross, A. J., Cooper, S. J. *et al.* Neurotransmitter receptors and monoamine metabolites in the brains of patients with Alzheimer-type dementia and depression, and suicides. *Neuropharmacology*, **23** (1984), 1561–1569.
49. Gross-Isseroff, R., Israeli, M. and Biegon, A. Autoradiographic analysis of [3H]desmethylimipramine binding in the human brain postmortem. *Brain Res.*, **456** (1988), 120–126.
50. Arora, R. C. and Meltzer, H. Y. Laterality and 3H-imipramine binding: studies in the frontal cortex of normal controls and suicide victims. *Biol. Psychiatry*, **29** (1991), 1016–1022.
51. Westwood, N. N., Scarcella, D. L. and Bryan-Lluka, L. J. Evidence for uptake1-mediated efflux of catecholamines from pulmonary endothelial cells of perfused lungs of rats. *Naunyn-Schmiedeberg's Arch. Pharmacol.*, **353** (1996), 528–535.
52. Ramamoorthy, S., Leibach, F. H., Mahesh, V. B. *et al.* Functional characterization and chromosomal localization of a cloned taurine transporter from human placenta. *Biochem. J.*, **300**: Pt 3 (1994), 893–900.
53. Kimelberg, H. K. and Pelton, E. W., 2nd. High-affinity uptake of [3H]norepinephrine by primary astrocyte cultures and its inhibition by tricyclic antidepressants. *J. Neurochem.*, **40** (1983), 1265–1270.
54. Inazu, M., Takeda, H. and Matsumiya, T. Functional expression of the norepinephrine transporter in cultured rat astrocytes. *J. Neurochem.*, **84** (2003), 136–144.
55. Apparsundaram, S., Moore, K. R., Malone, M. D., Hartzell, H. C. and Blakely, R. D. Molecular cloning and characterization of an L-epinephrine transporter from sympathetic ganglia of the bullfrog, *Rana catesbiana. J. Neurosci.*, **17** (1997), 2691–2702.
56. Iversen, L. L. Role of transmitter uptake mechanisms in synaptic neurotransmission. *Br. J. Pharmacol.*, **41** (1971), 571–591.

57. Callado, L. F. and Stamford, J. A. Spatiotemporal interaction of alpha(2) autoreceptors and noradrenaline transporters in the rat locus coeruleus: implications for volume transmission. *J. Neurochem.*, **74** (2000), 2350–2358.
58. Carboni, E., Tanda, G. L., Frau, R. and Di Chiara, G. Blockade of the noradrenaline carrier increases extracellular dopamine concentrations in the prefrontal cortex: evidence that dopamine is taken up in vivo by noradrenergic terminals. *J. Neurochem.*, **55** (1990), 1067–1070.
59. Mundorf, M. L., Joseph, J. D., Austin, C. M., Caron, M. G. and Wightman, R. M. Catecholamine release and uptake in the mouse prefrontal cortex. *J. Neurochem.*, **79** (2001), 130–142.
60. Moron, J. A., Brockington, A., Wise, R. A., Rocha, B. A. and Hope, B. T. Dopamine uptake through the norepinephrine transporter in brain regions with low levels of the dopamine transporter: evidence from knock-out mouse lines. *J. Neurosci.*, **22** (2002), 389–395.
61. Daws, L. C., Toney, G. M., Gerhardt, G. A. and Frazer, A. In vivo chronoamperometric measures of extracellular serotonin clearance in rat dorsal hippocampus: contribution of serotonin and norepinephrine transporters. *J. Pharmacol. Exp. Ther.*, **286** (1998), 967–976.
62. Schomig, E., Fischer, P., Schonfeld, C. L. and Trendelenburg, U. The extent of neuronal re-uptake of 3H-noradrenaline in isolated vasa deferentia and atria of the rat. *Naunyn-Schmiedeberg's Arch. Pharmacol.*, **340** (1989), 502–508.
63. Stjarne, L., Bao, J. X., Gonon, F. and Msghina, M. Nerve activity-dependent variations in clearance of released noradrenaline: regulatory roles for sympathetic neuromuscular transmission in rat tail artery. *Neuroscience*, **60** (1994), 1021–1038.
64. Kopin, I. J., Rundqvist, B., Friberg, P. *et al.* Different relationships of spillover to release of norepinephrine in human heart, kidneys, and forearm. *Am. J. Physiol.*, **275** (1998), R165–R173.
65. Glowinski, J. and Iversen, L. L. Regional studies of catecholamines in the rat brain. I. The disposition of [3H]norepinephrine, [3H]dopamine and [3H]dopa in various regions of the brain. *J. Neurochem.*, **13** (1966), 655–669.
66. Iversen, L. L. and Glowinski, J. Regional studies of catecholamines in the rat brain. II. Rate of turnover of catecholamines in various brain regions. *J. Neurochem.*, **13** (1966), 671–682.
67. Glowinski, J. and Iversen, L. Regional studies of catecholamines in the rat brain. 3. Subcellular distribution of endogenous and exogenous catecholamines in various brain regions. *Biochem. Pharmacol.*, **15** (1966), 977–987.
68. Glowinski, J., Axelrod, J. and Iversen, L. L. Regional studies of catecholamines in the rat brain. IV. Effects of drugs on the disposition and metabolism of H3-norepinephrine and H3-dopamine. *J. Pharmacol. Exp. Ther.*, **153** (1966), 30–41.
69. Iversen, L. L. and Glowinski, J. Regional differences in the rate of turnover of norepinephrine in the rat brain. *Nature*, **210** (1966), 1006–1008.
70. Snyder, S. H. and Coyle, J. T. Regional differences in H3-norepinephrine and H3-dopamine uptake into rat brain homogenates. *J. Pharmacol. Exp. Ther.*, **165** (1969), 78–86.
71. van Veldhuizen, M. J., Feenstra, M. G. and Boer, G. J. Regional differences in the in vivo regulation of the extracellular levels of noradrenaline and its metabolites in rat brain. *Brain Res.*, **635** (1994), 238–248.

72. Zahniser, N. R. and Doolen, S. Chronic and acute regulation of Na^+/Cl^- -dependent neurotransmitter transporters: drugs, substrates, presynaptic receptors, and signaling systems. *Pharmacol. Ther.*, **92** (2001), 21–55.
73. Xu, F., Gainetdinov, R. R., Wetsel, W. C. *et al.* Mice lacking the norepinephrine transporter are supersensitive to psychostimulants. *Nat. Neurosci.*, **3** (2000), 465–471.
74. Gainetdinov, R. R. and Caron, M. G. Monoamine transporters: from genes to behavior. *Annu. Rev. Pharmacol. Toxicol.*, **43** (2003), 261–284.
75. Sanders, J. D., Happe, H. K., Bylund, D. B. and Murrin, L. C. Development of the norepinephrine transporter in the rat CNS. *Neuroscience*, **130** (2005), 107–117.
76. Hall, F. S., Li, X. F., Sora, I. *et al.* Cocaine mechanisms: enhanced cocaine, fluoxetine and nisoxetine place preferences following monoamine transporter deletions. *Neuroscience*, **115** (2002), 153–161.
77. Mead, A. N., Rocha, B. A., Donovan, D. M. and Katz, J. L. Intravenous cocaine induced-activity and behavioural sensitization in norepinephrine-, but not dopamine-transporter knockout mice. *Eur. J. Neurosci.*, **16** (2002), 514–520.
78. Weinshenker, D., White, S. S., Javors, M. A., Palmiter, R. D. and Szot, P. Regulation of norepinephrine transporter abundance by catecholamines and desipramine in vivo. *Brain Res.*, **946** (2002), 239–246.
79. Miles, P. R., Mundorf, M. L. and Wightman, R. M. Release and uptake of catecholamines in the bed nucleus of the stria terminalis measured in the mouse brain slice. *Synapse*, **44** (2002), 188–197.
80. Owens, M. J., Morgan, W. N., Plott, S. J. and Nemeroff, C. B. Neurotransmitter receptor and transporter binding profile of antidepressants and their metabolites. *J. Pharmacol. Exp. Ther.*, **283** (1997), 1305–1322.
81. Richelson, E. Interactions of antidepressants with neurotransmitter transporters and receptors and their clinical relevance. *J. Clin. Psychiatry*, **64**: Suppl 13 (2003), 5–12.
82. Burrows, G. D., Maguire, K. P. and Norman, T. R. Antidepressant efficacy and tolerability of the selective norepinephrine reuptake inhibitor reboxetine: a review. *J. Clin. Psychiatry*, **59**: Suppl 14 (1998), 4–7.
83. Simpson, D. and Perry, C. M. Atomoxetine. *Paediatr. Drugs*, **5** (2003), 407–415; discussion 416–407.
84. Corman, S. L., Fedutes, B. A. and Culley, C. M. Atomoxetine: the first nonstimulant for the management of attention-deficit/hyperactivity disorder. *Am. J. Health Syst. Pharm.*, **61** (2004), 2391–2399.
85. Bymaster, F. P., Katner, J. S., Nelson, D. L. *et al.* Atomoxetine increases extracellular levels of norepinephrine and dopamine in prefrontal cortex of rat: a potential mechanism for efficacy in attention deficit/hyperactivity disorder. *Neuropsychopharmacology*, **27** (2002), 699–711.
86. Barker, E. L. and Blakely, R. D. Norepinephrine and serotonin transporters: molecular targets of antidepressant drugs. In *Psychopharmacology: The Fourth Generation of Progress*, ed. F. E. Bloom and D. J. Kupfer (New York: Raven Press, 1995), pp. 321–333.
87. Torres, G. E., Gainetdinov, R. R. and Caron, M. G. Plasma membrane monoamine transporters: structure, regulation and function. *Nat. Rev. Neurosci.*, **4** (2003), 13–25.

88. Bonisch, H., Fuchs, G. and Graefe, K. H. Sodium-dependence of the saturability of carrier-mediated noradrenaline efflux from noradrenergic neurones in the rat vas deferens. *Naunyn-Schmiedeberg's Arch. Pharmacol.*, **332** (1986), 131–134.
89. Rudnick, G. and Clark, J. From synapse to vesicle: the reuptake and storage of biogenic amine neurotransmitters. *Biochim. Biophys. Acta*, **1144** (1993), 249–263.
90. DeFelice, L. J. and Blakely, R. D. Pore models for transporters? *Biophys. J.*, **70** (1996), 579–580.
91. DeFelice, L. J. Transporter structure and mechanism. *Trends Neurosci.*, **27** (2004), 352–359.
92. Schwartz, J. W., Novarino, G., Piston, D. W. and DeFelice, L. J. Substrate binding stoichiometry and kinetics of the norepinephrine transporter. *J. Biol. Chem.*, **280** (2005), 19177–19184.
93. Blakely, R. D., Defelice, L. J. and Galli, A. Biogenic amine neurotransmitter transporters: just when you thought you knew them. *Physiology (Bethesda)*, **20** (2005), 225–231.
94. Galli, A., Blakely, R. D. and DeFelice, L. J. Norepinephrine transporters have channel modes of conduction. *Proc. Natl. Acad. Sci. U. S. A.*, **93** (1996), 8671–8676.
95. Kilic, F. and Rudnick, G. Oligomerization of serotonin transporter and its functional consequences. *Proc. Natl. Acad. Sci. U. S. A.*, **97** (2000), 3106–3111.
96. Torres, G. E., Carneiro, A., Seamans, K. *et al.* Oligomerization and trafficking of the human dopamine transporter. Mutational analysis identifies critical domains important for the functional expression of the transporter. *J. Biol. Chem.*, **278** (2003), 2731–2739.
97. Kocabas, A. M., Rudnick, G. and Kilic, F. Functional consequences of homo- but not hetero-oligomerization between transporters for the biogenic amine neurotransmitters. *J. Neurochem.*, **85** (2003), 1513–1520.
98. Hastrup, H., Karlin, A. and Javitch, J. A. Symmetrical dimer of the human dopamine transporter revealed by cross-linking Cys-306 at the extracellular end of the sixth transmembrane segment. *Proc. Natl. Acad. Sci. U. S. A.*, **98** (2001), 10055–10060.
99. Sucic, S., Paczkowski, F. A., Runkel, F., Bonisch, H. and Bryan-Lluka, L. J. Functional significance of a highly conserved glutamate residue of the human noradrenaline transporter. *J. Neurochem.*, **81** (2002), 344–354.
100. Sucic, S. and Bryan-Lluka, L. J. The role of the conserved GXXXRXG motif in the expression and function of the human norepinephrine transporter. *Brain Res. Mol. Brain Res.*, **108** (2002), 40–50.
101. Paczkowski, F. A. and Bryan-Lluka, L. J. Tyrosine residue 271 of the norepinephrine transporter is an important determinant of its pharmacology. *Brain Res. Mol. Brain Res.*, **97** (2001), 32–42.
102. Buck, K. J. and Amara, S. G. Chimeric dopamine-norepinephrine transporters delineate structural domains influencing selectivity for catecholamines and 1-methyl-4-phenylpyridinium. *Proc. Natl. Acad. Sci. U. S. A.*, **91** (1994), 12584–12588.
103. Giros, B., Wang, Y. M., Suter, S. *et al.* Delineation of discrete domains for substrate, cocaine, and tricyclic antidepressant interactions using chimeric dopamine-norepinephrine transporters. *J. Biol. Chem.*, **269** (1994), 15985–15988.
104. Bonisch, H., Runkel, F., Roubert, C., Giros, B. and Bruss, M. The human desipramine-sensitive noradrenaline transporter and the importance of defined amino acids for its function. *J. Auton. Pharmacol.*, **19** (1999), 327–333.

105. Roubert, C., Cox, P. J., Bruss, M. *et al.* Determination of residues in the norepinephrine transporter that are critical for tricyclic antidepressant affinity. *J. Biol. Chem.*, **276** (2001), 8254–8260.
106. Gu, H. H., Wu, X., Giros, B. *et al.* The NH(2)-terminus of norepinephrine transporter contains a basolateral localization signal for epithelial cells. *Mol. Biol. Cell*, **12** (2001), 3797–3807.
107. Bauman, P. A. and Blakely, R. D. Determinants within the C-terminus of the human norepinephrine transporter dictate transporter trafficking, stability, and activity. *Arch. Biochem. Biophys.*, **404** (2002), 80–91.
108. Apparsundaram, S., Galli, A., DeFelice, L. J., Hartzell, H. C. and Blakely, R. D. Acute regulation of norepinephrine transport: I. protein kinase C-linked muscarinic receptors influence transport capacity and transporter density in SK-N-SH cells. *J. Pharmacol. Exp. Ther.*, **287** (1998), 733–743.
109. Apparsundaram, S., Schroeter, S., Giovanetti, E. and Blakely, R. D. Acute regulation of norepinephrine transport: II. PKC-modulated surface expression of human norepinephrine transporter proteins. *J. Pharmacol. Exp. Ther.*, **287** (1998), 744–751.
110. Bauman, A. L., Apparsundaram, S., Ramamoorthy, S. *et al.* Cocaine and antidepressant-sensitive biogenic amine transporters exist in regulated complexes with protein phosphatase 2A. *J. Neurosci.*, **20** (2000), 7571–7578.
111. Nicklas, W. J. and Berl, S. Effects of cytochalasin B on uptake and release of putative transmitters by synaptosomes and on brain actomyosin-like protein. *Nature*, **247** (1974), 471–473.
112. Walker, J. E., Goodman, P., Jacobs, D. and Lewin, E. Uptake and release of norepinephrine by slices of rat cerebral cortex: effect of agents that increase cyclic AMP levels. *Neurology*, **28** (1978), 900–904.
113. Tas, P. W., Schomig, E. and Koschel, K. Phorbol ester interferes with [3H]norepinephrine accumulation into monolayers of PC 12 cells by stimulating the release of endogenous norepinephrine. *Eur. J. Neurosci.*, **2** (1990), 920–927.
114. Bunn, S. J., O'Brien, K. J., Boyd, T. L. and Powis, D. A. Pertussis toxin inhibits noradrenaline accumulation by bovine adrenal medullary chromaffin cells. *Naunyn-Schmiedeberg's Arch. Pharmacol.*, **346** (1992), 649–656.
115. Kaye, D. M., Wiviott, S. D., Kobzik, L., Kelly, R. A. and Smith, T. W. S-nitrosothiols inhibit neuronal norepinephrine transport. *Am. J. Physiol.*, **272** (1997), H875–H883.
116. Uchida, J., Kiuchi, Y., Ohno, M., Yura, A. and Oguchi, K. Ca(2+)-dependent enhancement of [3H]noradrenaline uptake in PC12 cells through calmodulin-dependent kinases. *Brain Res.*, **809** (1998), 155–164.
117. Haughey, H. M., Brown, J. M., Fleckenstein, A. E. and Hanson, G. R. Oxygen radicals differentially affect Na+/Cl(−)-dependent transporters. *Eur. J. Pharmacol.*, **379** (1999), 107–110.
118. Yang, H. and Raizada, M. K. Role of phosphatidylinositol 3-kinase in angiotensin II regulation of norepinephrine neuromodulation in brain neurons of the spontaneously hypertensive rat. *J. Neurosci.*, **19** (1999), 2413–2423.
119. Kaye, D. M., Gruskin, S., Smith, A. I. and Esler, M. D. Nitric oxide mediated modulation of norepinephrine transport: identification of a potential target for S-nitrosylation. *Br. J. Pharmacol.*, **130** (2000), 1060–1064.

120. Apparsundaram, S., Sung, U., Price, R. D. and Blakely, R. D. Trafficking-dependent and -independent pathways of neurotransmitter transporter regulation differentially involving p38 mitogen-activated protein kinase revealed in studies of insulin modulation of norepinephrine transport in SK-N-SH cells. *J. Pharmacol. Exp. Ther.*, **299** (2001), 666–677.
121. Bryan-Lluka, L. J., Paczkowski, F. A. and Bonisch, H. Effects of short- and long-term exposure to c-AMP and c-GMP on the noradrenaline transporter. *Neuropharmacology*, **40** (2001), 607–617.
122. Francis, S. C., Sunshine, C. and Kirk, K. L. Coordinate regulation of catecholamine uptake by rab3 and phosphoinositide 3-kinase. *J. Biol. Chem.*, **277** (2002), 7816–7823.
123. Jayanthi, L. D., Samuvel, D. J. and Ramamoorthy, S. Regulated internalization and phosphorylation of the native norepinephrine transporter in response to phorbol esters. Evidence for localization in lipid rafts and lipid raft-mediated internalization. *J. Biol. Chem.*, **279** (2004), 19315–19326.
124. Palaic, D. and Khairallah, P. A. Effect of angiotensin on uptake and release of norepinephrine by brain. *Biochem. Pharmacol.*, **16** (1967), 2291–2298.
125. Palaic, D. and Khairallah, P. A. Inhibition of noradrenaline uptake by angiotensin. *J. Pharm. Pharmacol.*, **19** (1967), 396–397.
126. Panisset, J. C. and Bourdois, P. Effect of angiotensin on the response to noradrenaline and sympathetic nerve stimulation, and on 3H-noradrenaline uptake in cat mesenteric blood vessels. *Can. J. Physiol. Pharmacol.*, **46** (1968), 125–131.
127. Janowsky, D. S., Davis, J. M., Fann, W. E. *et al.* Angiotensin effect on uptake of norepinephrine by synaptosomes. *Life Sci. I*, **11** (1972), 1–11.
128. Liao, J. C. and Zimmerman, B. G. Effect of angiotensin on uptake of 3H-norepinephrine in dog cutaneous arteries. *Proc. Soc. Exp. Biol. Med.*, **139** (1972), 216–219.
129. Sumners, C. and Raizada, M. K. Angiotensin II stimulates norepinephrine uptake in hypothalamus-brain stem neuronal cultures. *Am. J. Physiol.*, **250** (1986), C236–C244.
130. Suvannapura, A. and Levens, N. R. Norepinephrine uptake by rat jejunum: modulation by angiotensin II. *Am. J. Physiol.*, **254** (1988), G135–G141.
131. Fernandez, B. E., Dominguez, A. E., Vatta, M. S. *et al.* Atrial natriuretic peptide and angiotensin II interaction on noradrenaline uptake in the central nervous system. *Arch. Int. Pharmacodyn Ther.*, **307** (1990), 11–17.
132. Vatta, M. S., Bianciotti, L. G., Papouchado, M. L., Locatelli, A. S. and Fernandez, B. E. Effects of atrial natriuretic peptide and angiotensin III on the uptake and intracellular distribution of norepinephrine in medulla oblongata of the rat. *Comp. Biochem. Physiol. C*, **99** (1991), 293–297.
133. Vatta, M. S., Bianciotti, L. G., Locatelli, A. S., Papouchado, M. L., Fernandez, B. E. Monophasic and biphasic effects of angiotensin II and III on norepinephrine uptake and release in rat adrenal medulla. *Can. J. Physiol. Pharmacol.*, **70** (1992), 821–825.
134. Lu, D., Yu, K., Paddy, M. R., Rowland, N. E. and Raizada, M. K. Regulation of norepinephrine transport system by angiotensin II in neuronal cultures of normotensive and spontaneously hypertensive rat brains. *Endocrinology*, **137** (1996), 763–772.
135. Fernandez, B. E., Vatta, M. S. and Bianciotti, L. G. Comparative effects of bradykinin and atrial natriuretic factor on neuronal and non-neuronal noradrenaline uptake in the central nervous system of the rat. *Arch. Int. Physiol. Biochim. Biophys.*, **101** (1993), 337–340.

136. Goto, S., Egashira, T. and Yamanaka, Y. Further studies on the endogenous serotonin-uptake-inhibitor-like substances in the human cerebrospinal fluid. *Jpn. J. Pharmacol.*, **61** (1993), 51–56.
137. Backs, J., Bresch, E., Lutz, M., Kristen, A. V. and Haass, M. Endothelin-1 inhibits the neuronal norepinephrine transporter in hearts of male rats. *Cardiovasc. Res.*, **67** (2005), 283–290.
138. Bhagat, B., Burke, W. J. and Dhalla, N. S. Insulin-induced enhancement of uptake of noradrenaline in atrial strips. *Br. J. Pharmacol.*, **74** (1981), 325–332.
139. Figlewicz, D. P., Szot, P., Israel, P. A., Payne, C. and Dorsa, D. M. Insulin reduces norepinephrine transporter mRNA in vivo in rat locus coeruleus. *Brain Res.*, **602** (1993), 161–164.
140. Boyd, F. T., Jr., Clarke, D. W., Muther, T. F. and Raizada, M. K. Insulin receptors and insulin modulation of norepinephrine uptake in neuronal cultures from rat brain. *J. Biol. Chem.*, **260** (1985), 15 880–15 884.
141. Boyd, F. T., Jr., Clarke, D. W. and Raizada, M. K. Insulin inhibits specific norepinephrine uptake in neuronal cultures from rat brain. *Brain Res.*, **398** (1986), 1–5.
142. Raizada, M. K., Shemer, J., Judkins, J. H. *et al.* Insulin receptors in the brain: structural and physiological characterization. *Neurochem. Res.*, **13** (1988), 297–303.
143. Vatta, M. S., Bianciotti, L. G. and Fernandez, B. E. Influence of atrial natriuretic factor on uptake, intracellular distribution, and release of norepinephrine in rat adrenal medulla. *Can. J. Physiol. Pharmacol.*, **71** (1993), 195–200.
144. Vatta, M., Papouchado, M., Bianciotti, L., Locatelli, A. and Fernandez, B. Atrial natriuretic factor modifies noradrenaline release in a sodium-free medium. *J. Auton. Pharmacol.*, **14** (1994), 267–273.
145. Vatta, M. S., Presas, M., Bianciotti, L. G., Zarrabeitia, V. and Fernandez, B. E. B and C types natriuretic peptides modulate norepinephrine uptake and release in the rat hypothalamus. *Regul. Pept.*, **65** (1996), 175–184.
146. Cardinali, D. P., Nagle, C. A., Freire, F. and Rosner, J. M. Effects of melatonin on neurotransmitter uptake and release by synaptosome-rich homogenates of the rat hypothalamus. *Neuroendocrinology*, **18** (1975), 72–85.
147. Chuluyan, H. E., Rosenstein, R. E., Chang, S. M., Galvez, M. M. and Cardinali, D. P. Presynaptic effects of melatonin on norepinephrine release and uptake in rat pineal gland. *J. Pineal Res.*, **10** (1991), 165–173.
148. Balfour, D. J. Effects of nicotine on the uptake and retention of 14C-noradrenaline and 14C-5-hydroxytryptamine by rat brain homogenates. *Eur. J. Pharmacol.*, **23** (1973), 19–26.
149. Role, L. W. and Perlman, R. L. Catecholamine uptake into isolated adrenal chromaffin cells: inhibition of uptake by acetylcholine. *Neuroscience*, **10** (1983), 987–996.
150. Rosenstein, R. E., Chuluyan, H. E. and Cardinali, D. P. Presynaptic effects of gamma-aminobutyric acid on norepinephrine release and uptake in rat pineal gland. *J. Neural Transm. Gen. Sect.*, **82** (1990), 131–140.
151. Carmichael, F. J. and Israel, Y. In vitro inhibitory effects of narcotic analgesics and other psychotropic drugs on the active uptake of norepinephrine in mouse brain tissue. *J. Pharmacol. Exp. Ther.*, **186** (1973), 253–260.

152. Codd, E. E., Shank, R. P., Schupsky, J. J. and Raffa, R. B. Serotonin and norepinephrine uptake inhibiting activity of centrally acting analgesics: structural determinants and role in antinociception. *J. Pharmacol. Exp. Ther.*, **274** (1995), 1263–1270.
153. Ekesbo, A., Rydin, E., Torstenson, R. *et al.* Dopamine autoreceptor function is lost in advanced Parkinson's disease. *Neurology*, **52** (1999), 120–125.
154. Hendley, E. D., Whittemore, S. R., Chaffee, J. E. and Ehrlich, Y. H. Regulation of norepinephrine uptake by adenine nucleotides and divalent cations: role for extracellular protein phosphorylation. *J. Neurochem.*, **50** (1988), 263–273.
155. Hardwick, J. C., Ehrlich, Y. H. and Hendley, E. D. Extracellular ATP stimulates norepinephrine uptake in PC12 cells. *J. Neurochem.*, **53** (1989), 1512–1518.
156. Cao, C. J., Eldefrawi, A. T. and Eldefrawi, M. E. ATP-regulated neuronal catecholamine uptake: a new mechanism. *Life Sci.*, **47** (1990), 655–667.
157. Eshleman, A., Dunigan, C., Shamoo, A. and Eldefrawi, M. ATP enhances catecholamine uptake into PC12 cells. *Life Sci.*, **56** (1995), 1613–1621.
158. Imamura, M., Lander, H. M. and Levi, R. Activation of histamine H3-receptors inhibits carrier-mediated norepinephrine release during protracted myocardial ischemia. Comparison with adenosine A1-receptors and alpha2-adrenoceptors. *Circ. Res.*, **78** (1996), 475–481.
159. Hess, R. A., Bunick, D. and Bahr, J. Oestrogen, its receptors and function in the male reproductive tract – a review. *Mol. Cell. Endocrinol.*, **178** (2001), 29–38.
160. Janowsky, D. S. and Davis, J. M. Progesterone-estrogen effects on uptake and release of norepinephrine by synaptosomes. *Life Sci.*, **9** (1970), 525–531.
161. Endersby, C. A. and Wilson, C. The effect of ovarian steroids on the uptake of 3H-noradrenaline, 3H-dopamine and 3H-5-hydroxytryptamine by hypothalamic tissue in vitro. *Br. J. Pharmacol.*, **47** (1973), 647P-648P.
162. Nixon, R. L., Janowsky, D. S. and Davis, J. M. Effects of progesterone, beta-estradiol, and testosterone on the uptake and metabolism of 3H-norepinephrine, 3H-dopamine and 3H-serotonin in rat brain synaptosomes. *Res. Commun. Chem. Pathol. Pharmacol.*, **7** (1974), 233–236.
163. Pedroza-Garcia, E., Garcia-Bienere, W., Inglesi, J., Szerezewski, A. and Rosner, J. M. Interactions in the uptake between catecholamines and estradiol in the human uterus. *Am. J. Obstet. Gynecol.*, **122** (1975), 96–100.
164. Davis, B. F., Davis, C. F. and Halaris, A. E. Variations in the uptake of 3H-norepinephrine during the rat estrous cycle. *Life Sci.*, **21** (1977), 1387–1394.
165. Hamlet, M. A., Rorie, D. K. and Tyce, G. M. Effects of estradiol on release and disposition of norepinephrine from nerve endings. *Am. J. Physiol.*, **239** (1980), H450-H456.
166. Pau, K. Y., Lee, C. J., Cowles, A. *et al.* Possible involvement of norepinephrine transporter activity in the pulsatility of hypothalamic gonadotropin-releasing hormone release: influence of the gonad. *J. Neuroendocrinol.*, **10** (1998), 21–29.
167. Israel, Y., Carmichael, F. J. and Macdonald, J. A. Effects of ethanol on norepinephrine uptake and electrically stimulated release in brain tissue. *Ann. N. Y. Acad. Sci.*, **215** (1973), 38–48.
168. Thadani, P. V. and Truitt, E. B., Jr. Effect of acute ethanol or acetaldehyde administration on the uptake, release, metabolism and turnover rate of norepinephrine in rat brain. *Biochem. Pharmacol.*, **26** (1977), 1147–1150.

169. Lin, A. M., Bickford, P. C., Palmer, M. R. and Gerhardt, G. A. Ethanol inhibits the uptake of exogenous norepinephrine from the extracellular space of the rat cerebellum. *Neurosci. Lett.*, **164** (1993), 71–75.
170. Lin, A. M., Bickford, P. C., Palmer, M. R., Cline, E. J. and Gerhardt, G. A. Effects of ethanol and nomifensine on NE clearance in the cerebellum of young and aged Fischer 344 rats. *Brain Res.*, **756** (1997), 287–292.
171. Magyar, K. Uptake of biogenic amines and behaviour. *Act. Nerv. Super.* (Praha), **14** (1972), 303–304.
172. Welch, B. L., Hendley, E. D. and Turek, I. Norepinephrine uptake into cerebral cortical synaptosomes after one fight or electroconvulsive shock. *Science*, **183** (1974), 220–221.
173. Hendley, E. D. and Welch, B. L. Electroconvulsive shock: sustained decrease in norepinephrine uptake affinity in a reserpine model of depression. *Life Sci.*, **16** (1975), 45–54.
174. Weiss, J. M., Glazer, H. I., Pohorecky, L. A., Brick, J. and Miller, N. E. Effects of chronic exposure to stressors on avoidance-escape behavior and on brain norepinephrine. *Psychosom. Med.*, **37** (1975), 522–534.
175. Modigh, K. Long-term effects of electroconvulsive shock therapy on synthesis, turnover and uptake of brain monoamines. *Psychopharmacology* (Berl.), **49** (1976), 179–185.
176. Hendley, E. D., Burrows, G. H., Robinson, E. S., Heidenreich, K. A. and Bulman, C. A. Acute stress and the brain norepinephrine uptake mechanism in the rat. *Pharmacol. Biochem. Behav.*, **6** (1977), 197–202.
177. Moisset, B. Factors contributing to the modulation of norepinephrine uptake by synaptosomes from mouse brain cortex. *Brain Res.*, **121** (1977), 113–120.
178. Van Der Gugten, J. and Slangen, J. L. Norepinephrine uptake by hypothalamic tissue from the rat related to feeding. *Pharmacol. Biochem. Behav.*, **3** (1975), 855–860.
179. Coker, R. H., Krishna, M. G., Lacy, D. B., Allen, E. J. and Wasserman, D. H. Sympathetic drive to liver and nonhepatic splanchnic tissue during heavy exercise. J. Appl. Physiol., **82** (1997), 1244–1249.
180. Gillis, C. N. Altered cardiac retention of exogenous noradrenaline produced by stress in young rabbits. *Nature*, **207** (1965), 1302–1304.
181. Ungerer, M., Chlistalla, A. and Richardt, G. Upregulation of cardiac uptake 1 carrier in ischemic and nonischemic rat heart. *Circ. Res.*, **78** (1996), 1037–1043.
182. Rorie, D. K., Hunter, L. W. and Tyce, G. M. Dihydroxyphenylglycol as an index of neuronal uptake in dog saphenous vein. *Am. J. Physiol.*, **257** (1989), H1945–H1951.
183. Cubeddu, L., Barnes, E. M., Langer, S. Z. and Weiner, N. Release of norepinephrine and dopamine-beta-hydroxylase by nerve stimulation. I. Role of neuronal and extraneuronal uptake and of alpha presynaptic receptors. *J. Pharmacol. Exp. Ther.*, **190** (1974), 431–450.
184. Yamamoto, H. and Kirpekar, S. M. Effects of nerve stimulation on the uptake of norepinephrine by the perfused spleen of the cat. *Eur. J. Pharmacol.*, **17** (1972), 25–33.
185. Friedman, E. and Bhagat, B. Uptake of (3H)noradrenaline in the rat heart during increased sympathetic nervous activity associated with cold. *J. Pharm. Pharmacol.*, **20** (1968), 963–965.
186. Lee, C. M., Javitch, J. A. and Snyder, S. H. Recognition sites for norepinephrine uptake: regulation by neurotransmitter. *Science*, **220** (1983), 626–629.

187. Okada, K., Shinozuka, K., Shimoura, K. *et al.* Effects of reserpine on the content and uptake of dopamine and noradrenaline in rabbit arteries. *Clin. Exp. Pharmacol. Physiol.*, **20** (1993), 261–267.
188. Ramamoorthy, S. and Blakely, R. D. Phosphorylation and sequestration of serotonin transporters differentially modulated by psychostimulants. *Science*, **285** (1999), 763–766.
189. Saunders, C., Ferrer, J. V., Shi, L. *et al.* Amphetamine-induced loss of human dopamine transporter activity: an internalization-dependent and cocaine-sensitive mechanism. *Proc. Natl. Acad. Sci. U. S. A.*, **97** (2000), 6850–6855.
190. Bauer, M. E. and Tejani-Butt, S. M. Effects of repeated administration of desipramine or electroconvulsive shock on norepinephrine uptake sites measured by [3H]nisoxetine autoradiography. *Brain Res.*, **582** (1992), 208–214.
191. Szot, P., Ashliegh, E. A., Kohen, R. *et al.* Norepinephrine transporter mRNA is elevated in the locus coeruleus following short- and long-term desipramine treatment. *Brain Res.*, **618** (1993), 308–312.
192. Shores, M. M., Szot, P. and Veith, R. C. Desipramine-induced increase in norepinephrine transporter mRNA is not mediated via alpha 2 receptors. *Brain Res. Mol. Brain Res.*, **27** (1994), 337–341.
193. Cubells, J. F., Kim, K. S., Baker, H. *et al.* Differential in vivo regulation of mRNA encoding the norepinephrine transporter and tyrosine hydroxylase in rat adrenal medulla and locus ceruleus. *J. Neurochem.*, **65** (1995), 502–509.
194. Frazer, A. Pharmacology of antidepressants. *J. Clin. Psychopharmacol.*, **17**: Suppl 1 (1997), 2S-18S.
195. Frazer, A., Gerhardt, G. A. and Daws, L. C. New views of biogenic amine transporter function: implications for neuropsychopharmacology. *Int. J. Neuropsychopharmacol.*, **2** (1999), 305–320.
196. Shearman, L. P. and Meyer, J. S. Cocaine up-regulates norepinephrine transporter binding in the rat placenta. *Eur. J. Pharmacol.*, **386** (1999), 1–6.
197. Zhu, M. Y., Shamburger, S., Li, J. and Ordway, G. A. Regulation of the human norepinephrine transporter by cocaine and amphetamine. *J. Pharmacol. Exp. Ther.*, **295** (2000), 951–959.
198. Beveridge, T. J., Smith, H. R., Nader, M. A. and Porrino, L. J. Effects of chronic cocaine self-administration on norepinephrine transporters in the nonhuman primate brain. *Psychopharmacology* (*Berl.*), **120**, (2005), 781–788.
199. Hebert, C., Habimana, A., Elie, R. and Reader, T. A. Effects of chronic antidepressant treatments on 5-HT and NA transporters in rat brain: an autoradiographic study. *Neurochem. Int.*, **38** (2001), 63–74.
200. Benmansour, S., Altamirano, A. V., Jones, D. J. *et al.* Regulation of the norepinephrine transporter by chronic administration of antidepressants. *Biol. Psychiatry*, **55** (2004), 313–316.
201. Popoli, M., Brunello, N., Perez, J. and Racagni, G. Second messenger-regulated protein kinases in the brain: their functional role and the action of antidepressant drugs. *J. Neurochem.*, **74** (2000), 21–33.

202. Mori, S., Popoli, M., Brunello, N., Racagni, G. and Perez, J. Effect of reboxetine treatment on brain cAMP- and calcium/calmodulin-dependent protein kinases. *Neuropharmacology*, **40** (2001), 448–456.

203. Fuller, R. W. and Perry, K. W. Effect of uptake inhibitors on the depletion of brain norepinephrine and serotonin after alpha-methyl-m-tyrosine administration to rats. *Arch. Int. Pharmacodyn. Ther.*, **234** (1978), 229–235.

204. Swann, A. C., Duman, R. and Hewitt, L. Desipramine binding: relationship to central and sympathetic noradrenergic activity. *J. Neurochem.*, **44** (1985), 611–615.

205. Dong, E., Yatani, A., Mohan, A. and Liang, C. S. Myocardial beta-adrenoceptor downregulation by norepinephrine is linked to reduced norepinephrine uptake activity. *Eur. J. Pharmacol.*, **384** (1999), 17–24.

206. Koed, K. and Linnet, K. Opposing changes in serotonin and norepinephrine transporter mRNA levels after serotonin depletion. *Eur. Neuropsychopharmacol.*, **10** (2000), 501–509.

207. Cameron, O. G. and Smith, C. B. Comparison of acute and chronic lithium treatment on 3H-norepinephrine uptake by rat brain slices. *Psychopharmacology (Berl)*, **67** (1980), 81–85.

208. Ahluwalia, P. and Singhal, R. L. Effect of lithium treatment and withdrawal on uptake of noradrenaline into rat brain synaptosomes: a kinetic study. *Prog. Neuropsychopharmacol. Biol. Psychiatry*, **6** (1982), 339–342.

209. Figlewicz, D. P., Bentson, K. and Ocrant, I. The effect of insulin on norepinephrine uptake by PC12 cells. *Brain Res. Bull.*, **32** (1993), 425–431.

210. Figlewicz, D. P., Brot, M. D., McCall, A. L. and Szot, P. Diabetes causes differential changes in CNS noradrenergic and dopaminergic neurons in the rat: a molecular study. *Brain Res.*, **736** (1996), 54–60.

211. Zigmond, M. J., Chalmers, J. P., Simpson, J. R. and Wurtman, R. J. Effect of lateral hypothalamic lesions on uptake of norepinephrine by brain homogenates. *J. Pharmacol. Exp. Ther.*, **179** (1971), 20–28.

212. Vermes, I., Smelik, P. G. and Mulder, A. H. Effects of hypophysectomy, adrenalectomy and corticosterone treatment on uptake and release of putative central neurotransmitters by rat hypothalamic tissue in vitro. *Life Sci.*, **19** (1976), 1719–1725.

213. Shang, Y. and Dluzen, D. E. Castration increases nisoxetine-evoked norepinephrine levels in vivo within the olfactory bulb of male rats. *Neurosci. Lett.*, **328** (2002), 81–84.

214. Shaul, P. W. Regulation of endothelial nitric oxide synthase: location, location, location. *Annu. Rev. Physiol.*, **64** (2002), 749–774.

215. Hara, K., Minami, K., Ueno, S. *et al.* Up-regulation of noradrenaline transporter in response to prolonged exposure to ketamine. *Naunyn-Schmiedeberg's Arch. Pharmacol.*, **365** (2002), 406–412.

216. Szot, P., White, S. S. and Veith, R. C. Effect of pentylenetetrazol on the expression of tyrosine hydroxylase mRNA and norepinephrine and dopamine transporter mRNA. *Brain Res. Mol. Brain Res.*, **44** (1997), 46–54.

217. Chang, Y. Y. and Ho, I. K. Effects of acute and chronic administration of pentobarbital on catecholamine uptake in mouse brain synaptosomes. *Res. Commun. Chem. Pathol. Pharmacol.*, **23** (1979), 465–474.

218. Habecker, B. A., Grygielko, E. T., Huhtala, T. A., Foote, B. and Brooks, V. L. Ganglionic tyrosine hydroxylase and norepinephrine transporter are decreased by increased sodium chloride in vivo and in vitro. *Auton. Neurosci.*, **107** (2003), 85–98.
219. Torres, G. E., Yao, W. D., Mohn, A. R. *et al.* Functional interaction between monoamine plasma membrane transporters and the synaptic PDZ domain-containing protein PICK1. *Neuron*, **30** (2001), 121–134.
220. Sung, U., Jennings, J. L., Link, A. J. and Blakely, R. D. Proteomic analysis of human norepinephrine transporter complexes reveals associations with protein phosphatase 2A anchoring subunit and 14–3–3 proteins. *Biochem. Biophys. Res. Commun.*, **333** (2005), 671–678.
221. Sung, U., Apparsundaram, S., Galli, A. *et al.* A regulated interaction of syntaxin 1A with the antidepressant-sensitive norepinephrine transporter establishes catecholamine clearance capacity. *J. Neurosci.*, **23** (2003), 1697–1709.
222. Bruss, M., Kunz, J., Lingen, B. and Bonisch, H. Chromosomal mapping of the human gene for the tricyclic antidepressant-sensitive noradrenaline transporter. *Hum. Genet.*, **91** (1993), 278–280.
223. Gelernter, J., Kruger, S., Pakstis, A. J. *et al.* Assignment of the norepinephrine transporter protein (NET1) locus to chromosome 16. *Genomics*, **18** (1993), 690–692.
224. Kitayama, S., Kumagai, K., Morita, K. and Dohi, T. Identification and functional characterization of the novel isoforms of bovine norepinephrine transporter produced by alternative splicing. *Brain Res.*, **934** (2002), 152–156.
225. Kitayama, S., Morita, K. and Dohi, T. Functional characterization of the splicing variants of human norepinephrine transporter. *Neurosci. Lett.*, **312** (2001), 108–112.
226. Burton, L. D., Kippenberger, A. G., Lingen, B. *et al.* A variant of the bovine noradrenaline transporter reveals the importance of the C-terminal region for correct targeting to the membrane and functional expression. Biochem. J., **330**: Pt 2 (1998), 909–914.
227. Iwasa, H., Kurabayashi, M., Nagai, R., Nakamura, Y. and Tanaka, T. Genetic variations in five genes involved in the excitement of cardiomyocytes. *J. Hum. Genet.*, **46** (2001), 549–552.
228. Halushka, M. K., Fan, J. B., Bentley, K. *et al.* Patterns of single-nucleotide polymorphisms in candidate genes for blood-pressure homeostasis. *Nat. Genet.*, **22** (1999), 239–247.
229. Runkel, F., Bruss, M., Nothen, M. M. *et al.* Pharmacological properties of naturally occurring variants of the human norepinephrine transporter. *Pharmacogenetics*, **10** (2000), 397–405.
230. Hahn, M. K., Mazei-Robison, M. S. and Blakely, R. D. Single nucleotide polymorphisms in the human norepinephrine transporter gene affect expression, trafficking, antidepressant interaction, and protein kinase C regulation. *Mol. Pharmacol.*, **68** (2005), 457–466.
231. Ruoho, A. E. How the monoamine transporter garden grows. *Mol. Pharmacol.*, **68** (2005), 272–274.
232. Paczkowski, F. A., Bonisch, H. and Bryan-Lluka, L. J. Pharmacological properties of the naturally occurring Ala(457)Pro variant of the human norepinephrine transporter. *Pharmacogenetics*, **12** (2002), 165–173.
233. Hahn, M. K., Robertson, D. and Blakely, R. D. A mutation in the human norepinephrine transporter gene (SLC6A2) associated with orthostatic intolerance disrupts surface expression of mutant and wild-type transporters. *J. Neurosci.*, **23** (2003), 4470–4478.

234. Owen, D., Du, L., Bakish, D., Lapierre, Y. D. and Hrdina, P. D. Norepinephrine transporter gene polymorphism is not associated with susceptibility to major depression. *Psychiatry Res.*, **87** (1999), 1–5.

235. Stober, G., Hebebrand, J., Cichon, S. *et al.* Tourette syndrome and the norepinephrine transporter gene: results of a systematic mutation screening. *Am. J. Med. Genet.*, **88** (1999), 158–163.

236. Samochowiec, J., Kucharska-Mazur, J., Kaminski, R. *et al.* Norepinephrine transporter gene polymorphism is not associated with susceptibility to alcohol dependence. *Psychiatry Res.*, **111** (2002), 229–233.

237. Zill, P., Engel, R., Baghai, T. C. *et al.* Identification of a naturally occurring polymorphism in the promoter region of the norepinephrine transporter and analysis in major depression. *Neuropsychopharmacology*, **26** (2002), 489–493.

238. Ishida, Y., Shirokawa, T., Miyaishi, O., Komatsu, Y. and Isobe, K. Age-dependent changes in projections from locus coeruleus to hippocampus dentate gyrus and frontal cortex. *Eur. J. Neurosci.*, **12** (2000), 1263–1270.

239. Ishida, Y., Shirokawa, T., Miyaishi, O., Komatsu, Y. and Isobe, K. Age-dependent changes in noradrenergic innervations of the frontal cortex in F344 rats. *Neurobiol. Aging*, **22** (2001), 283–286.

240. de Avellar, M. C. and Markus, R. P. Neuronal uptake of noradrenaline in the rat isolated trachea: effect of ageing. *Comp. Biochem. Physiol. C*, **96** (1990), 287–290.

241. Zhao, X. H. and Nomura, Y. Age-related changes in uptake and release on L-[3H]noradrenaline in brain slices of senescence accelerated mouse. *Int. J. Dev. Neurosci.*, **8** (1990), 267–272.

242. Zaman, V., Li, Z., Middaugh, L. *et al.* The noradrenergic system of aged GDNF heterozygous mice. *Cell Transplant.*, **12** (2003), 291–303.

243. Morrow, L. A., Linares, O. A., Hill, T. J. *et al.* Age differences in the plasma clearance mechanisms for epinephrine and norepinephrine in humans. *J. Clin. Endocrinol. Metab.*, **65** (1987), 508–511.

244. Leineweber, K., Wangemann, T., Giessler, C. *et al.* Age-dependent changes of cardiac neuronal noradrenaline reuptake transporter (uptake1) in the human heart. *J. Am. Coll. Cardiol.*, **40** (2002), 1459.

245. Sumners, C., Phillips, M. I. and Raizada, M. K. Angiotensin II stimulates changes in the norepinephrine content of primary cultures of rat brain. *Neurosci. Lett.*, **36** (1983), 305–309.

246. Daly, R. N., Goldberg, P. B. and Roberts, J. Effects of age on neurotransmission at the cardiac sympathetic neuroeffector junction. *J. Pharmacol. Exp. Ther.*, **245** (1988), 798–803.

247. Snyder, D. L., Aloyo, V. J., Wang, W. and Roberts, J. Influence of age and dietary restriction on norepinephrine uptake into cardiac synaptosomes. *J. Cardiovasc. Pharmacol.*, **32** (1998), 896–901.

248. Kiyono, Y., Iida, Y., Kawashima, H. *et al.* Norepinephrine transporter density as a causative factor in alterations in MIBG myocardial uptake in NIDDM model rats. *Eur. J. Nucl. Med. Mol. Imaging*, **29** (2002), 999–1005.

249. Markus, R. P. and Werneck de Avellar, M. C. Effect of ageing on the number of neuronal noradrenaline uptake sites in the rat vas deferens. *Naunyn-Schmiedeberg's Arch. Pharmacol.*, **346** (1992), 405–409.
250. de Avellar, M. C., Kobashi, Y. L. and Markus, R. P. Age-related changes in neuronal uptake of noradrenaline. *Naunyn-Schmiedeberg's Arch. Pharmacol.*, **341** (1990), 295–300.
251. Salvatore, M. F., Apparsundaram, S. and Gerhardt, G. A. Decreased plasma membrane expression of striatal dopamine transporter in aging. *Neurobiol. Aging*, **24** (2003), 1147–1154.
252. Tejani-Butt, S. M., Yang, J. and Zaffar, H. Norepinephrine transporter sites are decreased in the locus coeruleus in Alzheimer's disease. *Brain Res.*, **631** (1993), 147–150.
253. Gesi, M., Soldani, P., Giorgi, F. S. *et al.* The role of the locus coeruleus in the development of Parkinson's disease. *Neurosci. Biobehav. Rev.*, **24** (2000), 655–668.
254. Biederman, J. and Spencer, T. Attention-deficit/hyperactivity disorder (ADHD) as a noradrenergic disorder. *Biol. Psychiatry*, **46** (1999), 1234–1242.
255. Friedman, J. I., Adler, D. N. and Davis, K. L. The role of norepinephrine in the pathophysiology of cognitive disorders: potential applications to the treatment of cognitive dysfunction in schizophrenia and Alzheimer's disease. *Biol. Psychiatry*, **46** (1999), 1243–1252.
256. Bobb, A. J., Addington, A. M., Sidransky, E. *et al.* Support for association between ADHD and two candidate genes: NET1 and DRD1. *Am. J. Med. Genet. B Neuropsychiatr. Genet.*, **134** (2005), 67–72.
257. Esler, M., Leonard, P., Kelleher, D. *et al.* Assessment of neuronal uptake of noradrenaline in humans: defective uptake in some patients with essential hypertension. *Clin. Exp. Pharmacol. Physiol.*, **7** (1980), 535–539.
258. Esler, M., Jackman, G., Bobik, A. *et al.* Norepinephrine kinetics in essential hypertension. Defective neuronal uptake of norepinephrine in some patients. *Hypertension*, **3** (1981), 149–156.
259. Oliverio, A. and Wang, H. H. Effects of acute heart failure and administration of ouabain on cardiac catecholamine uptake. *Acta Physiol. Scand.*, **66** (1966), 278–281.
260. Petch, M. C. and Nayler, W. G. Uptake of catecholamines by human cardiac muscle in vitro. *Br. Heart J.*, **41** (1979), 336–339.
261. Merlet, P., Dubois-Rande, J. L., Adnot, S. *et al.* Myocardial beta-adrenergic desensitization and neuronal norepinephrine uptake function in idiopathic dilated cardiomyopathy. *J. Cardiovasc. Pharmacol.*, **19** (1992), 10–16.
262. Rossowska, M. and Khachatrian, L. Effect of hypoxia on norepinephrine uptake and activity of (Na+–K+)–ATPase in synaptosomes. *Neuropatol. Pol.*, **21** (1983), 123–128.
263. Jaques, S., Jr., Tobes, M. C. and Sisson, J. C. Sodium dependency of uptake of norepinephrine and m-iodobenzylguanidine into cultured human pheochromocytoma cells: evidence for uptake-one. *Cancer Res.*, **47** (1987), 3920–3928.
264. Beal, M. F., Atuk, N. O., Westfall, T. C. and Turner, S. M. Catecholamine uptake, accumulation, and release in acute porphyria. *J. Clin. Invest.*, **60** (1977), 1141–1148.
265. Lahti, R. A. Antidepressant and antipsychotic agents. *Naturwissenschaften*, **66** (1979), 403–409.

266. Bassett, J. R. and Cairncross, K. D. Effect of stress on the uptake of 3H-norepinephrine into rat myocardium. *Pharmacol. Biochem. Behav.*, **4** (1976), 39–44.
267. Hayashi, H., Takagi, N., Kamimoto, N. and Takeo, S. Microsphere embolism-induced changes in presynaptic function of the cerebral cortex in rats. *Brain Res.*, **737** (1996), 64–70.
268. Gewitz, M. H. and Tait, M. S. Effects of hypoxia on norepinephrine uptake by developing rabbit lung. *Pediatr. Pharmacol. (New York)*, **4** (1984), 109–113.
269. Preda, I., Karpati, P. and Endroczi, E. Myocardial noradrenaline uptake after coronary occlusion in the rat. *Acta Physiol. Acad. Sci. Hung.*, **46** (1975), 99–106.
270. Kitagawa, H., Yamazaki, T., Akiyama, T. *et al.* Modulatory effects of ketamine on catecholamine efflux from in vivo cardiac sympathetic nerve endings in cats. *Neurosci. Lett.*, **324** (2002), 232–236.
271. Himura, Y., Felten, S. Y., Kashiki, M. *et al.* Cardiac noradrenergic nerve terminal abnormalities in dogs with experimental congestive heart failure. *Circulation*, **88** (1993), 1299–1309.
272. Herman, L. M., Dai, S., Hartman, N. G. and McNeill, J. H. Meta-iodobenzylguanidine uptake in the hypertensive-diabetic rat heart: a marker for myocardial dysfunction? *Can. J. Physiol. Pharmacol.*, **72** (1994), 1162–1167.
273. Kawai, H., Fan, T. H., Dong, E. *et al.* ACE inhibition improves cardiac NE uptake and attenuates sympathetic nerve terminal abnormalities in heart failure. *Am. J. Physiol.*, **277** (1999), H1609–H1617.
274. Rho, J. H. and Alexander, N. Altered in vitro uptake of norepinephrine by cardiovascular tissues of young spontaneously hypertensive rats. *Biochem. Pharmacol.*, **33** (1984), 3906–3909.
275. Krakoff, L. R., Ben-Ishay, D. and Mekler, J. Reduced sympathetic neuronal uptake (uptake1) in a genetic model of desoxycorticosterone-NaCl hypertension. *Proc. Soc. Exp. Biol. Med.*, **178** (1985), 240–245.
276. Sumners, C., Muther, T. F. and Raizada, M. K. Altered norepinephrine uptake in neuronal cultures from spontaneously hypertensive rat brain. *Am. J. Physiol.*, **248** (1985), C488–C497.
277. Koivunen, D. G. and Johnson, J. A. Effect on pressor and vascular responsiveness in rabbits of drugs that decrease norepinephrine uptake. *Proc. Soc. Exp. Biol. Med.*, **206** (1994), 375–383.
278. Stancheva, S., Georgiev, V., Alova, L. and Opitz, M. Effects of angiotensin II on biogenic monoamine uptake in the brain of nonhypoxic and hypoxic mice. *Methods Find Exp. Clin. Pharmacol.*, **19** (1997), 383–386.
279. Matsumoto, M., Togashi, H., Mori, K. *et al.* Characterization of endogenous serotonin-mediated regulation of dopamine release in the rat prefrontal cortex. *Eur. J. Pharmacol.*, **383** (1999), 39–48.
280. Michalak, A., Rose, C. and Butterworth, R. F. Loss of noradrenaline transporter sites in frontal cortex of rats with acute (ischemic) liver failure. *Neurochem. Int.*, **38** (2001), 25–30.
281. Karpati, P., Preda, I. and Endroczi, E. Effect of acidosis and noradrenaline infusion on 14C-noradrenaline uptake by the rat myocardium. *Acta Physiol. Acad. Sci. Hung.*, **45** (1974), 109–114.
282. Lindmar, R. and Loffelholz, K. Differential effects of hypothemia on neuronal efflux, release and uptake of noradrenaline. *Naunyn-Schmiedeberg's Arch. Pharmacol.*, **274** (1972), 410–414.

283. Ramasamy, S., Hodgson, W. C. and Ventura, S. Protein kinase C and the sub-sensitivity and sub-reactivity of the diabetic rat prostate gland to noradrenaline. *Eur. J. Pharmacol.*, **434** (2002), 151–161.

284. Ganguly, P. K., Beamish, R. E. and Dhalla, K. S., Innes, I. R. and Dhalla, N. S. Norepinephrine storage, distribution, and release in diabetic cardiomyopathy. *Am. J. Physiol.*, **252** (1987), E734–E739.

285. Szot, P., Reigel, C. E., White, S. S. and Veith, R. C. Alterations in mRNA expression of systems that regulate neurotransmitter synaptic content in seizure-naive genetically epilepsy-prone rat (GEPR): transporter proteins and rate-limiting synthesizing enzymes for norepinephrine, dopamine and serotonin. *Brain Res. Mol. Brain Res.*, **43** (1996), 233–245.

286. Schomig, A. Catecholamines in myocardial ischemia. Systemic and cardiac release. *Circulation*, **82** (1990), II 13–22.

287. Hatta, E., Yasuda, K. and Levi, R. Activation of histamine H3 receptors inhibits carrier-mediated norepinephrine release in a human model of protracted myocardial ischemia. *J. Pharmacol. Exp. Ther.*, **283** (1997), 494–500.

Part II

Norepinephrine and behavior

6

Role of the locus coeruleus-norepinephrine system in arousal and circadian regulation of the sleep–wake cycle

Gary Aston-Jones, Ph.D., Monica Gonzalez, and Scott Doran
University of Pennsylvania, Philadelphia, USA

Introduction

A variety of previous results indicate that the norepinephrine (NE) locus coeruleus (LC) system is integrally involved in regulation of sleep and waking. In particular, LC activation promotes wakefulness. This appears to be one of the major arousal systems in the brain. The LC was also found to be critical for rebound sleep following sleep deprivation, and increased sleep that occurs following a stressor. We recently identified a circuit from the suprachiasmatic nucleus (SCN) to the LC, in which the dorsomedial hypothalamus (DMH) serves as a relay. The functionality of this circuit was confirmed in our studies, showing that LC neurons have a circadian rhythm in their firing activity and that this circadian fluctuation in the LC requires an intact DMH. Other recent studies have also shown that lesions of the LC decrease the amplitude of the circadian rhythm in the sleep–wake cycle. These results indicate that the LC may be an important component in SCN efferent circuitry for driving circadian rhythms in sleep–wake cycles. Other recent results in our lab have revealed that light deprivation produces a profound loss of NE in the frontal cortex in rats, and this NE loss in dark-maintained animals is associated with decreased amplitude of the circadian sleep–wake rhythm. The dependence of the LC system on light for normal function has implications for clinical disorders such as seasonal affective disorder. Other disorders that are comorbid with sleep anomalies and associated with LC dysfunction are discussed.

Brain systems involved in sleep–wake regulation

Waking, slow-wave sleep and paradoxical sleep

Three naturally occurring, distinct vigilance states have been identified in all higher vertebrates (birds and mammals) according to cortical, muscular, autonomic, and

Brain Norepinephrine: Neurobiology and Therapeutics, ed. Gregory A. Ordway, Michael A. Schwartz and Alan Frazer. Published by Cambridge University Press.

behavioral activities: wakefulness (W), slow-wave sleep (SWS), and paradoxical sleep (PS). During W, individuals interact with the external environment using a great variety of volitional responses. Cortical activity as measured with the electroencephalogram (EEG) is dominated by high-frequency, low-amplitude signals, with a predominance of beta (20 to 30 Hz) and gamma (30 to 60 Hz) frequencies. Theta activity (5 to 8 Hz) appears during quiet waking and focused attention, and may facilitate learning during wakefulness.[1] Postural muscle tonus, as measured by the electromyogram (EMG), is variable depending on the posture or locomotor activity of the individual. Autonomic indices of W include an elevated cardiac rhythm and a rapid and irregular respiratory frequency.[2,3] During sleep onset and SWS, EEG-recorded cortical activity slows and becomes more synchronized until high-amplitude spindle waves (12 to 15 Hz) intermingle with increasingly dominant high-voltage delta waves (1 to 4 Hz) and slow waves (<1 Hz).[4,5] Both voluntary muscular and cardiorespiratory activities decrease, and the brain temperature reaches a nadir.

Paradoxical sleep (also referred to as rapid eye movement or REM sleep) is always preceded by SWS, except during narcolepsy, idiopathic hypersomnia, and some drug/alcohol withdrawal periods.[6,7] It is characterized by low-amplitude high-frequency desynchronized cortical EEG activity similar to EEG recordings during W.[2,8] In fact, the EEGs of W and PS are sometimes difficult, or impossible, to distinguish without other polygraphic signals, especially the EMG. Paradoxical sleep is also characterized by pronounced postural muscular atonia and body temperature dysregulation.[2,9] Periodic reemergence of high-frequency low-amplitude EEG activity during sleep with atonia inspired Jouvet to rename the previously described "Rapid Eye Movement" sleep of Aserinsky and Kleitman[10] as "Paradoxical Sleep," which he classified as a distinctive sleep state.[11–12] Other physiological indices of PS include penile erections, myoclonic twitches, irregular cardiorespiratory rhythms, and unique extracellular field potentials found in the pons, lateral geniculate nucleus, and occipital cortex (PGO waves).[2,4,13,14]

Networks involved in regulation of the sleep–wake cycle

Neuronal groups

Figure 6.1 illustrates the main neuronal groups involved in sleep–wake cycle regulation. Neurons associated with cortical activation during waking are concentrated in the pontine and midbrain central tegmentum, posterior hypothalamus, and basal forebrain (Figure 6.1a), primarily within the area described as the "reticular activating system" by Moruzzi and Magoun.[15] In particular, the noradrenergic LC system in the pons; the cholinergic neurons located in the brain stem reticular formation (pons and midbrain) and the basal forebrain (substantia innominata, diagonal band nuclei and septum); the glutamatergic neurons of the mesencephalic reticular

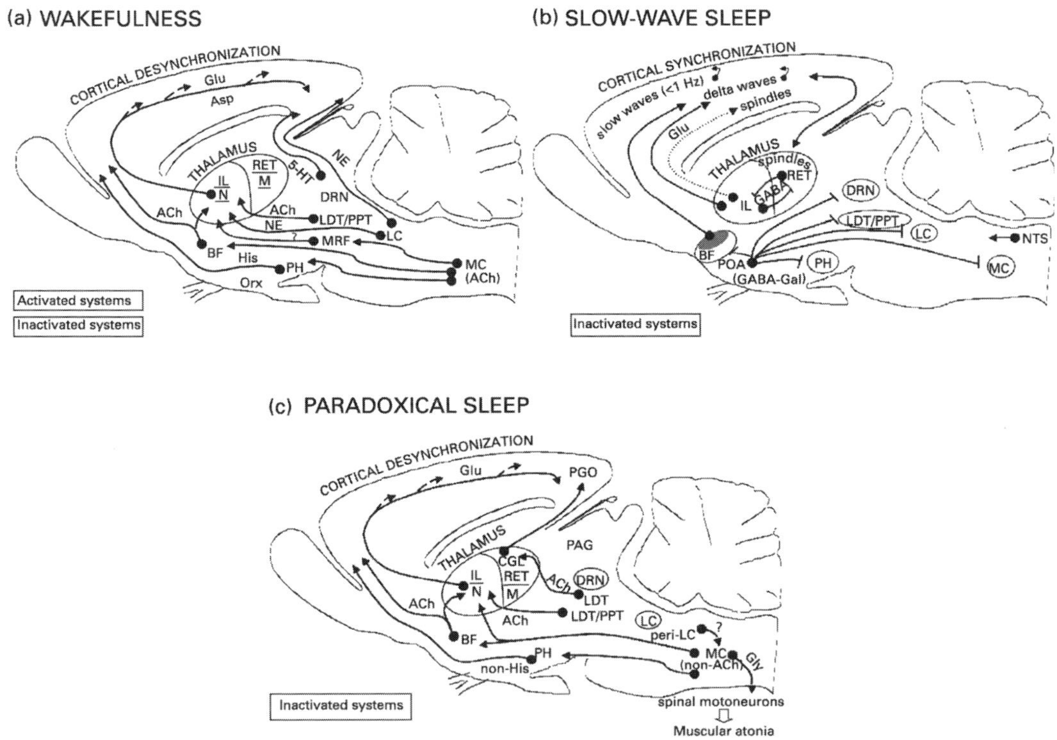

Figure 6.1 Simplified schema of the circuits involved in the vigilance states. MC, nucleus reticularis magnocellularis; LC, locus coeruleus; LDT/PPT, laterodorsal and peduculopontin tegmental nucleus; PAG, periaqueductal gray; NTS, nucleus tractus solitarius; DRN, dorsal raphe nucleus; PH, posterior hypothalamus; POA, anterior preoptic area; IL, intralaminar nucleus; RET, reticular nucleus; BF, basal forebrain; MRF, mesencephalic reticular formation; peri-LC, peri-LC area; PGO, ponto-geniculo-occipital (waves); CGL, lateral geniculate nucleus. Other abbreviations: Glu, glutamate; Asp, aspartate; Gly, glycine; ACh, acetylcholine; NE, norepinephrine; 5-HT, serotonin; His, histamine; Orx, orexin; GABA, γ-aminobutyric acid; Gal, galanin; N, nicotinic receptor, M, muscarinic receptor. Modified from Gonzalez (1997).[117]

formation; the dopaminergic neurons of substantia nigra and ventral tegmental area in the midbrain; the orexin neurons from the lateral and perifornical hypothalamus; the histamine neurons of the posterior hypothalamus; and serotonergic neurons (5-HT) of the dorsal raphe nucleus (DRN) are all prominently implicated in arousal and waking.[5,16–24] Figure 6.1b shows that the SWS-generating systems are concentrated within the midline brain stem, dorsolateral medullary reticular formation, the anterior hypothalamus-preoptic region and basal forebrain.[19] Pioneering studies done by Jouvet showed that rhombencephalic structures were responsible for generating PS.[12] Subsequent transection studies identified the region caudal to the midbrain and rostral to the spinal cord (primarily the lateral portion of the nucleus reticularis pontis oralis) to be the core network responsible for generating PS (Figure 6.1c).[14]

Mechanisms involved in sleep–waking cycle stage changes

Daily transitions into sleep from wakefulness appear to rely on circadian and sleep homeostatic cues from the hypothalamus that inhibit wakefulness-promoting activity from the brain stem. Tonic firing of NE-, 5-HT-, and histamine-containing neurons decrease during SWS and become practically silent before and during PS.[25–29] Active inhibition of this waking network to allow sleep is proposed to be mediated by sleep-generating systems including γ-aminobutyric acid (GABA) ergic and galaninergic neurons of the preoptic area, periaqueductal gray, prepositus hypoglossi, and the solitary tract nucleus.[23,30–35] A PS-*on* or executive system has been proposed to inhibit the monoaminergic waking systems (PS-*off* or permissive system) which is indispensable for the genesis and maintenance of PS.[36,37] This PS executive system is proposed to include cholinergic or cholinoceptive systems of the mesopontine tegmentum, the ventromedial medulla, and part of the nucleus reticularis parvocellularis (see Figure 6.1).[22,28,36,38–40]

LC involvement in sleep–wake regulation

Activity of LC neurons during sleep and waking

Tonic LC activity varies with behavioral state

The LC is the major source of NE in the brain, and the sole source of NE in the cerebral cortex, cerebellum, and hippocampus. Although the LC also contains non-NE neurons, which are particularly numerous in certain species such as cats and dogs, our focus will be on the NE neurons of this nucleus.

Spontaneous activity of NE neurons in the LC of rats and cats has been found to co vary with stages of the sleep–wake cycle, firing most rapidly during W, more slowly during SWS, and becoming virtually silent during PS.[25,41,42] State-dependent LC firing has also been found in monkeys, although some differences exist as described below.[43–45] Although average LC impulse activity co varies with arousal state, it is notable that LC firing changes markedly in anticipation of most state changes. Thus, for example, LC neurons decrease their activity during W in anticipation of SWS. In addition, LC neurons engaged in low tonic firing during SWS increase average spike rates dramatically just prior to a transition into W.[25,41] Locus coeruleus activity is also reported to increase just prior to the PS-to-W transition if arousal state is delineated by measuring changes in EMG.[41] This observation led to the view that LC activation was a primary signal causing the PS-to-W transition.[36] However, in our recordings, LC activation did not precede but, rather, followed the change in EEG activity during the PS-to-W transition.[25] This discrepancy results from the fact that EEG indices of this state transition precede the corresponding EMG transition. These findings also indicate that, although LC neurons may participate in producing most transitions between sleep–waking cycle stages, LC activity is not the primary signal terminating PS episodes.

Primate LC activity during sleep and waking

Very few reports are available that detail how LC activity varies across or within arousal states in primates. The only previous data recording LC activity during arousal state changes in primates reported distinct LC slowing from about 2 Hz during active waking to less than 1 Hz (often inactive) during naturally occurring drowsiness.[46] Locus coeruleus activity was very low or absent during PS, as was found in other species (described above).[45] More recent studies reveal that monkey LC activity is primarily associated with wakefulness, and varies less with different stages of drowsiness and sleep than in other species: during quiet wakefulness (defined by the presence of alpha waves in the EEG) impulse activity of the monkey LC neurons was not significantly different from that during either SWS or PS.[47] Furthermore, LC activity during SWS and PS was not significantly different. Thus, monkey LC activity may be less graded during different sleep–waking states than in rats or cats, and appears to be active primarily during periods of high cognitive engagement. However, at present this is based upon a limited data set, and additional study of LC activity during sleep and waking in the monkey is needed to confirm these initial results.

Monkey LC activity during W exhibits different modes of activity (including changes in the tonic levels of discharge as well as in phasic responses to task stimuli) that correspond to the current behavioral strategy.[48] These properties are described in detail in Chapter 7 of this volume and are not considered further here. However, it is noteworthy that this system appears to have functions in both state control and in behavioral performance within waking.

LC activity and cortical arousal

The midbrain lesion and stimulation experiments performed by Moruzzi and Magoun pointed to the relevance of the ascending tracts of the brain stem in arousal and established the idea of a "reticular activating system" located in the mesopontine tegmentum.[15] It is notable that the LC is located centrally within this arousal system. As reviewed above, state-dependent LC activity and the location of the LC within the reticular activating system are highly suggestive of a role for the LC in regulating arousal. However, this evidence was not definitive of a causal role of the LC in arousal. Experiments by Berridge and Foote[49] using direct stimulation of the LC provided evidence for such a causal relationship. This research showed that LC stimulation via direct microinfusion of cholinergic agonists desynchronized the normally synchronized EEG of halothane-anesthetized rats. This effect was shown to be NE-mediated as it was blocked by pretreatment with the β-adrenoceptor antagonist propranolol. Conversely, inhibition of LC activity via microinfusion of an α_2-agonist into the LC of more lightly anesthetized rats caused an increase in EEG slow-wave activity. Of importance, this effect required bilateral LC inactivation, indicating that even a small amount of LC activity was sufficient to prevent EEG slow waves.[50] Additional studies in unanesthetized rats indicated that the wake-promoting actions of the LC appear to involve NE actions at α_1- or β-receptors within the medial septum or medial preoptic area.[51]

Effects of locus coeruleus lesions on sleep and waking

Spontaneous Sleep

Because the NE-LC is part of the sleep–waking network and is thought to be a major arousal system (see Figure 6.1a), lesions of the LC would be expected to decrease waking and arousal. However, early studies of lesions of the LC did not find substantial changes in total amount of waking.[52–54] In addition, different authors have described an increase,[55] a decrease,[56,57] or no effect[53,54] on PS duration following LC lesions.[14,27] Inconsistent findings on this topic could be a consequence of differences in lesion research techniques, leading to inconsistent damage to the brain's aminergic nuclei between studies.

N-(2-chloroethyl)-*N*-ethyl-2-bromobenzylamine (DSP-4) is a tertiary haloalkylamine that induces long-term degeneration of noradrenergic fibers coming exclusively from the LC, with a concomitant depletion of NE levels in principal LC target areas.[58–63] Lesions produced by DSP-4 have been shown to produce minimal impact on any peripheral or central aminergic neurons other than the NE-LC system at 2 weeks post-DSP-4 administration.[61,64–66] Thus, DSP-4 is a useful tool to evaluate the specific role of the LC in sleep–waking mechanisms. Monti *et al.* were the first to evaluate the effects of DSP-4 on spontaneous sleep in rats.[55] However, that study was restricted to the light period, and measures were taken at 5 to 6 days after DSP-4 treatment, a time when the toxin also affects other amine systems that are potentially involved in the sleep–wake cycle (e.g., dopamine and 5-HT neurons).[59,65–67] Thus, the increase in PS observed by these authors could not be ascribed exclusively to a depletion of NE from the LC.

Later, Gonzalez and colleagues found that at one month after DSP-4 administration rats showed significantly decreased PS during the day (inactive period) and increased SWS during the night (active period).[68] These results indicated a differential involvement of the NE-LC in the SWS and PS mechanisms depending on the light–dark cycle. Subsequent circadian analyses of DSP-4 lesions confirmed and extended these findings, as described below.

Sleep Rebound

In 1960, Dement observed in humans an increase of both SWS and PS during the night following sleep deprivation.[69] This enhanced sleep duration above baseline level after sleep deprivation was designated "sleep rebound," and considered to be a component of normal sleep regulation. Since then, different sleep deprivation methods (water-tank, methylamphetamine administration, forced locomotion, disk-over-water method) have been used to produce enhanced PS to study the mechanism underlying PS.

The water-tank method (Wt) is the most widely used technique to sleep-deprive laboratory animals. This method consists of placing the animal on a small pedestal surrounded by water. At the onset of each PS period the loss of postural muscle tonus causes the animal to contact the water, which awakens him and prevents PS episodes. After one or two wettings, the animal avoids further water contact by periodic awakening and light SWS only, resulting in a total suppression of PS and a partial loss of deep SWS.[70] During sleep deprivation, cholinergic, peripheral and central monoaminergic systems are activated.[71–78] As most of these studies point out, the traditional Wt not only triggers a sleep rebound mechanism but also triggers peripheral and central responses similar to those elicited by stressors. For example, Wt treatment induces oxidative damage, immunosuppression, gastric ulcers, enlargement of the adrenals, and increases in corticosteroid levels.[73,79–81] Water-tank treatment also decreases the number of β-adrenoceptors in the brain and reduces both thymus and body weight.[63,73,80] Cerebral NE and 5-HT synthesis and turnover are enhanced, as is tyrosine hydroxylase (*TH*) gene expression in LC neurons.[74,76–78,81] Tyrosine hydroxylase activity and extrahypothalamic and hypothalamic corticotropin-releasing factor (CRF) levels also increase.[77,82]

As the NE-LC system is strongly activated by stress,[83–88] the stress inherent in most sleep deprivation methods must be taken into account when interpreting sleep deprivation results designed to study the role of the LC in spontaneous sleep mechanisms. In addition, a stressor stimulus without sleep loss, such as immobilization stress, produces a sleep increase on the following night, similar to that seen following sleep deprivation.[89–91] To characterize the role of the NE-LC in sleep rebound mechanisms after sleep deprivation, Gonzalez and colleagues evaluated the effect of NE-LC lesion by DSP-4 on sleep that was induced exclusively by stress alone.[89] One month after DSP-4 administration, neurotoxic NE-LC lesions reduced or eliminated the SWS and PS increases typically recorded after immobilization stress (Figure 6.2).[89] Thereafter, the same authors studied the effect of DSP-4 NE-LC lesions on sleep rebound following sleep deprivation. Ten hours of Wt treatment induced significant SWS and PS rebound during the first dark period following sleep deprivation in control animals, while DSP-4 treatment significantly attenuated both PS and SWS rebound (Figure 6.3).[90] Water-tank treatment did not induce sleep rebound in BALB/c mice, a genetic strain with 38% of the NE-LC

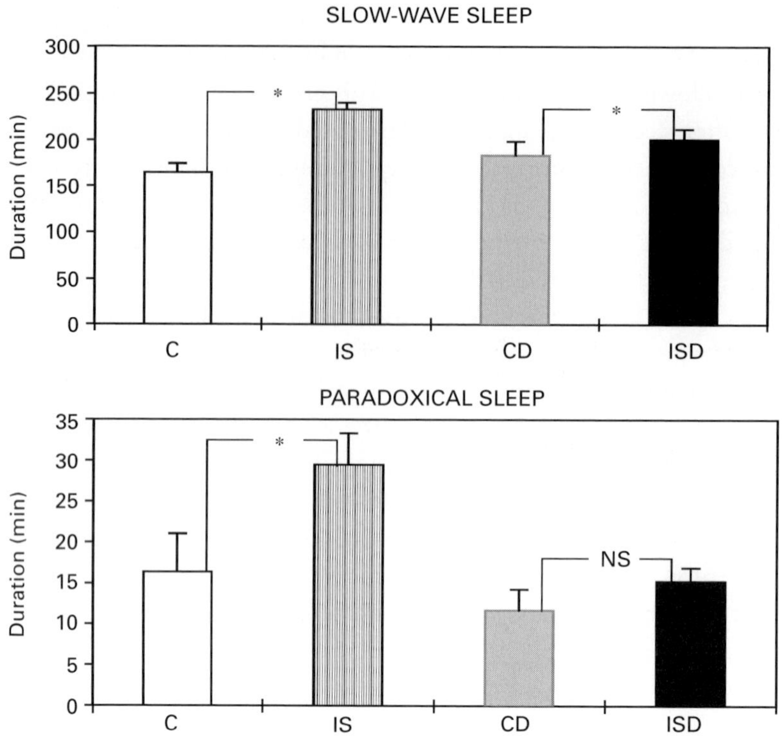

Figure 6.2 Vigilance states duration (10 h) following 1 h immobilization stress before (IS) and after (ISD) DSP-4 treatment. Control values before (C) and after (CD) DSP-4. Mean duration ± SEM; *$p < 0.05$; n = 10. From Gonzalez *et al.* (1995).[89]

perikarya missing and decreased density of cortical NE terminals.[92–94] Given the stressful nature of Wt treatment and the requirement for an intact NE-LC to provoke increased sleep after immobilization stress,[89] it is possible the NE-LC participates in sleep increases via stress responses triggered by sleep deprivation.[90] The view that sleep rebound is stress-related, and not necessarily a concomitant of sleep loss per se, is consistent with recent findings that the wake-enhancing drug modafinil increases waking without causing a subsequent sleep rebound.[95]

Role of corticotropin-releasing factor in noradrenergic regulation of sleep and waking

Anatomical, physiological and behavioral studies have shown that CRF plays a key role in the stress response. In particular, CRF is a key neurotransmitter involved in the NE-LC activation by certain stressors.[87,96,97] Immunohistochemical studies in the rat describe CRF-immunoreactive axons in the LC and peri-LC.[98–100]

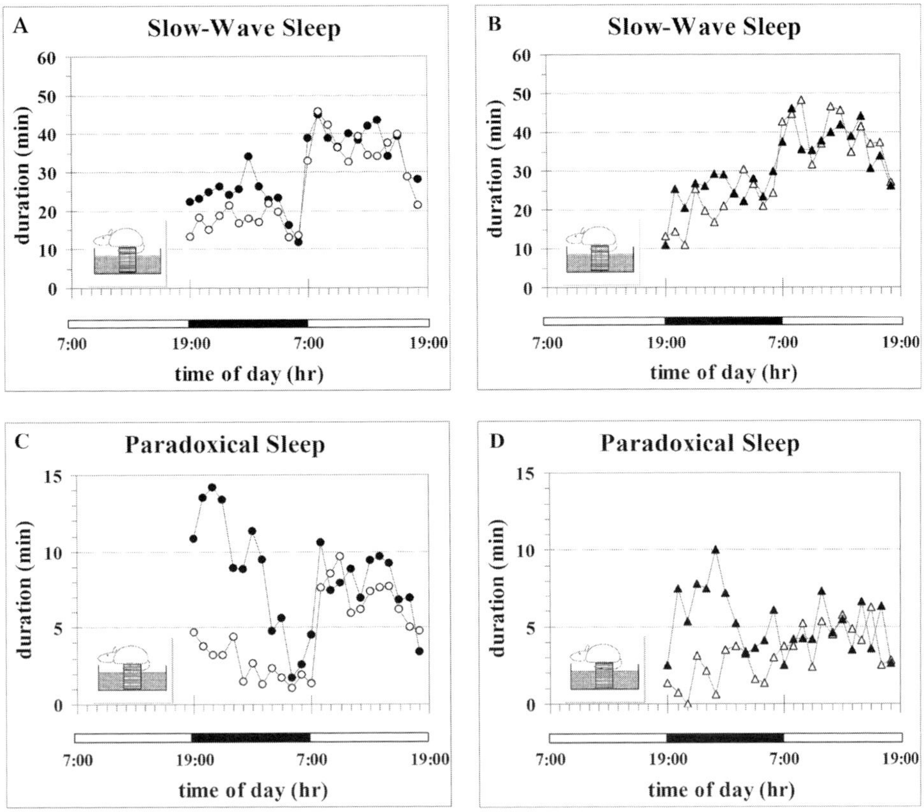

Figure 6.3 Sleep recordings after 10 h of sleep deprivation by the water-tank method. Ordinates: each value represents the mean time (in min) spent in SWS or PS every hour during 24 h of continuous recordings. (a,c) Before DSP-4: spontaneous sleep (open symbols), sleep following sleep deprivation (solid symbols). (b,d) 20 to 24 days after DSP-4: spontaneous sleep (open symbols), sleep following sleep deprivation (solid symbols). From Gonzalez *et al.* (1996).[90]

Neurophysiological experiments showed that activation of LC neurons by hypotensive stress is dependent upon CRF neurotransmission within the LC.[101] Also, biochemical studies reveal that both acute and chronic stress can double the concentration of CRF in the LC.[102] Finally, CRF application in the LC produces a number of stress-related effects in behavior and physiology. For example, local administration of CRF generates anxiogenic behavior, induces EEG activation, suppresses immune responses, and increases plasma corticosterone concentration.[103–106]

To evaluate the participation of CRF in sleep rebound, experiments were conducted using the specific CRF receptor antagonist, α-helical CRF (9–41) (ahCRF).[107,108] This compound has been shown to prevent behavioral stress responses caused by endogenous CRF activation while having no effect on unstressed animals.[86,96,109–114] Intracerebroventricular (ICV) injection of ahCRF

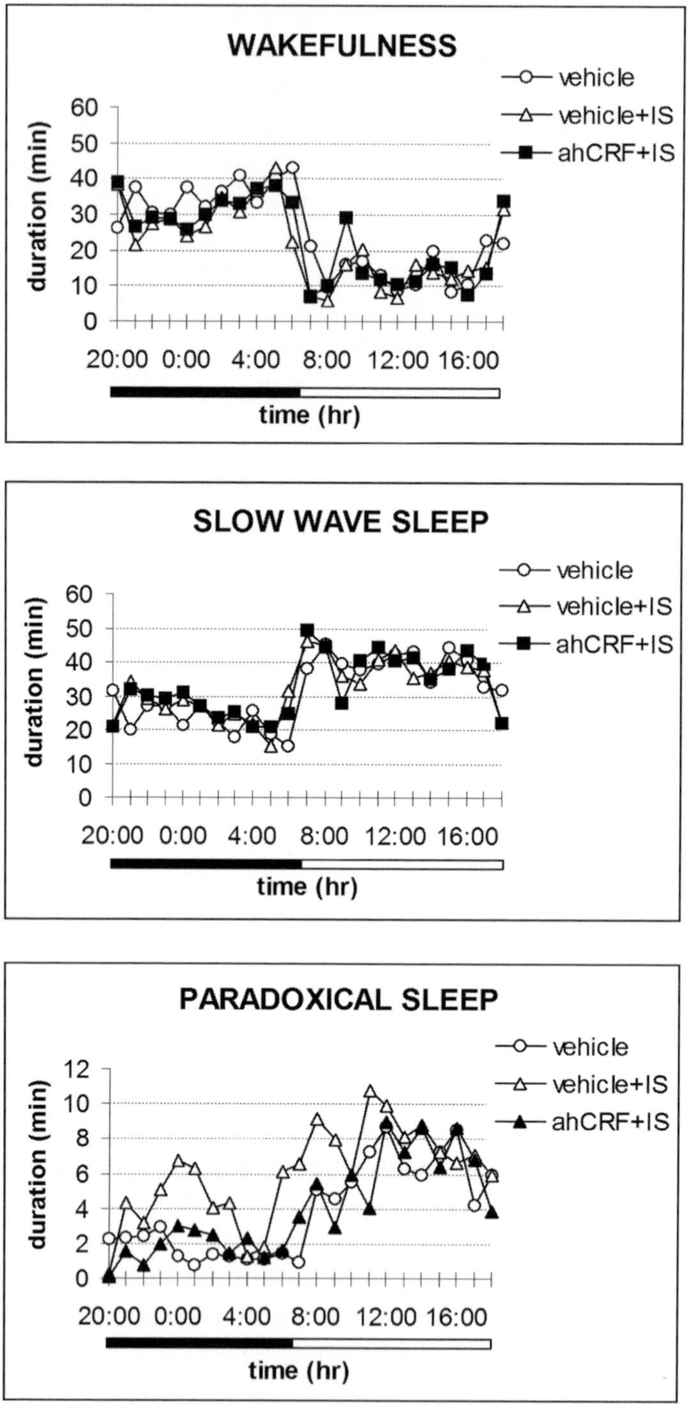

Figure 6.4 Changes in hourly mean duration of the vigilance states up to the next dark period following immobilization stress (IS). The groups were: distilled water (vehicle), distilled water after immobilization stress (vehicle + IS) or α-helical CRF (9–41) after immobilization stress (ahCRF + IS). From Gonzalez and Valatx (1997).[115]

abolished the PS increase typically induced by stress while stress-induced SWS increase was unchanged (Figure 6.4).[115] When administered ICV throughout 10 hours of sleep deprivation by Wt, ahCRF decreased PS rebound in rats without altering SWS rebound (Figure 6.5).[116] These results support the view that stress, acting via CRF activity, is a main factor inducing PS rebound following sleep deprivation. Moreover, ahCRF treatment was more effective than specific lesions of the NE-LC system in decreasing PS rebound, suggesting that neural systems activated in parallel to the NE-LC system by CRF, but spared from DSP-4 neurotoxicity, may be involved in PS rebound following sleep deprivation.[68,116] Because ahCRF had no effect on spontaneous vigilance states, these studies indicate that endogenous CRF is involved in the PS mechanism only under stressful conditions, and that the substrates involved in spontaneous PS differ from those in PS rebound.[115,117]

Brain regions showing the largest depletion of NE after DSP-4 administration include the neocortex, hippocampus and amygdala,[58] all of which are important inputs to the sleep-generating network (i.e., preoptic area, hypothalamus).[97,118] Sleep rebound could result from modulation of spontaneous sleep networks by structures related to stress but not directly involved in sleep mechanisms (i.e. the amygdala), implying that rebound sleep differs significantly in mechanism from spontaneous sleep.[115] Based on this hypothesis, Charifi and colleagues found that a local injection of DSP-4 in the central nucleus of the amygdala decreased sleep rebound induced by Wt without altering spontaneous sleep.[119]

Role of LC in circadian regulation of sleep and waking

Sleep and waking are expressed as a circadian rhythm that is under the control of the primary endogenous circadian pacemaker. The SCN appears to orchestrate diurnal variation of most physiological and behavioral variables in mammals.[120,121] Hypothalamic projections from the SCN appear to be prominently involved in the control of circadian sleep–waking rhythms, but precisely which projections and neurotransmitters exert circadian control is unclear. Arginine vasopressin (AVP)- and vasoactive intestinal peptide (VIP)-containing SCN efferents to the hypocretin/orexin arousal system of the posterior hypothalamus have been described in rats.[122] Inhibitory (GABAergic) and excitatory (glutamatergic) inputs from the SCN modulate the activity of ventromedial and ventrolateral preoptic areas, both known to be involved in thermoregulation and the induction and maintenance of sleep and waking.[123,124] Lesioning direct SCN targets, such as the ventral area of the subparaventricular zone, abolished the circadian rhythm of sleep.[125]

Indirect projection from the SCN to the LC

As part of our program of research into afferent regulation of the LC, we employed the transsynaptic retrograde tracer pseudorabies virus (PRV) to study indirect

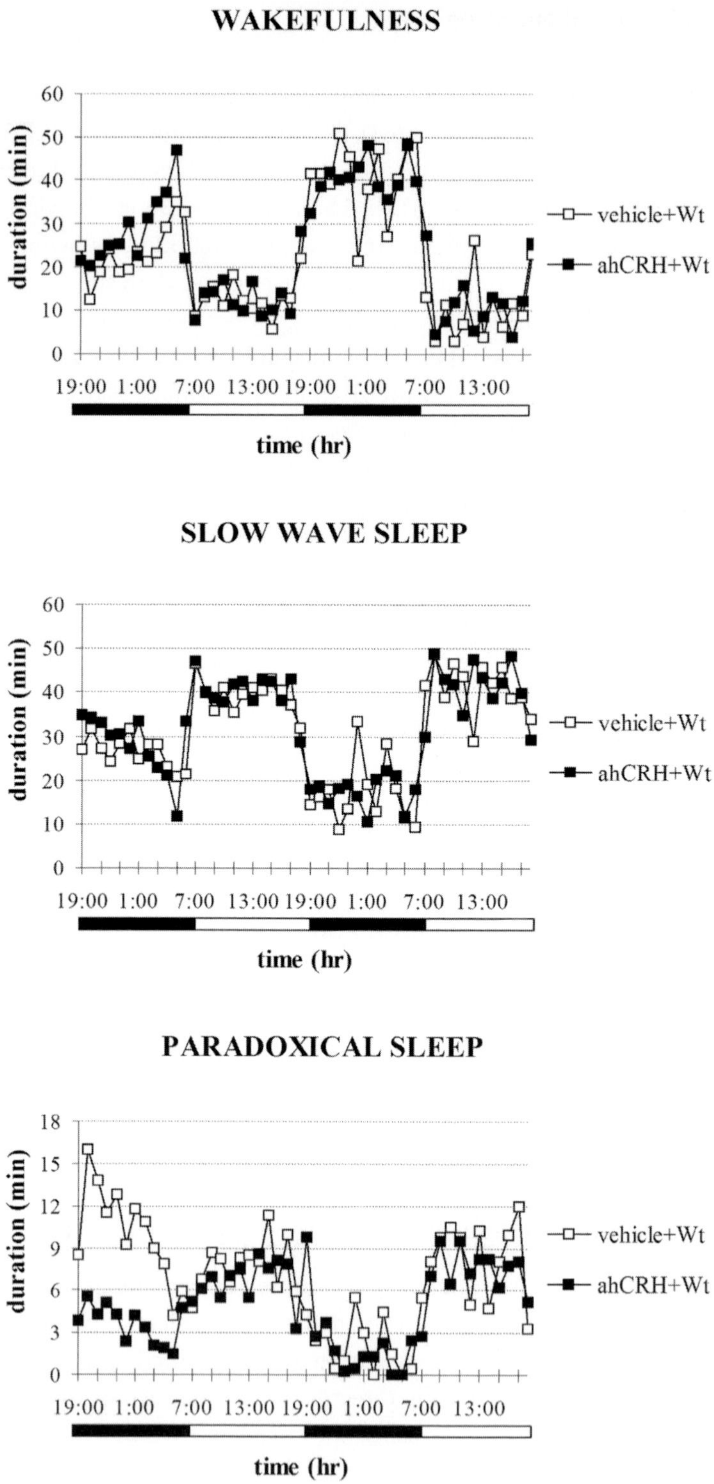

Figure 6.5 Hourly mean duration of sleep states up to 48 hours after the end of sleep deprivation. Rats were repeatedly treated during sleep deprivation with vehicle (vehicle+Wt) or α-helical CRF (9–41) (ahCRF+Wt). From Gonzalez and Valatx (1998).[116]

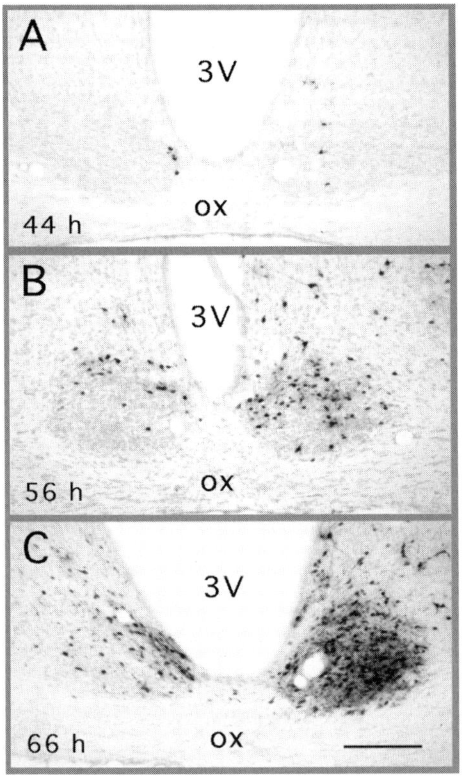

Figure 6.6 Transsynaptic retrograde labeling with PRV in the SCN. Frontal sections taken through the SCN at different survival times after PRV injection in the LC, as indicated. The number of labeled neurons in the SCN increased from 44 h (A) to 66 h of survival (C). Also, labeling was more prominent on the side ipsilateral to the LC injection site (right side in these photographs). Sections were counterstained with neutral red. 3V, third ventricle; Ox, optic chiasm. Bar = 200 μm. Taken from Aston-Jones *et al.* (2001).[129]

inputs to the LC.[126] We, and other researchers, have developed methods for using PRV with central injections to study chains of neurons connecting identified brain nuclei.[127,128] The Bartha strain of this virus is a very powerful tool in neuroanatomy because it moves only in a retrograde fashion and only across synapses to produce specific labeling of direct and indirect afferents to the nucleus injected. Injection of PRV into the LC resulted in strong labeling of the SCN after approximately 60 hours of survival, a time long after known direct inputs are well labeled (Figure 6.6).[129] This, in addition to previous studies and our own unpublished results, has shown that the SCN does not directly project to the LC. These findings indicate that the SCN is an indirect input to the LC. At 60 hours after PRV injection into the LC, the great majority of ipsilateral SCN neurons were labeled, indicating that the SCN is potentially a major indirect input to the LC. This finding prompted a series of double-labeling and lesion studies to identify the relay in the SCN–LC circuit. Results indicated that the DMH is a major relay in this circuit, as significantly fewer SCN cells were labeled with PRV following injection in the LC of animals with ibotenic acid-induced DMH lesions.[129]

Circadian rhythm in LC impulse activity

The above findings prompted the hypothesis that LC neurons may exhibit a circadian rhythm in their activity, a possibility that had not previously been tested. Although a prior report found a circadian rhythm in brain NE levels,[130] LC neurons vary their activity strongly with sleep and waking (as reviewed above), and therefore any changes in activity seen in behaving animals over the circadian cycle would be confounded by coincident changes in sleep and waking. To circumvent this possible confound, we recorded impulse activity of LC neurons from anesthetized animals taken from their active (dark cycle) or rest periods (light cycle). Anesthesia ensured that animals taken from either period were in a similar state of arousal when recordings were obtained, so that state-dependent changes in LC activity (e.g., waking vs. sleep) were eliminated. These experiments revealed that tonic LC neural activity was significantly faster during the active period compared to the rest period (Figure 6.7).[129] This was true even in animals maintained in continuous darkness in the absence of extrinsic circadian cues for whom active and rest periods were defined by spontaneous locomotor activity. This finding supports our hypothesis that activity in LC neurons reflects an endogenous circadian signal. Finally, we found that this rhythm was absent in animals with ibotenate lesions of the DMH, showing that the SCN–DMH–LC circuit defined with PRV tracing is a functional circuit for efferent transmission of circadian information.

LC lesions decrease circadian amplitude of sleep–waking cycle

As activity in the LC is associated with arousal, the finding of a circadian rhythm in LC activity indicated that the LC may contribute to circadian rhythms in the sleep–wake cycle. To test this hypothesis, we recently extended the previous study of effects of DSP-4-induced LC lesions on sleep and waking,[68] and analyzed the effect of specific lesions of the NE-LC system on the circadian rhythm of the sleep–waking cycle in rats. For this circadian study, we included cosinor analysis of results, as well as recordings while the animals were kept in continuous darkness (DD). Three weeks after DSP-4 administration, we recorded electrographic sleep signs by telemetry (EEG, EMG, temperature), and quantified the amount of W, SWS, and PS in lesioned versus nonlesioned control rats maintained in normal lighting conditions (12 hours:12 hours light–dark), or in DD. In the light–dark (LD) group, DSP-4 treatment caused a significant decrease in W and increase in sleep at night, while W increased and sleep slightly decreased during the day (Figure 6.8).

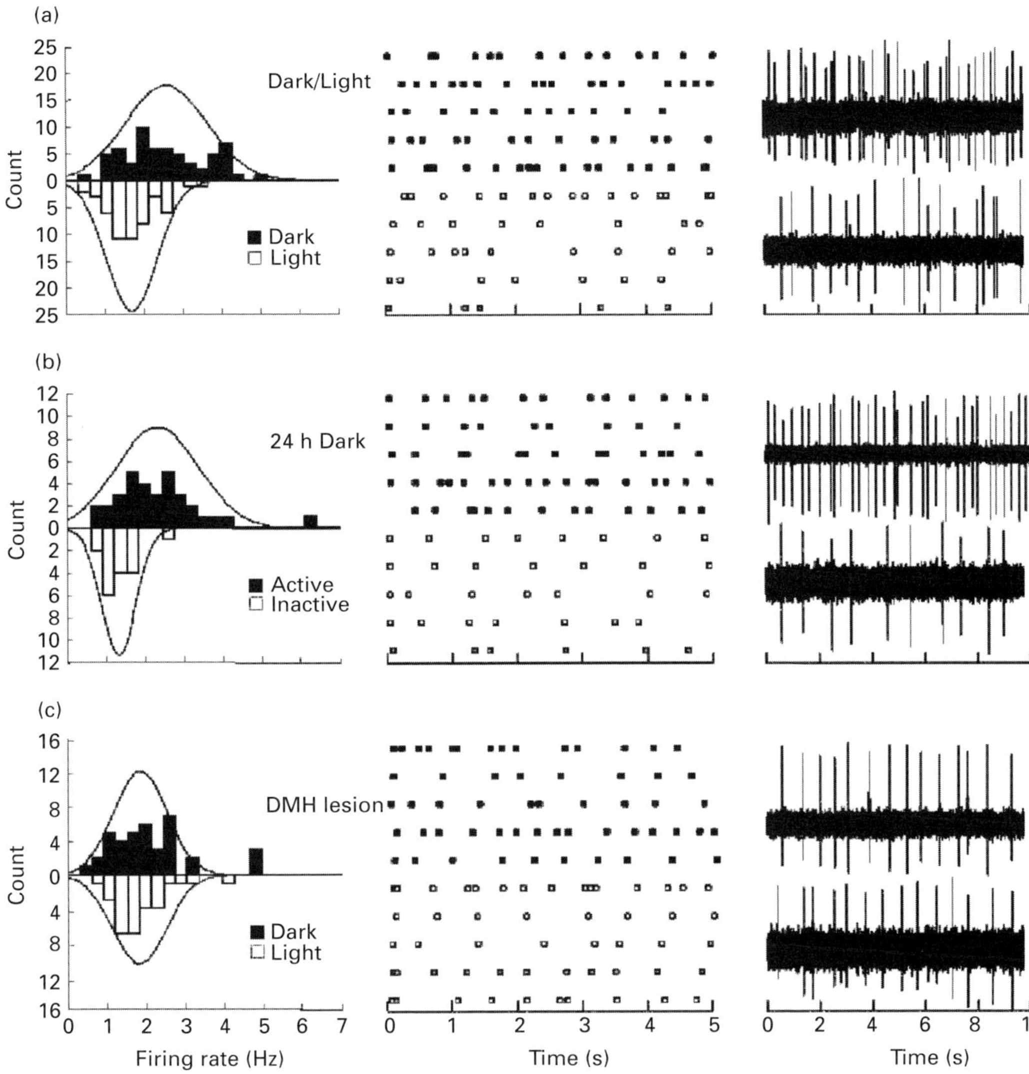

Figure 6.7 Histograms showing the distributions of LC firing rates during different epochs of the circadian cycle. Three paired groups of rats were maintained in either 12 h/12 h dark/light (panels a and c) or 24 h darkness (panel b). One paired group received bilateral ibotenic acid lesions of the DMH (panel c). Impulse activity was recorded in LC neurons during either the dark or light photoperiod (panels a and c) or the active or inactive epoch of the rat's circadian cycle in 24 h darkness (panel b). There was a significant difference in LC firing rates in animals taken from their dark vs. light periods (panel a), and from their active vs. inactive periods when maintained in continuous darkness (panel b; $p < 0.0001$, each). Lesions of the DMH eliminated the difference in LC firing rates during the dark vs. light photoperiods (panel c). The scale on the y-axis only corresponds to the histogram bars. The solid lines on each of the graphs represent best-fit normal distribution curves for the histogram data. A one sample Kolmogorov-Smirnov test revealed that in all cases the data fit a normal distribution with a > 0.725 probability. The following is a tabulation of the number of animals and LC neurons used to calculate each histogram. Panel a – dark: n = 52 neurons from 10 animals; light: n = 61 neurons from 12 animals. Panel b – active: n = 17 neurons from 3 animals; inactive: n = 33 neurons from 3 animals. Panel c – dark: n = 35 neurons from 6 animals; light: n = 30 neurons from 5 animals. Taken from Aston-Jones *et al.* (2001).[129]

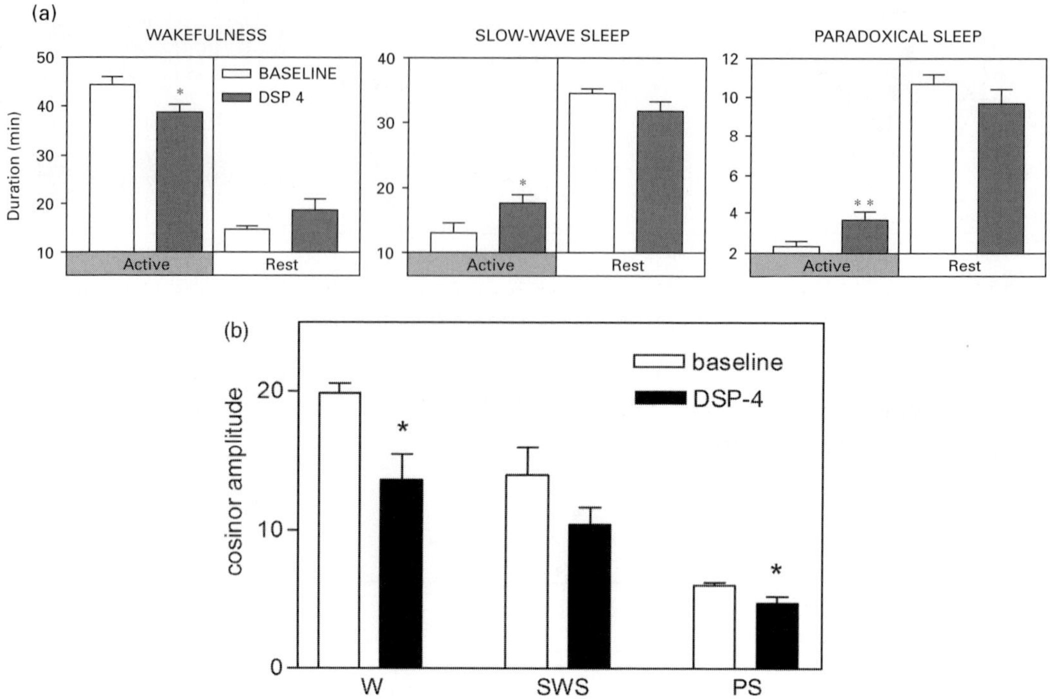

Figure 6.8 (a) Hourly mean duration (in min) of each vigilance state during the active period (night) and rest period (day) under 12:12 LD, n = 6%. (b) Cosinor analysis of effect of DSP-4 lesion on circadian amplitude of the sleep–wake cycle. Note that the amplitude is decreased for each sleep–wake stage in rats kept under 12:12 light-dark, n = 6. Values represent the mean ± SEM of 3 consecuitive 24 h periods (* $p < 0.05$, ** $p < 0.01$). W, wakefulness; SWS, slow-wave sleep; PS, paradoxica sleep.

We hypothesize that less W occurs in the active period of lesioned rats because there is less LC influence to drive arousal from the SCN. We also propose that the decreased sleep (increased waking) observed in the rest period reflects a decreased homeostatic sleep need due to decreased W during the preceding active period. Decreased W/increased sleep during the night coupled with increased W/decreased sleep during the day produced a flattening of the circadian rhythm of sleep and W expression following LC lesioning.[131] These results confirmed that LC lesions reduce the amplitude of the circadian rhythm of the sleep–wake cycle, as expected if LC activity contributed to waking in a circadian fashion. Interestingly, although there was a significant decrease in circadian amplitude of the sleep–waking rhythm, there was no overall change in the absolute amounts of sleep or waking in the lesioned animals. This may explain the minimal or inconsistent effects of LC lesions on total amounts of sleep and waking in previous studies that did not employ a circadian analysis. Another important finding was that sleep onset, and body and

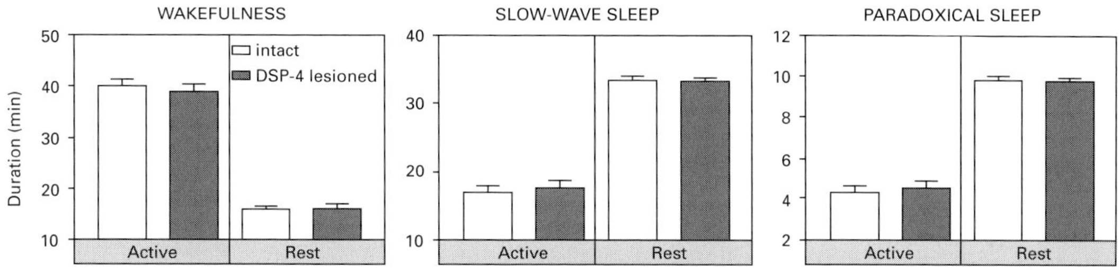

Figure 6.9 Hourly mean duration (in min) of each vigilance state during the active period and rest period under constant darkness; n = 7; vertical bars = SEM.

brain temperature were not affected by DSP-4 treatment in either LD or DD animals compared to controls.[68,131]

Light deprivation-induced loss of cortical NE

Contrary to the above findings for decreased amplitude of the sleep–wake rhythm after LC lesions in animals maintained in LD conditions, animals maintained in DD conditions did not exhibit a significant alteration of the sleep–wake rhythm following LC lesions, as compared to nonlesioned rats maintained in DD (Figure 6.9).[131] This was initially puzzling, as loss of a circadian signal following LC lesions should have resulted in decreased sleep–wake cycle rhythm amplitudes in DD conditions. However, analysis of NE fibers revealed that control (nonlesioned) DD-maintained animals exhibited substantially reduced NE fiber staining density in the frontal cortex when compared to control LD animals. This finding was confirmed with high-performance liquid chromatography analysis, which showed a decrease in frontal cortex NE of 45%.[132] We hypothesize that LC lesions in the DD animals were ineffective in modifying the circadian rhythm of sleep and waking because the control (nonlesioned) DD-maintained animals had less cortical NE, reducing the apparent effect of lesions by reducing the functional contrast with DD-lesioned subjects. Indeed, we observed that DD control rats (nonlesioned) exhibited reduced amplitude in the waking and sleep rhythms, as previously reported.[133] Furthermore, this reduced amplitude is indistinguishable from that observed in lesioned animals recorded during LD. In addition to offering a possible explanation for the lack of effect of LC lesions in the DD group, the finding that light deprivation caused substantial decreases in NE neurotransmission in the frontal cortex raises a number of important questions. Does this change reflect degeneration of LC efferent fibers or only a decrease in dopamine-β-hydroxylase and NE in otherwise intact fibers? Are there behavioral changes (aside from sleep–wake rhythms) associated with this change in NE projections? For example, do light-deprived animals exhibit a depressive behavioral phenotype? If so, this might correspond to clinical problems such as seasonal affective disorder. Finally, can these anatomical and associated behavioral changes be reversed by light exposure or other treatments? These questions are being actively pursued in our laboratory at present. In any case, these results indicate that LC-NE projections depend upon light for maintenance of normal function.

Clinical perspectives: LC participation in sleep alterations due to clinical pathologies

Many psychiatric disorders are prompted by, or include, stress responses. This may be important for understanding the neurobiology of corresponding sleep disturbances, because LC neurons are strongly activated by several physiological and visceral stressors.[134–139] As described above, activation of the NE-LC system increases wakefulness and arousal. We hypothesize that hyperactivity of the LC via enhanced stress responses may help to explain sleep disruptions emerging secondary to many clinical conditions, some of which are detailed below. Long-term up-regulation in NE production, release, or receptor regulation could contribute to sleep disruptions simply by increasing the drive towards wakefulness.

Depression

Disordered sleep is a cardinal symptom of both unipolar and bipolar clinical depression required for clinical assessment of chronic depression according to the Diagnostic and Statistical Manual of Mental Disorders (DSM-IV; 1994). Patients with major depressive disorder (MDD) typically report insomnia, a claim supported by polysomnographic recordings showing increased W, decreased SWS, and increased PS in people with depression compared to age-matched controls.[140,141] Increased W and light sleep typically replaces SWS in people with MDD. A sleep EEG taken during MDD contains significant amounts of high-frequency (10 to 35 Hz) EEG intruding into normally slow-wave (0.5 to 4 Hz) EEG during sleep.[140,142] Paradoxical sleep abnormalities during MDD include shortened time to PS from sleep onset (i.e., reduced PS latency), increased total PS time, and an increase in the phasic eye movements and muscle twitches characteristic of active PS.[143,144] Findings of persistent insomnia, reductions in SWS, high-frequency sleep EEG, and increases in PS support a hypothesis that generalized physiologic activation (heart rate, body temperature, metabolism, mental activity) is a source of disordered sleep during MDD. Disordered sleep, especially involving PS changes, is predictive of depression relapse, recurrence, and increased risk of suicide.[145]

Conversely, insomnia itself is a significant risk factor for the development of depression and is prodromal in people at risk for depression.[146,147] This bidirectional relationship between depression and sleep has long suggested that intrasleep variables may provide diagnostic benefit. Altered PS during MDD (primarily short

PS latency) is so prevalent that it was once considered a reliable trait of people prone to depression.[148] Unfortunately, shortened PS latency is also found in several other disorders, including narcolepsy, borderline personality schizophrenia, and eating disorders,[141] limiting diagnostic use of PS measures in depression. Sleep disruptions are found during the chronic phase of numerous psychiatric disorders, which also limits their diagnostic utility but strongly suggests common disruptions of the neural mechanisms responsible for sleep expression.

NE and depression

A large body of literature has implicated the LC-NE system in depression. This view arises, in part, because several clinically effective antidepressants have actions, sometimes specifically, on the NE system (including the new selective NE reuptake blockers). However, measurements of NE and its metabolites in depressed individuals have been inconsistent, and therefore inconclusive, regarding alterations in the NE of depressed individuals.[149] This may reflect NE involvement in some subtypes of depression but not others. For example, one study found that cerebrospinal fluid (CSF) levels of 3-methoxy-4-hydroxyphenylglycol (MHPG) were higher among patients who were cortisol nonsuppressors than among either patients who were cortisol suppressors or controls. Also, urinary outputs of NE and normetanephrine were significantly higher among patients who were cortisol nonsuppressors than among controls. These results suggest that dysregulation of the noradrenergic system and hypothalamic–pituitary–adrenal axis occur together in a subgroup of depressed patients,[150] consistent with a previous hypothesis that depression is associated with dysregulation of the NE-LC system.[151] These data also imply that depression associated with stress disorders may be particularly associated with NE-LC dysregulation. Given the role of the NE-LC in sleep regulation, such dysregulation may account for some of the sleep disturbances in depression.

Stress, depression, CRF, NE-LC system and sleep

There are several links between stress, CRF, the NE-LC system and depression. Behavioral alarm and stress response were reduced in rhesus monkeys given a CRF_1 receptor antagonist.[152] Both patients with MDD and animal models of depression show increased levels of circulating cortisol and CRF compared to nonstressed controls.[153] Hyperactivity of CRF-containing neurons appears to account for several symptoms of depression, including disrupted sleep, blunted daily temperature fluctuation, and altered growth hormone-releasing hormone to CRF ratio.[154,155] Administration of CRF in normal adults increases W and reduces SWS but not PS.[156] Stress, depression, and sleep may interact via the NE-LC system because effects of stress on sleep have been shown (as described above) to involve the NE-LC system, probably via CRF.[68,117] Moreover, the LC receives a CRF input and CRF potently activates LC neurons.[88,99,139] It seems possible that alterations in the CRF influence on LC neurons may underlie sleep alterations associated with depression as a by-product of stress responses. Altered sleep associated with depression may involve specific stress mechanisms and LC functioning distinct from those involved in normal sleep regulation, as indicated in animal studies of the role of NE in stress effects on sleep.[68,117]

Sleep deprivation and depression

Sleep deprivation itself is a surprisingly potent strategy for alleviating symptoms of depression. Total sleep deprivation, partial sleep deprivation, and PS deprivation all improve mood, albeit transiently, especially for patients with increasingly severe depression.[157] Partial sleep deprivation is most successful when enforced for the second half of the night and/or when PS is preferentially reduced. Mood elevation due to any type of sleep limitation only occurs for about 50% of patients tested and typically lasts only until the next sleep period.[158] Given the links between NE, depression, and sleep, it seems possible that the LC could be involved in the antidepressant effect of sleep deprivation. Consistent with this possibility, Schrieber *et al.* found significant increases in plasma NE after sleep deprivation in healthy adults compared to baseline.[159] Likewise, patients with MDD have increased MHPG after sleep deprivation, indicating increased NE neurotransmission.[160]

Several methodologies of neuroimaging have recently been reviewed in order to better understand the functional neurology of the antidepressant response to sleep deprivation. Gillin *et al.*[158] found, in general, that patients whose moods improved after either total sleep deprivation or partial sleep deprivation tend to have higher brain activity in the ventral anterior cingulate cortex (ACC), a localized increase in brain activity that normalized after sleep deprivation treatment. Both the ventral ACC and the orbital frontal regions were most often recognized as modified by sleep deprivation if researchers were careful to sort the data into "responders" vs. "nonresponders" (>50% improvement on Hamilton Depression (HAM-D) ratings). Both the ACC and orbital frontal cortices are strongly innervated by the NE-LC system in primates, and our recent data (see Chapter 7 of this volume) also indicate that both are prominent afferents to the primate LC.[161–163] Thus, dysfunctions in these areas could affect sleep and depression, in part, via interactions with the NE-LC system.

Anxiety

Abundant symptomatic similarities exist between diseases of excessive anxiety and major depression. As in depression, both types of insomnia (difficulty initiating sleep and difficulty staying asleep) are reported to persist during several anxiety disorders, e.g. panic attacks, attention-deficit/hyperactivity disorder (ADHD), posttraumatic stress disorder (PTSD), and obsessive-compulsive disorder (OCD). Generalized anxiety disorder (GAD) is comorbid in patients diagnosed first with depression in an estimated 58% of cases.[164] Subjectively, 50% to 70% of patients with anxiety tend to report disturbed, unsatisfying sleep with morning drowsiness.[165] Objectively, the polysomnographic study with the largest number of patients with GAD reported decreased total sleep time, decreased sleep efficiency (i.e. more time awake once sleep was attained), and more early morning awakenings.[166] Selective 5-HT uptake inhibitors are as effective in treating GAD and other disorders of excessive anxiety, as they are in treating depression.[167] Similar neurobiologic disruptions in NE, 5-HT, and CRF systems are thought to contribute to both depression and anxiety. A review by Ressler and Nemeroff[149] suggests that down-regulation of 5-HT and up-regulation of NE activity may occur in both anxiety and depression. According to their model, the NE-LC system primarily sensitizes the amygdala and hippocampus to increase the salience of, and conditioned responding to, stressful or fearful stimuli. Although Ressler and Nemeroff note that both 5-HT and NE dysregulation likely contribute to depression, NE is more directly implicated in the fear and stress responses contributing to persistent clinical anxiety.

A noradrenergic contribution to GAD has been difficult to ascertain because GAD is often comorbid with depression and other anxiety disorders.[168] Conditioned anxiety disorders (PTSD and panic disorder) are marked by increased peripheral sympathetic arousal. Work in animals reveals that LC-NE neuronal activity often correlates strongly with sympathetic activity, probably reflecting the major inputs to the LC from the medulla.[169–172] Thus, these disorders may also be associated with elevated LC activity.[167] Ordway[173] summarized the neurology of mood disorders to suggest that stress-induced CRF hyperactivation of the LC and ventral tegmental area (VTA) alters the balance of those monoamines with 5-HT. In addition, experimentally stressed macacque monkeys show higher CRF and depressed cortisol in the CSF while being hyperresponsive to yohimbine.[174] Higher levels of CRF are also reported for patients with panic disorders and PTSD. Thus, chronic NE-LC hyperactivity via elevated CRF inputs may help sustain GAD and related disorders, and this LC hyperactivity may occur primarily as a conditioned response to stress that does not naturally ameliorate.

Postmortem evidence for changes in the human LC is not as available for clinical anxiety as for depression. However, there is an abundance of data in animal models linking LC activation to stress, pain, learned helplessness, hyperarousal, and nearly all face- and construct-valid animal measures of human anxiety.[175–178] Activation of the amygdala has been linked to CRF-mediated LC hyperactivation as an explanation for the link between conditioned anxiety, hypothalamic activation, and subsequent NE-LC up-regulation.[179]

Posttraumatic Stress Disorder (PTSD)

Posttraumatic stress disorder is an anxiety disorder arising from the conditioned emotional or injury-based responses to a singular or repeated salient event.[180] Sleep disturbances are a common symptom reported by people who experience an acutely stressful episode. Polysomnographic assessment of patients with PTSD has produced a literature of mixed results. The symptoms of PTSD – recurrent anxiety, increased startle reflex, and sleep problems – do not occur in all who experience a traumatic event. Prevalence of PTSD varies widely, depending on the age of the subject, the severity of the event, and other unrecognized factors.[181] Furthermore, comorbidity with other psychiatric, putatively emergent, disorders (e.g., depression) has been reported to be as high as 85%. A wide variety of events can cause PTSD, and the heterogeneity of responses make subject selection in studies so difficult that generalities about their sleep are difficult to support. For example, Hefez *et al.*[182] found that some PTSD patients have decreased sleep efficiency, but that the sleep of other patients is not different from age-matched controls. Lavie reviewed the literature on sleep in PTSD patients to summarize that "patients with PTSD appear to have deeper sleep and lower rates of dream recall than normal persons."[183]

Locus coeruleus activity itself cannot be measured in people with PTSD, but there is evidence for disruptions in the NE-LC system contributing to the symptomatology of PTSD. Excessive startle reflexes and behavioral hypervigilance have been suggested to result from an increase in NE production from the LC, which may be a response to decreased NE receptor sensitivity following an acute stress event. Norepinephrine levels in the CSF of PTSD patients have been evaluated using an indwelling catheter to allow for serial recordings without sampling stress. Geracioto *et al.* reported higher CSF norepinephrine levels across 6 hours of wakefulness in men with PTSD as compared to healthy controls.[184] Notably, CSF NE correlated with PTSD symptom severity while plasma NE levels did not. Increased chronic NE production in PTSD patients can explain increased arousal and disturbed sleep by virtue of increasing the drive to wakefulness via LC hyperactivity. Circadian variation of NE levels has not been seen when measured in people with PTSD or depression compared to control subjects.[185,186] The tendency towards wakefulness at all times of day, not dysregulation of sleep state control, appears to arise from LC hyperactivity that increases nightmares and the perception of poor sleep during clinical PTSD.

The neurobiology of PTSD suggests a functional connection between behavioral hyperresponsivity, the NE-LC system, and stress-evoked CRF increases. Heim and Nemeroff[187] suggest that early trauma chronically increases CRF reactivity leading to subsequent supersensitivity to stressful events. Chronic stress in animals has been shown to increase NE levels in the hippocampus and amygdala, which may lead to increased memory consolidation during periods of stress.[188] Conditioned anxiety responses to specific triggers (in cases of PTSD or panic disorder), or generalized to many triggers, would thereby help maintain NE hyperactivity in the central nervous system, leading secondarily to sleep disruptions of excessive arousal.

Attention-deficit/hyperactivity disorder (ADHD)

Impulsive, hyperactive, and inattentive behavior identifies a (typically young) person as having ADHD. These same types of behavior are seen in nonhuman primates during the tonic mode of activity in monkey LC neurons (as described in Chapter 7 of this volume). This finding is consistent with the possibility that this mode of LC activity contributes to the disorder. Treatment of ADHD with amphetamines is effective clinically but is not sufficient to explain underlying mechanisms.

Arousal state dysregulation, primarily nocturnal insomnia and wakeful hyperactivity, are common in as many as 75% of young people with ADHD, according to parental reports. Polysomnography studies of children with ADHD are few in number but shortened REM latency[189,190] and an increase in motor activity[191,192] are most commonly reported. Comorbidity between ADHD and nocturnal motor activity has been seen in several studies of ADHD, with a strong trend towards excessive sleep movements, including clinical levels of period limb movements in sleep (PLMS). Sleep instability measured with actigraphy in 38 students with ADHD found greater sleep onset variability and shorter average sleep duration than age-matched controls.[193]

Solanto surveyed existing data to surmise that ADHD, like anxiety, occurs due to dysregulation of more than one catecholamine to manifest the full range of cognitive and behavioral signs of ADHD, specifically implicating both NE and dopamine (DA).[194,195] Up-regulated NE release may contribute to both the sleep and behavioral symptomatology of ADHD, as described above for anxiety. In vivo microdialysis finds larger increases in local NE levels than local DA levels in response to amphetamine-like stimulants, further suggesting that increased NE levels contribute to both distractible behavior and more disrupted sleep.[196] However, it has proved difficult to specify the role of NE changes in ADHD. Reports of NE levels measured via metabolites in ADHD patients are so mixed as to be unusable for understanding its role in this disorder.[197–203] Amphetamines, the primary drug treatment for ADHD, activate the NE-LC system but also alter dopaminergic activity in the VTA. Dopamine transporter knockout mice, normally hyperactive and impulsive, show reduced hyperactivity following methylphenidate and dextroamphetamine, behavioral responses similar to the behavioral responses seen for patients taking these medications.[204] Because DA transporter knockout mice do not have DA transporters, these drugs must operate through another neurotransmitter system. The LC is a good candidate, but confirmation requires further examination.

Aging

Aging represents a "natural disorder" that typically causes an increase in sleep problems, including significant loss of SWS and PS, increased awakenings, and less total sleep time.[205] Czeisler and colleagues suggest that aging disrupts the ability to maintain sleep and reduces the sleep response (homeostatic sleep response) to increased waking periods.[206,207] The dominant theory at this point is that comorbid health problems conspire to disrupt sleep but that the elderly may retain their capacity for good sleep if external health and lifestyle sleep impediments are eliminated. However, older subjects measured in the "forced desynchrony" protocols showed significantly worse sleep than younger subjects when they were forced to sleep at nonoptimal circadian periods.[207]

The symptomology of aging suggests a gradual diminution of NE-LC basal activity and reactivity. Aged primates and humans show a profound loss of LC neurons (particularly in Alzheimer's Disease),[208–210] and aged monkeys have forebrain catecholamine depletion.[211] Memory problems that emerge with age in rhesus monkeys can be temporarily offset by administration of α_2-agonists,[212,213] but sleep effects of this treatment have not been studied. One possibility is that sleep–wake state instability results from a loss of wake-drive promotion due to LC insufficiency. Locus coeruleus insufficiency, in contrast to other clinical disorders reviewed here, could reduce daily sleep drive secondary to a reduction in active wakefulness. In contrast to these signs of decreased LC function, one set of studies found that aged rats produced higher levels of plasma catecholamines during both acute and chronic stress than did young controls,[214–216] implying that elevated NE could contribute to sleep problems in the elderly. However, much more work is needed in this area before conclusions can be drawn regarding an LC role in sleep problems of the elderly.

Insomnia

Two types of insomnia exist: difficulty falling asleep or difficulty staying asleep. Together they result in chronic sleep problems reported by one-third to one-half of all Americans at some time in their lives. Subjective insomnia is often overstated when compared with polysomnographic signs of sleep, but clearly at least 10% of people have great trouble falling asleep or staying asleep, and this disruption can last for months at a time.[217] Furthermore, although insomniacs sleep less and report being sleepier, they do not tend to fall asleep faster if given multiple chances to sleep across the day.

The neurobiology of insomnia remains a mystery even though several therapeutic strategies find good success treating insomnia in its variants. Benzodiazepines (transient insomnia), cognitive behavioral therapy (chronic insomnia), phototherapy (seasonal insomnia), and stimulus control therapy (conditioned insomnia) are capable of providing both subjective and objective sleep improvements. Few of these pharmacobehavioral strategies to treat insomnia have received adequate basic neurobiology research.

Notably, abundant clinical evidence suggests that most chronic insomnia develops subsequent to stress-related transient insomnia.[218] People with insomnia have been shown to perceive minor stressful events as more difficult to cope with than people with the same number of stressful events who have normal sleep.[219] Nocturnal cortisone levels are higher in insomniacs than controls suggesting that insomnia results from increased hypothalamic–pituitary axis activity.[220] Increased hypothalamic – pituitary–adrenal activation, as in the other disorders, is capable of increasing both central and peripheral release of epinephrine and NE. In addition, one study found that circulating levels of NE were higher in insomniacs than patients with depression or age-matched controls,[221] while another found that circulating levels of NE correlated with sleep disturbance frequency.[222] Furthermore, levels of nocturnal catecholamine correlated with the amount of sleep disruption found for insomniacs but not for patients with depression.[221] In a recent clinical review, Hauri[223] notes that signs of hyperarousal can be recorded in patients with primary insomnia: increased body temperature, high-frequency EEG during sleep, and metabolism, and they tend to be "sensation avoiders," suggesting they feel generally overstimulated all the time. Physiologic overstimulation coupled with stress responses may be adequate to stimulate stress-sensitive input to the LC and cause NE hyperactivity to result in sleep disruption as described earlier in this chapter for other disorders.

Conclusions

Arousal state maintenance and sleep–wake regulation clearly involves the NE-LC system. Participation of the LC system in several clinical disorders implies that this brain nucleus may contribute to the sleep disturbances associated with those disorders, as outlined above. One theme that emerges from our review of this clinical literature is that sleep disturbances in psychiatric disorders are often associated with enhanced stress response. This could be significant because the LC plays a large role in such responses. Thus, CRF release in the LC, and consequent LC activation during stress appears to be sufficient to evoke both sleep disturbance and aberrant PS expression found in many clinical mood disorders. However, CRF is not the only messenger to communicate stress-related information to the LC. For example, glutamate inputs from the medulla are also involved in LC activation by somatic or visceral stressors.[224–226] Whatever the input that drives the LC, increased NE output consequent to stress may participate in sleep disturbances associated with a variety of psychiatric problems. Further work along these lines may yield new approaches to treating these sleep disturbances and sleep improvement may, in turn, facilitate clinical resolution of the primary psychiatric disorder.

REFERENCES

1. Seager, M. A., Johnson, L. D., Chabot, E. S., Asaka, Y. and Berry, S. D. Oscillatory brain states and learning: impact of hippocampal theta-contingent training. *Proc. Natl. Acad. Sci. U. S. A.*, **99** (2002), 1616–1620.
2. Michel, F., Klein, M., Jouvet, D. and Valatx, J. L. Etude polygraphique du sommeil chez le rat. *C. R. Seances Soc. Biol. Fil.*, **12** (1961), 2389–2391.
3. Timo-Iaria, C., Negrao, N., Schmidek, W. R. *et al.* Phases and states of sleep in the rat. *Physiol. Behav.*, **5** (1970), 1057–1062.
4. Cespuglio, R., Calvo, J. M., Musolino, R. and Valatx, J. L. Phasic activity in rats. *Physiol. Behav.*, **19** (1977), 589–596.
5. Steriade, M., McCormick, D. A. and Sejnowski, T. J. Thalamocortical oscillations in the sleeping and aroused brain. *Science*, **262** (1993), 679–685.
6. Obermeyer, W. H. and Benca, R. M. Effects of drugs on sleep. *Neurol. Clin.*, **14** (1996), 827–840.
7. Thompson, P. M., Gillin, J. C., Golshan, S. and Irwin, M. Polygraphic sleep measures differentiate alcoholics and stimulant abusers during short-term abstinence. *Biol. Psychiatry*, **38** (1995), 831–836.
8. Borbely, A. A., Tobler, I. and Hanagasioglu, M. Effect of sleep deprivation on sleep and EEG power spectra in the rat. *Behav. Brain Res.*, **14** (1984), 171–182.
9. Parmeggiani, P. L. Interaction between sleep and thermoregulation. *Waking Sleep*, **1** (1977), 123–132.
10. Aserinsky, E. and Kleitman, W. Regularly occurring periods of eye motility and concomitant phenomena during sleep. *Science*, **118** (1953), 273–274.
11. Jouvet, M., Michel, F. and Courjon, J. Sur un stade d'activité électrique cérébrale rapide au cours du sommeil physiologique. *C. R. Seances Soc. Biol. Fil.*, **153** (1959), 1024–1028.
12. Jouvet, M. Recherches sur les structures nerveuses and les mécanismes responsables des différentes phases du sommeil physiologique. *Arch. Ital. Biol.*, **100** (1962), 125–206.
13. Schmidt, M. H., Valatx, J. L., Schmidt, H. S., Wauquier, A. and Jouvet, M. Experimental evidence of penile erections during paradoxical sleep in the rat. *NeuroReport*, **5** (1994), 561–564.
14. Siegel, J. M. Brainstem mechanisms generating REM sleep. In *Principles and Practice of Sleep Medicine*, ed. M. H. Kryer, T. Roth and W. C. Dement. (London: Saunders Company, 1994), pp. 125–144.
15. Moruzzi, G. and Magoun, H. W. Brain stem reticular formation and activation of the EEG. *Electroencephalogr. Clin. Neurophysiol.*, **1** (1949), 455–473.
16. Aston-Jones, G., Chiang, C. and Alexinsky, T. Discharge of noradrenergic locus coeruleus neurons in behaving rats and monkeys suggests a role in vigilance. *Prog. Brain Res.*, **88** (1991), 501–520.

17. Denoyer, M., Sallanon, M., Kitahama, K., Aubert, C. and Jouvet, M. Reversibility of para-chlorophenylalanine-induced insomnia by intrahypothalamic microinjection of L-5-hydroxytryptophan. *Neuroscience*, **28** (1989), 83–94.
18. Hungs, M. and Mignot, E. Hypocretin/orexin, sleep and narcolepsy. *Bioessays*, **23** (2001), 397–408.
19. Jones, B. E. Arousal systems. *Front. Biosci.*, **8** (2003), S438–S451.
20. Lin, J. S., Sakai, K. and Jouvet, M. Hypothalamo-preoptic histaminergic projections in sleep-wake control in the cat. *Eur. J. Neurosci.*, **6** (1994), 618–625.
21. McCormick, D. A. Neurotransmitter actions in the thalamus and cerebral cortex. *J. Clin. Neurophysiol.*, **9** (1992), 212–223.
22. Pace-Schott, E. F. and Hobson, J. A. The neurobiology of sleep: genetics, cellular physiology and subcortical networks. *Nat. Rev. Neurosci.*, **3** (2002), 591–605.
23. Sallanon, M., Denoyer, M., Kitahama, K. *et al.* Long-lasting insomnia induced by preoptic neuron lesions and its transient reversal by muscimol injection into the posterior hypothalamus in the cat. *Neuroscience*, **32** (1989), 669–683.
24. Semba, K. Multiple output pathways of the basal forebrain: organization, chemical heterogeneity, and roles in vigilance. *Behav. Brain Res.*, **115** (2000), 117–141.
25. Aston-Jones, G. and Bloom, F. E. Activity of norepinephrine-containing locus coeruleus neurons in behaving rats anticipates fluctuations in the sleep-waking cycle. *J. Neurosci.*, **1** (1981), 876–886.
26. Jacobs, B. L. and Azmitia, E. C. Structure and function of the brain serotonin system. *Physiol. Rev.*, **72** (1992), 165–299.
27. Jones, B. E. Paradoxical sleep and its chemical/structural substrates in the brain. *Neuroscience*, **40** (1991), 637–656.
28. Sakai, K. Neurons responsible for paradoxical sleep. In *Sleep: Neurotransmitters and Neuromodulators*, ed. W. A. Monti, J. M. Gaillard, and M. Radulovacki. (New York: Raven Press, 1985), pp. 405–429.
29. Vanni-Mercier, G., Gigout, S., Debilly, G. and Lin, J. S. Waking selective neurons in the posterior hypothalamus and their response to histamine H3-receptor ligands: an electrophysiological study in freely moving cats. *Behav. Brain Res.*, **144** (2003), 227–241.
30. Gervasoni, D., Darracq, L., Fort, P. *et al.* Electrophysiological evidence that noradrenergic neurons of the rat locus coeruleus are tonically inhibited by GABA during sleep. *Eur. J. Neurosci.*, **10** (1998), 964–970.
31. Gervasoni, D., Peyron, C., Rampon, C. *et al.* Role and origin of the GABAergic innervation of dorsal raphe serotonergic neurons. *J. Neurosci.*, **20** (2000), 4217–4225.
32. Kaur, S., Saxena, R. N. and Mallick, B. N. GABAergic neurons in prepositus hypoglossi regulate REM sleep by its action on locus coeruleus in freely moving rats. *Synapse*, **42** (2001), 141–150.
33. Lu, J., Bjorkum, A. A., Xu, M. *et al.* Selective activation of the extended ventrolateral preoptic nucleus during rapid eye movement sleep. *J. Neurosci.*, **22** (2002), 4568–4576.
34. Nitz, D. and Siegel, J. M. GABA release in the locus coeruleus as a function of sleep/wake state. *Neuroscience*, **78** (1997), 795–801.

35. Sherin, J. E., Elmquist, J. K., Torrealba, F. and Saper, C. B. Innervation of histaminergic tuberomammillary neurons by GABAergic and galaninergic neurons in the ventrolateral preoptic nucleus of the rat. *J. Neurosci.*, **18** (1998), 4705–4721.
36. McCarley, R. W. and Hobson, J. A. Neuronal excitability modulation over the sleep cycle: a structural and mathematical model. *Science*, **189** (1975), 58–60.
37. Sakai, K., Sastre, J. P., Kanamori, N. and Jouvet, M. State specific neurons in the ponto-medullary reticular formation with special reference to the postural atonia during paradoxical sleep in the cat. In *Brain Mechanisms of Perceptual Awareness and Purposeful Behavior*, ed. O. Pompeiano and C. Ajmone Marsan. (New York: Raven Press, 1981), pp. 405–429.
38. Chase, M. H. and Morales, F. R. The atonia and myoclonia of active (REM) sleep. *Annu. Rev. Psychol.*, **41** (1990), 557–584.
39. Sakai, K. Executive mechanisms of paradoxical sleep. *Arch. Ital. Biol.*, **126** (1988), 239–257.
40. Sakai, K. and Koyama, Y. Are there cholinergic and non-cholinergic paradoxical sleep-on neurones in the pons? *NeuroReport*, **7** (1996), 2449–2453.
41. Hobson, J. A., McCarley, R. W. and Wyzinski, P. W. Sleep cycle oscillation: reciprocal discharge by two brainstem neuronal groups. *Science*, **189** (1975), 55–58.
42. Rasmussen, K., Morilak, D. A. and Jacobs, B. L. Single unit activity of locus coeruleus neurons in the freely moving cat. I. During naturalistic behaviors and in response to simple and complex stimuli. *Brain Res.*, **371** (1986), 324–334.
43. Foote, S. L., Aston-Jones, G. and Bloom, F. E. Impulse activity of locus coeruleus neurons in awake rats and monkeys is a function of sensory stimulation and arousal. *Proc. Natl. Acad. Sci. U. S. A.*, **77** (1980), 3033–3037.
44. Rajkowski, J., Kubiak, P., Ivanova, S. and Aston-Jones, G. State related activity and reactivity of locus coeruleus neurons in behaving monkeys. In *Catecholamines: Bridging Basic Science with Clinical Medicine*, ed. D. Goldstein, G. Eisenhofer and R. McCarty. (New York: Academic Press, 1998), pp. 740–744.
45. Rajkowski, J., Silakov, V., Ivanova, S. and Aston-Jones, G. Locus coeruleus neurons in monkey are quiescent in paradoxical sleep. *Soc. Neurosci. Abstr.*, **23** (1997), 2130.
46. Rajkowski, J., Kubiak, P. and Aston-Jones, G. Locus coeruleus activity in monkey: phasic and tonic changes are associated with altered vigilance. *Brain Res. Bull.*, **35** (1994), 607–616.
47. Doran, S., Rajkowski, J. and Aston-Jones, G. Locus coeruleus neurons in primate are active during waking but do not co-vary with arousal state. Activity across states of arousal. *Soc. Neurosci. Abstr.*, (2003), Program No. 618.14.
48. Aston-Jones, G., Rajkowski, J. and Cohen, J. Locus coeruleus and regulation of behavioral flexibility and attention. *Prog. Brain Res.*, **126** (2000), 165–182.
49. Berridge, C. W. and Foote, S. L. Effects of locus coeruleus activation on electroencephalographic activity in the neocortex and hippocampus. *J. Neurosci.*, **11** (1991), 3135–3145.
50. Berridge, C. W., Page, M. E., Valentino, R. J. and Foote, S. L. Effects of locus coeruleus inactivation on electroencephalographic activity in neocortex and hippocampus. *Neuroscience*, **55** (1993), 381–393.
51. Berridge, C. W., Isaac, S. O. and Espana, R. A. Additive wake-promoting actions of medial basal forebrain noradrenergic α_1- and β-receptor stimulation. *Behav. Neurosci.*, **117** (2003), 350–359.

52. Jones, B. E., Harper, S. T. and Halaris, A. E. Effects of locus coeruleus lesions upon cerebral monoamine content, sleep-wakefulness states and the response to amphetamine in the cat. *Brain Res.*, **124** (1977), 473–496.
53. Lidbrink, P. The effect of lesions of ascending noradrenaline pathways on sleep and waking in the rat. *Brain Res.*, **74** (1974), 19–40.
54. Roussel, B., Pujol, J. F. and Jouvet, M. Effects of lesions in the pontine tegmentum on the sleep stages in the rat. *Arch. Ital. Biol.*, **114** (1976), 188–209.
55. Monti, J. M., D'Angelo, L., Jantos, H., Barbeito, L. and Abo, V. Effect of DSP-4, a noradrenergic neurotoxin, on sleep and wakefulness and sensitivity to drugs acting on adrenergic receptors in the rat. *Sleep*, **11** (1988), 370–377.
56. Laguzzi, R., Petitjean, F., Pujol, J. E. and Jouvet, M. Effects of intraventricular injection of 6-hydroxydopamine on sleep states and cerebral monoamines in cats. *C. R. Seances Soc. Biol. Fil.*, **165** (1971), 1649–1653.
57. Matsuyama, S., Coindet, J. and Mouret, J. Intracisternal 6-hydroxydopamine and sleep in the rat. *Brain Res.*, **57** (1973), 85–95.
58. Fritschy, J. M. and Grzanna, R. Immunohistochemical analysis of the neurotoxic effects of DSP-4 identifies two populations of noradrenergic axon terminals. *Neuroscience*, **30** (1989), 181–197.
59. Jonsson, G., Hallman, H., Ponzio, F. and Ross, S. DSP4 (N-(2-chloroethyl)-N-ethyl-2-bromobenzylamine) – a useful denervation tool for central and peripheral noradrenaline neurons. *Eur. J. Pharmacol.*, **72** (1981), 173–188.
60. Hallman, H., Sundstrom, E. and Jonsson, G. Effects of the noradrenaline neurotoxin DSP 4 on monoamine neurons and their transmitter turnover in rat CNS. *J. Neural. Transm.*, **60** (1984), 89–102.
61. Logue, M. P., Growdon, J. H., Coviella, I. L. and Wurtman, R. J. Differential effects of DSP-4 administration on regional brain norepinephrine turnover in rats. *Life Sci.*, **37** (1985), 403–409.
62. Lyons, W. E., Fritschy, J. M. and Grzanna, R. The noradrenergic neurotoxin DSP-4 eliminates the coeruleospinal projection but spares projections of the A5 and A7 groups to the ventral horn of the rat spinal cord. *J. Neurosci.*, **9** (1989), 1481–1489.
63. Mogilnicka, E., Przewlocka, B., Van Luijtelaar, E. L., Klimek, V. and Coenen, A. M. Effects of REM sleep deprivation on central alpha 1- and β-adrenoceptors in rat brain. *Pharmacol. Biochem. Behav.*, **25** (1986), 329–332.
64. Dooley, D. J., Bittiger, H., Hauser, K. L., Bischoff, S. F. and Waldmeier, P. C. Alteration of central alpha 2- and beta-adrenergic receptors in the rat after DSP-4, a selective noradrenergic neurotoxin. *Neuroscience*, **9** (1983), 889–898.
65. Fety, R., Misere, V., Lambas-Senas, L. and Renaud, B. Central and peripheral changes in catecholamine-synthesizing enzyme activities after systemic administration of the neurotoxin DSP-4. *Eur. J. Pharmacol.*, **124** (1986), 197–202.
66. Theron, C. N., de Villiers, A. S. and Taljaard, J. J. Effects of DSP-4 on monoamine and monoamine metabolite levels and on beta adrenoceptor binding kinetics in rat brain at different times after administration, *Neurochem. Res.*, **18** (1993), 1321–1327.
67. Jaim-Etcheverry, G. and Zieher, L. M. DSP-4: a novel compound with neurotoxic effects on noradrenergic neurons of adult and developing rats. *Brain Res.*, **188** (1980), 513–523.

68. Gonzalez, M. M., Debilly, G. and Valatx, J. L. Noradrenaline neurotoxin DSP-4 effects on sleep and brain temperature in the rat. *Neurosci. Lett.*, **248** (1998), 93–96.
69. Dement, W. C. The effect of dream deprivation. *Science*, **131** (1960), 1705–1707.
70. Mouret, J., Pujol, J. F. and Kiyono, S. Paradoxical sleep rebound in the rat. Effects of physical procedures involved in intracisternal injection. *Brain Res.*, **15** (1969), 501–506.
71. Asikainen, M., Deboer, T., Porkka-Heiskanen, T., Stenberg, D. and Tobler, I. Sleep deprivation increases brain serotonin turnover in the Djungarian hamster. *Neurosci. Lett.*, **198** (1995), 21–24.
72. Cramer, H., Tagliamonte, A., Tagliamonte, P., Perez-Cruet, J. and Gessa, G. L. Stimulation of brain serotonin turnover by paradoxical sleep deprivation in intact and hypophysectomized rats. *Brain Res.*, **54** (1973), 372–375.
73. Devi, R. S., Maheswari, K. S. and Namasivayam, A. Immunity and REM sleep deprivation. *Med. Sci. Res.*, **22** (1994), 753–755.
74. Hery, F., Pujol, J. F., Lopez, M., Macon, J. and Glowinski, J. Increased synthesis and utilization of serotonin in the central nervous system of the rat during paradoxical sleep deprivation. *Brain Res.*, **21** (1970), 391–403.
75. Mallick, B. N. and Thakkar, M. Short-term REM sleep deprivation increases acetylcholinesterase activity in the medulla of rats. *Neurosci. Lett.*, **130** (1991), 221–224.
76. Pujol, J. F., Mouret, J., Jouvet, M. and Glowinski, J. Increased turnover of cerebral norepinephrine during rebound of paradoxical sleep in the rat. *Science*, **159** (1968), 112–114.
77. Sinha, A. K., Ciaranello, R. D., Dement, W. C. and Barchas, J. D. Tyrosine hydroxylase activity in rat brain following "REM" sleep deprivation. *J. Neurochem.*, **20** (1973), 1289–1290.
78. Stern, W. C., Miller, F. P., Cox, R. H. and Maickel, R. P. Brain norepinephrine and serotonin levels following REM sleep deprivation in the rat. *Psychopharmacologia*, **22** (1971), 50–55.
79. D'Almeida, V., Hipolide, D. C., Azzalis, L. A. *et al.* Absence of oxidative stress following paradoxical sleep deprivation in rats. *Neurosci. Lett.*, **235** (1997), 25–28.
80. Kovalzon, V. M. and Tsibulsky, V. L. REM-sleep deprivation, stress and emotional behavior in rats. *Behav. Brain Res.*, **14** (1984), 235–245.
81. Porkka-Heiskanen, T., Smith, S. E., Taira, T. *et al.* Noradrenergic activity in rat brain during rapid eye movement sleep deprivation and rebound sleep. *Am. J. Physiol.*, **268** (1995), R1456–R1463.
82. Fadda, P. and Fratta, W. Stress-induced sleep deprivation modifies corticotropin releasing factor (CRF) levels and CRF binding in rat brain and pituitary. *Pharmacol. Res.*, **35** (1997), 443–446.
83. Abercrombie, E. D. and Jacobs, B. L. Single unit response of noradrenergic neurons in locus coeruleus of freely moving cats I. Acutely presented stressful and nonstressful stimuli. *J. Neurosci.*, **7** (1987), 2837–2843.
84. Bremner, J. D., Krystal, J. H., Southwick, S. M. and Charney, D. S. Noradrenergic mechanisms in stress and anxiety: I. Preclinical studies. *Synapse*, **23** (1996), 28–38.
85. Emoto, H., Tanaka, M., Koga, C. *et al.* Corticotropin-releasing factor activates the noradrenergic neuron system in the rat brain. *Pharmacol. Biochem. Behav.*, **45** (1993), 419–422.

86. Shimizu, N., Nakane, H., Hori, T. and Hayashi, Y. CRF receptor antagonist attenuates stress-induced noradrenaline release in the medial prefrontal cortex of rats. *Brain Res.*, **654** (1994), 145–148.
87. Stanford, S. C. Central noradrenergic neurones and stress. *Pharmacol. Ther.*, **68** (1995), 297–342.
88. Valentino, R. J., Foote, S. L. and Aston-Jones, G. Corticotropin-releasing factor activates noradrenergic neurons of the locus coeruleus. *Brain Res.*, **270** (1983), 363–367.
89. Gonzalez, M. M., Debilly, G., Valatx, J. L. and Jouvet, M. Sleep increase after immobilization stress: role of the noradrenergic locus coeruleus system in the rat. *Neurosci. Lett.*, **202** (1995), 5–8.
90. Gonzalez, M. M., Valatx, J. L. and Debilly, G. Role of the locus coeruleus in the sleep rebound following two different sleep deprivation methods in the rat. *Brain Res.*, **740** (1996), 215–226.
91. Rampin, C., Cespuglio, R., Chastrette, N. and Jouvet, M. Immobilisation stress induces a paradoxical sleep rebound in rat. *Neurosci. Lett.*, **126** (1991), 113–118.
92. Kitahama, K. and Valatx, J. L. Strain differences in amphetamine sensitivity in mice. II. Overcompensation of paradoxical sleep after deprivation in two C57 strains. *Psychopharmacology (Berl)*, **66** (1979), 291–295.
93. Touret, M., Valatx, J. L. and Jouvet, M. The locus coeruleus: a quantitative and genetic study in mice. *Brain Res.*, **250** (1982), 353–357.
94. Berger, B., Herve, D., Dolphin, A. *et al.* Genetically determined differences in noradrenergic input to the brain cortex: a histochemical and biochemical study in two inbred strains of mice. *Neuroscience*, **4** (1979), 877–888.
95. Edgar, D. M. and Seidel, W. F. Modafinil induces wakefulness without intensifying motor activity or subsequent rebound hypersomnolence in the rat. *J. Pharmacol. Exp. Ther.*, **283** (1997), 757–769.
96. Dunn, A. J. and Berridge, C. W. Physiological and behavioral responses to corticotropin-releasing factor administration: is CRF a mediator of anxiety or stress responses? *Brain Res. Brain Res. Rev.*, **15** (1990), 71–100.
97. Owens, M. J. and Nemeroff, C. B. Physiology and pharmacology of corticotropin-releasing factor. *Pharmacol. Rev.*, **43** (1991), 425–473.
98. Merchenthaler, I. Corticotropin releasing factor (CRF)-like immunoreactivity in the rat central nervous system. Extrahypothalamic distribution. *Peptides*, **5** (1984), 53–69.
99. Valentino, R. J., Page, M., Van Bockstaele, E. and Aston-Jones, G. Corticotropin-releasing factor innervation of the locus coeruleus region: distribution of fibers and sources of input. *Neuroscience*, **48** (1992), 689–705.
100. Van Bockstaele, E. J., Colago, E. E. and Valentino, R. J. Corticotropin-releasing factor-containing axon terminals synapse onto catecholamine dendrites and may presynaptically modulate other afferents in the rostral pole of the nucleus locus coeruleus in the rat brain. *J. Comp. Neurol.*, **364** (1996), 523–534.
101. Valentino, R., Page, M. and Curtis, A. Activation of noradrenergic locus coeruleus neurons by hemodynamic stress is due to local release of corticotropin-releasing factor. *Brain Res.*, **555** (1991), 25–34.

102. Chappell, P. B., Smith, M. A., Kilts, C. D. *et al.* Alterations in corticotropin-releasing factor-like immunoreactivity in discrete rat brain regions after acute and chronic stress. *J. Neurosci.*, **6** (1986), 2908–2914.
103. Butler, P. D., Weiss, J. M., Stout, J. C. and Nemeroff, C. B. Corticotropin-releasing factor produces fear-enhancing and behavioral activating effects following infusion into the locus coeruleus. *J. Neurosci.*, **10** (1990), 176–183.
104. Curtis, A. L., Lechner, S. M., Pavcovich, L. A. and Valentino, R. J. Activation of the locus coeruleus noradrenergic system by intracoerulear microinfusion of corticotropin-releasing factor: effects on discharge rate, cortical norepinephrine levels and cortical electroencephalographic activity. *J. Pharmacol. Exp. Ther.*, **281** (1997), 163–172.
105. Rassnick, S., Sved, A. F. and Rabin, B. S. Locus coeruleus stimulation by corticotropin-releasing hormone suppresses in vitro cellular immune responses. *J. Neurosci.*, **14** (1994), 6033–6040.
106. Butler, P., Weiss, J., Stout, J. and Nemeroff, C. B. Corticotropin-releasing factor produces fear-enhancing and behavioral activating effects following infusion into the LC. *J. Neurosci.*, **10** (1990), 176–183.
107. Rivier, C. L. and Plotsky, P. M. Mediation by corticotropin releasing factor (CRF) of adenohypophysial hormone secretion. *Annu. Rev. Physiol.*, **48** (1986), 475–494.
108. Rivier, J., Rivier, C. and Vale, W. Synthetic competitive antagonists of corticotropin-releasing factor: effect on ACTH secretion in the rat. *Science*, **224** (1984), 889–891.
109. Britton, K. T., Lee, G., Vale, W., Rivier, J. and Koob, G. F. Corticotropin releasing factor (CRF) receptor antagonist blocks activating and 'anxiogenic' actions of CRF in the rat. *Brain Res.*, **369** (1986), 303–306.
110. Emoto, H., Koga, C., Ishii, H. *et al.* A CRF antagonist attenuates stress-induced increases in NA turnover in extended brain regions in rats. *Brain Res.*, **627** (1993), 171–176.
111. Heinrichs, S. C., Menzaghi, F., Pich, E. M. *et al.* Anti-stress action of a corticotropin-releasing factor antagonist on behavioral reactivity to stressors of varying type and intensity. *Neuropsychopharmacology*, **11** (1994), 179–186.
112. Krahn, D. D., Gosnell, B. A., Grace, M. and Levine, A. S. CRF antagonist partially reverses CRF- and stress-induced effects on feeding. *Brain Res. Bull.*, **17** (1986), 285–289.
113. Diamant, M. and de Wied, D. Autonomic and behavioral effects of centrally administered corticotropin-releasing factor in rats. *Endocrinology*, **129** (1991), 446–454.
114. Morimoto, A., Nakamori, T., Morimoto, K., Tan, N. and Murakami, N. The central role of corticotrophin-releasing factor (CRF-41) in psychological stress in rats. *J. Physiol.*, **460** (1993), 221–229.
115. Gonzalez, M. C. and Valatx, J. L. Effect of intracerebroventricular administration of alpha-helical CRH (9–41) on the sleep/waking cycle in rats under normal conditions or after subjection to an acute stressful stimulus. *J. Sleep Res.*, **6** (1997), 164–170.
116. Gonzalez, M. M. and Valatx, J. L. Involvement of stress in the sleep rebound mechanism induced by sleep deprivation in the rat: use of alpha-helical CRH (9–41). *Behav. Pharmacol.*, **9** (1998), 655–662.
117. Gonzalez, M. C. Role du locus coeruleus dans les mecanismes du rebond de sommeil chez le rat. Thèse de Neurosciences. Lyon: Universite Claude Bernard (1997).

118. Kupfermann, I. Hypothalamus and limbic system: peptidergique neurons, homeostasis, and emotional behavior. In *Principles of Neural Science*, ed. E. R. Kandel and J. H. Schwartz. (New York: Elsevier, 1991), pp. 735–749.
119. Charifi, C., Paut-Pagano, L., Debilly, G. *et al.* Effect of noradrenergic denervation of the amygdala upon recovery after sleep deprivation in the rat. *Neurosci. Lett.*, **287** (2000), 41–44.
120. Dijk, D. J. and Edgar, D. M. Circadian and homeostatic control of wakefulness. In *Regulation of Sleep and Circadian Rhythms*, ed. F. W. Turek and P. C. Zee. (Basel: Marcel Dekker, 1999), pp. 111–147.
121. Rusak, B., Guido, M. and Semba, K. (2002) Immediate-early gene expression in the analysis of circadian rhythms and sleep. In *Handbook of Chemical Neuroanatomy*, ed. L. Kaczmarek and H. A. Robertson. New York: Marcel Dekker, Inc., pp. 147–170.
122. Abrahamson, E. E., Leak, R. K. and Moore, R. Y. The suprachiasmatic nucleus projects to posterior hypothalamic arousal systems. *NeuroReport*, **12** (2001), 435–440.
123. Sun, X., Rusak, B. and Semba, K. Electrophysiology and pharmacology of projections from the suprachiasmatic nucleus to the ventromedial preoptic area in rat. *Neuroscience*, **98** (2000), 715–728.
124. Sun, X., Whitefield, S., Rusak, B. and Semba, K. Electrophysiological analysis of suprachiasmatic nucleus projections to the ventrolateral preoptic area in the rat. *Eur. J. Neurosci.*, **14** (2001), 1257–1274.
125. Lu, J., Zhang, Y. H., Chou, T. C. *et al.* Contrasting effects of ibotenate lesions of the parventricular nucleus and subparaventricular zone on sleep-wake cycle and temperature regulation. *J. Neurosci.*, **21** (2001), 4864–4874.
126. Aston-Jones, G., Akaoka, H., Charlety, P. and Chouvet, G. Serotonin selectively attenuates glutamate-evoked activation of locus coeruleus neurons in vivo. *J. Neurosci.*, **11** (1991), 760–769.
127. Aston-Jones, G. and Card, J. P. Use of Pseudorabies virus to delineate multisynaptic circuits in brain: opportunities and limitations. *J. Neurosci. Methods*, **103** (2000), 51–61.
128. Chen, S., Ming, X., Miselis, R. and Aston-Jones, G. Characterization of transsynaptic tracing with central application of Pseudorabies virus. *Brain Res.*, **838** (1999), 171–183.
129. Aston-Jones, G., Chen, S., Zhu, Y. and Oshinsky, M. A neural circuit for circadian regulation of arousal. *Nat. Neurosci.*, **4** (2001), 732–738.
130. Semba, J. I., Toru, M. and Mataga, N. Twenty-four hour rhythms of norepinephrine and serotonin in nucleus suprachiasmaticus, raphe nuclei, and locus coeruleus in the rat. *Sleep*, **7** (1984), 211–218.
131. Gonzalez, M. C., Lu, W. and Aston-Jones, G. Role of the noradrenergic locus coeruleus system in sleep and waking circadian factors. *Soc. Neurosci. Abstr.*, (2002), Program No. 871.11.
132. Gonzalez, M. M., Zhu, Y. and Aston-Jones, G. Decrease in locus coeruleus-noradrenergic fiber staining under long-term constant darkness. *Soc. Neurosci. Abstr.*, (2003), Program No. 930.17.
133. Kawakami, M., Yamaoka, S. and Yamaguchi, T. Influence of light and hormones upon circadian rhythm of EEG slow wave and paradoxical sleep. In *Advances in Climatic Physiology*, ed. S. Itoh, K. Ogata and H. Yoshimura. (Tokyo: Igaken Shoin, 1972), pp. 349–366.

134. Elam, M., Yao, T., Svensson, T. H. and Thoren, P. Regulation of locus coeruleus neurons and splanchnic, sympathetic nerves by cardiovascular afferents. *Brain Res.*, **290** (1984), 281–287.
135. Elam, M., Yao, T., Thorén, P. and Svensson, T. H. Hypercapnia and hypoxia: chemoreceptor-mediated control of locus coeruleus neurons and splanchnic, sympathetic nerves. *Brain Res.*, **222** (1981), 373–381.
136. Morilak, D. A., Fornal, C. A. and Jacobs, B. L. Effects of physiological manipulations on locus coeruleus neuronal activity in freely moving cats. I. Thermoregulatory challenge. *Brain Res.*, **422** (1987), 17–23.
137. Morilak, D. A., Fornal, C. A. and Jacobs, B. L. Effects of physiological manipulations on locus coeruleus neuronal activity in freely moving cats. II. Cardiovascular challenge. *Brain Res.*, **422** (1987), 24–31.
138. Morilak, D. A., Fornal, C. A. and Jacobs, B. L. Effects of physiological manipulations on locus coeruleus neuronal activity in freely moving cats. III. Glucoregulatory challenge. *Brain Res.*, **422** (1987), 32–39.
139. Valentino, R. J., Curtis, A. L., Page, M. E., Pavcovich, L. A. and Florin-Lechner, S. M. Activation of the locus coeruleus brain noradrenergic system during stress: circuitry, consequences, and regulation. *Adv. Pharmacol.*, **42** (1998), 781–784.
140. Armitage, R., Trivedi, M., Hoffmann, R. and Rush, A. J. Relationship between objective and subjective sleep measures in depressed patients and healthy controls. *Depress. Anxiety*, **5** (1997), 97–102.
141. Benca, R., Obermeyer, W., Thisted, R. and Gillin, J. Sleep and psychiatric disorders: a meta-analysis. *Arch. Gen. Psychiatry*, **49** (1992), 651–668.
142. Armitage, R., Calhoun, J. S., Rush, A. J. and Roffwarg, H. P. Comparison of the delta EEG in the first and second non-REM periods in depressed adults and normal controls. *Psychiatry Res.*, **41** (1992), 65–72.
143. Foster, F. G., Kupfer, D. J., Coble, P. and McPartland, R. J. Rapid eye movement sleep density. An objective indicator in severe medical-depressive syndromes. *Arch. Gen. Psychiatry*, **33** (1976), 1119–1123.
144. Kupfer, D. J., Ulrich, R. F., Coble, P. A. *et al.* Electroencephalographic sleep of younger depressives. Comparison with normals. *Arch. Gen. Psychiatry*, **42** (1985), 806–810.
145. Holsboer-Trachsler, E. and Seifritz, E. Sleep in depression and sleep deprivation: a brief conceptual review. *World J. Biol. Psychiatry*, **1** (2000), 180–186.
146. Ford, D. E. and Kamerow, D. B. Epidemiologic study of sleep disturbances and psychiatric disorders. An opportunity for prevention? *JAMA* **262** (1989), 1479–1484.
147. Lustberg, L. and Reynolds, C. F. Depression and insomnia: questions of cause and effect. *Sleep Med. Rev.*, **4** (2000), 253–262.
148. Giles, D. E., Roffwarg, H. P., Rush, A. J. and Guzick, D. S. Age-adjusted threshold values for reduced REM latency in unipolar depression using ROC analysis. *Biol. Psychiatry*, **27** (1990), 841–853.
149. Ressler, K. J. and Nemeroff, C. B. Role of norepinephrine in the pathophysiology and treatment of mood disorders. *Biol. Psychiatry*, **46** (1999), 1219–1233.
150. Roy, A., Pickar, D., De Jong, J., Karoum, F. and Linnoila, M. Norepinephrine and its metabolites in cerebrospinal fluid, plasma, and urine. Relationship to hypothalamic–pituitary–adrenal axis function in depression. *Arch. Gen. Psychiatry*, **45** (1988), 849–857.

151. Siever, L. J. and Davis, K. L. Overview: toward a dysregulation hypothesis of depression. *Am. J. Psychiatry*, **142** (1985), 1017–1031.
152. Habib, K. E., Weld, K. P., Rice, K. C. *et al.* Oral administration of a corticotropin-releasing hormone receptor antagonist significantly attenuates behavioral, neuroendocrine, and autonomic responses to stress in primates. *Proc. Natl. Acad. Sci. U. S. A.*, **97** (2000), 6079–6084.
153. Holsboer, F. Corticotropin-releasing hormone modulators and depression. *Curr. Opin. Invest. Drugs*, **4** (2003), 46–50.
154. Krueger, J. M. and Majde, J. A. Humoral links between sleep and the immune system: research issues. *Ann. N. Y. Acad. Sci.*, **992** (2003), 9–20.
155. Steiger, A. Sleep and endocrinology, *J. Intern. Med.*, **254** (2003), 13–22.
156. Tsuchiyama, Y., Uchimura, N., Sakamoto, T., Maeda, H. and Kotorii, T. Effects of hCRH on sleep and body temperature rhythms. *Psychiatry Clin. Neurosci.*, **49** (1995), 299–304.
157. Benca, R. Mood disorders. In *Principles and Practices of Sleep Medicine*, ed. M. H. Kryger, T. Roth and W. C. Dement. (Philadelphia: W. B. Saunders Company, 2000), pp. 1140–1158.
158. Gillin, J. C., Buchsbaum, M., Wu, J., Clark, C. and Bunney, W., Jr. Sleep deprivation as a model experimental antidepressant treatment: findings from functional brain imaging. *Depress. Anxiety*, **14** (2001), 37–49.
159. Schreiber, W., Opper, C., Dickhaus, B. *et al.* Alterations of blood platelet MAO-B activity and LSD-binding in humans after sleep deprivation and recovery sleep. *J. Psychiatr. Res.*, **31** (1997), 323–331.
160. Muller, H. U., Riemann, D., Berger, M. and Muller, W. E. The influence of total sleep deprivation on urinary excretion of catecholamine metabolites in major depression. *Acta Psychiatr. Scand.*, **88** (1993), 16–20.
161. Lewis, D. A. and Morrison, J. H. Noradrenergic innervation of monkey prefrontal cortex: a dopamine-beta-hydroxylase immunohistochemical study. *J. Comp. Neurol.*, **282** (1989), 317–330.
162. Aston-Jones, G., Rajkowski, J., Lu, W. *et al.* Prominent projections from the orbital prefrontal cortex to the locus coeruleus in monkey. *Soc. Neurosci. Abstr.*, (2002), Program No. 86.9.
163. Rajkowski, J., Lu, W., Zhu, Y., Cohen, J. and Aston-Jones, G. Prominent projections from the anterior cingulate cortex to the locus coeruleus in Rhesus monkey. *Soc. Neurosci. Abstr.*, (2000), Program No. 838.15.
164. Kessler, R. C., Berglund, P. A., Dewit, D. J. *et al.* Distinguishing generalized anxiety disorder from major depression: prevalence and impairment from current pure and comorbid disorders in the US and Ontario. *Int. J. Methods Psychiatr. Res.*, **11** (2002), 99–111.
165. Gillin, J. C., Seifritz, E., Zoltoski, R. and Salin-Pascual, R. J. Basic science of sleep. In *Comprehensive Textbook of Psychiatry*, ed. B. J. Sadock. (Philadelphia: Lippincott Williams & Wilkins, 2000), pp. 199–208.
166. Saletu-Zyhlarz, G., Saletu, B., Anderer, P. *et al.* Nonorganic insomnia in generalized anxiety disorder. 1. Controlled studies on sleep, awakening and daytime vigilance utilizing polysomnography and EEG mapping. *Neuropsychobiology*, **36** (1997), 117–129.
167. Charney, D. and Drevets, W. The neurobiological basis of anxiety disorders. In *Neuropsychopharmacology: The Fifth Generation of Progress*, ed. K. L. Davis, D. Charney, J. T. Coyle and C. Neroff. (Philadelphia: Lippincott Williams & Wilkins, 2002), pp. 901–930.

168. Weissman, M. M. Panic and generalized anxiety: are they separate disorders? *J. Psychiatr. Res.*, **24** (1990), 157–162.

169. Aston-Jones, G., Ennis, M., Pieribone, V. A., Nickell, W. T. and Shipley, M. T. The brain nucleus locus coeruleus: restricted afferent control of a broad efferent network. *Science*, **234** (1986), 734–737.

170. Aston-Jones, G., Shipley, M. T., Chouvet, G. *et al.* Afferent regulation of locus coeruleus neurons: anatomy, physiology and pharmacology. *Prog. Brain Res.*, **88** (1991), 47–75.

171. Elam, M., Svensson, T. H. and Thorén, P. Locus coeruleus neurons and sympathetic nerves: activation by cutaneous sensory afferents. *Brain Res.*, **366** (1986), 254–261.

172. Elam, M., Thorén, P. and Svensson, T. H. Locus coeruleus neurons and sympathetic nerves: activation by visceral afferents. *Brain Res.*, **375** (1986), 117–125.

173. Ordway, G. A., Klimek, V. and Mann, J. J. Neurocircuitry of mood disorders. In *Neuropsychopharmacology: The Fifth Generation of Progress*, ed. K. L. Davis, D. Charney, J. T. Coyle and C. Neroff. (Philadelphia: Lippincott Williams & Wilkins, 2002), pp. 1051–1064.

174. Coplan, J. D., Andrews, M. W., Rosenblum, L. A. *et al.* Persistent elevations of cerebrospinal fluid concentrations of corticotropin-releasing factor in adult nonhuman primates exposed to early-life stressors: implications for the pathophysiology of mood and anxiety disorders. *Proc. Natl. Acad. Sci. U. S. A.*, **93** (1996), 1619–1623.

175. Barros, M. and Tomaz, C. Non-human primate models for investigating fear and anxiety. *Neurosci. Biobehav. Rev.*, **26** (2002), 187–201.

176. Davis, M. Anatomic and physiologic substrates of emotion in an animal model. *J. Clin. Neurophysiol.*, **15** (1998), 378–387.

177. Harris, J. C. Experimental animal modeling of depression and anxiety. *Psychiatr. Clin. North Am.*, **12** (1989), 815–836.

178. McKinney, W. T., Jr. and Bunney, W. E., Jr. Animal model of depression. I. Review of evidence: implications for research. *Arch. Gen. Psychiatry*, **21** (1969), 240–248.

179. Koob, G. F. and Heinrichs, S. C. A role for corticotropin releasing factor and urocortin in behavioral responses to stressors. *Brain Res.*, **848** (1999), 141–152.

180. Newport, D. J. and Nemeroff, C. B. Neurobiology of posttraumatic stress disorder. *Curr. Opin. Neurobiol.*, **10** (2000), 211–218.

181. Stein, D. J. Anxiety and stress disorders: course over the lifetime. In *Neuropsychopharmacology: The Fifth Generation of Progress*, ed. K. L. Davis, D. Charney, J. T. Coyle and C. Neroff. (Philadelphia: Lippincott Williams & Wilkins, 2002), pp. 859–866.

182. Hefez, A., Metz, L. and Lavie, P. Long-term effects of extreme situational stress on sleep and dreaming. Am. J. Psychiatry, **144** (1987), 344–347.

183. Lavie, P. Sleep disturbances in the wake of traumatic events. *New Engl. J. Med.*, **345** (2001), 1825–1832.

184. Geracioti, T. D., Jr., Baker, D. G., Ekhator, N. N. *et al.* CSF norepinephrine concentrations in posttraumatic stress disorder. *Am. J. Psychiatry*, **158** (2001), 1227–1230.

185. Mellman, T. A., Kumar, A., Kulick-Bell, R., Kumar, M. and Nolan, B. Nocturnal/daytime urine noradrenergic measures and sleep in combat-related PTSD. *Biol. Psychiatry*, **38** (1995), 174–179.

186. Yehuda, R., Southwick, S., Giller, E. L., Ma, X. and Mason, J. W. Urinary catecholamine excretion and severity of PTSD symptoms in Vietnam combat veterans. *J. Nerv. Ment. Dis.*, **180** (1992), 321–325.
187. Heim, C. and Nemeroff, C. B. Neurobiology of early life stress: clinical studies. *Semin. Clin. Neuropsychiatry*, **7** (2002), 147–159.
188. Roozendaal, B. Stress and memory: opposing effects of glucocorticoids on memory consolidation and memory retrieval. *Neurobiol. Learn. Mem.*, **78** (2002), 578–595.
189. Khan, A. U. Sleep REM latency in hyperkinetic boys. *Am. J. Psychiatry*, **139** (1982), 1358–1360.
190. O'Brien, L. M., Ivanenko, A., Crabtree, V. M. *et al.* Sleep disturbances in children with attention deficit hyperactivity disorder. *Pediatr. Res.*, **54** (2003), 237–243.
191. Picchietti, D. L., England, S. J., Walters, A. S., Willis, K. and Verrico, T. Periodic limb movement disorder and restless legs syndrome in children with attention-deficit hyperactivity disorder. *J. Child Neurol.*, **13** (1998), 588–594.
192. Picchietti, D. L., Underwood, D. J., Farris, W. A. *et al.* Further studies on periodic limb movement disorder and restless legs syndrome in children with attention-deficit hyperactivity disorder. *Mov. Disord.*, **14** (1999), 1000–1007.
193. Gruber, R., Sadeh, A. and Raviv, A. Instability of sleep patterns in children with attention-deficit/hyperactivity disorder. *J. Am. Acad. Child Adolesc. Psychiatry*, **39** (2000), 495–501.
194. Solanto, M. V. Clinical psychopharmacology of AD/HD: implications for animal models. *Neurosci. Biobehav. Rev.*, **24** (2000), 27–30.
195. Solanto, M. V. Dopamine dysfunction in AD/HD: integrating clinical and basic neuroscience research. *Behav. Brain Res.*, **130** (2002), 65–71.
196. Berridge, C. W. and Waterhouse B. D. The locus coeruleus-noradrenergic system: modulation of behavioral state and state-dependent cognitive processes. *Brain Res. Brain Res. Rev.*, **42** (2003), 33–84.
197. Hanna, G. L., Ornitz, E. M. and Hariharan, M. Urinary catecholamine excretion and behavioral differences in ADHD and normal boys. *J. Child Adolesc. Psychopharmacol.*, **6** (1996), 63–73.
198. Oades, R. D., Daniels, R. and Rascher, W. Plasma neuropeptide-Y levels, monoamine metabolism, electrolyte excretion and drinking behavior in children with attention-deficit hyperactivity disorder. *Psychiatry Res.*, **80** (1998), 177–186.
199. Pliszka, S. R., Maas, J. W., Javors, M. A., Rogeness, G. A. and Baker, J. Urinary catecholamines in attention-deficit hyperactivity disorder with and without comorbid anxiety. *J. Am. Acad. Child Adolesc. Psychiatry*, **33** (1994), 1165–1173.
200. Shekim, W. O., Bylund, D. B., Frankel, F. *et al.* Platelet alpha 2-adrenergic receptor binding to 3H-yohimbine and personality variations in normals. *Psychiatry Res.*, **32** (1990), 125–134.
201. Shekim, W. O., Bylund, D. B., Hodges, K. *et al.* Platelet alpha 2-adrenergic receptor binding and the effects of d-amphetamine in boys with attention deficit hyperactivity disorder. *Neuropsychobiology*, **29** (1994), 120–124.
202. Shekim, W. O., Sinclair, E., Glaser, R. *et al.* Norepinephrine and dopamine metabolites and educational variables in boys with attention deficit disorder and hyperactivity. *J. Child Neurol.*, **2** (1987), 50–56.

203. Spivak, B., Vered, Y., Yoran-Hegesh, R. *et al.* Circulatory levels of catecholamines, serotonin and lipids in attention deficit hyperactivity disorder. *Acta Psychiatr. Scand.*, **99** (1999), 300–304.
204. Trinh, J. V., Nehrenberg, D. L., Jacobsen, J. P., Caron, M. G. and Wetsel, W. C. Differential psychostimulant-induced activation of neural circuits in dopamine transporter knockout and wild type mice. *Neuroscience*, **118** (2003), 297–310.
205. Nofzinger, E. A. and Keshavan, M. Sleep disturbances associated with neuropsychiatric disease. In *Neuropsychopharmacology: The Fifth Generation of Progress*, ed. K. L. Davis, D. Charney, J. T. Coyle and C. Neroff. (Philadelphia: Lippincott Williams & Wilkins, 2002), pp. 1945–1959.
206. Czeisler, C. A., Dumont, M., Duffy, J. F. *et al.* Association of sleep–wake habits in older people with changes in output of circadian pacemaker. *Lancet*, **340** (1992), 933–936.
207. Dijk, D. J., Duffy, J. F., Riel, E., Shanahan, T. L. and Czeisler, C. A. Ageing and the circadian and homeostatic regulation of human sleep during forced desynchrony of rest, melatonin and temperature rhythms. *J. Physiol.*, **516** (1999), 611–627.
208. Iwanaga, K., Yamada, M., Wakabayashi, K., Ikuta, F. and Takahashi, H. A newly discovered age-related synaptic change in the human locus coeruleus: morphometric and ultrastructural studies. *Acta Neuropathol. (Berl.)*, **91** (1996), 337–342.
209. Manaye, K. F., McIntire, D. D., Mann, D. M. and German, D. C. Locus coeruleus cell loss in the aging human brain: a non-random process. *J. Comp. Neurol.*, **358** (1995), 79–87.
210. Tejani-Butt, S. M., Yang, J. and Zaffar, H. Norepinephrine transporter sites are decreased in the locus coeruleus in Alzheimer's disease. *Brain Res.*, **631** (1993), 147–150.
211. Coull, J. T. Pharmacological manipulations of the alpha 2-noradrenergic system. Effects on cognition. [Review]. *Drugs Aging*, **5** (1994), 116–126.
212. Arnsten, A. F., Cai, J. X. and Goldman-Rakic, P. S. The alpha-2 adrenergic agonist guanfacine improves memory in aged monkeys without sedative or hypotensive side effects: evidence for alpha-2 receptor subtypes. *J. Neurosci.*, **8** (1988), 4287–4298.
213. Arnsten, A. F. T. and Goldman-Rakic, P. S. Alpha2-adrenergic mechanisms in prefrontal cortex associated with cognitive decline in aged nonhuman primates. *Science*, **230** (1985), 1273–1276.
214. Mabry, T. R., Gold, P. E. and McCarty, R. Age-related changes in plasma catecholamine and glucose responses of F-344 rats to a single footshock as used in inhibitory avoidance training. *Neurobiol. Learn. Mem.*, **64** (1995), 146–155.
215. Mabry, T. R., Gold, P. E. and McCarty, R. Age-related changes in plasma catecholamine responses to acute swim stress. *Neurobiol. Learn. Mem.*, **63** (1995), 260–268.
216. Mabry, T. R., Gold, P. E. and McCarty, R. Age-related changes in plasma catecholamine responses to chronic intermittent stress. *Physiol. Behav.*, **58** (1995), 49–56.
217. Buysse, D. J. and Dorsey, C. M. Current and experimental therapeutics of insomnia. In *Neuropsychopharmacology: The Fifth Generation of Progress*, ed. K. L. Davis, D. Charney, J. T. Coyle and C. Neroff. (Philadelphia: Lippincott Williams & Wilkins, 2002), pp. 1931–1944.
218. Roehrs, T. A., Zorick, F. and Roth, T. Transient and short-term insomnias. In *Principles and Practices of Sleep Medicine*, ed. M. H. Kryger, T. Roth and W. C. Dement. (Philadelphia: W. B. Saunders Company, 2000), pp. 624–632.

219. Morin, C. M., Rodrigue, S. and Ivers, H. Role of stress, arousal, and coping skills in primary insomnia. *Psychosom. Med.*, **65** (2003), 259–267.

220. Rodenbeck, A. and Hajak, G. Neuroendocrine dysregulation in primary insomnia. *Rev. Neurol. (Paris)*, **157** (2001), S57-S61.

221. Vgontzas, A. N., Tsigos, C., Bixler, E. O. *et al.* Chronic insomnia and activity of the stress system: a preliminary study. *J. Psychosom. Res.*, **45** (1998), 1–31.

222. Irwin, M., Clark, C., Kennedy, B., Christian Gillin, J. and Ziegler, M. Nocturnal catecholamines and immune function in insomniacs, depressed patients, and control subjects. *Brain Behav. Immun.*, **17** (2003), 365–372.

223. Hauri, P. Primary insomnia. In *Principles and Practices of Sleep Medicine*, ed. M. H. Kryger, T. Roth and W. C. Dement. (Philadelphia: W. B. Saunders Company, 2000), pp. 633–639.

224. Chiang, C. and Aston-Jones, G. Response of locus coeruleus neurons to footshock stimulation is mediated by neurons in the ventrolateral medulla. *Neuroscience*, **53** (1993), 705–715.

225. Lechner, S. M., Curtis, A. L., Brons, R. and Valentino, R. J. Locus coeruleus activation by colon distention: role of corticotropin-releasing factor and excitatory amino acids. *Brain Res.*, **756** (1997), 114–124.

226. Page, M., Akaoka, H., Aston-Jones, G. and Valentino, R. Bladder distension activates locus coeruleus neurons by an excitatory amino acid mechanism. *Neuroscience*, **51** (1992), 555–563.

7

The locus coeruleus and regulation of behavioral flexibility and attention: clinical implications

Gary Aston-Jones, Michiyo Iba, Edwin Clayton, Janusz Rajkowski
University of Pennsylvania, Philadelphia, USA

Jonathan Cohen
Princeton University, Princeton, USA

Introduction

It has been proposed that the locus coeruleus (LC) regulates nonspecific arousal and thereby may participate in a wide range of functions. Our work indicates that, while the LC may indeed play an important role in arousal, it has more specific effects on behavior and may regulate cortical mechanisms involved in selective attention and task performance. In one study, we recorded impulse activity of LC neurons in monkeys performing a visual discrimination task. Phasic and tonic firing characteristics of LC neurons varied in close relation to task performance. Phasically, LC neurons were selectively activated by target cues and not by other task events, including behavioral responses. The target-elicited LC responses were limited to periods of good performance, when tonic firing rates were at an intermediate level ($\sim$ 1 to 2 spikes/s). Higher levels of tonic activity were associated with few or no phasic LC responses, and poor task performance. Direct manipulations of LC activity via local microinfusions yielded behavioral results consistent with the above recordings. A computational model was constructed to explore mechanisms that underlie these patterns of LC activity and their relationship to task performance.[1] This model revealed that electrotonic coupling among LC neurons can provide a mechanism for regulating the pattern of LC activity between two modes of functioning, which may in turn regulate task performance. In one mode (high electrotonic coupling, resulting in intermediate levels of tonic LC activity and robust phasic responses to task-defined target stimuli), LC responses facilitate the processing of target stimuli while responses to distractors are reduced. In the

Brain Norepinephrine: Neurobiology and Therapeutics, ed. Gregory A. Ordway, Michael A. Schwartz and Alan Frazer. Published by Cambridge University Press.

other mode (low coupling, resulting in high tonic LC activity and reduced phasic responding), task performance worsens and nonspecific responding to distractors (e.g., false alarms) is increased. This leads us to hypothesize that the LC may play a role in modulating attentional state, by favoring the processing of task-defined stimuli (selective attention) in one mode, versus a broader sampling of stimuli in the environment (scanning attention) in the other. Additional experiments using a forced-choice task allowed us to analyze in detail LC phasic responses with respect to behavioral responses. This revealed that LC responses are more tightly linked temporally to the behavioral response than to the sensory stimulus. This, plus the lack of LC response on omission error trial or during behavioral responding out of the task context, leads us to hypothesize that LC responses are driven by decision processes rather than by sensory or motor activities per se. We propose that LC response activity then serves to facilitate behavioral responding once a decision is made concerning the approriate response. Other of our recent studies reveal that inputs from the orbital and anterior cingulate cortices innervate the monkey LC. Such inputs are likely sources for the selective LC phasic responses, and for inputs that regulate electrotonic coupling among LC cells. Overall, these results indicate that, during waking, the LC may participate in a system that alters attentional mode to facilitate selective attention vs. scanning attention. This in turn suggests a more specific role for the LC in regulating cognitive function than earlier hypotheses limiting its role to simple arousal.

The ability to respond selectively to certain aspects of the environment, and filter out others that are irrelevant or disruptive to the current behavioral plan, is critical for goal-directed behavior. At the same time, behavior must be flexible and adaptive, so as to quickly adjust to imperative or unexpected events. Thus, successful behavior in both animals and humans requires the capacity for both selective responding in a stable environment, and rapid adaptive responding in a changing environment. This capacity represents a fundamental regulation of attention between two states: selective/focused attention vs. scanning/labile attentiveness. This interplay between focused vs. flexible responding is a fundamental property of advanced nervous systems. However, these interactions are poorly understood. Knowledge of the mechanisms that regulate this interplay would not only advance our understanding of normal behavior, but also of disruptions in behavior associated with a variety of psychopathological disorders, such as schizophrenia, attention-deficit disorder, and depression, in which overly focused or overly labile behaviors are commonly observed.

Our recent work on the locus coeruleus-norepinephrine (LC-NE) system has led us to hypothesize that this system plays a central role in regulating this balance between focused vs. flexible responding, or selective vs. scanning attention. The present chapter reviews some of the most salient previous work on the LC system

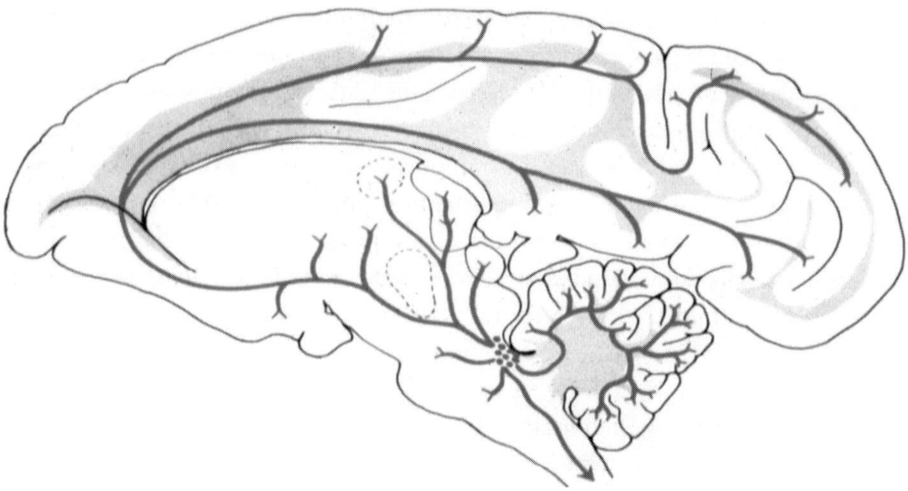

Figure 7.1 Sagittal diagram of a monkey brain illustrating broad projection network from LC neurons. A similar projection system exists in human.

relevant to understanding its role in cognitive activity and attention. The chapter then describes recent neurophysiology in behaving monkeys and modeling work aimed at understanding the mechanisms by which this neuromodulatory brain system operates and regulates behavior.

Background

Global efferent projections

Locus coeruleus-norepinephrine neurons give rise to a global network of efferent projections.[2–5] Notably, the LC projects throughout the cerebral cortex, hippocampus, thalamus, midbrain, brain stem, cerebellum and spinal cord (Figure 7.1).[6,7] It is also noteworthy, however, that the LC exhibits substantial regional and laminar specificity in its efferent projections.[4] In particular, brain areas that are associated with attentional processing (e.g., parietal cortex, pulvinar nucleus, and superior colliculus) receive a particularly dense LC-NE innervation.[8] Recent studies also reveal that LC terminals make conventional synapse-like appositions with postsynaptic specializations on target neurons rather than contacts onto blood vessels.[9–12] Although this indicates that NE is released from LC terminals at discrete sites of synaptic contact, it does not rule out the possibility of a more diffuse, paracrine-like mode of neurotransmission in some areas as well.

NE modulates activity of neurons in LC target areas

Early studies found that iontophoretic NE inhibited basal activity of cerebellar and hippocampal neurons.[13,14] However, subsequent experiments by Foote, Segal, and colleagues found that NE decreased baseline or spontaneous impulse activity to a greater extent than activity evoked by afferent or sensory stimulation.[15,16] Furthermore, in many cases NE was found to augment evoked responses (either excitatory or inhibitory) while decreasing spontaneous activity of the same neuron.[17–19] Such relative enhancement of responses to strong inputs relative to low-level or basal activity has been found in several LC target areas including cerebral cortex, hippocampus, midbrain, thalamus, and spinal cord.[7,20,21] These findings are consistent with neural modeling work hypothesizing that NE acts to enhance signal-to-noise ratios in target systems.[22] As described below, this modulatory property of NE may be a key mechanism for its role in the attentional state.

Tonic LC activity varies with behavioral state

The role of the LC-NE system in regulation of sleep and waking is presented in more detail in Chapter 6 of this volume, and these properties are only briefly summarized here. Spontaneous activity of identified NE neurons in the rat LC covaried consistently with stages of the sleep–waking cycle, firing most rapidly during waking, more slowly during slow-wave sleep, and becoming virtually silent during paradoxical sleep.[23] These results are similar to findings from other labs for cat LC neurons,[24,25] and we have observed similar discharge properties in monkeys[26–28] (see Chapter 6 of this volume). We also found that LC activity is decreased during certain behaviors in aroused waking. When rats and monkeys either groomed or drank a glucose solution, LC impulse activity is decreased compared to other epochs of similar electroencephalogram (EEG) arousal.[23,29,30] These findings indicate that LC activity is reduced not only for periods of low arousal (drowsiness or sleep), but also moderately so during certain behaviors (grooming and consumption) when animals are in active waking but engaged in automatic or very stable behavior and inattentive to most extrapersonal environmental stimuli.

LC neurons exhibit polymodal sensory responsiveness

In addition to the above fluctuations in LC tonic activity, LC neurons in unanesthetized rats and monkeys were phasically responsive to conspicuous environmental stimuli in many modalities.[25,26,31] Notably, stimuli that elicited large LC responses in either rats or monkeys also typically disrupted ongoing behavior and evoked a behavioral orienting response. The same stimuli did not disrupt behavior if they elicited small LC responses. Thus, there was a strong correspondence in rats and monkeys between sensory-evoked LC impulse activity and behavioral disruption and reorientation.[26,29,31]

Other studies have revealed strong phasic and tonic activation of LC neurons by stressors. Stimuli such as sciatic nerve activation or other painful events strongly activate LC cells.[32,33] In the awake monkey, other stressors such as air puff,[29] or a variety of environmental[34,35] or physiological stressors[36–38] also activate LC neurons. In addition, LC neurons are activated by the stress hormone, corticotropin-releasing factor (CRF), which mediates the response of LC cells to certain physiological stressors such as hypotension.[39,40]

Recent results recording LC neurons in behaving monkeys

Although these previous findings for LC neurons were consistent with a role in arousal and attention, they did not specify what such a role might be or the mechanisms by which the LC might contribute. Therefore, we recorded LC neuronal activity in monkeys during performance of a visual discrimination task that demanded focused attention for optimal performance. The results of our studies are described below for changes in tonic activity and phasic responses, both of which vary in close relation to task performance. We will further describe modeling studies that simulate the patterns of LC activity during task performance, and that also indicate that altered electrotonic coupling among LC neurons may play a pivotal role in the function of this system. Finally, we will consider this new perspective for implications in clinical disorders, with a focus on attention-deficit/hyperactivity disorder, stress disorders such as posttraumatic stress disorder, and in learning and memory.

Target detection task

This behavioral task was described in our other publications,[30,41,42] and it is only briefly summarized here. During training and recording sessions, animals assumed a natural squatting position in a modified primate chair, facing a color video monitor.

Monkeys were trained to perform a visual target detection task. The animal initiated each trial by depressing a lever, after which a small rectangular spot (fix spot) appeared in the center of the monitor. The monkey was required to stably foveate this spot, after which the fix spot was extinguished and a conditioned stimulus (CS) was presented at the fix spot location. Conditioned stimuli were either a vertical or horizontal bar: one orientation was used as the target (CS+) stimulus and the other as the distractor (nontarget, or CS–) stimulus. The animal's task was to release the lever immediately following a target stimulus to receive a juice reward, but to continue holding the lever down following distractor stimuli. Incorrect releases to distractor stimuli resulted in a 3-second time-out. Target stimuli occurred on 10% to 20% of trials and were semirandomly dispersed among distractor stimuli. Intertrial intervals averaged 1.6 seconds. Performance was measured in terms of hits (correct releases to target stimuli), misses (incorrect nonreleases to target stimuli), rejections (correct nonreleases to distractor stimuli), and false alarms (incorrect releases to distractor stimuli).

Phasic activation of LC neurons by meaningful stimuli

As in the rat, the monkey LC is composed almost entirely of NE-containing neurons.[3] Norepinephrine-locus coeruleus neurons were identified by electrophysiologic characteristics, as previously described.[42]

Responses of LC neurons during this task were surprisingly selective. As illustrated for one example in Figure 7.2, nearly all of the >300 LC neurons recorded to date were phasically activated preferentially by target stimuli but only weakly, or not at all, by distractor stimuli; lever release outside of the task elicited no response.[42] In addition, juice delivery following target stimuli and correct behavioral responses evoked no LC response. Some LC neurons were weakly inhibited by distractor stimuli.

Contrary to some concepts of the LC system as a nonspecific, slowly acting system, the latencies of LC responses to targets were surprisingly short (~100 ms onset), and preceded behavioral responses by ~200 ms. Moreover, the latencies of response for LC neurons and lever releases were significantly correlated over trials, so that shorter LC responses were associated with shorter behavioral responses to the same cues.[42] These findings indicate that LC target responses could facilitate behavioral responses to target stimuli.

Locus coeruleus responses to target stimuli did not depend on which stimulus was chosen to be the target within a particular session (vertical or horizontal rectangle). Recordings during reversal training confirmed that LC responses to targets were independent of sensory attributes. Reversal of cue meaning caused LC cells to reverse responsiveness, and become selectively activated by the new target stimulus and lose responsiveness to the old target.[41] Thus, phasic LC activation in this task is specifically related to the meaning of the stimuli, not to their physical attributes.

In these studies, the target stimulus was infrequently presented (20% of trials). To test whether the selective target response reflected the relative novelty of the target, an infrequent distractor stimulus (square) was included in additional experiments. Of the three distinct stimuli presented in these experiments (20% vertical targets, 20% square distractors, and 60% horizontal distractors), only target stimuli elicited phasic excitatory responses in LC neurons.[43] This indicates that frequency was not the determining factor for LC activation. Instead, in this task LC cells are specifically phasically activated by attended stimuli that signal the availability of reward. Although there is a close correspondence between such LC responses and reward prediction,[44] LC neurons are also phasically activated by stimuli that elicit attentional demands but are not associated with reward, e.g., novel or highly salient unconditioned stimuli.[29] Therefore, we conclude that these responses reflect attention-related functions of the LC system (as described in more detail below).

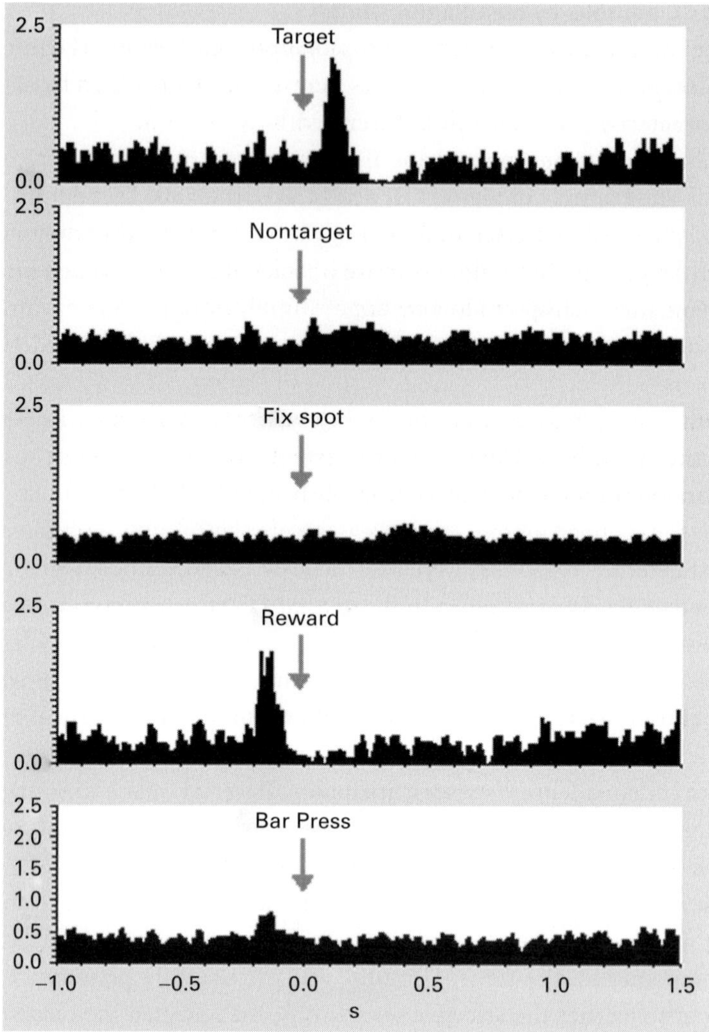

Figure 7.2 Peri-event time histograms (PETHs) for a typical individual LC neuron in response to various events during performance of the discrimination task. PETHs are each accumulated for 100 sweeps of activity in this neuron synchronized with different task events, as indicated. Note the selective activation by target stimuli. The activation seen before reward presentation or bar release is due to activation by target cues. The ordinate represents spike counts accumulated per 10 ms bin (from Aston-Jones *et al.* [1999]).[162]

Fluctuations in tonic LC activity during task performance

In addition to these phasic responses, we also noted that levels of tonic LC activity varied during alert task performance. In some of our recordings, LC neurons changed levels of tonic discharge several times, between intermediate and elevated rates (e.g., Figure 7.3). These different levels of tonic LC discharge were closely associated with differences in performance on the task. As shown in Figure 7.3, periods of elevated tonic LC activity were consistently accompanied by more frequent false alarm errors but no changes in misses or hits.[1,45] Analyses using signal detection measures revealed that during periods of elevated LC activity the animal's ability to discriminate targets from distractors (δ factor) and its criterion for responding (ß factor) both decreased.[42,46] Thus, it appears that during elevated LC activity the animal is less focused on task stimuli (making it more difficult to discriminate target from nontarget stimuli), displaying increased distractibility, and a greater tendency to respond to nontarget stimuli (lower response criterion).

This interpretation is supported by additional analyses of LC activity and attentional focus. In brief, less frequent foveation of the fix spot (required at the onset of each trial), and increased "scanning eye movements," were associated with epochs of elevated LC activity. Conversely, consistent foveation of the fix spot occurred during periods of intermediate tonic LC discharge.[30,46,47] This inverse relationship between visual fixation and LC activity was highly significant. In contrast, there was no consistent relationship between LC activity and simple eye position or direction of eye movement. These results suggest that focused attention on this task is highest during periods of intermediate LC activity and lower with elevated LC discharge.

While the above results are intriguing, they do not establish whether these alterations in LC activity are causative of, simply correlated with, or result from the changes in attention. We have gathered preliminary evidence using microinfusions into the monkey LC. This information supports the view that the LC plays a causal role in these behaviors. We used the α_2-adrenoceptor agonist clonidine to decrease tonic LC activity, and the muscarinic cholinergic agonist pilocarpine to stimulate tonic LC discharge. In brief, we have found that microinfusion of clonidine into the LC of a monkey exhibiting hyperactive behavior and poor task performance significantly decreased tonic LC activity and improved performance by decreasing false alarm and omission errors. In contrast, during error-free performance in other monkeys, activation of LC neurons by local microinjection of the muscarinic cholinergic agonist pilocarpine interrupted task behavior.[48]

Relationship of tonic to phasic LC activity

Periods of elevated tonic LC activity were also consistently associated with decreased phasic responsiveness of LC neurons to target stimuli in all 27 LC cells examined in three monkeys (Figure 7.3). In other words, the phasic activation of LC neurons typically seen for target stimuli (described above) occurred selectively during epochs of intermediate tonic LC discharge and excellent behavioral performance. Conversely, elevated tonic LC discharge corresponds to both decreased attentional performance and decreased phasic activation of LC neurons by target stimuli. Moreover, phasic LC responses to target stimuli were also suppressed in the few instances when task

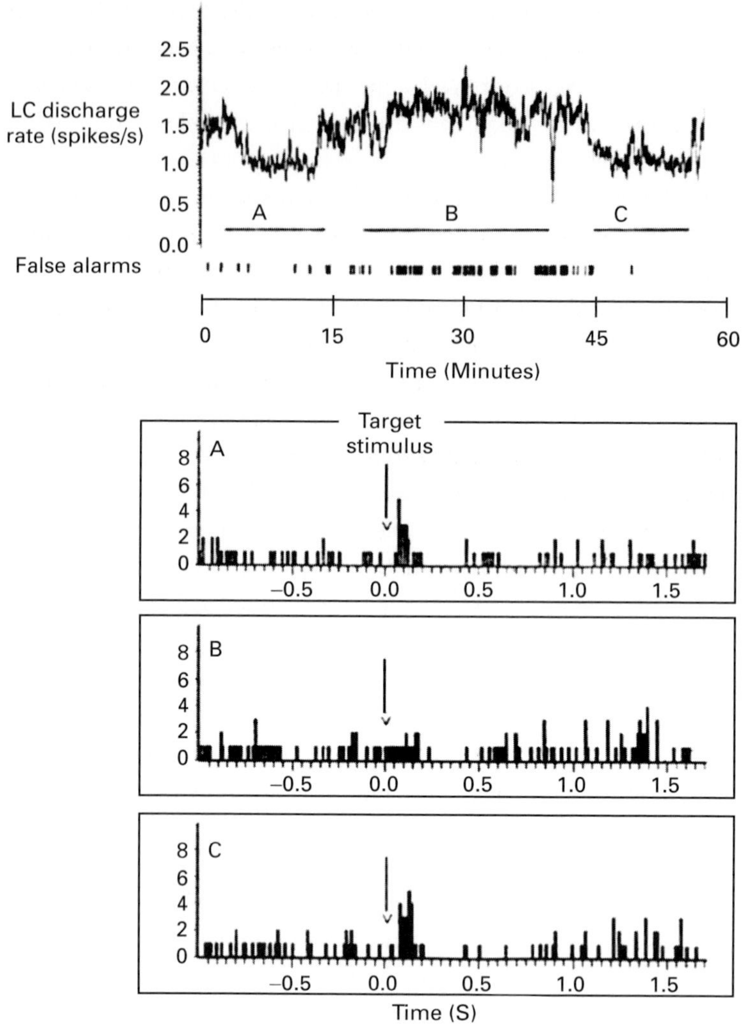

Figure 7.3 Upper panel: tonic impulse activity of a typical monkey LC neuron varies with the level of performance of the target detection task. Note that higher tonic LC activity is associated with worse performance (more false alarm errors, epoch B) than during lower levels of LC tonic activity (epochs A and C). Lower panel: PETHs of activity of this same LC neuron in response to target stimuli that elicited hit responses during epochs marked A, B or C in the upper panel. Note phasic responses to targets during periods of good performance and moderate LC tonic activity (epochs A and C) which are absent during poor performance associated with higher tonic LC activity (epoch B).

performance continued during drowsiness and very low LC tonic activity. These findings indicate that phasic-evoked responses are closely related to intermediate tonic discharge levels of monkey LC neurons and focused attentional performance.

Task difficulty

The above studies led us to hypothesize that LC phasic activation was a response to stimuli that had been identified as a target. This predicts that situations requiring increased processing time to identify a target would also produce longer-latency LC responses. In fact, this was observed during task reversal, where LC latencies and response times (RTs) increased in parallel. To further test this idea, we examined LC responses as a function of discrimination difficulty in the target detection task. We had predicted that more difficult discriminations would require more processing time, and should be associated with longer LC latencies and RTs. Discrimination was made more difficult by making target and nontarget stimuli more similar, i.e., more "square-like." As we predicted, these experiments revealed that during difficult discrimination conditions the latencies of LC phasic responses increased in proportion to the behavioral RTs.[43] Using population and response-locked vs. stimulus-locked histogram analysis, we also found that LC responses were more closely aligned temporally with behavioral responses than with sensory stimuli. This finding was also true for LC responses during a forced-choice task (described below, p. 211), and has led us to hypothesize that LC responses occur immediately postdecision and serve to facilitate impending behavioral responses (see below).

Interpretation of results – LC neurophysiology experiments

Overall, these results indicate that there are three modes of LC activity corresponding to different levels of performance on this task that require focused attention. In the *hypoactive mode* (1), LC neurons exhibit very little activity either tonically or in response to task events. This mode is associated with drowsiness and inattention to the task or other external stimuli. In the *phasic mode* (2), LC cells exhibit phasic activation selectively for stimuli that are discriminated as targets but only a moderate level of tonic discharge. This mode of LC activity is consistently associated with excellent performance on the task with few errors and high visual fixation. In the *tonic mode* (3), LC cells fail to phasically respond to any task stimuli, but exhibit higher levels of ongoing tonic activity. This mode corresponds to poor performance on this task, with many false alarm errors, less consistent fixation of fix spot stimuli, and more scanning eye movements.

The results of these studies, while corroborating certain aspects of the relationship we proposed between LC activity and vigilance, paint a more complex picture than our previous hypotheses suggested.[20,31,49] At very low levels of LC activity, the animal failed to engage sufficiently to perform the task, apparently because of low

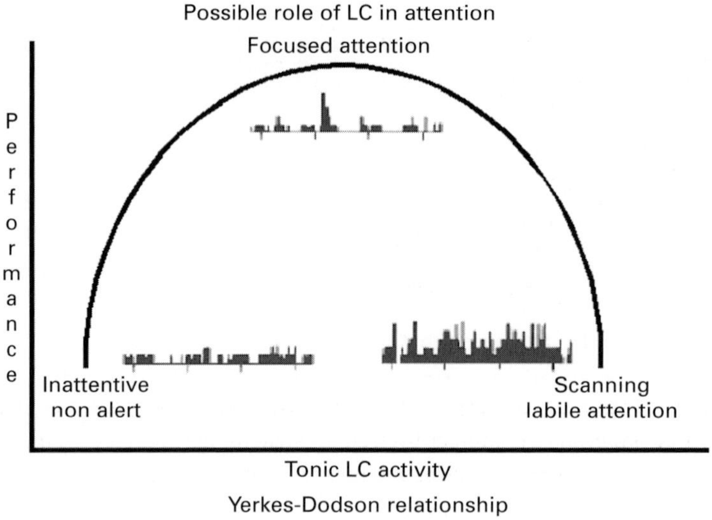

Figure 7.4 Inverted-U relationship between LC tonic activity and performance on the discrimination task. Performance is poor at very low levels of LC tonic discharge because animals are drowsy and nonalert.[23,26] Performance on this task that requires focused attention is optimal with moderate LC tonic activity and phasic LC responses to target stimuli (phasic LC mode). Performance is poor at high levels of tonic LC activity (tonic mode) because of scanning, labile attentiveness, which is incompatible with this task. Note that the tonic LC mode would be optimal for tasks (contexts) that require high behavioral flexibility rather than focused attention. In this view, the LC regulates the balance between focused and flexible behavior (from Aston-Jones *et al.* [1999]).[162]

arousal and drowsiness. At an intermediate level of tonic LC activity with phasic responses to targets, arousal and performance increased. However, at the highest tonic LC activities, performance *decreased*, a finding that is inconsistent with earlier predictions. These new results for an inverted-U relationship between the level of tonic LC discharge and task performance resemble the classic Yerkes-Dodson relationship (Figure 7.4) that has frequently been observed between arousal and performance.[50,51]

Computational modeling: simulation of LC activity and task performance

The data described above pose several questions about the mechanisms underlying LC activity and its relationship to behavioral performance. Two in particular stand out. (1) What mechanisms underlie the difference between phasic and tonic modes of LC firing during alert task performance? (2) What mechanisms explain the relationship between each of these modes and the pattern of behavioral performance with which it was associated? Contained within these questions are more detailed

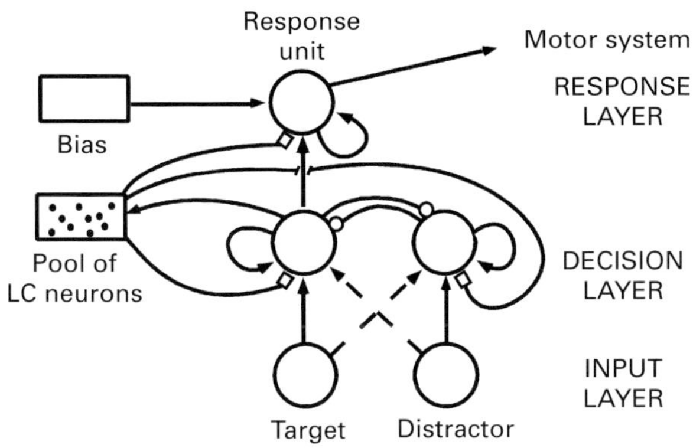

Figure 7.5 Architecture of the model of task performance. Arrows represent excitatory links, small circles represent inhibition, and the squares from LC projections represent the modulatory influence of NE on target units. There is a moderate positive bias on the response unit which captures the observation that monkeys in this task make many false alarms but very few misses (from Aston-Jones *et al.* [1994],[42] Usher *et al.* [1999]).[1]

ones. For example, how is it that a reduction of false alarms during periods of good performance was not accompanied by an increase in response latency, as might have been predicted by a simple increase in response threshold? To address these questions, we developed a simulation model of LC function and its influence on performance in the target detection task. The model is described in detail by Usher *et al.* and outlined briefly below.[1] This analysis deals only with LC activity during alert performance, i.e., the tonic and phasic modes of LC activity; the hypoactive mode is presumed to involve other mechanisms related to internal vegetative function and homeostatic needs (e.g., sleep drive). As illustrated in Figure 7.5, the model is a hybrid with two components: a simple stimulus discrimination network that simulated performance in the behavioral task, and a detailed model of LC neuronal activity. The former was the simplest network model capable of performing the behavioral task and was used to examine the influence of LC activity on performance. The model of LC was significantly more elaborate and biologically realistic, permitting examination of the neural mechanisms that might be responsible for its different modes of functioning.

Stimulus discrimination network

The task model consisted of a small number of units, each of which represented cell assemblies supporting stimulus or response representations necessary for performing the task. Thus, there were two input units (for the target and distractor stimuli), two decision units, and one response unit (Figure 7.5). Only the target decision unit was connected to the response unit, based on the assumption that the animal was overtrained to respond to the target but not the distractor. Connections between units in different processing layers were excitatory (information flow), while those within a layer were inhibitory (competition) and activity of units was subject to small random variations (noise). Finally, the physiological effect of NE was simulated as an increase in the gain parameter of the activation function of processing units. The activation function is a sigmoidal curve that describes the output of each unit for any input; a gain increase is produced by increasing the steepness of this curve. Elsewhere,[22] we have argued that this is consistent with a large body of data concerning the ability of NE to enhance the responsiveness of target neurons to other inputs, sometimes referred to as increasing the signal-to-noise ratio of target cell activity.[3,7,16,19]

LC model

This consisted of a population of 250 spiking neurons, each of which was a leaky integrate-and-fire cell that exhibited temporal dynamics similar to those obtained in detailed compartmental models.[52,53] Each LC cell received input from the target decision unit, as well as noise, which was responsible for a weak spontaneous firing rate of about 1 Hz (as observed in vivo).[26,29,54] Locus coeruleus cells in this model interacted with each other in two ways. First, lateral inhibition simulated the effect of local NE release.[55,56] Second, we included electrotonic coupling among LC cells, which simulated such coupling found empirically.[57–61] As described below and illustrated in Figure 7.6, modulation of electrotonic coupling among LC neurons alone in our model was able to capture the full set of neurophysiological and behavioral findings observed in the monkey experiments described above.

Electrotonic coupling

Several studies indicate that LC neurons may be coupled by electrotonic junctions. In neonatal rats, Christie and colleagues[57,58] demonstrated robust electrotonic and dye coupling among LC neurons. However, in these studies such coupling appeared to be transient and was much less frequently observed with these techniques in slices from older rats. While it is possible that coupling is expressed only transiently in the LC during development, it is also possible that coupling persists but becomes weaker and more distributed over distant dendrites, and thus more difficult to demonstrate in adulthood by these conventional methods. Evidence for weak coupling in adult LC has been reported by Williams and colleagues,[59,60] and we have accumulated similar evidence in slice studies.[62] These results are consistent with the possibility that, in adulthood, coupling among LC neurons occurs on distal dendrites, so that virtually all LC neurons may be weakly coupled to all other LC cells in a syncitium-like arrangement. Our computational model of the LC predicts that

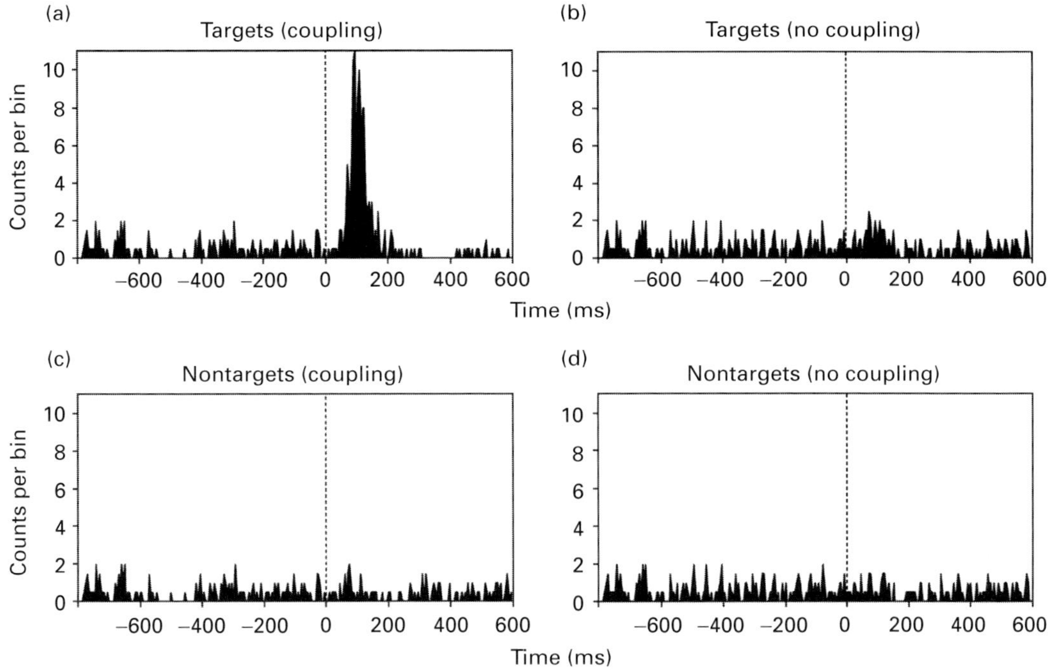

Figure 7.6 Poststimulus time histograms (PSTHs) for the simulated data. (a, b) Response to targets. (c, d) Response to distractors. (a, c) Coupling among LC neurons. (b, d) No coupling among LC neurons. PSTHs are normalized for 100 trials, as for the empirical data (see Figure 7.3). Note that coupling reduces tonic (baseline) LC activity but increases phasic (transient) response to target stimuli, capturing the phasic mode of LC neurons in our recordings (Figure 7.2 and Figure 7.3 panels A and C). Taken from Usher *et al.* (1999).[1]

weak, ubiquitous coupling among LC neurons is sufficient to substantially regulate LC tonic and phasic activity.[1] Our model also predicts that coupling among LC neurons is dynamic and can be altered by neural inputs, as has been reported for electrotonic coupling in other central nervous system structures.[63,64]

Simulation results

The model explained the phasic and tonic modes of LC firing in terms of differences in the degree of electrotonic coupling within the LC. High coupling caused stronger, more synchronized (phasic) activation of the LC in response to the target input due to the distribution of the voltage in spiking cells to the remainder of the population across electrotonic links (Figure 7.6). At the same time, coupling reduced spontaneous firing (which results from noise randomly injected into each unit) by averaging uncorrelated noise among the population (shunting the depolarizing effect of noise to other coupled neurons), resulting in an overall reduction in tonic activity. Thus, our model revealed that changes in phasic and tonic firing properties may be inversely related to one another, and governed by the same mechanism: electrotonic coupling among LC units. The model also provides an explanation of the behavioral effects of LC. As noted above, we simulated the physiological effect of the LC on targets by increasing the gain of the activation function of processing units in the task model. Thus, reduced coupling that produces high tonic LC activity causes higher responsivity of units in the task model, and thus a greater number of false alarms (triggered by intrinsic noise). Conversely, increased coupling leads to a reduction of tonic activity and an increase in phasic activity. The reduction in tonic activity produces a reduction of the overall responsivity of units in the task model, and therefore a concomitant reduction in false alarms. On target stimulus trials during high coupling, however, this reduced responsivity is compensated for by an increased phasic LC response to the target, which in turn produces a temporary increase in responsivity and therefore a quick response.

Thus, an increase in electrotonic coupling is able to produce an increase in the accuracy of performance (reduced false alarms), without a cost in the response time to targets. These results are described in detail in our publication.[1]

Interpretation of results: modeling experiments

Our model offers several new insights into LC function and its influence on behavior. First, it predicts that the phasic and tonic properties of LC firing may be inversely related to one another by means of a single parameter, electrotonic coupling among LC neurons. The model also predicts that an increase in tonic LC activity, and the concomitant reduction of phasic activity, will impair performance under certain circumstances. This is consistent with empirical observations in the vigilance task described above. While this mode of LC function impaired performance in the vigilance task, we believe that, in fact, it may serve an important adaptive function under certain circumstances. By increasing responsivity to all stimuli, it may provide a mechanism for insuring behavioral flexibility. This suggests a general theory concerning the psychological function of the LC-NE system, as elaborated below.

More recent work using a more biophysically detailed model of the LC has extended these results, showing that altering the excitatory drive to LC neurons may be sufficient to cause a shift between phasic and tonic modes of activity without a change in the amount of coupling.[65] This possibility is consistent with other recent modeling and physiological work showing the importance of excitatory drive on patterns of activity in LC neurons.[66] These results also show that mechanisms, in addition to alterations in coupling, may underlie transitions between LC modes.

Decision-related activation of monkey LC neurons

As described above, the phasic activation of LC neurons during the target detection task was not dependent upon physical stimulus properties. This activation readily switched to a previously nonrewarded nontarget stimulus that was made after task reversal.[41] Given that the same stimuli elicited different LC responses, depending upon whether they demanded a behavioral response, these findings indicate that LC activation by targets is not simply a response to the sensory features of the stimuli but instead reflects stimulus meaning and its consequence for behavior. However, it remained unclear if the phasic LC response was more closely related to stimulus processing or to decision formation and the initiation of the corresponding behavioral response.

We extended our analysis of LC activity in monkeys and addressed this issue by recording monkey LC activity during performance in a forced-choice discrimination task.[67] This task required the subject to discriminate between different arrays of visual cues for a juice reward. This more complex task required a behavioral response on 80% to 100% of trials and produced greater variability in reaction times than previously found in our target detection studies. In order to more fully explore the role of the LC in decision making processes, these characteristics allowed us to conduct a detailed examination of the timing of LC phasic activity in relation to stimulus presentation and behavioral response.

Forced-choice task

The monkey was trained to continuously depress two pedals and foveate a centrally located fixation spot on a computer monitor 60 cm away for at least 500 to 750 ms. After successful foveation, the animal was presented with an array of stimuli consisting of a central cue (bracket, > or <) that was flanked by similar or opposite brackets. The monkey was required to make a behavioral response by releasing the lever that corresponded to the directional orientation of the central cue (e.g., “<<<<<” or “>><>>” both required release of the left lever). Correct responses were reinforced with a reward of a drop of juice. Incorrect responses ended the trial and no reward was available until the following trial.

Phasic activation of the LC preceded the behavioral response

As seen in Figure 7.7, individual peri-event time histograms (PETHs) of LC activity synchronized with stimulus onset (stimulus-locked PETHs) indicated a stronger LC response on trials in which a correct behavioral response was made versus when the animal responded incorrectly. However, examination of LC activity time-locked to the behavioral response (response-locked PETHs) revealed that the magnitudes

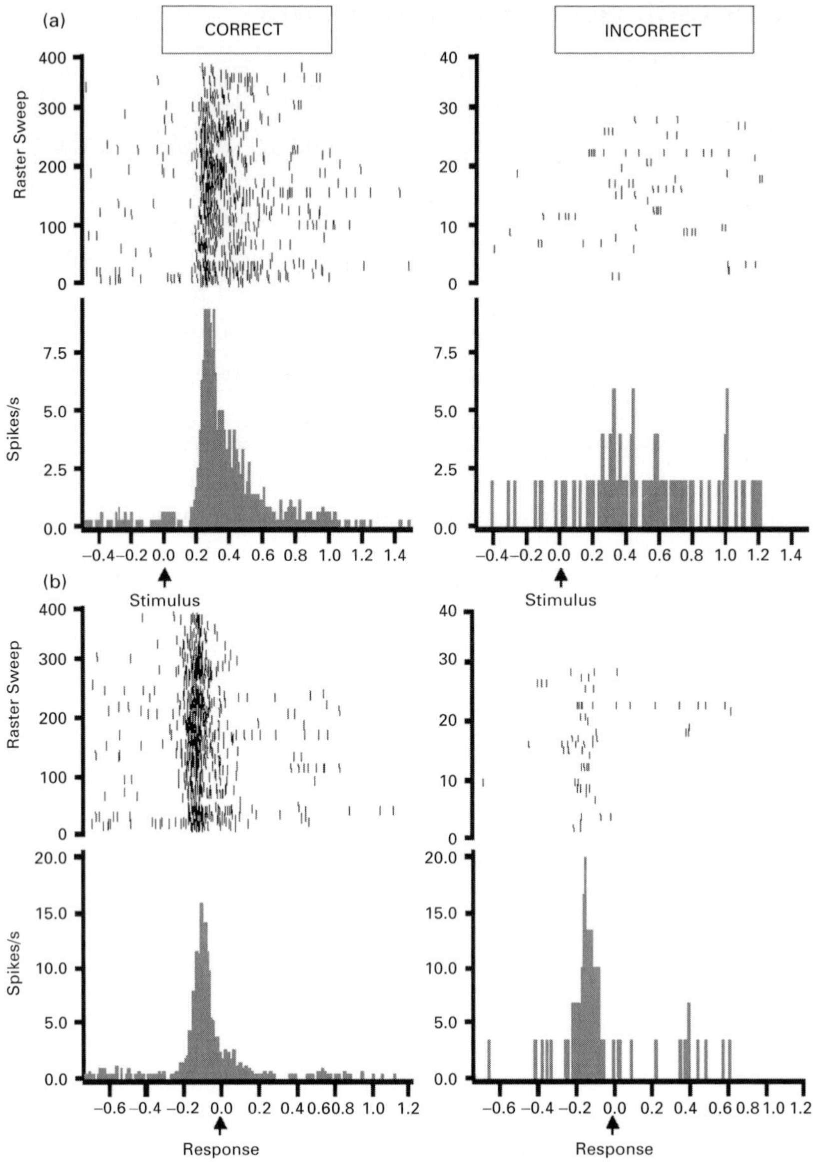

Figure 7.7 Peri-event time histogram and raster plots of a typical single LC neuron illustrating response to task cues and LC activity immediately preceding the behavioral response for correct and incorrect trials in the forced-choice task. Stimulus-locked correct and incorrect PETHs and rasters are in Figure 7.7a. Response-locked PETHs and rasters for correct and incorrect trials are located beneath the corresponding stimulus-locked analysis in Figure 7.7b. The same LC cell was used in each analysis. Note that LC activity was much more strongly aligned with the behavioral response than with stimulus presentation. This result is most evident when comparing stimulus-locked vs. response-locked rasters and PETHs for incorrect trials. Taken from Clayton *et al.* [2004].[67]

of phasic LC responses were similar for both correct and incorrect trials (Figure 7.7). Population PETHs of correct vs. incorrect responses (Figure 7.8) further demonstrated that the LC response was more tightly related to lever release than to presentation of the stimulus array. These response-locked PETHs revealed that LC responses to forced-choice stimuli consistently preceded behavioral responses by about 0.23 seconds. In addition, it is important to note that lever releases that occurred when the animal was not engaged in task performance (i.e., when no stimuli were present and the animal was not foveating a fix spot), and trials in which the animal viewed the stimulus but failed to make a behavioral response (omission errors), were not associated with phasic LC activation.

These recent findings demonstrated that the LC response is more closely linked with behavioral responding than with the sensory stimuli. Furthermore, no LC response was found on omission trials even though the monkey looked at the stimulus. The lack of an LC response on omission trials was not due to decreased foveation of the stimulus nor simply to fatigue, as omissions did not occur preferentially toward the end of recording sessions. We hypothesize that omission trials reflect attentional lapses, so that despite foveation of the stimulus, the animal was not sufficiently engaged in the task to produce a behavioral response. Furthermore, there were no LC responses associated with errant bar releases between trials, indicating that LC responses are not strictly motor or premotor in nature. Together, these findings indicate that the LC phasic response is driven by the outcome of the decision process rather than by the appearance of the stimulus or by the motor act itself. According to this view, variability of the LC response in relation to stimulus onset is attributable to variability in the duration of the decision process, consistent with existing models of the dynamics of decision processes in two-alternative forced-choice tasks.[68–70] We hypothesize that this decision-related activation of LC neurons serves to facilitate behavioral responses once the decision for responding has been made.

Influence of prefrontal cortex on activity of LC neurons

The above results raise two important questions concerning inputs that regulate LC neuronal activity: (1) what input mediates the selective response to target stimuli, and (2) what input regulates coupling among LC cells? Responses to target stimuli appear to reflect highly pre processed information impinging on the LC. Associational cortical areas process such information, and would be logical possible sources of the selective response to targets in the LC. In particular, the orbital frontal cortex (OC) has been found to represent the relative value of, or preference for, exogenous stimuli.[71–73] Thus, the OC appears to process information about stimuli and identify those stimuli of most value at the time. This is the type of information that

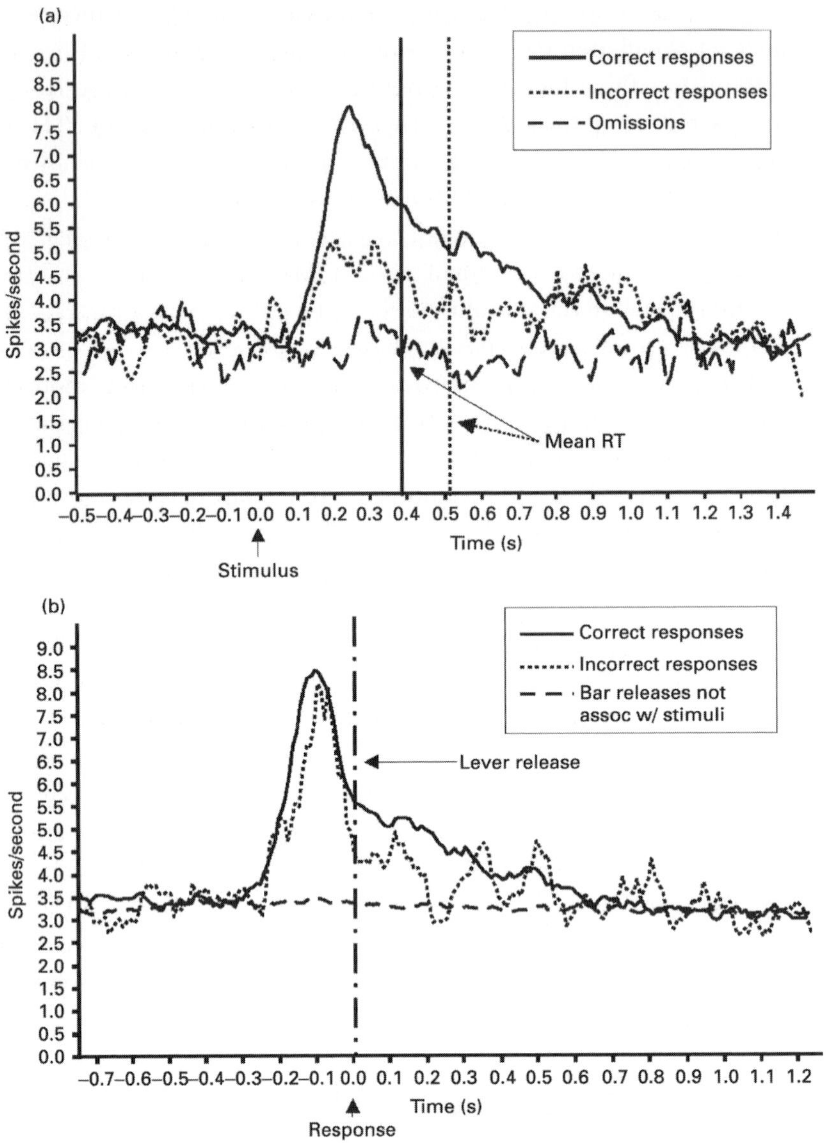

Figure 7.8 Stimulus- and response-locked population PETHs showing LC responses to cues for trials yielding correct or incorrect behavioral responses. (a) Stimulus-locked population PETHs showing LC responses to cues (presented at time 0) for trials yielding correct or incorrect behavioral responses. Note that the LC response magnitude is larger on correct compared to incorrect trials in this analysis ($p < 0.01$; 6243 and 584 trials, respectively). In addition, no fluctuation in LC activity was detected on omission trials in which no behavioral response was made (498 trials). Vertical lines indicate the mean behavioral RTs for correct and incorrect responses. (b) The difference in magnitude between correct and incorrect trials is no longer evident on response-locked population PETHs. It is also important to note that no LC response was observed prior to or following lever releases not associated with stimuli presentation (dashed line; 1480 trials (from Clayton *et al.* [2004]).[67]

seems to be reflected in phasic LC responses, indicating the OC to be involved in these responses. Consistent with this possibility, in recent studies we found retrogradely labeled neurons in the OC in both rats and monkeys after injections into the LC area, especially when injections included the peri-LC dendritic zone (Figure 7.9).[74,75]

Regulation of coupling among LC neurons may also involve a cortical projection. Recent findings indicate that the anterior cingulate cortex (ACC) in humans becomes activated by task difficulty and response conflict.[76,77] These findings led to the proposal that the ACC may be important for monitoring performance and signaling appropriate brain areas when behavior needs to be adjusted to better meet task demands. This is the type of system that would be appropriate for regulating coupling among LC neurons: when performance is good (as judged by reward frequency), coupling among LC neurons would remain high and might even increase in response to occasional conflict (to maintain good performance). However, if the task becomes more difficult and performance degrades, or the motivation for the current task decreases so that its intrinsic reward value is low, then the ACC may cause coupling among LC neurons to decrease. This would produce a tonic mode of LC activity and facilitate shifting to another strategy or pursuing a different reward. It is noteworthy that, consistent with this possibility, we have recently found that the medial prefrontal cortex (PFC) in the rat strongly innervates the peri-LC zone of extranuclear LC dendrites.[75,78] Other of our recent studies reveal a prominent ACC projection to LC in monkeys as well (Figure 7.9).[79] In addition, using electrophysiology we found that stimulation of the rat medial PFC (ACC area) activates noradrenergic LC neurons via a glutamate input (Figure 7.10).[80,81]

Our recent retrograde tracing studies in monkeys indicate that the OC and ACC are the two most prominent inputs to the LC from the PFC. Other studies have also reported inputs to the LC area from the dorsolateral PFC.[82] We are currently conducting additional studies to test the hypothesis that these PFC projections are importantly involved in the phasic LC responses during cognitive tasks (described above) as well as different modes of LC activity.

Discussion

Although evidence for electrotonic coupling in the adult LC exists, our analysis is independent of the specific mechanism that mediates coupling among LC cells. Elsewhere, coupling among neurons has been shown to be mediated by alterations in γ-aminobutyric acid interneuron activity,[83] and it is possible that a combined altered input from specific afferents mediates the change in coupling observed among LC cells.

It is also possible that the different modes of LC activity are produced by mechanisms other than alterations in coupling. Recent theoretical work with a

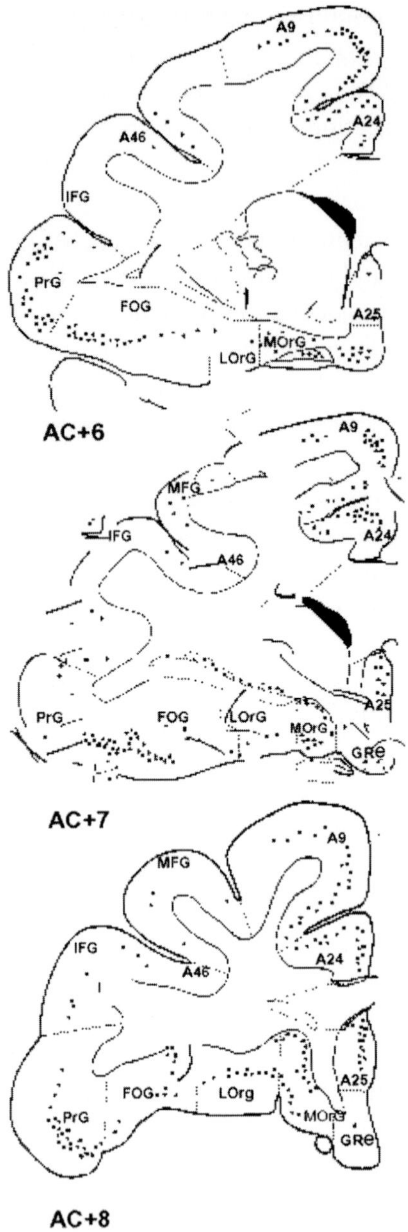

Figure 7.9 Plots of retrogradely labeled neurons in the prefrontal cortex of a monkey after injection of the retrograde tracer cholera toxin b subunit (Ctb) into the LC. Note that the caudolateral orbital and anterior cingulate cortices contained large numbers of neurons that were labeled, indicating that these are major inputs to the LC in monkey. Abbreviations: A9, area 9; A24, area 24; A25, area 25; A46, area 46; AC, anterior commissure; FOG, frontal orbital gyrus; GRe, gyrus rectus; IFG, inferior frontal gyrus; LOrG, lateral orbital gyrus; MFG, middle central gyrus; MOrG, medial orbital gyrus; PrG, precentral gyrus. Neurons are plotted on illustrations taken from Martin and Brown (1997).[163]

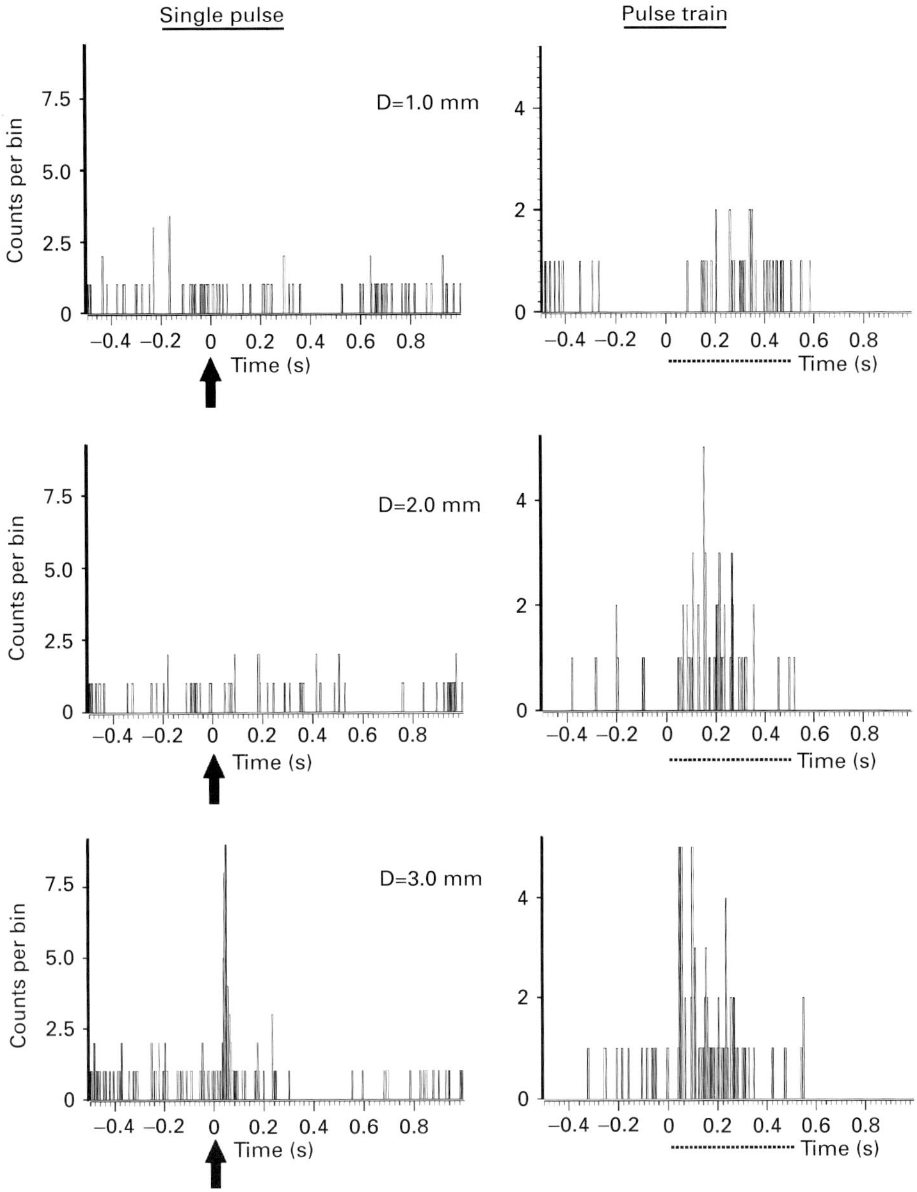

Figure 7.10 Responses to electrical stimulation of the dorsomedial prefrontal cortex in a typical rat LC neuron. Electrical stimulation was given at three different depths below the surface of the brain (1, 2, 3 mm) in the cortex, corresponding approximately to superficial, middle and deep cortical locations along the same stimulation electrode tracks in the FR2 region. In the left PETHs single-pulse stimulation is designated by arrows. In the right PETHs train stimulation (20 Hz, 0.5 s) was given during the epoch designated by small dots. Bin width in each PETH was 5 ms. Taken from Jodo *et al.* (1998).[81]

biophysically based model of the LC indicates that phasic and tonic modes of activity may result from different levels of excitatory drive to the LC with no change in coupling.[65] Additional work is needed to determine the cellular/molecular mechanism underlying different modes of activity in the LC.

A new theory of LC function

These empirical and computational modeling studies have led us to a new hypothesis of LC function. At the neurophysiological level, we propose that the LC governs the responsivity of its target neuronal assemblies to their afferent inputs, consistent with previous data.[21] This can manifest overtly as the likelihood that a stimulus will elicit a behavioral response, or have internal consequences, such as on attentional selection. Thus, the influence that LC-evoked changes in responsivity has on performance will depend upon the current task demands (e.g., whether they engage behavioral and/or attentional responses). It will also depend on the magnitude and pattern of LC discharge. With very low tonic and phasic LC activity (hypoactive mode), systems modulated by LC input will be relatively unresponsive. This level of activity is important for facilitating sleep and unresponsiveness to external stimuli. In addition, persistent hypoactive LC function may lead to clinical dysfunction such as depression, as discussed above. In the phasic LC mode (intermediate tonic activity plus stimulus-associated phasic activity), systems that receive LC input will become responsive to specific task-relevant stimuli. As the levels of tonic activity increase further (tonic LC mode, with a concomitant reduction in phasic activity), increases in responsivity will occur to a broader class of stimuli, ultimately to the point that behavior (and attention) become relatively indiscriminate and labile. We argue that these different modes of LC activity each have adaptive advantages under different environmental circumstances. The phasic mode (intermediate tonic activity) may support cortically driven behaviors optimized to specific stable environments (e.g., tasks requiring focused attention). While the behavioral lability associated with the higher tonic activity would impair performance under such circumstances, it may be adaptive in unpredictable or changing environments. That is to say, the variability associated with high tonic LC activity may be critical to behavioral flexibility and responsiveness to unexpected events. Note that persistence in either of these modes may facilitate clinical disorders such as autism (phasic mode) or attention-deficit/hyperactivity disorder (ADHD) (tonic mode), as discussed below.

Relationship of LC function to attention

Previous views held that phasic LC activation functioned to regulate sensory processing. Indeed, in rat the LC preferentially innervates sensory-related areas in the brain stem,[84] and in several species salient sensory stimuli reliably activate these neurons.[6,20,25,54] Other studies showed that LC stimulation or NE application increased the responsiveness of neurons in sensory regions to other inputs, supporting the view that LC activation could regulate behavior by altering sensory processing in distributed networks.[85,86] However, our recent findings in the primate demonstrate that the LC response is more closely linked with the decision to respond than the sensory event. In addition, no observable LC response was noted on omissions trials in which no behavioral response was made to rewardable cues. Neither was there LC activation preceding errant bar releases between trials. Together, these findings indicate that the LC phasic response is driven by the outcome of the decision process rather than by the appearance of the stimulus or by the motor act if it is not task-related. Furthermore, these recent results, taken together with previous empirical and modeling studies that address primate LC responses, suggest a revision of traditional views of LC function. These findings provide support for a novel theory that the LC plays an important and specific role in facilitating the influence of decision processes on behavioral response. This theory may impact not only our understanding of the neural bases of normal cognitive function but also disturbances of cognitive function thought to involve the LC-NE system.

Clinical implications

The LC-noradrenergic system is implicated in a number of psychiatric conditions. Given its widespread cortical projections and its reported role in cognition and behavior, it is an ideal candidate for exploration in the etiology and treatment of diseases of the mind. In the following section, we attempt to interpret the putative role of the LC mode of activity in a variety of clinical disorders.

Hypertonic LC activity: attention-deficit/hyperactivity disorder and posttraumatic stress disorder

Attention-deficit hyperactivity disorder (ADHD)

The firing patterns of LC neurons in behaving monkeys indicate that this system plays an important role in attention and performance.[1,54,87] In particular, one mode of LC activity, characterized by elevated tonic discharge, corresponds to poor performance on a task that requires focused attention. In addition, drugs that treat ADHD including stimulants (such as methylphenidate) and clonidine improve performance of ADHD subjects in a similar task.[88] These pharmaceutical agents decrease tonic LC activity and facilitate a shift to the phasic discharge mode (reviewed above). These and other results have led us to propose that the tonic mode of LC activity promotes high behavioral flexibility and disables focused or selective attention.[1,87] This view also implies that attentional disorders may be associated with LC dysregulation in which the proper mode of activity is not engaged adaptively for the context at hand. Specifically, there are several parallels between behaviors in monkeys during the tonic mode of LC activity and symptoms of ADHD, including irritability, poor focused attentiveness, and a high false alarm rate in continuous performance tasks. These findings indicate that the LC may play an important role in ADHD. Consequently, drugs that modulate LC mode, or switching between LC modes, could be helpful in treating this disorder. In fact, methylphenidate, the most common treatment for ADHD, decreases LC activity[89] while clonidine (which we found to switch LC from the tonic to phasic mode; described above) is also useful in treating ADHD.[90]

Posttraumatic stress disorder (PTSD)

As with ADHD, symptoms of posttraumatic stress disorder (PTSD) include sleep disturbances, difficulty with concentration, and modification of startle responses.[91,92] These symptoms are seen in the majority of PTSD cases and are associated with recall of the traumatic event or by "flashbacks" stimulated by new but reminiscent environmental stimuli. Our results showing phasic LC activation in association with behavioral responding to conditioned, meaningful stimuli in waking primates suggest that sensory and arousal disturbances of PTSD may be associated with hyperresponsive LC activity. This indicates that drugs that decrease tonic LC discharge rate and increase phasic LC responses to current goal-relevant stimuli may improve PTSD symptoms.

Decreased activity within the serotonergic system has also been associated with PTSD and ADHD.[93,94] The LC receives a dense innervation of serotonergic fibers and terminals in both the rodent and primate, and serotonin attenuates glutamate-evoked phasic activity of LC neurons.[95–98] These findings suggest that decreased activity within the serotonergic system may induce hypertonic or hyperresponsive LC activity, which could contribute to the attentional deficits associated with ADHD or PTSD. This may also offer some insight into why selective serotonin uptake inhibitors are effective in some patients with these disorders.

Hyperphasic LC activity: autism

Patients with autism are clinically defined by abnormalities of social interaction, disordered language, and repetitive and stereotyped patterns of movements and behaviors.[99,100] In addition, autistics show deficits in the ability to shift attention to new stimuli.[101,102] These data indicate that, in at least some ways, autism resembles a persistent, highly focused attentive state in which the patient does not adaptively disengage attention to adjust behavior to new or compelling stimuli.

As reviewed above, we propose that focused attention is facilitated by the phasic mode of activity in LC neurons. When in the phasic mode, LC cells are transiently activated selectively by task-relevant stimuli. The monkey exhibits enhanced task performance indicating focused attention and decreased distractibility. It therefore seems possible that, in patients with autism, LC neurons are in a persistent "hyperphasic" mode, so that an individual would have difficulty shifting attention to new stimuli. This hypothesis is supported by recent findings demonstrating that the NE reuptake inhibitor venlafaxine suppresses LC neuronal activity,[103] and is also an effective treatment for attention-impairment symptoms associated with autism.[104] These findings suggest that the hypervigilance associated with autism may be treated with pharmacological agents targeting the LC-noradrenergic system. Future clinical trials examining the efficacy of drugs effective at shifting LC activity from the phasic to the tonic mode may also prove fruitful in treating autistic patients with attention disorders.

Hypotonic LC activity: depression

As we have discussed, the LC is strongly linked to the two major indices used to measure emotional responses, autonomic activity and the EEG.[105] The strong projection to the LC from the nucleus paragigantocellularis (PGi), a major sympathoexcitatory brain region, links the LC to autonomic activity.[106] Emotionally charged stimuli that elicit sympathetic activation do so in part by activating PGi neurons. The parallel connections from the PGi to both spinal sympathetic areas and the LC may form the basis for the finding that stimuli that cause sympathetic activation typically also activate the LC. The activation of the LC, with the consequent increase in responsivity of receiving units throughout the brain, may be a key element in promoting emotional activation.

The present analysis indicates that a chronically hyperactive LC system may give rise to some symptoms of manic-depressive disorder, including sleeplessness and impulsivity.[107] On the other hand, chronically decreased LC function may be associated with limited emotionality and flat affect, a common characteristic of a subpopulation of depressed patients. It is noteworthy in this regard that alterations in receptors and biochemical parameters of LC neurons have been reported in brains of suicide victims.[108–111] In addition, inappropriate LC hyperresponsiveness to stimuli may participate in the exaggerated stimulus-responsivity and emotionality seen in patients with stress or anxiety disorders.[112] Thus, in a variety of ways, dysregulation of the LC system may be associated with clinical manifestations with a substantial emotional component. Independent clinical work has indicated that the LC system may be dysregulated in depression and other disorders.[107] It is also noteworthy that several studies have found reduced numbers of LC neurons in patients suffering from dementia associated with Parkinson's or Alzheimer's diseases.[113–117]

The majority of pharmacological treatments for depression attempt to regulate monoamine neuronal activity and/or release. A recent review article by Harro and Oreland[118] hypothesizes that LC dysfunction is the principal cause of depression. This hypothetical etiology takes place through the disruption of widespread noradrenergic projections and impaired interactions with other neurotransmitter systems that may also be involved in severe depression. Other studies support the contention that the LC is implicated in depression, demonstrating that suicidal patients with severe depression had 23% fewer LC neurons and 38% lower density of LC neurons than controls.[119] These observations are consistent with those of other workers finding evidence of depleted brain NE and diminished LC function in suicide victims.[109,120–123] Such hypothesized overstimulation of the LC with consequent depletion of function may reflect chronic stress, which is known to activate the LC system (as discussed above) and has long been linked to depression. Thus, a predominant theory of the noradrenergic system's role in depression is that there is reduced LC function in target systems. This hypothesis is supported by pharmacological studies demonstrating that the NE reuptake inhibitor reboxetine is effective in the treatment of depression[122] and deficits in cognitive function associated with depression via its ability to potentiate tonic NE release in target structures of the LC.[124–127]

These findings suggest the hypothesis that enhancing tonic noradrenergic activity would alleviate depression and the associated impairments in cognitive ability. Our results in behaving monkeys support this hypothesis. We found that when the monkeys were drowsy, fatigued, or apparently unmotivated to perform the target detection task, tonic neuronal activity in the LC decreased, as did phasic response magnitudes (left side of inverted-U curve, Figure 7.4). Therefore, in monkeys, decreased tonic and phasic LC activity were associated with lethargy and poor cognitive ability similar to what is observed in clinically depressed humans. These findings further support the contention that depression is associated with a hypoactive mode of LC activity.

Dysregulated LC phasic activity: schizophrenia

Schizophrenia is proposed by many to be primarily an attentional disorder.[128–131] Schizophrenics typically exhibit poor performance in the Stroop color-word task in a manner that suggests a selective attention deficit.[128,132] They also show impairments in tasks that require covert shifts of attention or sustained attention.[133,134] These results have led investigators to hypothesize that schizophrenics have difficulty using context information appropriately, and that their volitional attention is impaired.[135] The role for LC in selective attention outlined above suggests the possibility that attentional deficits in schizophrenia could involve deficits in engaging and maintaining the phasic mode of LC activity to support selective attention.

Pharmacologic treatments that normalize LC modes of activity might therefore be beneficial in schizophrenia.

Schizophrenic patients characteristically exhibit poor working memory (WM) performance.[136] Evidence indicates that this deficit could be due, in part, to noradrenergic dysfunction within the PFC. Postmortem analysis of schizophrenics with severe cognitive impairments revealed that these patients tended to have significantly reduced NE levels in the PFC relative to other schizophrenic patients without memory deficits.[137] Noradrenergic activation of the PFC is considered an important component in the regulation of WM processes. Previous studies have examined the possibility of reinstating normal WM function in young, catecholamine-depleted monkeys or aged, cognitively impaired nonhuman primates with decreased levels of prefrontal NE levels. Findings indicate that WM can be improved in these animals by peripheral injection of α_2-adrenergic agonists such as clonidine.[138] Additionally, some schizophrenic patients administered clonidine or other α_2-adrenergic agonists such as guanfacine have demonstrated an improvement in cognitive performance.[139,140] Taken together, these findings suggest that the cognitive impairment observed in schizophrenic patients may be due in part to dysfunction of the noradrenergic innervation of the PFC.

Cognitive function and LC activity may also be significantly impaired in schizophrenic patients as a result of altered input from higher cortical regions that have dysregulated activity in this disorder. We have recently found that the LC receives prominent projections from the ACC and OFC.[74,79] The ACC is a key component in the regulation of cognitive conflict,[76,141] while the OFC is crucially involved in decision making and the motivational control of goal-directed behaviors.[73,142,143] Schizophrenic patients show impaired development or decreased volume and/or neuronal activity in both the ACC and OFC relative to normal controls.[144–149] In addition, these patients also performed poorly on tasks sensitive to ACC or OFC damage.[150–152] These findings suggest that in schizophrenic patients, decreased neuronal input from hypoactive regions such as the ACC and OFC may negatively influence the LC's ability to respond to environmental cues, which may subsequently lead to impaired cognitive performance.

Loss of LC neurons: dementias of Alzheimer's and Parkinson's diseases

Although it is widely known that Alzheimer's and Parkinson's diseases are related to cholinergic and dopaminergic systems, respectively, substantial evidence also indicates dysfunction of the LC-NE system in these disorders. A common symptom of these diseases is cognitive decline. Several lines of studies have found that patients suffering from dementia with Alzheimer's and Parkinson's diseases have reduced numbers of LC neurons.[113–117] Levels of NE are decreased in the brains of aged monkeys and humans, with particularly pronounced loss in Alzheimer's disease.[153–155] Our own studies for LC function in behaving monkeys (reviewed above) reveal that the LC system plays an important role in attention and cognitive performance. In addition, experimentally induced loss of LC neurons produces impairments in attention, learning, and behavior.[156,157] Together, these findings indicate that the LC system is important for normal cognitive function, and its decline contributes to cognitive dysfunctions including the dementias found in Alzheimer's and Parkinson's diseases.

Sleep disorders

Many of the above conditions are comorbid with sleep disorders, and there is strong evidence that the LC plays an important role in regulation of arousal, sleep, and waking. Therefore, it is possible that dysfunction of the LC is important in sleep anomalies associated with any of these clinical disorders (as reviewed above and in Chapter 6 of this volume).

Regulation of LC mode: implications for new pharmacotherapies

The present results indicate that modes of phasic vs. tonic LC activity may participate in important behavioral functions as well as several mental dysfunctions. Our modeling and neurophysiology work in monkeys predicts that these modes may be produced by degrees of electrotonic coupling among LC neurons, and that such couplings may be modulated by inputs (as yet unidentified) to LC cells. Such modulated coupling has been studied in the retina, where coupling between horizontal cells is rapidly (within seconds) reduced in response to a dopamine input in a cAMP–protein kinase A-dependent manner.[158,159] These considerations suggest that sites may exist that selectively regulate coupling among central neurons such as the LC and could be pharmacologically targeted. New agents that increase coupling among LC cells could be a new means of treating disorders such as ADHD. Coupling has been found commonly among many brain neurons during development, and recent evidence indicates that it persists in adulthood not only in the LC but also in midbrain dopamine neurons and in the striatum, among other sites.[160,161] Our results indicate that such coupling may regulate important functional properties of these brain neurons, and that this coupling is therefore a new potential target for drug development for a variety of mental disorders.

Other recent modeling work reveals that phasic and tonic modes of LC activity may result simply from different levels of excitatory drive to LC neurons, independent of changes in coupling.[65] This reveals that multiple mechanisms may operate in regulating LC mode. This finding also indicates that pharmacologic agents are already at hand which are known to have overall excitatory or inhibitory effects on LC neurons. These agents may have clinical efficacy due, in part, to changes that they evoke in the modes of LC activity.

Our analysis indicates that drugs that alter modes of LC activity may be useful in treating various mental disorders. Our work also suggests that abnormal LC activity could contribute to the initial etiology of such disorders. Thus, such symptoms as those seen in ADHD could originate in part from a hypertonic/subphasic LC system. Additional animal work in which the modes of LC activity might be changed by genetic or other early intervention would be very important in testing this basis for different clinical maladies. It also seems likely that, regardless of whether altered LC activity plays a role in disease etiology, several mental disorders could be improved by treatments that would adjust modes of LC activity appropriately.

ACKNOWLEDGEMENTS

This work was supported by PHS grants MH55309, MH58480, MH62196 and MH12817.

REFERENCES

1. Usher, M., Cohen, J. D., Servan-Schreiber, D. *et al.* The role of locus coeruleus in the regulation of cognitive performance. *Science*, **283** (1999), 549–554.
2. Dahlstrom, A. and Fuxe, K. Evidence for the existence of monoamine-containing neurons in the central nervous system. I. Demonstration of monoamines in the cell bodies of brain stem neurons. *Acta Physiol. Scand.*, **62** (1964), 5–55.
3. Freedman, R., Foote, S. L. and Bloom, F. E. Histochemical characterization of a neocortical projection of the nucleus locus coeruleus in the squirrel monkey. *J. Comp. Neurol.*, **164** (1975), 209–231.
4. Morrison, J. H., Foote, S. L., O'Connor, D. and Bloom, F. E. Laminar, tangential and regional organization of the noradrenergic innervation of the monkey cortex: dopamine-beta-hydroxylase immunohistochemistry. *Brain Res. Bull.*, **9** (1982), 309–319.
5. Ungerstedt, U. Stereotaxic mapping of the monoamine pathways in the rat brain. *Acta Physiol. Scand. Suppl.*, **367** (1971), 1–48.
6. Aston-Jones, G., Foote, S. L. and Bloom, F. E. Anatomy and physiology of locus coeruleus neurons: functional implications. In *Norepinephrine: Frontiers of Clinical Neuroscience*, Vol. 2, ed. M. Ziegler and C. R. Lake. (Baltimore: Williams and Wilkins, 1984), pp. 92–116.
7. Foote, S. L., Bloom, F. E. and Aston-Jones, G. Nucleus locus coeruleus: new evidence of anatomical and physiological specificity. *Physiol. Rev.*, **63** (1983), 844–914.
8. Morrison, J. and Foote, S. Noradrenergic and serotonergic innervation of cortical, thalamic and tectal visual structures in old and new world monkeys. *J. Comp. Neurol.*, **243** (1986), 117–128.
9. Koda, L. Y., Schulman, J. A. and Bloom, F. E. Ultrastructural identification of noradrenergic terminals in the rat hippocampus: unilateral destruction of the locus coeruleus with 6-hydroxydopamine. *Brain Res.*, **145** (1978), 190–195.

10. Olschowka, J. A., Molliver, M. E., Grzanna, R., Rice, F. L. and Coyle, J. T. Ultrastructural demonstration of noradrenergic synapses in the rat central nervous system by dopamine-beta-hydroxylase immunocytochemistry. *J. Histochem. Cytochem.*, **29** (1981), 271–280.
11. Papadopoulos, G. C., Parnavelas, J. G. and Buijs, R. Monoaminergic fibers form conventional synapses in the cerebral cortex. *Neurosci. Lett.*, **76** (1987), 275–279.
12. Papadopoulos, G. C., Parnavelas, J. G. and Buijs, R. M. Light and electron microscopic immunocytochemical analysis of the noradrenaline innervation of the rat visual cortex. *J. Neurocytol.*, **18** (1989), 1–10.
13. Hoffer, B. J., Siggins, G. R., Oliver, A. P. and Bloom, F. E. Activation of the pathway from locus coeruleus to rat cerebellar Purkinje neurons: pharmacological evidence of noradrenergic central inhibition. *J. Pharmacol. Exp. Ther.*, **184** (1973), 553–569.
14. Segal, M. and Bloom, F. E. The action of norepinephrine in the rat hippocampus. I. Iontophoretic studies. *Brain Res.*, **72** (1974), 79–97.
15. Foote, S. L., Freedman, R. and Oliver, A. P. Effects of putative neurotransmitters on neuronal activity in monkey auditory cortex. *Brain Res.*, **86** (1975), 229–242.
16. Segal, M. and Bloom, F. E. The action of norepinephrine in the rat hippocampus. IV. The effects of locus coeruleus stimulation on evoked hippocampal unit activity. *Brain Res.*, **107** (1976), 513–525.
17. Waterhouse, B. D., Moises, H. C. and Woodward, D. J. Noradrenergic modulation of somatosensory cortical neuronal responses to iontophoretically applied putative neurotransmitters. *Exp. Neurol.*, **69** (1980), 30–49.
18. Waterhouse, B. D., Moises, H. C., Yeh, H. H., Geller, H. M. and Woodward, D. J. Comparison of norepinephrine- and benzodiazepine-induced augmentation of Purkinje cell responses to gamma-aminobutyric acid (GABA). *J. Pharmacol. Exp. Ther.*, **228** (1984), 257–267.
19. Waterhouse, B. D. and Woodward, D. J. Interaction of norepinephrine with cerebrocortical activity evoked by stimulation of somatosensory afferent pathways in the rat. *Exp. Neurol.*, **67** (1980), 11–34.
20. Aston-Jones, G. and Alexinsky, T. Discharge of noradrenergic locus coeruleus neurons in behaving rats and monkeys suggests a role in vigilance. In *Progress in Brain Research*, ed. O. Pompeiano. (Amsterdam: Elsevier Science Publishers BV, 1991), pp. 501–520.
21. Berridge, C. W. and Waterhouse, B. D. The locus coeruleus-noradrenergic system: modulation of behavioral state and state-dependent cognitive processes. *Brain Res. Brain Res. Rev.*, **42** (2003), 33–84.
22. Servan-Schreiber, D., Printz, H. and Cohen, J. D. A network model of catecholamine effects gain signal to noise ratio and behavior. *Science*, **249** (1990), 892–895.
23. Aston-Jones, G. and Bloom, F. E. Activity of norepinephrine-containing locus coeruleus neurons in behaving rats anticipates fluctuations in the sleep waking cycle. *J. Neurosci.*, **1** (1981), 876–886.
24. Hobson, J. A., McCarley, R. W. and Wyzinski, P. W. Sleep cycle oscillation: reciprocal discharge by two brainstem neuronal groups. *Science*, **189** (1975), 55–58.
25. Rasmussen, K., Morilak, D. A. and Jacobs, B. L. Single unit activity of locus coeruleus neurons in the freely moving cat. I. During naturalistic behaviors and in response to simple and complex stimuli. *Brain Res.*, **371** (1986), 324–334.

26. Foote, S. L., Aston-Jones, G. and Bloom, F. E. Impulse activity of locus coeruleus neurons in awake rats and monkeys is a function of sensory stimulation and arousal. *Proc. Natl. Acad. Sci. U. S. A.*, **77** (1980), 3033–3037.
27. Rajkowski, J., Kubiak, P., Ivanova, S. and Aston-Jones, G. State related activity and reactivity of locus coeruleus neurons in behaving monkeys. In: *Catecholamines: Bridging Basic Science with Clinical Medicine*, ed. D. Goldstein, G. Eisenhofer and R. McCarty. (New York: Academic Press, 1998), pp. 740–744.
28. Rajkowski, J., Silakov, V., Ivanova, S. and Aston-Jones, G. Locus coeruleus neurons in monkey are quiescent in paradoxical sleep. *Soc. Neurosci. Abstr.*, **23** (1997), 2130.
29. Grant, S. J., Aston-Jones, G. and Redmond, D. E. J. Responses of primate locus coeruleus neurons to simple and complex sensory stimuli. *Brain Res. Bull.*, **21** (1988), 401–410.
30. Rajkowski, J., Kubiak, P. and Aston-Jones, G. Locus coeruleus activity in monkey: phasic and tonic changes are associated with altered vigilance. *Brain Res. Bull.*, **35** (1994), 607–616.
31. Aston-Jones, G. and Bloom, F. E. Norepinephrine-containing locus coeruleus neurons in behaving rats exhibit pronounced responses to non-noxious environmental stimuli. *J. Neurosci.*, **1** (1981), 887–900.
32. Chiang, C. and Aston-Jones, G. Response of locus coeruleus neurons to footshock stimulation is mediated by neurons in the ventrolateral medulla. *Neuroscience.*, **53** (1993), 705–715.
33. Ennis, M. and Aston-Jones, G. Activation of locus coeruleus from nucleus paragigantocellularis: a new excitatory amino acid pathway in brain. *J. Neurosci.*, **8** (1988), 3644–3657.
34. Abercrombie, E. and Jacobs, B. Single unit response of noradrenergic neurons in locus coeruleus of freely moving cats. I. Adaptation to chronically presented stressful stimuli. *J. Neurosci.*, **7** (1987), 2844–2848.
35. Abercrombie, E. D. and Jacobs, B. L. Single unit response of noradrenergic neurons in locus coeruleus of freely moving cats I. Acutely presented stressful and nonstressful stimuli. *J. Neurosci.*, **7** (1987), 2837–2843.
36. Morilak, D. A., Fornal, C. A. and Jacobs, B. L. Effects of physiological manipulations on locus coeruleus neuronal activity in freely moving cats. I. Thermoregulatory challenge. *Brain Res.*, **422** (1987), 17–23.
37. Morilak, D. A., Fornal, C. A. and Jacobs, B. L. Effects of physiological manipulations on locus coeruleus neuronal activity in freely moving cats. II. Cardiovascular challenge. *Brain Res.*, **422** (1987), 24–31.
38. Morilak, D. A., Fornal, C. A. and Jacobs, B. L. Effects of physiological manipulations on locus coeruleus neuronal activity in freely moving cats. III. Glucoregulatory challenge. *Brain Res.*, **422** (1987), 32–39.
39. Valentino, R. J., Foote, S. L. and Aston-Jones, G. Corticotropin-releasing factor activates noradrenergic neurons of the locus coeruleus. *Brain Res.*, **270** (1983), 363–367.
40. Valentino, R., Page, M. and Curtis, A. Activation of noradrenergic locus coeruleus neurons by hemodynamic stress is due to local release of corticotropin-releasing factor. *Brain Res.*, **555** (1991), 25–34.

41. Aston-Jones, G., Rajkowski, J. and Kubiak, P. Conditioned responses of monkey locus coeruleus neurons anticipate acquisition of discriminative behavior in a vigilance task. *Neuroscience.*, **80** (1997), 697–715.
42. Aston-Jones, G., Rajkowski, J., Kubiak, P. and Alexinsky, T. Locus coeruleus neurons in the monkey are selectively activated by attended stimuli in a vigilance task. *J. Neurosci.*, **14** (1994), 4467–4480.
43. Rajkowski, J., Majczynski, H., Clayton, E. C. and Aston-Jones, G. Activation of monkey locus coeruleus neurons varies with difficulty and behavioral performance in a target detection task. *J. Neurophysiol.*, **92:1** (2004), 361–371.
44. Rajkowski, J., Kubiak, P., Ivanova, S. and Aston-Jones, G. LC neurons in monkey are selectively activated by the earliest predictors of reward. *Soc. Neurosci. Abstr.*, **22** (1996), 2063.
45. Kubiak, P., Rajkowski, J. and Aston-Jones, G. Behavioral performance and sensory responsiveness of LC neurons in a vigilance task varies with tonic LC discharge rate. *Soc. Neurosci. Abstr.*, **18** (1992), 538.
46. Rajkowski, J., Kubiak, P. and Aston-Jones, G. Activity of locus coeruleus neurons in behaving monkeys varies with focused attention: short- and long-term changes. *Soc. Neurosci. Abstr.*, **18** (1992), 538.
47. Rajkowski, J., Kubiak, P. and Aston-Jones, G. Correlations between locus coeruleus (LC) neural activity, pupil diameter and behavior in monkey support a role of LC in attention. *Soc. Neurosci. Abstr.*, **19** (1993), 974.
48. Ivanova, S., Rajkowski, J., Silakov, V., Watanabe, T. and Aston-Jones, G. Local chemomanipulations of locus coeruleus (LC) activity in monkeys alter cortical event-related potentials (E. R. Ps) and task performance. *Soc. Neurosci. Abstr.*, **23** (1997), 1587.
49. Aston-Jones, G. Behavioral functions of locus coeruleus derived from cellular attributes. *Physiol. Psychol.*, **13** (1985), 118–126.
50. Easterbrook, J. A. The effect of emotion on cue utilization and the organization of behavior. *Psychol. Rev.*, **66** (1959), 183–201.
51. Parasuraman, R. The psychobiology of sustained attention. In *Sustained Attention in Human Performance*, ed. J. S. Warm. (New York: John Wiley & Sons, 1984), pp. 61–101.
52. Knight, B. Dynamics of encoding in a population of neurons. *J. Gen. Physiol.*, **59** (1972), 734–766.
53. Bernander, O., Koch, C. and Usher, M. The effect of synchronized inputs at the single neuron level. *Neural Comput.*, **6** (1994), 622–641.
54. Aston-Jones, G., Valentino, R. J., Van Bockstaele, E. and Meyerson, A. Locus coeruleus, stress, and PTSD: neurobiological and clinical parallels. In *Catecholamine Function in Post-Traumatic Stress Disorder*, ed. M. Murburg. (Washington, DC: American Psychiatric Press, 1994), pp. 17–62.
55. Aghajanian, G. K., Cedarbaum, J. M. and Wang, R. Y. Evidence for norepinephrine-mediated collateral inhibition of locus coeruleus neurons. *Brain Res.*, **136** (1977), 570–577.
56. Ennis, M. and Aston-Jones, G. Evidence for self- and neighbor-mediated postactivation inhibition of locus coeruleus neurons. *Brain Res.*, **374** (1986), 299–305.
57. Christie, M. J. and Jelinek, H. F. Dye-coupling among neurons of the rat locus coeruleus during postnatal development. *Neuroscience*, **56** (1993), 129–137.

58. Christie, M. J., Williams, J. T. and North, R. A. Electrical coupling synchronizes subthreshold activity in locus coeruleus neurons in vitro from neonatal rats. *J. Neurosci.*, **9** (1989), 3584–3589.
59. Ishimatsu, M. and Williams, J. T. Synchronous activity in locus coeruleus results from dendritic interactions in pericoerulear regions. *J. Neurosci.*, **16** (1996), 5196–5204.
60. Travagli, R. A., Dunwiddie, T. V. and Williams, J. T. Opioid inhibition in locus coeruleus. *J. Neurophysiol.*, **74** (1995), 519–528.
61. Travagli, R. A., Wessendorf, M., Dunwiddie, T. V. and Williams, J. T. The opioid-induced current in rat locus coeruleus: electrotonic considerations. *Soc. Neurosci. Abstr.*, **20** (1994), 1734.
62. Ivanov, A. Y. and Aston-Jones, G. Intrinsic mechanisms regulating spontaneous discharge and subthreshold oscillations in locus coeruleus (LC) neurons in vitro. *Soc. Neurosci. Abstr.*, **21** (1995), 633.
63. Velazquez, J. L., Han, D. and Carlen, P. L. Neurotransmitter modulation of gap junctional communication in the rat hippocampus. *Eur. J. Neurosci.*, **9** (1997), 2522–2531.
64. Gladwell, S. J. and Jefferys, J. G. Second messenger modulation of electrotonic coupling between region CA3 pyramidal cell axons in the rat hippocampus. *Neurosci. Lett.*, **300** (2001), 1–4.
65. Brown, E., Moehlis, J., Holmes, P. *et al.* The influence of spike rate and stimulus duration on noradrenergic neurons. *J. Comput. Neurosci.*, **17** (2004), 13–29.
66. Alvarez, V. A., Chow, C. C., Van Bockstaele, E. J. and Williams, J. T. Frequency-dependent synchrony in locus coeruleus: role of electrotonic coupling. *Proc. Natl. Acad. Sci. U. S. A.*, **99** (2002), 4032–4036.
67. Clayton, E. C., Rajkowski, J., Cohen, J. D. and Aston-Jones, G. Phasic activation of monkey locus coeruleus neurons by simple decisions in a forced-choice task. *J. Neurosci.*, **24** (2004), 9914–9920.
68. Ratcliff, R. A theory of memory retrieval. *Psychol. Rev.*, **85** (1978), 59–108.
69. Ratcliff, R. and Rouder, J. N. Modeling response times for two-choice decisions: a theory of memory retrieval. *Psychol. Sci.*, **9** (1998), 347–356.
70. Usher, M. and McClelland, J. L. The time course of perceptual choice: the leaky, competing accumulator model. *Psychol. Rev.*, **108** (2001), 550–592.
71. Breiter, H. C., Aharon, I., Kahneman, D., Dale, A. and Shizgal, P. Functional imaging of neural responses to expectancy and experience of monetary gains and losses. *Neuron.*, **30** (2001), 619–639.
72. Rolls, E. T. The orbitofrontal cortex and reward. *Cereb. Cortex*, **10** (2000), 284–294.
73. Tremblay, L. and Schultz, W. Relative reward preference in primate orbitofrontal cortex. *Nature*, **398** (1999), 704–708.
74. Aston-Jones, G., Rajkowski, J., Lu, W. *et al.* Prominent projections from the orbital prefrontal cortex to the locus coeruleus in monkey. *Soc. Neurosci. Abstr.*, **28** (2002), 86–89.
75. Zhu, Y. and Aston-Jones, G. The medial prefrontal cortex prominently innervates a peri-locus coeruleus dendritic zone in rat. *Soc. Neurosci. Abstr.*, **22** (1996), 601.
76. Botvinick, M., Nystrom, L. E., Fissell, K., Carter, C. S. and Cohen, J. D. Conflict monitoring versus selection-for-action in anterior cingulate cortex. *Nature*, **402** (1999), 179–181.

77. Carter, C. S., Braver, T. S., Barch, D. M. *et al.* Anterior cingulate cortex, error detection, and the online monitoring of performance. *Science*, **280** (1998), 747–749.
78. Shipley, M. T., Fu, L., Ennis, M., Liu, W. and Aston-Jones, G. Dendrites of locus coeruleus neurons extend preferentially into two pericoerulear zones. *J. Comp. Neurol.*, **365** (1996), 56–68.
79. Rajkowski, J., Lu, W., Zhu, Y., Cohen, J. and Aston-Jones, G. Prominent projections from the anterior cingulate cortex to the locus coeruleus in Rhesus monkey. *Soc. Neurosci. Abstr.*, **25** (2000), Program No. 838.15.
80. Jodo, E. and Aston-Jones, G. Activation of locus coeruleus by prefrontal cortex is mediated by excitatory amino acid inputs. *Brain Res.*, **768** (1997), 327–332.
81. Jodo, E., Chiang, C. and Aston-Jones, G. Potent excitatory influence of prefrontal cortex activity on noradrenergic locus coeruleus neurons. *Neuroscience*, **83** (1998), 63–80.
82. Arnsten, A. F. and Goldman, R. P. Selective prefrontal cortical projections to the region of the locus coeruleus and raphe nuclei in the rhesus monkey. *Brain Res.*, **306** (1984), 9–18.
83. Traub, R., Whittington, M., Colling, S., Buzsaki, G. and Jefferys, J. Analysis of gamma rhythms in the rat hippocampus in vitro and in vivo. *J. Physiol.*, **493** (1996), 471–484.
84. Levitt, P., Moore, R. Y. Origin and organization of brainstem catecholamine innervation in the rat. *J. Comp Neurol.*, **186** (1979), 505–528.
85. Devilbiss, D. M. and Waterhouse, B. D. Norepinephrine exhibits two distinct profiles of action on sensory cortical neuron responses to excitatory synaptic stimuli. *Synapse*, **37** (2000), 273–282.
86. Waterhouse, B. D., Mouradian, R., Sessler, F. M. and Lin, R. C. Differential modulatory effects of norepinephrine on synaptically driven responses of layer V barrel field cortical neurons. *Brain Res.*, **868** (2000), 39–47.
87. Aston-Jones, G., Rajkowski, J. and Cohen, J. Locus coeruleus and regulation of behavioral flexibility and attention. *Prog. Brain Res.*, **126** (2000), 165–182.
88. Solanto, M. V. Neuropsychopharmacological mechanisms of stimulant drug action in attention-deficit hyperactivity disorder: a review and integration. *Behav. Brain Res.*, **94** (1998), 127–152.
89. Ishimatsu, M., Kidani, Y., Tsuda, A. and Akasu, T. Effects of methylphenidate on the membrane potential and current in neurons of the rat locus coeruleus. *J. Neurophysiol.*, **87** (2002), 1206–1212.
90. Connor, D., Barkley, R. and Davis, H. A pilot study of methylphenidate, clonidine, or the combination in ADHD comorbid with aggressive oppositional defiant or conduct disorder. *Clin. Pediatr.*, **39** (2000), 15–25.
91. Pitman, R. K., Orr, S. P., Shalev, A. Y., Metzger, L. J. and Mellman, T. A. Psychophysiological alterations in post-traumatic stress disorder. *Semin. Clin. Neuropsychiatry*, **4** (1999), 234–241.
92. Southwick, S. M., Paige, S., Morgan, C. A. III *et al.* Neurotransmitter alterations in PTSD: catecholamines and serotonin. *Semin. Clin. Neuropsychiatry*, **4** (1999), 242–248.
93. Donnelly, C. L. Pharmacologic treatment approaches for children and adolescents with posttraumatic stress disorder. *Child Adolesc. Psychiatr. Clin. N. Am.*, **12** (2003), 251–269.

94. Davids, E., Zhang, K., Kula, N. S., Tarazi, F. I. and Baldessarini, R. J. Effects of norepinephrine and serotonin transporter inhibitors on hyperactivity induced by neonatal 6-hydroxydopamine lesioning in rats. *J. Pharmacol. Exp. Ther.*, **301** (2002), 1097–1102.
95. Pickel, V. M., Joh, T. H. and Reis, D. J. A serotonergic innervation of noradrenergic neurons in nucleus locus coeruleus: demonstration by immunocytochemical localization of the transmitter specific enzymes tyrosine and tryptophan hydroxylase. *Brain Res.*, **131** (1977), 197–214.
96. Steinbusch, H. W. M. Serotonin-immunoreactive neurons and their targets in the CNS. In '*Classical Transmitters and Transmitter Receptors in the CNS*, Part II, ed. M. J. Kuhar. (Amsterdam: Elsevier Science Publishers BV, 1984), pp. 68–118.
97. Aston-Jones, G., Akaoka, H., Charlety, P. and Chouvet, G. Serotonin selectively attenuates glutamate-evoked activation of noradrenergic locus coeruleus neurons. *J. Neurosci.*, **11** (1991), 760–769.
98. Bobker, D. H. and Williams, J. T. Serotonin agonists inhibit synaptic potentials in the rat locus coeruleus in vitro via 5-hydroxytryptamine1A and 5-hydroxytryptamine1B receptors. *J. Pharmacol. Exp. Ther.*, **250** (1989), 37–43.
99. Friedman, S. D., Shaw, D. W., Artru, A. A. *et al.* Regional brain chemical alterations in young children with autism spectrum disorder. *Neurology*, **60** (2003), 100–107.
100. Wassink, T. H., Sutcliffe, J. S., Vieland, V. J. and Piven, J. The molecular and cellular genetics of autism. In *Neuropsychopharmacology. The Fifth Generation of Progress*, ed. C. Nemeroff. (Philadelphia: Lippincott Williams and Wilkins, 2002), pp. 549–563.
101. Mann, T. and Walker, P. Autism and a deficit in broadening the spread of visual attention. *J. Child Psychol. Psychiatry*, **44** (2003), 274–284.
102. Wainwright-Sharp, J. and Bryson, S. Visual orienting deficits in high-functioning people with autism. *J. Autism Dev. Disord.*, **23** (1993), 1–13.
103. Beique, J., de Montigny, C., Blier, P. and Debonnel, G. Effects of sustained administration of the serotonin and norepinephrine reuptake inhibitor venlafaxine: I. In vivo electrophysiological studies in the rat. *Neuropharmacology*, **39** (2000), 1800–1812.
104. Hollander, E., Kaplan, A., Cartwright, C. and Reichman, D. Venlafaxine in children, adolescents, and young adults with autism spectrum disorders: an open retrospective clinical report. *J. Child Neurol.*, **15** (2000), 132–135.
105. Aston-Jones, G., Rajkowski, J., Kubiak, P., Valentino, R. and Shipley, M. Role of the locus coeruleus in emotional activation. *Prog. Brain Res.*, **107** (1996), 379–402.
106. Aston-Jones, G., Ennis, M., Pieribone, V. A., Nickell, W. T. and Shipley, M. T. The brain nucleus locus coeruleus: restricted afferent control of a broad efferent network. *Science*, **234** (1986), 734–737.
107. Siever, L. J. and Davis, K. L. Overview: toward a dysregulation hypothesis of depression. *Am. J. Psychiatry*, **142** (1985), 1017–1031.
108. Biegon, A. and Fieldust, S. Reduced tyrosine hydroxylase immunoreactivity in locus coeruleus of suicide victims. *Synapse*, **10** (1992), 79–82.
109. Ordway, G. A., Smith, K. S. and Haycock, J. W. Elevated tyrosine hydroxylase in the locus coeruleus of suicide victims. *J. Neurochem.*, **62** (1994), 680–685.

110. Ordway, G. A., Widdowson, P. S., Smith, K. S. and Halaris, A. Agonist binding to alpha 2-adrenoceptors is elevated in the locus coeruleus from victims of suicide. *J. Neurochem.*, **63** (1994b), 617–624.

111. Widdowson, P. S., Ordway, G. A. and Halaris, A. E. Reduced neuropeptide Y concentrations in suicide brain. *J. Neurochem.*, **59** (1992), 73–80.

112. Aston-Jones, G., Valentino, R. J., Van Bockstaele, E. and Meyerson, A. Locus coeruleus, stress, and PTSD: neurobiological and clinical parallels. In *Catecholamine Function in Post-Traumatic Stress Disorder*, ed. M. Murburg. (Washington, DC, American Psychiatric Press, 1994c), pp. 17–62.

113. Chan-Palay, V. Alterations in the locus coeruleus in dementias of Alzheimer's and Parkinson's disease. *Prog. Brain Res.*, **88** (1991), 625–630.

114. Chan-Palay, V. Depression and dementia in Parkinson's disease. Catecholamine changes in the locus coeruleus, a basis for therapy. *Adv. Neurol.*, **60** (1993), 438–446.

115. German, D. C., Manaye, K. F., White, C. L. D. *et al.* Disease-specific patterns of locus coeruleus cell loss. *Ann. Neurol.*, **32** (1992), 667–676.

116. Hoogendijk, W. J., Pool, C. W., Troost, D., van Zwieten, E. and Swaab, D. F. Image analyser-assisted morphometry of the locus coeruleus in Alzheimer's disease, Parkinson's disease and amyotrophic lateral sclerosis. *Brain*, **118** (1995), 131–143.

117. Zweig, R. M., Cardillo, J. E., Cohen, M., Giere, S. and Hedreen, J. C. The locus coeruleus and dementia in Parkinson's disease. *Neurology*, **43** (1993), 986–991.

118. Harro, J. and Oreland, L. Depression as a spreading adjustment disorder of monoaminergic neurons: a case for primary implication of the locus coeruleus. *Brain Res. Brain Res. Rev.*, **38** (2001), 79–128.

119. Arango, V., Underwood, M. D. and Mann, J. Fewer pigmented locus coeruleus neurons in suicide victims: preliminary results. *Biol. Psychiatry*, **39** (1996), 112–120.

120. Bissette, G., Klimek, V., Pan, J., Stockmeier, C. and Ordway, G. Elevated concentrations of CRF in the locus coeruleus of depressed subjects. *Neuropsychopharmacology*, **28** (2003), 1328–1335.

121. Ordway, G. A. Pathophysiology of the locus coeruleus in suicide. *Ann. N. Y. Acad. Sci.*, **836** (1997), 233–252.

122. Ordway, G. A., Schenk, J., Stockmeier, C. A., May, W. and Klimek, V. Elevated agonist binding to alpha2-adrenoceptors in the locus coeruleus in major depression. *Biol. Psychiatry*, **53** (2003), 315–323.

123. Zhu, M. Y., Klimek, V., Dilley, G. E. *et al.* Elevated levels of tyrosine hydroxylase in the locus coeruleus in major depression. *Biol. Psychiatry*, **46** (1999), 1275–1286.

124. Montgomery, S., Ferguson, J. M. and Schwartz, G. E. The antidepressant efficacy of reboxetine in patients with severe depression. *J. Clin. Psychopharmacol.*, **23** (2003), 45–50.

125. Ferguson, J. M., Wesnes, K. A. and Schwartz, G. E. Reboxetine versus paroxetine versus placebo: effects on cognitive functioning in depressed patients. *Int. Clin. Psychopharmacol.*, **18** (2003), 9–14.

126. Invernizzi, R. W., Parini, S., Sacchetti, G. *et al.* Chronic treatment with reboxetine by osmotic pumps facilitates its effect on extracellular noradrenaline and may desensitize alpha(2)-adrenoceptors in the prefrontal cortex. *Br. J. Pharmacol.*, **132** (2001), 183–188.

127. Sacchetti, G., Bernini, M., Bianchetti, A. *et al.* Studies on the acute and chronic effects of reboxetine on extracellular noradrenaline and other monoamines in the rat brain. *Br. J. Pharmacol.*, **128** (1999), 1332–1338.
128. Barch, D. M. and Carter, C. S. Selective attention in schizophrenia: relationship to verbal working memory. *Schizophr Res.*, **33** (1998), 53–61.
129. Braff, D. L. Information processing and attention dysfunctions in schizophrenia. *Schizophr. Bull.*, **19** (1993), 233–259.
130. Hagh-Shenas, H., Toobai, S. and Makaremi, A. Selective, sustained, and shift in attention in patients with diagnoses of schizophrenia. *Percept. Mot. Skills*, **95** (2002), 1087–1095.
131. Perlstein, W. M., Carter, C. S., Noll, D. C. and Cohen, J. D. Relation of prefrontal cortex dysfunction to working memory and symptoms in schizophrenia. *Am. J. Psychiatry*, **158** (2001), 1105–1113.
132. Perlstein, W. M., Carter, C. S., Barch, D. M. and Baird, J. W. The Stroop task and attention deficits in schizophrenia: a critical evaluation of card and single-trial Stroop methodologies. *Neuropsychology*, **12** (1998), 414–425.
133. Gold, J. M., Randolph, C., Coppola, R. *et al.* Visual orienting in schizophrenia. *Schizophr. Res.*, **7** (1992), 203–209.
134. Green, M. F., Nuechterlein, K. H. and Gaier, D. J. Sustained and selective attention in schizophrenia. *Prog. Exp. Pers. Psychopathol. Res.*, **15** (1992), 290–313.
135. Braver, T. S., Barch, D. M. and Cohen, J. D. Cognition and control in schizophrenia: a computational model of dopamine and prefrontal function. *Biol. Psychiatry*, **46** (1999), 312–328.
136. Fleming, K., Goldberg, T. E., Binks, S. *et al.* Visuospatial working memory in patients with schizophrenia. *Biol. Psychiatry*, **41** (1997), 43–49.
137. Bridge, T. P., Kleinman, J. E., Karoum, F. and Wyatt, R. J. Postmortem central catecholamines and antemortem cognitive impairment in elderly schizophrenics and controls. *Neuropsychobiology*, **14** (1985), 57–61.
138. Arnsten, A. F. T. and Goldman-Rakic, P. S. Alpha2-adrenergic mechanisms in prefrontal cortex associated with cognitive decline in aged nonhuman primates. *Science*, **230** (1985), 1273–1276.
139. Fields, R. B., Van Kammen, D. P., Peters, J. L. *et al.* Clonidine improves memory function in schizophrenia independently from change in psychosis. Preliminary findings. *Schizophr. Res.*, **1** (1988), 417–423.
140. Friedman, J. I., Adler, D. N., Temporini, H. D. *et al.* Guanfacine treatment of cognitive impairment in schizophrenia. *Neuropsychopharmacology*, **25** (2001), 402–409.
141. Botvinick, M. M., Braver, T. S., Barch, D. M., Carter, C. S. and Cohen, J. D. Conflict monitoring and cognitive control. *Psychol Rev.*, **108** (2001), 624–652.
142. Hikosaka, K. and Watanabe, M. Delay activity of orbital and lateral prefrontal neurons of the monkey varying with different rewards. *Cereb. Cortex*, **10** (2000), 263–271.
143. Tremblay, L. and Schultz, W. Modifications of reward expectation-related neuronal activity during learning in primate orbitofrontal cortex. *J. Neurophysiol.*, **83** (2000), 1877–1885.
144. Carter, C. S., MacDonald, A. W. 3rd, Ross, L. L. and Stenger, V. A. Anterior cingulate cortex activity and impaired self-monitoring of performance in patients with schizophrenia: an event-related fMRI study. *Am. J. Psychiatry*, **158** (2001), 1423–1428.

145. Gallinat, J., Mulert, C., Bajbouj, M. *et al.* Frontal and temporal dysfunction of auditory stimulus processing in schizophrenia. *Neuroimage*, **17** (2002), 110–127.
146. Laurens, K. R., Ngan, E. T., Bates, A. T., Kiehl, K. A. and Liddle, P. F. Rostral anterior cingulate cortex dysfunction during error processing in schizophrenia. *Brain*, **126** (2003), 610–622.
147. Pantelis, C., Velakoulis, D., McGorry, P. D. *et al.* Neuroanatomical abnormalities before and after onset of psychosis: a cross-sectional and longitudinal MRI comparison. *Lancet*, **361** (2003), 281–288.
148. Convit, A., Wolf, O. T., de Leon, M. J. *et al.* Volumetric analysis of the pre-frontal regions: findings in aging and schizophrenia. *Psychiatry Res.*, **107** (2001), 61–73.
149. Sanfilipo, M., Lafargue, T., Rusinek, H. *et al.* Volumetric measure of the frontal and temporal lobe regions in schizophrenia: relationship to negative symptoms. *Arch. Gen. Psychiatry*, **57** (2000), 471–480.
150. Nordahl, T. E., Carter, C. S., Salo, R. E. *et al.* Anterior cingulate metabolism correlates with stroop errors in paranoid schizophrenia patients. *Neuropsychopharmacology*, **25** (2001), 139–148.
151. Yucel, M., Volker, C., Collie, A. *et al.* Impairments of response conflict monitoring and resolution in schizophrenia. *Psychol. Med.*, **32** (2002), 1251–1260.
152. Hutton, S. B., Murphy, F. C., Joyce, E. M. *et al.* Decision making deficits in patients with first-episode and chronic schizophrenia. *Schizophr. Res.*, **55** (2002), 249–257.
153. Goldman-Rakic, P. and Brown, R. Regional changes of monoamines in cerebral cortex and subcortical structures of aging rhesus monkeys. *Neuroscience*, **6** (1981), 177–187.
154. Chan-Palay, V. and Asan, E. Quantitation of catecholamine neurons in the locus coeruleus in human brains of normal young and older adults and in depression. *J. Comp. Neurol.*, **287** (1989), 357–372.
155. Adolfsson, R., Gottfries, C. G., Roos, B. E. and Winblad, B. Changes in the brain catecholamines in patients with dementia of Alzheimer type. *Br. J. Psychiatry*, **135** (1979), 216–223.
156. Anlezark, G. M., Crow, T. J. and Greenway, A. P. Impaired learning and decreased cortical norepinephrine after bilateral locus coeruleus lesions. *Science*, **181** (1973), 682–684.
157. Selden, N. R., Robbins, T. W. and Everitt, B. J. Enhanced behavioral conditioning to context and impaired behavioral and neuroendocrine responses to conditioned stimuli following coeruleocortical noradrenergic lesions: support for an attentional hypothesis of central noradrenergic function. *J. Neurosci.*, **10** (1990), 531–539.
158. McMahon, D. and Brown, D. Modulation of gap-junction channel gating at zebrafish retinal electrical synapses. *J. Neurophysiol.*, **72** (1994), 2257–2268.
159. McMahon, D. G. and Mattson, M. P. Horizontal cell electrical coupling in the giant danio: synaptic modulation by dopamine and synaptic maintenance by calcium. *Brain Res.*, **718** (1996), 89–96.
160. Grace, A. A. and Bunney, B. S. Intracellular and extracellular electrophysiology of nigral dopaminergic neurons–3. Evidence for electrotonic coupling. *Neuroscience*, **10** (1983), 333–348.

161. O'Donnell, P. and Grace, A. A. Cortical afferents modulate striatal gap junction permeability via nitric oxide. *Neuroscience*, **76** (1997), 1–5.
162. Aston-Jones, G., Rajkowski, J. and Cohen, J. Role of locus coeruleus in attention and behavioral flexibility. *Biol. Psychiatry*, **46** (1999), 1309–1320.
163. Martin, R. F. and Bowden, D. M. *Template Atlas of the Primate Brain.* (Seattle: Primate Information Center, University of Washington, 1997).

8

Norepinephrine and long-term memory function

Benno Roozendaal Ph.D

Center for the Neurobiology of Learning and Memory, and Department of Neurobiology and Behavior, University of California, Irvine, USA

Introduction

This chapter reviews research investigating the role of norepinephrine in modulating the consolidation of long-term memory. The focus is primarily on noradrenergic activation-influenced storage of lasting explicit (episodic/declarative) memory. The hypothesis that norepinephrine plays a role in learning and memory emerged more than three decades ago when Kety[1] suggested that adrenergic catecholamines released by certain emotional states may serve "to reinforce and consolidate new and significant sensory patterns in the neocortex." The initial experiments investigating this general hypothesis examined the effects of a large variety of sympathomimetics or antiadrenergic agents on learning and memory. Systemic injections of amphetamine were found to facilitate memory[2] and, in contrast, depletion of norepinephrine by tyrosine hydroxylase inhibition with α-methyl-p-tyrosine (α-MPT), by 6-hydroxydopamine or by dopamine-β-hydroxylase inhibition with diethyldithiocarbamate (DDC) usually impaired learning performance. Similar impairing effects were obtained with systemic injections of the noradrenergic toxin *N*-(2-chloroethyl)-*N*-ethyl-2-bromobenzylamine (DSP-4) or disruption of vesicular storage of catecholamines with reserpine. Although the use of these nonspecific antiadrenergic agents clearly implicated norepinephrine in learning and memory, it remained unclear whether the effects involved peripheral or central mechanisms.

The findings of Gold and van Buskirk[3] were among the first to suggest the involvement of central norepinephrine in memory. They showed that inhibitory avoidance training decreased brain norepinephrine levels under conditions that resulted in good retention, suggesting an enhanced release and metabolism of this neuromodulator during training. In support of this view, Haycock *et al.*[4] showed

Brain Norepinephrine: Neurobiology and Therapeutics, ed. Gregory A. Ordway, Michael A. Schwartz and Alan Frazer. Published by Cambridge University Press.

that intraventricular infusions of norepinephrine facilitated retention, thus providing evidence for the view that central norepinephrine function modulates memory. Other experiments[5] reported that intracerebroventricular administration of DDC impaired inhibitory avoidance retention when administered posttraining and that concurrent infusions of norepinephrine into the ventricles or systemic injections of norepinephrine or epinephrine blocked the retention impairment induced by administration of DDC.

Although these early studies implicated central norepinephrine in memory processes, they did not investigate whether the effects involved activation of specific brain regions or whether the effects involved a widespread activation of the brain. The findings of many subsequent studies have strongly suggested that noradrenergic activation of the amygdala, arising from catecholaminergic cell bodies in the nucleus of the solitary tract and the locus coeruleus (LC), is critically involved in memory consolidation. Norepinephrine release in the amygdala is enhanced during emotionally arousing training experiences and this noradrenergic response is essential for inducing memory enhancement.[6] Recent work involving the use of adrenoceptor agonists and antagonists has provided support for the notion of adrenergic involvement in emotionally influenced memory consolidation. Undoubtedly there are many other influences on memory consolidation provided by adrenomedullary and adrenocortical stress hormones, opioid peptidergic, γ-aminobutyric acid (GABA) ergic, cholinergic, and other neurotransmitter systems, as well as local circuit neurons. Although memory modulation can be achieved by manipulation of any of these components, there are many interactions between the amygdala norepinephrine system and these neuromodulatory influences. Peripheral stress hormones, released during emotionally arousing training conditions, enhance memory consolidation by facilitating norepinephrine release and synaptic mechanisms in the amygdala. The effects of several other neurotransmitter systems on memory consolidation are also mediated by an alteration of noradrenergic activity in the amygdala. There are also multiple, often interacting, brain systems involved in memory consolidation processes. Our findings suggest that noradrenergic activation of the amygdala, or more precisely the basolateral complex of the amygdala, serves to modulate neuroplasticity and memory consolidation in other brain regions. Evidence further indicates that norepinephrine also affects memory consolidation within these other brain regions. Studies using infusions of norepinephrine into several forebrain regions, including the hippocampus, the entorhinal cortex, and the medial prefrontal cortex, indicate that these brain regions contribute to mediating norepinephrine effects on memory consolidation. Finally, recent findings are discussed indicating that noradrenergic activation modulates long-term formation of memory for emotionally arousing or traumatic experiences in human subjects.

Role of adrenergic catecholamines in memory consolidation

Müller and Pilzecker[7] were the first to propose that memory traces are initially fragile after learning and become consolidated over time. Guided by this view, numerous studies have shown that treatments such as electrical brain stimulation or drug injections given during a critical time shortly after training can alter the consolidation of lasting memory traces. The neuropharmacological analysis of memory processes is complicated by the fact that drugs can indirectly influence learning and memory by affecting sensorimotor, arousal, or attentional processes. Thus it is of critical importance to be able to distinguish the effects of the treatments on the inferred process of memory from the effects of the treatments on performance during training. The use of posttraining treatments to alter the activity of the brain shortly after training has provided an effective technique for dealing with this issue.[8] With the use of this paradigm, brain functioning is experimentally altered only after learning has occurred, thus, affecting memory consolidation in a retrograde fashion. With most treatments used, the acute effects of the treatments are short-lasting and do not directly affect brain function when retention is tested a day or longer after the learning.

The observation that some experiences are remembered vividly, whereas others are not remembered at all, or at best very remotely, led to the suggestion that there may be an endogenous mechanism that serves to regulate the strength of the memories.[9] Strong memories are often based on experiences that are meaningful or emotionally arousing. Arousal is a somewhat vague concept covering a range of behaviors from wakefulness, attention, and vigilance to stress and fear, as well as motivational factors, such as hunger. In an early effort to offer a neurobiological account of the influence of emotional arousal on memory, Livingston[10] suggested that stimulation of the limbic system and brain stem reticular formation might promote the storage of recently activated brain events by initiating a "neurohormonal influence [favoring] future repetitions of the same neural activities." Although the specific details of more recent experimental findings and theoretical interpretations differ in many ways from those early views offered by Livingston and Kety,[1,10] they are consistent with their general hypotheses. There is now extensive evidence supporting the hypothesis that the central noradrenergic response to an experience is a determining factor in establishing a brain state that promotes the storage of recently acquired information.

Amphetamine effects on memory consolidation

Gerard[11] may have been the first to suggest that the adrenomedullary stress hormone epinephrine might have enhancing effects on memory consolidation. Gerard proposed that as "epinephrine . . . is released in vivid emotional experiences, such an intense adventure should be highly memorable." Early support for this suggestion came from experiments using systemic injections of amphetamines. In addition to amphetamine-induced effects on dopaminergic mechanisms, amphetamine increases the release of adrenergic catecholamines from peripheral storage sites. Numerous studies have shown that amphetamine enhances memory when administered systemically either shortly before or shortly after training. Enhancing effects of amphetamine have been observed in a variety of tasks such as inhibitory avoidance, active avoidance, and appetitive discrimination.[12] Moreover, the fact that amphetamine enhances memory when administered posttraining supports the view that amphetamine enhances retention by influencing memory consolidation processes. The effects of amphetamine administration on memory consolidation follow an inverted-U-shaped dose-response relationship. Moderate doses of amphetamine enhance retention performance, whereas lower doses or higher doses are less effective. This biphasic response pattern is characteristic for almost every drug tested, but the reason (or reasons) for this pattern is not known. It is possible that this is a pharmacological artifact due to a skewed stimulation of a single modulatory system, rather than a coordinated activation of all systems involved in memory consolidation. Otherwise, this finding may reflect that higher doses of amphetamine, or other drugs, may simultaneously activate other, antagonizing mechanisms to protect against overactivation. Another consequence of this inverted-U-shaped dose-response relationship on memory consolidation is that the drug effects also depend on the training-induced emotional response as well as the training intensity. For example, if rats are trained on the inhibitory avoidance task with a high-intensity footshock, this experience alone would result in good retention in the control animals, and amphetamine administration may induce relative memory impairment. Therefore, when it is stated that a drug enhances memory consolidation, this means that this effect is observed in animals given a moderate amount of training.

As amphetamine readily passes the blood–brain barrier, other experiments investigated whether the memory-modulating effects of amphetamine involved influences on peripheral or central catecholamines. Posttraining systemic injections of 4-OH-amphetamine (a derivative of amphetamine that does not pass the blood–brain barrier) also enhanced inhibitory avoidance retention.[13] These findings provide support for the view that amphetamine administration enhances memory consolidation by stimulating catecholamine release from peripheral sites. The effects of systemic injections of amphetamine and 4-OH-amphetamine do not seem to involve peripheral sympathetic neurons, the primary source of peripheral norepinephrine, as sympathetic denervation induced 24 hours before training with 6-hydroxydopamine did not attenuate the memory-enhancing effects of either drug.[13] In contrast, adrenal demedullation, i.e., elimination of peripheral epinephrine, blocked the effects of both amphetamine and 4-OH-amphetamine on memory for active avoidance and inhibitory avoidance. Moreover, Williams *et al.*[14] showed that the peripherally acting β-adrenoceptor antagonist sotalol blocked the memory-enhancing effects of systemic 4-OH-amphetamine. These findings provide strong support for the view that amphetamine influences memory consolidation, at least in part, through effects involving the release of peripheral epinephrine from the adrenal medulla.

Peripheral epinephrine and norepinephrine effects on memory consolidation

Systemic administration of epinephrine or norepinephrine also enhances retention in a wide variety of tasks, including inhibitory avoidance, multitrial avoidance, a one-trial appetitive task, and an aversively motivated discrimination task.[15,16] Epinephrine is effective when given immediately after training and moderate doses produce greatest enhancement, whereas larger doses are less effective and may even impair retention. Comparable to the effects described above for amphetamine, the memory-modulating effect of epinephrine administered posttraining appears to vary with the training intensity: doses of epinephrine that enhance memory when administered following a mild training footshock impair memory when administered shortly after a high footshock.[17] The effect is comparable to that produced by increasing the dose of epinephrine beyond an optimal level when a low footshock is used for training. Retention enhancement induced by epinephrine or norepinephrine is blocked by injection of the β-adrenoceptor antagonist propranolol, a drug that readily enters the brain, as well as by sotalol, a β-adrenoceptor antagonist that does not enter the brain. Posttraining administration of β-adrenoceptor agonists that enter the brain, including dipivefrin and clenbuterol, also enhance memory consolidation. The memory enhancement induced by dipivefrin and clenbuterol is blocked by propranolol but not by sotalol.[18] Such findings indicate that the effects of systemic injections of epinephrine and norepinephrine on memory consolidation are initiated by an activation of peripheral β-adrenoceptors, but that the effects ultimately must involve activation of β-adrenoceptors within the brain.

Role of brainstem noradrenergic cell groups in memory consolidation

A large number of studies have suggested that the vagus nerve may play a critical role in mediating the peripheral effects of epinephrine and norepinephrine on brain systems involved in memory consolidation. Electrical stimulation of vagal afferents immediately after training enhances later retention performance in an inverted-U-shaped fashion.[19] Thus, the memory enhancement induced by vagal stimulation shows a striking similarity with that observed after peripheral injections of epinephrine. Systemically administered or released adrenergic compounds may activate vagal afferents by binding to peripheral β-adrenoceptors, and the vagus nerve may serve to influence pontine and limbic structures by relaying input regarding the peripheral endocrine/autonomic milieu to the central nervous system. Peripheral vagal afferents project to the nucleus of the solitary tract (NTS). Studies using triple fluorescence labeling have revealed that norepinephrine is the major neurotransmitter in projection neurons from the NTS. The cell bodies of these neurons were detected in the A2 catecholamine region of the NTS.[20] The fact that the NTS receives peripheral input and conveys information to forebrain structures involved in learning and memory suggests that central noradrenergic neurons arising in the NTS may mediate the effects of peripheral physiological influences on memory consolidation. This implication is supported by evidence that infusions of the local anesthetic lidocaine into the NTS posttraining impair inhibitory avoidance retention. Moreover, injections of lidocaine into the NTS block the memory-enhancing effects of posttraining systemic injections of epinephrine.[21]

Electron microscopic studies have revealed that the NTS not only contains noradrenergic cell bodies, but also an abundance of noradrenergic terminals and receptors. Thus, norepinephrine may play an important role in influencing NTS activity during memory formation. To examine this hypothesis, Williams and colleagues trained rats in a Y-maze discrimination task followed immediately afterward by an infusion of the β-adrenoceptor agonist clenbuterol or a vehicle control solution administered into the NTS. On the retention test, groups receiving clenbuterol showed significantly improved performance.[22] These findings demonstrate that memory consolidation processes can be improved by posttraining activation of noradrenergic receptors in the NTS. The effects of clenbuterol may be mediated by facilitating norepinephrine output from terminals that innervate the NTS as opposed to any specific actions of this compound on the A2 catecholamine neurons in the NTS. This effect occurs because these cells are activated only by the binding of norepinephrine to postsynaptic α_1-adrenoceptors. The authors further reported that posttraining intra-NTS injections of epinephrine, which has a high affinity for α_1-adrenoceptors, also enhanced retention performance on the Y-maze task. Furthermore, there is evidence that other neuromodulators may affect memory consolidation by influencing noradrenergic neurons in the NTS. Immediate posttraining infusions of glutamate or the glucocorticoid receptor agonist RU 28362 administered into the NTS dose-dependently enhanced memory consolidation.[23,24] The finding that a blockade of β-adrenoceptors in the amygdala prevented the memory enhancement induced by either drug provides evidence that intra-NTS glutamate or glucocorticoid infusions enhance memory consolidation by affecting noradrenergic mechanisms to limbic forebrain structures.

Noradrenergic projections originating in the NTS innervate forebrain structures involved in learning and memory, including the amygdala. However, these structures may also influence norepinephrine release via projections to the nucleus paragigantocellularis in the lower medulla, which projects to the LC. The LC noradrenergic system is viewed as a broadly projecting system. Locus coeruleus efferent projections are widespread, including projections to the amygdala, the hippocampus, and the prefrontal cortex, and single LC neurons send divergent projections to target cells of very different functions.[25] Approximately 85% of noradrenergic fibers innervating the amygdala arrive from the LC, suggesting that manipulation of LC activity, like that of the NTS, should induce marked effects on memory consolidation. The LC receives a very restricted set of input from two rostral medullary nuclei (i.e., the nucleus paragigantocellularis and the prepositus hypoglossi). Glutamate infused into the nucleus paragigantocellularis immediately after training increases LC activity and induces retention enhancement, whereas inactivation of the nucleus paragigantocellularis with local infusions of the GABAergic agonist muscimol induces memory impairment.[21] Many neurotransmitters are involved in activation of the LC, including epinephrine, glutamate, corticotropin-releasing factor, GABA and estrogens.[25] The release rate of these various neurotransmitters is modified by arousal and stressful stimuli and may affect memory consolidation processes by influencing noradrenergic activation of LC neurons. Suppression of LC activity with local infusions of the α_2-adrenoceptor agonist clonidine induces dose-dependent retention impairment.[26] This effect is ameliorated by either pretraining infusions of yohimbine, an α_2-adrenoceptor antagonist, into the LC or by posttraining infusions of norepinephrine into the lateral ventricles. When clonidine is infused into the LC it also blocks memory enhancement induced by systemic injections of epinephrine. This finding supports the view that noradrenergic cell groups in both the NTS and the LC are actively involved in mediating epinephrine effects on memory consolidation.

Pharmacological and genetic evidence of central norepinephrine in regulating long-term memory consolidation

Many of the findings of drug administration studies are consistent with a role for central norepinephrine in memory consolidation.[27] As noted above (p. 236), depletion of norepinephrine with the norepinephrine-specific toxin DSP-4 impairs memory consolidation on a variety of tasks,[28] and administration of the centrally acting β-adrenoceptor antagonist propranolol impairs memory consolidation.[29,30] Activation of central β-adrenoceptors with systemic administration of centrally acting β-adrenoceptor agonists such as dipivefrin and clenbuterol dose-dependently enhances memory consolidation.[18,31] These effects are blocked by administration of centrally acting β-adrenoceptor antagonists, indicating an activation of central receptors for norepinephrine. Several findings support the hypothesis that norepinephrine plays a role in the modulation of long-term memory formation. Propranolol injected shortly after inhibitory avoidance training had no effect on performance tested either 2 or 6 hours later.[30] Furthermore, injections of propranolol before water-maze spatial training did not affect acquisition performance, but selectively impaired retention performance tested a day or more after training.

Neurochemical studies performed on genetically selected rat lines revealed a higher noradrenergic response to acute stress in the LC in the Maudsley reactive strain of rats than in Maudsley nonreactive rats. The behavioral profile of the Maudsley reactive rats is similar to rats treated with drugs that enhance noradrenergic function in that they show superior long-term memory performance.[32] Obviously, it is difficult or impossible to conclude that the behavioral differences in these psychogenetically selected breeding lines are caused selectively by increased noradrenergic function, as these rats no doubt differ in a wide spectrum of ways. A recent genetic study reported finding an involvement of norepinephrine in long-term memory in mice heterozygous for a mutation in the gene encoding tyrosine hydroxylase, the first and rate-limiting step in the biosynthesis of catecholamines.[33] The mice exhibited reduced tyrosine hydroxylase activity and also had a moderate reduction in norepinephrine accumulation and release in brain regions. The most important finding was that these mice displayed deficits in long-term memory formation of learning evaluated with three distinct behavioral paradigms, including active avoidance, cued fear conditioning, and conditioned taste aversion. The mice, however, were not impaired in short-term memory. Furthermore, these memory deficits were restored by posttraining stimulation of norepinephrine activity.

Many studies have examined the actions of norepinephrine or adrenoceptor agonists and antagonists on memory consolidation processes in specific brain regions. These findings are discussed below. Because most studies have examined those effects on memory consolidation in the amygdala, this brain region is discussed extensively and used to exemplify norepinephrine actions on memory consolidation as well as its interactions with other neurotransmitter systems. As the amygdala is not the only brain region regulating norepinephrine effects on memory consolidation, the major findings of norepinephrine influences in other brain regions are also discussed.

Norepinephrine in the amygdala and memory consolidation

Extensive evidence indicates that norepinephrine affects memory consolidation by influences involving the amygdala, and that the effects of peripheral epinephrine on memory involve noradrenergic activation of the amygdala. For example, systemic injections of epinephrine in doses that produce memory enhancement also induce norepinephrine release in the amygdala.[34] Experiments using the expression of immediate early genes as a marker for activated neurons reported that vagal nerve stimulation increases the number of amygdala neurons expressing *c-fos*. Electrophysiological findings indicate that the firing rate of amygdala neurons is increased following electrical stimulation of either the vagus nerve[35] or the NTS.[35,36] Lesions of the amygdala block the memory-modulatory effects of systemic administration of epinephrine and several other neuromodulators.[6] Depletion of amygdala norepinephrine with local injections of the norepinephrine-specific neurotoxin DSP-4 blocks systemic epinephrine effects on memory consolidation.[37] Posttraining intra-amygdala infusions of the β-adrenoceptor antagonist propranolol also block the memory enhancement induced by systemic administration of epinephrine,[38,39] and posttraining infusions of norepinephrine into the amygdala attenuate retention impairment induced by adrenal demedullation.[39] Gallagher and colleagues[40] found that β-adrenoceptor antagonism within the amygdala immediately after inhibitory avoidance training impaired retention performance, and that concurrent administration of norepinephrine attenuated the impairment. Additionally, posttraining infusions of norepinephrine or the β-adrenoceptor agonist clenbuterol administered into the amygdala induced a dose-dependent enhancement of retention.[31,41] Considered together, these findings strongly suggest that noradrenergic activation in the amygdala plays a critical role in mediating epinephrine-induced memory enhancement.

Selective involvement of the basolateral complex of the amygdala

The findings of a series of experiments have indicated that the basolateral complex of the amygdala (BLA) is selectively involved in mediating noradrenergic influences on memory consolidation. These findings are consistent with extensive evidence indicating an involvement of the BLA in mediating the memory-modulatory effects of drugs influencing other neurotransmitter systems as well.[6] Posttraining infusions of drugs affecting several classes of receptors administered into the BLA, but not the adjacent central nucleus of the amygdala, enhanced or impaired memory consolidation. Selective microinfusions of norepinephrine or the β-adrenoceptor agonist clenbuterol into the BLA enhanced inhibitory avoidance retention.[42,43] Posttraining intra-BLA infusions of norepinephrine also enhanced memory for contextual fear conditioning in a Y-maze apparatus.[44] In this task, rats received several footshocks in a closed arm of the Y-maze apparatus without having the opportunity to escape from the aversive stimulation. On the retention test performed 24 hours later, rats were placed in another arm that was not previously associated with shock. They were allowed to freely explore the entire maze without further footshock delivery. Rats given norepinephrine after footshock stimulation showed superior memory as compared to rats receiving a vehicle control solution, as assessed both by an increased avoidance of the place where they received footshock as well as a facilitated expression of freezing behavior. Moreover, studies using strains of rats that differ in their tendencies to freeze in response to fearful stimulation indicated that posttraining intra-amygdala infusions of norepinephrine enhanced memory of an aversive event (mild shock delivered through a probe inserted into the cage). However, the expression of the enhanced memory depended on the rats' ability to freeze. In the strain of rats that normally displayed freezing (i.e., Roman low-avoidance rats), the posttraining norepinephrine infusions enhanced conditioned freezing. In the other strain (i.e., Roman high-avoidance rats) the posttraining norepinephrine did not enhance freezing but, instead, enhanced other natural defensive responses (i.e., burying and biting of the probe) typically displayed by rats of this strain.[45] Such findings indicate that amygdala norepinephrine has a comparable role in regulating memory consolidation for different kinds of fear conditioning training, including inhibitory avoidance. Although many researchers have focused their investigations of the amygdala on its role in the acquisition and expression of fear conditioning, the amygdala noradrenergic system influences memory consolidation of many types of training. For example, immediate posttraining infusions of norepinephrine or β-adrenergic agents into the BLA or the amygdala complex also modulate memory for changes in reward magnitude,[46] discrimination learning,[47] spatial learning in a water maze,[48] and conditioned taste aversion.[49]

Stress- and training-induced norepinephrine release in the amygdala and memory consolidation

In an early study examining the relationship between brain norepinephrine content and memory, Gold and van Buskirk reported evidence suggesting that noradrenergic activity in the brain was increased in rats trained on an inhibitory avoidance task or given memory-enhancing doses of epinephrine immediately after training.[3] Recent studies using in vivo microdialysis and high-performance liquid chromatography (HPLC) indicate that norepinephrine is normally released within the amygdala during training on aversively motivated tasks. Several studies report that stress induced by prolonged immobilization or tail pinch increases amygdala norepinephrine levels.[50] As is shown in Figure 8.1, a single, brief footshock stimulation of the kind typically used in inhibitory avoidance training induced norepinephrine release within the amygdala that varied directly with the stimulus intensity.[51,52] In recent experiments, this microdialysis technique was used to observe ongoing fluctuations in amygdala norepinephrine levels in individual rats as they acquired and consolidated new, emotionally influenced memories. With this method, it is possible to relate the training-induced increase in norepinephrine levels to subsequent retention performance within individual animals.[53] For this experiment, rats were trained on an inhibitory avoidance task, in which they learned to avoid the place in the apparatus where they receive footshock. A surprising finding was that the

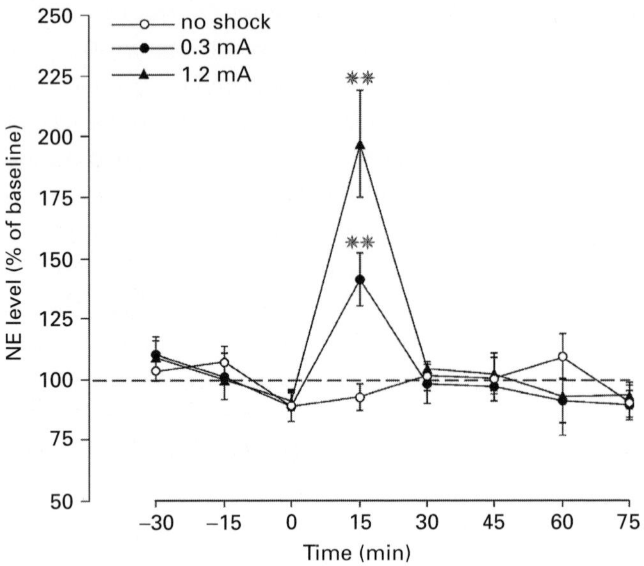

Figure 8.1 Effects of low- and high-intensity footshock on norepinephrine levels in the amygdala as assessed by in vivo microdialysis and high-performance liquid chromatography. Rats were given a single 3-s footshock at an intensity of 0.3 or 1.2 mA. Microdialysis samples were collected at 15-minute intervals throughout the session. Norepinephrine levels are represented as mean (± SEM) of basal levels prior to footshock $^{**}p < 0.01$ as compared with the no footshock group. (From Quirarte *et al.* [1998], with permission.)[52]

mean percentage and duration of the increase in amygdala norepinephrine levels assessed after inhibitory avoidance training greatly exceeded that seen in rats receiving a footshock of the same intensity and duration in a familiar holding cage.[51,53] The critical difference between animals that were trained and those that received shock in a holding cage was the exposure to the novel context of an inhibitory avoidance apparatus. Therefore, the pattern of amygdala norepinephrine release appears to be influenced by context. However, novelty alone cannot explain the enhanced increase in norepinephrine levels, as no marked increase in norepinephrine levels was found in rats that were exposed to the novel inhibitory avoidance apparatus in the absence of footshock. Therefore, the amplitude as well as the duration of norepinephrine release in the amygdala may determine norepinephrine effects on memory enhancement.

Further analysis examined the relationship between training-induced release of amygdala norepinephrine and subsequent retention performance of individual animals. As shown in Figure 8.2, amygdala norepinephrine levels assessed during and immediately after training were highly related to inhibitory avoidance performance

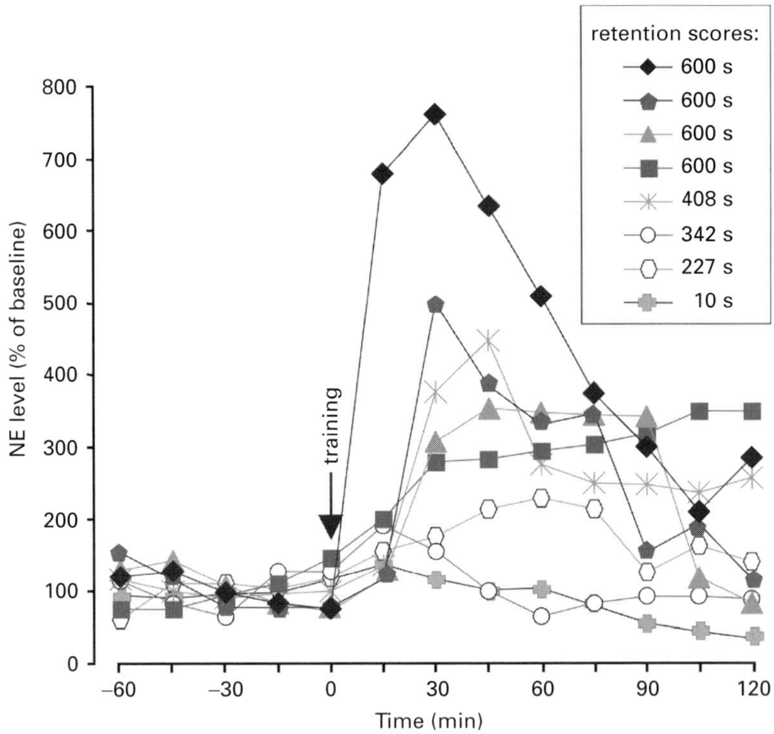

Figure 8.2 Norepinephrine levels in the amygdala in individual animals following inhibitory avoidance training. Percent of baseline norepinephrine following inhibitory avoidance training is shown for each individual rat. The key notes retention score on the following day. Amygdala norepinephrine levels correlate with 24-hour retention performance. Correlation values for the first five posttraining samples varied from +0.75 to +0.92. (From McIntyre *et al.* [2002], with permission.)[54]

tested 24 hours after the training. Those rats that exhibited the best retention (i.e., did not enter the shock compartment on the testing day) were also the rats in which the increase in amygdala norepinephrine levels in the samples following training was greatest.

Involvement of β-adrenoceptors and α-adrenoceptors in mediating norepinephrine effects on memory consolidation

The evidence reviewed above indicates that intra-amygdala infusions of β-adrenoceptor agonists dose-dependently enhanced memory consolidation, whereas infusions of β-adrenoceptor antagonists impaired memory consolidation. The amygdala also contains high densities of both α_1-adrenoceptor and α_2-adrenoceptor subtypes. α_1-adrenoceptors are found at the postsynaptic membrane,

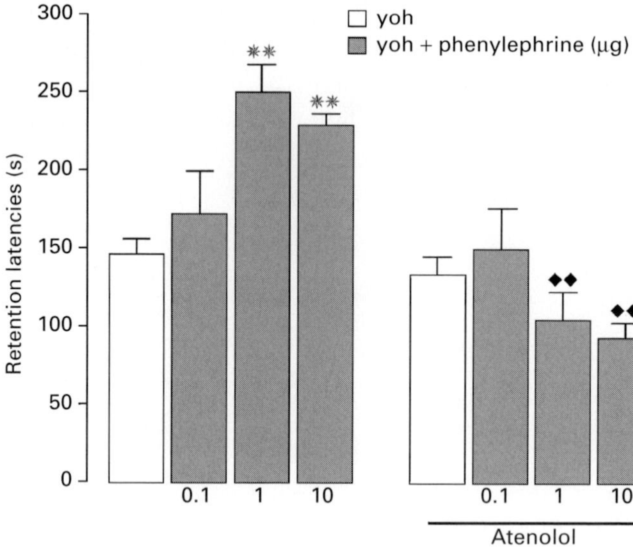

Figure 8.3 Step-through latencies (mean + SEM) in seconds for a 48-hour inhibitory avoidance test. Effects of immediate posttraining concurrent infusions of the nonselective α-adrenoceptor agonist phenylephrine (0.1, 1 or 10 μg) with the selective α_2-adrenoceptor antagonist yohimbine (yoh; 0.2 μg) into the basolateral amygdala alone or together with the β_1-adrenoceptor antagonist atenolol (1.0 μg). $^{**}p < 0.01$ as compared with the yohimbine group; ♦♦$p < 0.01$ as compared with the corresponding yohimbine + phenylephrine groups. (From Ferry *et al.* [1999], with permission.)[54]

whereas the α_2-adrenoceptor in the amygdala may be predominantly an autoreceptor and found at presynaptic sites. Recent findings indicate that the effects of norepinephrine on memory consolidation also involve α-adrenoceptors within the BLA in a close interaction with β-adrenergic mechanisms. In our experiments, activation of α-adrenoceptors in the BLA with posttraining infusions of the nonselective agonist phenylephrine induced a complex pattern of effects. Although none of the doses induced significant effects, low doses of phenylephrine tended to impair inhibitory avoidance retention, whereas higher doses did not show this tendency.[54] Although phenylephrine has typically been regarded as a selective α_1-adrenoceptor agonist, there is evidence that phenylephrine also stimulates prejunctional α_2-adrenoceptors. Therefore, the complex dose-response effects of phenylephrine are due most likely to a combined activation of α_1- and α_2-adrenoceptors. Concurrent infusions of the α_2-adrenoceptor antagonist yohimbine with phenylephrine enhanced retention (Figure 8.3). That is, yohimbine reversed the tendency of phenylephrine to impair retention, suggesting that presynaptic α_2-adrenoceptor activation impairs memory consolidation. This hypothesis is consistent with evidence that activation of presynaptic α_2-adrenoceptors blocks norepinephrine release. Moreover, immediate posttraining inactivation of postsynaptic α_1-adrenoceptors with the selective antagonist prazosin impaired retention, and selective activation of α_1-adrenoceptors by infusing phenylephrine together with the α_2-adrenoceptor antagonist yohimbine dose-dependently enhanced inhibitory avoidance retention.

The α_1-adrenergic influence on memory seems to be mediated by an interaction with β-adrenoceptors within the BLA. As is shown in Figure 8.3, posttraining intra-BLA infusions of the β-adrenoceptor antagonist atenolol block the memory enhancement induced by selective activation of α_1-adrenoceptors.[54] These findings indicate that the enhancing effects of amygdala α_1-adrenoceptor activation on memory consolidation require a coactivation of β-adrenoceptors. This idea was further investigated in another experiment. As is shown in Figure 8.4a, intra-BLA infusions of a low dose of the α_1-adrenoceptor antagonist prazosin did not impair retention performance alone but shifted the dose-response effects of the β-adrenoceptor agonist clenbuterol to the right when both drugs were infused together into the BLA after training.[42] That is, in the case of α_1-adrenoceptor blockade, a 1000 times higher dose of the β-adrenoceptor agonist was required to induce comparable memory enhancement. These results suggest that α_1-adrenergic activity in the BLA facilitates the effects of β-adrenergic activation on memory formation. Activation of β-adrenoceptors affects adenylate cyclase to stimulate intracellular cyclic adenosine monophosphate (cAMP) and cAMP-dependent protein kinase (PKA). Other findings indicate that intra-BLA infusions of the synthetic cAMP analog 8-bromo-cAMP enhance retention in a manner similar to that of clenbuterol. However, as prazosin did not alter the memory-enhancing effects induced by 8-bromo-cAMP (Figure 8.4b), cAMP must act downstream from the locus of interaction between α_1-adrenoceptors and the β-adrenoceptor–cAMP pathway.[42] These findings are consistent with pharmacological evidence suggesting that β-adrenoceptors modulate memory consolidation by a direct coupling to adenylate cyclase. The α_1-adrenoceptor is not coupled to this enzyme, but may act indirectly on this process by influencing the β-adrenergic-induced synthesis of cAMP. There is evidence that α_1-adrenoceptor activation influences the β–adrenoceptor-cAMP/PKA pathway through phospholipase A_2 effects on adenylate cyclase.[55]

Electrophysiological and molecular events in the amygdala induced by noradrenergic activation

The evidence that memory impairment induced by intra-amygdala infusions of propranolol is reversed by glutamate infusions indicates that noradrenergic stimulation may activate glutamatergic neurotransmission in the BLA.[56] Furthermore,

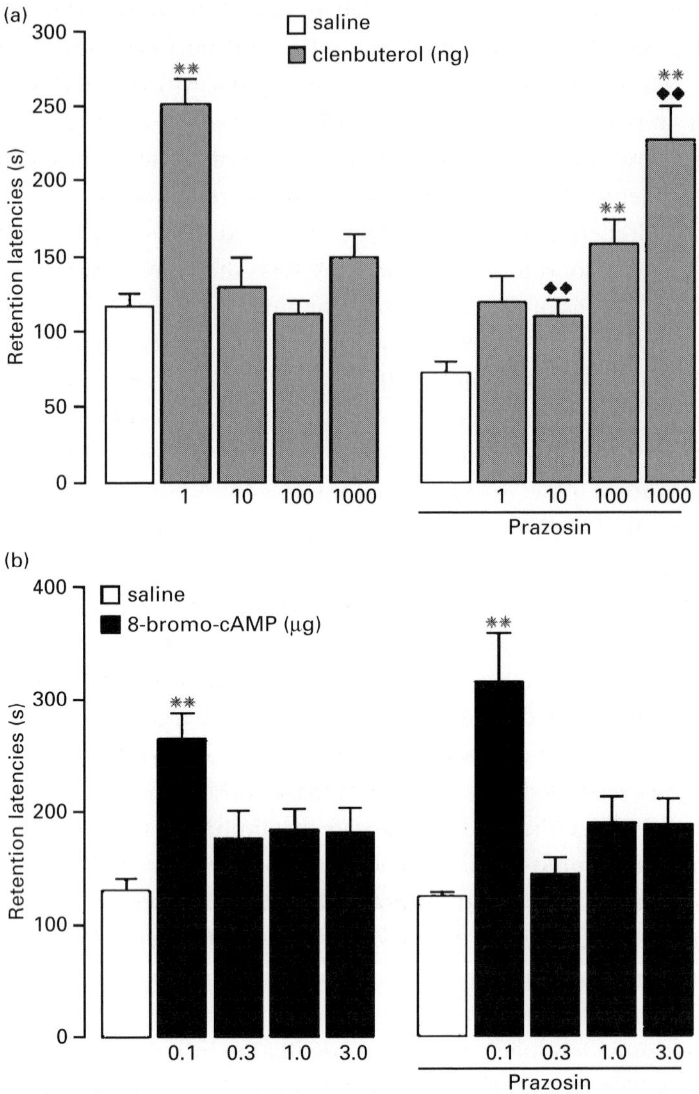

Figure 8.4 Step-through latencies (mean + SEM) in seconds for a 48-hour inhibitory avoidance test. Effects of immediate posttraining infusions of the α_1-adrenoceptor antagonist prazosin (0.1 μg) into the basolateral amygdala on the dose-response effects of the concurrently administered β-adrenoceptor agonist clenbuterol (1, 10, 100, or 1000 ng; a) or the synthetic cAMP analog 8-bromo-cAMP (0.1, 0.3, 1.0, or 3.0 μg; b). $^{**}p < 0.01$ as compared with the corresponding saline group; ♦♦$p < 0.01$ as compared with the corresponding clenbuterol group. (From Ferry *et al.* [1999], with permission.)[42]

findings of electrophysiological studies indicate that norepinephrine modulates *N*-methyl-D-aspartate (NMDA)-dependent neurotransmission in BLA pyramidal neurons in vitro.[57] Consistent with the behavioral evidence, pharmacological analysis revealed that the noradrenergic modulation of neurotransmission in the BLA can be dissected into a β-adrenoceptor-mediated excitation and a predominantly α_2-adrenoceptor-mediated inhibition. Application of the β-adrenoceptor agonist isoproterenol results in a long-lasting enhancement of the amplitude of excitatory postsynaptic potentials.[58,59] This effect appears to involve β-adrenoceptor influences on cAMP formation, as application of the direct adenylate cyclase activator forskolin induces a very similar facilitation of neuronal plasticity. Furthermore, blockade of PKA prevents late-phase long-term potentiation (L-LTP) in the lateral amygdala.[60] The effects are very similar to those found in studies of consolidation of fear-based memories. Auditory-cue fear conditioning is attenuated by administration of a PKA inhibitor directly into the BLA,[61] and fear memory consolidation is impaired in transgenic mice that overexpress an inhibitory isoform of PKA.[62] Protein Kinase A activation results in phosphorylation of the transcription factor cAMP-response element-binding protein (CREB), and manipulation of CREB activity levels in the amygdala has been reported to influence long-term memory formation for aversive conditioning.[63] Thus, norepinephrine effects on cAMP and PKA accumulation may activate the CREB signaling pathway to induce long-term changes in amygdala neurons. Norepinephrine effects on cAMP formation also induce activation of mitogen-activated protein kinases (MAPK).[64] A coordinated activation of CREB and MAPK in the amygdala is considered critical for long-term neuronal plasticity and memory consolidation.[60] Finally, norepinephrine effects on amygdala neurons and glial cells may include nonspecific influences on metabolic activity, such as increased glucose uptake and glycogenolysis, as well as on regional blood flow. Stimulation of brain β_2-adrenoceptors induces vasodilatation and hence may increase supply of glucose and other nutrients. Extensive evidence supports a role for glucose in memory consolidation. However, it is currently unknown whether norepinephrine effects on metabolic activity in the amygdala contribute to memory consolidation, as the memory impairment induced by intra-amygdala infusions of propranolol is not reversed by glucose administration.[56] Chapters 2 and 3 provide a more detailed description of the many actions of norepinephrine on cellular signaling.

Role of norepinephrine in the amygdala in regulating memory processes in other brain regions

Findings of many experiments have indicated that after memories have been consolidated, the BLA is no longer involved in the expression of long-term memories.[6] For example, lesions of the BLA or drug infusions into the BLA one day to one month after training do not block inhibitory avoidance retention performance. This evidence that the BLA may not be a site of long-term memory storage suggests that BLA activity influences other neural systems during the consolidation phase of memory storage. These findings may at first appear to conflict with those discussed above indicating that norepinephrine induces long-lasting changes in BLA neuroplasticity through an activation of intracellular signal transduction pathways. However, several findings indicate that training-induced norepinephrine release in the BLA is necessary for regulating memory consolidation processes in other brain regions. One opinion holds that the amygdala norepinephrine system participates in the consolidation of memory through its influence on efferent neural systems. Support for this view includes the findings that lesions of the stria terminalis, a major output pathway of the amygdala, block memory enhancement induced by posttraining injections of epinephrine,[65] as well as the enhancement induced by intra-amygdala infusions of norepinephrine.[39] For a detailed summary of BLA influences on other brain regions in regulating memory consolidation see McGaugh's review.[66]

Studies of the amygdala's influences on different forms of memory have provided further evidence that the BLA noradrenergic system influences other neural systems during consolidation of long-term memories. For example, posttraining injections of amphetamine into either the hippocampus or the caudate nucleus selectively enhance memory for the hippocampus-dependent water-maze spatial task or the caudate nucleus-dependent cued version of the water-maze task, respectively.[67] Amphetamine actions on memory consolidation are mediated, in part, through influences on noradrenergic mechanisms. Importantly, amphetamine infused posttraining into the amygdala enhances memory for both tasks. Furthermore, inactivation of the hippocampus or the caudate nucleus, but *not* the amygdala, during retention testing selectively attenuates amphetamine-induced memory enhancement.[68] These findings indicate that although posttraining activation of the amygdala induces memory enhancement, the memory traces for these types of training are not permanently stored in the amygdala. Recent experiments have also indicated a role for β-adrenoceptor activation within the BLA in modulating memory consolidation in the hippocampus. Posttraining intrahippocampal infusions of a glucocorticoid receptor agonist enhanced retention performance on an inhibitory avoidance task, but infusions of the β-adrenoceptor antagonist atenolol into the BLA blocked this memory enhancement.[69] These findings indicate a direct involvement of BLA β-adrenoceptors in the facilitation of hippocampal processing of newly acquired information.

Electrophysiological findings are consistent with a role of the amygdala noradrenergic system in influencing hippocampal processes underlying neuroplasticity and cognition. High-frequency stimulation of the BLA has been reported to facilitate population spike long-term potentiation (LTP) in the dentate gyrus of the hippocampus.[70] This potentiation depends on activation of β-adrenoceptors in the BLA.[71] Intra-BLA infusion of the β-adrenoceptor antagonist propranolol did not affect baseline neuronal activity in the hippocampus but blocked dentate gyrus LTP induced by high-frequency stimulation of the perforant path. Moreover, dentate gyrus LTP is reinforced by appetitive or aversive stimuli, or BLA stimulation, administered within 30 minutes before or after perforant path tetanization. Beta-adrenoceptor inactivation blocks this LTP reinforcement.[72] Other recent findings indicate that electrical stimulation of the BLA immediately preceding perforant path tetanization facilitates dentate gyrus LTP, but that depletion of norepinephrine with the noradrenergic toxin DSP-4 blocks the priming effect of BLA stimulation on hippocampal plasticity.[73]

Other evidence is also consistent with a role for β-adrenoceptors in the amygdala in regulating memory consolidation processes in a close relationship with other brain regions. Memory enhancement produced by activation of BLA β-adrenoceptors appears to depend on intact projections from the nucleus basalis magnocellularis (NBM) to the neocortex. Selective lesions of corticopetal NBM cholinergic projections with 192 IgG-saporin block inhibitory avoidance memory enhancement induced by intra-BLA infusions of norepinephrine.[43] As 192 IgG-saporin lesions do not deplete acetylcholine in the BLA, noradrenergic activation of the BLA may modulate memory consolidation processes in the neocortex by influencing the activity of corticopetal NBM neurons or, alternatively, through a coincident interaction in the neocortex of BLA and NBM efferents.

Interactions of the amygdala noradrenergic system with other neuromodulatory systems on memory consolidation

Glucocorticoid hormones

Noradrenergic activation in the amygdala is critical for other neuromodulatory influences on memory consolidation. Numerous studies in mice, rats, and chicks have shown that corticosterone (the main glucocorticoid in these species), or the synthetic ligand dexamethasone, enhances memory consolidation when administered immediately after training on a variety of aversively or appetitively motivated tasks.[74] The effects rely selectively on an activation of the low-affinity glucocorticoid receptor. Considerable evidence indicates that acute glucocorticoid effects on memory consolidation are mediated, at least in part, by influences involving the BLA. Lesions of the BLA block the enhancing effects of posttraining dexamethasone administration on memory for inhibitory avoidance training, and infusions

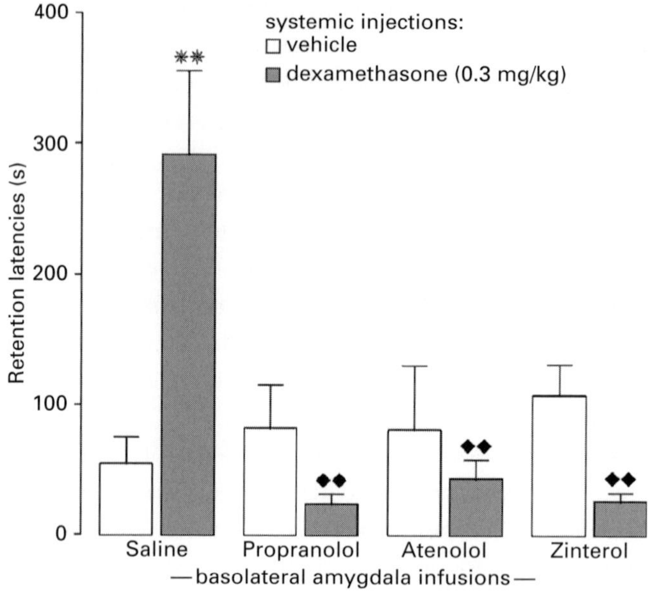

Figure 8.5 Step-through latencies (mean + SEM) in seconds for a 48-hour inhibitory avoidance test. Effects of pretraining infusions of either the nonselective β-adrenoceptor antagonist propranolol (0.5 μg), the β_1-adrenoceptor antagonist atenolol (0.5 μg), or the β_2-adrenoceptor antagonist zinterol (0.5 μg) into the basolateral amygdala on the effects of immediate post training injections of the synthetic glucocorticoid dexamethasone (0.3 mg/kg, s.c.). $^{**}p < 0.01$ as compared with the corresponding vehicle group; $^{\blacklozenge\blacklozenge}p < 0.01$ as compared with the saline-dexamethasone group. (From Quirarte *et al.* [1997], with permission.)[75]

of glucocorticoid receptor agonists or antagonists administered selectively into the BLA enhance or impair memory consolidation, respectively. Recent findings indicate that the enhancing effects of glucocorticoids on memory consolidation depend on the integrity of the amygdala noradrenergic system. As is shown in Figure 8.5, microinfusions of the nonselective β-adrenoceptor antagonist propranolol administered into the BLA 10 minutes before training blocked the enhancing effects of posttraining systemic injections of dexamethasone on memory for inhibitory avoidance training.[75] These effects appear to involve both β_1-adrenoceptors and β_2-adrenoceptors, as pretraining infusions of selective antagonists for either receptor subtype (i.e., atenolol or zinterol, respectively) also blocked dexamethasone effects on memory consolidation.

Atenolol or the PKA inhibitor Rp-cAMPS administered into the BLA also blocked memory enhancement induced by infusions of a glucocorticoid receptor agonist into the BLA.[76] These findings indicate that glucocorticoid effects on memory consolidation require activation of the β-adrenoceptor–cAMP/PKA pathway in the BLA. Other findings suggest that glucocorticoids enhance memory consolidation, in a permissive fashion, by potentiating β-adrenoceptor–cAMP/PKA efficacy in the BLA.[76] As noted above, posttraining intra-BLA infusions of the β-adrenoceptor agonist clenbuterol or the cAMP analog 8-bromo-cAMP enhance memory consolidation in a dose-dependent fashion.[31,41,42] As is shown in Figure 8.6a, the glucocorticoid receptor antagonist RU 38486 infused into the BLA shortly before training shifted the dose-response effects of clenbuterol such that a much higher dose of clenbuterol was required to induce memory enhancement.[76] In contrast, the glucocorticoid receptor antagonist did not modify the dose-response effects of 8-bromo-cAMP (Figure 8.6c), indicating that cAMP acts in the BLA downstream from the locus of interaction of glucocorticoids with the β-adrenoceptor–cAMP/PKA pathway.

This selective effect of glucocorticoid receptor blockade on β-adrenoceptor-, but not cAMP-, induced memory enhancement is very similar to that obtained with α_1-adrenoceptor blockade in the BLA and suggests that glucocorticoid receptor and α_1-adrenoceptor activation may influence the β-adrenoceptor–cAMP/PKA pathway through a common mechanism.[42] Furthermore, as reported for the glucocorticoid receptor agonist, the memory-enhancing effects of α_1-adrenoceptor activation in the BLA are blocked by concurrent infusions of atenolol.[54] There is evidence that both glucocorticoids and α_1-adrenoceptor activation influence the β-adrenoceptor–cAMP/PKA pathway through phospholipase A_2 effects on adenylate cyclase.[55] The finding that the glucocorticoid receptor antagonist blocked memory enhancement induced by α_1-adrenoceptor activation in the BLA (Figure 8.6b) is congruent with those of neurochemical studies in cerebral cortex indicating that glucocorticoids influence the α_1-adrenoceptor-mediated augmentation of β-adrenoceptor–cAMP efficacy.[77]

In addition to interacting with the noradrenergic signaling pathway at a postsynaptic level, other studies have shown that glucocorticoids influence noradrenergic function by altering levels of available norepinephrine. As discussed above (p. 242), the BLA receives noradrenergic innervation from brain stem noradrenergic cell groups in the LC and the NTS. The cell bodies of the noradrenergic neurons in the brain stem that project to the BLA contain among the highest densities of glucocorticoid receptors in the mammalian brain.[78] Glucocorticoid influences on noradrenergic neurons have been examined extensively in catecholaminergic neurons in the LC. Consistent with a permissive role, glucocorticoids activate noradrenergic neurons in the LC during emotionally arousing situations, but they do not seem to affect levels of tyrosine hydroxylase or norepinephrine synthesis rate in the LC of nonstressed rodents.[79] Glucocorticoids may affect the release of norepinephrine from its terminals via a genomic action on the activity or density of presynaptic α_2-adrenoceptors.[80] However, presynaptic influences on the release of

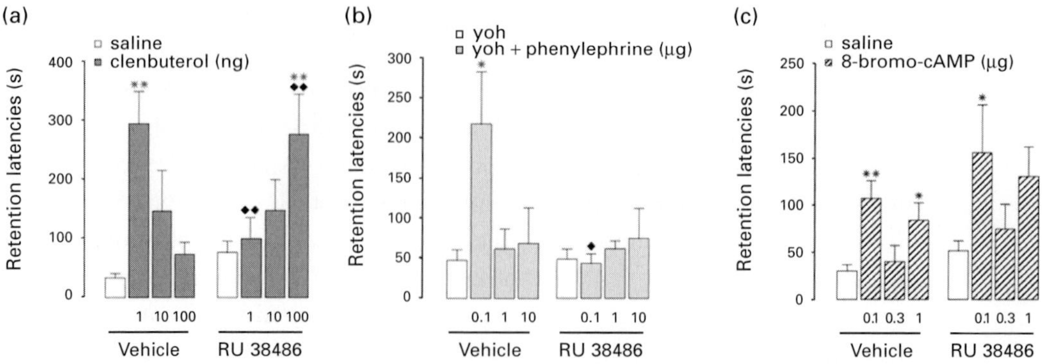

Figure 8.6 Step-through latencies (mean + SEM) in seconds on a 48-hour inhibitory avoidance test. Effects of pretraining infusions into the basolateral amygdala of the glucocorticoid receptor antagonist RU 38486 (1 ng) on the dose-response effects of immediate posttraining infusions into the basolateral amygdala of the β-adrenoceptor agonist clenbuterol (1, 10, or 100 ng; a), the nonselective α-adrenoceptor agonist phenylephrine (0.1, 1, or 10 μg) together with the selective α_2-adrenoceptor antagonist yohimbine (yoh; 0.2 μg; b), or the synthetic cAMP analog 8-bromo-cAMP (0.1, 0.3 or 1 μg; c). *$p < 0.05$; **$p < 0.01$ compared with the corresponding control group; ♦$p < 0.05$; ♦♦$p < 0.01$ compared with the corresponding vehicle group. (From Roozendaal *et al.* [2002], with permission.)[76]

norepinephrine appear not to be exclusively genomic. A microdialysis study found evidence for enhanced release of norepinephrine in the prefrontal cortex within 15 minutes after systemic administration of corticosterone.[81] Our findings, indicating that posttraining infusions of the glucocorticoid receptor agonist RU 28362 administered into the NTS modulate memory consolidation, provide further evidence that glucocorticoids may activate the noradrenergic system at the brain stem level.[24] Further, as is shown in Figure 8.7, inactivation of β-adrenoceptors in the BLA with atenolol blocked the memory enhancement induced by intra-NTS infusions of the glucocorticoid receptor agonist. Therefore, acute posttraining glucocorticoid receptor activation of noradrenergic soma in the NTS (and most likely the LC) may increase the release of norepinephrine in the amygdala. Lastly, glucocorticoids may increase norepinephrine levels via an extraneuronal mechanism. Norepinephrine is rapidly taken up by glial cells after its release into the synaptic cleft through a specific catecholamine-transporter protein. Corticosterone is a potent inhibitor of this catecholamine uptake through a rapid, nongenomic action.[82]

Opioid peptidergic system

The noradrenergic system of the amygdala is also involved in mediating other neuromodulatory influences on memory, including those of the opioid peptidergic system. Posttraining systemic injections of the opioid peptide β-endorphin and

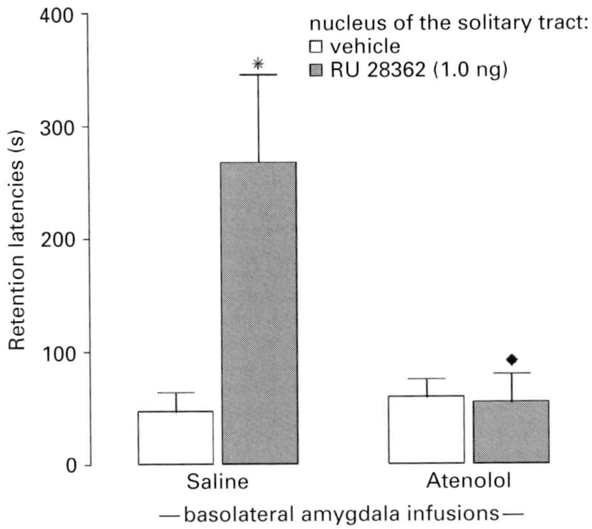

Figure 8.7 Step-through latencies (mean + SEM) in seconds for a 48-hour inhibitory avoidance test. Effects of pretraining infusions of the β_1-adrenoceptor antagonist atenolol (0.5 μg) into the basolateral amygdala on the effects of immediate posttraining infusions of the specific glucocorticoid receptor agonist RU 28362 (1.0 ng) into the nucleus of the solitary tract. $^*p <$ 0.05 as compared with the corresponding vehicle group; $^{\blacklozenge}p < 0.05$ as compared with the saline-RU 28362 group. (From Roozendaal *et al.* [1999], with permission.)[24]

opiates generally impair memory, while the opioid peptidergic antagonist naloxone enhances memory.[83,12] Lesions of the amygdala block opioid peptidergic influences on memory consolidation.[6] Importantly, 6-hydroxydopamine-induced lesions of the dorsal noradrenergic forebrain bundle also block the memory-enhancing effects of systemic or intra-amygdala infusions of naloxone.[84] The memory-enhancing effects of naloxone are also blocked by systemic injections of the β-adrenoceptor antagonist propranolol. In contrast, naloxone-induced memory enhancement is not blocked by the peripherally acting β-adrenoceptor antagonist sotalol, suggesting that opioids interact with central noradrenergic mechanisms. Subsequent studies indicated that the amygdala is a locus of interaction between opioids and the noradrenergic system in influencing memory consolidation. Intra-amygdala infusions of the nonselective β-adrenoceptor antagonist propranolol, the selective β_1-adrenoceptor antagonist atenolol, or the selective β_2-adrenoceptor antagonist zinterol all blocked the memory enhancement induced by systemic injections of naloxone.[85] In contrast, systemic or intra-amygdala injections of α_1- or α_2-adrenoceptor antagonists did not block naloxone effects on memory consolidation. These findings are consistent with extensive evidence suggesting a direct role of β-adrenoceptors in memory consolidation and a modulating role of

α_1-adrenoceptors. In experiments using both inhibitory avoidance and water-maze spatial tasks, intra-amygdala infusions of the β-adrenoceptor agonist clenbuterol blocked the memory-impairing effect of β-endorphin administered concurrently. Moreover, low and otherwise ineffective doses of β-endorphin and propranolol impaired memory when infused concurrently into the amygdala.[86] Findings of studies using in vivo microdialysis techniques have shown that opioids or opioid peptidergic antagonists influence the release of norepinephrine in the amygdala. As is shown in Figure 8.8, systemic injections of memory-enhancing doses of the opioid peptidergic antagonist naloxone significantly increased norepinephrine levels in the amygdala in nonstressed rats and potentiated the increase induced by footshock.[52] In contrast, systemic injections of β-endorphin, which impair memory consolidation, reduced norepinephrine levels in the amygdala of nonstressed rats and completely blocked the footshock-induced increase in norepinephrine levels.[52]

GABAergic system

Studies using various training tasks have demonstrated that posttraining systemic injections of GABAergic compounds modulate memory. The GABAergic antagonist picrotoxin enhances memory, whereas the GABAergic agonist muscimol impairs it.[12] As was found with adrenergic, glucocorticoid, and opioid peptidergic influences, lesions of the amygdala block GABAergic influences on memory.[6] Additionally, and most importantly, intra-amygdala infusions of propranolol block GABAergic influences on memory consolidation. Studies using in vivo microdialysis and HPLC techniques have shown that systemic injections of the memory-enhancing GABAergic antagonist picrotoxin produce a large and prolonged increase in norepinephrine levels in the amygdala, even in the absence of footshock; whereas the memory-impairing GABAergic agonist muscimol decreases norepinephrine levels below baseline.[87] Other evidence indicates that the GABAergic system also interacts with the noradrenergic system at a postsynaptic level. Gamma-aminobutyric acid receptor activation may influence the effects of β-adrenoceptor activation on cAMP accumulation, possibly through a similar mechanism as described for α_1-adrenoceptors and glucocorticoid receptors.

Cholinergic system

The cholinergic system also interacts with the amygdala noradrenergic system, but the mode of interaction appears to be different from that described above for glucocorticoid, opioid peptidergic and GABAergic systems.[6] There is extensive evidence that activation of muscarinic receptors within the BLA is a requirement for norepinephrine-induced memory enhancement. Systemic injections of the muscarinic cholinergic antagonist atropine attenuate the memory-enhancing

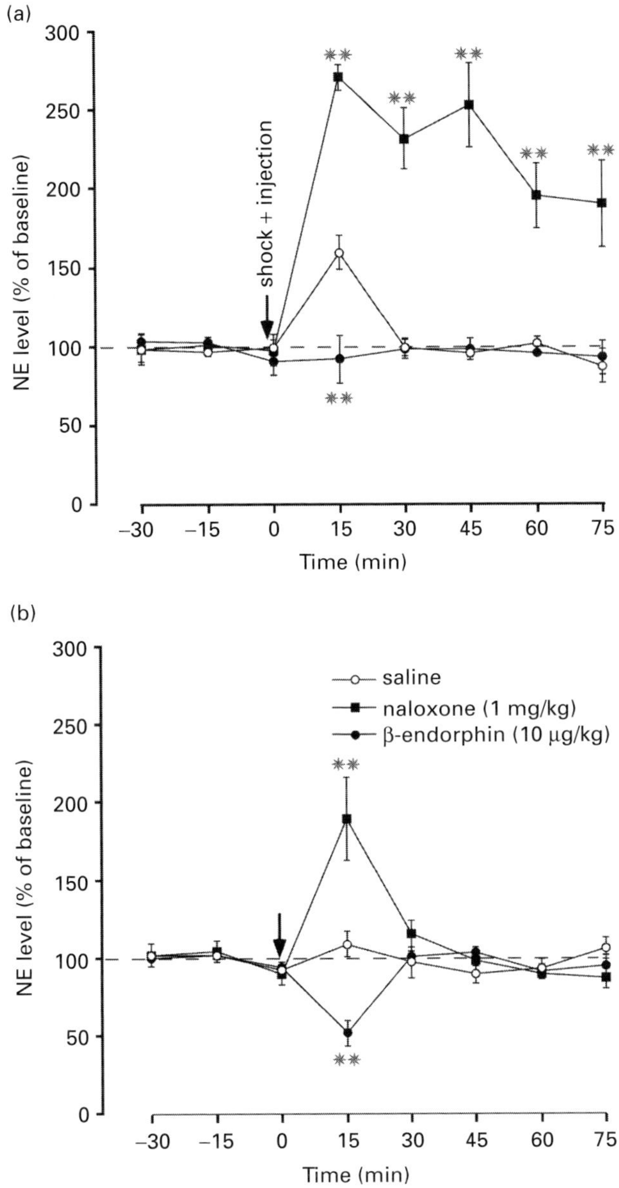

Figure 8.8 Effects of systemic injections of the opioid peptidergic agonist β-endorphin (10 μg/kg; intraperitoneal [ip]) and the antagonist naloxone (1 mg/kg, ip) on norepinephrine (NE) levels in the amygdala when administered immediately after footshock (0.55 mA, 1 s; a), or without prior footshock (b). Microdialysis samples were collected at 15-minute intervals throughout the session. Norepinephrine levels are presented as means (±SEM) of basal levels prior to drug administration or footshock. $^{**}p < 0.01$ as compared to saline group. (From Quirarte *et al.* [1998], with permission.)[52]

effects of the β-adrenoceptor agonist clenbuterol as well as that of epinephrine.[88] Intra-amygdala infusions of propranolol do not attenuate memory enhancement induced by systemic injections of the cholinergic agonist oxotremorine. However, a subeffective dose of atropine infused into the amygdala prevents the memory-enhancing effects of oxotremorine as well as those of the β-adrenoceptor agonist clenbuterol.[88] Recent neuroanatomical findings indicate that cholinergic fibers contact noradrenergic terminals in the BLA. Noradrenergic terminals also receive input from other (noncholinergic) terminals and form synapses with dendrites or spines of intrinsic cholinergic neurons in the BLA.[89] Taken together, these findings suggest that converging neuromodulatory systems influence noradrenergic and cholinergic activation in the BLA. This evidence strongly supports the hypothesis that noradrenergic activation of the BLA serves as a bridge between peripheral and central arousal systems and brain regions involved in memory consolidation. A schematic model of the role of norepinephrine in the BLA in regulating memory consolidation is represented in Figure 8.9.

Effects of norepinephrine in other brain regions on memory consolidation

Hippocampus/entorhinal cortex

The hippocampal formation (hippocampus and entorhinal cortex) is also involved in mediating norepinephrine effects on memory consolidation. This norepinephrine arises exclusively from norepinephrine-containing cell groups in the LC. As in the amygdala, norepinephrine is released in the hippocampus during stress[90] or following training on emotionally arousing tasks.[91] Furthermore, evidence suggests an involvement of noradrenergic mechanisms in the hippocampus in neuroplasticity. Considerable evidence indicates that norepinephrine increases the signal-to-noise ratio by inhibiting neuronal background firing while enhancing responses to significant stimuli.[25] Norepinephrine applied exogenously or released endogenously can initiate both a short- and a long-term potentiation of the dentate gyrus response to perforant path input. Triggering of the potentiated response depends on β-adrenoceptor activation and does not require a high-frequency stimulus.[92] Other evidence indicates that β-adrenoceptor activation selectively potentiates excitatory postsynaptic potentials and population spikes evoked by stimulation of the medial perforant pathway, but depresses responses to lateral perforant pathway stimulation.[93] Furthermore, an increase in LC activity can initiate both short- and long-term potentiation in the perforant path response, although a reduction in LC activity does not alter baseline perforant path responses. These signal-enhancing effects of LC stimulation on dentate gyrus population spikes are blocked by administration of a β-, but not an α-adrenoceptor antagonist, and

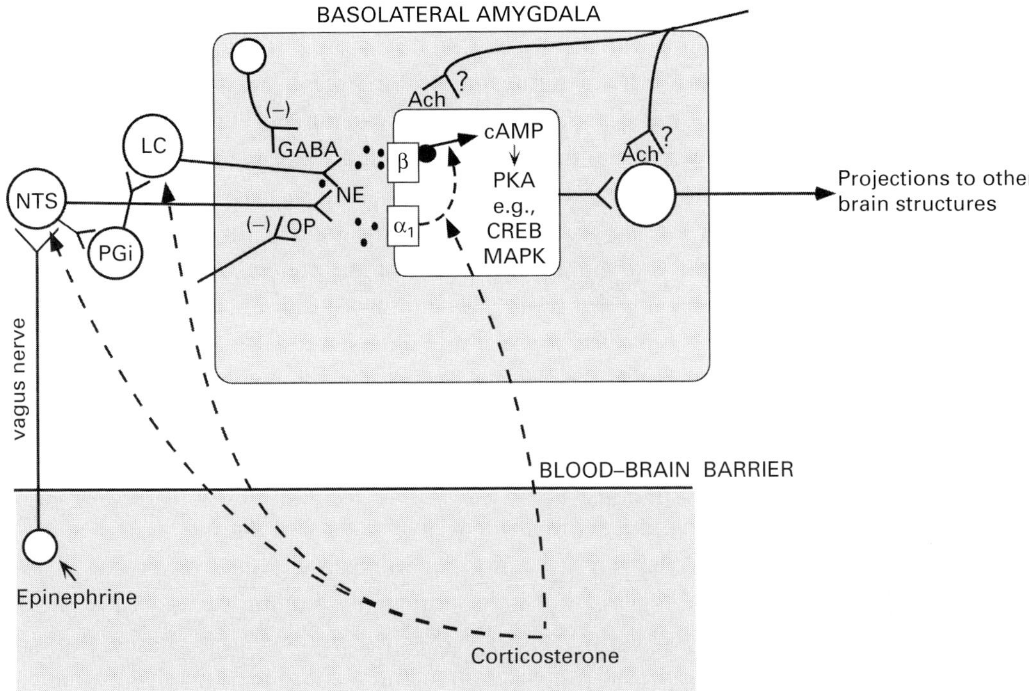

Figure 8.9 Schematic model summarizing the role of the noradrenergic system of the basolateral amygdala in memory consolidation. Norepinephrine (NE) is released in the basolateral amygdala following training in aversively motivated tasks and binds to both β-adrenoceptors and α_1-adrenoceptors at postsynaptic sites. The β-adrenoceptor is coupled directly to adenylate cyclase to stimulate cAMP formation. The α_1-adrenoceptor modulates the response induced by β-adrenoceptor stimulation. Intracellular cAMP can initiate a cascade of molecular events in the basolateral amygdala. The memory-modulatory effects of several other neuromodulatory influences, including those of epinephrine, glucocorticoid, opioid peptidergic, and GABAergic systems, are mediated by converging influences on the noradrenergic system of the basolateral amygdala. Drug interactions with the noradrenergic system can occur at both presynaptic and postsynaptic loci. These noradrenergic effects in the basolateral amygdala are required for regulating memory consolidation in other brain regions. α_1, α_1-adrenoceptor; ACh, acetylcholine; β, β-adrenoceptor; cAMP, cyclic adenosine monophosphate; CREB, cAMP-response element-binding protein; GABA, γ-aminobutyric acid; LC, locus coeruleus; MAPK, mitogen-activated protein kinase; NTS, nucleus of the solitary tract; OP, opioid peptide; PGi, nucleus paragigantocellularis; PKA, protein kinase A. (From McGaugh *et al.* [2000];[6] Roozendaal *et al.* [2000], with permission.)[74]

are further abolished by pretreatment of the animals with the catecholamine neurotoxin 6-hydroxydopamine.[94]

Other evidence indicates that infusions of norepinephrine directly into the hippocampus enhance memory consolidation.[95] Norepinephrine enhances memory for inhibitory avoidance training when infused into the hippocampus immediately after or 3 to 6 hours after training. However, infusions of norepinephrine into the hippocampus administered 1.5 hours after training do not enhance retention. Infusions of the β-adrenoceptor antagonist timolol administered into the hippocampus impair memory consolidation when infused 3 or 6 hours after training.[95] These findings suggest that norepinephrine in the hippocampus appears to be involved in two temporally separated processes of memory formation. Norepinephrine also induces memory enhancement when infused into the entorhinal cortex or the posterior parietal cortex up to 6 hours after inhibitory avoidance training.[96] Conversely, inactivation of β-adrenoceptors with local infusions of the antagonist timolol into the entorhinal cortex induces impairment of long-term memory.

Although there is relatively little information regarding the involvement of specific adrenoceptor subtypes in the hippocampus and the entorhinal cortex in memory consolidation, biochemical studies performed on these brain regions strongly suggest that norepinephrine influences on memory consolidation rely on an interaction between both β-adrenoceptors and α-adrenoceptors. Furthermore, a novel β_3-adrenoceptor was described in the rat hippocampus.[97] Although its function in the rat has not been investigated, findings of studies in chicks suggest that this receptor subtype may also play a role in memory consolidation. The effects of norepinephrine on long-term memory formation in the hippocampus and the entorhinal cortex also appear to be mediated through adrenoceptor actions on the cAMP/PKA signal transduction pathway. Infusions of the synthetic cAMP analog 8-bromo-cAMP, or the adenylate cyclase activator forskolin into the hippocampus or the entorhinal cortex, also enhance memory consolidation when infused immediately, 3, or 6 hours after training; whereas the PKA inhibitor KT5720 impairs inhibitory avoidance retention performance.[96,97] These findings correlate with the findings of increased cAMP, PKA, and CREB phosphorylation levels at 3 to 6 hours from training in rat hippocampus.

In a manner comparable to that found with the amygdala, the noradrenergic system of the hippocampus is critically involved in regulating the modulatory effects of several other neurotransmitters on memory consolidation. Benzodiazepines (such as midazolam) given systemically to rats before inhibitory avoidance training produce anterograde amnesia. However, infusions of norepinephrine or 8-bromo-cAMP into the hippocampus (as well as the entorhinal cortex and parietal cortex) 3 hours after training reversed this midazolam-induced amnesia.[98] These findings suggest that benzodiazepine (and GABAergic) effects on memory consolidation are mediated through converging actions on the noradrenergic system within these brain regions. There are many other examples of studies examining interactions of hippocampal norepinephrine with various neurotransmitter systems. As discussed above, intrahippocampal infusions of amphetamine enhance memory consolidation on a variety of tasks.[68,69] Infusion of the β-adrenoceptor antagonist propranolol, at a dose that does not affect retention alone, antagonizes the memory-enhancing effect of amphetamine.[99] Along with this memory-enhancing effect, amphetamine also elevates the level of norepinephrine release in the hippocampus. These results support the hypothesis that amphetamine enhances memory consolidation, in part, through facilitation of hippocampal norepinephrine release. Similar effects were observed with the stress-activated neuropeptide corticotropin-releasing factor. Corticotropin-releasing factor may affect neuroplasticity in the hippocampus, depending on the noradrenergic system.[100] Corticotropin-releasing factor–noradrenergic interactions also play a direct role in memory consolidation. The β-adrenoceptor antagonist propranolol or the noradrenergic toxin DSP-4 block memory enhancement induced by corticotropin-releasing factor infusions into the hippocampus.[100] These findings suggest that corticotropin-releasing factor, through a presynaptic facilitation mechanism, stimulates the release of norepinephrine in the hippocampus. Moreover, both the β-adrenoceptor and the corticotropin-releasing factor receptor are coupled to the adenylate cyclase system, providing an additional potential locus of interaction at the postsynaptic level. The hippocampal noradrenergic system also interacts with the cholinergic system in modulating memory consolidation. It has been reported that partial lesions of the fornix, which deprives the hippocampus of cholinergic input, results in severe impairments on spatial memory tasks.[101] This manipulation, in addition, induces functional changes in the hippocampal noradrenergic system. Norepinephrine turnover in these animals is increased and quantitative autoradiography techniques revealed significantly lower levels of β-adrenoceptors in both the dorsal and ventral hippocampus of fornix-lesioned rats. Furthermore, α_2-adrenoceptors were significantly decreased in the dorsal region of the hippocampus, but increased in the ventral regions. These findings indicate that after disruption of cholinergic innervation of the hippocampus with a fornix transsection, the noradrenergic system is reorganized. This reorganization may contribute to some of the cognitive impairments seen after fornix lesions. Consistent with these observations (suggesting an interaction of the hippocampal noradrenergic system with the cholinergic system) is the finding that noradrenergic depletion with systemic injections of the toxin DSP-4 potentiates impairment of working memory produced by intrahippocampal infusions of subeffective doses of the muscarinic antagonist scopolamine.[102] Other evidence indicating interactions between noradrenergic and cholinergic receptor signaling comes from studies investigating drug effects on hippocampal LTP. Low doses of the β-adrenoceptor agonist isoproterenol and the cholinergic receptor agonist carbachol do not affect the induction of LTP in the hippocampal CA1 region when applied individually, but they dramatically facilitate LTP induction when coapplied.[103] Furthermore, several other neuromodulators, including glucocorticoids, estrogens, and neuropeptides such as vasopressin have been reported to influence hippocampal noradrenergic function in the modulation of long-term neuroplasticity and memory consolidation.

Medial prefrontal cortex

Most studies examining the role of the noradrenergic system in the medial prefrontal cortex evaluated its effects on working memory performance.[104] Norepinephrine effects in the medial prefrontal cortex on working memory appear to be opposite to the effects on long-term memory consolidation, in that stress levels of norepinephrine induce working memory impairments, whereas basal levels of norepinephrine improve working memory. These dual effects appear to be mediated by a dose-dependent activation of α-adrenoceptor subtypes. The α_2-adrenoceptor agonist guanfacine improves working memory, an effect reversed by the α_2-adrenoceptor antagonist idazoxan. Alternately, infusions of the α_1-adrenoceptor agonist cirazoline, or cAMP activation with Sp-cAMPS impair working memory performance. These findings are discussed further in Chapter 14. The role of norepinephrine in the prefrontal cortex in long-term memory consolidation has not been investigated extensively, but it has been reported that immediate posttraining infusions of norepinephrine into the medial prefrontal cortex enhance 24-hour inhibitory avoidance retention performance.[26] Thus these findings would be consistent with the general hypothesis suggesting that noradrenergic activation enhances long-term memory formation. It is currently unknown whether these norepinephrine effects in the prefrontal cortex on memory consolidation are mediated by the same neural mechanisms as those influencing working memory performance.

Insular cortex

The insular cortex, which processes visceral information, is critically involved in conditioned taste aversion and is therefore also known as the "gustatory cortex." In this task, a novel taste is paired with a gastric malaise induced by an intraperitoneal administration of lithium chloride, resulting in a long-term avoidance of this taste. Several drug administration studies have implicated central norepinephrine in memory for conditioned taste aversion. Mice heterozygous for a mutation in the gene encoding tyrosine hydroxylase also displayed deficits in conditioned taste aversion.[33] Although some of these effects may be due to actions of norepinephrine elsewhere in the brain (i.e., the BLA),[49] norepinephrine also directly acts in the insular cortex in modulating conditioned taste aversion memory. It has been reported that posttraining infusions of the cAMP analog 8-bromo-cAMP into the insular cortex enhance memory for conditioned taste aversion[105] and that infusions of the β-adrenoceptor antagonist propranolol into the insular cortex block long-term conditioned taste aversion memory.[106] Noradrenergic function in the insular cortex, however, does not seem to be restricted to conditioned taste aversion, as post-training infusions of 8-bromo-cAMP into the insular cortex also enhance memory for inhibitory avoidance training.[105]

Norepinephrine effects on memory in human subjects

Several studies indicate that activation of the noradrenergic system in humans is required for the enhancing effect of arousal (either emotionally or physically induced) on memory. For example, propranolol attenuates the enhanced long-term memory associated with emotional information without affecting memory for emotionally arousing neutral information.[107] Because of the selectivity for emotionally salient information, the impairing effect of propranolol cannot easily be attributed to potential effects of the drug on attention, sedation, or emotional reactions to the material.[107] Van Stegeren and colleagues[108] investigated whether the impairing effects of propranolol are due to actions at peripheral or central β-adrenoceptors. Subjects in this study received placebo or propranolol before viewing either a relatively emotionally arousing story, or a closely matched but more emotionally neutral story (the same story used by Cahill *et al.*[107]). An additional group received nadolol (a peripherally acting β-adrenoceptor antagonist). Propranolol, but not nadolol, impaired memory for the emotionally arousing story, strongly suggesting an involvement of central noradrenergic mechanisms in emotionally influenced memory. Metoprolol, a selective β_1-adrenoceptor antagonist, administered before viewing an emotionally arousing narrated slide show also impaired retention performance assessed one week later. In contrast, activation of central noradrenergic mechanisms with yohombine (which stimulates noradrenergic activity via blockade of α_2-adrenergic inhibitory autoreceptors) induced significantly improved retention performance for this information.[109] In another study yohimbine was given to subjects shortly after viewing the series of slides depicting an emotionally arousing story.[110] In addition, blood samples were drawn for determining plasma free 3-methoxy-4-hydroxyphenylglycol (MHPG) levels, a major metabolite of norepinephrine. During the surprise retention test one week later, yohimbine did not induce significant effects on retention performance. However, linear regression revealed a significant effect of MHPG levels during encoding on memory score for the group as a whole (i.e., both after yohimbine and placebo treatment).

As in rats, adrenergic stimulation can act in a retrograde manner to enhance memory in humans. Soetens *et al.*[111] reported that amphetamine given to subjects immediately after learning lists of words enhanced long-term memory for the words. Findings from another recent study[112] indicated that administration of epinephrine enhanced memory consolidation. Immediately after viewing a series of 21 slides, subjects received a 3 minute intravenous infusion of either saline or epinephrine. A surprise memory test one week later revealed that epinephrine produced a dose-dependent enhancement of memory for the initial slides in the series. These authors further reported a higher emotional (i.e., electrodermal) response to the primacy slides. These findings support the view that endogenous stress hormones modulate memory consolidation for experiences that induce their release. In support of this view, a recent study reported that the total dose of epinephrine administered to patients treated in intensive care for traumatic injury or disease was, among many measures assessed, the single best predictor for the number of traumatic memories reported months later.[113]

Findings suggest that the release of adrenergic stress hormones regulates the strength of long-term memory. These findings, as well as the evidence that administration of the β-adrenoceptor blocker propranolol attenuates the enhanced long-term memory induced by emotionally arousing information, also suggest that overactivation of this modulatory system may contribute to the development of chronic stress symptoms such as posttraumatic stress disorder (PTSD).[114] In support of this suggestion, Pitman *et al.*[115] reported that chronic administration of a peripherally acting β-blocker to individuals who had experienced trauma reduced the subsequent incidence of PTSD assessed weeks after the traumatic experience.

Summary

The evidence summarized here indicates that norepinephrine is critically involved in regulating the consolidation of long-term memory formation. Emotional arousal or traumatic experiences stimulate norepinephrine-producing cells in brain stem regions of the NTS and the LC and induce the release of norepinephrine in forebrain structures, including the BLA, the hippocampus, and the medial prefrontal cortex. Norepinephrine affects neuroplasticity and molecular events to induce long-term changes in cellular properties, even if the memory traces are not permanently stored here. Several other neuromodulators affect memory consolidation through converging influences on the central noradrenergic system in these brain regions. This norepinephrine-activated mechanism serves the adaptive function of influencing the consolidation of strong, long-lasting memories of emotionally significant experiences. However, overactivation of this system may contribute to the development of chronic stress symptoms such as PTSD.

ACKNOWLEDGEMENTS

The author thanks Drs. James L. McGaugh and Christa K. McIntyre for helpful comments on an earlier draft of the manuscript. Research supported by US Public Health Service Grant MH12526 from National Institute of Mental Health.

REFERENCES

1. Kety, S. Brain catecholamines, affective states and memory. In *The Chemistry of Mood, Motivation and Memory*, ed. J. L. McGaugh. (New York: Plenum Press, 1972), pp. 65–80.
2. Doty, B. A. and Doty, L. A. Facilitative effects of amphetamine on avoidance conditioning in relation to age and problem difficulty. *Psychopharmacologia*, **9** (1966), 234–241.
3. Gold, P. E. and van Buskirk, R. Posttraining brain norepinephrine concentrations: correlation with retention performance of avoidance training and with peripheral epinephrine modulation of memory processing. *Behav. Biol.*, **25** (1978), 509–520.
4. Haycock, J. W., van Buskirk, R. and McGaugh, J. L. Effects of diethyldithiocarbamate on memory processes in mice: a neurobiological analysis. *Behav. Biol.*, **20** (1977), 281–310.
5. Meligeni, J. A., Ledergerber, S. A. and McGaugh, J. L. Norepinephrine attenuation of amnesia produced by diethyldithiocarbamate. *Brain Res.*, **149** (1978), 155–164.
6. McGaugh, J. L., Ferry, B., Vazdarjanova, A. and Roozendaal, B. Amygdala: role in modulation of memory storage. In *The Amygdala*, 2nd ed. J. P. Aggletyon., (New York), (Oxford University Press, 2000), pp. 391–423.
7. Müller, G. E. and Pilzecker, A. Experimentelle Beitrage zur lehre vom Gedachtniss. *Zeitschrift für Psychologie*, **1** (1900), 1–288.
8. McGaugh, J. L. Time-dependent processes in memory storage. *Science*, **153** (1966), 1351–1358.
9. Gold, P. E and McGaugh, J. L. A single-trace, two process view of memory storage processes. In *Short-Term Memory*, ed. D. and J. A. Deutsch. (New York: Academic Press, 1975), pp. 355–378.
10. Livingston, R. B. Reinforcement: a study program. In *The Neurosciences*, ed. G. C. Quarton, T. Melnechuk and F. O. Schmitt. (New York: Rockefeller University Press, 1967), pp. 568–576.
11. Gerard, R. W. The fixation of experience. In *Brain Mechanisms and Learning*, ed. A. Fessard, R. W. Gerard and J. Konorski. (Springfield: Chars C. Thomas, 1961), pp. 21–35.
12. McGaugh, J. L. Involvement of hormonal and neuromodulatory systems in the regulation of memory storage. *Annu. Rev. Neurosci.*, **12** (1989), 255–287.
13. Martinez, J. L. J., Vasquez, B. J., Rigter, H. *et al.* Attenuation of amphetamine-induced enhancement of learning by adrenal demedullation. *Brain. Res.*, **195** (1980), 433–443.
14. Williams, C. L., Packard, M. G. and McGaugh, J. L. Amphetamine facilitation of win-shift radial maze retention: the involvement of peripheral adrenergic and central dopaminergic systems. *Psychobiology*, **22** (1994), 141–148.

15. Gold, P. E. and van Buskirk, R. Facilitation of time-dependent memory processes with posttrial epinephrine injections. *Behav. Biol.*, **13** (1975), 145–153.
16. Gold, P. E. and van Buskirk, R. Effects of posttrial hormone injections on memory processes. *Horm. Behav.*, **7** (1976), 509–517.
17. Gold, P. E., van Buskirk, R. and Haycock, J. Effects of posttraining epinephrine injections on retention of avoidance training in mice. *Behav. Biol.*, **20** (1977), 197–207.
18. Introini-Collison, I. B., Saghafi, D., Novack, G. and McGaugh, J. L. Memory enhancing effects of posttraining dipivefrin and epinephrine: involvement of peripheral and central adrenergic receptors. *Brain. Res.*, **572** (1992), 81–86.
19. Clark, K. B, Krahl, S. E, Smith, D. C. and Jensen, R. A. Post-training unilateral vagal stimulation enhances retention performance in the rat. *Neurobiol. Learn. Mem.*, **63** (1995), 213–216.
20. Petrov, T., Krukoff, T. L. and Jharmandas, J. H. Branching projections of catecholaminergic brainstem neurons to the paraventricular hypothalamic nucleus and the central nucleus of the amygdala in the rat. *Brain. Res.*, **609** (1993), 81–92.
21. Williams, C. L. and Clayton, E. C. Contribution of brainstem structures in modulating memory storage processes. In *Memory Consolidation: Essays in Honor of James L. McGaugh*, ed. P. E. Gold and W. T. Greenough (Washington: American Psychological Association, 2001), pp. 141–163.
22. Williams, C. L., Men, D. and Clayton, E. C. The effects of noradrenergic activation of the nucleus tractus solitarius on memory and in potentiating norepinephrine release in the amygdala. *Behav. Neurosci.*, **114** (2000), 1131–1144.
23. Miyashita, T. and Williams, C. L. Glutamatergic neurotransmission in the nucleus of the solitary tract modulates memory through influences on amygdala noradrenergic systems. *Behav. Neurosci.*, **116** (2002), 13–21.
24. Roozendaal, B., Williams, C. L. and McGaugh, J. L. Glucocorticoid receptor activation in the rat nucleus of the solitary tract facilitates memory consolidation: involvement of the basolateral amygdala. *Eur. J. Neurosci.*, **11** (1999), 1317–1323.
25. Valentino, R. J. and Aston-Jones, G. Physiological and anatomical determinants of locus coeruleus discharge. In *Psychopharmacology: The Fourth Generation of Progress*, ed. F. E. Bloom and D. J. Kipfer (New York: Raven, 1995), pp. 373–385.
26. Liang, K. C. Epinephrine modulation of memory: amygdala activation and regulation of long-term memory storage. In *Memory Consolidation: Essays in Honor of James L. McGaugh*, ed. P. E Gold and W. T. Greenough (Washington D. C: American Psychological Association, 2001), pp. 165–183.
27. Ferry, B., Roozendaal, B. and McGaugh, J. L. Role of norepinephrine in mediating stress hormone regulation of long-term memory storage: a critical involvement of the amygdala. *Biol. Psychiatry*, **46** (1999), 1140–1152.
28. Archer, T., Soderberg, U., Ross, S. B. and Jonsson, G. Role of olfactory bulbectomy and DSP-4 treatment in avoidance learning in the rat. *Behav. Neurosci*, **98** (1984), 496–505.
29. Cahill, L., Pham, C. A. and Setlow, B. Impaired memory consolidation in rats produced with β-adrenergic blockade. *Neurobiol. Learn. Mem.*, **74** (2000), 259–266.

30. Cohen, R. P. and Hamburg, M. D. Evidence for adrenergic neurons in a memory access pathway. *Pharmacol. Biochem. Behav.*, **3** (1975), 519–523.
31. Introini-Collison, I. B., Miyazaki, B. and McGaugh, J. L. Involvement of the amygdala in the memory-enhancing effects of clenbuterol. *Psychopharmacology*, **104** (1991), 541–544.
32. Sara, S. J., Devauges, V., Biegon, A. and Blizard, D. A. The Maudsley rat strains as a probe to investigate noradrenergic–cholinergic interaction in cognitive function. *J. Physiol. Paris*, **88** (1994), 337–345.
33. Kobayashi, K. and Kobayashi, T. Genetic evidence for noradrenergic control of long-term memory consolidation. *Brain Dev.*, **23** (2001), 16–23.
34. Williams, C. L., Men, D., Clayton, E. C. and Gold, P. E. Norepinephrine release in the amygdala following systemic injection of epinephrine or escapable foot shock: contribution of the nucleus of the solitary tract. *Behav. Neurosci.*, **112** (1998), 1414–1422.
35. Radna, R. J. and MacLean, P. D. Vagal elicitation of respiratory-type and other unit responses in basal limbic structures of squirrel monkeys. *Brain Res.*, **213** (1981), 45–61.
36. Rogers, R. C. and Fryman, D. L. Direct connections between the central nucleus of the amygdala and the nucleus of the solitary tract: an electrophysiological study in the rat. *J. Auton. Nerv. Syst.*, **22** (1988), 83–87.
37. Liang, K. C. Pretraining infusions of DSP-4 into the amygdala impaired retention in the inhibitory avoidance task: involvement of norepinephrine but not serotonin in memory facilitation. *Chin. J. Physiol.*, **41** (1998), 223–233.
38. Liang, K. C., Chen, L. L. and Huang, T. E. The role of amygdala norepinephrine in memory formation: involvement of memory enhancing effect of peripheral epinephrine. *Chin. J. Physiol.*, **38** (1995), 81–91.
39. Liang, K. C., Juler, R. and McGaugh, J. L. Modulating effects of posttraining epinephrine on memory: involvement of the amygdala noradrenergic system. *Brain Res.*, **368** (1986), 125–133.
40. Gallagher, M., Kapp, B. S., Musty, R. E. and Driscoll, P. A. Memory formation: evidence for a specific neurochemical system in the amygdala. *Science*, **198** (1977), 423–425.
41. Liang, K. C., McGaugh, J. L. and Yao, H. Y. Involvement of amygdala pathways in the influence of post-training intra-amygdala norepinephrine and peripheral epinephrine on memory storage. *Brain Res.*, **508** (1990), 225–233.
42. Ferry, B., Roozendaal, B. and McGaugh, J. L. Basolateral amygdala noradrenergic influences on memory storage are mediated by an interaction between β- and α1-adrenoceptors. *J. Neurosci.*, **19** (1999), 5119–5123.
43. Power, A. E., Thal, L. J. and McGaugh, J. L. Lesions of the nucleus basalis magnocellularis induced by 192 IgG-saporin block memory enhancement with posttraining norepinephrine in the basolateral amygdala. *Proc. Natl. Acad. Sci. U. S. A.*, **99** (2002), 2315–2319.
44. LaLumiere, R. T., Buen, T. V. and McGaugh, J. L. Post-training intra-basolateral amygdala infusions of norepinephrine enhance consolidation of memory for contextual fear conditioning. *J. Neurosci.*, **23** (2003), 6754–6758.
45. Roozendaal, B., Koolhaas, J. M. and Bohus, B. Posttraining norepinephrine infusion into the central amygdala differentially enhances later retention in Roman high-avoidance and low-avoidance rats. *Behav. Neurosci.*, **107** (1993), 575–579.

46. Salinas, J. A., Introini-Collison, I. B., Dalmaz, C. and McGaugh, J. L. Posttraining intraamygdala infusions of oxotremorine and propranolol modulate storage of memory for reductions in reward magnitude. *Neurobiol. Learn. Mem.*, **68** (1997), 51–59.
47. Introini-Collison, I. B., Nagahara, A. H. and McGaugh, J. L. Memory enhancement with intra-amygdala post-training naloxone is blocked by concurrent administration of propranolol. *Brain Res.*, **476** (1989), 94–101.
48. Hatfield, T. and McGaugh, J. L. Norepinephrine infused into the basolateral amygdala post-training enhances retention in a spatial water maze task. *Neurobiol. Learn. Mem.*, **91** (1999), 232–239.
49. Miranda, M. I., LaLumiere, R. T., Buen, T. V., Bermudez-Rattoni, F. and McGaugh, J. L. Blockade of noradrenergic receptors in the basolateral amygdala impairs taste memory. *Eur. J. Neurosci.*, **18** (2003), 2605–2610.
50. Tanaka, T., Yokoo, H., Mizoguchi, K. *et al.* Noradrenaline release in the rat amygdala is increased by stress: studies with intracerebral microdialysis. *Brain Res.*, **544** (1991), 174–176.
51. Galvez, R., Mesches, M. and McGaugh, J. L. Norepinephrine release in the amygdala in response to footshock stimulation. *Neurobiol. Learn. Mem.*, **66** (1996), 253–257.
52. Quirarte, G. L., Galvez, R., Roozendaal, B. and McGaugh, J. L. Norepinephrine release in the amygdala in response to footshock and opioid peptidergic drugs. *Brain Res.*, **808** (1998), 134–140.
53. McIntyre, C. K., Hatfield, T. and McGaugh, J. L. Norepinephrine levels in the amygdala following inhibitory avoidance training predict retention score. *Eur. J. Neurosci.*, **16** (2002), 1223–1226.
54. Ferry, B., Roozendaal, B. and McGaugh, J. L. Involvement of α_1-adrenoceptors in the basolateral amygdala in modulation of memory storage. *Eur. J. Pharmacol.*, **372** (1999), 9–16.
55. Duman, R. S., Karbon, E. W., Harrington, C. and Enna, S. J. An examination of the involvement of phospholipase A_2 and C in the α-adrenergic and γ-aminobutyric acid receptor modulation of cyclic AMP accumulation in rat brain slices. *J. Neurochem.*, **47** (1986), 800–810.
56. Lennartz, R. C., Hellems, K. L., Mook, E. R. and Gold, P. E. Inhibitory avoidance impairments induced by intra-amygdala propranolol are reversed by glutamate but not glucose. *Behav. Neurosci.*, **110** (1996), 1033–1039.
57. Ferry, B., Magistretti, P. J. and Pralong, E. Noradrenaline modulates glutamate-mediated neurotransmission in the rat basolateral amygdala in vitro. *Eur. J. Neurosci.*, **9** (1997), 1356–1364.
58. Huang, C. C., Lin, C. H. and Gean, P. W. Potentiation of N-methyl-D-aspartate currents by isoproterenol in the acutely dissociated rat amygdalar neurons. *Neurosci. Lett.*, **253** (1998) 9–12.
59. Wang, S. J., Huang, C. C., Hsu, K. S. *et al.* Blockade of isoproterenol-induced synaptic potentiation by tetra-9-aminoacridine in the rat amygdala. *Neurosci. Lett.*, **214** (1996), 87–90.
60. Huang, Y. Y., Martin, K. C. and Kandel, E. R. Both protein kinase A and mitogen-activated protein kinase are required in the amygdala for the macromolecular synthesis-dependent late phase of long-term potentiation. *J. Neurosci.*, **20** (2000), 6317–6325.

61. Schafe, G. E. and LeDoux, J. E. Memory consolidation of auditory pavlovian fear conditioning requires protein synthesis and protein kinase A in the amygdala. *J. Neurosci.*, **20** (2000), RC96(1–5).
62. Abel, T., Nguyen, P. V., Bard, M. *et al.* Genetic demonstration of a role for PKA in the late phase of LTP and in hippocampus-based long-term memory. *Cell*, **88** (1997), 615–626.
63. Lamprecht, R., Hazui, S. and Dudai, Y. cAMP response element-binding protein in the amygdala is required for long- but not short-term conditioned taste aversion memory. *J. Neurosci.*, **17** (1997), 8443–8450.
64. Winder, D. G., Martin, K. C., Muzzio, I. A. *et al.* ERK plays a regulatory role in induction of LTP by theta frequency stimulation and its modulation by β-adrenergic receptors. *Neuron*, **24** (1999), 715–726.
65. Liang, K. C. and McGaugh, J. L. Lesions of the stria terminalis attenuate the enhancing effect of posttraining epinephrine on retention of an inhibitory avoidance response. *Behav. Brain Res.*, **9** (1983), 49–58.
66. McGaugh, J. L. Memory consolidation and the amygdala: a systems perspective. *Trends Neurosci.*, **25** (2002), 456–461.
67. Packard, M. G., Cahill, L. and McGaugh, J. L. Amygdala modulation of hippocampal-dependent and caudate nucleus-dependent memory processes. *Proc. Natl. Acad. Sci. U. S. A.*, **91** (1994), 8477–8481.
68. Packard, M. G. and Teather, L. Amygdala modulation of multiple memory systems: hippocampus and caudate-putamen. *Neurobiol. Learn. Mem.*, **69** (1998), 163–203.
69. Roozendaal, B., Nguyen, B. T., Power, A. E. and McGaugh, J. L. Basolateral amygdala noradrenergic influence enables enhancement of memory consolidation induced by hippocampal glucocorticoid receptor activation. *Proc. Natl. Acad. Sci. U. S. A.*, **96** (1999), 11 642–11 647.
70. Ikegaya, Y., Saito, H. and Abe, K. High-frequency stimulation of the basolateral amygdala facilitates the induction of long-term potentiation in the dentate gyrus in vivo. *Neurosci. Res.*, **22** (1995), 203–207.
71. Ikegaya, Y., Nakanishi, K., Saito, H. and Abe, K. Amygdala β-noradrenergic influence on hippocampal long-term potentiation *in vivo*. *NeuroReport*, **8** (1997), 3143–3146.
72. Frey, S., Bergado-Rosado, J., Seidenbecher, T., Pape, H. C., Frey, J. U. Reinforcement of early long-term potentiation (early-LTP) in dentate gyrus by stimulation of the basolateral amygdala: heterosynaptic induction mechanisms of late-LTP. *J. Neurosci.*, **21** (2001), 3697–3703.
73. Akirav, I. and Richter-Levin, G. Mechanisms of amygdala modulation of hippocampal plasticity. *J. Neurosci.*, **22** (2002), 9912–9921.
74. Roozendaal, B. Glucocorticoids and the regulation of memory consolidation. *Psychoneuroendocrinology*, **25** (2000), 213–238.
75. Quirarte, G. L., Roozendaal, B. and McGaugh, J. L. Glucocorticoid enhancement of memory storage involves noradrenergic activation in the basolateral amygdala. *Proc. Natl. Acad. Sci. U. S. A.*, **94** (1997), 14 048–14 053.
76. Roozendaal, B., Quirarte, G. L. and McGaugh, J. L. Glucocorticoids interact with the basolateral amygdala β-adrenoceptor–cAMP/PKA system in influencing memory consolidation. *Eur. J. Neurosci.*, **15** (2002), 553–560.

77. Stone, E. A., McEwen, B. S., Herrera, A. S. and Carr, K. D. Regulation of α- and β-components of noradrenergic cyclic AMP response in cortical slices. *Eur. J. Pharmacol.*, **141** (1987), 347–356.
78. Härfstrand, A., Fuxe, K., Cintra, A. *et al.* Glucocorticoid receptor immunoreactivity in monoaminergic neurons of rat brain. *Proc. Natl. Acad. Sci. U. S. A.*, **83** (1987), 9779–9783.
79. McEwen, B. S. Glucocorticoid–biogenic amine interaction in relation to mood and behavior. *Biochem. Pharmacol.*, **36** (1987), 1755–1763.
80. Jhanwar-Uniyal, M. and Leibowitz, S. F. Impact of circulating corticosterone on α_1- and α_2-noradrenergic receptors in discrete brain areas. *Brain Res.*, **368** (1986), 404–408.
81. Thomas, D. N., Post, R. M. and Pert, A. Central and systemic corticosterone differentially affect dopamine and norepinephrine in the frontal cortex of the awake freely moving rat. *Ann. N. Y. Acad. Sci.*, **746** (1994), 467–469.
82. Grundemann, D., Schechinger, B., Rappold, G. A. and Schöming, E. Molecular identification of the corticosterone-sensitive extraneuronal catecholamine transporter. *Nat. Neurosci.*, **1** (1998), 349–351.
83. Gallagher, M., Kapp, B. S., McNall, C. L. and Pascoe, J. P. Opiate effects in the amygdala central nucleus on heart rate conditioning in rabbits. *Pharmacol. Biochem. Behav.*, **4** (1981), 497–505.
84. Gallagher, M., Rapp, P. R. and Fanelli, R. J. Opiate antagonist facilitation of time-dependent memory processes: dependence upon intact norepinephrine function. *Brain Res.*, **347** (1985), 284–290.
85. McGaugh, J. L., Introini-Collison, I. B. and Nagahara, A. H. Memory-enhancing effects of posttraining naloxone: involvement of β-noradrenergic influences in the amygdaloid complex. *Brain Res.*, **446** (1988), 37–49.
86. Introini-Collison, I. B., Ford, L. and McGaugh, J. L. Memory impairment induced by intra-amygdala β-endorphin is mediated by noradrenergic influences. *Neurobiol. Learn. Mem.*, **63** (1995), 200–205.
87. Hatfield, T., Spanis, C. and McGaugh, J. L. Response of amygdalar norepinephrine to footshock and GABAergic drugs using in vivo microdialysis and HPLC. *Brain Res.*, **835** (1999), 340–345.
88. Dalmaz, C., Introini-Collison, I. B. and McGaugh, J. L. Noradrenergic and cholinergic interactions in the amygdala and the modulation of memory storage. *Behav. Brain Res.*, **58** (1993), 167–174.
89. Li, R., Nishijo, H., Wang, Q. *et al.* Light and electron microscopic study of cholinergic and noradrenergic elements in the basolateral nucleus of the rat amygdala: evidence for interactions between the two systems. *J. Comp. Neurol.*, **439** (2001), 411–425.
90. Abercrombie, E. D, Nisenbaum, L. K. and Zigmond, M. J. Impact of acute and chronic stress on the release and synthesis of norepinephrine in the brain: microdialysis studies in behaving animals. In *Stress: Neuroendocrine and Molecular Approaches*, ed. R. Kvetnansky, R. McCarty and J. Axelrod. (Philadelphia: Gordon & Breach, 1992), pp. 29–42.
91. Men, D., McCarty, R. and Gold, P. E. Enhanced release of norepinephrine in rat hippocampus during spontaneous alternation tests. *Neurobiol. Learn. Mem.*, **71** (1999), 289–300.

92. Harley, C. Noradrenergic and locus coeruleus modulation of the perforant path-evoked potential in the rat dentate gyrus supports a role for the locus coeruleus in attentional and memory processes. *Prog. Brain Res.*, **88** (1991), 307–321.
93. Dahl, D. and Sarvey, J. M. Norepinephrine induces pathway-specific long-lasting potentiation and depression in the hippocampal dentate gyrus. *Proc. Natl. Acad. Sci. U. S. A.*, **86** (1989), 4776–4780.
94. Washburn, M. and Moises, H. C. Electrophysiological correlates of presynaptic α_2-adrenoceptor-mediated inhibition of norepinephrine release at locus coeruleus synapses in dentate gyrus. *J. Neurosci.*, **9** (1989), 2131–2140.
95. Bevilaqua, L., Ardenghi, P., Schröder, N. *et al.* Drugs acting upon the cyclic adenosine monophosphate/protein kinase A signalling pathway modulate memory consolidation when given late after training into rat hippocampus but not amygdala. *Behav. Pharmacol.*, **8** (1997), 331–338.
96. Ardenghi, P., Barros, D., Izquierdo, L. A. *et al.* Late and prolonged post-training memory modulation in entorhinal and parietal cortex by drugs acting on the cAMP/protein kinase A signalling pathway. *Behav. Pharmacol.*, **8** (1997), 745–751.
97. Summers, R. J., Papaioannou, M., Harris, S. and Evans, B. A. Expression of β_3-adrenoceptors in rat brain *Br. J. Pharmacol.*, **116** (1995), 2547–2548.
98. Barros, D. M., Izquierdo, L. A., Quevedo, J. *et al.* Interaction between midazolam-induced anterograde amnesia and memory enhancement by treatments given hours later in hippocampus, entorhinal cortex or posterior parietal cortex. *Behav. Pharmacol.*, **9** (1998), 163–167.
99. Lee, E. H. and Ma, Y. L. Amphetamine enhances memory retention and facilitates norepinephrine release from the hippocampus in rats. *Brain Res. Bull.*, **37** (1995), 411–416.
100. Lee, E. H., Lee, C. P., Wang, H. I. and Lin, W. R. Hippocampal CRF, NE, and NMDA system interactions in memory processing in the rat. *Synapse*, **14** (1993), 144–153.
101. Dyon-Laurent, C., Romand, S., Biegon, A. and Sara, S. Functional reorganization of the noradrenergic system after partial fornix section: a behavioral and autoradiographic study. *Exp. Brain Res.*, **96** (1993), 203–211.
102. Ohno, M., Yoshimatsu, A., Kobayashi, M. and Watanabe, S. Noradrenergic DSP-4 lesions aggravate impairment of working memory produced by hippocampal muscarinic blockade in rats. *Pharmacol. Biochem. Behav.*, **57** (1997), 257–261.
103. Watabe, A. M., Zaki, P. A. and O'Dell, T. J. Coactivation of β-adrenergic and cholinergic receptors enhances the induction of long-term potentiation and synergistically activates mitogen-activated protein kinase in the hippocampal CA1 region. *J. Neurosci.*, **20** (2000), 5924–5931.
104. Arnsten, A. F. Stress impairs prefrontal cortical function in rats and monkeys: Role of dopamine D1 and norepinephrine alpha-1 receptor mechanisms. *Prog. Brain Res.*, **126** (2000), 183–192.
105. Miranda, M. I. and McGaugh, J. L. Enhancement of inhibitory avoidance and conditioned taste aversion memory with insular cortex infusions of 8-Br-cAMP: involvement of the basolateral amygdala. *Learn. Mem.*, **11** (2004), 312–317.

106. Berman, D. E., Hazvi, S., Neduva, V. and Dudai, Y. The role of identified neurotransmitter systems in the response of insular cortex to unfamiliar taste: activation of ERK1–2 and formation of a memory trace. *J. Neurosci.*, **20** (2000), 7017–7023.
107. Cahill, L., Prins, B., Weber, M. and McGaugh, J. L. Beta-adrenergic activation and memory for emotional events. *Nature*, **371** (1994), 702–704.
108. van Stegeren, A., Everaerd, W., Cahill, L. and van Gooren, L. Memory for emotional events: differential effects of centrally versus peripherally acting β-blocking agents. *Psychopharmacology*, **138** (1998), 305–310.
109. O'Carrol, R. E., Drysdale, E., Cahill, L., Shajahan, P. and Ebmeier, K. P. Stimulation of the noradrenergic system enhances and blockade reduces memory for emotional material in man. *Psychol. Med.*, **29** (1999), 1083–1088.
110. Southwick, S. M., Davis, M., Horner, B. *et al.* Relationship of enhanced norepinephrine activity during memory consolidation to enhanced long-term memory in humans. *Am. J. Psychiatry*, **159** (2002), 1429–1432.
111. Soetens, E., Casaer, S., D'Hooge, R. and Hueting, J. E. Effect of amphetamine on long-term retention of verbal material. *Psychopharmacology*, **119** (1995), 155–162.
112. Cahill, L. and Alkire, M. T. Epinephrine enhancement of human memory consolidation: interaction with arousal at encoding. *Neurobiol. Learn. Mem.*, **79** (2003),2:194–198.
113. Schelling, G., Richter, M., Roozendaal, B. *et al.* Exposure to high stress in the intensive care unit may have negative effects on health-related quality-of-life outcomes after cardiac surgery. *Crit. Care. Med.*, **31** (2003), 1971–1980.
114. Pitman, R. K. Post-traumatic stress disorder, hormones, and memory. *Biol. Psychiatry*, **26** (1989), 221–223.
115. Pitman, R., Sanders, K. M., Zusman, R. M. *et al.* Pilot study of secondary prevention of posttraumatic stress disorder with propranolol. *Biol. Psychiatry*, **51** (2002), 189–192.

9

Norepinephrine and stress

David A. Morilak, Ph.D.

University of Texas Health Science Center at San Antonio, San Antonio, USA

Introduction

Stress is a major factor in the development of many psychopathological conditions, and the dysregulation of noradrenergic neurotransmission has been implicated in a variety of stress-related neuropsychiatric diseases, such as depression, posttraumatic stress disorder (PTSD), and other anxiety disorders.[1–4] Further, the brain noradrenergic system is a key regulatory target for psychotherapeutic drugs used to treat such disorders. Thus, understanding the role of brain norepinephrine (NE) in short-term coping and the response to acute stress, as well as long-term adaptation and restoration in the face of chronic or repeated stress, may be important for understanding the neurobiological processes involved in the development of or vulnerability to such pathological states and the processes involved in their effective treatment.

The brain noradrenergic neurotransmitter system, originating in the locus coeruleus (LC) and other cell groups in the medulla and pons, innervates essentially the entire neuraxis, the notable exception being the neostriatum (see Chapter 1). Such a widespread and divergent innervation pattern arises from a relatively small number of cells concentrated in a few regions of the hindbrain. This system is positioned ideally to enact a global "state-change" function and influence the operating characteristics of the entire nervous system under conditions of elevated noradrenergic activity. That the system indeed operated in this manner was supported by electrophysiologic experiments in the 1980s and early 1990s, which showed the effect of NE on target neurons to be primarily modulatory in nature. Rather than inducing simple inhibition or excitation, NE alters the "signal-to-noise ratio" of evoked activity, both excitatory and inhibitory, making synaptic transmission in target circuits more effective.[5] Such modulatory effects have been described for NE in many brain circuits, and have been shown to be mediated, via different transduction mechanisms, by both β- and α_1-adrenergic receptors.[6–9]

Brain Norepinephrine: Neurobiology and Therapeutics, ed. Gregory A. Ordway, Michael A. Schwartz and Alan Frazer. Published by Cambridge University Press.

In this chapter, the function of brain norepinephrine in the context of stress will be considered from a more global, systemic perspective. This approach represents the bias of the author. It is certainly possible, for purposes of experimental access, control, and interpretation, to study in isolation a variety of specific behavioral and physiological responses that may be related to noradrenergic activity in specific pathways, acting on specific receptors in specific brain target regions. Indeed, much of the work conducted in the author's laboratory, some of which is described in this chapter, is the result of such a reductionist approach. However, in so doing, it is sometimes easy to lose sight of the fact that these responses rarely, if ever, occur in such isolation. Rather, due to its divergent anatomical organization, originating in relatively few noradrenergic neurons, the effects of norepinephrine are likely to be exerted globally throughout the brain in contexts in which these cells are activated. The individual responses mediated in each of these regions thus represent components of more complex, integrated processes of physiologic and behavioral regulation, and the overall function of the brain noradrenergic system would therefore appear to be greater than the sum of these individual parts. Such a global and systemic view of noradrenergic function is consistent both with its anatomical organization and with its modulatory postsynaptic cellular effects.

A useful analogy can be drawn with the peripheral sympathetic nervous system, the activation of which elicits a myriad of responses mediated by specific receptors and effectors. Any one of these can be understood in terms of its unique effect, e.g., mydriasis increases visual sensitivity, cardiovascular responses divert blood to skeletal muscle and increase aerobic capacity, and other changes facilitate energy mobilization and increased catabolic activity. However, the adaptational significance and the essential role that each of these responses serves in the survival of the individual and survival of the species can be fully appreciated only when the system is considered as a whole, as an integrated entity mediating the well-known "fight-or-flight" response to stress. We view the brain noradrenergic system in much the same way. Although studying noradrenergic effects in specific brain regions is important to understanding how this system does what it does, recognizing the significance of these responses lies in the fact that together they comprise an integrated and systematic modulatory process. Such a global, integrated view would seem the key to understanding the role of brain norepinephrine in the response and adaptation to stress.

Elevation of tonic noradrenergic neurotransmission with arousal

To understand the behavioral function of the brain noradrenergic system at such a global level, it is useful to differentiate first between tonic, basal, or steady-state noradrenergic activity on one hand, and phasic, or stimulus-evoked, reactivity on the other. The cellular effects, even the overt behavioral effects of NE, may be similar in these two modes of activity, but the pattern, duration and context in which they are elicited are distinct. Tonic, or basal, noradrenergic activity has been closely related to the overall state of behavioral arousal, which may be conceptualized as the degree to which an organism is engaged with or responsive to its external environment.[10,11] (For further information see Chapters 5 and 6.) Tonic electrical activity of noradrenergic neurons is highest during an active, alert state, and decreases steadily as the organism passes through states of progressively decreasing arousal, from quiet waking to drowsy to slow-wave sleep, until these neurons are completely silent during rapid eye movement sleep.[10,11] Tonic noradrenergic activity is associated with many behavioral and physiological indices of arousal, such as locomotion, eye movements, electroencephalogram desynchronization and overt behavioral activity.[13–14] Blockade of postsynaptic α_1-adrenergic receptors induces a state of sedation and behavioral inactivity,[16,17] and more specifically, blockade of α_1-receptors in the prefrontal cortex antagonizes stimulant-induced locomotor hyperactivity.[18] This suggests that the relationship of tonic noradrenergic activity to arousal is not only correlative, but that the modulatory influence of NE contributes to the behavioral expression of at least some aspects of arousal.

Thus, tonic noradrenergic activity, particularly in regions such as the prefrontal cortex, may maintain or facilitate neuronal processes involved in directing attention to or interacting with the external environment, thereby promoting a state of arousal or alertness. In a similar way, the phasic activation of noradrenergic transmission to even higher levels, in response to temporally defined environmental stimuli, may likewise facilitate acute behavioral reactivity to those stimuli. Noradrenergic neurons are excited briefly by punctate sensory stimuli of several modalities.[10,14] As the behavioral response to the stimulus habituates, so does the response of noradrenergic neurons,[14,19,20] suggesting that the "salience" of the stimulus is an important factor in evoking a noradrenergic response. This has been elaborated in a series of studies addressing a role for NE in facilitating neuronal processes involved in vigilance and attention, the cognitive aspects of arousal.[19,20] This is considered in detail in Chapter 5 and 6, and will not be discussed further here, except to note that the same concepts pertaining to arousal may not differ qualitatively from those pertaining to acute stress: they may differ only in terms of context, magnitude, and stimulus-dependence. An acute stressor, by definition, is a salient stimulus, and increased alertness and vigilance may be viewed not only as cognitive components of a state of elevated arousal, but also as a phasically evoked cognitive response to acute stress.

Phasic activation of noradrenergic neurotransmission in response to acute stress

Before considering the role of NE in the acute response to stress, it is first useful to specify objectively what is meant by "stress." At a conceptual level, stress can be defined as any threat, either real or perceived, to the homeostasis and well-being of an organism. Within this definition, two broad types of stressors can be differentiated. The first, termed a systemic or physiologic stressor, is a real and imminent physical threat to health and well-being, with the physical characteristics of the stimulus itself placing immediate restorative response demands on the organism (e.g., injury with loss of blood). In this case, conscious awareness or perception of the stimulus by the organism is not requisite for "stress" to be present. The second type of stress-inducing stimulus, called a processive or psychogenic stressor, depends upon perception and cognitive processing of the stimulus to confer upon it a stressful quality (e.g., the threat posed by potential loss of a job). In this case, the relative severity of the stressor and its impact upon the individual, or even whether it is considered a stressor at all, can vary between individuals, as it is based upon the interpretation of the stimulus and the stimulus context.

It has also been useful, for purposes of quantifying the severity of a stressor, to define stress in terms of the physiological response it elicits, for instance by measuring activation of the hypothalamic–pituitary–adrenal (HPA) axis. In this regard, the differentiation of systemic and processive stressors is of more than academic interest. Although stimuli of both broad categories robustly and acutely activate both HPA and noradrenergic responses, the neural pathways by which these stimuli are processed in the brain, and by which they activate stress-responsive systems including the ascending noradrenergic system and the HPA axis, are quite distinct.[21,22]

Both electrophysiological and neurochemical studies (i.e., in vivo microdialysis) have shown that the brain noradrenergic system is phasically and robustly activated by a diverse array of acutely stressful stimuli, both systemic and processive, including, among others, immobilization, loud noise, immune challenge, electric shock, hypoglycemia, hypotension, cold exposure, bladder distension and forced activity.[24–29] Whereas the brief bursts of electrical activity elicited by punctate, innocuous stimuli occur over a range of hundreds of milliseconds, phasic activation of noradrenergic neurotransmission by acutely stressful stimuli is much longer lasting, of the order of seconds to minutes or hours, depending on the stimulus, often outlasting the duration of the stimulus itself, and correlated temporally with peripheral physiological indicators of the stress response.[11] Stimulus-specific and transmitter-specific afferent pathways inducing phasic activation of the LC in response to different types of stimuli have been identified. For example, an ascending glutamatergic projection from the rostral medulla mediates activation in response to somatic sensory stimulation,[31] whereas a descending corticotropin-releasing factor (CRF) pathway arising from the central amygdala and the pontine Barrington's nucleus mediates activation of the LC by acute hypotensive stress.[30,32] Other studies also suggest that processive stressors activate the lateral tegmental noradrenergic neurons in the hindbrain, as well as the LC, via descending projections from the amygdala.[33]

Role of norepinephrine in stress-induced activation of the HPA axis

The potential role of NE as a central mediator of stress-induced activation of the HPA axis has been controversial, and remains unclear to this day. Ascending noradrenergic afferents, primarily from the A1 and A2 medullary noradrenergic cell groups, innervate the parvocellular component of the hypothalamic paraventricular nucleus (PVN), and make synaptic contact with dendrites of neurons that synthesize CRF, the final common output for HPA activation.[34] Exogenous administration of adrenergic receptor agonists, in particular agonists acting at α_1-adrenergic receptors, given either intracerebroventricularly or directly into the PVN, stimulates CRH secretion and activation of the HPA axis.[35,36] Similarly, electrical activation of the ascending ventral medullary noradrenergic pathway elicits both NE release in the PVN and pituitary secretion of adrenocorticotropic hormone (ACTH).[37] Given the multimodal responsivity of the noradrenergic system to acute stress, it has been observed that exogenously administered or exogenously activated NE neurotransmission in the PVN can stimulate the HPA axis. These observations support the possibility of a potential role for endogenous NE in mediating stress-induced HPA activation. However, a considerable array of approaches to the investigation of the potential mediating role for endogenous NE has yielded inconsistent and contradictory findings. Abolition of endogenous noradrenergic innervation of the PVN, or blockade of adrenergic receptors in the PVN, has been shown to have little or no effect on the HPA response to a number of acutely stressful stimuli of many classes, including hypoglycemia, osmotic stress, systemic histamine, footshock, restraint stress or exposure to a novel environment.[39–41] By contrast, similar interventions have produced a variety of effects, ranging from partial attenuation to near-complete blockade of CRF and/or ACTH responses to an equally diverse set of stimuli including restraint stress, ether stress, and immune stress.[35,39,40,43] Results of a study comparing the effects of neurotoxic lesions of noradrenergic afferents innervating the PVN with the effects of local adrenergic antagonist administration on ACTH secretion suggested that, whereas the integrity of the ascending noradrenergic pathway innervating the PVN appears to be essential for stress-induced activation of the HPA axis, it may, in fact, not be NE, but another, as of yet unidentified, neurotransmitter substance (presumably colocalized and coreleased with NE from those same afferent fibers) that actually mediates the response.[37] A similar dissociation has been observed more recently using the induction of c-Fos expression in PVN as a marker of HPA activation, as well as activation of other forebrain limbic regions that may secondarily modulate components of the HPA axis, such as the medial amygdala.[44]

Thus, taken as a whole, the bulk of evidence suggests the following conclusions: that the noradrenergic innervation of the PVN represents but one of several convergent and redundant stress-responsive pathways regulating activation of the HPA axis; that the influence of the noradrenergic input is context- and stressor-specific; that this specificity may be reflected by a differential influence of NE on subpopulations of PVN neurons synthesizing CRF- or other HPA-related secretagogues; and that NE acts in the PVN, as it does elsewhere, more as a facilitating modulatory influence than as a primary mediating mechanism in the activation of the HPA axis.[45]

Modulatory effects of stress-induced noradrenergic neurotransmission on neuroendocrine and behavioral responses to acute stress

Given the widespread anatomical organization of the noradrenergic system, the modulatory cellular effects of NE, and the robust activation of noradrenergic neurotransmission by a host of acute stressors, it seems logical to hypothesize that stress-induced NE release may serve to enhance a broad array of behavioral, cognitive, autonomic, and neuroendocrine response capabilities, mediated by a number of distinct neural circuits throughout the brain. For example, activation of noradrenergic transmission, either pharmacologically or by acute stress exposure, facilitates a simple sensorimotor reflex[46,47] and also enhances more complex sensorimotor behaviors, such as the acoustic startle response.[48] At an even higher level of complexity, elevating noradrenergic transmission by NE reuptake inhibition specifically reduces immobility and promotes active escape behaviors, such as struggling and climbing, in the forced-swim test.[49,50]

Acute stress-induced release of NE has been shown to occur in specific regions of the limbic forebrain that are intimately involved in behavioral, affective, and neuroendocrine responses to stress (Figure 9.1), including, the lateral bed nucleus of the stria terminalis (BSTL) and the central nucleus of the amygdala (CeA).[24,51–53] In a recent series of studies conducted in our laboratory, we addressed directly the potential modulatory influence of stress-induced NE release in these regions in adult male Sprague–Dawley rats using two well-validated tests of anxiety-like behavior: the elevated plus-maze and the social interaction test. Behavioral reactivity was defined operationally as a change in behavior induced after a brief (5 minute) exposure to immobilization as compared to the behavior of unstressed control rats. Adrenergic receptors in the BSTL or CeA were blocked by local microinjection of selective antagonists immediately prior to acute immobilization stress.[24,54]

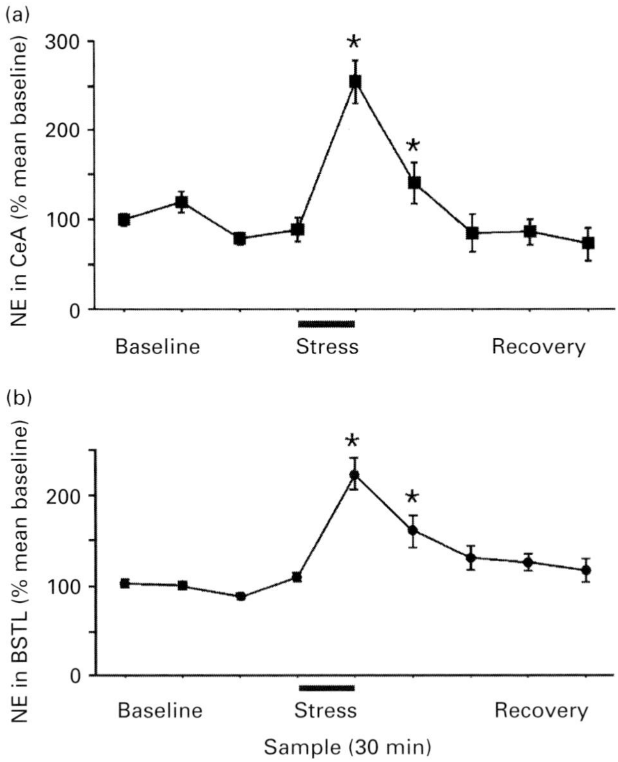

Figure 9.1 Acute immobilization stress induced a significant phasic increase in NE release measured by microdialysis in the CeA (a) and BSTL (b) of adult male Sprague–Dawley rats. * $p < 0.05$ compared to baseline samples. Adapted from Cecchi *et al.* (2002a);[24] Khoshbouei *et al.* (2002),[51] and reproduced with permission from Morilak *et al.* (2003).[77]

In these experiments, the behavioral response to acute immobilization stress was seen as a reduction in social behavior on the social interaction test, and a specific reduction in open-arm exploration on the elevated plus-maze (Figure 9.2). Behavioral response was evaluated independent of any changes in locomotion, which were not seen. Blockade of α_1-adrenergic receptors in the CeA attenuated the stress-induced reduction in social behavior, but β-antagonists had no effect, suggesting that stress-induced NE release acted specifically on α_1-receptors in CeA to facilitate the behavioral response seen on the social interaction test. By contrast, adrenergic receptor blockade in the CeA had no effect on the stress-induced reduction in open-arm exploration on the elevated plus-maze (Figure 9.2). The reverse was true when adrenergic antagonists were administered into the BSTL. Here, blockade of either α_1- or β-receptors attenuated the reduction in open-arm exploration on the plus-maze, but had no effect on the stress-induced reduction in social behavior.

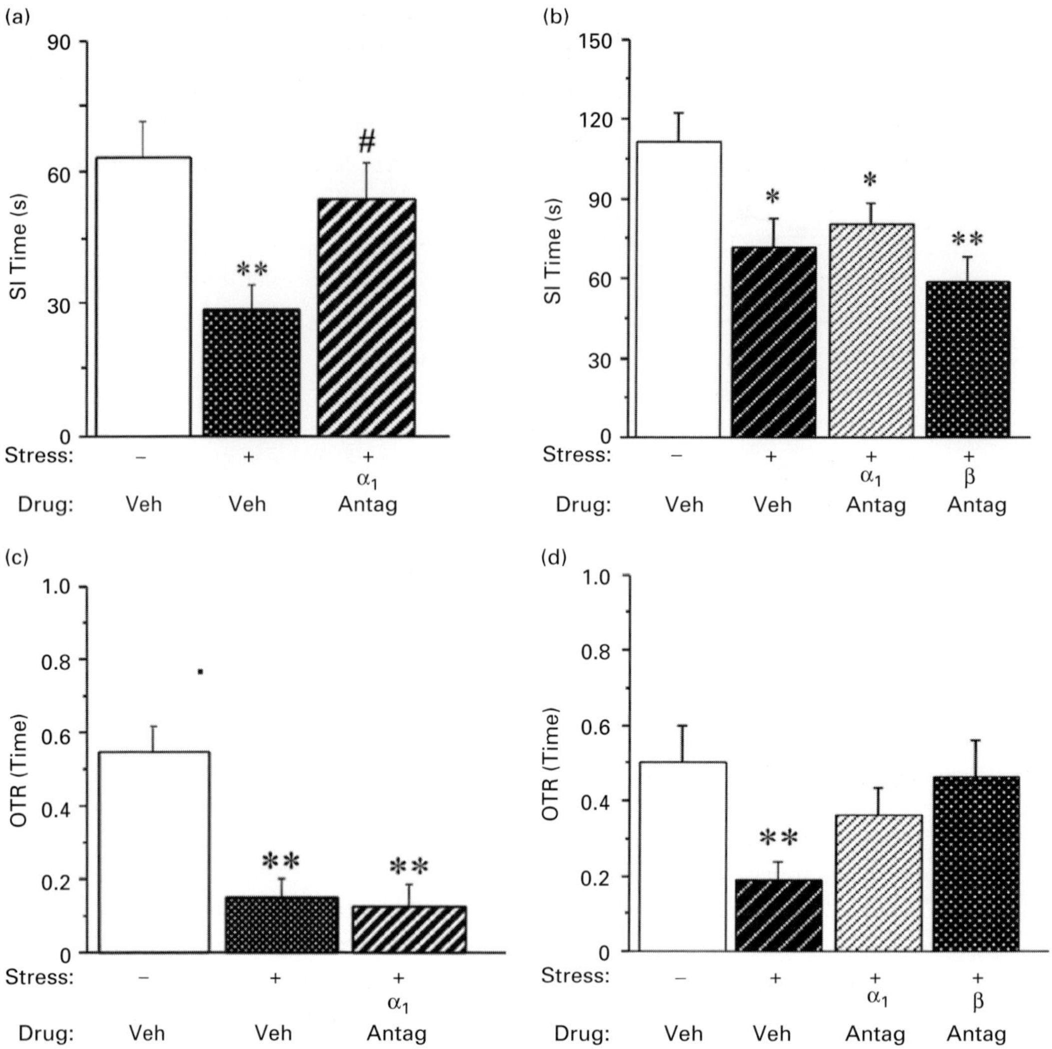

Figure 9.2 Behavioral stress modulatory effects of NE in the CeA (a and c) or BSTL (b and d). Administration of the α_1-antagonist benoxathian (2.0 nmole) into CeA blocked the acute immobilization stress-induced reduction in social behavior seen on the social interaction (SI) test (a), but had no effect on the stress-induced reduction in open-arm exploration on the plus-maze (c). By contrast, either benoxathian or a cocktail of β-antagonists (betaxolol + ICI 118,551, 1 nmole each) microinjected into the BSTL significantly blocked the stress-induced reduction in open-arm exploration (d), but had no effect on the reduction in social behavior (b). OTR, open-to-total ratio; $^*p < 0.05$ compared to unstressed vehicle controls; $^{**}p < 0.01$ compared to unstressed vehicle controls, $^{\#}p < 0.05$ compared to vehicle-treated stressed group. Adapted from Cecchi *et al.* (2002a); [24] Cecchi *et al.* 2002b,[54] and reproduced with permission from Morilak *et al.* (2003).[77]

We found in these same studies that blockade of α_1-adrenergic receptors in the BSTL attenuated the elevation of ACTH secretion seen in response to acute immobilization stress by about 40% compared to vehicle-injected control rats. This finding is also consistent with a facilitating modulatory influence, rather than a primary mediating role, for NE in stress-induced activation of the HPA axis in these extrahypothalamic limbic brain regions.[24] By contrast, administration of β-receptor antagonists had no effect in the BSTL, and similar to the region-specificity seen with the acute behavioral responses, administration of either α_1- or β-receptor antagonists into CeA had no effects on stress-induced ACTH secretion.[54]

Thus, stress-induced activation of NE neurotransmission in both the CeA and BSTL facilitated the behavioral and neuroendocrine responses to acute immobilization stress, but did so in a region-specific, receptor-specific and response-specific fashion.[24,54] Importantly, none of the antagonists in either region had any effect on baseline ACTH levels or on unstressed behavioral activity, suggesting that it was specifically the phasic, stress-activated noradrenergic neurotransmission that facilitated these acute stress-evoked responses. Tonic, basal noradrenergic neurotransmission, however, had little effect on baseline, unstimulated activity mediated by these neuroendocrine and behavioral response circuits.

The noradrenergic system as a potential substrate for stress vulnerability?

An inability to appropriately initiate or regulate aspects of the stress response has been proposed as a potentially critical factor in the pathophysiology of various stress-related disorders,[55,56] and dysregulation of noradrenergic neurotransmission has been implicated in stress-related psychiatric diseases such as depression, PTSD, and other anxiety disorders.[3,4,57] However, although stress is a precipitating factor in a number of psychiatric disorders, it is also clear that not all individuals exposed to an equivalent level of stress develop pathological consequences. Even for PTSD, in which a clear connection exists between an identifiable trauma and subsequent development of inappropriate responses to neutral or mild stimuli, not all survivors of the same trauma exhibit the disorder. Thus, individual vulnerability factors must also be involved, for instance a genetic predisposition or an increased susceptibility to stress as a consequence of prior experience. Therefore, the degree and nature of any stress-related psychopathology must result from an interaction of this predisposition with exposure to a sufficiently sensitizing environmental stimulus. We have conducted a series of studies recently to examine this interaction, and to address the possibility that endogenous differences in the reactivity or regulation of the brain noradrenergic system might contribute to a predisposition to stress vulnerability, using rat strains that are defined by differences in their stress response and their vulnerability to certain stress-related illnesses. The inbred Wistar-Kyoto (WKY) rat strain in particular has been proposed as a model of stress susceptibility, as they develop stress-induced gastric ulcers more readily than their parent Wistar strain, and they also show exaggerated HPA responses to stress.[58,59] We thus examined potential differences in noradrenergic responsivity to acute stress, as well as stress-induced noradrenergic modulatory influence, in the inbred WKY rat compared to the outbred Sprague–Dawley control strain, using approaches similar to those described above.[60,61]

In the first of these experiments,[60] WKY rats showed reduced baseline exploratory activity, depressed baseline startle reactivity, and less baseline social behavior than Sprague–Dawley rats. Thus, in this study as well as in the literature, WKY rats tend to exhibit passivity, immobility, and an overall deficit of baseline behavioral activity, perhaps suggestive of a lower level of baseline arousal. Similarly, in comparison with controls, WKY rats also exhibited a lack of fear potentiation of the startle response, and a lack of acute stress-induced reductions in both social behavior and open-arm exploratory behavior on the plus-maze. These findings indicate an attenuation of acute behavioral stress responsivity as well (Figure 9.3). For comparison, another inbred rat strain, the Lewis rat, was also examined in this experiment. Although the baseline scores on several of the measures were also low in Lewis rats, comparable to the basal values in WKY rats, they still exhibited a degree of acute stress responsivity that was similar to that seen in the control Sprague–Dawley rats, suggesting that the deficit observed in the WKY rats was not attributable to a floor effect (Figure 9.3). Along with these behavioral observations, and consistent with the previous pharmacological studies described above, microdialysis experiments revealed a corresponding attenuation of phasic stress-induced NE release in the BSTL of WKY rats. Thus, the reduced basal activity and blunted acute behavioral stress responsivity seen in WKY rats were associated with deficient brain noradrenergic reactivity (Figure 9.4).

We hypothesized that such a deficit in noradrenergic modulatory function might alter the ability of WKY rats to cope with stress, which may explain not only their increased susceptibility to stress-related pathology, but also their exaggerated neuroendocrine responses. Thus, despite the facilitation of HPA activity shown previously for NE in the BSTL,[24] these data would suggest that the exaggerated HPA responses to acute stress seen in WKY rats are not attributable to elevated noradrenergic facilitation, but rather may be secondary to a deficit of adequate coping capability, resulting in an increase in the relative "impact" of stress. Consistent with this interpretation, adrenergic antagonist administration in BSTL failed to reduce stress-induced ACTH secretion in WKY rats as it did in controls[61] (see Figure 9.6).

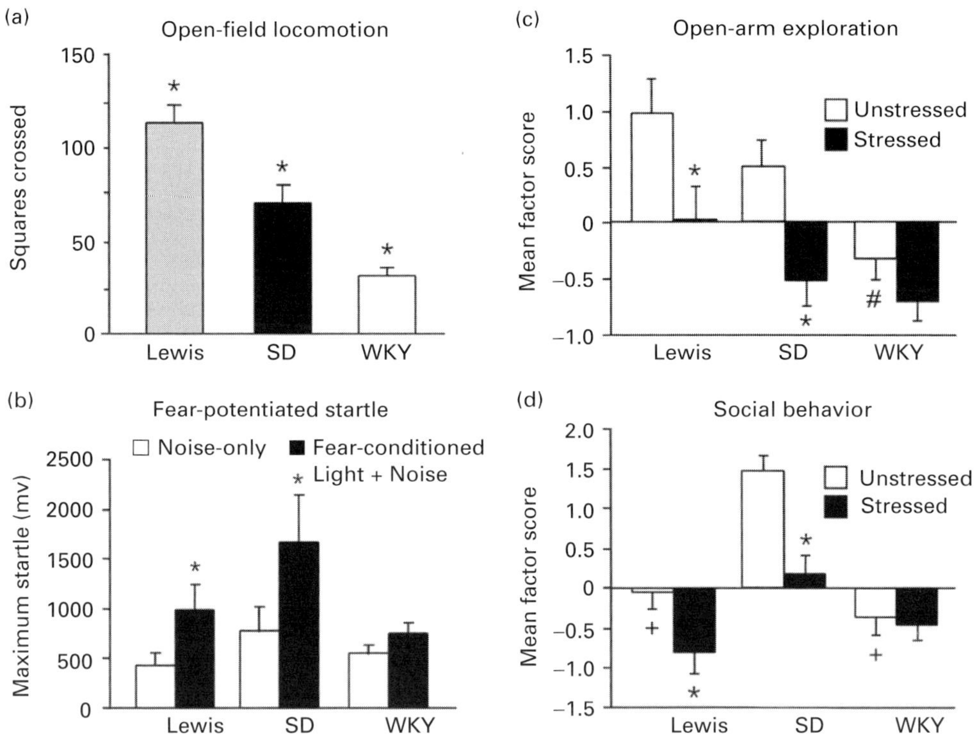

Figure 9.3 Differences in behavioral activity and acute behavioral stress reactivity in WKY, Sprague–Dawley (SD), and for comparison, Lewis rats. (a) Baseline locomotor activity in a novel open field. WKY rats showed the lowest level of locomotor activity, Lewis were highest, and SD rats were intermediate (*$p < 0.05$ all comparisons significant). (b) Fear-potentiated startle reactivity. WKY rats showed reduced baseline startle reactivity compared to SD rats and failed to exhibit any potentiation of the startle response magnitude by a fear-conditioned light stimulus. By contrast, despite a comparably low baseline startle response, the Lewis rats nonetheless showed a degree of fear-potentiation that was similar to that observed in SD rats. *$p < 0.05$ compared to noise-only startle trials within strain. (c and d) Acute behavioral stress responsivity on the elevated plus-maze and social interaction tests, both of which are related to phasic activation of noradrenergic transmission by stress (cf. Figure 9.2). Dependent measures represent factor scores derived from a principal components extraction and factor analysis. Acute immobilization stress induced a reduction in both open-arm exploration and social behavior in SD rats but not WKY rats. Note that Lewis rats showed low baseline social interaction similar to that of WKY rats. However, acute stress still induced a reduction in the Lewis rats, suggesting that the failure to see an effect in WKY rats was not attributable to a floor effect. *$p < 0.05$ compared to unstressed control rats of the same strain; $^{+}p < 0.05$ compared to unstressed SD rats; $_{\#}\, p < 0.05$ compared to unstressed Lewis rats. Adapted and modified from Pardon *et al.* (2002).[60]

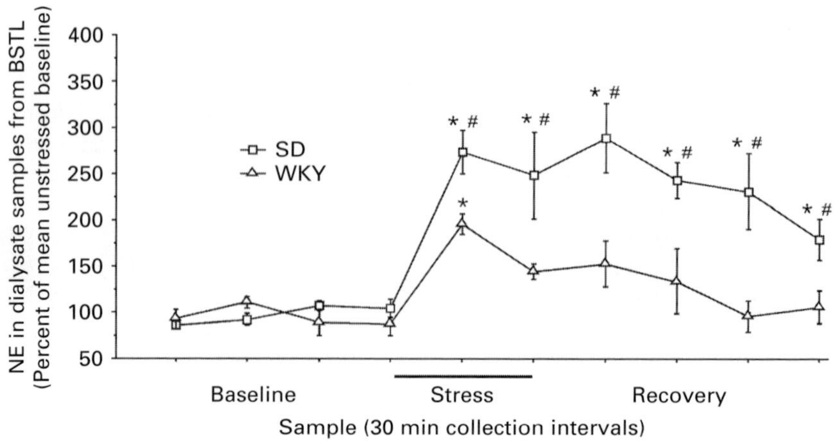

Figure 9.4 Acute immobilization stress-induced activation of NE release in BSTL, measured by microdialysis, was significantly less in WKY rats than in Sprague–Dawley rats. There were no strain differences in the mean unstressed baseline levels of NE in dialysate samples collected before stress. $^{*}p < 0.05$ compared to unstressed baseline values from the same strain; $^{\#}p < 0.05$ difference between strains at the same time point. Adapted and modified from Pardon *et al.* (2002).[60]

Norepinephrine and the interaction of stress vulnerability and environmental sensitization

Chronic intermittent exposure to cold has been shown to sensitize subsequent activation of the HPA axis by a novel acute stressor.[62] More severe and prolonged cold exposure has also been shown to enhance excitation of electrophysiological activity of noradrenergic neurons in the LC,[63] and acute stress-induced activation of NE release in forebrain terminal regions.[64,65] Thus, using a mild cold stress protocol (7 days, 4 hours/day, 4 °C) to avoid maximal effects and perhaps better reveal any potential strain differences, we next compared the sensitivity of WKY and Sprague–Dawley rats to the sensitizing influence of chronic intermittent cold exposure on neuroendocrine and noradrenergic stress reactivity, and also on noradrenergic facilitation of the HPA response in the BSTL. We found that this relatively mild chronic cold treatment enhanced the subsequent acute immobilization stress-induced release of NE in the BSTL of WKY rats but not Sprague–Dawley rats (Figure 9.5). Chronic cold also potentiated activation of the HPA axis by acute immobilization stress in both strains, although to a greater extent in WKY rats (Figure 9.6). It was remarkable, however, that after cold-sensitization, the ACTH response to acute stress in WKY rats was now attenuated by microinjection of the α_1-antagonist benoxathian into the BSTL, even more so than in Sprague–Dawley rats, such that the ACTH response of WKY rats after benoxathian was not different

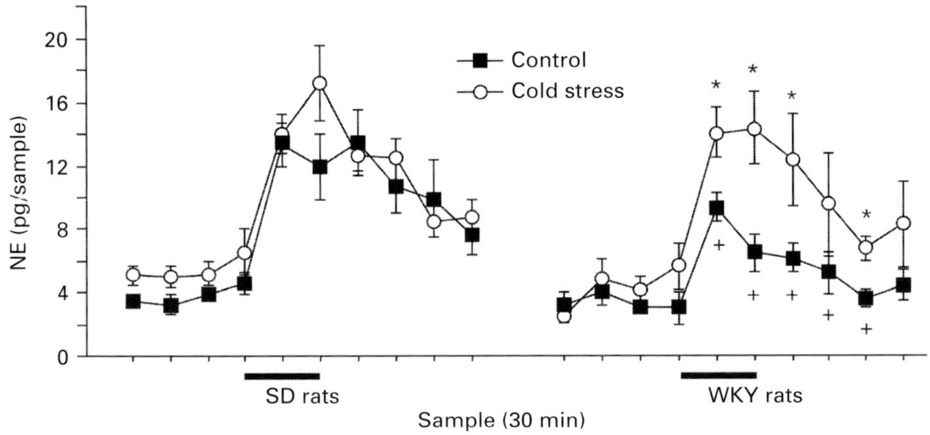

Figure 9.5 As seen in Figure 9.4, the phasic activation of NE release induced by acute immobilization stress, as measured by microdialysis, was attenuated in the BSTL of unsensitized WKY rats compared to unsensitized Sprague Dawley (SD) rats (black squares, compare left and right panels). By contrast, after one-week chronic intermittent cold exposure (white circles), the acute immobilization stress-induced activation of NE release in the BSTL of WKY rats was significantly enhanced relative to their unsensitized controls (right panel). No such sensitizing effect was seen in SD rats. $^*p < 0.05$ compared to control rats of the same strain measured at the same time point; $^+p < 0.05$ control WKY rats compared to control SD rats at the same time point. Reproduced with permission from Pardon *et al.* (2003).[61]

from that of Sprague–Dawley rats (Figure 9.6). Thus, chronic cold exposure not only sensitized the stress-induced release of NE in the BSTL of WKY rats, but it also restored noradrenergic facilitation of their already-elevated HPA response. Such functional sensitization of a previously deficient facilitatory system may thus be one mechanism whereby exposure to repeated or severe stress may induce pathologic dysregulation of the stress response in susceptible individuals.

Pharmacologic regulation of tonic and phasic noradrenergic neurotransmission as a mechanism for antidepressant and anxiolytic effects of chronic NE reuptake blockade

Whether or not primary dysregulation of noradrenergic function is a factor in the etiology or manifestation of stress-related psychiatric disorders, noradrenergic neurotransmission is a regulatory target of many of the drugs used to treat such disorders. Although selective serotonin reuptake inhibitors (SSRIs) have become preeminent among antidepressant (AD) drugs, it is clear that alterations in noradrenergic neurotransmission are important in the actions of older tricyclic ADs, and of a newer generation of dual uptake inhibitors and selective NE reuptake

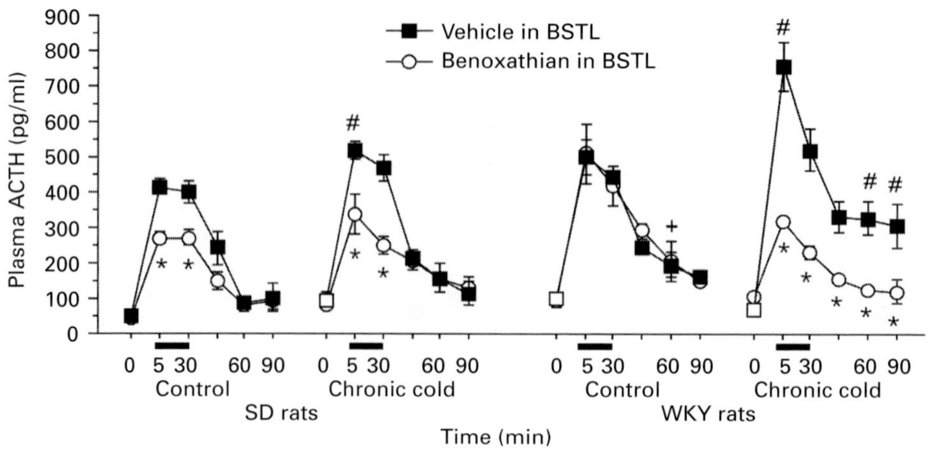

Figure 9.6 Administration of the α_1-receptor antagonist benoxathian into the BSTL attenuated acute stress-induced ACTH secretion in unsensitized control Sprague–Dawley (SD) rats (panel 1), but had no effect in unsensitized WKY rats (panel 3), illustrating the deficit of noradrenergic modulatory influence in this strain. Chronic intermittent cold exposure enhanced the stress-induced elevation in plasma ACTH in both WKY and SD rats. However, whereas the response of SD rats was enhanced only at 5 min, both the magnitude and duration of the ACTH response were enhanced in WKY rats (black squares, compare the difference in panels 1 and 2 to that in panels 3 and 4). Further, after chronic cold exposure, benoxathian microinjected into the BSTL significantly attenuated the acute ACTH stress response in WKY rats (panel 4), indicating that cold exposure had sensitized the previously deficient noradrenergic facilitation of acute HPA stress reactivity in this strain. $^{*}p < 0.05$ benoxathian-treated rats compared to vehicle-treated rats of the same strain and cold condition measured at the same time point; $^{+}p < 0.05$ vehicle-treated control WKY rats compared to vehicle-treated control SD rats at the same time point; $^{\#}p < 0.05$ vehicle-treated cold-exposed rats compared to vehicle-treated control rats of the same strain at the same time point. Reproduced with permission from Pardon *et al.* (2003).[61]

inhibitors (NRIs). These drugs are as effective as SSRIs in alleviating depressed mood, social withdrawal, and other inhibitory symptoms of depression. In addition, anxiety is a prominent component of depression, and these drugs are also effective at alleviating anxiety and other symptoms of depression that may be indicative of behavioral or emotional activation (aggression, agitation, psychomotor tension, distress, etc). There is some limited evidence that they may be effective in treating certain anxiety disorders as well.[66–68]

The level of tonic, steady-state noradrenergic activity has been implicated in promoting behavioral states of arousal and vigilance, and so it seems logical that enhancing such a function could improve the inhibitory symptoms of depression. However, given the role of phasically activated noradrenergic transmission in promoting acute anxiety-like behavioral reactivity to stress, it is less clear how enhancing NE transmission could contribute to the anxiolytic effects of AD drugs. More specifically, how can the selective pharmacological enhancement of noradrenergic neurotransmission improve both depressive and anxiety-related symptoms?

To address this dual behavioral effect, we have recently begun testing the hypothesis that chronic NE reuptake blockade may differentially regulate tonic noradrenergic transmission and phasic, stress-activated noradrenergic transmission. A possible mechanism by which these different modes of neurotransmission may be regulated differentially is through the actions of inhibitory α_2-adrenergic autoreceptors. As extracellular levels of a transmitter increase, it acts not only on postsynaptic receptors but also on somatodendritic and terminal autoreceptors to decrease the activity and excitability of the cell and to inhibit the further release of transmitter. This action counteracts, in part, the elevated extracellular concentration caused by reuptake blockade. In the serotonergic system, chronic SSRI treatment induces a time-dependent desensitization of somatodendritic serotonin autoreceptors, and this is thought to enhance serotonergic transmission via a gradual restoration of baseline firing rate and serotonin release in the face of ever-increasing extracellular levels.[69] By contrast, in the noradrenergic system, although there is some evidence that chronic NE reuptake blockade may induce a degree of desensitization,[70,71] it nonetheless appears that inhibitory α_2-autoreceptors remain largely functional. Tonic electrical activity of noradrenergic neurons remains inhibited,[71–73] and phasic excitation of noradrenergic activity, for instance by acute hypotensive stress, is attenuated.[74]

Evidence that noradrenergic autoreceptor activity is maintained after chronic reuptake blockade is seen after acute autoreceptor antagonist administration. In this case, if autoreceptors were desensitized, not only should agonist effects be diminished, but so should the effects of antagonists that block endogenous transmitter activity.[75] However, it has been shown in rats treated chronically with a NRI that systemic administration of an autoreceptor antagonist elevates the activity of NE neurons even more than it does after acute reuptake blockade,[76] and induces a robust elevation of cortical NE release (see Figure 9.7).[72,77]

Both the preclinical data described above,[24,54,60] as well as clinical investigations seem to suggest that it is not simply an absolute elevation of NE levels per se, but rather an increased volatility or reactivity of the noradrenergic system that is most closely associated with stress-induced anxiety-like behaviors, and also with clinical anxiety disorders.[57] Thus, chronic reuptake blockade may exert a dual influence on noradrenergic neurotransmission, in effect clamping noradrenergic neurotransmission at an elevated but less volatile level, both increasing arousal and reducing anxiety-like behavioral reactivity to stress. The tonic elevation of extracellular NE levels produced by chronic reuptake blockade may thus contribute to an alleviation

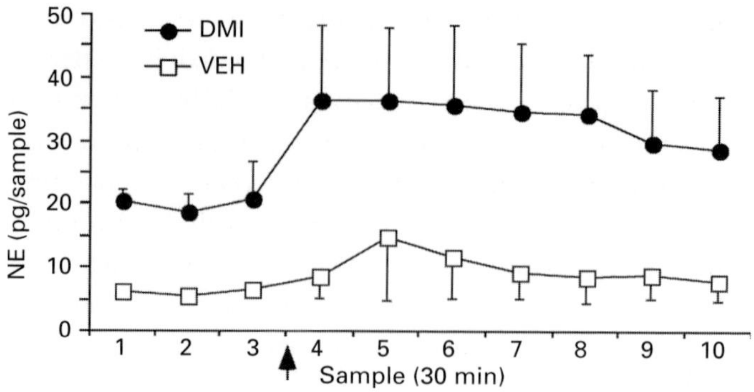

Figure 9.7 Administration of a low dose of the α_2-autoreceptor antagonist yohimbine (0.5 mg/kg, intraperitoneal, arrow) robustly elevated extracellular NE collected by microdialysis in medial prefrontal cortex of rats treated with chronic desipramine (DMI; 15 mg/kg/day) as compared to rats treated chronically with vehicle, reflecting a persistent functional inhibitory influence of α_2-autoreceptors.

of the inhibitory or arousal-related symptoms of depression, such as psychomotor retardation, lack of interest, fatigue, languor, depressed mood, etc. At the same time, the reduction of acute stress-induced phasic activation of noradrenergic transmission, resulting from persistent activation of inhibitory autoreceptors, may attenuate acute anxiety-like behavioral reactivity. In preliminary experiments (and unpublished observations), we have obtained evidence in support of the hypothesis, as chronic treatment with a selective NRI attenuated acute stress-induced reductions in social behavior and open-arm exploratory behavior on the elevated plus-maze.[78] This obviously remains an area of intense and ongoing investigation.

Summary and conclusions

It has long been understood that the brain noradrenergic system participates in promoting arousal and is activated by stress, but its precise role in these behavioral states has not been easy to identify. We are now beginning to understand how the widespread and divergent anatomical organization of this system, together with the cellular modulatory effects exerted by NE on its postsynaptic targets, can enact a global state-change function, enhancing the operating characteristics of a number of diverse response-specific circuits throughout the brain. We have speculated that elevations in the tonic activity of this system may contribute to functional changes related to arousal, and that the phasic activation of noradrenergic neurotransmission in stress acutely facilitates neuroendocrine, autonomic, behavioral and cognitive components of the integrated stress response. Given this, it is possible that dysregulation of this modulatory system may represent a potential substrate for the interaction between environmental sensitization and individual vulnerability leading to stress-related psychopathology. Further, whether or not a primary dysfunction of noradrenergic modulatory activity is a factor in the development of specific disorders, the pharmacological regulation of noradrenergic modulatory function represents a powerful and important mechanism whereby psychotherapeutic drug intervention can alleviate many of the symptoms of stress-related psychiatric illnesses such as depression, PTSD, and other anxiety disorders.

ACKNOWLEDGEMENTS

Research conducted in the author's laboratory was supported by research grants from the National Institute of Mental Health (MH60118 and MH53851), and from the Veterans Administration and Department of Defense. The author would like to acknowledge the contributions made to the original research described in this chapter by the following individuals: Dr. Marco Cecchi, Ms. April Garcia, Dr. Habibeh Khoshbouei, Dr. Shuaike Ma, Dr. Marie-Christine Pardon, Ms. Lauren Phillips, Dr. Scott Sands and Dr. Amelia Williams.

REFERENCES

1. Charney, D. S., Woods, S. W., Nagy, L. M. *et al.* Noradrenergic function in panic disorder. *J. Clin. Psychiatry*, **51** (1990), 5–11.
2. Pitman, R. K., Orr, S. P. and Shalev, A. Y. Once bitten, twice shy: beyond the conditioning model of PTSD. *Biol. Psychiatry*, **33** (1993), 145–146.
3. Schatzberg, A. F. and Schildkraut, J. J. Recent studies on norepinephrine systems in mood disorders. In *Psychopharmacology: The Fourth Generation of Progress*, ed. F. E. Bloom and D. J. Kupfer. (New York: Raven Press Ltd, 1995), pp. 911–920.
4. Southwick, S. M., Krystal, J. H., Morgan, C. A. *et al.* Abnormal noradrenergic function in posttraumatic stress disorder. *Arch. Gen. Psychiatry*, **50** (1993), 266–274.
5. Woodward, D. J., Moises, H. C., Waterhouse, B. D., Yeh, H. H. and Cheun, J. E.. Modulatory actions of norepinephrine on neural circuits. *Adv. Exp. Med. Biol.*, **287** (1991), 193–208.
6. Aghajanian, G. K. and Rogawski, M. A. The physiological role of α-adrenoceptors in the CNS: new concepts from single-cell studies. *Trends Pharmacol. Sci.*, **4** (1983), 315–317.
7. Jiang, M., Griff, E. R., Ennis, M., Zimmer, L. A. and Shipley, M. T. Activation of locus coeruleus enhances the responses of olfactory bulb mitral cells to weak olfactory nerve input. *J. Neurosci.*, **16** (1996), 6319–6329.
8. Waterhouse, B. D., Azizi, S. A., Burne, R. A. and Woodward, D. J. Modulation of rat cortical area 17 neuronal responses to moving visual stimuli during norepinephrine and serotonin microiontophoresis. *Brain Res.*, **514** (1990), 276–292.

9. Woodward, D. J., Moises, H. C., Waterhouse, B. D., Yeh, H. H. and Cheun, J. E. The cerebellar norepinephrine system: inhibition, modulation, and gating. *Prog. Brain Res.*, **88** (1991), 331–341.
10. Aston-Jones, G., Chiang, C. and Alexinsky, T. Discharge of noradrenergic locus coeruleus neurons in behaving rats and monkeys suggests a role in vigilance. *Prog. Brain Res.*, **88** (1991), 501–520.
11. Jacobs, B. L., Abercrombie, E. D., Fornal, C. A. *et al.* Single-unit and physiological analyses of brain norepinephrine function in behaving animals. *Prog. Brain Res.*, **88** (1991), 159–165.
12. Berridge, C. W. and Foote, S. L. Effects of locus coeruleus activation on electroencephalographic activity in neocortex and hippocampus. *J. Neurosci.*, **11** (1991), 3135–3145.
13. Page, M. E., Berridge, C. W., Foote, S. L. and Valentino, R. J. Corticotropin-releasing factor in the locus coeruleus mediates EEG activation associated with hypotensive stress. *Neurosci. Lett.*, **164** (1993), 81–84.
14. Rasmussen, K., Morilak, D. A. and Jacobs, B. L. Single unit activity of locus coeruleus neurons in the freely moving cat. I. During naturalistic behaviors and in response to simple and complex stimuli. *Brain Res.*, **371** (1986), 324–334.
15. Rosario, L. A. and Abercrombie, E. D. Individual differences in behavioral reactivity: correlation with stress-induced norepinephrine efflux in the hippocampus of Sprague-Dawley rats. *Brain Res. Bull.*, **48** (1999), 595–602.
16. Hilakivi, I. and Leppavuori, A. Effects of methoxamine, an alpha-1 adrenoceptor agonist, and prazosin, an alpha-1 antagonist, on the stages of the sleep–waking cycle in the cat. *Acta Physiol. Scand.*, **120** (1984), 363–372.
17. Hilakivi-Clarke, L. A., Turkka, J., Lister, R. G. and Linnoila, M. Effects of early postnatal handling on brain β-adrenoceptors and behavior in tests related to stress. *Brain Res.*, **542** (1991), 286–292.
18. Blanc, G., Trovero, F., Vezina, P. *et al.* Blockade of prefronto-cortical α1-adrenergic receptors prevents locomotor hyperactivity induced by subcortical D-amphetamine injection. *Eur. J. Neurosci.*, **6** (1994), 293–298.
19. Aston-Jones, G., Rajkowski, J. and Cohen, J. Role of locus coeruleus in attention and behavioral flexibility. *Biol. Psychiatry*, **46** (1999), 1309–1320.
20. Aston-Jones, G., Rajkowski, J. and Cohen, J. Locus coeruleus and regulation of behavioral flexibility and attention. *Prog. Brain Res.*, **126** (2000), 165–182.
21. Dayas, C. V., Buller, K. M., Crane, J. W. and Day, T. A. Stressor categorization: acute physical and psychological stressors elicit distinctive recruitment patterns in the amygdala and in medullary noradrenergic cell groups. *Eur. J. Neurosci.*, **14** (2001), 1143–1152.
22. Sawchenko, P. E., Li, H. Y. and Ericsson, A. Circuits and mechanisms governing hypothalamic responses to stress: a tale of two paradigms. *Prog. Brain Res.*, **122** (2000), 61–78.
23. Abercrombie, E. D. and Jacobs, B. L. Single-unit response of noradrenergic neurons in the locus coeruleus of freely moving cats. I. Acutely presented stressful and nonstressful stimuli. *J. Neurosci.*, **7** (1987), 2837–2843.
24. Cecchi, M., Khoshbouei, H., Javors, M. and Morilak, D. A. Modulatory effects of norepinephrine in the lateral bed nucleus of the stria terminalis on behavioral and neuroendocrine responses to acute stress. *Neuroscience*, **112** (2002a), 13–21.

25. Morilak, D. A., Fornal, C. A. and Jacobs, B. L. Effects of physiological manipulations on locus coeruleus neuronal activity in freely moving cats. II. Cardiovascular challenge. *Brain Res.*, **422** (1987), 24–31.
26. Morilak, D. A., Fornal, C. A. and Jacobs, B. L. Effects of physiological manipulations on locus coeruleus neuronal activity in freely moving cats. III. Glucoregulatory challenge. *Brain Res.*, **422** (1987), 32–39.
27. Pacak, K., Palkovits, M., Kopin, I. J. and Goldstein, D. S. Stress-induced norepinephrine release in the hypothalamic paraventricular nucleus and pituitary-adrenocortical and sympathoadrenal activity: in vivo microdialysis studies. *Front. Neuroendocrinol.*, **16** (1995), 89–150.
28. Page, M. E., Akaoka, H., Aston-Jones, G. and Valentino, R. J. Bladder distension activates noradrenergic locus coeruleus neurons by an excitatory amino acid mechanism. *Neuroscience*, **51** (1992), 555–563.
29. Svensson, T. H. Peripheral, autonomic regulation of locus coeruleus noradrenergic neurons in brain: putative implications for psychiatry and psychopharmacology. *Psychopharmacology (Berl.)*, **92** (1987), 1–7.
30. Valentino, R. J., Foote, S. L. and Page, M. E. The locus coeruleus as a site for integrating corticotropin-releasing factor and noradrenergic mediation of stress responses. *Ann. N. Y. Acad. Sci.*, **97** (1993), 173–188.
31. Ennis, M. and Aston-Jones, G. Activation of locus coeruleus from nucleus paragigantocellularis: a new excitatory amino acid pathway in brain. *J. Neurosci.*, **8** (1988), 3644–3657.
32. Valentino, R. J., Page, M. E. and Curtis, A. L. Activation of noradrenergic locus coeruleus neurons by hemodynamic stress is due to local release of corticotropin-releasing factor. *Brain Res.*, **555** (1991), 25–34.
33. Dayas, C. V. and Day, T. A. Opposing roles for medial and central amygdala in the initiation of noradrenergic cell responses to a psychological stressor. *Eur. J. Neurosci.*, **15** (2001), 1712–1718.
34. Liposits, Z., Sherman, D., Phelix, C. and Paull, W. K. A combined light and electron microscopic immunocytochemical method for simultaneous localization of multiple tissue antigens: tyrosine hydroxylase immunoreactive innervation of corticotropin releasing factor synthesizing neurons in the paraventricular nucleus of the rat. *Histochemistry*, **85** (1986), 95–106.
35. Kiss, A. and Aguilkara, G. Participation of α1-adrenergic receptors in the secretion of hypothalamic corticotropin-releasing hormone during stress. *Neuroendocrinology*, **56** (1992), 153–160.
36. Plotsky, P. M. Facilitation of immunoreactive corticotropin-releasing factor secretion into the hypophysial-portal circulation after activation of catecholaminergic pathways or central norepinephrine injection. *Endocrinology*, **121** (1987), 924–930.
37. Gartside, S. E., Suaud-Chagny, M. F. and Tappaz, M. Evidence that activation of the hypothalamo–pituitary–adrenal axis by electrical stimulation of the noradrenergic A1 group is not mediated by noradrenaline. *Neuroendocrinology*, **62** (1995), 2–12.
38. Castagne, V., Rivet, J. M. and Mormede, P. The integrity of the ventral noradrenergic bundle (VNAB) is not necessary for a normal neuroendocrine stress response. *Brain Res.*, **511** (1990), 349–352.

39. Ericsson, A., Kovacs, K. J. and Sawchenko, P. E. A functional anatomical analysis of central pathways subserving the effects of interleukin-1 on stress-related neuroendocrine neurons. *J. Neurosci.*, **14** (1994), 897–913.
40. Gaillet, S., Lachuer, J., Malaval, F., Assenmacher, I. and Szafarczyk, A. The involvement of noradrenergic ascending pathways in the stress-induced activation of ACTH and corticosterone secretions is dependent on the nature of stressors. *Exp. Brain Res.*, **87** (1991), 173–180.
41. Li, H. Y., Ericsson, A. and Sawchenko, P. E. Distinct mechanisms underlie activation of hypothalamic neurosecretory neurons and their medullary catecholaminergic afferents in categorically different stress paradigms. *Proc. Natl. Acad. Sci. USA*, **93** (1996), 2359–2364.
42. Williams, A. M. and Morilak, D. A. α1B adrenoceptors in rat paraventricular nucleus overlap with, but do not mediate, the induction of c-Fos expression by osmotic or restraint stress. *Neuroscience*, **76** (1997), 901–913.
43. Szafarczyk, A., Malaval, F., Laurent, A., Gibaud, R. and Assenmacher, I. Further evidence for a central stimulatory action of catecholamines on adrenocorticotropin release in the rat. *Endocrinology*, **121** (1987), 883–892.
44. Dayas, C. V., Buller, K. M. and Day, T. A. Medullary neurones regulate hypothalamic corticotropin-releasing factor cell responses to an emotional stressor. *Neuroscience*, **105** (2001), 707–719.
45. Sands, S. A. and Morilak, D. A. Expression of α1D adrenergic receptor mR. N. A in oxytocin- and corticotropin-releasing hormone-synthesizing neurons in the rat paraventricular nucleus. *Neuroscience*, **91** (1999), 639–649.
46. Morilak, D. A. and Jacobs, B. L. Noradrenergic modulation of sensorimotor processes in intact rats: the masseteric reflex as a model system. *J. Neurosci.*, **5** (1985), 1300–1306.
47. Stafford, I. L. and Jacobs, B. L. Noradrenergic modulation of the masseteric reflex in behaving cats. II. Physiological studies. *J. Neurosci.*, **10** (1990), 99–107.
48. Davies, M., Astrachan, D., Kehne, J., Commissaris, R. and Gallager, D. Catecholamine modulation of sensorimotor reactivity measured with acoustic startle. In *Catecholamines: Part B: Neuropharmacology and Central Nervous System – Theoretical Aspects*, ed. E. Usdin, A. Carlsson, A. Dahlström and J. Engel. (New York: Alan R. Liss Inc, 1984), pp. 245–258.
49. Detke, M. J., Rickels, M. and Lucki, I. Active behaviors in the rat forced swimming test differentially produced by serotonergic and noradrenergic antidepressants. *Psychopharmacology*, **121** (1995), 66–72.
50. Kitada, Y., Miyauchi, T., Kanazawa, Y., Nakamichi, H. and Satoh, S. Involvement of α- and β1-adrenergic mechanisms in the immobility-reducing action of desipramine in the forced swimming test. *Neuropharmacology*, **22** (1983), 1055–1060.
51. Khoshbouei, H., Cecchi, M., Dove, S., Javors, M. and Morilak, D. A. Behavioral reactivity to stress: amplification of stress-induced noradrenergic activation elicits a galanin-mediated anxiolytic effect in central amygdala. *Pharmacol. Biochem. Behav.*, **71** (2002), 407–417.
52. Pacak, K., McCarty, R., Palkovits, M., Kopin, I. J. and Goldstein, D. S. Effects of immobilization on in vivo release of norepinephrine in the bed nucleus of the stria terminalis in conscious rats. *Brain Res.*, **688** (1995), 242–246.

53. Pacak, K., Palkovits, M., Kvetnansky, R. *et al.* Effects of single or repeated immobilization on release of norepinephrine and its metabolites in the central nucleus of the amygdala in conscious rats. *Neuroendocrinology*, **57** (1993), 623–633.
54. Cecchi, M., Khoshbouei, H. and Morilak, D. A. Modulatory effects of norepinephrine, acting on α1-receptors in the central nucleus of the amygdala, on behavioral and neuroendocrine responses to acute immobilization stress. *Neuropharmacology*, **43** (2002b), 1139–1147.
55. Gold, P. W. and Chrousos, G. P. The endocrinology of melancholic and atypical depression: relation to neurocircuitry and somatic consequences. *Proc. Assoc. Am. Physicians*, **111** (1999), 22–34.
56. Johnson, E. O., Kamilaris, T. C., Chrousos, G. P. and Gold, P. W. Mechanisms of stress: a dynamic overview of hormonal and behavioral homeostasis. *Neurosci. Biobehav. Rev.*, **16** (1992), 115–130.
57. Sullivan, G. M., Coplan, J. D., Kent, J. M. and Gorman, J. M. The noradrenergic system in pathological anxiety: a focus on panic with relevance to generalized anxiety and phobias. *Biol. Psychiatry*, **46** (1999), 1205–1218.
58. Redei, E., Pare, W. P., Aird, F. and Kluczynski, J. Strain differences in hypothalamic–pituitary–adrenal activity and stress ulcer. *Am. J. Physiol.*, **266** (1994), R353–R360.
59. Pare, W. P. and Redei, E. Depressive behavior and stress ulcer in Wistar Kyoto rats. *J. Physiol.*, **87** (1993), 229–238.
60. Pardon, M. C., Gould, G. G., Garcia, A. *et al.* Stress reactivity of the brain noradrenergic system in three rat strains differing in their neuroendocrine and behavioral responses to stress: implications for susceptibility to stress-related neuropsychiatric disorders. *Neuroscience*, **115** (2002), 229–242.
61. Pardo, M. C., Ma, S. and Morilak, D. A. Chronic cold stress sensitizes brain noradrenergic reactivity and noradrenergic facilitation of the HPA stress response in Wistar Kyoto rats. *Brain Res.*, **971** (2003), 55–65.
62. Bhatnagar, S. and Dallman, M. Neuroanatomical basis for facilitation of hypothalamic–pituitary–adrenal responses to a novel stressor after chronic stress. *Neuroscience*, **84** (1998), 1025–1039.
63. Mana, M. J. and Grace, A. A. Chronic cold stress alters the basal and evoked electrophysiological activity of rat locus coeruleus neurons. *Neuroscience*, **81** (1997), 1055–1064.
64. Jedema, H. P., Sved, A. F., Zigmond, M. J. and Finlay, J. M. Sensitization of norepinephrine release in medial prefrontal cortex: effect of different chronic stress protocols. *Brain Res.*, **830** (1999), 211–217.
65. Nisenbaum, L. K., Zigmond, M. J., Sved, A. J. and Abercrombie, E. D. Prior exposure to chronic stress results in enhanced synthesis and release of hippocampal norepinephrine in response to a novel stressor. *J. Neurosci.*, **11** (1991), 1478–1484.
66. Ferguson, J. M., Mendels, J. and Schwart, G. E. Effects of reboxetine on Hamilton Depression Rating Scale factors from randomized, placebo-controlled trials in major depression. *Int. Clin. Psychopharmacol.*, **17** (2000), 45–51.
67. Nelson, J. C. A review of the efficacy of serotonergic and noradrenergic reuptake inhibitors for treatment of major depression. *Biol. Psychiatry*, **46** (1999), 1301–1308.

68. Versani, M., Cassano, G., Perugi, G. *et al.* Reboxetine, a selective norepinephrine reuptake inhibitor, is an effective and well-tolerated treatment for panic disorder. *J. Clin. Psychiatry.*, **63** (2002), 31–37.

69. Piñeyro, G. and Blier, P. Autoregulation of serotonin neurons: role in antidepressant drug action. *Pharmacol. Rev.*, **51** (1999), 533–591.

70. Sacchetti, G., Bernini, M., Gobbi, M. *et al.* Chronic treatment with desipramine facilitates its effect on extracellular noradrenaline in the rat hippocampus: studies on the role of presynaptic α2-adrenoceptors. *Naunyn-Schmiedeberg's Arch. Pharmacol.*, **363** (2001), 66–72.

71. Svensson, T. H. Effect of chronic treatment with tricyclic antidepressant drugs on identified brain noradrenergic and serotonergic neurons. *Acta Psychiatr. Scand.*, **61**: 280 (1980), 121–131.

72. Linnér, L., Arborelius, L., Nomikos, G. G., Bertilsson, L. and Svensson, T. H. Locus coeruleus neuronal activity and noradrenaline availability in the frontal cortex of rats chronically treated with imipramine: effects of α2-adrenoceptor blockade. *Biol. Psychiatry*, **46** (1999), 766–774.

73. Szabo, S. T. and Blier, P. Effect of the selective noradrenergic reuptake inhibitor reboxetine on the firing activity of noradrenaline and serotonin neurons. *Eur. J. Neurosci.*, **13** (2001), 2077–2087.

74. Valentino, R. J., Curtis, A. L., Parris, D. G. and Wehby, R. G. Antidepressant actions on brain noradrenergic neurons. *J. Pharmacol. Exp. Ther.*, **253** (1990), 833–840.

75. Meana, J. J., Herrera-Marschitz, M., Goiny, M. and Silveira, R. Modulation of catecholamine release by α2-adrenoceptors and I1-imidazoline receptors in rat brain. *Brain Res.*, **744** (1997), 216–226.

76. Svensson, T. H. Brain noradrenaline and the mechanisms of action of antidepressant drugs. *Acta Psychiatr. Scand.*, **101**: 402 (2000), 18–27.

77. Morilak, D. A., Cecchi, M. and Khoshbouei, H. Interactions of norepinephrine and galanin in the central amygdala and lateral bed nucleus of the stria terminalis modulate the behavioral response to acute stress. *Life Sci.*, **73** (2003), 715–726.

78. Morilak, D. A., Gould, G. and Pardon, M. C. Changes in baseline behavior and acute stress reactivity on the interaction test and elevated plus-maze following chronic reboxetine treatment. *Soc. Neurosci. Abstr.*, **27** (2001), program No. 572.14.

Part III

The biology of norepinephrine in CNS pathology

10

Animal models of psychopathology: focus on norepinephrine

Leonie A. M. Welberg, Ph.D. and Paul M. Plotsky, Ph.D.

Emory University School of Medicine, Atlanta, USA

Introduction: an overview

Primary functions of norepinephrine during brain development

Catecholamines are essential to normal neuronal development and thus can be detected from early ontogenesis onwards.[1] The cellular environment influences which type of neurotransmitter (e.g. adrenergic or cholinergic) a neuron will produce. Glucocorticoids, for example, can affect the differentiation of neural crest cells in an adrenergic direction, at least in vitro.[2] Norepinephrine (NE) is found in early cortical structures, and is probably required for appropriate migration and apoptosis of neurons, as well as for laminar formation and glial development.[3,4] The crucial importance of NE in development has been shown by Thomas *et al.* These researchers produced mice completely lacking NE production due to the absence of the gene for dopamine-β-hydroxylase, and showed that this state led to death in utero from day 11 of gestation onwards.[5,6] A small percentage of homozygous pups born to heterozygous mothers did survive, probably due to transplacental passage of maternal NE.[5] In other cases, fetuses could only be rescued by adding an NE precursor that did not require conversion by dopamine-β-hydroxylase to the maternal drinking water during pregnancy,[5] indicating that the presence of NE during development is essential for survival.

Primary functions of norepinephrine in the mature CNS

Norepinephrine neurons in the central nervous system (CNS) originate mainly from two cell groups: the lateral tegmentum area sends fibers through the ventral ascending noradrenergic bundle innervating the hypothalamic, basal telencephalic, and preoptic areas. Alternately, the locus coeruleus (LC) neurons project through the dorsal ascending noradrenergic bundle to the thalamus, hippocampus, cerebellum, and neocortex. The NE system interacts with the corticotropin-releasing

Brain Norepinephrine: Neurobiology and Therapeutics, ed. Gregory A. Ordway, Michael A. Schwartz and Alan Frazer. Published by Cambridge University Press.

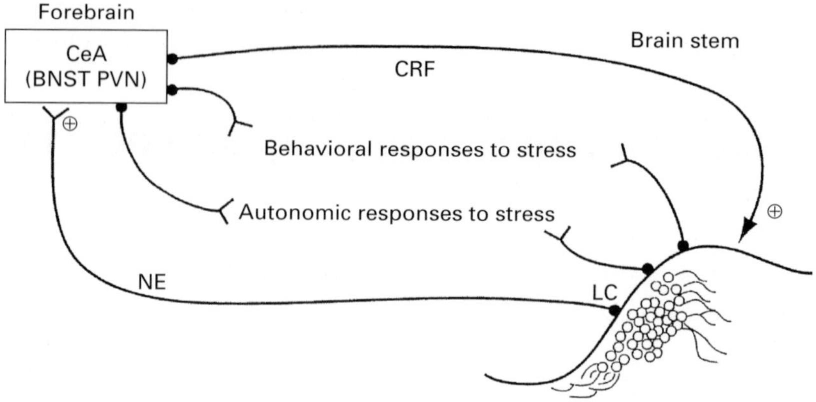

Figure 10.1 Interaction between CRF and NE systems within the CNS. BNST, bed nucleus of the stria terminalis; CeA, central nucleus of the amygdala; CRF, corticotropin-releasing factor; LC, locus coeruleus; NE, norepinephrine; PVN, paraventricular nucleus of the hypothalamus. Reprinted from Koob, G. F. Corticotropin-releasing factor, norepinephrine, and stress, *Biol. Psychiatry*, **46**, 1167–1180, Copyright (1999), with permission from the Society for Biological Psychiatry.

factor (CRF) system, an important regulator of stress and anxiety: CRF neurons from the bed nucleus of the stria terminalis (BNST) and the central nucleus of the amygdala (CeA) innervate the LC and vice versa, causing fear and stress to increase both neuronal activity of the LC and NE turnover in the brain. A diagrammatic representation of the interaction between the CRF and NE systems is shown in Figure 10.1. Activation of the LC, by stress or artificially by electrical stimulation, produces a state of heightened arousal. As such, central NE pathways play an important role in controlling arousal, vigilance, and alertness, and, consequently, learning and memory.[7]

Changes and inconsistencies in the noradrenergic system in mood disorders

The majority of research investigating the role of NE in mood disorders has focused on depression. Although early studies with reserpine suggested that depression results from central NE depletion, controversy still exists over whether the NE system is actually hyper- or hypoactive in depression. Many effective antidepressants serve to increase NE levels in the synapse, indicating that a deficit in NE levels, at least in part, underlies the depressive symptomatology. However, there is also evidence that NE may actually be elevated in depression.[8,9] It has been proposed that the state of the central NE system may depend on the specific subtype of depression.[10] Similarly, specific abnormalities in the NE system appear to underlie different subgroups of anxiety and panic disorders. All subtypes display some level of increased arousal and autonomic activity in which noradrenergic dysregulation plays a major role.[11] For example, administration of the α_2-adrenergic receptor antagonist yohimbine can evoke a panic attack in people with panic disorder, as well as in around 30% of posttraumatic stress disorder (PTSD) patients. The α_2-receptor agonist clonidine on the other hand, induces a blunted growth hormone response in patients from most anxiety disorder subtypes compared to healthy controls. Subjects with social phobia show elevated plasma NE concentrations compared to people with panic disorder and healthy controls. Another anxiety-type form of psychopathology, PTSD, has also shown evidence for hyperactivity of the NE system. Studies of PTSD are performed most often in combat veterans and women who have suffered mental, physical, or sexual abuse early in life. There are some indications of a hyperactive noradrenergic system in PTSD patients under baseline conditions, such as elevations in plasma NE levels or 24-hour urine NE concentrations,[12] but these findings are not consistent. Rather, the NE system appears to be hyperreactive in PTSD patients. For example, war veterans suffering from PTSD respond to hearing recorded combat sounds with an increase in NE, whereas non-PTSD veterans do not.[12] Administration of the α_2-adrenergic receptor antagonist yohimbine causes panic attacks, flashbacks, and evidence of increased autonomic activity in a large percentage of combat veterans with PTSD. Moreover, PTSD patients respond with a greater release of NE to yohimbine administration than controls.

Since activation of the LC and the concomitant release of NE throughout the brain play a major role in controlling arousal, selective attention, and alertness, it is reasonable to suggest that disturbances in the NE system are likely to underlie the pathophysiology of attention-deficit/hyperactive disorder (ADHD), a disorder characterized by hyperactivity and concentration problems.[13] However, the exact role of NE in ADHD is unclear. There may be an elevated tonic activation of the LC and peripheral NE neurons in ADHD patients, resulting in an inefficient stress-induced activation of those systems. However, ADHD does not appear to result simply from a hyperactive NE system, but rather from a distortion in the interaction between NE, epinephrine, and dopamine.[14]

Considerations for the development of animal models

Criteria for a satisfactory animal model of disease

The original and now widely accepted criteria for animal models of human psychopathology were developed by McKinney and Bunney,[15] who proposed that a satisfactory animal model should be similar to the human disorder under study, with regard to symptomatology, etiology, underlying pathophysiological mechanisms, and sensitivity to treatment. Willner[16] has attempted to summarize the above criteria as three *validities* that together determine the overall strength of an animal model. The first of these is *face validity*, representing the extent to which the behavior and/or physiology of an animal model resembles that in the human disorder. For example, an animal model of depression should show at least a number of the symptoms that characterize major depression in humans, such as psychomotor retardation (loss of activity), lack of grooming, anhedonia (lack of pleasure from acts that would normally produce pleasure), loss of reward-stimulated behaviors, substantial body weight loss or gain, hypothalamic–pituitary–adrenal (HPA) axis activity alterations, changes in sleep pattern, and cognitive disturbances. The second of Willner's validities is *predictive validity*, that is, the degree to which the animal model is able to predict clinical efficacy of drugs without false positives or false negatives. Third, *construct validity*, signifies whether the animal model and the human disorder have similar underlying neurobiological mechanisms.

Two additional criteria for accurate animal models of disease were later added by Koob *et al.*[7] A good *etiological validity* means that the animal model and human disease have the same cause. Finally, the *convergent validity* shows the degree to which the animal model correlates with other animal models of the same disorder.

NE system alterations in different animal models of human psychopathology

In the next sections we will discuss NE system alterations observed in some common and less common animal models of human psychopathology (Table 10.1). Four different types of models will be distinguished: homologous models typically have good face, construct, and etiological validities. Second, "animal assay" models[17] are behavioral tests with a good predictive validity, but they are limited in the other criteria listed above, and are used mainly to screen new drugs. Next, in "breeding-to-phenotype" models, animals from the high and low ends of the behavioral spectrum in a particular test are selectively bred to create two different lines. These models are often used to discriminate genetic from environmental influences on behavior (cf. Francis *et al.*[18]). Finally, recently developed models based on selective gene manipulations use mice in which a particular gene, the product of which is thought to be involved in a psychopathological process, is either knocked out or overexpressed.

Homologous models

Perinatal models of psychopathology

The paradigms listed below were not necessarily developed with the intention of providing models for affective or anxiety disorders, but rather with the purpose of studying the long-term effects of environmental injury early in life. The CNS has proven to be especially sensitive to such insults, suggesting that vulnerability to

Table 10.1 Overview of behavioral and NE system changes in different animal models. Listed effects are in rats unless stated otherwise.

Developmental models	*Phenotype*	*NE* activity	*NE* synthesis/content	*NE* turnover	α_1- *receptors*	α_2-*receptors*	β-*receptors*
Prenatal stress	anxiety ↑ HPA responsivity ↑ learning/memory ↓ depression-like behavior	↑ (monkeys)	↓ LC, cortex	↑ LC, cortex ↓ POA (transient)	↓ cortex	↓ hypothalamus	↓ cortex
Prenatal dexamethasone	anxiety ↑ HPA responsivity ↑ depression-like behavior	?	↓ hippocampus, cortex	↓ midbrain, brain stem	?	↑ midbrain, brain stem	?
Maternal separation	anxiety ↑ HPA responsivity ↑ depression-like behavior cognitive functioning ↓ drug sensitivity ↑	↑ (monkeys, in response to separation)	↑ hippocampus (females), PVN, accumbens	?	?	= PVN	?
Neonatal handling	anxiety ↓ efficient HPA response, learning/memory ↑	?	↓ PVN	?	= PVN, LC, NTS	= PVN ↑ LC, NTS	?
Early malnutrition	anxiety ↓ social behavior ↓ reward responsiveness ↑	↑ cortex	↑ whole brain	↓ cortex, cerebellum	?	↓ cortex	↓ cortex
Isolation rearing	behavioral stress sensitivity ↑ HPA alterations	↓ hippocampus	↓ CSF (monkeys)	?	?	↑ hippocampus, = cortex	?

(*cont.*)

Table 10.1 *(cont.)*

Developmental models	*Phenotype*	*NE* activity	*NE* synthesis/content	*NE* turnover	α_1- *receptors*	α_2-*receptors*	β-*receptors*
Neonatal infection	behavioral, emotional and learning alterations?	? ↑ frontal cortex, hippocampus, hypothalamus, cerebellum, SN, raphe (Borna virus) ↓ occipital cortex = hippocampus, hypothalamus, olfactory bulb (Vesicular stomatitis virus) ↓ spinal cord, cerebellum, cortex (cytomegalovirus, mice)			?	?	?
Adult models							
Learned helplessness	depression-like behavior loss of reward sleep disturbances anorexia grooming ↓	Frontal cortex/ hippocampal NE levels protective	= hippocampus	?	?	?	↑ hippocampus
CVS/CMS	open-field activity ↓ anhedonia	?	= hippocampus, = hypothalamus	?	?	↑ sensitivity, ↓ amygdala	↑↓ cortex, ↓ hippocampus, amygdala, hypothalamus, ↓ β_1 amygdala
Olfactory bulbectomy	muricide, hyperactivity, ↓ sexual behavior ↓ neophobia ↓ learning	?	↓ lateral hypothalamus, amygdala, midbrain ↑ hypothalamus	?	↑ hippocampus amygdala	↑ forebrain	↑ hippocampusa, amygdala

Breeding-to-phenotype models							
Flinders (FSL vs. FRL)	high vs. low responsiveness to ↑ cholinergic activity	?	↑ PFC, raphe, accumbens, hippocampus = hypothalamus	?	?	?	?
Maudsley (MR vs. MNRA)	high vs. low level of "emotionality"	Increased release in LC after acute stress, lower after chronic stress	?	?	?	↓ LC	?
Roman (RHA vs RLA)	active *vs* passive coping behavior	?	= cortex, hypothalamus, pons/medulla	?	?	?	?
Gene manipulation models							
DBH $^{-/-}$ *mice*	normal learning, memory ↓ maternal care ↓						
NET $^{-/-}$ *mice*	antidepressant-like effect						
Alpha$_{1B}$ *receptor* $^{-/-}$ *mice*	stimulant sensitivity ↓ reward sensitivity ↓						
Alpha$_2$*-receptor* $^{-/-}$ *mice*	anxiety ($\alpha_{2A}{}^{-/-}$) antidepressant, HPA sensitivity ↓ ($\alpha_{2C}{}^{-/-}$)						
Galanin-overexpressing mice	= basal anxiety						

CVS/CMS, chronic variable stress/chronic mild stress; PFC, prefrontal cortex; SN, substantia nigra.

psychopathology in adulthood may be increased by adverse events occurring during periods of rapid brain development. Consequently, some of these paradigms have reasonable face, construct, and etiological validities as models for adult psychopathology. In the models mentioned in this section, we focus on changes in the central NE system resulting from early life events.

Prenatal stress

Stress during pregnancy has been linked to psychopathology and behavioral disorders in children.[19–21] Studies in nonhuman primates have shown that prepartum stress may cause fetal loss and, in the case of live births, may trigger impairments in attention, motor development, and exaggerated stress responses.[22–27] In rodent models, prenatal stress has consistently been shown to result in HPA hyperactivity and anxiety-like behavior in adulthood, sometimes in combination with "behavioral depression."[28,29] When pregnant animals are stressed relatively early in gestation, severe and permanent effects on growth and survival occur.[27,30] However, prenatal stress in rats is usually applied in the latter part of pregnancy. Thus prenatal stress influences differentiation processes, rather than the cruder process of growth, although some research points to potential effects at this level as well.[30] Subtle differences in study outcomes of prenatal stress are likely the result of differences in timing, predictability, and controllability of the stressors applied, in sex, and age of tested offspring, and importantly, in postnatal rearing conditions.

Several studies have shown alterations in the NE system resulting from exposure to prenatal stress. Specifically, NE turnover appears to be affected,[31] with the direction of the change depending upon the brain region. For example, Takahashi *et al.*[32] showed that male offspring from dams exposed to uncontrollable shock during pregnancy had lower NE and higher 3-methoxy-4-hydroxyphenylglycol (MHPG) levels in the cortex and LC compared to control rats. This increase then resulted in an increased MHPG/NE ratio in these areas. In contrast, reduced NE turnover and elevated NE content have been found in the preoptic area of prenatally stressed male rats, in combination with a reduced stress-induced depletion of hypothalamic NE content.[33,34] Peters subjected pregnant females to the relatively mild stressor of daily saline injections and found reductions in cortical levels of α_1- and β-adrenoceptors and NE content in the offspring, as well as permanently reduced α_1-adrenoceptor binding in the hypothalamus.[35,36] These changes were transient, suggesting that prenatal stress altered the development of the noradrenergic system. Clearly, effects of prenatal stress on the central NE system are highly region-specific. However, a general increase in activity of the central NE system was observed in monkeys: offspring from stressed mothers showed higher concentrations of MHPG in cerebrospinal fluid (CSF), with a similar trend for NE.[37] In another study, prenatal stress appeared to have no effect on the basal level of sympathetic nervous system activity, indicated by equal concentrations of NE and its metabolites in plasma of prenatally stressed and control rats.[38] Stress-induced sympathetic activity, however, was increased in prenatally stressed rats. These animals showed higher levels of plasma NE after footshock stress. Surprisingly, no studies have been undertaken to investigate whether administration of antidepressant drugs to prenatally stressed animals can reverse or modify any of the behavioral or biochemical effects of prenatal stress. One study administered α_2-adrenoceptor antagonists to prenatally stressed male rats and found no difference in sensitivity to these drugs compared to unstressed rats.[39] Thus, although prenatal stress clearly causes some changes in the developing NE system, it is unclear whether these changes underlie the anxious phenotype of these animals.

Prenatal glucocorticoid exposure

Many of the permanent effects of prenatal stress on behavior and brain biochemistry can be replicated by prenatal exposure to glucocorticoids.[29] Indeed, low-dose (0.05 mg/kg) injections of the synthetic glucocorticoid dexamethasone on gestational days 17 to 19 accelerated the developmental pattern of NE turnover in the midbrain and brain stem in rats, resulting in an early "peak" in NE turnover and reduced turnover later in life.[40] Muneoka *et al.*,[41] using the same treatment, also found a reduced NE content in the hippocampus from 3 weeks of age onwards, and in the neocortex at 14 weeks. In addition, exposure to the same low dose of dexamethasone enhanced presynaptic α_2-adrenoceptor binding in midbrain, brain stem, and cerebellum.[42] It appears that at low doses, glucocorticoids accelerate differentiation of NE neurons, whereas at higher doses (0.8 mg/kg) a more generalized growth-suppressing effect is seen.[42] The studies mentioned above measured activity of the NE system under basal conditions, and it is not known whether the central NE response to stress is also altered in these animals, although one study showed that prenatal dexamethasone exposure tended to increase plasma NE levels both after fasting and after stress in adult rats.[43] It is not unlikely that permanent alterations in basal NE content and turnover in prenatal-glucocorticoid exposed rats play a role in the behavioral deficits observed in these rats,[44] but this remains to be verified. Reversal of the prenatal-glucocorticoid exposure "phenotype" by administration of NE uptake inhibitors or adrenergic receptor antagonists to these rats in adulthood would suggest a direct involvement of the NE system in this phenotype, but such studies have not yet been undertaken.

Maternal separation and neonatal handling

Separation of pups from their dams in the preweaning period has long-term effects on stress reactivity. Rats that were maternally separated for relatively long (3 to 24 hour) periods display as adults increased anxiety-like behavior, increased alcohol preference and sensitivity to opiates, and have deficits in cognitive functioning.[45–49] Some of the behavioral effects of maternal separation can be reversed in adulthood by chronic treatment with the selective serotonin reuptake inhibitor (SSRI) paroxetine,[45] indicating that this model may provide some predictive validity in addition to face, construct, and etiological validities. Although the most consistent effect of extended periods of maternal separation is a hyperresponsiveness of the HPA axis,[48,49] which is often attributed to a decrease in mineralocorticoid and glucocorticoid receptor levels in the hippocampus, few studies have investigated the contribution of the LC-NE system in HPA regulation of maternally separated rats. Dent *et al.*[50] showed that rats normally respond to a saline injection with an increase in tyrosine hydroxylase (TH) mRNA in the LC on postnatal day 18 but not earlier in development, i.e. not until after the so-called stress-hyporesponsive period. However, this response is accelerated if rats undergo a single 24-hour separation from the dam on day 11. Thus, the separation amplified the response to a subsequent stressor. In the long term, repeated maternal separation (10 times, 6-hour periods, randomly done between postnatal days 5 and 10) has been shown to increase NE levels in the dorsal hippocampus in female, but not male, rats,[51] although no changes were seen in the prefrontal cortex. Maternal separation paradigms have also been tested in primates: Clarke[52] compared juvenile rhesus monkeys that were raised by their mothers with monkeys raised with peers only, and showed that maternal deprivation reduced CSF NE in response to social separation later in life as compared to peer-deprivation. However, this study did not contain a "normal" control group of monkeys raised with both mothers and peers, leaving it unclear whether maternal separation in itself increases or decreases the CSF NE content. Nevertheless, it appears that maternal separation in rhesus macaques raises NE levels at least acutely. One study compared infant monkeys separated, but housed adjacent to their mothers, for 4 days to monkeys that were completely isolated during the same period. The results showed that the monkeys who were completely isolated displayed greater and longer NE responses as measured by NE metabolite levels in CSF in response to the separation.[53]

In contrast to long-term separation, short (3 to 15 minutes) daily maternal separations, also called neonatal handling, have the opposite effect on HPA activity and anxiety-like behavior.[54,55] Thus, Liu *et al.*[56] compared maternally separated (3 hours daily, postnatal day 2 to 14) offspring with handled offspring (15 minutes daily, postnatal day 2 to 14) and with undisturbed offspring in adulthood. Liu's study found that maternally separated rats had elevated basal NE levels and an increased NE response to stress in the paraventricular nucleus of the hypothalamus (PVN), whereas handled rats showed a decreased response compared to control (undisturbed) rats. Levels of α_2-adrenoceptors in the PVN were not affected by either paradigm. However, in the LC and nucleus solitary tract, handled rats showed increased α_2-receptor densities compared to both undisturbed and maternally separated offspring. As noradrenergic neurons from these areas project to the PVN and thus play a role in the regulation of the HPA response to stress, these results suggest that an increased level of autoinhibition in handled rats may result in the efficient turn-off of the stress response commonly observed in such rats.[48] Levels of α_1-adrenoceptors in these regions were not affected by maternal separation or handling.[56]

Recently it has become apparent that the quality of maternal care upon reunion, rather than separation from the dam per se, is the main determinant of these outcomes. Indeed, offspring from mothers that naturally show a high level of licking, grooming, and arched-back nursing[57] have increased α_2-receptor levels in the LC compared to offspring from mothers that show less maternal care. The rats with higher quality maternal care also resemble handled rats in their lower levels of both anxiety-like behavior and HPA reactivity, suggesting that these changes may be caused by a decreased LC-NE activity. In summary, prolonged maternal separation results in a hyperactive central NE system later in life with a subsequent exaggerated hormonal stress response and increased anxiety-like behavior. In contrast, high-quality maternal care, whether spontaneous or induced by short-term daily separations from the pups, appears to ensure a more efficient stress response and improved cognitive functioning in offspring. Interestingly, NE plays a crucial role in the development of maternal behavior itself. Mice lacking the gene for dopamine-β-hydroxylase failed to retrieve pups, and this deficit could be restored by administering an alternative NE precursor that does not require dopamine-β-hydroxylase, shortly before parturition.[58] It is tempting to suggest that a dam's own early life experiences may, via changes in central NE systems resulting from these experiences, in turn determine her behavior towards her pups.

Early malnutrition

Studies have shown that early malnutrition in children is reflected by poorer IQ levels and cognitive functioning, as well as behavioral problems during their school years (although there do not appear to be any specific cognitive deficits).[59] In rats, both nutrition and protein restriction prenatally and preweaning have profound effects on brain development, with permanent consequences for behavior later in life.[60,61] The precise effects crucially depend upon the timing of the malnourishment.[62] Manipulations used to achieve perinatal malnourishment in these paradigms include increasing litter size to 18 pups; nipple ligation in the dam; reduction of maternal caloric intake during pregnancy and/or lactation; and, most commonly applied, protein reduction in the maternal diet (6% to 8% casein diet instead of 16% to 25%). The latter form of malnutrition is a subtle one that does not always lead to obvious effects on body or brain weight in the offspring.

However, as discussed below, this restriction has pronounced effects on the central NE system.

Behavioral effects of prenatal protein reduction include decreased social behavior in juvenile rats,[63] reduced anxiety-like behavior in adult female offspring,[64] alterations in sleep patterns,[65] and an increased responsiveness to reward.[60] Decreased nutrition during lactation also reduces anxiety in adulthood[66] and decreases the "anticonflict" effects of anxiolytics,[67] whereas Rotta *et al.*[68] showed that preweaning protein restriction blocks the anxiogenic effect of naltrexone on rat exploratory behavior. Campbell and Bedi[69] failed to show effects of preweaning malnourishment on learning and memory in a water-maze task. In addition, no effects were seen in other hippocampus-associated learning tasks[70] or on social behavior in adulthood,[71] suggesting that prenatal malnourishment affects specific neuronal pathways or systems, rather than causing general impairments in brain function.

Prenatal malnourishment produces more pronounced behavioral effects than neonatal malnourishment. This corresponds with the fact that neuronal morphology is altered after prenatal malnutrition, regardless of the postnatal maternal diet.[72] In many "prenatal undernutrition" studies, it is important to note that newborn pups were adopted to dams that had received a normal diet during pregnancy, in order to avoid the possibility that the effects of malnutrition on the pregnant dam would continue into the lactation period. Massaro *et al.*[73] showed that adoption of prenatally underfed pups to well-fed mothers increases the level of maternal behavior by the adoptive dams; this may affect, or even reverse, possible effects of prenatal malnourishment. Such modifying effects of adoption have been shown at least after prenatal stress.[74] Maternal food restriction both during and after pregnancy (until weaning) has permanent effects on offspring HPA activity,[75] with reduced plasma corticosteroid-binding globulin, elevated free corticosterone concentrations, and a faster termination of the stress response. These effects appear to depend on undernutrition-induced increases in maternal corticosterone levels during pregnancy.[76]

There is a wealth of evidence suggesting that early life malnutrition results in increased noradrenergic activity in the brain.[77] Thus, increased NE synthesis,[78,79] higher NE content in various brain regions,[80] reduced NE turnover,[81,82] and (possibly compensatory) reductions in α- and β-adrenoceptor levels[83] have been observed in brains of adult rats exposed to perinatal malnutrition. Undernourished rats also showed a deficit in down-regulation of cortical and hippocampal β-adrenoceptors in response to desipramine,[84] although their capacity to up-regulate these receptors after propranalol administration was unaffected. Furthermore, brains of perinatally malnourished rats showed elevated K^+-induced NE release in the prefrontal and occipital cortex,[72,85] but only if the protein restriction was continued into the preweaning period.[72,86] Protein deprivation from day 14 of gestation until day 40 of life increased LC activity and reduced sensitivity of α_2-autoreceptors in this area, indicating insufficient autoinhibition of LC activity by NE. Repeated administration of the antidepressant desipramine normalized these deficits,[87] although it did not completely reverse the malnutrition-induced immobility in the forced-swim test.[88]

Importantly, using clonidine to reduce the release of NE during gestation or lactation prevents (to some extent) the effects of early malnutrition.[89,90] This finding suggests that prenatal malnutrition increases central NE activity from early development onwards, and that this affects the normal organizational role of NE on neuronal growth and differentiation.

Isolation rearing

In their natural environment, most mammals are reared by their mothers, though they also have contact with littermates or other peers. Thus, an animal's earliest social environment consists of mother and peers. Depriving an infant of either of these two sources of contact has proven to have severe consequences lasting into adulthood. The effects of maternal separation have been discussed above. Peer contact deprivation has also received some attention in research, in primates as well as in rats.

Social isolation in infant nonhuman primates can result in disturbed behavior later in life, such as self-injurious behavior[91] and exaggerated behavioral responses to stress.[92,93] Furthermore, it alters HPA activity: Clarke *et al.*[94] showed that monkeys reared without peers have higher plasma levels of cortisol and adrenocorticotropic hormone (ACTH) than peer-reared (but mother-deprived) monkeys, basally as well as in response to social stress and to cage transfer.[95] Nevertheless, both groups are equally sensitive to the effects on cortisol and ACTH of chronic treatment with the SSRI fluoxetine or with the NE reuptake inhibitor desipramine.[94] It is important to note that the above studies did not contain a control group of monkeys raised in the presence of mother *and* peers, rendering it impossible to interpret from these results whether peer-deprivation leads to increased HPA activity or whether maternal deprivation blunts this activity.

A few studies have investigated how the NE system is affected by social deprivation early in life. Kraemer *et al.*[96] compared rhesus monkey infants raised by their mothers who from 9 months onwards either lived with peers (social rearing) or alone (peer-deprived rearing) with infants that had lived in individual cages from birth onwards (social isolation). Social isolation consistently resulted in lower CSF NE levels than those found in socially reared monkeys, which in turn had lower NE levels than peer-deprived monkeys. Furthermore, infant peer-deprived monkeys had lower basal CSF NE and MHPG levels than monkeys raised with peers but without their mothers, but displayed a greater NE increase in response to a social stressor than peer-reared infants.[52] Later in life, peer-deprived juvenile monkeys show higher CSF NE levels than monkeys that were artificially reared in the presence of peers. Norepinephrine responses to an injection with the NE reuptake inhibitor desipramine were similar in both groups,[97] suggesting that differences in NE reuptake do not account for the deprivation-induced effects on NE levels in the CSF. Similar studies were performed by Higley *et al.*[98,99] who showed permanently decreased levels of both MHPG in CSF and cortisol in plasma in monkeys that were peer-deprived compared to monkeys that were mother-deprived for the first 6 months of life. However, as stated earlier, without a control group of animals raised in the presence of both mother and peers, it is difficult to say which responses can be considered "abnormal" or "inefficient." Interestingly, the alterations in activity of the NE and HPA systems consistently observed after early social isolation are not reflected by changes in the noradrenergic input into the PVN.[100]

Isolation rearing in rats starts after weaning, when rats are housed individually rather than in groups of five or six. Social isolation induces behavioral disturbances, especially in stressful situations, such as locomotor hyperactivity,[101] anxiety-like behavior on the elevated plus-maze,[102] learning impairments,[103] and disruptions in prepulse inhibition of acoustic startle.[104] These behavioral changes are reflected by alterations in the central NE system. For example, Fulford *et al.*[105] revealed a social isolation-induced hypersensitivity to clonidine injections, suggesting elevated levels of presynaptic α_2-adrenoceptors in isolation-reared rats. Indeed, isolated rats expressed more α_2-receptors in the hippocampus (but not in cortex) than socially reared rats. This was later confirmed by in vitro experiments[106] in which the hippocampus of isolated rats proved more sensitive to the effects of the α_2-receptor antagonist idazoxan on NE release. The same researchers then used microdialysis to show the functional effects of increased α_2-receptor levels. Although social isolation did not affect basal hippocampal NE release, it decreased K^+-induced NE release. Moreover, there was an increased inhibition by clonidine of K^+-induced NE release and a decreased stress-induced NE release in the hippocampus of isolated rats.[107] The hippocampus is a relatively late-developing structure, and therefore may be especially sensitive to postweaning manipulations. The altered central NE state in isolated animals also became apparent after central NE depletion using the neurotoxin *N*-(2-chloroethyl)-*N*-ethyl-2-bromobenzylamine (DSP-4), in that isolated animals had different behavioral responses to this depletion compared to control rats.[108,109] Taken together, these animal studies indicate that social isolation early in life appears to reduce central NE activity, possibly via increased autoinhibition through elevated numbers of presynaptic α_2-adrenoceptors, with a more pronounced reduction of NE activity after more severe deprivation.

The few studies that investigated effects of antidepressant administration on isolation-induced deficits in rats have confirmed some predictive validity for social isolation as a model for human psychiatric disorders. For example, Heritch *et al.*[110] showed that postweaning-isolated rats displayed increased immobility in a forced-swim test and had elevated brain NE levels, both of which could be normalized by desipramine treatment during the isolation procedure. In another study, postweaning isolation produced hyperactivity in an open field compared to group-housed rats. Hyperactivity was reduced by previous exposure to restraint stress. Chronic treatment with imipramine during isolation prevented the effects of restraint on locomotor activity in isolated, but not group-housed, rats.[111] Imipramine treatment of peer-deprived monkeys also significantly improved their behavior.[112]

Neonatal infection

Prenatal or early postnatal infection has been implicated in the etiology of some neurodevelopmental disorders, notably schizophrenia.[113] Neonatal infection with the Borna Disease virus has been used in rats as a model for autism spectrum disorders,[114] on account of the alterations in emotional and learning behavior displayed by Borna virus-infected animals as well as hippocampal and cerebellar damage.[114] In adulthood, neonatally Borna virus-infected rats show increased levels of NE in frontal cortex, hippocampus, hypothalamus, cerebellum, substantia nigra and raphe nuclei. However, the magnitude of these increases can be strain dependent.[115,116] In contrast, Mohammed *et al.*[117] infected female rats through the nose on postnatal day 11 or 12 with the vesicular stomatitis virus and 5 months later observed a reduction of NE levels in the occipital cortex. However, NE reduction was not observed in other cortical areas, hippocampus, hypothalamus, or olfactory bulb. These animals showed hyperactivity in an open field and disturbed learning behavior in a water-maze test. The authors argued that the behavioral effects were likely to be due to infection-induced alterations in the serotonin system, which were much more pronounced than the effects on the NE system. An intraventricular infection with cytomegalovirus[118] in mice on postnatal day 10 resulted in moderately reduced NE levels in spinal cord, cerebellum, and cortex 5 days later, though no permanent effects on the NE system could be detected. Similar transient effects were found in levels of dopamine and serotonin. As with other "adverse early life event" models, the timing of the viral infection appeared to be an important factor in producing the eventual outcome, as did the route of administration. Clearly, those parts of the CNS that undergo maturation or differentiation at the time of infection are most vulnerable to the insult.

Plasticity

The findings cited above clearly show that the sensitive window for environmental effects on CNS development is rather broad. Events that occur during a time of neuronal organization and specialization (prenatally in the rat) as well as insults taking place during rapid growth of the brain or specific structures in it (such as the hippocampus and the cerebellum early postnatally in the rat) can have enduring effects on behavior and on hormone and neurotransmitter systems, including the NE system. In both cases, however, some plasticity is apparent even after the most sensitive window. There is some room for reversibility of effects that result from adverse early life events, provided that the environment is optimal. Increased maternal care (after prenatal stress[74]) or nutritional rehabilitation (after early malnutrition) can modify the effects of such events. An example of interaction between effects on the CNS of different environmental events was given by Bedi *et al.*[119] Their research showed that postnatal malnutrition only adversely affected synaptic density in the visual cortex if pups were fed and raised artificially, i.e. were maternally deprived: pups raised by undernourished dams in fact had a higher synaptic density than pups nurtured by normally nourished dams. Whether such interactions also exist at the level of the NE system remains to be examined.

Adult models of psychopathology

Learned helplessness

The term "learned helplessness" was first introduced by Seligman and Maier,[120] who observed that, after exposure to uncontrollable or inescapable stress (usually shock), rats lose the motivation and/or ability to avoid that stressor (or cues associated with it) during subsequent exposures. These rats also behave differently in other tests. This phenomenon does not happen if the stressor is escapable or controllable. The paradigm is thought to model aspects of human depression, as animals exposed to it display symptoms such as reduced food intake and body weight, behavioral suppression, decreased grooming and play behavior, changes in sleep patterns, loss of reward, and cognitive disturbances.[17] However, an important drawback of the learned-helplessness model is that effects are very short-lasting.[121]

The role of NE in this paradigm has been intensively investigated. Inescapable shock reduces noradrenergic neurotransmission in the LC.[122] This suggests that central NE depletion may underlie the development of learned helplessness, potentially giving the model some construct validity. However, Soubrie *et al.*[123] showed that pharmacological lesioning of the hippocampal NE innervation did not prevent the development of learned helplessness after exposure to inescapable shock, but it did decrease the effect of imipramine in a subsequent anxiety test. Moreover, microinjections of the tricyclic desipramine into the hippocampus, but not other areas, prevented the development of learned helplessness, presumably by increasing local NE concentrations.[124] Thus, it appears that depletion of hippocampal NE does not underlie the development of learned helplessness, but that increased levels of NE in this structure protect against it. This underlines the possibility that the locus of action of antidepressants is not necessarily the same as the locus of the original deficit. Frontal cortex NE neurotransmission appears to play a similar protective role, as shown by Sagen *et al.*,[125] who increased NE levels in this area in rats by transplanting adrenal medullary tissue. This prevented the induction of learned helplessness. Similarly, Dougherty *et al.*[126] showed that NE microinjections into the frontal cortex as well as chronic imipramine treatment decreased immobility in a forced-swim test, which is thought to represent "behavioral despair" and thus can also be considered a form of learned helplessness. Adrenal medullary transplants had the same effect in this test, and it was prevented by acute α- (phentolamine) or β- (propranolol) adrenergic antagonist injections. Interestingly, when Plaznik *et al.*[127] injected NE into the nucleus accumbens, it acutely increased struggling behavior in the forced-swim test, an effect that could be antagonized by the α-adrenoceptor antagonist phentolamine, but not with the β-adrenoceptor antagonist propranolol. Petty *et al.*[128,129] showed that rats displaying learned helplessness after inescapable tail-shock stress had lower K^+-induced hippocampal NE release after a subsequent shuttle box test than rats that did not display helplessness. Taken together, these studies indicate that an increase of NE levels in the brain, directly or through the actions of antidepressants, protect against learned helplessness. However, both the mechanism underlying the induction of learned helplessness and the possible role of NE depletion in this process remain to be elucidated.

Chronic variable stress/chronic mild stress

In this model for human depression, animals are exposed to a regime of several relatively mild stressors in an unpredictable fashion for 2 to 3 weeks. Initial studies involving this paradigm were performed by Roth and Katz.[130] Animals show reduced activity when they are subsequently tested in an open field, and this reduction can be reversed by administration of different types of antidepressants as well as by electroconvulsive therapy.[131,132] As such, the model has good levels of both face and predictive, and possibly etiological, validity. Chronically stressed rats are also prone to anhedonia as tested in a sucrose-consumption test (although there is some debate as to whether chronic stress affects sucrose preference or merely total intake[133,134]), and this can be counteracted by antidepressant treatment.[134–136] Although chronic stress-induced anhedonia is probably a reflection of alterations in the dopaminergic system,[17] the chronic variable stress protocol also produces changes in the noradrenergic system. Thus, an up-regulation of cortical β-adrenoceptors has been observed in chronically stressed rats, and could be prevented by imipramine administered during (part of) the stress protocol.[137,138] Imipramine-induced restoration of sucrose intake and of β-adrenoceptor levels are not necessarily correlated,[136,138] suggesting, as stated above, that changes in the β-adrenergic receptor system probably do not underlie the anhedonic effect of exposure to chronic variable stress.

Not all studies show consistent effects of chronic variable stress. For example, after 7 weeks of chronic variable stress, mice displayed reduced sucrose intake; however, they did not show any alterations in hypothalamic or hippocampal NE content. Moreover, chronic imipramine treatment affected neither behavior nor NE levels in these mice.[139] A variable stress paradigm employed by Areso and Frazer[140] showed a reduction of β_1-, but not β_2-adrenoceptors exclusively in the amygdala. The authors explained the discrepancy with other studies such as that by Molina *et al.*[137] by the fact that their method of detection distinguished between β_1- and β_2-adrenoceptors, whereas more commonly used methods do not. In addition, inconsistencies between studies may be due to strain differences: it has been shown that strains with different sensitivities to stress also respond differently to chronic variable stress, including responses measured on the NE system level.[141] Furthermore, the choice of stressors applied within the paradigm (mild vs. strong; psychological vs. physical) may be of importance. Interestingly, Harro *et al.*[142] showed that chronic variable stress has an antidepressant-like effect in group-housed rats in that it decreased immobility in a forced-swim test and also prevented the effects on β-adrenoceptors induced by pharmacological denervation of LC projections. This result stresses the important modulating influences of housing conditions on outcomes of behavioral tests. Indeed, it has been suggested that isolation housing should not be considered a chronic stressor in itself, but that the lack of the daily mild stress associated with social housing conditions may hypersensitize rats to subsequent stressors[143] and have anxiogenic effects.[144]

Olfactory bulbectomy

Olfactory bulbectomy in rats causes retrograde degeneration of several limbic regions, including amygdala, stria terminalis, and hippocampus, and results in many behavioral changes such as increased levels of muricide, novelty-induced hyperactivity, reduced sexual behavior, reduced neophobia, increased extinction of conditioned taste aversion, and impaired learning. However, it does not change behavior in a classic forced-swim test.[145] Many of these behavioral changes do not resemble "depressive" behavior, and the strength of this model lies mainly in the fact that bulbectomy-induced behavioral alterations are reversed by chronic antidepressant treatment. Thus, it offers good predictive validity.[121] It is thought that changes in the NE system underlie many of the behavioral consequences of olfactory bulbectomy. Indeed, many studies have shown that both NE levels and turnover in the brain are reduced after olfactory bulbectomy.[146] For example, a decreased NE release from the lateral, but not the ventromedial, hypothalamus has been observed,[147] and Leonard *et al.* found reduced levels of NE in amygdaloid cortex and midbrain structures that could be reversed by the antidepressant mianserin.[148,149] However, a number of studies showed an increase in NE levels in some brain areas following olfactory bulbectomy. For example, in some studies bulbectomy induced a higher NE content in the hypothalamus,[150,151] irrespective of whether animals displayed muricidal behavior,[151] whereas Iwasaki *et al.*[152] found elevated NE levels in the medial amygdala and both ventromedial and lateral hypothalamus, but only in muricidal bulbectomized rats. Other changes in the central NE system detected after olfactory bulbectomy include increased levels of α_2-adrenoceptors in forebrain,[153] and an increased density of α_1- and β-adrenoceptors in amygdala and hippocampus.[145] Administration of α_2-, but not α_1-antagonists suppressed muricide in bulbectomized rats,[153] whereas the tricyclic desipramine also suppressed muricide and normalized NE levels.[152]

Animal assay models

This section describes how drugs known to modulate NE neurotransmission alter the behavior of otherwise normal animals in tests for depression or anxiety.

Assay models

Forced-swim test

This often-used test was initially developed as a model for depression, but has severely limited face and construct validities.[17,121] Predictive validity for this test is good, although it reveals a relatively high number of false positive and false negative results. The test is based on the premise that when a rat is placed in a cylinder filled with water, it will, after an initial period of vigorous struggling (presumably in an attempt to escape), either adopt an immobile position or keep itself above water by calmly swimming. When the test is repeated 24 hours later, the rat will display immobility sooner than during first exposure. Antidepressant drugs administered before the second exposure generally reduce the time spent immobile. The role of the NE system in these behaviors has received much attention in previous years. Drugs that increase central NE levels generally reduce passive behavior in this test. Microinjections of desipramine directly into the hippocampus also reduced floating in forced-swim test.[154] Moreover, increasing local NE levels in the frontal cortex by transplantation of adrenal medullary tissue reduced immobility when rats were tested 6 weeks after surgery. Systemic imipramine treatment had a similar effect, and both could be reversed with α- or β-adrenoceptor antagonist injections.[126] Weiss *et al.*[155,156] showed that chronic administration of tricyclics (desipramine, imipramine) and the monoamine oxidase inhibitor phenelzine to rats that naturally display low levels of struggling in the swim test specifically increased active behavior without significantly affecting immobility levels. Other researchers have also shown that selective NE reuptake inhibitors increase struggling,[157] whereas SSRIs increase calm swimming in some studies but not all.[157,158]

Elevated plus-maze

This popular screening test for anxiolytic drugs is based on an approach/avoidance response. A rat is placed in the center of a cross-shaped platform with two walled ("closed") arms and two unwalled, exposed ("open") arms. It is assumed that anxious rats will spend more time on the closed arms of the maze than exploring the open arms. Stress generally reduces, while anxiolytic drugs increase, the number of entries into and the time spent on the open arms.[7,159] Modulation of the central NE system greatly affects plus-maze behavior. In general, increasing NE levels (directly or indirectly and in different regions of the brain) has anxiogenic effects when measured in a plus-maze test. For example, antagonizing the stress-induced NE increase in the BNST with either an α_1-antagonist or with a combination of β_1- and β_2-antagonists prevented the reduced open-arm exploration normally observed after stress, although it did not affect behavior in a social-interaction test.[160] When antisense oligonucleotides directed against α_{2A}-adrenoreceptor mRNA was injected into the LC, there was decreased anxiety in a plus-maze test.[161,162] Yohimbine injections, which are thought to increase NE levels by antagonizing the inhibitory α_2-receptor, had anxiogenic effects.[163] However, these effects could only be reversed by systemic administration of the dopamine receptor agonist apomorphine and by clonidine, but not by selective α_2- or α_1-adrenoceptor agonists or a β-adrenoreceptor antagonist, indicating that activation of the NE system alone does not underlie the anxiogenic effect of yohimbine in this test. When very high NE levels are present, galanin may be released to "buffer" the anxiogenic actions of NE in the central amygdala.[164]

Open-field test

This test of exploratory behavior in a large novel environment without escape or hiding possibilities is commonly used in combination with the elevated plus-maze test. When changes in anxiety levels have been detected in a plus-maze test, animals are often exposed to the open-field test to ensure that the observed differences were not due to changes in general locomotor activity. However, in most cases the open field will represent a novel environment for the rat or mouse and will likely be anxiogenic. Therefore the open-field test should be considered as an alternative anxiety test. Several parameters can be measured in this paradigm: total locomotion, entries into the center of the field, defecation and rearing. Interpretation of these parameters, however, has been inconsistent. High open-field activity has been thought to reflect a high level of escape-seeking behavior, whereas in an alternative interpretation, a low level of exploratory behavior in this field is an indication of anxiety. For an excellent review regarding these matters, see Ramos and Mormede.[165] In any case, NE-modulating drugs do not appear to specifically affect open-field behavior. Thus, acute administration of selective NE reuptake inhibitors[166,167] or imipramine in mice had no effect on locomotion in an open field, although the latter did show an anxiolytic effect in a light/dark-box test. In contrast, Cervo and Samanin[168] showed that acute clonidine injections in rats reduced immobility in a forced-swim test and also decreased activity in an open field. These effects were reversed with administration of an α_2-, but not an α_1-adrenoceptor antagonist. Furthermore, pharmacological depletion of NE in the LC could not prevent the actions of clonidine, indicating that clonidine probably acted on α_2-receptors outside the LC to alter open-field activity.

Social interaction test

In this test, one rat is placed in a novel cage with another, unfamiliar rat, and the time spent in active social investigation is recorded and used as a measure of "nonfearfulness." Fearful rats in this test will show aggressive, freezing, or flight behavior. Sajdyk *et al.*[169] showed that NE levels in the dorsomedial hypothalamus do not increase after exposure to a social interaction test. Central NE in return also does not appear to play a major role in behavior in this test, as has been shown in a number of studies in which the central NE system had been modulated. Thus, an acute injection of MCI-225, which is a selective NE reuptake inhibitor with additional serotonin 5-HT_3-receptor antagonism, increased social interaction to unfamiliar partners. Although this effect was blocked by a 5-HT_3-agonist, imipramine had no effect, suggesting that alterations in the serotonin rather than the NE system modify behavior in this test at least in acute manipulations.[170] Similarly, Cecchi *et al.*[160] showed that microinjections into the BNST of both α_1- and $\beta_{1/2}$-antagonists had no effect in a social interaction test. In contrast, adrenergic antagonists injected into the central amygdala did reduce social interaction responses,[171] indicating that any influence of NE on stress-induced social behavior is receptor and site specific. Lesioning of the NE terminals in the PVN with 6-hydroxydopamine (6-OHDA) did not alter social interaction scores, although it did result in more vigorous contact.[172] Interestingly, acute injections of venlafaxine, a reuptake inhibitor of both serotonin and NE, resulted in reduced aggression, whereas chronic administration increased aggression and reduced flight behavior.[173] This corresponds with old data showing that lesions of the NE innervation from LC to cortex using 6-OHDA had no effect on social behavior in an interaction test, but again increased aggressive behavior such as "boxing" and "wrestling."[174] In view of these findings, the NE system does not appear to play a major role in anxiety-induced reductions in social behavior. This suggests that the social interaction test may not be the recommended paradigm for screening noradrenergic drugs for possible anxiolytic properties.

Operant conflict test

This widely used anxiety test evokes an approach/avoidance response in the subject.[7] It combines a positive reinforcement (such as food) with a negative stimulus (such as shock), thus creating a conflict situation. However, because a number of known anxiolytic drugs produce false negative results in this test, it has limited predictive validity. Moreover, the NE system does not appear to play a major role in behavior in the operant conflict test. For example, Beaufour *et al.*[175] assessed the efficacy of several different antidepressants for potential anxiolytic effects in this test. Only desipramine had (some) anxiolytic properties. Other antidepressants, including imipramine, did not. Curiously, anxiolytic actions were observed after discontinuation of these drugs. Lesions of the LC or its projections neither affected operant conflict behavior nor did they prevent the anxiolytic actions of ethanol or benzodiazepines. This finding suggests that NE projections from the LC are not involved in anxiety measured in this test.[176] The operant conflict test thus appears not to be a sensitive method for screening drugs that affect the central NE system.

Models based on selective breeding to phenotype

Breeding-to-phenotype models

Although selective breeding is usually applied to discern genetic from environmental influences on behavior, it is possible that the "bred phenotype" nevertheless reflects environmental effects: females from one end of the behavioral spectrum may differ from females from the other end in uterine characteristics (i.e. hormone levels to which the fetus will be exposed), responses to stress during pregnancy or lactation (again changing hormone levels such as corticosterone and testosterone), and maternal behavior.[177] Embryonic transfer and cross-fostering studies are needed to verify that rat lines "bred to phenotype" are indeed the result of genetic influences on behavior. An important recent study in mice showed that at least some aspects of the phenotypical differences between mouse strains are determined by their in utero environments and quality of maternal care, rather than by their genetic backgrounds.[18]

Flinders sensitive line

Overstreet *et al.*[178] selected rats for responsiveness to the effects of increased cholinergic activity (by administering the anticholinesterase diisopropyl fluorophosphate; DFP). Thus, the Flinders sensitive line (FSL) of rats reacts to both mild stressors and cholinergic agonists with greater behavioral depression than the so-called Flinders resistant line (FRL), and displays other depression-like behaviors such as locomotor retardation, alterations in sleep, cognitive defects, and anhedonia.[179,180] The FSL also displays lower plasma ACTH concentrations and lower CRF levels in the median eminence, LC, and prefrontal cortex.[181] A host of neuropeptides are altered in FSL, notably dopamine and neuropeptide Y, but thus far only one study has investigated the NE system in this rat line.[182] It was found that levels of NE in FSL are 2 to 3 times higher in prefrontal cortex, medial raphe nuclei, nucleus accumbens, and hippocampus (but not hypothalamus) compared to levels in normal Sprague–Dawley rats. Chronic treatment with several tricyclics affecting NE reuptake normalized NE levels in FSL rats, as well as their normally increased immobility in a forced-swim test, but did not affect these parameters in control Sprague–Dawleys, and only moderately in FRL.[182,183] Interestingly, chronic administration of an anticholinesterase did not alter swim test behavior,[183] indicating that changes in noradrenergic rather than cholinergic systems underlie the behavioral suppression in the forced-swim test in this rat line.

Maudsley line

The Maudsley rat lines were initially selected by Broadhurst[184] for high and low defecation in an open-field paradigm. They were termed Maudsley Reactive (MR) and Maudsley Non-Reactive (MNRA) rats, respectively, and were thought to represent rat lines with high and low levels of "emotionality." This was confirmed in several behavioral tests,[185,186] although learning did not appear to be different between the two strains.[185,187] However, some experiments yielded inconsistent results, and it has been called into question whether present-day MR and MNRA rats truly represent differences in general susceptibility to anxiety.[187,188] Maintenance and crossbreeding in different laboratories may have resulted in changes in the phenotypes of MR and MNRA lines. Nevertheless, MR and MNRA have shown differences in central noradrenergic activity, both under acute and chronic stress conditions. For example, MR have increased NE release in the LC in response to acute stress compared to MNRA, associated with reduced α_2-adrenoceptor binding in this region and lower behavioral response to clonidine.[189,190] In contrast, after exposure to chronic stress, the increase in TH content of the LC was greater in MNRA, as were peripheral NE levels, indicating a higher sympathetic tone in chronically stressed MNRA rats.[191] These data suggest differences in adaptation to chronic stress between MR and MNRA. Antidepressant drugs have not been tested in MR and MNRA rats, but the two strains differ in their response to administration of anxiolytics.[192]

Roman high- vs. low-avoidance rats

Roman high- and low-avoidance rats were selected for extreme differences in behaviors in an active, two-way avoidance task and were consequently bred by Driscoll *et al.*[193] Roman high-avoidance (RHA/Verh) rats show a rapid acquisition of avoidance behavior in this test (active coping), whereas the low-avoidance (RLA/Verh) do not learn this (passive coping, freezing), and it is thought that this reflects differences in emotionality rather than in learning. The central amygdala is involved in the expression of coping strategies, and microinjections of NE into the CeA increase the level of passive coping in RLA/Verh when tested for retention in a defensive burying paradigm. Microinjections of NE also evoke active coping (burying) in the RHA/Verh rats.[194] These findings suggest that differences in amygdala NE do not underlie behavioral differences between RHA/Verh and RLA/Verh rats but merely accentuate them. Driscoll *et al.*[195] showed that RHA/Verh and RLA/Verh did not differ in NE content of cortex, hypothalamus, and pons/medulla under basal or stress conditions, but had different serotonergic responses to stress. Surprisingly, to date, no studies have been undertaken to test whether chronic administration of anxiolytics or antidepressants can reduce the behavioral differences between the two rat lines.

Models based on selective gene manipulations

Dopamine-β-hydroxylase knockout mice

Mice in which the gene for dopamine-β-hydroxylase (DBH) is absent generally die in utero from day 11 of gestation onwards.[5,6] When born to heterozygous ($DBH^{+/-}$) mothers, about 10% of $DBH^{-/-}$ pups survive, probably due to transplacental passage of maternal NE.[5] Fetuses could also be rescued by supplying dams during pregnancy with drinking water containing an NE precursor that did not require conversion by dopamine-β-hydroxylase.[5] In adulthood, $DBH^{-/-}$ mice show normal general locomotor activity but display disturbed motor functioning when tested in a water-maze or on a rotarod.[6] Their learning abilities appear relatively normal, but memory function is reduced.[6] Interestingly, female $DBH^{-/-}$ mice show impaired maternal care of pups, particularly with regard to retrieval behavior,[58] and restoration of NE levels in $DBH^{-/-}$ dams normalizes these deficits.[6,58,196] Considering the role of maternal behavior in determining vulnerability to anxiety- and depression-like behavior in rodents later in life (see above subsection "Maternal separation and neonatal handling"), it is possible that some aspects of the $DBH^{-/-}$ phenotype may be the result of substandard maternal care.

Norepinephrine transporter knockout mice

Mice lacking the NE transporter (NET) have an anxiolytic phenotype. Xu *et al.*[197] showed that $NET^{-/-}$ mice display reduced immobility in a forced-swim test as well as in a tail-suspension test, and faster habituation to an open field. Behaviorally, $NET^{-/-}$ mice resembled wild-type mice treated with the tricyclic desipramine, whereas this antidepressant had no effect in $NET^{-/-}$ mice themselves. Extracellular NE levels were elevated in $NET^{-/-}$ mice, but tissue contents in LC projection areas such as frontal cortex, hippocampus, and cerebellum were >50% lower, although there were indications that NE synthesis was increased. Dopaminergic function was also altered in $NET^{-/-}$ mice, suggesting that modulation of the dopamine system may play a role in the action of antidepressant drugs. When exposed to chronic social stress, $NET^{-/-}$ mice responded in a manner similar to wild-type mice, i.e. they displayed weight loss and more defensive behavior. When tested in a forced-swim test, however, chronically stressed $NET^{-/-}$ mice showed more struggling, which to the authors suggested that enhanced NE activity may have prevented "depressed" behavior in specific circumstances from becoming more generalized.[198] Important alterations in the dopamine system were also reported in $NET^{-/-}$ mice; they displayed dopamine receptor supersensitivity resulting in increased responsiveness to amphetamine and cocaine.[197]

α_{1B}-adrenergic receptor knockout and overexpressing mice

In mice lacking the α_{1B}-adrenoceptor, psychostimulants evoke diminished hyperactivity and have decreased rewarding properties. Although psychostimulants act by increasing dopaminergic transmission, the dopamine system is unaltered in these knockout mice. This indicates a possible role for the NE system and, more specifically, for α_{1B}-receptors in sensitivity to drug addiction.[199] To date it is not known whether α_{1B}-receptor knockout mice display alterations in anxiety- or depression-like behavior. Overexpression of the α_{1B}-adrenoceptor has been described in one study, which mainly focused on the peripheral effects of this receptor: mice overexpressing the α_{1B}-receptor had lower plasma levels of NE, epinephrine and cortisol,[200] although it is unclear why cortisol rather than corticosterone was measured as an indication of HPA axis activity. Awaiting detailed behavioral characterization of α_{1B}-receptor transgenic mice, it is clear that they display at least a disturbed regulation of the autonomic nervous system.

α_2-adrenergic receptor knockout and overexpressing mice

Hein and colleagues have developed knockout mice that specifically lack either the α_{2A}-, the α_{2B}-, or the α_{2C}-adrenergic receptors.[201] Their results suggest that both α_{2A} and α_{2C} function as presynaptic inhibitory receptors regulating NE neurotransmission.[202] Detailed studies have shown that α_{2A}-receptors modulate transmitter release at high stimulation frequencies, whereas the α_{2C}-receptor, which has a much more restricted distribution than the α_{2A}-receptor, does so only at lower levels of neuronal activity. Both low- and high-frequency regulation seems to be physiologically important, as mice lacking both α_{2A}- and α_{2C}-receptor subtypes have elevated plasma NE concentrations.[203]

Behaviorally, mice lacking the α_{2A}-receptor display increased anxiety in several tests, as well as abnormal circadian activity patterns. They also showed elevated NE turnover in the brain without effects on dopamine or serotonin levels.[204] This study confirmed an earlier one by Schramm *et al.*[205] Schramm observed increased force-swim test immobility in α_{2A}-receptor knockout mice that were insensitive to imipramine administration. In addition, stress-induced anxiety was increased in these mice as measured in open-field and light-/dark-box tests. These observations suggest that activation of α_{2A}-receptors, via inhibition of the LC-NE system, protects against hyperactivity in noradrenergic neurotransmission, and thus has anxiolytic properties.

As mentioned above, knockout studies have shown that α_{2A}- and α_{2C}-adrenoreceptors both act as presynaptic receptors inhibiting NE release in the CNS.[202] However, they appear to have different roles in the regulation of behavior. Thus, Sallinen *et al.*[206] showed that α_{2C}-knockout mice display reduced immobility in a forced-swim test, similar to wild-type mice treated with antidepressants. Correspondingly, overexpression of the α_{2C}-receptor increased immobility in the forced-swim test. In addition, mice lacking the α_{2C}-receptor showed attenuated stress-induced elevation of plasma corticosterone, whereas overexpression of α_{2C}-receptors was linked with increased corticosterone levels after repeated stress. The α_{2C}-adrenoreceptor also plays a role in regulating dopamine and serotonin, but not NE turnover. From these results it appears that α_{2A}- and α_{2C}-adrenoreceptors work in conjunction, but may be activated at different times and in different brain regions.

Galanin-overexpressing mice

The neuropeptide galanin is coexpressed in NE neurons and is found in limbic brain areas such as the LC and amygdala. It appears to inhibit the actions of only high levels of NE in LC neurons by reducing their firing rate, as well as that of raphe serotonin neurons. As such, galanin is one of the peptide modulators of anxiety. Galanin overexpression in itself does not alter basal extracellular NE levels, at least not in the hippocampus.[207] Moreover, it does not alter basal anxiety-like behavior in plus-maze, open-field, or light/dark exploration tests. Administration of yohimbine increases anxiety in wild-type mice, but has no effect in galanin-overexpressing mice in the light/dark test, suggesting that the overexpressed galanin only springs into action in the presence of high levels of NE, such as exist after yohimbine administration, and may not do so under basal conditions.[208] This idea corresponds with data obtained by Khoshbouei *et al.*,[164] who showed that intra-amygdala administration of galanin had no effect on anxiety in previously restrained rats, but only in restrained rats that had also received a yohimbine injection.

Conclusions and avenues for future research

The involvement of NE in the etiology and/or pathophysiology of mood disorders was first postulated over 40 years ago. As reviewed above, NE plays a role in neural development. In adult animals, central NE pathways play important roles in arousal, vigilance, alertness, learning, and memory. It is therefore reasonable to hypothesize that gene polymorphisms in enzymes of the NE biosynthetic pathway, adrenergic receptor subtypes, or intracellular signaling cascades might have a developmentally significant impact on vulnerability to psychopathologies. Similarly, extreme environmental adversity, especially if chronic, at any time from the prenatal period through adulthood, may alter the set-point of stress-responsive neurocircuitry, such as the NE system. This may lead to psychopathologies.

Animal models, while imperfect, provide advantages on a number of fronts. However, they must be carefully developed and used judiciously. A useful approach is as follows: compile a list of diagnostic criteria from the Diagnostic and Statistical Manual (DSM-IV) for the target disease; survey the clinical literature for any behavioral, physiological, and postmortem findings that have been widely replicated; then review the animal literature for the appropriate species. Among the factors to consider are: (1) how to translate human symptomatology or etiologic factors to a particular animal species; (2) the overall ethological validity of housing conditions, nature of the manipulations, types of testing, and age at insult and testing for the selected species; (3) the difficulty of mapping outcome onto a species-specific developmental trajectory across species; (4) potential sex-specific effects; (5) the selection of appropriate comparison or control groups; (6) whether the disease or disease characteristics to be modeled are really categorical or on a continuum; and (7) criteria for model validity. Despite these potential complications, even restricted animal models can make potentially important contributions to our understanding of processes underlying human psychopathology.

It is imperative to be clear in presenting and discussing the benefits and limitations of various animal models. In particular, animal assay models should be distinguished from other types of animal models, as their sole purpose is to screen potential therapeutic compounds under development. These models are designed to mimic the efficacy of known therapeutic agents as a standard against which new compounds can be tested. As such, these animal models may or may not exhibit any characteristics of any psychiatric disease. One drawback of models based on the concept of pharmacological isomorphism is that they may not be useful for identifying new pharmacological agents with novel mechanisms of action or novel target structures.

Models developed by selective breeding of animals to a particular phenotype, as well as those developed by directly manipulating selected genes, suffer from "background" effects. Thus, although a particular phenotype or gene is the focus of these studies, other genes and phenotypes may be carried through the generations that have subtle or not-so-subtle effects on the phenotype of interest. In addition, transgenic studies continue to suffer from the limited use of inducible knockouts or knockins and issues of targeting gene manipulations to particular sets of cells. The end result is that the animals that survive to birth and beyond have activated a huge (and unidentified) cascade of compensatory mechanisms that complicate interpretation of the outcomes. Finally, it is naïve to isolate "nature" and "nurture" as if they are independent variables. These facets are inextricably linked in a continuous and complex interaction throughout development and life.

Future directions

Both "top-down" and "bottom-up" approaches to elucidating the role(s) of NE in psychopathology remain important. In fact, this type of cross validation of assumptions underlying clinical etiology and diagnosis as well as assumptions underlying various animal models can only be strengthened by further delineation of factors that contribute to these psychiatric disorders. To this end, future approaches to studies in this area will likely include: (1) clinical studies of subtypes of PTSD, depression, anxiety and attention-deficit disorder, with a particular focus upon symptom clusters, patterns of comorbidities and responsivity to the efficacy of agents with distinct mechanisms of action; (2) "finger printing" well-defined clinical populations in terms of history (family stability, medical illnesses, pre pubertal abuse or neglect, etc.) and single-nucleotide polymorphisms (SNP); (3) multi modal imaging studies combining structure (magnetic resonance imaging [MRI]), function (fMRI), and the receptor or neurotransmitter density and function (positron emission tomography) during active disease states and during remissions as well as in high-risk populations prior to expression of disease; (4) comprehensive electrophysiological studies of signal processing using sensory-evoked potentials across sensory domains; (5) multimodal human and animal studies of the dynamics of brain responses to acute and chronic antidepressant treatment; and (6) comprehensive postmortem analyses of morphometric, biochemical, binding and gene expression characteristics in medically well-characterized individuals and well-matched controls. No doubt, these suggestions do not cover the wealth of studies currently possible or studies that may be possible in the future with advances in technology. None-the-less, methods combining behavioral, molecular and genetic approaches should permit a functional dissection of the role of NE neurocircuits in the etiology and pathology of psychiatric disorders. This will hopefully allow the contributions of the NE system to be placed in the context of the many other factors affecting vulnerability to and expression of psychopathology.

REFERENCES

1. Pendleton, R. G., Rasheed, A., Roychowdhury, R. and Hillman, R. A new role for catecholamines: ontogenesis. *Trends Pharmacol. Sci.*, **19** (1998), 248–251.
2. Smith, J. and Fauquet, M. Glucocorticoids stimulate adrenergic differentiation in cultures of migrating and premigratory neural crest. *J. Neurosci.*, **4** (1984), 2160–2172.
3. Herlenius, E. and Lagercrantz, H. Neurotransmitters and neuromodulators during early human development. *Early Hum. Dev.*, **65** (2001), 21–37.
4. Naqui, S. Z., Harris, B. S., Thomaidou, D. and Parnavelas, J. G. The noradrenergic system influences the fate of Cajal-Retzius cells in the developing cerebral cortex. *Brain Res. Dev. Brain Res.*, **113** (1999), 75–82.
5. Thomas, S. A., Matsumoto, A. M. and Palmiter, R. D. Noradrenaline is essential for mouse fetal development. *Nature*, **374** (1995), 643–646.
6. Thomas, S. A. and Palmiter, R. D. Disruption of the dopamine beta-hydroxylase gene in mice suggests roles for norepinephrine in motor function, learning, and memory. *Behav. Neurosci.*, **111** (1997), 579–589.
7. Koob, G. F., Heinrichs, S. C. and Britton, K. Animal models of anxiety disorders. In *Textbook of Psychopharmacology*, ed. A. F. Schatzberg and C. B. Nemeroff. (Washington, DC: American Psychiatric Press, 1998), pp. 133–144.
8. Kelly, C. B. and Cooper, S. J. Differences and variability in plasma noradrenaline between depressive and anxiety disorders. *J. Psychopharmacol.*, **12** (1998), 161–167.
9. Roy, A., Pickar, D., Linnoila, M. and Potter, W. Z. Plasma norepinephrine level in affective disorders. Relationship to melancholia. *Arch. Gen. Psychiatry*, **42** (1985), 1181–1185.
10. Gold, P. W. and Chrousos, G. P. Organization of the stress system and its dysregulation in melancholic and atypical depression: high vs low CRH/NE states. *Mol. Psychiatry*, **7** (2002), 254–275.
11. Sullivan, G. M., Coplan, J. D., Kent, J. M. and Gorman, J. M. The noradrenergic system in pathological anxiety: a focus on panic with relevance to generalized anxiety and phobias. *Biol. Psychiatry*, **46** (1999), 1205–1218.
12. Southwick, S. M., Bremner, J. D., Rasmusson, A. *et al.* Role of norepinephrine in the pathophysiology and treatment of posttraumatic stress disorder. *Biol. Psychiatry*, **46** (1999), 1192–1204.
13. Biederman, J. and Spencer, T. Attention-deficit/hyperactivity disorder (ADHD) as a noradrenergic disorder. *Biol. Psychiatry*, **46** (1999), 1234–1242.
14. Pliszka, S. R., McCracken, J. T. and Maas, J. W. Catecholamines in attention-deficit hyperactivity disorder: current perspectives. *J. Am. Acad. Child Adolesc. Psychiatry*, **35** (1996), 264–272.

15. McKinney, W. T., Jr. and Bunney, W. E., Jr. Animal model of depression. I. Review of evidence: implications for research. *Arch. Gen. Psychiatry*, **21** (1969), 240–248.
16. Willner, P. The validity of animal models of depression. *Psychopharmacology*, **83** (1984), 1–16.
17. Weiss, J. M. and Kilts, C. D. Animal models of depression and schizophrenia. In *Textbook of Psychopharmacology*, ed. A. F. Schatzberg and C. B. Nemeroff. (Washington, DC: American Psychiatric Press, 1998), pp. 89–131.
18. Francis, D. D., Szegda, K., Campbell, G., Martin, W. D. and Insel, T. R. Epigenetic sources of behavioral differences in mice. *Nat. Neurosci.*, **6** (2003), 445–446.
19. Huttunen, M. O. and Niskanen, P. Prenatal loss of father and psychiatric disorders. *Arch. Gen. Psychiatry*, **35** (1978), 429–431.
20. Kofman, O. The role of prenatal stress in the etiology of developmental behavioral disorders. *Neurosci. Biobehav. Rev.*, **26** (2002), 457–470.
21. Ward, A. J. Prenatal stress and childhood psychopathology. *Child. Psychiatry. Hum. Dev.*, **22** (1991), 97–110.
22. Sackett, G. P. Pregnancy outcome following jet transport stress in nonhuman primates. *J. Med. Primatol.*, **10** (1981), 149–154.
23. Wilmut, I., Sales, D. I. and Ashworth, C. J. Maternal and embryonic factors associated with prenatal loss in mammals. *J. Reprod. Fertil.*, **76** (1986), 851–864.
24. Clarke, A. S., Wittwer, D. J., Abbott, D. H. and Schneider, M. L. Long-term effects of prenatal stress on HPA axis activity in juvenile rhesus monkeys. *Dev. Psychobiol.*, **27** (1994), 257–269.
25. Gorman, J. M., Mathew, S. and Coplan, J. Neurobiology of early life stress: nonhuman primate models. *Semin. Clin. Neuropsychiatry*, **7** (2002), 96–103.
26. Schneider, M. L., Coe, C. L. and Lubach, G. R. Endocrine activation mimics the adverse effects of prenatal stress on the neuromotor development of the infant primate. *Dev. Psychobiol.*, **25** (1992), 427–439.
27. Schneider, M. L., Roughton, E. C., Koehler, A. J. and Lubach, G. R. Growth and development following prenatal stress exposure in primates: an examination of ontogenetic vulnerability. *Child Dev.*, **70** (1999), 263–274.
28. Weinstock, M. Does prenatal stress impair coping and regulation of hypothalamic–pituitary–adrenal axis? *Neurosci. Biobehav. Rev.*, **21** (1997), 1–10.
29. Welberg, L. A. and Seckl, J. R. Prenatal stress, glucocorticoids and the programming of the brain. *J. Neuroendocrinol.*, **13** (2001), 113–128.
30. Lordi, B., Protais, P., Mellier, D. and Caston, J. Acute stress in pregnant rats: effects on growth rate, learning, and memory capabilities of the offspring. *Physiol. Behav.*, **62** (1997), 1087–1092.
31. Huttunen, M. O. Persistent alteration of turnover of brain noradrenaline in the offspring of rats subjected to stress during pregnancy. *Nature*, **230** (1971), 53–55.
32. Takahashi, L. K., Turner, J. G. and Kalin, N. H. Prenatal stress alters brain catecholaminergic activity and potentiates stress-induced behavior in adult rats. *Brain. Res.*, **574** (1992), 131–137.
33. Reznikov, A. G. and Nosenko, N. D. Catecholamines in steroid-dependent brain development. *J. Steroid Biochem. Mol. Biol.*, **53** (1995), 349–353.

34. Reznikov, A. G., Nosenko, N. D. and Tarasenko, L. V. Prenatal stress and glucocorticoid effects on the developing gender-related brain. *J. Steroid Biochem. Mol. Biol.*, **69** (1999), 109–115.
35. Peters, D. A. Prenatal stress: effects on brain biogenic amine and plasma corticosterone levels. *Pharmacol. Biochem. Behav.*, **17** (1982), 721–725.
36. Peters, D. A. Prenatal stress: effect on development of rat brain adrenergic receptors. *Pharmacol. Biochem. Behav.*, **21** (1984), 417–422.
37. Schneider, M. L., Clarke, A. S., Kraemer, G. W. *et al.* Prenatal stress alters brain biogenic amine levels in primates. *Dev. Psychopathol.*, **10** (1998), 427–440.
38. Weinstock, M., Poltyrev, T., Schorer-Apelbaum, D., Men, D. and McCarty, R. Effect of prenatal stress on plasma corticosterone and catecholamines in response to footshock in rats. *Physiol. Behav.*, **64** (1998), 439–444.
39. White, D. A. and Birkle, D. L. The differential effects of prenatal stress in rats on the acoustic startle reflex under baseline conditions and in response to anxiogenic drugs. *Psychopharmacology (Berl.)*, **154** (2001), 169–176.
40. Slotkin, T. A., Lappi, S. E., McCook, E. C. *et al.* Glucocorticoids and the development of neuronal function: effects of prenatal dexamethasone exposure on central noradrenergic activity. *Biol. Neonate*, **61** (1992), 326–336.
41. Muneoka, K., Mikuni, M., Ogawa, T. *et al.* Prenatal dexamethasone exposure alters brain monoamine metabolism and adrenocortical response in rat offspring. *Am. J. Physiol.*, **273** (1997), R1669–R1675.
42. Slotkin, T. A., Lappi, S. E., Tayyeb, M. I. and Seidler, F. J. Dose-dependent glucocorticoid effects on noradrenergic synaptogenesis in rat brain: ontogeny of [3H]desmethylimipramine binding sites after fetal exposure to dexamethasone. *Res. Commun. Chem. Pathol. Pharmacol.*, **73** (1991), 3–19.
43. Nyirenda, M. J., Welberg, L. A. and Seckl, J. R. Programming hyperglycaemia in the rat through prenatal exposure to glucocorticoids: fetal effect or maternal influence? *J. Endocrinol.*, **170** (2001), 653–660.
44. Welberg, L. A., Seckl, J. R. and Holmes, M. C. Prenatal glucocorticoid programming of brain corticosteroid receptors and corticotrophin-releasing hormone: possible implications for behavior. *Neuroscience*, **104** (2001), 71–79.
45. Huot, R. L., Thrivikraman, K. V., Meaney, M. J. and Plotsky, P. M. Development of adult ethanol preference and anxiety as a consequence of neonatal maternal separation in Long Evans rats and reversal with antidepressant treatment. *Psychopharmacology (Berl.)*, **158** (2001), 366–373.
46. Kalinichev, M., Easterling, K. W. and Holtzman, S. G. Early neonatal experience of Long-Evans rats results in long-lasting changes in reactivity to a novel environment and morphine-induced sensitization and tolerance. *Neuropsychopharmacology*, **27** (2002), 518–533.
47. Kalinichev, M., Easterling, K. W., Plotsky, P. M. and Holtzman, S. G. Long-lasting changes in stress-induced corticosterone response and anxiety-like behaviors as a consequence of neonatal maternal separation in Long-Evans rats. *Pharmacol. Biochem. Behav.*, **73** (2002), 131–140.

48. Ladd, C. O., Huot, R. L., Thrivikraman, K. V. *et al.* Long-term behavioral and neuroendocrine adaptations to adverse early experience. *Prog. Brain Res.*, **122** (2000), 81–103.
49. Rots, N. Y., de Jong, J., Workel, J. O. *et al.* Neonatal maternally deprived rats have as adults elevated basal pituitary–adrenal activity and enhanced susceptibility to apomorphine. *J. Neuroendocrinol.*, **8** (1996), 501–506.
50. Dent, G. W., Smith, M. A. and Levine, S. Stress-induced alterations in locus coeruleus gene expression during ontogeny. *Brain Res. Dev. Brain. Res.*, **127** (2001), 23–30.
51. Matthews, K., Dalley, J. W., Matthews, C., Tsai, T. H. and Robbins, T. W. Periodic maternal separation of neonatal rats produces region- and gender-specific effects on biogenic amine content in postmortem adult brain. *Synapse*, **40** (2001), 1–10.
52. Clarke, A. S., Hedeker, D. R., Ebert, M. H. *et al.* Rearing experience and biogenic amine activity in infant rhesus monkeys. *Biol. Psychiatry*, **40** (1996), 338–352.
53. Bayart, F., Hayashi, K. T., Faull, K. F., Barchas, J. D. and Levine, S. Influence of maternal proximity on behavioral and physiological responses to separation in infant rhesus monkeys (*Macaca mulatta*). *Behav. Neurosci.*, **104** (1990), 98–107.
54. Denenberg, V. H. Evolution proposes and ontogeny disposes. *Brain Lang.*, **73** (2000), 274–296.
55. Meaney, M. J., Bhatnagar, S., Diorio, J. *et al.* Molecular basis for the development of individual differences in the hypothalamic–pituitary–adrenal stress response. *Cell. Mol. Neurobiol.*, **13** (1993), 321–347.
56. Liu, D., Caldji, C., Sharma, S., Plotsky, P. M. and Meaney, M. J. Influence of neonatal rearing conditions on stress-induced adrenocorticotropin responses and norepinephrine release in the hypothalamic paraventricular nucleus. *J. Neuroendocrinol.*, **12** (2000), 5–12.
57. Caldji, C., Tannenbaum, B., Sharma, S. *et al.* Maternal care during infancy regulates the development of neural systems mediating the expression of fearfulness in the rat. *Proc. Natl. Acad. Sci. U. S. A.*, **95** (1998), 5335–5340.
58. Thomas, S. A. and Palmiter, R. D. Impaired maternal behavior in mice lacking norepinephrine and epinephrine. *Cell*, **91** (1997), 583–592.
59. Grantham-McGregor, S. A review of studies of the effect of severe malnutrition on mental development. *J. Nutr.*, **125** (1995), 2233S–2238S.
60. Tonkiss, J., Shukitt-Hale, B., Formica, R. N., Rocco, F. J. and Galler, J. R. Prenatal protein malnutrition alters response to reward in adult rats. *Physiol. Behav.*, **48** (1990), 675–680.
61. Morgane, P. J., Austin-LaFrance, R., Bronzino, J. *et al.* Prenatal malnutrition and development of the brain. *Neurosci. Biobehav. Rev.*, **17** (1993), 91–128.
62. Levitsky, D. A. and Strupp, B. J. Malnutrition and the brain: changing concepts, changing concerns. *J. Nutr.*, **125** (1995), 2212S–2220S.
63. Almeida, S. S., Tonkiss, J. and Galler, J. R. Prenatal protein malnutrition affects the social interactions of juvenile rats. *Physiol. Behav.*, **60** (1996), 197–201.
64. Almeida, S. S., Tonkiss, J. and Galler, J. R. Prenatal protein malnutrition affects exploratory behavior of female rats in the elevated plus-maze test. *Physiol. Behav.*, **60** (1996), 675–680.

65. Datta, S., Patterson, E. H., Vincitore, M. *et al.* Prenatal protein malnourished rats show changes in sleep/wake behavior as adults. *J. Sleep Res.*, **9** (2000), 71–79.
66. Almeida, S. S., de Oliveira, L. M. and Graeff, F. G. Early life protein malnutrition changes exploration of the elevated plus-maze and reactivity to anxiolytics. *Psychopharmacology*, **103** (1991), 513–518.
67. Almeida, S. S., de Oliveira, L. M. and Graeff, F. G. Decreased reactivity to anxiolytics caused by early protein malnutrition in rats. *Pharmacol. Biochem. Behav.*, **36** (1990), 997–1000.
68. Rotta, F. T., Mello, C. F., Rocha, J. B. and Souza, D. O. Early undernutrition blocks the effect of naltrexone on rat exploratory behavior. *Braz. J. Med. Biol. Res.*, **21** (1988), 305–312.
69. Campbell, L. F. and Bedi, K. S. The effects of undernutrition during early life on spatial learning. *Physiol. Behav.*, **45** (1989), 883–890.
70. Hall, R. D. Is hippocampal function in the adult rat impaired by early protein or protein-calorie deficiencies? *Dev. Psychobiol.*, **16** (1983), 395–411.
71. Schenck, P. E., Slob, A. K. and van der Werff ten, Bosch, J. J. Locomotor activity and social behavior of old rats after preweaning undernutrition. *Dev. Psychobiol.*, **11** (1978), 205–212.
72. Soto-Moyano, R., Fernandez, V., Sanhueza, M. *et al.* Effects of mild protein prenatal malnutrition and subsequent postnatal nutritional rehabilitation on noradrenaline release and neuronal density in the rat occipital cortex. *Brain Res. Dev. Brain Res.*, **116** (1999), 51–58.
73. Massaro, T. F., Levitsky, D. A. and Barnes, R. H. Protein malnutrition induced during gestation: its effect on pup development and maternal behavior. *Dev. Psychobiol.*, **10** (1977), 339–345.
74. Maccari, S., Piazza, P. V., Kabbaj, M. *et al.* Adoption reverses the long-term impairment in glucocorticoid feedback induced by prenatal stress. *J. Neurosci.*, **15** (1995), 110–116.
75. Lesage, J., Dufourny, L., Laborie, C. *et al.* Perinatal malnutrition programs sympathoadrenal and hypothalamic–pituitary–adrenal axis responsiveness to restraint stress in adult male rats. *J. Neuroendocrinol.*, **14** (2002), 135–143.
76. Lesage, J., Blondeau, B., Grino, M., Breant, B. and Dupouy, J. P. Maternal undernutrition during late gestation induces fetal overexposure to glucocorticoids and intrauterine growth retardation, and disturbs the hypothalamo–pituitary–adrenal axis in the newborn rat. *Endocrinology*, **142** (2001), 1692–1702.
77. Soto-Moyano, R., Belmar, J., Perez, H., Ruiz, S. and Hernandez, A. Central noradrenergic hyperactivity early in life: a hypothesis on the origin of morpho-functional brain disorders induced by malnutrition. *Biol. Res.*, **28** (1995), 105–111.
78. Marichich, E. S., Molina, V. A. and Orsingher, O. A. Persistent changes in central catecholaminergic system after recovery of perinatally undernourished rats. *J. Nutr.*, **109** (1979), 1045–1050.
79. Shoemaker, W. J. and Wurtman, R. J. Perinatal undernutrition: accumulation of catecholamines in rat brain. *Science*, **171** (1971), 1017–1019.
80. Stern, W. C., Morgane, P. J., Miller, M. and Resnick, O. Protein malnutrition in rats: response of brain amines and behavior to foot shock stress. *Exp. Neurol.*, **47** (1975), 56–67.
81. Burns, E. M. and Brown, K. B. Perinatal malnutrition: effects on brain norepinephrine content. *Brain Res. Bull.*, **2** (1977), 313–316.

82. Seidler, F. J., Bell, J. M. and Slotkin, T. A. Undernutrition and overnutrition in the neonatal rat: long-term effects on noradrenergic pathways in brain regions. *Pediatr. Res.*, **27** (1990), 191–197.
83. Keller, E. A., Munaro, N. I. and Orsingher, O. A. Perinatal undernutrition reduces alpha and beta adrenergic receptor binding in adult rat brain. *Science*, **215** (1982), 1269–1270.
84. Keller, E. A., Cuadra, G. R., Molina, V. A. and Orsingher, O. A. Perinatal undernutrition affects brain modulatory capacity of beta-adrenergic receptors in adult rats. *J. Nutr.*, **120** (1990), 305–308.
85. Soto-Moyano, R., Hernandez, A., Perez, H. *et al.* Early malnutrition and changes in the induced release of noradrenaline in the prefrontal cortex of adult rats. *Int. J. Neurosci.*, **37** (1987), 93–102.
86. Resnick, O. and Morgane, P. J. Animal models for small-for-gestational-age (SGA) neonates and infants-at-risk (IAR). *Brain Res.*, **312** (1983), 221–225.
87. Nasif, F. J., Ramirez, O. A., Cuadra, G. R. and Orsingher, O. A. Increased neuronal activity in locus coeruleus from adult rats undernourished at perinatal age: its reversal by desipramine. *Life Sci.*, **69** (2001), 2551–2559.
88. Molina, V. A., Keller, E. A. and Orsingher, O. A. Reduced anti-immobility effect of repeated desipramine (DMI) treatment in adult rats undernourished at perinatal age. *Pharmacol. Biochem. Behav.*, **26** (1987), 417–419.
89. Soto-Moyano, R., Hernandez, A., Perez, H. *et al.* Clonidine treatment during gestation prevents functional deficits induced by prenatal malnutrition in the rat visual cortex. *Int. J. Neurosci.*, **76** (1994), 237–248.
90. Soto-Moyano, R., Hernandez, A., Perez, H. *et al.* Effect of clonidine early in life on brain morphofunctional deficits induced by neonatal malnutrition in the rat. *Int. J. Neurosci.*, **49** (1989), 103–113.
91. Kraemer, G. W. and Clarke, A. S. The behavioral neurobiology of self-injurious behavior in rhesus monkeys. *Prog. Neuropsychopharmacol. Biol. Psychiatry*, **14** (1990), S141–S168.
92. Capitanio, J. P., Rasmussen, K. L., Snyder, D. S., Laudenslager, M. and Reite, M. Long-term follow-up of previously separated pigtail macaques: group and individual differences in response to novel situations. *J. Child Psychol. Psychiatry*, **27** (1986), 531–538.
93. Kalin, N. H. and Carnes, M. Biological correlates of attachment bond disruption in humans and nonhuman primates. *Prog. Neuropsychopharmacol. Biol. Psychiatry*, **8** (1984), 459–469.
94. Clarke, A. S., Kraemer, G. W. and Kupfer, D. J. Effects of rearing condition on HPA axis response to fluoxetine and desipramine treatment over repeated social separations in young rhesus monkeys. *Psychiatry Res.*, **79** (1998), 91–104.
95. Clarke, A. S. Social rearing effects on HPA axis activity over early development and in response to stress in rhesus monkeys. *Dev. Psychobiol.*, **26** (1993), 433–446.
96. Kraemer, G. W., Ebert, M. H., Schmidt, D. E. McKinney, W. T. A longitudinal study of the effect of different social rearing conditions on cerebrospinal fluid norepinephrine and biogenic amine metabolites in rhesus monkeys. *Neuropsychopharmacology*, **2** (1989), 175–189.

97. Clarke, A. S., Ebert, M. H., Schmidt, D. E., McKinney, W. T. and Kraemer, G. W. Biogenic amine activity in response to fluoxetine and desipramine in differentially reared rhesus monkeys. *Biol. Psychiatry*, **46** (1999), 221–228.
98. Higley, J. D., Suomi, S. J. and Linnoila, M. CSF monoamine metabolite concentrations vary according to age, rearing, and sex, and are influenced by the stressor of social separation in rhesus monkeys. *Psychopharmacology*, **103** (1991), 551–556.
99. Higley, J. D., Suomi, S. J. and Linnoila, M. A longitudinal assessment of CSF monoamine metabolite and plasma cortisol concentrations in young rhesus monkeys. *Biol. Psychiatry*, **32** (1992), 127–145.
100. Ginsberg, S. D., Hof, P. R., McKinney, W. T. and Morrison, J. H. The noradrenergic innervation density of the monkey paraventricular nucleus is not altered by early social deprivation. *Neurosci. Lett.*, **158** (1993), 130–134.
101. Sahakian, B. J., Robbins, T. W., Morgan, M. J. and Iversen, S. D. The effects of psychomotor stimulants on stereotypy and locomotor activity in socially-deprived and control rats. *Brain Res.*, **84** (1975), 195–205.
102. Wright, I. K., Upton, N. and Marsden, C. A. Resocialisation of isolation-reared rats does not alter their anxiogenic profile on the elevated X-maze model of anxiety. *Physiol. Behav.*, **50** (1991), 1129–1132.
103. Jones, G. H., Marsden, C. A. and Robbins, T. W. Behavioral rigidity and rule-learning deficits following isolation-rearing in the rat: neurochemical correlates. *Behav. Brain Res.*, **43** (1991), 35–50.
104. Geyer, M. A., Wilkinson, L. S., Humby, T. and Robbins, T. W. Isolation rearing of rats produces a deficit in prepulse inhibition of acoustic startle similar to that in schizophrenia. *Biol. Psychiatry*, **34** (1993), 361–372.
105. Fulford, A. J., Butler, S., Heal, D. J., Kendall, D. A. and Marsden, C. A. Evidence for altered alpha 2-adrenoceptor function following isolation-rearing in the rat. *Psychopharmacology (Berl.)*, **116** (1994), 183–190.
106. Fulford, A. J. and Marsden, C. A. Effect of isolation-rearing on noradrenaline release in rat hypothalamus and hippocampus in vitro. *Brain Res.*, **748** (1997), 93–99.
107. Fulford, A. J. and Marsden, C. A. Social isolation in the rat enhances alpha 2-autoreceptor function in the hippocampus in vivo. *Neuroscience*, **77** (1997), 57–64.
108. Lapiz, M. D., Mateo, Y., Durkin, S., Parker, T. and Marsden, C. A. Effects of central noradrenaline depletion by the selective neurotoxin DSP-4 on the behavior of the isolated rat in the elevated plus maze and water maze. *Psychopharmacology (Berl.)*, **155** (2001), 251–259.
109. Lapiz, M. D., Mateo, Y., Parker, T. and Marsden, C. Effects of noradrenaline depletion in the brain on response on novelty in isolation-reared rats. *Psychopharmacology (Berl.)*, **152** (2000), 312–320.
110. Heritch, A. J., Henderson, K. and Westfall, T. C. Effects of social isolation on brain catecholamines and forced swimming in rats: prevention by antidepressant treatment. *J. Psychiatr. Res.*, **24** (1990), 251–258.
111. Plaznik, A., Palejko, W., Stefanski, R. and Kostowski, W. Open field behavior of rats reared in different social conditions: the effects of stress and imipramine. *Pol. J. Pharmacol.*, **45** (1993), 243–252.

112. Suomi, S. J., Seaman, S. F., Lewis, J. K., DeLizio, R. D. and McKinney, W. T., Jr. Effects of imipramine treatment of separation-induced social disorders in rhesus monkeys. *Arch. Gen. Psychiatry*, **35** (1978), 321–325.
113. Pearce, B. D. Schizophrenia and viral infection during neurodevelopment: a focus on mechanisms. *Mol. Psychiatry*, **6** (2001), 634–646.
114. Pletnikov, M. V., Moran, T. H. and Carbone, K. M. Borna disease virus infection of the neonatal rat: developmental brain injury model of autism spectrum disorders. *Front. Biosci.*, **7** (2002), d593–d607.
115. Pletnikov, M. V., Rubin, S. A., Schwartz, G. J., Carbone, K. M. and Moran, T. H. Effects of neonatal rat Borna disease virus (BDV) infection on the postnatal development of the brain monoaminergic systems. *Brain Res. Dev. Brain Res.*, **119** (2000), 179–185.
116. Pletnikov, M. V., Rubin, S. A., Vogel, M. W., Moran, T. H. and Carbone, K. M. Effects of genetic background on neonatal Borna disease virus infection-induced neurodevelopmental damage. II. Neurochemical alterations and responses to pharmacological treatments. *Brain Res.*, **944** (2002), 108–123.
117. Mohammed, A. K., Magnusson, O., Maehlen, J. *et al.* Behavioral deficits and serotonin depletion in adult rats after transient infant nasal viral infection. *Neuroscience*, **35** (1990), 355–363.
118. O'Kusky, J. R., Boyes, B. E., Walker, D. G. and McGeer, E. G. Cytomegalovirus infection of the developing brain alters catecholamine and indoleamine metabolism. *Brain Res.*, **559** (1991), 322–330.
119. Bedi, K. S., Massey, R. F. and Smart, J. L. Neuronal and synaptic measurements in the visual cortex of adult rats after undernutrition during normal or artificial rearing. *J. Comp. Neurol.*, **289** (1989), 89–98.
120. Seligman, M. E. and Maier, S. F. Failure to escape traumatic shock. *J. Exp. Psychol.*, **74** (1967), 1–9.
121. Willner, P. Animal models of depression: an overview. *Pharmacol. Ther.*, **45** (1990), 425–455.
122. Pavcovich, L. A. and Ramirez, O. A. Effects of chronic desipramine administration on the locus coeruleus neuronal activity in the learned helplessness paradigm. *Brain Res. Bull.*, **32** (1993), 83–86.
123. Soubrie, P., Martin, P., el Mestikawy, S. and Hamon, M. Delayed behavioral response to antidepressant drugs following selective damage to the hippocampal noradrenergic innervation in rats. *Brain Res.*, **437** (1987), 323–331.
124. Sherman, A. D. and Petty, F. Additivity of neurochemical changes in learned helplessness and imipramine. *Behav. Neural Biol.*, **35** (1982), 344–353.
125. Sagen, J., Sortwell, C. E. and Pappas, G. D. Monoaminergic neural transplants prevent learned helplessness in a rat depression model. *Biol. Psychiatry.*, **28** (1990), 1037–1048.
126. Dougherty, D. D., Sortwell, C. E. and Sagen, J. Pharmacologic specificity of antidepressive activity by monoaminergic neural transplants. *Psychopharmacology (Berl.)*, **118** (1995), 10–18.
127. Plaznik, A., Danysz, W. and Kostowski, W. A stimulatory effect of intraaccumbens injections of noradrenaline on the behavior of rats in the forced swim test. *Psychopharmacology (Berl.)*, **87** (1985), 119–123.

128. Petty, F., Chae, Y., Kramer, G., Jordan, S. and Wilson, L. Learned helplessness sensitizes hippocampal norepinephrine to mild restress. *Biol. Psychiatry*, **35** (1994), 903–908.
129. Petty, F., Kramer, G., Wilson, L. and Chae, Y. L. Learned helplessness and in vivo hippocampal norepinephrine release. *Pharmacol. Biochem. Behav.*, **46** (1993), 231–235.
130. Roth, K. A. and Katz, R. J. Further studies on a novel animal model of depression: therapeutic effects of a tricyclic antidepressant. *Neurosci. Biobehav. Rev.*, **5** (1981), 253–258.
131. Katz, R. J. Animal model of depression: effects of electroconvulsive shock therapy. *Neurosci. Biobehav. Rev.*, **5** (1981), 273–277.
132. Katz, R. J. and Sibel, M. Animal model of depression: tests of three structurally and pharmacologically novel antidepressant compounds. *Pharmacol. Biochem. Behav.*, **16** (1982), 973–977.
133. Forbes, N. F., Stewart, C. A., Matthews, K. and Reid, I. C. Chronic mild stress and sucrose consumption: validity as a model of depression. *Physiol. Behav.*, **60** (1996), 1481–1484.
134. Willner, P., Towell, A., Sampson, D., Sophokleous, S. and Muscat, R. Reduction of sucrose preference by chronic unpredictable mild stress, and its restoration by a tricyclic antidepressant. *Psychopharmacology*, **93** (1987), 358–364.
135. Katz, R. J. Animal model of depression: pharmacological sensitivity of a hedonic deficit. *Pharmacol. Biochem. Behav.*, **16** (1982), 965–968.
136. Papp, M., Nalepa, I., Antkiewicz-Michaluk, L. and Sanchez, C. Behavioral and biochemical studies of citalopram and WAY 100635 in rat chronic mild stress model. *Pharmacol. Biochem. Behav.*, **72** (2002), 465–474.
137. Molina, V. A., Volosin, M., Cancela, L. *et al.* Effect of chronic variable stress on monoamine receptors: influence of imipramine administration. *Pharmacol. Biochem. Behav.*, **35** (1990), 335–340.
138. Papp, M., Klimek, V. and Willner, P. Effects of imipramine on serotonergic and beta-adrenergic receptor binding in a realistic animal model of depression. *Psychopharmacology (Berl.)*, **114** (1994), 309–314.
139. Azpiroz, A., Fano, E., Garmendia, L. *et al.* Effects of chronic mild stress (CMS) and imipramine administration, on spleen mononuclear cell proliferative response, serum corticosterone level and brain norepinephrine content in male mice. *Psychoneuroendocrinology*, **24** (1999), 345–361.
140. Areso, M. P. and Frazer, A. Effect of repeated administration of novel stressors on central beta adrenoceptors. *J. Neural. Transm. Gen. Sect.*, **86** (1991), 229–235.
141. Tejani-Butt, S. M., Pare, W. P. and Yang, J. Effect of repeated novel stressors on depressive behavior and brain norepinephrine receptor system in Sprague-Dawley and Wistar Kyoto (WKY) rats. *Brain Res.*, **649** (1994), 27–35.
142. Harro, J., Haidkind, R., Harro, M. *et al.* Chronic mild unpredictable stress after noradrenergic denervation: attenuation of behavioral and biochemical effects of DSP-4 treatment. *Eur. Neuropsychopharmacol.*, **10** (1999), 5–16.
143. Niesink, R. J. and van Ree, J. M. Involvement of the pituitary–adrenal axis in socio-behavioral disturbances after short-term isolation. *Physiol. Behav.*, **30** (1983), 825–830.
144. Haller, J. and Halasz, J. Mild social stress abolishes the effects of isolation on anxiety and chlordiazepoxide reactivity. *Psychopharmacology (Berl.)*, **144** (1999), 311–315.

145. Kelly, J. P., Wrynn, A. S. and Leonard, B. E. The olfactory bulbectomized rat as a model of depression: an update. *Pharmacol. Ther.*, **74** (1997), 299–316.
146. van Riezen, H. and Leonard, B. E. Effects of psychotropic drugs on the behavior and neurochemistry of olfactory bulbectomized rats. *Pharmacol. Ther.*, **47** (1990), 21–34.
147. Iwasaki, K., Fujiwara, M. and Ueki, S. In vivo changes in brain catecholamine release from rat hypothalamus following olfactory bulbectomy. *Pharmacol. Biochem. Behav.*, **34** (1989), 879–885.
148. Leonard, B. E. and O'Connor, W. T. The effect of chronically administered mianserin, 8-hydroxymianserin and desmethylmianserin on the 'open field' behavior and brain noradrenaline metabolism in the olfactory bulbectomized rat. *Neuropsychobiology*, **18** (1987), 118–121.
149. O'Connor, W. T. and Leonard, B. E. Effect of chronic administration of the 6-aza analogue of mianserin (Org. 3770) and its enantiomers on behavior and changes in noradrenaline metabolism of olfactory-bulbectomized rats in the "open field" apparatus. *Neuropharmacology*, **25** (1986), 267–270.
150. Tonnaer, J. A., Rigter, H., Versteeg, D. H. and Nickolson, V. J. Changes in rat brain norepinephrine levels and turnover after olfactory bulbectomy. *Brain Res. Bull.*, **5** (1980), 683–686.
151. Yoshimura, H. and Ueki, S. Regional changes in brain norepinephrine content in relation to mouse-killing behavior by rats. *Brain Res. Bull.*, **7** (1981), 151–155.
152. Iwasaki, K., Fujiwara, M., Shibata, S. and Ueki, S. Changes in brain catecholamine levels following olfactory bulbectomy and the effect of acute and chronic administration of desipramine in rats. *Pharmacol. Biochem. Behav.*, **24** (1986), 1715–1719.
153. Hong, K. W., Lee, W. S. and Rhim, B. Y. Role of central alpha 2-adrenoceptors on the development of muricidal behavior in olfactory bulbectomized rats: effect of alpha 2-adrenoceptor antagonists. *Physiol. Behav.*, **39** (1987), 535–539.
154. Kostowski, W., Plaznik, A. and Danysz, W. The role of the locus coeruleus-limbic noradrenergic transmission in the action of antidepressant drugs. *Psychopharmacol. Bull.*, **22** (1986), 512–522.
155. Weiss, J. M., Cierpial, M. A. and West, C. H. Selective breeding of rats for high and low motor activity in a swim test: toward a new animal model of depression. *Pharmacol. Biochem. Behav.*, **61** (1998), 49–66.
156. West, C. H. and Weiss, J. M. Effects of antidepressant drugs on rats bred for low activity in the swim test. *Pharmacol. Biochem. Behav.*, **61** (1998), 67–79.
157. Kelliher, P., Kelly, J. P., Leonard, B. E. and Sanchez, C. Effects of acute and chronic administration of selective monoamine re-uptake inhibitors in the rat forced swim test. *Psychoneuroendocrinology*, **28** (2003), 332–347.
158. Detke, M. J., Rickels, M. and Lucki, I. Active behaviors in the rat forced swimming test differentially produced by serotonergic and noradrenergic antidepressants. *Psychopharmacology (Berl.)*, **121** (1995), 66–72.
159. File, S. E. Recent developments in anxiety, stress, and depression. *Pharmacol. Biochem. Behav.*, **54** (1996), 3–12.

160. Cecchi, M., Khoshbouei, H., Javors, M. and Morilak, D. A. Modulatory effects of norepinephrine in the lateral bed nucleus of the stria terminalis on behavioral and neuroendocrine responses to acute stress. *Neuroscience*, **112** (2002), 13–21.
161. Shishkina, G. T., Kalinina, T. S., Sournina, N. Y. and Dygalo, N. N. Effects of antisense to the (alpha)2A-adrenoceptors administered into the region of the locus coeruleus on behaviors in plus-maze and sexual behavior tests in sham-operated and castrated male rats. *J. Neurosci.*, **21** (2001), 726–731.
162. Shishkina, G. T., Kalinina, T. S., Sournina, N. Y. *et al.* Effects of antisense oligodeoxynucleotide to the alpha2A-adrenoceptors on the plasma corticosterone level and on elevated plus-maze behavior in rats. *Psychoneuroendocrinology*, **27** (2002), 593–601.
163. Johnston, A. L. and File, S. E. Yohimbine's anxiogenic action: evidence for noradrenergic and dopaminergic sites. *Pharmacol. Biochem. Behav.*, **32** (1989), 151–156.
164. Khoshbouei, H., Cecchi, M., Dove, S., Javors, M. and Morilak, D. A. Behavioral reactivity to stress: amplification of stress-induced noradrenergic activation elicits a galanin-mediated anxiolytic effect in central amygdala. *Pharmacol. Biochem. Behav.*, **71** (2002), 407–417.
165. Ramos, A. and Mormede, P. Stress and emotionality: a multidimensional and genetic approach. *Neurosci. Biobehav. Rev.*, **22** (1998), 33–57.
166. Brocco, M., Dekeyne, A., Veiga, S., Girardon, S. and Millan, M. J. Induction of hyperlocomotion in mice exposed to a novel environment by inhibition of serotonin reuptake: a pharmacological characterization of diverse classes of antidepressant agents. *Pharmacol. Biochem. Behav.*, **71** (2002), 667–680.
167. de Angelis, L. Experimental anxiety and antidepressant drugs: the effects of moclobemide, a selective reversible MAO-A inhibitor, fluoxetine and imipramine in mice. *Naunyn-Schmiedeberg's Arch. Pharmacol.*, **354** (1996), 379–383.
168. Cervo, L. and Samanin, R. Clonidine causes antidepressant-like effects in rats by activating alpha 2-adrenoceptors outside the locus coeruleus. *Eur. J. Pharmacol.*, **193** (1991), 309–313.
169. Sajdyk, T. J., Katner, J. S. and Shekhar, A. Monoamines in the dorsomedial hypothalamus of rats following exposure to different tests of "anxiety". *Prog. Neuropsychopharmacol. Biol. Psychiatry*, **21** (1997), 193–209.
170. Eguchi, J., Inomata, Y. and Saito, K. The anxiolytic-like effect of MCI-225, a selective NA reuptake inhibitor with 5-HT3 receptor antagonism. *Pharmacol. Biochem. Behav.*, **68** (2001), 677–683.
171. Cecchi, M., Khoshbouei, H. and Morilak, D. A. Effects of stress-induced norepinephrine release in central amygdala on ACTH secretion and behavioral reactivity to stress. *Soc. Neurosci. Abstr.* (2001), Program No. 858.16.
172. Kellogg, C. K., Inglefield, J. R., Taylor, M. K. and Pleger, G. L. Importance of hypothalamic function to stressor-induced responsiveness of the G. A. B. A. A receptor in the cerebral cortex: a non-corticosterone influence. *Brain Res.*, **609** (1993), 244–252.
173. Mitchell, P. J. and Fletcher, A. Venlafaxine exhibits pre-clinical antidepressant activity in the resident-intruder social interaction paradigm. *Neuropharmacology*, **32** (1993), 1001–1009.

174. Crow, T. J., Deakin, J. F., File, S. E., Longden, A. and Wendlandt, S. The locus coeruleus noradrenergic system – evidence against a role in attention, habituation, anxiety and motor activity. *Brain Res.*, **155** (1978), 249–261.

175. Beaufour, C. C., Ballon, N., Le Bihan, C., Hamon, M. and Thiebot, M. H. Effects of chronic antidepressants in an operant conflict procedure of anxiety in the rat. *Pharmacol. Biochem. Behav.*, **62** (1999), 591–599.

176. Koob, G. F., Thatcher-Britton, K., Britton, D. R., Roberts, D. C. and Bloom, F. E. Destruction of the locus coeruleus or the dorsal NE bundle does not alter the release of punished responding by ethanol and chlordiazepoxide. *Physiol. Behav.*, **33** (1984), 479–485.

177. Francis, D., Diorio, J., Liu, D. and Meaney, M. J. Nongenomic transmission across generations of maternal behavior and stress responses in the rat. *Science*, **286** (1999), 1155–1158.

178. Overstreet, D. H. Selective breeding for increased cholinergic function: development of a new animal model of depression. *Biol. Psychiatry*, **21** (1986), 49–58.

179. Overstreet, D. H. The Flinders sensitive line rats: a genetic animal model of depression. *Neurosci. Biobehav. Rev.*, **17** (1993), 51–68.

180. Overstreet, D. H., Pucilowski, O., Rezvani, A. H. and Janowsky, D. S. Administration of antidepressants, diazepam and psychomotor stimulants further confirms the utility of Flinders Sensitive Line rats as an animal model of depression. *Psychopharmacology (Berl.)*, **121** (1995), 27–37.

181. Owens, M. J., Overstreet, D. H., Knight, D. L. *et al.* Alterations in the hypothalamic–pituitary–adrenal axis in a proposed animal model of depression with genetic muscarinic supersensitivity. *Neuropsychopharmacology*, **4** (1991), 87–93.

182. Zangen, A., Overstreet, D. H. and Yadid, G. Increased catecholamine levels in specific brain regions of a rat model of depression: normalization by chronic antidepressant treatment. *Brain Res.*, **824** (1999), 243–250.

183. Schiller, G. D., Pucilowski, O., Wienicke, C. and Overstreet, D. H. Immobility-reducing effects of antidepressants in a genetic animal model of depression. *Brain Res. Bull.*, **28** (1992), 821–823.

184. Broadhurst, P. L. The Maudsley reactive and nonreactive strains of rats: a survey. *Behav. Genet.*, **5** (1975), 299–319.

185. Berrettini, W. H., Harris, N., Ferraro, T. N. and Vogel, W. H. Maudsley reactive and non-reactive rats differ in exploratory behavior but not in learning. *Psychiatr. Genet.*, **4** (1994), 91–94.

186. Commissaris, R. L., Verbanac, J. S., Markovska, V. L., Altman, H. J. and Hill, T. J. Anxiety-like and depression-like behavior in Maudsley reactive (MR) and non-reactive (NMRA) rats. *Prog. Neuropsychopharmacol. Biol. Psychiatry*, **20** (1996), 491–501.

187. Overstreet, D. H., Rezvani, A. H. and Janowsky, D. S. Maudsley reactive and non-reactive rats differ only in some tasks reflecting emotionality. *Physiol. Behav.*, **52** (1992), 149–152.

188. Paterson, A., Whiting, P. J., Gray, J. A., Flint, J. and Dawson, G. R. Lack of consistent behavioral effects of Maudsley reactive and non-reactive rats in a number of animal tests of anxiety and activity. *Psychopharmacology (Berl.)*, **154** (2001), 336–342.

189. Sara, S. J., Devauges, V. and Biegon, A. Maudsley rat strains, selected for differences in emotional responses, differ in behavioral response to clonidine and in [125I]clonidine binding in the locus coeruleus. *Behav. Brain Res.*, **57** (1993), 101–104.
190. Sara, S. J., Devauges, V., Biegon, A. and Blizard, D. A. The Maudsley rat strains as a probe to investigate noradrenergic–cholinergic interaction in cognitive function. *J. Physiol Paris*, **88** (1994), 337–345.
191. Blizard, D. A., Freedman, L. S. and Liang, B. Genetic variation, chronic stress, and the central and peripheral noradrenergic systems. *Am. J. Physiol.*, **245** (1983), R600–R605.
192. Commissaris, R. L., McCloskey, T. C., Harrington, G. M. and Altman, H. J. MR/Har and MNRA/Har Maudsley rat strains: differential response to chlordiazepoxide in a conflict task. *Pharmacol. Biochem. Behav.*, **32** (1989), 801–805.
193. Driscoll, P., Escorihuela, R. M., Fernandez-Teruel, A. *et al.* Genetic selection and differential stress responses. The Roman lines/strains of rats. *Ann. N. Y. Acad. Sci.*, **851** (1998), 501–510.
194. Roozendaal, B., Koolhaas, J. M. and Bohus, B. Posttraining norepinephrine infusion into the central amygdala differentially enhances later retention in Roman high-avoidance and low-avoidance rats. *Behav. Neurosci.*, **107** (1993), 575–579.
195. Driscoll, P., Dedek, J., Martin, J. R. and Zivkovic, B. Two-way avoidance and acute shock stress induced alterations of regional noradrenergic, dopaminergic and serotonergic activity in Roman high- and low-avoidance rats. *Life Sci.*, **33** (1983), 1719–1725.
196. Thomas, S. A. and Palmiter, R. D. Examining adrenergic roles in development, physiology, and behavior through targeted disruption of the mouse dopamine beta-hydroxylase gene. *Adv. Pharmacol.*, **42** (1998), 57–60.
197. Xu, F., Gainetdinov, R. R., Wetsel, W. C. *et al.* Mice lacking the norepinephrine transporter are supersensitive to psychostimulants. *Nat. Neurosci.*, **3** (2000), 465–471.
198. Haller, J., Bakos, N., Rodriguiz, R. M. *et al.* Behavioral responses to social stress in noradrenaline transporter knockout mice: effects on social behavior and depression. *Brain Res. Bull.*, **58** (2002), 279–284.
199. Drouin, C., Darracq, L., Trovero, F. *et al.* Alpha1b-adrenergic receptors control locomotor and rewarding effects of psychostimulants and opiates. *J. Neurosci.*, **22** (2002), 2873–2884.
200. Zuscik, M. J., Chalothorn, D., Hellard, D. *et al.* Hypotension, autonomic failure, and cardiac hypertrophy in transgenic mice overexpressing the alpha 1B-adrenergic receptor. *J. Biol. Chem.*, **276** (2001), 13 738–13 743.
201. Philipp, M., Brede, M. and Hein, L. Physiological significance of alpha(2)-adrenergic receptor subtype diversity: one receptor is not enough. *Am. J. Physiol. Regul. Integr. Comp. Physiol.*, **283** (2002), R287–R295.
202. Bucheler, M. M., Hadamek, K. and Hein, L. Two alpha(2)-adrenergic receptor subtypes, alpha(2A) and alpha(2C), inhibit transmitter release in the brain of gene-targeted mice. *Neuroscience*, **109** (2002), 819–826.
203. Hein, L., Altman, J. D. and Kobilka, B. K. Two functionally distinct alpha2-adrenergic receptors regulate sympathetic neurotransmission. *Nature*, **402** (1999), 181–184.
204. Lahdesmaki, J., Sallinen, J., MacDonald, E. *et al.* Behavioral and neurochemical characterization of alpha(2A)-adrenergic receptor knockout mice. *Neuroscience*, **113** (2002), 289–299.

205. Schramm, N. L., McDonald, M. P. and Limbird, L. E. The alpha(2a)-adrenergic receptor plays a protective role in mouse behavioral models of depression and anxiety. *J. Neurosci.*, **21** (2001), 4875–4882.

206. Sallinen, J., Haapalinna, A., MacDonald, E. *et al.* Genetic alteration of the alpha2-adrenoceptor subtype c in mice affects the development of behavioral despair and stress-induced increases in plasma corticosterone levels. *Mol. Psychiatry*, **4** (1999), 443–452.

207. Kehr, J., Yoshitake, T., Wang, F. H. *et al.* Microdialysis in freely moving mice: determination of acetylcholine, serotonin and noradrenaline release in galanin transgenic mice. *J. Neurosci. Methods*, **109** (2001), 71–80.

208. Holmes, A., Yang, R. J. and Crawley, J. N. Evaluation of an anxiety-related phenotype in galanin overexpressing transgenic mice. *J. Mol. Neurosci.*, **18** (2002), 151–165.

11

Neuropathology of central norepinephrine in psychiatric disorders: postmortem research

Gregory A. Ordway, Ph.D.
East Tennessee State University, Johnson City, USA

The postmortem human brain as a tool to study central nervous system disease

Abnormalities in noradrenergic transmission are likely to play a role in behavioral expressions of a number of psychiatric and neurological disorders. The extent to which these abnormalities are pathognomonic, or even principal pathological features contributing to the illness, remains debatable. Interest in the potential for pathological abnormalities in central norepinephrine in central nervous system (CNS) disorders derives from the three general observations: (1) disruption of behaviors known to be heavily influenced by noradrenergic transmission that are associated with the illness; (2) demonstration that pharmacological manipulation of noradrenergic transmission can precipitate, modify, or alleviate symptoms of these disorders; and (3) certain CNS disorders are characterized pathologically by a loss of noradrenergic neurons in the brain. Research on the pathology of central noradrenergic systems in CNS diseases and their relationship to behavioral alterations utilizes a variety of techniques, most of which are technically indirect, given that we currently are unable to directly measure noradrenergic neuron activity, noradrenergic receptor signaling, or norepinephrine release in vivo in living humans. In vivo imaging methods now permit investigators to measure occupancy of certain receptors, but application of these methods specifically to noradrenergic proteins, such as receptors, enzymes or transporters, has been limited.

One method to study the role of norepinephrine in the CNS disorders is to utilize postmortem brain tissue from subjects with a given psychiatric or neurological condition. Success with the use of postmortem brain tissue in unraveling neuropathology is most evident for Parkinson's disease and Alzheimer's disease, although relationships between pathology and behaviors continue to be the most difficult to elucidate. In contrast to neurological disorders, postmortem human

Brain Norepinephrine: Neurobiology and Therapeutics, ed. Gregory A. Ordway, Michael A. Schwartz and Alan Frazer. Published by Cambridge University Press.

brain research has revealed only a few reproducible neuropathologies for psychiatric illnesses. Despite the relative lower success of postmortem human brain research of psychiatric disease, the use of this tissue still retains tremendous promise and funding institutions continue to encourage the use of postmortem brain tissues in research on psychiatric as well as neurological diseases.

Research utilizing human postmortem brain tissue to study the noradrenergic system has focused largely upon depressive disorders, schizophrenia, suicide, Parkinson's and Alzheimer's disease. There has been far less study of other psychiatric and neurological disorders. The focused interest on norepinephrine in depressive disorders comes from the vast amount of laboratory animal and human clinical literature that implicates a role of norepinephrine in depression. Noradrenergic pathology has also been of considerable interest in the study of Alzheimer's disease and Parkinson's disease because both of these disorders are characterized by a loss of central noradrenergic neurons. Noradrenergic dysfunction is linked to affective states that accompany these neurological disorders, although drugs that modulate noradrenergic transmission have had limited success in alleviating symptoms of either of these latter disorders. The role of norepinephrine in neurological disorders is considered in Chapter 15.

Shortcomings of postmortem brain research

There are many reasons why the knowledge yield from postmortem research of CNS diseases is lower than what one might anticipate. While the method allows the researcher to directly investigate neurotransmitter systems in the brains of humans suffering from specific illnesses, it provides information at only a single time point along the continuum of disease progression or remission for a given subject. Often, the point at which the researcher studies the brain is at the end of the disease process, i.e after the subject has died as a direct result or indirect result of the illness. In the case of psychiatric illnesses, it is common for the researcher to obtain access to the brain after the patient terminates his or her own life (suicide). Occasionally, researchers can gain access to brains from individuals with psychiatric illnesses who die of natural causes, but in many of these cases, the psychiatric illness is historical rather than active at the time of death. These issues continue to challenge the overall utility of postmortem research.

Other inherent weaknesses of postmortem research have been reviewed recently.[1,2,3] These include lack of appropriate control subjects being matched for age, gender, time between death in tissue preparation, as well as a lack of thorough toxicological examination, psychiatric and medical history, knowledge of psychiatric state at the time of death, medication and illicit drug history, social drug use (tobacco, alcohol, caffeine), and verification of the anatomical location of tissue collected. Within the past decade, the National Institutes of Health has invested money in the development of high-quality postmortem brain collections, the funding for which has required peer review and high standards for quality. In addition, a recent Request for Applications (RFA) from the National Institute of Mental Health demonstrates a renewed interest in utilizing postmortem brain tissues in mental health research, particularly in conjunction with state of the art genomic and proteomic technologies.

Strengths of postmortem brain research

Despite the shortcomings and difficulties associated with postmortem brain research, one should not overlook the enormous power of using postmortem brain tissue to study psychiatric and neurological disease. It will be many years into the future before in vivo imaging techniques can match the anatomical resolution or the breadth of study of in vitro postmortem brain research. Anatomical resolution may be the greatest advantage of postmortem research relative to other clinical research endeavors involving live human subjects. Many recent postmortem studies have combined modern biotechnological methods with microscopic resolution in order to study the neurochemistry and neuroanatomy of psychiatric and neurological illnesses. Relatively recent advances in laser microscopy affords researchers unparalleled anatomical resolution for biomolecular measures, whereby cells of a single neurochemical type can be microscopically dissected from brain sections using a laser and collected for analyses. A second advance in technology that makes postmortem studies exciting, if not exclusive, is the use of microarray technology to examine the expression of thousands of genes from a cytoarchitectonically defined and microscopically or macroscopically dissected brain region. These methods, combined with rigorous retrospective psychiatric autopsy methods to carefully define the behaviors of study subjects, have only recently begun to be applied to postmortem brain research, and many preliminary data from many ongoing studies have only appeared in abstract form. Nevertheless, the potential for a revolution in brain science exists and researchers are hopeful that application of these modern methods to postmortem human brain tissue will greatly advance our understanding of psychiatric and neurological diseases and improve their treatment.

Noradrenergic systems and presynaptic vs. postsynaptic markers

In general, research designed to elucidate potential neuropathological features of the central noradrenergic system can be broadly classified into studies of presynaptic and postsynaptic markers. Postsynaptic markers are those proteins with which norepinephrine, once released into the synapse, interacts directly, or signaling proteins along biochemical cascades resulting from norepinephrine receptor activation. Typical postsynaptic markers of noradrenergic neurons are the receptors for norepinephrine, including the α_1 (A, B, C), α_2 (A, B, C), β_1- and β_2-adrenergic receptors. Presynaptic markers studied in postmortem tissues include many different proteins that are either expressed solely on noradrenergic neurons, or proteins that are expressed more widely but are studied in a way to confirm their location on noradrenergic neurons. Studies of noradrenergic neurons themselves, either their projections or cell bodies, as identified through immunostaining or histological staining methods are considered presynaptic markers in this review.

In searching for alterations in postsynaptic markers, investigators typically assume that if there is altered noradrenergic transmission at the synapse, then postsynaptic receptor sensitivity or density will be altered, essentially as a form of adaptation. This assumption is based on animal studies demonstrating that depletion of norepinephrine or destruction of noradrenergic neurons results in an up-regulation of adrenergic receptors. In contrast, increases in activation of noradrenergic receptors, by excess agonist or activator, results in the opposite adaptation, i.e. down-regulation or a reduction in the amount of noradrenergic receptors. Hence, measurement of postmortem postsynaptic markers has the potential to uncover either decreases or increases in antemortem noradrenergic activity. It is also possible that an alteration at the postsynaptic receptor may in fact be the primary deficit in noradrenergic transmission, i.e. a deficit in the noradrenergic receptor may alter transmission postsynaptically via otherwise normal noradrenergic neuronal activity.

There are numerous presynaptic markers that have been studied in postmortem research of the central noradrenergic system. Several of these include proteins that are expressed only by, or predominately by, noradrenergic neurons. These include dopamine- β-hydroxylase, which converts dopamine to norepinephrine in noradrenergic neurons, and the norepinephrine transporter, which is responsible for terminating the action of norepinephrine in the synapse by reuptake. Other studies of presynaptic markers utilize anatomical identification of noradrenergic neurons, combined with immunostaining or radioligand binding to confirm the location of the marker on noradrenergic neurons. Cell bodies of neurons that synthesize norepinephrine in the brain are extremely limited in their distribution in the CNS, residing in the pontine-medullary region. The two major groups of noradrenergic cell bodies are the locus coeruleus (LC) and the lateral tegmental nucleus. These cells can be easily identified in the human because they contain the natural pigment, neuromelanin.

Postmortem studies of the noradrenergic system in depression

The popular postulate that norepinephrine plays a role in the neuropathology of depression is strongly based in the pharmacological actions of drugs that can either alleviate or precipitate the symptoms of depression. Drugs that alleviate depression in humans are those that have been shown to, using laboratory animals, increase the synaptic availability of norepinephrine and/or serotonin. In contrast, drugs that can precipitate depression in humans are drugs that have been shown to deplete norepinephrine and/or serotonin. With regard to norepinephrine, the strongest evidence for its role in depression comes from neurotransmitter depletion studies in living humans. These clinical studies are reviewed in Chapter 12. It is worth noting the salient findings from this research here. Depletion of catecholamines (norepinephrine, dopamine) in normal human subjects does not precipitate depression.[4] However, depleting catecholamines in subjects who are euthymic and who have a history of depression precipitates depressive symptoms.[5] These studies demonstrate that catecholamines, in particular norepinephrine, play an important role in the etiology of depression. However, they also demonstrate that catecholamine depletion by itself, for example in human subjects lacking a history of depression, is insufficient to precipitate depression. Such findings demonstrate that other factors including environment, genes, and perhaps development play a role as well. Depletion of catecholamines in subjects who are actively taking noradrenergic antidepressant drugs and who are euthymic also precipitates depression.[6] This finding demonstrates the importance of norepinephrine in the therapeutic action of noradrenergic antidepressant drugs. These more recent findings in human subjects, combined with the vast literature demonstrating the noradrenergic actions of antidepressant drugs in animals, provide a strong basis for investigating the putative role of norepinephrine in the neuropathology of depression in postmortem brain studies.

With regard to postmortem studies of psychiatric disorders, early studies often used suicide victims with little information regarding the psychiatric status of the subject at the time of death. While researchers reporting results of these studies did not often use suicide as a synonym for depression, other investigators used data from these studies to support theories about the biological basis of depression. To some extent, the assumption that suicide and depression are highly related has merit, although detailed retrospective studies demonstrate that the assumption has major limitations. Approximately 80% of suicide victims suffered some depressive symptoms prior to suicide. However, only about 50% of suicide victims suffered from a depressive mood disorder prior to death. The remaining suicide victims suffered from any of a variety of other psychiatric disorders, particularly alcoholism or schizophrenia.[7] Hence, a high degree of caution is required in evaluating studies using postmortem brain tissue from suicide victims when psychiatric and medical histories are not known.

Many studies have been performed using postmortem tissues from psychiatrically uncharacterized suicide victims. These studies may provide clues to the biological basis of suicide itself, although this is a highly complex issue since most, if not all, suicides suffer from a psychiatric disorder. The reader is referred to

Gross-Isseroff *et al.*[8] for a more thorough review. For the purpose of this review, studies of psychiatrically uncharacterized suicide victims are not considered because of the difficulty in interpreting findings.

Presynaptic markers in depression

Presynaptic markers of brain noradrenergic neurons have been studied in cortex, hippocampus, amygdala, and LC, although this research has limitations. This author knows of no study of dopamine-β-hydroxylase in major depressive subjects postmortem. Alpha$_2$-adrenergic receptors are often considered a presynaptic marker because these receptors are found on noradrenergic processes and are noradrenergic autoreceptors. Garcia-Sevilla *et al.* have consistently reported an elevation of agonist binding to these receptors in depressed suicide victims.[9–12] However, it should be noted that most α_2-adrenergic receptors in noradrenergic projection areas like the cortex are postsynaptic, not presynaptic, based on laboratory animal studies. Lesion of noradrenergic neurons in rats results in an increase in α_2-adrenergic receptor binding, rather than a decrease that would be expected if most of the α_2-adrenergic receptors were on the noradrenergic terminals.[13,14] Regardless of the issue of location, an increase in α_2-adrenergic receptor binding in depressed suicides is suggestive of reduced synaptic concentrations of norepinephrine, since these receptors are known to upregulate in response to norepinephrine depletion.

As mentioned above, the cellular source of norepinephrine represents a valuable source of information about the "presynaptic" side of noradrenergic transmission. Of the two cellular sources of norepinephrine, the LC is particularly suited for postmortem studies because of the ease of its identification and its compactness. This group of the cells lies on the floor of the fourth ventricle and can be seen by the naked eye at autopsy because of its distinct natural color (as a result of containing neuromelanin). There is a body of evidence from laboratory animal studies that implicate the LC in the pathophysiology of depression and in the mechanism of action of antidepressant drugs. Locus coeruleus activity and subsequent release of brain norepinephrine are associated with many behavioral responses that are disrupted in depression, such as attention, memory, and sleep (see Chapters 6 to 9). In rats, behavioral disturbances similar to human depression that are associated with the inescapable shock paradigm are paralleled by reductions of norepinephrine in the LC.[15] Lesion of the LC in the rat abolishes behavioral[16] and biochemical effects[17] of antidepressant drug administration. In addition, the LC contains a very high density of antidepressant drug binding sites, including norepinephrine transporters,[18,19] monoamine oxidase A (MAO-A),[20] α_2-adrenergic receptors,[21] and a moderate density of serotonin transporters.[22,23] The high density of antidepressant binding sites in the LC provides an explanation for why all antidepressant drugs, regardless of their chemical class, modulate LC neurochemistry and electrophysiology.[24–26]

Only a few laboratories have studied the LC in depressive subjects, and most of this work has been performed in the author's laboratory. Our work began with the study of the LC in brains from psychiatrically uncharacterized suicide victims, where we observed an increase in agonist binding to α_2-adrenergic receptors[27] and an increase in tyrosine hydroxylase (TH),[28] which is the rate-limiting enzyme in the synthesis of norepinephrine. Later we obtained LC tissues from psychiatrically characterized suicide victims and readdressed these two proteins, as well as a number of other proteins expressed in the LC. It is important to note that in all these subsequent studies, we utilized brain tissues from subjects whose psychiatric status was rigorously characterized, including control subjects as well as psychiatrically ill subjects. The psychiatric autopsy method used provides a highly reliable antemortem diagnosis and readers are referred to Stockmeier[3] for a description of this method. The presence of antidepressant drugs in the postmortem toxicology is also an exclusion criterion in these studies because we do not want to misinterpret the effects of antidepressant drugs on brain biology as pathophysiological underpinnings of the disorder.

In the LC of major depressive subjects, we found elevated agonist binding to α_2-adrenergic receptors,[29] and elevated TH[30] as compared to psychiatrically normal control subjects. These findings were essentially identical to those described above for psychiatrically uncharacterized suicide victims. In addition, major depressive subjects were characterized by reduced radioligand binding to norepinephrine transporters in the LC, as compared to normal control subjects.[31] In contrast to these findings, no differences were observed in radioligand binding to MAO-A[32] or serotonin transporters in the LC of major depressive subjects.[33] Other laboratories have investigated TH in the LC in major depression as well. Biegon and Fieldust[34] reported lower levels of TH-immunoreactivity over individual cells of the LC of suicide victims as compared to control subjects. Half of the suicide victims had a history of depression ($n = 3$) in this study, and none had antidepressant drugs in the postmortem toxicology. Methodologically, the study by Biegon and Fieldust differs considerably from our work. They reported relative density of TH staining over individual LC cell bodies using immunocytochemistry, whereas our studies of the LC measured TH in the entire LC region including cell bodies and dendrites using quantitative transfer immunoblotting or Western blotting. In addition, it is worth noting that our sample size ($n = 13$) was considerably larger than that of Beigon and Fieldust ($n = 3$ depressives). Our findings of an elevation of TH protein in the LC of suicide and major depressive subjects was recently repeated in an additional set of eight pairs of major depressive and control subjects (Ordway and Szebeni, unpublished).

The number of neuromelanin-containing cells in the LC of major depressive subjects has been reported to be normal,[29–31] as we have reported, or reduced.[35]

It is worth noting that the study by Baumann *et al.*[35] found that there was a reduction in the number of TH-positive neurons in the LC of depressed subjects who died of natural causes. However, there was no difference in TH-positive neurons comparing depressed suicides to controls, as we have observed counting neuromelanin-containing cells. Also, Baumann *et al.* did not observe differences in neuromelanin-containing cells in depressed suicide or depressed non suicide subjects, hinting that the difference between controls and depressed suicides may be in the number of TH-positive, non neuromelanin-containing LC neurons. Another very important difference between our work and Baumann *et al.* is that half of their depressed subjects were bipolar depressed, whereas all of our study subjects had unipolar major depression. Together, these studies suggest that while there do not appear to be changes in neuromelanin-containing cells in major depressive suicides, a reduction in TH immunostaining of cells may occur in certain depressive subjects that do not commit suicide.

In order to interpret postmortem findings of neurochemical pathology in the LC vis-à-vis the biology of depression, one must consider the laboratory animal research where manipulation of LC biochemistry either pharmacologically or behaviorally has been reported. Interestingly, the three salient neurochemical findings in major depression from our laboratory (elevated binding to α_2-adrenergic receptors, reduced norepinephrine transporter, and elevated TH) can be produced experimentally in rats by pharmacologically depleting norepinephrine. However, not all of these changes have been demonstrated to occur in the LC of the rat. Depletion of norepinephrine with reserpine or 6-hydroxydopamine increases TH expression in the LC.[36,37] Depletion of norepinephrine also upregulates α_2 adrenergic receptors[13,14] and downregulates the norepinephrine transporter,[38] although effects of depletion on these proteins has not been tested specifically in the LC. Reserpine treatment does modestly reduce norepinephrine transporter (*NET*) gene expression in the rat LC.[37] Hence, there is considerable evidence that norepinephrine depletion in rats produces abnormalities similar to those observed in the human LC in major depressive subjects (see Table 11.1).

Norepinephrine depletion in the LC can also be produced by chronic stress of animals.[15] Uncontrollable shock, a stress-based animal model of depression, increases the release of norepinephrine[39] and reduces norepinephrine stores.[15] In addition, stress activates the LC and repeated stress exposure increases TH in the LC of rats,[36] similar to the increase of TH observed in depressed suicide victims. Repeated stress has also been shown to increase α_2-adrenergic receptors in the brain stem[40] and to decrease radioligand binding to the norepinephrine transporter in rats.[41] Hence, like pharmacological depletion of norepinephrine, chronic stress, which activates the LC, can produce effects on the rat LC that are similar to the abnormalities in the LC of major depressive human subjects.

Table 11.1 Summary of neuropathological findings in the locus coeruleus (LC) from postmortem studies of depression in humans and relevant experimental animal studies.

In the human LC: finding in major depression	In the rat LC: effect of NE depletion	In the rat LC: effect of stress	In the rat LC: effect of chronic antidepressant
$\uparrow$ tyrosine hydroxylase	$\uparrow$ tyrosine hydroxylase	$\uparrow$ tyrosine hydroxylase	$\downarrow$ tyrosine hydroxylase
$\uparrow$ α_2-adrenergic receptor	$\uparrow$ α_2-adrenergic receptor	$\uparrow$ α_2-adrenergic receptor	$\downarrow$ α_2-adrenergic receptor
$\uparrow$ corticotropin-releasing factor	–	$\uparrow$ corticotropin-releasing-factor	–
$\uparrow$ NR2C subunit (NMDA receptor)	–	–	–
$\downarrow$ norepinephrine transporter	$\downarrow$ norepinephrine transporter	$\downarrow$ norepinephrine transporter	$\leftrightarrow$ norepinephrine transporter
$\downarrow$ nitric oxide synthase	–	–	–
$\leftrightarrow$ monoamine oxidase A	–	–	–
$\leftrightarrow$ monoamine oxidase B	–	–	–
$\leftrightarrow$ serotonin transporter	–	–	–

The symbol $\leftrightarrow$ indicates no change. Dashes indicate that the protein has not been measured in the LC. NMDA, *N*-methyl-D-aspartate.

Recently, we proposed that postmortem biochemical changes in the LC reflect central noradrenergic overdrive, i.e. an over active LC, premortem[30,42] and that this overdrive leads to a synaptic deficiency of norepinephrine. This hypothesis is based on the laboratory animal studies discussed above, and also is supported by the opposing effects of antidepressant drugs on LC activity and biochemistry (see Table 11.1). For example, repeated treatment of rats with antidepressants from different pharmacological classes, but not with non-antidepressant drugs, down-regulates LC TH,[24] with exception.[43,44] Antidepressant treatment blocks elevation of TH or TH mRNA in the rat LC that is produced by stress or reserpine-induced norepinephrine depletion.[45,46] Repeated antidepressant treatment also downregulates α_2-adrenergic receptors in the rat LC.[47] Unlike effects on TH and α_2-adrenergic receptors, antidepressant treatment appears to have no effect on the norepinephrine transporter in the LC.[48] Overall, these pre-clinical findings demonstrate that repeated antidepressant treatment downregulates TH and α_2-adrenergic receptors, proteins showing elevated levels in the LC of human major depressives. Given the hypothesis that LC activity may be elevated in depression, recent studies demonstrate that repeated treatment of rats with antidepressant of many pharmacological classes (including serotonin uptake inhibitors) reduces LC activity.[25,26] Hence, animal data strongly support the contention that drugs produce antidepressant effects, at least in part, by reducing demand for norepinephrine (reducing biochemical measures of demand such as the biosynthetic enzyme for norepinephrine) and reducing LC activity.

As mentioned above, stress activates the LC and the relationship between stress and depression is well established, i.e. stress is a common precipitating factor of depression and suicide. The possibility of central noradrenergic overactivity in depression draws attention to stress-sensitive brain circuits that regulate LC activity. Numerous neurotransmitters are implicated in mediating the actions of stress on the brain. Several of these stress-sensitive neurotransmitters are released from terminals innervating the immediate region of the LC, and thereby regulate LC activity, e.g. corticotropin releasing factor (CRF), substance P, glutamate, serotonin and γ-aminobutyric acid (GABA). It is very intriguing that each of these are major inputs to the LC, most are robustly modulated by stress, and each have been strongly implicated in the pathobiology of depression (see reference 42 for review). It is noteworthy that antagonists of CRF, substance P, and glutamate receptors appear to have antidepressant activity in humans,[49] although none of these compounds have yet received Food and Drug Administration (FDA) approval for the treatment of depression.

With regard to possible disruption of presynaptic noradrenergic activity by altered inputs to the LC in depression, the author's laboratory has initiated a systematic evaluation of these inputs to the LC. A modest, but significant, elevation of levels of CRF-immunoreactivity in the LC has been reported comparing major depressive subjects to control subjects.[50] Interestingly, chronic stress elevates CRF-immunoreactivity in the rat LC.[51] Alterations in the subunit composition of the *N*-methyl-D-aspartate (NMDA) receptor, specifically elevated NR2C subunit levels, and reductions in neuronal nitric oxide synthase, an intracellular mediator of NMDA receptor activation, have been found in the LC of depressive subjects compared to control subjects.[52,53] Although the effects of stress and norepinephrine depletion on NMDA receptor expression in the LC has not been studied in the rat, exposure to repeated unpredictable stress has been shown to increase levels of NR2 subunit in the rat ventral tegmental area.[54] In addition, stress increases NR2 mRNA levels in the rat hippocampus.[55] These findings draw attention to the need for further study of the relationship between stress, depression, and NMDA receptor signaling in the LC.

In contrast to glutamatergic and CRF inputs, no change in serotonin transporter binding in the LC of depressed subjects has been reported[33] and preliminary evidence of no change in radioligand binding to $GABA_A$ receptors in the LC of depressives has been reported.[56] Hence, there is initial evidence of altered CRF and glutamate input to the LC in depression. The putative antidepressant activity of antagonists of CRF and glutamate receptors may be at least partially mediated by antagonizing excitatory input to the LC.

Postsynaptic markers in depression

Postsynaptic markers of noradrenergic neurons include receptors for norepinephrine and their signaling cascades. The receptors for norepinephrine that have been studied in postmortem brains from depressed subjects include the α_2-, β_1-, and β_2-adrenergic receptors. Alpha$_1$-adrenergic receptors have been studied in brains from suicide victims, but these studies have utilized psychiatrically uncharacterized subjects.

α_2-adrenergic receptors

Numerous investigators have reported results of radioligand binding studies of α_2-adrenergic receptors, most of which have utilized prefrontal cortex. Three studies from two independent laboratories report no change in α_2-receptor binding,[57–59] while one laboratory consistently reports elevated agonist binding to α_2-adrenergic receptors in the prefrontal cortex of depressed suicide victims.[9–12] This latter group of investigators has reported selective increases in the high-affinity state of the receptor, as identified in agonist competition studies using an antagonist radioligand, or using an agonist radioligand. In contrast, these investigators did not observe changes in antagonist binding to α_2-adrenergic receptors in the same subjects. Interestingly, a similar observation, change in agonist but not antagonist binding, has been reported for α_2-adrenergic receptors in the LC.[27] Garcia-Sevilla *et al.*[60] have explored this finding further by using quantitative Western blotting to measure amounts of α_{2A}-adrenergic receptor-immunoreactivity, as well as α_{2A}-receptor-associated signaling proteins, in prefrontal cortex from depressed suicide subjects. Elevated levels of immunoreactivities for α_{2A}-adrenergic receptor protein, $G_{\alpha 1/2}$ protein, and G protein receptor kinase$_{2/3}$ were observed in depressed suicides as compared to control subjects. Since $G_{\alpha 1/2}$ protein and G protein receptor kinase$_{2/3}$ lie downstream of α_{2A}-receptor activation, these data suggest the existence of a supersensitive α_{2A}-adrenergic receptor signaling system in depression. G proteins are transducers of many different G-protein-coupled receptors. Hence, an up-regulation of G_{α} protein could mean that signal transduction via many different receptors may be affected in depression. To test this, Gonzalez-Maeso *et al.*[12] studied G protein receptor signaling via multiple G_{α}-coupled receptors, including α_{2A}-adrenergic, serotonin 5-HT_{1A}, μ-opioid, $GABA_B$, and muscarinic cholinergic receptors. Only α_{2A}-adrenergic receptor-stimulated G_{alphai} was increased in suicide victims, while activation via the other receptors, or direct activation of G, was normal. These latter findings demonstrate supersensitivity specific to the α_2-adrenergic receptor. These authors argue that this receptor is the presynaptic autoreceptor on noradrenergic neurons and that a supersensitivity of this receptor may contribute to reduced noradrenergic neurotransmission, or more classically, a deficiency of norepinephrine. It should be noted however, that a deficiency of norepinephrine can cause supersensitivity of α_2-adrenergic receptors,[13] as evidenced by the effect of 6-hydroxydopamine lesions of norepinephrine neurons in rats. In addition, monoamine depletion by treatment of rats with reserpine not only upregulates agonist binding to α_2-adrenergic receptors, but also up-regulates $G_{alphai1/2}$ protein,[61] similar to the increase in these proteins observed in depressed suicide victims. Hence, it remains unknown whether the supersensitivity of brain α_2-adrenergic receptors is a consequence of a deficiency that has its cause elsewhere, or whether supersensitivity of these receptors itself is causal. The available literature suggests that α_2-adrenergic receptor sensitivity in depression is secondary to a deficiency of norepinephrine in the synapse.

β-adrenergic receptors

The study of β-adrenergic receptors in depression has received the most attention of all noradrenergic proteins. Unfortunately, there is little agreement in the literature with respect to consistent alterations in these receptors in the brain in depressive disorders. Interest in β-adrenergic receptors was stimulated by the findings of Mann *et al.*[62] and Biegon and Israeli[63] in the 1980s, where elevating radioligand binding to β-adrenergic receptors was demonstrated in the prefrontal cortex from brains of psychiatrically uncharacterized suicide victims. Other investigators have not been able to reproduce these findings.[64,65] Measurement of β-adrenergic receptors in brain tissues from depressed or affective disorder (including bipolar depressed, schizoaffective, and depression) subjects have also yielded mixed results, including no change in binding in the hippocampus[59] or frontal cortex[57,58] or reductions in binding in the cortex[66] and hippocampus. It is difficult to draw a conclusion from these mixed findings. Further research on β-adrenergic receptors in postmortem brains from depressed subjects seems unwarranted, unless a considerably different approach is undertaken to the study of the relationship between these receptors and abnormal behavior.

Postmortem studies of the noradrenergic system in schizophrenia

Postmortem studies investigating the pathophysiological basis of schizophrenia have focused largely on the dopaminergic system. This "bias" is a result of the fact that antagonists of dopamine D_2-receptors alleviate many of the symptoms of schizophrenia, as opposed to definitive evidence of pathology in brain dopamine in the disorder. The fact that D_2-receptor antagonism is effective treatment for psychoses led many researchers to consider that schizophrenia is a disorder of overactive dopamine systems, particularly mesolimbic dopamine, in the brain. Despite a strong focus on dopamine, there have been a number of interesting studies examining a possible involvement of norepinephine in schizophrenia, including studies of noradrenergic indices in postmortem brains from schizophrenic subjects. The reader is referred to a recent review by Yamamoto and Hornykiewicz[67] for a detailed consideration of norepinephrine in schizophrenia. With regard to postmortem studies of schizophrenia, it is worth noting that, as opposed to studies on depression, it is difficult to discount the potential effects of chronic psychotherapeutic drug treatment on measurements made postmortem for schizophrenic subjects. Most brains from schizophrenics have long histories of antipsychotic drug use and this must be considered when interpreting the findings of postmortem pathology.

Presynaptic markers in schizophrenia

An early study on the noradrenergic system in schizophrenia was that of Wise and Stein,[68] who demonstrated a reduction in dopamine-β-hydroxylase activity in the pons-medulla, diencephalon, and hippocampus from chronically institutionalized subjects with schizophrenia as compared to control subjects who died as a result of natural or accidental causes. Interestingly, these researchers demonstrated that chronic antipsychotic administration to rats increased, not decreased, dopamine-β-hydroxylase activity in the brain, suggesting that the reductions in the enzyme in schizophrenics was not a result of chronic antipsychotic drug administration. Catechol-O-methyltransferase activity was also significantly decreased in the diencephalon of schizophrenics. In contrast, no change in MAO activity was observed in the same schizophrenics, as compared to the control subjects. Later, Wyatt *et al.*[69] were unable to reproduce these findings. It is generally accepted now that enzyme activities undergo significant postmortem decay and this factor complicates interpretation of data. This may partially explain the disparate findings of these two studies.

Numerous investigators have measured norepinephrine, or its metabolite 3-methoxy-4-hydroxyphenylglycol (MHPG), in postmortem brain tissues.[70–74] The majority of studies have demonstrated elevations of norepinephrine in schizophrenia, although the data are not consistent for the area of the brain. The possibility of elevated norepinephrine in schizophrenia is also modestly supported by studies of norepinephrine and MHPG in the cerebrospinal fluid (CSF) of schizophrenic patients (see reference [67] for discussion). Again, for CSF studies the data have not been consistent between laboratories. Elevated CSF norepinephrine appears to be most robustly observed in patients with positive symptoms of schizophrenia, or the type I syndrome.

Recently, Friedman *et al.*[75] proposed that norepinephrine deficits, rather than increases, may contribute to cognitive dysfunction in schizophrenia. They provided the argument that studies of norepinephrine and MHPG in postmortem brains from schizophrenics, as cited above, did not consider antemortem cognitive function. In contrast, Bridge *et al.*[76] demonstrated that cognitive impairment in schizophrenics was associated with increased noradrenergic turnover in the nucleus accumbens, i.e. reduced norepinephrine and increased MHPG. Friedman *et al.*[75] demonstrated significantly reduced norepinephrine concentrations in the frontal cortex comparing cognitively impaired schizophrenics (Clinical Dementia Rating, CDR ≥ 1) to schizophrenics having low levels of cognitive impairment (CDR, 0 and 0.5). The postmortem findings of elevated and low norepinephrine in schizophrenia seem contradictory on the surface. However, it is conceivable that both conditions can exist, either as a result of regional differences within the brain, or differences associated with specific behavioral disturbance, i.e. psychosis or impaired cognition.

Postsynaptic markers in schizophrenia

Several laboratories have measured receptors for norepinephrine in postmortem brain tissue from schizophrenic subjects. One of the earliest studies was that of Bennett *et al.*[77] who found no differences in radioligand binding to β-adrenergic receptors in frontal cortices from schizophrenic subjects, while reductions in serotonin receptors were observed. Joyce *et al.*[78] reported significant increases in the amount of β_1-, but not β_2-, adrenergic receptors in subregions of the human striatum, including the caudate, nucleus accumbens septi, and the putamen, from schizophrenic subjects compared to control subjects. In addition, the pattern of β-adrenergic receptor expression in the striatum was altered in schizophrenics as compared to controls. In contrast to the striatum, other areas of the schizophrenic brain, including cortex, hippocampus, thalamus, and amygdala did not exhibit significant changes in β- adrenergic receptors. The author's laboratory found significant reductions in β_1-, but not β_2-, adrenergic receptors in the right hippocampus of schizophrenic subjects as compared to control subjects or to major depressive subjects.[59] Joyce *et al.*[78] also observed a comparable percentage lower binding in the right hippocampus of schizophrenics, as compared to control subjects, although this difference did not reach statistical significance. Overall, the data are less than compelling, although the reduction of β_1-adrenergic receptors in the hippocampus of schizophrenic subjects suggests an involvement of norepinephrine in the pathobiology of schizophrenia.

Two groups have measured α_2-adrenergic receptors in postmortem brains from schizophrenic subjects. The author's laboratory reported no differences in radiolabeled-agonist binding to α_2-adrenergic receptors in several hippocampal subregions from schizophrenics compared to control subjects.[59] Dean[79] found no differences in radiolabeled-antagonist binding to α_2 adrenergic receptors in the homogenates of prefrontal cortex of schizophrenics as compared to control subjects. These are the only two studies of α_2-adrenergic receptors in postmortem tissues from schizophrenic subjects. The findings do little to warrant additional investigation of α_2-adrenergic receptors in the pathophysiology of schizophrenia.

Summary

Overall, there is a considerable clinical literature demonstrating pathology of the noradrenergic system in schizophrenia. In addition, clozapine, considered a prototype of atypical antipsychotic drugs, is a high-affinity antagonist at α_1-and α_2-adrenergic receptors, and these actions may contribute to the uniqueness of this compound in the treatment of schizophrenia. However, few published studies have been conducted to investigate the noradrenergic system in the postmortem brain from schizophrenic brains. The small amount of work that has been performed in this regard demonstrates some interesting findings, but it is difficult to draw conclusions because of the lack of sufficient investigation. Clearly, there is enough clinical and preclinical data to justify further study of the noradrenergic system in schizophrenia.

Conclusions

Postmortem studies have yielded many important clues regarding the involvement of noradrenergic neurons in depression. A considerable cadre of postmortem evidence suggests that there is elevated central noradrenergic (LC) activity in depression (see Figure 11.1). This elevated activity may drive an increased demand

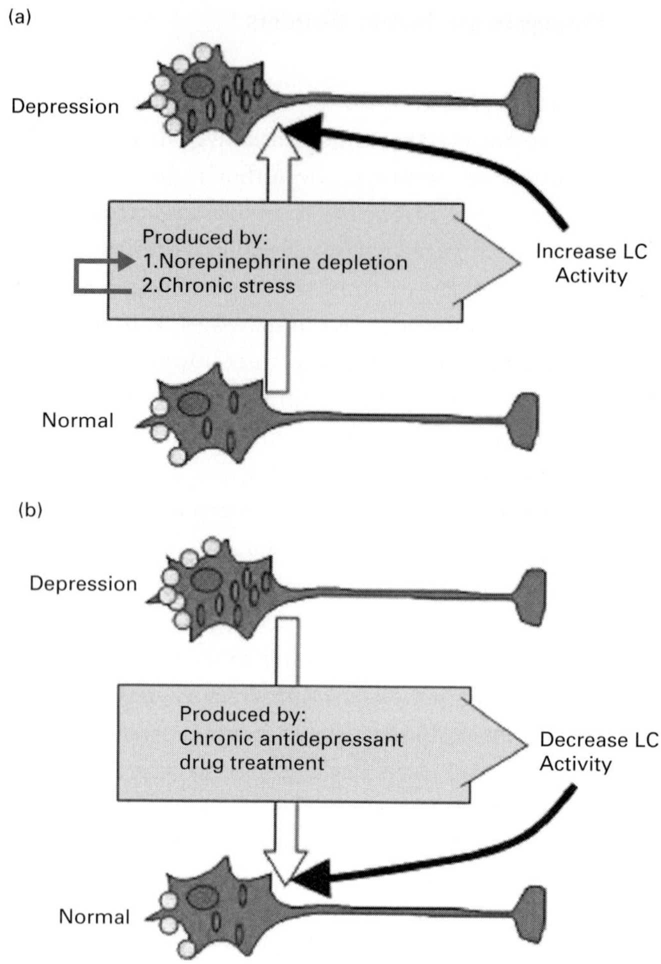

Figure 11.1 The hypothetical role of the activity and biochemistry of noradrenergic locus coeruleus (LC) neurons in the biology of depression, based on an integration of postmortem findings in depressed humans and laboratory animal studies. In (a) the schematic neurons represent LC neurons, where an elevation of tyrosine hydroxylase (ovals) and α_2-adrenergic receptors (small circles) have been observed postmortem in depressed subjects relative to normal control subjects. These alterations in the LC can be produced experimentally in rats by depleting norepinephrine or by chronic stress (which also deplete LC norepinephrine, as indicated by the dark grey arrow). Norepinephrine depletions and chronic stress elevate LC activity, presumably driving an increase in demand for norepinephrine, which results in an increase in the production of the rate-limiting enzyme in norepinephrine biosynthesis (tyrosine hydroxylase). Presumably, if norepinephrine supply is unable to meet the demand (as may occur in depression) for the neurotransmitter, a relative deficiency of norepinephrine occurs and α_2-adrenergic receptors are up-regulated and there is a further drive for an increase in synthesizing capacity. In (b), the schematic neurons depict the effect of repeated treatment of rats with antidepressant drugs on the expression of tyrosine hydroxylase and α_2-adrenergic receptors in the LC. Repeated antidepressant treatments (regardless of the pharmacological class and including selective serotonin uptake inhibitors) reduce the activity of LC neurons and thereby reduce the demand for norepinephrine. This reduction in demand is met by a reduction in the expression of the biosynthetic enzyme tyrosine hydroxylase.

for norepinephrine, which may not be met and may result in a relative synaptic deficiency of norepinephrine. Abnormal levels of several key noradrenergic proteins in postmortem noradrenergic neurons from depressive subjects support this hypothesis. Experimental studies of stress and norepinephrine depletion in laboratory animals also supports the relevance of central noradrenergic dysfunction in depression, as well as other stress-related disorders. In addition, there is growing evidence from postmortem studies of elevated excitatory input (CRF, glutamate) to central noradrenergic neurons in depression. Given that most, if not all, antidepressants dampen the activity of central noradrenergic neurons, and many simultaneously elevate synaptic norepinephrine, it is reasonable to conclude that at least some therapeutic behavioral effects of antidepressant drugs are achieved by their corrective actions on central noradrenergic activity. Novel antidepressant drugs that can block stress-sensitive excitatory inputs to central noradrenergic neurons, or that can enhance inhibitory inputs to these neurons, may prove to be useful in the treatment of depression.

Schizophrenia research using postmortem tissues has focused largely on the dopaminergic system, although disrupted noradrenergic transmission likely plays a role in this disorder as well. The most recent data regarding norepinephrine and schizophrenia relates to cognitive dysfunction resulting from schizophrenia-related disruption of noradrenergic transmission.

The field of postmortem research of psychiatric disorders will be benefited in the future through the use of gene expression and proteomic microarrays in combination with laser capture microscopy. This combination of methods is an elegant confluence of a hypothesis-generating approach (microarrays examining large numbers of genes without bias about which will be affected) with the traditional hypothesis-driven approach (e.g. laser capture microscopy to collect only noradenergic neurons for assay). Such a combined approach holds the promise of elucidating a more complete picture of the role of norepinephrine in psychiatric disorders. In addition, the field will likely be benefited by a change in the experimental focus from searching for relationships between postmortem biology and psychiatric diagnoses, disorders which are inherently heterogeneous with respect to behavioral symptoms, to the examination of potential relationships between biology and specific symptoms or symptom clusters that are associated with the psychiatric diagnoses.

ACKNOWLEDGEMENTS

This work was supported in part by grants from the National Institute for Mental Health (MH/AG 02031, MH46692, MH63187).

REFERENCES

1. Ordway, G. A. and Klimek, V. Noradrenergic pathology in psychiatric disorders: postmortem studies. *CNS Spectr.*, **6** (2001), 697–703.
2. Lewis, D. A. The human brain revisited: opportunities and challenges in postmortem studies of psychiatric disorders. *Neuropsychopharmacology*, **26**: 2 (2002), 143–154.
3. Stockmeier, C. A. Involvement of serotonin in depression: evidence from postmortem and imaging studies of serotonin receptors and the serotonin transporter. *J. Psychiatr. Res.*, **37**: 5 (2003), 357–373.
4. Salomon, R. M., Miller, H. L., Krystal, J. H., Heninger, G. R. and Charney, D. S. Lack of behavioral effects of monoamine depletion in healthy subjects. *Biol. Psychiatry*, **41**: 1 (1997) 58–64.
5. Berman, R. M., Narasimhan, M., Miller, H. L. *et al.* Transient depressive relapse induced by catecholamine depletion: potential phenotypic vulnerability marker? *Arch. Gen. Psychiatry*, **56**: 5 (1999), 395–403.
6. Charney, D. S. Monoamine dysfunction and the pathophysiology and treatment of depression. *J. Clin. Psychiatry*, **59**: Suppl. 14 (1998), 11–14 .
7. Rich, C. L., Young, D., Fowler, R. C. San Diego suicide study. *Arch. Gen. Psychiatry*, **43** (1986), 577–582.
8. Gross-Isseroff, R., Biegon, A., Voet, H. and Weizman, A. The suicide brain: a review of postmortem receptor/transporter binding studies. *Neurosci. Biobehav. Rev.*, **22**: 5 (1998), 653–661.
9. Meana, J. J. and Garcia-Sevilla, J. A. Increased alpha 2-adrenoceptor density in the frontal cortex of depressed suicide victims. *J. Neural Transm.*, **70**: 3–4 (1987), 377–381.
10. Gonzalez, A. M., Pascual, J., Meana, J. J. *et al.* Autoradiographic demonstration of increased alpha 2-adrenoceptor agonist binding sites in the hippocampus and frontal cortex of depressed suicide victims. *J. Neurochem.*, **63**: 1 (1994), 256–265.
11. Callado, L. F., Meana, J. J., Grijalba, B. *et al.* Selective increase of alpha2A-adrenoceptor agonist binding sites in brains of depressed suicide victims. *J. Neurochem.*, **70**: 3 (1998), 1114–1123.
12. Gonzalez-Maeso, J., Rodriguez-Puertas, R., Meana, J. J., Garcia-Sevilla, J. A. and Guimon, J. Neurotransmitter receptor-mediated activation of G-proteins in brains of suicide victims with mood disorders: selective supersensitivity of alpha(2A)-adrenoceptors. *Mol. Psychiatry*, **7**: 7 (2002), 755–767.
13. U'Prichard, D. C., Bechtel, W. D., Rouot, B. M. and Snyder, S. H. Multiple apparent alpha-noradrenergic receptor binding sites in rat brain: effect of 6-hydroxydopamine. *Mol. Pharmacol.*, **16** (1979), 47–60.
14. Gross, G., Gothert, M., Glapa, U., Engel, G. and Schumann, H. J. Lesioning of serotoninergic and noradrenergic nerve fibres of the rat brain does not decrease binding of 3H-clonidine and 3H-rauwolscine to cortical membranes. *Naunyn-Schmiedeberg's Arch. Pharmacol.*, **328** (1985), 229–235.
15. Weiss, J. M., Goodman, P. A., Losito, B. G. *et al.* Behavioral depression produced by an uncontrollable stressor: relationship to norepinephrine, dopamine, and serotonin levels in various regions of rat brain. *Brain. Res. Rev.*, **3** (1981), 167–205.

16. Danysz, W., Kostowski, W. and Hauptmann, M. Evidence for the locus coeruleus involvement in desipramine action in animal models of depression. *Pol. J. Pharmacol. Pharm.*, **37**: 6 (1985), 855–864.
17. Wolfe, B. B., Harden, T. K., Sporn, J. R. and Molinoff, P. B. Presynaptic modulation of beta adrenergic receptors in rat cerebral cortex after treatment with antidepressants. *J. Pharmacol. Exp. Ther.*, **207**: 2 (1978), 446–457.
18. Biegon, A. and Rainbow, T. C. Localization and characterization of [3H]desmethylimipramine binding sites in rat brain by quantitative autoradiography. *J. Neurosci.*, **3**: 5 (1983), 1069–1076.
19. Ordway, G. A., Stockmeier, C. A., Cason, G. W. and Klimek, V. Pharmacology and distribution of norepinephrine transporters in the human locus coeruleus and raphe nuclei. *J. Neurosci.*, **17** (1997), 1710–1719.
20. Richards, J. G., Saura, J., Ulrich, J. and Da Prada, M. Molecular neuroanatomy of monoamine oxidases in human brainstem. *Psychopharmacology (Berl.)*, **106**: Suppl. (1992), S21–S23.
21. Klimek, V. and Ordway, G. A. Distribution of alpha2-adrenoceptors in the human locus coeruleus. *Brain Res.*, **741** (1996), 263–274.
22. Cortes, R., Soriano, E., Pazos, A., Probst, A. and Palacios, J. M. Autoradiography of antidepressant binding sites in the human brain: localization using [3H]imipramine and [3H]paroxetine. *Neuroscience*, **27**: 2 (1988), 473–496.
23. Klimek, V., Stockmeier, C. A., Overholser, J. C. *et al.* Serotonin transporter and MAO-B levels in monoamine nuclei of the human brain stem are normal in major depression. *J. Psychiatr. Res.*, **37** (2003), 387–397.
24. Nestler, E. J., McMahon, A., Sabban, E. L., Tallman, J. F. and Duman, R. S. Chronic antidepressant administration decreases the expression of tyrosine hydroxylase in the rat locus coeruleus. *Proc. Natl. Acad. Sci. U. S. A.*, **87**: 19 (1990), 7522–7526.
25. Szabo, S. T., de Montigny, C. and Blier, P. Progressive attenuation of the firing activity of locus coeruleus noradrenergic neurons by sustained administration of selective serotonin reuptake inhibitors. *Int. J. Neuropsychopharmacol.*, **3**: 1 (2000), 1–11.
26. Grant, M. M. and Weiss, J. M. Effects of chronic antidepressant drug administration and electroconvulsive shock on locus coeruleus electrophysiologic activity. *Biol. Psychiatry*, **49**: 2 (2001), 117–129.
27. Ordway, G. A., Widdowson, P. S., Smith, K. and Halaris, A. E. Agonist binding to alpha2 adrenoceptors is elevated in the locus coeruleus from victims of suicide. *J. Neurochem.*, **63** (1994), 617–624.
28. Ordway, G. A., Streator-Smith, K. and Haycock, J. W. Elevated tyrosine hydroxylase in the locus coeruleus of suicide victims. *J. Neurochem.*, **62** (1994), 680–685.
29. Ordway, G. A., Schenk, J., Meltzer, H. Y. *et al.* Elevated agonist binding to alpha2-adrenoceptors in the locus coeruleus in major depression. *Biol. Psychiatry*, **53**: (2003), 315–323.
30. Zhu, M.-Y., Klimek, V., Dilley, G. E. *et al.* Elevated levels of tyrosine hydroxylase in the locus coeruleus in major depression. *Biol. Psychiatry*, **46** (1999), 1275–1286.
31. Klimek, V., Stockmeier, C. A., Overholser, J. *et al.* Reduced levels of norepinephrine transporters in the locus coeruleus in major depression. *J. Neurosci.*, **17** (1997), 8451–8458.

32. Ordway, G. A., Farley, J. T., Dilley, G. E. *et al.* Quantitative distribution of monoamine oxidase A in brainstem monoamine nuclei is normal in major depression. *Brain Res.*, **847** (1999), 71–79.
33. Klimek, V., Stockmeier, C. A., Overholser, J. C. *et al.* Serotonin transporter and MAO-B levels in monoamine nuclei of the human brain stem are normal in major depression. *J. Psychiatr. Res.*, **37** (2003), 387–397.
34. Biegon, A. and Fieldust, S. Reduced tyrosine hydroxylase immunoreactivity in locus coeruleus of suicide victims. *Synapse*, **10** (1992), 79–82.
35. Baumann, B., Danos, P., Diekmann, S. *et al.* Tyrosine hydroxylase immunoreactivity in the locus coeruleus is reduced in depressed non-suicidal patients but normal in depressed suicide patients. *Eur. Arch. Psychiatry Clin. Neurosci.*, **249**: 4 (1999), 212–219.
36. Melia, K. R., Rasmussen, K., Terwilliger, R. Z. *et al.* Coordinate regulation of the cyclic AMP system with firing rate and expression of tyrosine hydroxylase in the rat locus coeruleus: effects of chronic stress and drug treatments. *J. Neurochem.*, **58** (1992), 494–502.
37. Cubells, J. F., Kim, K. S., Baker, H. *et al.* Differential in vivo regulation of mRNA encoding the norepinephrine transporter and tyrosine hydroxylase in rat adrenal medulla and locus coeruleus. *J. Neurochem.*, **65** (1995), 502–509.
38. Lee, C.-M., Javitch, J. A. and Snyder, S. H. Recognition sites for norepinephrine uptake: regulation by neurotransmitter. *Science*, **220** (1983), 626–629.
39. Weiss, J. M., Glazer, H. I. and Pohorecky, L. A. Coping behavior and neurochemical changes: an alternative explanation for the original "learned helplessness" experiments. In *Animal Models in Human Psychobiology*, ed. S. Kling (New York: Plenum, 1976) pp. 133–145.
40. Torda, T., Kvetnansky, R. and Petrikova, M. Effect of repeated immobilization stress on central and peripheral adrenoceptors in rats. *Endocrinol. Exp.*, **19**: 3 (1985), 157–163.
41. Zafar, H. M., Pare, W. P. and Tejani-Butt, S. M. Effect of acute or repeated stress on behavior and brain norepinephrine system in Wistar-Kyoto (WKY) rats. *Brain Res. Bull.*, **44**: 3 (1997), 289–295.
42. Ordway, G. A., Klimek, V. and Mann, J. J. Neurocircuitry of mood disorder. In *Neuropsychopharmacology: The Fifth Generation of Progress*, ed. K. L. Davis, D. Charney, J. T. Coyle and C. Nemeroff. (Philadelphia: Lippincott Williams & Wilkins, 2002) pp. 1051–1064.
43. Brady, L. S., Gold, P. W., Herkenham, M., Lynn, A. B. and Whitfield, H. J. The antidepressants fluoxetine, idazoxan, and phenelzine alter corticotropin-releasing hormone and tyrosine hydroxylase mRNA levels in rat brain: therapeutic implications. *Brain Res.*, **572** (1992), 117–125.
44. Ordway, G. A. and Szebeni, K. Effect of olanzapine on tyrosine hydroxylase in the rat locus coeruleus. *Int. J. Neuropsychopharmacol.*, **7** (2004), 1–7.
45. Schultzberg, M., Austin, M. C., Crawley, J. N. and Paul, S. M. Repeated administration of desmethylimipramine blocks reserpine-induced increase in tyrosine hydroxylase mRNA in locus coeruleus neurons of the rat. *Mol. Brain Res.*, **10** (1991), 307–314.
46. Melia, K. R., Nestler, E. J. and Duman, R. S. Chronic imipramine treatment normalizes levels of tyrosine hydroxylase in the locus coeruleus of chronically stressed rats. *Psychopharmacology (Berl.)*, **108**:1–2 (1992), 23–26.

47. Kovachich, G. B., Frazer, A. and Aronson, C. E. Effect of chronic administration of antidepressants on alpha 2-adrenoceptors in the locus coeruleus and its projection fields in rat brain determined by quantitative autoradiography. *Neuropsychopharmacology*, **8** (1993), 57–65.
48. Bauer, M. E. and Tejani-Butt, S. M. Effects of repeated administration of desipramine or electroconvulsive shock on norepinephrine uptake sites measured by [3H]nisoxetine autoradiography. *Brain Res.*, **582** (1992), 208–214.
49. Holmes, A., Heilig, M., Rupniak, N. M., Steckler, T. and Griebel, G. Neuropeptide systems as novel therapeutic targets for depression and anxiety disorders. *Trends Pharmacol. Sci.*, **24**: 11 (2003), 580–588.
50. Bissette, G., Klimek, V., Pan, J., Stockmeier, C. A. and Ordway, G. A. Elevated concentrations of CRF in the locus coeruleus of depressed subjects. *Neuropsychopharmacology*, **28** (2003), 1328–1335.
51. Chappell, P. B., Smith, M. A., Kilts, C. D. *et al.* Alterations in corticotropin-releasing factor-like immunoreactivity in discrete rat brain regions after acute and chronic stress. *J. Neurosci.*, **6** (1986), 2908–2914.
52. Karolewicz, B., Szebeni, K., Stockmeier, C. A. *et al.* Low nNOS protein in the locus coeruleus in major depression. *J. Neurochem.*, **91** (2004), 1057–1066.
53. Karolewicz, B., Stockmeier, C. A. and Ordway, G. A. Elevated levels of the NR2C subunit of the NMDA receptor in the locus coeruleus in depression. *Neuropsychopharmacology*, **30** (2005), 1557–1567.
54. Fitzgerald, L. W., Ortiz, J., Hamedani, A. G. and Nestler, E. J. Drugs of abuse and stress increase the expression of GluR1 and NMDAR1 glutamate receptor subunits in the rat ventral tegmental area: common adaptations among cross-sensitizing agents. *J. Neurosci.*, **16** (1996), 274–282.
55. Bartanusz, V., Aubry, J. M., Pagliusi, S. *et al.* Stress-induced changes in messenger RNA levels of N-methyl-D-aspartate and AMPA receptor subunits in selected regions of the rat hippocampus and hypothalamus. *Neuroscience*, **66** (1995), 247–252.
56. Zhu, H., Karolewicz, B., Nail, E. *et al.* Analysis of $GABA_A$ receptors in the locus coeruleus in major depression. *Soc. Neurosci. Abstr.*, (2004), program No. 799. 15.
57. Crow, T. J., Cross, A. J., Cooper, S. J. *et al.* Neurotransmitter receptors and monoamine metabolites in the brains of patients with Alzheimer-type dementia and depression, and suicides. *Neuropharmacology*, **23** (1984), 1561–1569.
58. Ferrier, I. N., McKeith, I. G., Cross, A. J. *et al.* Postmortem neurochemical studies in depression. *Ann. N. Y. Acad. Sci.*, **487** (1986), 128–142.
59. Klimek, V., Rajkowska, G., Luker, S. N. *et al.* Brain noradrenergic receptors in major depression and schizophrenia. *Neuropsychopharmacology*, **21** (1999), 69–81.
60. Garcia-Sevilla, J. A., Escriba, P. V., Ozaita, A. *et al.* Up-regulation of immunolabeled alpha2A-adrenoceptors, Gi coupling proteins, and regulatory receptor kinases in the prefrontal cortex of depressed suicides. *J. Neurochem.*, **72**: 1 (1999), 282–291.
61. Ribas, C., Miralles, A., Busquets, X. and Garcia-Sevilla, J. A. Brain alpha(2)-adrenoceptors in monoamine-depleted rats: increased receptor density, G coupling proteins, receptor turnover and receptor mRNA. *Br. J. Pharmacol.*, **132**: 7 (2001), 1467–1476.

62. Mann, J. J., Stanley, M., McBride, P. A. and McEwen, B. S. Increased serotonin2 and beta-adrenergic receptor binding in the frontal cortices of suicide victims. *Arch. Gen. Psychiatry*, **43**: 10 (1986), 954–959.
63. Biegon, A. and Israeli, M. Regionally selective increases in beta-adrenergic receptor density in the brains of suicide victims. *Brain Res.*, **442**: 1 (1988), 199–203.
64. Stockmeier, C. A. and Meltzer, H. Y. Beta-adrenergic receptor binding in frontal cortex of suicide victims. *Biol. Psychiatry*, **29**: 2 (1991), 183–191.
65. Little, K. Y., Clark, T. B., Ranc, J. and Duncan G. E. Beta-adrenergic receptor binding in frontal cortex from suicide victims. *Biol. Psychiatry*, **34**: 9 (1993), 596–605.
66. De Paermentier, F., Cheetham, S. C., Crompton, M. R., Katona, C. L. and Horton, R. W. Brain beta-adrenoceptor binding sites in antidepressant-free depressed suicide victims. *Brain Res.*, **525**: 1 (1990), 171–77.
67. Yamamoto, K. and Hornykiewicz, O. Proposal for a noradrenaline hypothesis of schizophrenia. *Prog. Neuropsychopharmacol. Biol. Psychiatry*, **28**: 5 (2004), 913–922.
68. Wise, C. D. and Stein, L. Dopamine beta-hydroxylase deficits in the brains of schizophrenic patients. *Science*, **181**: 97 (1973), 344–347.
69. Wyatt, R. J., Schwartz, M. A., Erdelyi, E. and Barchas, J. D. Dopamine beta-hydroxylase activity in brains of chronic schizophrenic patients. *Science*, **187**: 4174 (1975), 368–370.
70. Farley, I. J., Price, K. S., McCullough, E. *et al.* Norepinephrine in chronic paranoid schizophrenia: above-normal levels in limbic forebrain. *Science*, **200**: 4340 (1978), 456–458.
71. Winblad, B., Bucht, G., Gottfries, C. G. and Roos, B. E. Monoamines and monoamine metabolites in brains from demented schizophrenics. *Acta Psychiatr. Scand.*, **60**: 1 (1979), 17–28.
72. Crow, T. J., Baker, H. F., Cross, A. J. *et al.* Monoamine mechanisms in chronic schizophrenia: post-mortem neurochemical findings. *Br. J. Psychiatry*, **134** (1979), 249–256.
73. Bird, E. D., Spokes, E. G. and Iversen, L. L. Dopamine and noradrenaline in post-mortem brain in Huntington's disease and schizophrenic illness. *Acta Psychiatr. Scand. Suppl.*, **280** (1980), 63–73.
74. Bridge, T. P., Kleinman, J. E., Karoum, F. and Wyatt, R. J. Postmortem central catecholamines and antemortem cognitive impairment in elderly schizophrenics and controls. *Neuropsychobiology*, **14**: 2 (1985), 57–61.
75. Friedman, J. I., Temporini, H. and Davis, K. L. Pharmacologic strategies for augmenting cognitive performance in schizophrenia. *Biol. Psychiatry*, **45**: 1 (1999), 1–16.
76. Bridge, T. P., Kleinman, J. E., Soldo, B. J. and Karoum, F. Central catecholamines, cognitive impairment, and affective state in elderly schizophrenics and controls. *Biol. Psychiatry*, **22**: 2 (1987), 139–147.
77. Bennett, J. P. Jr., Enna, S. J., Bylund, D. B. *et al.* Neurotransmitter receptors in frontal cortex of schizophrenics. *Arch. Gen. Psychiatry*, **36**: 9 (1979), 927–934.
78. Joyce, J. N., Lexow, N., Kim, S. J. *et al.* Distribution of beta-adrenergic receptor subtypes in human post-mortem brain: alterations in limbic regions of schizophrenics. *Synapse*, **10**: 3 (1992), 228–246.
79. Dean, B. [3H]RX 821002 in human dorsolateral prefrontal cortex: no changes in postmortem tissue from subjects with schizophrenia. *Psychiatry Res.*, **119**: 1–2 (2003), 25–31.

12

Norepinephrine in mood disorders

Pedro L. Delgado, M.D.
University of Texas Health Science Center, San Antonio, USA

Cristinel Coconcea, M.D.
University Hospitals of Cleveland and Case Western Reserve University School of Medicine, Cleveland, USA

Introduction

Research interest in the role of norepinephrine (NE) in the pathophysiology of mood and anxiety disorders and the therapeutic effects of antidepressant medications has waxed and waned over the past 50 years. After an intense round of research in the 1960s and 1970s, interest in NE decreased substantially in the decade between 1980 and 1990. This was due to various factors, including the introduction of selective serotonin (5-HT) reuptake inhibitors (SSRIs), the failure to find evidence of a NE deficiency in depressed or anxious patients, and the disappointing results of studies of NE receptor agonists and antagonists as treatment for major depression.

The reawakening of interest in the role of NE in depression and antidepressant effects was fueled by results from neurotransmitter depletion studies[1–7] and the clinical development of selective NE reuptake inhibitors (NRIs) such as reboxetine. Neurotransmitter depletion studies showed that the therapeutic effects of SSRIs could be transiently reversed by rapid depletion of 5-HT but not by depletion of NE.[1–3,7] Conversely, the therapeutic effects of the NRI (desipramine) could be transiently reversed by depletion of NE but not by depletion of 5-HT.[1–3,7] These key studies strongly suggest that antidepressants do not simply work by enhancing neurotransmission through a single final monoamine common pathway.

Unfortunately, progress has been slowed by the lack of available methods for direct measurement of central nervous system (CNS) NE function in living humans. Many strategies have been used over the past 40 years to try to evaluate CNS NE function, including measurement of NE and/or NE metabolites in urine, blood, and cerebrospinal fluid (CSF), neuroendocrine and neurotransmitter depletion challenge studies, and neuroimaging studies. Although the evidence suggests that some of these strategies are reflective of central NE function, none has proven to be adequate.

Brain Norepinephrine: Neurobiology and Therapeutics, ed. Gregory A. Ordway, Michael A. Schwartz and Alan Frazer. Published by Cambridge University Press.

This chapter will review data from selected human studies investigating the role of NE in the pathophysiology of major depression and the mechanisms underlying antidepressant action.

Peripheral and central markers

The study of metabolic pathways for NE suggests the use of plasma, urine, and CSF levels of NE and NE's main metabolite 3-methoxy-4-hydroxyphenylglycol (MHPG) as markers. Subsequent work showed that there were moderate correlations between CSF, plasma, and urine measures of NE.[8] Additionally, manic patients and depressed patients with "high" anxiety had elevated CSF NE compared with "low" anxiety depressed patients and neurological controls.[9] However, there is more variability in NE than in MHPG in the CSF and the origin of NE is more likely to derive from spinal neurons than from cells originating in the locus coeruleus and other rostral-projecting NE neurons.[9,10] Movement or exercise also affects NE and MHPG. It is therefore very important to address potentially small methodological differences when assessing these markers.[11]

Levels of MHPG have been measured in the urine, plasma, and CSF of patients with depression, but no clear correlations have been found between these levels and depressive symptoms. Increased, decreased and unchanged urinary concentrations of MHPG have all been reported in patients with depression.[12–16] While it is clear that NE and MHPG in CSF, blood, and urine derive from NE neurons, it remains unclear to what extent, and under which circumstances, they may provide useful information on the status of brain NE systems of interest in regard to mood and anxiety disorders.

Platelets and lymphocytes

Many studies have investigated NE receptors on platelets and lymphocytes in hope that abnormalities inherent to the NE system might be paralleled in the peripheral systems. One consistent finding was that, as in the CNS, presynaptic α_2-adrenoceptors on peripheral NE neurons also play an important physiological role in the regulation of NE release.[17] In addition, some differences in NE measures between depressed patients and controls have been described, as well as an increase in the affinity and density of α_2-adrenergic receptors on isolated human platelets from depressed patients.[18]

Peripheral α_2-adrenergic receptors located on blood platelets have been extensively investigated with a variety of radiolabeled α_2-adrenergic receptor agonist and antagonist probes. Platelet α_2-adrenergic receptor number as measured by [^{3}H]clonidine,[19] [^{3}H]dihydroergocryptine,[20] [^{3}H]epinephrine, and

[^{3}H]para-aminoclonidine[18] binding have been reported to be increased compared to healthy subjects. However, other studies utilizing [^{3}H]clonidine,[21] [^{3}H]dihydro-ergocryptine,[22] [^{3}H]yohimbine and [^{3}H]rauwolscine[23] have revealed no significant difference between healthy subjects and depressed patients.

Studies of peripheral α_2-adrenergic receptor responsiveness have been used to assess the above conflicting results in receptor number. However, these studies have also been contradictory, with some studies noting no difference between patients and controls and other studies noting a blunted cyclic adenosine monophosphate (cAMP) response.[23,24] Antidepressant drug treatment and electroconvulsive therapy are also associated with a decrease in the density of α_2-adrenergic receptors and in the affinity of ligands at these receptors in platelets from depressed patients.[25]

β-adrenergic receptors are present on blood lymphocytes. Measurement of these receptors has also yielded mixed results with some authors reporting decreased numbers[26] and others reporting no difference[27,28] between depressed patients and healthy subjects. Alternately, the cAMP response to stimulation of β-adrenergic receptors by β-adrenergic agonists has led to consistent reports of a blunted response in depressed patients when compared to healthy subjects.[26,28–31]

Neuroendocrine challenges

While there was considerable optimism in the 1980s that the neuroendocrine challenge strategy might uncover the biological origin of mood and anxiety disorders, towards the end of the decade the optimism faded. In part, this was due to the lack of specificity emerging as neuroendocrine tests were studied in increasingly broad patient groups. Additionally, as the multitude of environmental, clinical, and demographic variables that could affect the results of these tests became known, the complexity of controlling for so many variables increased, requiring either prohibitively large samples or patient samples that had been so highly selected that it was questionable if the findings could be generalized.

However, this approach has produced one of the most rigorously studied neuroendocrine challenges in psychiatry: the growth hormone response to clonidine. Clonidine, an α_2-adrenergic receptor agonist, causes an increase in the secretion of growth hormone by the anterior pituitary, and a decrease in the plasma MHPG, blood pressure, and sedation.[32] The release of growth hormone by α_2-adrenergic receptor agonists is mediated by the activation of postsynaptic α_2-adrenergic receptors within the hypothalamus, exerting a stimulatory effect on the secretion of growth hormone-releasing hormone (GHRH).[33] In fact regulation of growth hormone secretion includes many other neurotransmitters, including 5-HT and dopamine (DA). For example, depletion of both 5-HT and NE in rats leads to a weaker growth hormone response to clonidine than depletion of NE alone.[34]

One of the most consistent findings from research investigation of growth hormone response to clonidine in humans is that this response is blunted in drug-free depressed patients.[32,35,36] Acute administration of desipramine also causes an increase in the release of growth hormone. Depressed patients also show a blunted growth hormone response to desipramine, supporting the hypothesis that aspects of NE function may be different in depressed individuals.[37–39]

Unfortunately, a blunted growth hormone response to clonidine is not specific to depression, as it is also seen in patients with panic disorder,[40] generalized anxiety disorder,[41] social phobia,[42] obsessive-compulsive disorder (OCD),[43] alcohol dependence,[44] Gilles de la Tourette syndrome,[29] and schizoaffective disorder.[45] In alcoholic subjects with concurrent major depression, the blunted growth hormone response to clonidine was present during retesting, even though depression and anxiety symptoms had remitted.[46] By contrast, a study in 24 children (aged 7 to 14) with anxiety disorders and 15 age-matched control children did not find evidence of blunting, rather clonidine-stimulated growth hormone secretion was increased in the anxious children relative to the controls.[47]

In a study of patients with OCD, the growth hormone response to GHRH was normal while the response to clonidine was blunted compared with healthy controls.[43] This finding suggests that the blunted response is not simply due to a deficit at the pituitary level. There is a high correlation between the release of growth hormone by GHRH and that caused by clonidine.[48] Also, a variety of factors have been shown to affect the growth hormone response to clonidine. Some antidepressants appear to cause a diminution of the response.[42,49] These effects have been observed to last as long as 6 months after the exposure to desipramine in healthy subjects.[50] Weight loss increases the growth hormone response to clonidine[51] and menstruation and recent alcohol use reduce the response.[42,52]

Neurotransmitter depletion challenges: catecholamine depletion

Neurotransmitter depletion in humans has developed into a highly productive research tool. There are currently two methods in use: pharmacological inhibition of synthesis or dietary manipulations resulting in inhibition of synthesis. Both approaches will be described.

Pharmacological inhibition of synthesis

It is possible to specifically enhance or block a particular step in the synthetic pathways for monoamines. Through chemical modification of various precursors, it is possible to synthesize reversible inhibitors of some of the key rate-limiting enzymes. A classical example is α-methyl-p-tyrosine (AMPT), which reversibly inhibits tyrosine hydroxylase, the rate-limiting enzyme in the synthesis of NE and

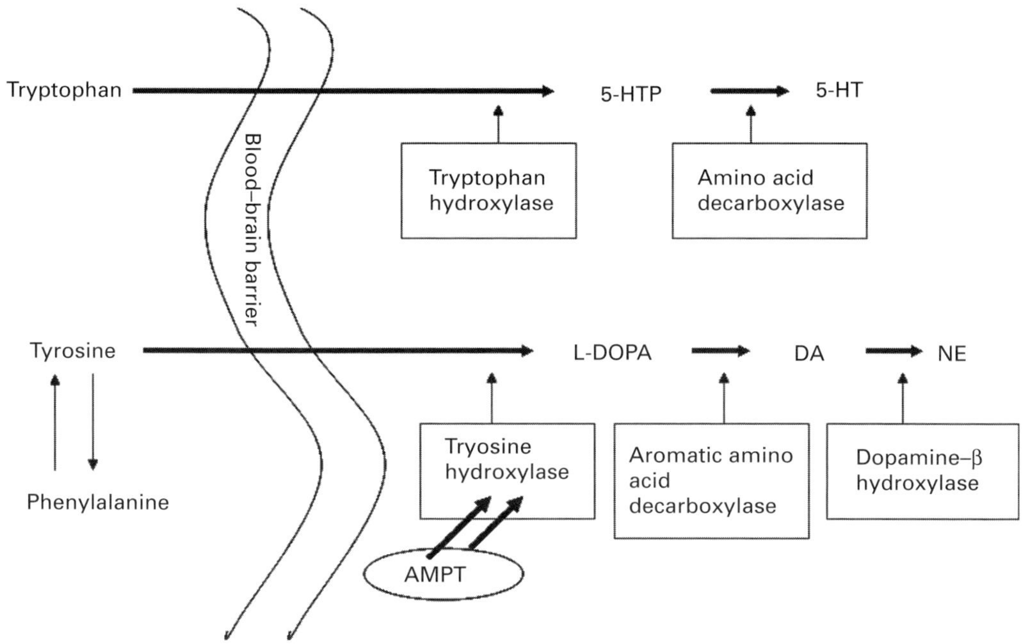

Figure 12.1 Metabolic pathways for serotonin (5-HT), dopamine (DA), and norepinephrine (NE). AMPT, α-methyl-p-tyrosine; L-DOPA, L-dihydroxyphenylalanine; 5-HTP, 5-hydroxytryptophan.

DA. Figure 12.1 shows the metabolic pathways for synthesis of NE and 5-HT. Alpha-methyl-p-tyrosine treatment in man (600 to 4000 mg/day) decreases urinary excretion of catecholamine metabolites by up to 75%.[53] A 3-gm/day dose of AMPT reduces urinary MHPG by 70% and CSF levels of the DA metabolite, homovanillic acid (HVA) by 61% with no change in the 5-HT metabolite 5-hydroxyindoleacetic acid (5-HIAA).[54] Maximum reduction of catecholamine metabolites during AMPT treatment occurs within two to three days of initiation of treatment and returns to normal within three to four days after withdrawal of the drug.[53,54]

Dietary manipulations resulting in inhibition of synthesis

Another approach to neurotransmitter depletion in humans is based on the fact that mammals are not capable of synthesizing eight key amino acids. Several of these amino acids serve as precursors for the synthesis of monoamine neurotransmitters. For example, tryptophan is the precursor for 5-HT synthesis and tyrosine for NE and DA synthesis.[55–57] It has been possible to cause a rapid depletion of any of these amino acids in humans by restricting their intake. Additionally, certain amino acids that are large and chemically neutral require transport across the blood–brain barrier in order to pass from the serum to the brain. The large neutral amino acids (LNAAs) share a common transporter and they compete for it with each other.

This means that if there is a large excess of any one or more of these LNAAs, then transport of the others is greatly reduced. By combining dietary restriction of an essential precursor amino acid with the ingestion of a large amount of competing LNAAs, investigators have successfully caused robust depletion of 5-HT[1,3] or NE/DA.[55,58] Because phenylalanine and tyrosine can be interconverted in the body and both are essential amino acids, depletion of NE and DA requires depletion of both phenylalanine and tyrosine[55–57] while depletion of 5-HT requires only tryptophan depletion.[1] Both of these paradigms involve oral consumption of about 100 gm of a solution of 14 to 15 amino acids that do not contain any of the amino acid(s) targeted for depletion. Serum levels of the targeted amino acid plummet following ingestion of the imbalanced amino acid solution, with peak serum depletion (typically ranging from 60% to 80%) occurring within 7 hours.[1,57] Because limited quantities of monoamines are available in CNS neurons, peak depletion of the targeted monoamine lags by an additional 2 hours. Because of the relative safety and tolerability, precursor amino acid depletion has emerged as the most commonly used method for monoamine depletion.

Results

Initial studies using AMPT to deplete catecholamines were poorly controlled and difficult to interpret. No significant changes in mood were reported even after a few months of administration of AMPT to patients with general medical illness.[53] Most nonpsychiatric patients and healthy subjects demonstrate little behavioral change other than sedation during AMPT treatment with rebound insomnia frequently seen for one to two days after AMPT discontinuation.[53,54,59] In an open treatment trial of AMPT in patients with essential hypertension, 6 of 20 hypertensive patients had a history of a previous depressive episode. Three of these six became agitated on AMPT, requiring drug discontinuation, although none of the subjects without a history of depression experienced depressive symptoms.[53] Alpha-methyl-p-tyrosine was also reported to decrease craving for opiates and amphetamines in a small case series,[60] decrease tic movements in Tourette's syndrome,[61] reduce oral tardive dyskinesia,[62] and potentiate antipsychotic efficacy in schizophrenia.[63] In none of these studies were depressive symptoms reported.[60–62] In a double-blind trial, AMPT reduced manic symptoms in five of seven bipolar patients in the manic phase, but two had an increase in manic symptoms.[54,64] In the same study, three of four psychotic depressed patients became more depressed during AMPT treatment.[54,64] In three depressed patients who had a therapeutic response to imipramine, AMPT had no significant effect on the antidepressant response.[23] In one of these patients, AMPT was started prior to initiation of antidepressant treatment and that patient went on to have a therapeutic response to imipramine.[23] The inconsistent results of these studies and the shift towards a focus on 5-HT in research into the pathophysiology of depression led to a more than 20-year hiatus in human research with catecholamine depletion.

The first carefully controlled studies utilizing AMPT to investigate the role of catecholamines in major depression and antidepressant action began in the early 1990s.[2] This research was a direct outgrowth of research showing that depressed patients having been successfully treated with and continuing to take the NE-selective antidepressant desipramine were generally not affected by tryptophan/5-HT depletion, while those taking SSRIs experienced a transient depressive relapse.[1] A logical conclusion from these findings was that if 5-HT depletion did not disrupt the therapeutic response to a NE reuptake inhibitor, then catecholamine depletion should.

Previous studies had simply given open-label AMPT to people over varying lengths of time.[53,54,60,64] Data from these studies also showed that AMPT had moderate sedating properties. Therefore, in order to devise an adequate placebo-controlled paradigm, an "active" placebo was chosen: diphenhydramine. Based on the known time-course of AMPT's effects on catecholamines, it was estimated that three consecutive days of AMPT exposure at 3 gm/day was going to be required to induce behavioral changes. Therefore, the testing paradigm chosen involved two test periods, 1 week apart. Each test period was five days long: a pretest day, three depletion days, and a follow-up day. The "active" placebo control was implemented by administering AMPT 1 gm three times per day for three days during one test period and diphenhydramine 50 mg three times per day on the alternate test period. Assignment to AMPT or diphenhydramine was double-blind and the sequence of testing was randomly assigned. Patients continued to take their regular dose of antidepressant medication throughout testing. However, the severity of depression seen in the first patient studied was so pronounced by the middle of the second day of AMPT exposure that the protocol was shortened to a maximum of five 1-gm doses of AMPT over 36 hours and diphenhydramine was decreased to five 25-mg doses over 36 hours. The outline of the final schedule utilized is depicted in Table 12.1.[2,7]

The first controlled human study to assess the effects of AMPT in antidepressant-treated depressed patients' action was published in 1993[2] and a second random assignment study was published in 1996.[7] The results of these studies show that the most prominent effects of AMPT are in NE-selective antidepressant-treated depressed patients.[2,7,65] Thus, patients who achieved and maintained a treatment response to desipramine were more likely to have a temporary return of depressive symptoms during exposure to AMPT 3 gm/day than those successfully treated with fluoxetine or sertraline.[2,7] These findings are the mirror image of the results with the use of rapid plasma tryptophan depletion, a method for reducing brain 5-HT synthesis.[1,3] In those studies, tryptophan depletion was more likely to cause a

Table 12.1 Catecholamine depletion schedule.

Baseline day	
09.00	Ratings of mood, plasma MHPG, HVA
Depletion day 1	
09.00	Ratings of mood, plasma MHPG, HVA
09.15; 12.00; 15.00	AMPT 1 g or diphenhydramine 25 mg
16.00	Ratings of mood
Depletion day 2	
09.00	Ratings of mood, plasma MHPG, HVA
09.15; 12.00	AMPT 1 g or diphenhydramine 25 mg
16.00	Ratings of mood
Follow-up day	
09.00	Ratings of mood, plasma MHPG, HVA

MHPG, 3-methoxy-4-hydroxyphenylglycol; HVA, homovanillic acid; AMPT, α-methyl-p-tyrosine.

temporary return of depressive symptoms in depressed patients successfully treated with and taking fluoxetine than those treated with and taking desipramine.[1,3]

On the other hand, symptomatic untreated depressed patients did not worsen during either exposure to AMPT 3 gm/day,[65] tryptophan/5-HT depletion,[66] or a simultaneous combination of AMPT and tryptophan/5-HT depletion.[67] Severity of depression did not alter the behavioral response to either type of depletion: patients did not become more depressed, whether mildly ill or severely ill. Further to this point, clinically significant depressive symptoms have not been elicited in studies administering AMPT to normal volunteers,[59,68,69] patients with OCD,[70] bipolar disorder in stable long-term remission with lithium treatment,[71] schizophrenia,[62,63] Tourette's syndrome,[61] or amphetamine-dependent subjects.[60] These results suggest that the pathophysiology of depression is unlikely to be understood by focusing on monoamine neurotransmission alone.[4,6]

In contrast to the failure to induce clinically significant depressive symptoms during exposure to AMPT, tyrosine/phenylalanine depletion, or tryptophan depletion in untreated depressives and healthy subjects, there are prominent mood responses in people with a history of prior depression or a family history of recurrent affective illness.[57,72–75] Additionally, healthy women report moderate negative mood states and irritability during phenylalanine/tyrosine depletion[57] or tryptophan depletion.[76–78] Further, patients homozygous for the ll polymorphism of the 5-HT transporter promoter were more likely to experience depressive symptoms during tryptophan depletion.[74] Table 12.2 summarizes the results of studies with tryptophan depletion or catecholamine depletion in different populations of people.

Table 12.2 Results of neurotransmitter depletion.

Subjects	Depletion type		
	NE/DA depletion	5-HT depletion	Simultaneous NE/DA/5-HT depletion
SSRI-treated MDE	++	++++	No data
NRI-treated MDE	++++	++	No data
Untreated MDE	+/−	+/−	+/−
Nondepressed history positive	+++	+++	No data
OCD	+/−	+/−	No data
Cocaine dependent	+/−	+/−	No data
Alcohol dependent	+/−	+/−	No data
Healthy controls	+	+	+/−

NE, norepinephrine; DA, dopamine; 5-HT, serotonin; SSRI, selective serotonin reuptake inhibitor; NRI, norepinephrine reuptake inhibitor; MDE, major depressive episode; OCD, obsessive-compulsive disorder. Nondepressed history positive, subjects not meeting DSM-IV criteria for major depression prior to testing who have had at least one prior MDE. Symbols: +/−, no reported mood effects; +, mild nonclinical depressive symptoms; ++, clinically significant depressive symptoms in 25% or less; +++, clinically significant depressive symptoms in 50% or less; ++++, clinically significant depressive symptoms in more than 75%.

No single study has attempted to compare the individual symptoms induced by catecholamine depletion to tryptophan depletion. A meta-analysis of previously published data with tryptophan and catecholamine depletion revealed important differences in the symptoms that emerge during each type of depletion.[79] At the date of the review, there were a total of 123 depressed patients on and off antidepressants (44 catecholamine depletion and 79 tryptophan depletion) who participated in one of several prior studies[3,7,65,72] including a subset of 11 patients who had both tryptophan and catecholamine depletion.[80] Changes from baseline during depletion were assessed with the Hamilton Depression (HAM-D) scale. Both tryptophan and catecholamine depletion led to statistically significant increases in total HAM-D score, and on the following subitems: depressed mood; loss of interest; retardation; agitation; psychic and somatic anxiety; loss of energy; loss of sexual interest; decreased appetite; decreased concentration; and helplessness, hopelessness, and worthlessness. The increases from baseline in item score for "impaired concentration," "loss of interest," and "retardation" was significantly greater for catecholamine than tryptophan depletion. The magnitude of increase in somatic anxiety was greater for tryptophan than catecholamine depletion. Table 12.3 shows these changes. Overall, it appears that the symptoms evoked by NE

Table 12.3 Maximum increase in item score on the Hamilton Depression Rating Scale (Ham-D) during depletion

Symptom	NE/DA depletion	5-HT depletion	*p value
Loss of energy	1.93	1.59	<.001
Decreased concentration	1.73	1.13	<.001
Loss of interest	1.55	0.98	<.001
Retardation	0.55	0.33	<.001
Depressed mood	0.70	0.60	NS
Somatic anxiety	0.53	0.84	<.001

NE, norepinephrine; DA, dopamine; 5-HT, serotonin.
*Symptoms listed are based on the items of the HAM-D that significantly changed from baseline during depletion based on analysis of variance. p value refers to results of t-test of mean change from baseline in item score during NE/DA and 5-HT depletion. NS, item score increased equally for both depletions.

depletion were more similar than different to those evoked by 5-HT depletion. As might have been expected, NE depletion impaired cognitive symptoms more than 5-HT depletion, while 5-HT depletion led to more somatic anxiety. The similarities and differences in symptoms suggest that NE and 5-HT share considerable overlap in modulating brain areas mediating the symptoms of depression but have less of an overlap with some symptoms of anxiety and cognition. This may have relevance to putative differences between 5-HT-selective, NE-selective, and dual-action antidepressants.

Relevance of treatment research to the biology of depression

The results from clinical trials can also be used to indirectly test biological theories of mood disorders. However, it is important to keep in mind that even when a treatment helps, this does not necessarily help us understand the pathophysiology of an illness. The results are simply confirmatory. For example, the fact that topical corticosteroids alleviate rashes does not show that rashes are due to a corticosteroid deficiency or that rashes are due to an imbalance in endogenous steroid systems.

Since the early 1950s, when monoamine oxidase (MAO) inhibitors and imipramine were serendipitously discovered to have therapeutic effects in the treatment of depression,[81] and imipramine was found to be effective for panic attacks in patients with agoraphobia,[82] a steady stream of new medications for depression have become available for clinical use. The first generations of these drugs were simply modifications of the original tricyclic or MAO inhibitor compounds. Gradually, as our understanding of the pharmacology of these compounds has evolved, newer drugs have been created with specific neurochemical effects in mind. The newer generation compounds have been designed to potently enhance NE and/or 5-HT neurotransmission without anticholinergic, antihistaminergic, or antiadrenergic properties.

All available antidepressants increase monoamine neurotransmission by blocking reuptake or inhibiting metabolism of NE and/or 5-HT, or else blocking receptors that either secondarily increase signal transduction through other receptors, increase neuronal firing rate, or stimulate release of NE and/or 5-HT.[83] These initial effects on monoamines are well characterized, and there are several converging lines of evidence supporting the view that they are essential for antidepressant efficacy. First and foremost is the ability to reverse antidepressant effects in living patients by depletion of the neurotransmitter most affected by the drug which they are taking (see Table 12.3). These data have been extensively reviewed in a prior section of this chapter (see p. 369). Second is the fact that drugs synthesized to selectively block 5-HT (i.e., selective serotonin reuptake inhibitors, or SSRIs) or NE reuptake (i.e., norepinephrine reuptake inhibitors, or NRIs) are effective in the treatment of major depression. Either one of these properties is sufficient, though neither of these properties is necessary to convey therapeutic antidepressant effects. Furthermore, the effects on histamine, acetylcholine, γ-aminobutyric acid (GABA), or ion channel blockade seen with older antidepressants have been eliminated in selective agents with no apparent reduction in efficacy. Thus it is reasonable to conclude that although such effects might contribute directly or indirectly to antidepressant efficacy in some patients, they are not either sufficient or necessary for antidepressant action.[83]

We now have a large number of selective 5-HT and NE reuptake inhibitors, some of which are listed in Table 12.4. The evolution of SSRIs has been in the direction of increasing selectivity and potency. The recently developed escitalopram is the most potent and selective SSRI available, and has demonstrated equal or greater efficacy compared to its slightly less selective and potent parent compound, citalopram.[84]

Although the ability to engineer these medications has sharpened, our understanding of the clinical differences between SSRIs and NRIs with regard to the symptom profiles that each improves is less clear. Nelson[85] published a comprehensive review of 16 studies comparing an NRI to an SSRI in patients with major depression. Most of the agents listed in Table 12.4 were used, and 1563 patients were involved in the studies. Response rates were similar for the SSRIs and NRIs (65% and 60%, respectively), and there were no consistent findings across studies when baseline symptoms that predict response were examined. Some research, however, has shown that baseline anxiety predicts preferential response to SSRIs versus NRIs.[86,87]

Table 12.4 Binding affinity of common antidepressants.

Drug	NE (nM/l)	5-HT (nM/l)	NE/5-HT ratio
Reboxetine[20]	11.0	440.0	0.03
Atomoxetine[20]	5.0	77.0	0.06
Desipramine[105]	13.7	75.5	0.18
Amitriptyline[105]	13.8	9.4	1.47
Nefazodone[105]	586.5	339.5	1.73
Imipramine[105]	15.5	5.0	3.10
Duloxetine[19]	5.6	0.7	8.00
Venlafaxine[105]	1668.0	13.5	123.56
Paroxetine[96]	45.0	0.1	450.00
Fluoxetine[96]	599.0	1.1	544.55
Fluvoxamine[96]	1427.0	2.3	620.43
Sertraline[96]	714.0	0.3	2380.00
Citalopram[96]	6190.0	1.6	3868.75
Escitalopram[96]	7841.0	1.1	7128.18

The topic of whether NRIs are as effective for anxiety symptoms and disorders as SSRIs continues to be an important focus of debate. Discussion is fueled in part by the lack of efficacy of NRIs for OCD[88–90] and posttraumatic stress disorder,[91–93] as well as a generally held perception that SSRIs are more effective in depressed patients with moderate to severe anxiety symptoms.[94] Arguing against a significant class difference are the studies showing NRIs to be effective in panic disorder[95–98] and a recent study showing that baseline Hamilton Anxiety Rating Scale score did not predict response to either sertraline or bupropion during treatment of depression.[99]

Role of NE in mood disorders and their treatment

The assumptions underlying early models for the neurobiology of depression were greatly influenced by the discovery of the CNS DA deficiency in patients suffering from Parkinson's disease[100–103] and by the remarkable therapeutic effects of the DA precursor, L-dihydroxyphenylalanine (L-DOPA).[104] It was in this historical context that the therapeutic and pharmacological effects of antidepressant drugs were discovered and the first neurobiological theories of depression and anxiety disorders were proposed. The research of the past 30 years suggests that new, more complex models are in order.

While there has been substantial progress in elucidating many of the normal functions of NE in the brain, there continues to be lack of consensus about NE's role in illness. The locus coeruleus appears to be a critical component in the brain's "attention" system.[105] Norepinephrine modulation of the prefrontal cortex explains one aspect of how we modulate our focus on rapidly changing external events.[106–108] In the periphery, NE serves to ready the body for a potential challenge in the classical "flight-or-fight" paradigm. Central nervous system NE also serves to modify the storage of emotion-laden memory, perhaps conveying aspects of emotional significance to memory of events.[109–111] Chronic activation by NE causes adaptive changes in NE neurons and postsynaptic brain areas, possibly contributing to the symptoms of stress.[112–114] Increasing NE release in healthy people sometimes results in mild anxiety.[59,68,115] Decreasing it can cause a mild decrease in attention, fatigue, and irritability.[2,23,57,59–61,65,67,68,75,103]

Many studies have identified differences in measures of NE between patients with major depression or anxiety disorders and healthy controls. One of the most consistent findings has been the blunted growth hormone response to clonidine in drug-free depressed patients (see p. 365). But this finding is highly nonspecific, being present in a wide variety of psychiatric conditions other than major depression. Few of the other markers have been comparably researched in multiple patient populations, but the studies that have been published tend to show a comparable lack of specificity for major depression.

An important consideration in reviewing studies of the NE system is to take into account the possible confounding nonspecific effects of stress on observed differences between groups. Almost all prior studies investigating NE function in depressed patients used healthy controls as a comparison. The NE system is highly stress responsive, and adaptive changes to chronic stress include many of the differences from controls identified in patient populations.[112–114,116–118] Therefore, rather than using healthy controls, studies investigating the NE system in clinical samples should be using "stressed" healthy controls or comparing individuals experiencing similar degrees of stress with and without major depression. This would help to clarify whether observed differences in NE reflect a nonspecific effect of stress or the underpinnings of a psychiatric disorder.

Neurotransmitter depletion studies suggest that diminished NE is not sufficient to cause major depression. The fact that NE depletion does not lead to depressive symptoms in healthy controls, nonantidepressant-treated clinical samples, or in depressed patients taking SSRIs[2,23,57,59–65,68] suggests that NE dysfunction is unlikely to be causal. However, the results of NE depletion studies confirm the importance of NE to the antidepressant response of NRIs relative to SSRIs.[2,7]

Biological vulnerability to major depression is heterogeneous, with many factors converging to cause the constellation of symptoms making up the syndrome. However, the brain systems underlying alterations in specific emotions, cognitions, and somatic processes are likely to be similar across patients. Considerable evidence supports the idea that specific brain networks and circuits are responsible for emotional experience, cognition, and somatic function.[119] While our current understanding of the role and function of these circuits and networks remains primitive, a rough outline is beginning to emerge.[120] Research suggests that the limbic system and areas of the frontal cortex are central to these functions of the brain. Recent data suggest that depression may be associated with localized reductions in brain morphology in the frontal cortex.[91] Therefore, it may be more appropriate to consider major depression as a neuroanatomical disease involving a constellation of factors that lead to cellular dysfunction in critical brain circuits regulating emotion and cognition.

Most antidepressant drugs are comparably effective across patients. This suggests that the drugs are causing widespread effects in many brain regions rather than simply reversing a common neurochemical abnormality. While it is well established that there are considerable interactions between NE and 5-HT neurons in the brain, it is also clear that there can be independent effects of changing neurotransmission in either system.[1–3,7] This may explain why the pharmacological profile of the antidepressant drug that the patient is taking is so important to the effects of monoamine depletion. While NE and 5-HT systems innervate similar brain regions, the pattern can involve different layers of the cortex, and both common and dissimilar effects are observed in intracellular responses.[73,113,121] It is still unclear whether the separate effects of increasing NE and 5-HT could be additive in causing antidepressant action, or if perhaps they are two separate and distinct pathways to the same endpoint.[122]

The bulk of available evidence suggests that increasing monoamines is the initial, necessary step for antidepressant action. The reasons as to why this strategy does not rapidly improve depression are still unclear, although the bulk of the evidence points to the likelihood that delayed improvment may be due to increases in neurogenesis or neuronal arborization that follow monoamine receptor activation.[113,114,123]

Conclusion

When considering the possible role of NE in mood disorders, it is clear that much more work is needed to elucidate the role of NE in normal brain function, normal stress responses, and mental disorders. It is also clear that NE has an important role to play in the initial steps of the mechanism of antidepressant action. Future work will benefit from a broader perspective, incorporating research from the cognitive neurosciences, systems neuroscience, molecular and cellular research as well as neuropsychopharmacology.

REFERENCES

1. Delgado, P. L., Charney, D. S., Price, L. H. *et al.* Serotonin function and the mechanism of antidepressant action. Reversal of antidepressant-induced remission by rapid depletion of plasma tryptophan. *Arch. Gen. Psychiatry*, **47** (1990), 411–418.
2. Delgado, P. L., Miller, H. L., Salomon, R. M. *et al.* Monoamines and the mechanism of antidepressant action: effects of catecholamine depletion on mood of patients treated with antidepressants. *Psychopharmacol. Bull.*, **29** (1993), 389–396.
3. Delgado, P. L., Miller, H. M., Salomon, R. M. *et al.* Tryptophan depletion challenge in depressed patients treated with desipramine or fluoxetine: implications for the role of serotonin in the mechanism of antidepressant action. *Biol. Psychiatry*, **46** (1999), 212–220.
4. Delgado, P. L., Moreno, F. A., Potter, R. and Gelenberg, A. J. Norepinephrine and serotonin in antidepressant action: evidence from neurotransmitter depletion studies. In *Antidepressant Therapy at the Dawn of the Third Millennium*, ed. M. Briley. (London: Martin Dunitz LTD, 1997), pp. 141–163.
5. Delgado, P. L., Price, L. H., Miller, H. M. *et al.* Rapid serotonin depletion as a provocative challenge test for patients with major depression: relevance to antidepressant action and the neurobiology of depression. *Psychopharmacol. Bull.*, **27** (1991), 321–330.
6. Delgado, P. L. Are there differences between the role of monoamine systems in the pathophysiology of depression and the mechanisms of antidepressant drug action: evidence from neurotransmitter depletion studies. *Eur. Neuropsychopharmacol.*, **7** (1997), S77.
7. Miller, H. L., Delgado, P. L., Salomon, R. M. *et al.* Clinical and biochemical effects of catecholamine depletion on antidepressant-induced remission of depression. *Arch. Gen. Psychiatry*, **53** (1996), 117–128.
8. Roy, A., Pickar, D., De Jong, J., Karoum, F. and Linnoila, M. Norepinephrine and its metabolites in cerebrospinal fluid, plasma, and urine. *Arch. Gen. Psychiatry*, **45** (1988), 849–857.
9. Post, R. M., Lake, C. R., Jimerson, D. C. *et al.* Cerebrospinal fluid norepinephrine in affective illness. *Am. J. Psychiatry*, **135** (1978), 907–912.
10. Ziegler, M. G., Wood, J. H., Lake, R. and Kopin, I. J. Norepinephrine and 3-methoxy-4-hydroxyphenyl glycol gradients in human cerebrospinal fluid. *Am. J. Psychiatry*, **134** (1977), 565–568.
11. Schatzberg, A. F. and Schildkraut, J. J. Recent studies on norepinephrine systems in mood disorders. In *Psychopharmacology: The Fourth Generation of Progress*, ed. F. E. Bloom and D. J. Kupfer. (New York: Raven Press, 1995), pp. 911–920.
12. Potter, W. Z., Muscettola, G. and Goodwin, F. K. Sources of variance in clinical studies in MHPG. In *MHPG: Basic Mechanisms and Basic Psychopathology*, ed. J. W. Maas. (New York: Academic Press, 1983), pp. 145–165.
13. Maas, J. W., Fawcett, J. A. and Dekirmenjiam, H. Catecholamine metabolism, depressive illness and drug response. *Arch. Gen. Psychiatry*, **26** (1972), 252–262.
14. Schatzberg, A. F. and Rothschild, A. J. Psychotic major depression: should it be included as a distinct syndrome in DSM-IV? *Am. J. Psychiatry*, **149** (1992), 733–745.

15. Schildkraut, J. J., Orsulak, P. J., Schatzberg, A. F. *et al.* Toward a biochemical classification of depressive disorders. I. Differences in urinary excretion of MHPG and other catecholamine metabolites in clinically defined subtypes of depressions. *Arch. Gen. Psychiatry*, **35** (1978), 1427–1433.
16. Muscettola, G., Potter, W. Z., Pickar, D. and Goodwin, F. K. Urinary 3-methoxy-4-hydroxyphenyl-glycol and major affective disorders. A replication and new findings. *Arch. Gen. Psychiatry*, **41** (1984), 337–342.
17. Langer, S. Z. Presynaptic regulation of the release of catecholamines. *Pharmacol. Rev.*, **32** (1981), 337–362.
18. Piletz, J. E., Halaris, A., Saran, A. and Marler, M. Elevated ^{3}H-para-aminoclonidine binding to platelet purified plasma membranes from depressed patients. *Neuropsychopharmacology*, **3** (1990), 201–210.
19. Garcia-Sevilla, J. A., Zis, A. P., Hollingsworth, P. J., Greden, J. F. and Smith, C. B. Platelet alpha 2-adrenergic receptors in major depressive disorder. Binding of tritiated clonidine before and after tricyclic antidepressant drug treatment. *Arch. Gen. Psychiatry*, **38** (1981), 1327–1333.
20. Siever, L. J., Kafka, M. S., Insel, T. R., Lake, C. R. and Murphy, D. L. Effect of clorgyline administration on human platelet alpha-adrenergic receptor binding and platelet cyclic AMP responses. *Psychiatry Res.*, **9** (1984), 37–44.
21. Georgotas, A., Schweitzer, J., McCue, R. E., Armour, M. and Friedhoff, A. J. Clinical and treatment effects on ^{3}H-clonidine and ^{3}H-imipramine binding in elderly depressed patients. *Life Sci.*, **40** (1987), 2137–2143.
22. Wood, K. and Coppen, A. Alpha2 adrenergic receptors in depression. *Lancet*, **1** (1982), 1121–1122.
23. Siever, L. J. Role of noradrenergic mechanisms in the etiology of the affective disorders. In *Psychopharmacology: The Third Generation of Progress*, ed. H. T. Meltzer. (New York: Raven Press, 1987), pp. 493–504.
24. Kindler, S. and Lerer, B. Norepinephrine and depression: a reappraisal. In *The Biologic Basis of Psychiatric Treatment. Progress in Basic and Clinical Pharmacology*, ed. R. Pohl and S. Gershon. (Basel: Karger, 1990), pp. 120–141.
25. García-Sevilla, J. A., Padró, D., Giralt T., Guimón, J. and Areso, P. α_2-Adrenoreceptor-mediated inhibition of platelet adenylate cyclase and induction of aggregation in major depression: effect of long-term cyclic antidepressant drug treatment. *Arch. Gen. Psychiatry*, **147** (1990), 125–132.
26. Extein, I., Tallman, J., Smith, C. C. and Goodwin, F. K. Changes in lymphocyte beta-adrenergic receptors in depression and mania. *Psychiatry Res.*, **1** (1979), 191–197.
27. Cooper, S. J., Kelly, J. G. and King, D. J. Adrenergic receptors in depression: effects of electroconvulsive therapy. *Br. J. Psychiatry*, **147** (1985), 23–29.
28. Mann, J. J., Brown, R. P., Halper, J. P. *et al.* Reduced sensitivity of lymphocyte beta-adrenergic receptors in patients with endogenous depression and psychomotor agitation. *New Engl. J. Med.*, **313** (1985), 715–720.
29. Mueller, N., Putz, A., Klages, U. *et al.* Blunted growth hormone response to clonidine in Gilles de la Tourette syndrome. *Psychoneuroendocrinology*, **19** (1994), 335–341.

30. Pandey, G. N., Sudershan, P. and Davis, J. M. Beta adrenergic receptor function in depression and the effect of antidepressant drugs. *Acta Pharmacol. Toxicol.*, **56** (1985), 66–79.
31. Ebstein, R. P., Lerer, B., Shapira, B. *et al.* Cyclic AMP second messenger amplification in depression. *Br. J. Psychiatry*, **152** (1988), 665–669.
32. Charney, D. S., Heninger, G. R., Sternberg, D. E. *et al.* Adrenergic receptor sensitivity in depression: effects of clonidine in depressed patients and healthy subjects. *Arch. Gen. Psychiatry*, **39** (1982), 290–294.
33. Frohman, L. A. and Jansson, J. O. Growth hormone releasing hormone. *Endocrine*, **7** (1986), 223–253.
34. Soderpalm, B., Andersson, L., Carlsson, M., Modigh, K. and Eriksson, E. Serotonergic influence on the growth hormone response to clonidine in the rat. *J. Neural Transm.*, **69** (1987), 105–114.
35. Matussek, N., Ackenheil, M., Hippius, H. *et al.* Effects of clonidine on growth hormone release in psychiatric patients and controls. *Psychiatry Res.*, **2** (1980), 25–36.
36. Ansseau, M., Von Frenckell, R. Cerfontaine, R. *et al.* Blunted response of growth hormone to clonidine and apomorphine in endogenous depression. *Br. J. Psychiatry*, **153** (1988), 65–71.
37. Meesters, P., Kerkhofs, M., Charles, G. *et al.* Growth hormone response after desipramine in depressive illness. *Eur. Arch. Psychiatry Neurol. Sci.*, **235** (1985), 140–142.
38. Laakmann, G., Zygan, K., Schoen, H. W. *et al.* Effects of receptor blockers (methysergide, propranolol, phentolamine, yohimbine, and prazosin) on the desipramine-induced pituitary hormone stimulation in humans. Part I: Growth hormone. *Psychoneuroendocrinology*, **11** (1986), 447–461.
39. Asnis, G. M., Lemus, C. Z. and Halbreich, U. The desipramine cortisol test – a selective noradrenergic challenge (relationship to other cortisol tests in depressives and normals). *Psychopharmacol. Bull.*, **22** (1986), 571–578.
40. Coplan, J. D., Pine, D. S., Papp, L. A. and Gorman, J. M. A view on noradrenergic, hypothalamic–pituitary–adrenal axis and extrahypothalamic corticotropin-releasing factor function in anxiety and affective disorders: the reduced growth hormone response to clonidine. *Psychopharmacol. Bull.*, **33** (1997), 193–204.
41. Abelson, J. L., Glitz, D., Cameron, O. G. *et al.* Blunted growth hormone response to clonidine in patients with generalized anxiety disorder. *Arch. Gen. Psychiatry*, **48** (1991), 157–162.
42. Siever, L. J., Trestman, R. L., Coccaro, E. F. *et al.* The growth hormone response to clonidine in acute and remitted depressed male patients. *Neuropsychopharmacology*, **6** (1999), 165–177.
43. Brambilla, F., Perma, G., Bellodi, L. *et al.* Noradrenergic receptor sensitivity in obsessive-compulsive disorders: I. Growth hormone response to clonidine stimulation. *Psychiatry Res.*, **69** (1997), 155–162.
44. Berggren, U., Fahlke, C., Norrby, A., Zachrisson, O. and Balldin, J. Subsensitive alpha-2-adrenoceptor function in male alcohol-dependent individuals during 6 months of abstinence. *Drug Alcohol Depend.*, **57** (2000), 255–260.

45. Mokrani, M. C., Duval, F., Diep, T. S., Bailey, P. E. and Macher, J. P. Multihormonal response to clonidine in patients with affective and psychotic symptoms. *Psychoneuroendocrinology*, **25** (2000), 741–752.
46. Fahlke, C., Berggren, U., Lundborg, C. and Balldin, J. Psychopathology in alcohol withdrawal: Relationship to alpha-sub-2-adrenoceptor function. *Alcohol Alcohol.*, **34** (1999), 750–759.
47. Sallee, F. R., Richman, H., Ethurman, G. *et al.* Clonidine challenge in childhood anxiety disorder. *J. Am. Acad. Child Adolesc. Psychiatry*, **37** (1998), 655–662.
48. Eriksson, E., Balldin, J., Lindstedt, G. and Modigh, K. Growth hormone response to the alpha$_2$-adrenoreceptor agonist guanfacine and to growth hormone releasing hormone in depressed patients and controls. *Psychiatry Res.*, **26** (1988), 59–67.
49. Schittecatte, M., Charles, G., Machowski, R. and Wilmotte, J. Tricyclic wash-out and growth hormone response to clonidine. *Br. J. Psychiatry*, **154** (1989), 858–863.
50. Schittecatte, M., Charles, G., Nefve, C. *et al.* Long-term downregulation of central adreno-receptor function by desipramine treatment: a clonidine study in normal subjects. *Biol. Psychiatry*, **31** (1992), 856–858.
51. Goodwin, G. M., Fairburn, C. G. and Cowen, P. The effects of dieting and weight loss on neuroendocrine responses to tryptophan, clonidine, and apomorphine in volunteers: implications for neuroendocrine investigations in depression. *Arch. Gen. Psychiatry*, **44** (1987), 952–957.
52. Matussek, N., Ackenheil, M. and Herz, M. The dependence of the clonidine growth hormone test on alcohol drinking habits and the menstrual cycle. *Psychoneuroendocrinology*, **9** (1984), 173–177.
53. Engelman, K., Horwitz, D., Jequier, E. and Sjoerdsma, A. Biochemical and pharmacologic effects of α-methyltyrosine in man. *J. Clin. Invest.*, **47** (1968), 577–594.
54. Bunney, W. E., Brodie, H. K. H., Murphy, D. L. and Goodwin, F. K. Studies of alpha-methyl-para-tyrosine, L-dopa, and L-tryptophan in depression and mania. *Am. J. Psychiatry*, **127** (1971), 872–881.
55. Biggio, G., Porceddu, M. L. and Gessa, G. L. Decrease of homovanillic, dihydroxyphenyl-acetic acid and cyclic-adenosine-3', 5'-monophosphate content in the rat caudate nucleus induced by the acute administration of an amino acid mixture lacking tyrosine and phenyl-alanine. *J. Neurochem.*, **26** (1976), 1253–1255.
56. Moja, E. A., Lucini, V., Beneditti, F. and Lucca, A. Decrease in plasma phenylalanine and tyrosine after phenylalanine-tyrosine free amino acid solutions in man. *Life Sci.*, **58** (1996), 2389–2395.
57. Leyton, M., Young, S. N., Pihl, R. O. *et al.* Effects on mood of acute phenylalanine/tyrosine depletion in healthy women. *Neuropsychopharmacology*, **22** (2000), 52–63.
58. Fernstrom, J. D., Wurtman, R. J., Hammarstrom-Wiklund, B. *et al.* Diurnal variations in plasma concentrations of tryptophan, tyrosine, and other neutral amino acids. Effects of dietary protein intake. *Am. J. Clin. Nutr.*, **32** (1979), 1912–1922.
59. McCann, U. D., Thorne, D., Hall, M. *et al.* The effects of L-dihydroxyphenylalanine on alertness and mood in α-methyl-para-tyrosine-treated healthy humans: further evidence for the role of catecholamines in arousal and anxiety. *Neuropsychopharmacology*, **13** (1995), 41–52.

60. Pozuelo, J. Suppression of craving and withdrawal in humans addicted to narcotics or amphetamines by administration of alpha-methyl-para-tyrosine (AMPT) and 5-butylpicolinic acid (fusaric acid). *Cleve. Clin. Q.*, **43** (1976), 89–94.
61. Sweet, R. D., Bruun, R., Shapiro, E. and Shapiro, A. K. Presynaptic catecholamine antagonists as treatment for Tourette syndrome. *Arch. Gen. Psychiatry*, **31** (1974), 857–861.
62. Gerlach, J. and Thorsen, K. The movement pattern of oral tardive dyskinesia in relation to anticholinergic and antidopaminergic treatment. *Int. Pharmacopsychiatry*, **11** (1976), 1–7.
63. Carlsson, A., Persoon, T., Roose, B.-E. and Walinder, J. Potentiation of phenothiazines by α-methyltyrosine in treatment of chronic schizophrenia. *J. Neural Transm.*, **33** (1972), 83–90.
64. Brodie, H. K. H., Murphy, D. L., Goodwin, F. K. and Bunney, W. E. Catecholamines and mania: the effect of alpha-methyl-para-tyrosine on manic behavior and catecholamine metabolism. *Clin. Pharmacol. Ther.*, **12** (1970), 218–224.
65. Miller, H. L., Delgado, P. L., Salomon, R. M., Heninger, G. R. and Charney, D. S. Effects of alpha-methyl-para-tyrosine (AMPT) in drug-free depressed patients. *Neuropsychopharmacology*, **14** (1996), 151–157.
66. Delgado, P. L., Price, L. H., Aghajanian, G. K. *et al.* Serotonin and the neurobiology of depression: effects of tryptophan depletion in drug-free depressed patients. *Arch. Gen. Psychiatry*, **51** (1994), 865–874.
67. Berman, R. M., Sanacora, G., Anand, A. *et al.* Monoamine depletion in unmedicated depressed subjects. *Biol. Psychiatry*, **51** (2002), 469–473.
68. McCann, U. D., Penetar, D. M., Shaham, Y. *et al.* Effects of catecholamine depletion on alertness and mood in rested and sleep deprived normal volunteers. *Neuropsychopharmacology*, **8** (1993), 345–356.
69. Krahn, L. E., Lin, S. C., Klee, G. G. *et al.* The effect of presynaptic catecholamine depletion on 6-hydroxymelatonin sulfate: a double-blind study of α-methyl-para-tyrosine. *Eur. Neuropsychopharmacol.*, **9** (1999), 61–66.
70. Longhurst, J. G., Carpenter, L. L., Epperson, C. N., Price, L. H. and McDougle, C. J. Effects of catecholamine depletion with AMPT (alpha-methyl-para-tyrosine) in obsessive-compulsive disorder. *Biol. Psychiatry*, **46** (1999), 573–576.
71. Anand, A., Darnell, A., Miller, H. L. *et al.* Effect of catecholamine depletion on lithium-induced long-term remission of bipolar disorder. *Biol. Psychiatry*, **45** (1999), 972–978.
72. Moreno, F. A., McKnight, K., Gelenberg, A. J. *et al.* Tryptophan depletion and depressive vulnerability. *Biol. Psychiatry*, **46** (1999), 498–505.
73. Benkelfat, C., Seletti, E., Mark, A., Dean, P. and Palmour, R. M. Mood-lowering effects of tryptophan depletion: enhanced susceptibility in young men at risk for major affective disorders. *Arch. Gen. Psychiatry*, **51** (1994), 687–697.
74. Moreno, F. A., Rowe, D. C., Kaiser, B. *et al.* Association between a serotonin transporter promoter region polymorphism and mood response during tryptophan depletion. *Mol. Psychiatry*, **7** (2002), 213–216.
75. Berman, R. M., Narasimhan, M., Miller, H. L. *et al.* Transient depressive relapse induced by catecholamine depletion: potential phenotypic vulnerability marker? *Arch. Gen. Psychiatry*, **56** (1999), 395–403.

76. Smith, K. A., Clifford, E. M., Hockney, R. A. and Clark, D. M. Effect of tryptophan depletion on mood in male and female volunteers: a pilot study. *Hum. Psychopharmacol.*, **12** (1997), 111–117.
77. Ellenbogen, M. A., Young, S. N., Dean, P., Palmour, R. M. and Benkelfat, C. Mood response to acute tryptophan depletion in healthy volunteers: sex differences and temporal stability. *Neuropsychopharmacology*, **15** (1996), 465–474.
78. Ellenbogen, M. A., Young, S. N., Dean, P., Palmour, R. M. and Benkelfat, C. Acute tryptophan depletion in healthy young women with a family history of major affective disorder. *Psychol. Med.*, **29** (1999), 35–46.
79. Delgado, P. L. and Moreno, F. A. Noradrenaline dysfunction in depression: implications for treatment. Presented at symposium titled, *Noradrenaline in Depression*, 22nd Annual Meeting of the CINP, Brussels, Belgium, July 10, 2000.
80. Delgado, P. L., Moreno, F. A., Onate, L. and Gelenberg, A. J. Sequential catecholamine and serotonin depletion in mirtazapine-treated depressed patients. *Int. J. Neuropsychopharmacol.*, **5** (2002), 63–66.
81. Loomer, H. P., Saunders, J. C. and Kline, N. S. A clinical and pharmacodynamic evaluation of iproniazid as a psychic energizer. *Psychiatr. Res. Rep. Am. Psychiatr. Assoc.*, **8** (1957), 129–141.
82. Klein, D. F. and Fink, M. Psychiatric reaction patterns of imipramine. *Am. J. Psychiatry*, **119** (1962), 432–438.
83. Bolden-Watson, C. and Richelson, E. Blockade by newly-developed antidepressants of biogenic amine uptake into rat brain synaptosomes. *Life Sci.*, **52** (1993), 1023–1029.
84. Gorman, J. M., Korotzer, A. and Su, G. Efficacy comparison of escitalopram and citalopram in the treatment of major depressive disorder: pooled analysis of placebo-controlled trials. *CNS Spectr.* **7**: 4 Suppl. 1 (2002), 40–44.
85. Nelson, J. C. A review of the efficacy of serotonergic and noradrenergic reuptake inhibitors for treatment of major depression. *Biol. Psychiatry*, **46** (1999), 1301–1308.
86. Aberg-Wistedt, A. A double-blind study of zimelidine, a serotonin uptake inhibitor, and desipramine, a noradrenaline uptake inhibitor, in endogenous depression. I. Clinical findings. *Acta Psychiatr. Scand.*, **66** (1982), 50–65.
87. Tyrer, P. J., Lee, I., Edwards, J. G. *et al.* Prognostic factors determining response to antidepressant drugs in psychiatric out-patients and general practice. *J. Affect Disord.*, **2** (1980), 149–156.
88. Goodman, W. K., Price, L. H., Delgado, P. L. *et al.* Specificity of serotonin reuptake inhibitors in the treatment of obsessive-compulsive disorder. Comparison of fluvoxamine and desipramine. *Arch. Gen. Psychiatry*, **47** (1990), 577–585.
89. Leonard, B. E. The comparative pharmacology of new antidepressants. *J. Clin. Psychiatry*, **54** (1993), 3–15.
90. Lelliott, P. T. and Monteiro, W. O. Drug treatment of obsessive-compulsive disorder. *Drugs*, **31** (1986), 75–80.
91. Reist, C., Kauffmann, C. D., Haier, R. J. *et al.* A controlled trial of desipramine in 18 men with posttraumatic stress disorder. *Am. J. Psychiatry*, **146** (1989), 513–516.

92. Dow, B. and Kline, N. Antidepressant treatment of posttraumatic stress disorder and major depression in veterans. *Ann. Clin. Psychiatry*, **9** (1997), 1–5.
93. Canive, J. M., Clark, R. D., Calais, L. A., Qualls, C. and Tuason, V. B. Bupropion treatment in veterans with posttraumatic stress disorder: an open study. *J. Clin. Psychopharmacol.*, **18** (1998), 379–383.
94. Blackwell, B. Side effects of antidepressant drugs. In *The American Psychiatric Association Annual Review*, ed. A. J. Frances. (Washington, DC: American Psychiatric Press, 1987), pp. 724–725.
95. Kalus, O., Asnis, G. M., Robinson, E. *et al.* Desipramine treatment in panic disorder. *J. Affect Disord.*, **21** (1991), 239–244.
96. Lydiard, B. R., Morton, W. A., Emmanuel, N. P. *et al.* Preliminary report: placebo-controlled, double-blind study of the clinical and metabolic effects of desipramine in panic disorder. *Psychopharmacol. Bull.*, **29** (1993), 183–188.
97. Villarreal, G. Desipramine plasma levels and treatment response in panic disorder. *Can. J. Psychiatry*, **40** (1995), 110–111.
98. Versiani, M., Cassano, G., Perugi, G. *et al.* Reboxetine, a selective norepinephrine reuptake inhibitor, is an effective and well-tolerated treatment for panic disorder. *J. Clin. Psychiatry*, **63** (2002), 31–37.
99. Rush, A. J., Batey, S. R., Donahue, R. M. J. *et al.* Does pretreatment anxiety predict response to either bupropion SR or sertraline? *J. Affect. Disord.*, **64** (2001), 81–87.
100. Hornykiewicz, O. Dopamine (3-hydroxytyramine) and brain function. *Pharmacol. Rev.*, **18** (1966), 925–964.
101. Ehringer, H. and Hornykiewicz, O. Verteilung von Noradrenalin and Dopamin (3-Hydroxytyramin) im Gehirn des Menschen und ihr Verhalten bei Erkrankungen des estrapyramidalen Systems. *Klin. Wschr.*, **38** (1960), 1236–1239.
102. Schildkraut, J. J. The catecholamine hypothesis of affective disorders: a review of supporting evidence. *Am. J. Psychiatry*, **122** (1965), 509–522.
103. Bunney, W. E., Jr. and Davis, J. M. Norepinephrine in depressive reactions: a review. *Arch. Gen. Psychiatry*, **13** (1965), 483–494.
104. Cotzias, G. C., Van Woert, M. H. and Schiffer, L. M. Aromatic amino acids and modification of parkinsonism. *N. Engl. J. Med.*, **276** (1967), 374–379.
105. Aston-Jones, G., Rajkowski, J. and Cohen, J. Role of locus coeruleus in attention and behavioral flexibility. *Biol. Psychiatry*, **46** (1999), 1309–1320.
106. Carli, M., Robbins, T. W., Evenden, J. L. and Everitt, B. J. Effects of lesions to ascending noradrenergic neurons on performance of a 5-choice serial reaction task in rats. Implications for theories of dorsal noradrenergic bundle functions based on selective attention and arousal. *Behav. Brain Res.*, **9** (1983), 361–380.
107. Devauges, V. and Sara, S. J. Activation of the noradrenergic system facilitates an attention shift in the rat. *Behav. Brain Res.*, **39** (1990), 19–29.
108. Friedman, J. I., Adler, D. N. and Davis, K. The role of norepinephrine in the pathophysiology of cognitive disorders: potential applications to the treatment of cognitive dysfunction in schizophrenia and Alzheimer's Disease. *Biol. Psychiatry*, **46** (1999), 1243–1252.

109. Cahill, L., Pham, C. A. and Setlow, B. Impaired memory consolidation in rats produced with β-adrenergic blockade. *Neurobiol. Learn. Mem.*, **74** (2000), 259–266.

110. Cahill, L., Prins, B., Weber, M. and McGaugh, J. L. β-Adrenergic activation and memory for emotional events. *Nature*, **371** (1994), 702–704.

111. Van Stegeren, A. H., Everaerd, W., Cahill, L., McGaugh, J. L. and Gooren, L. J. Memory for emotional events: differential effects of centrally versus peripherally acting beta-blocking agents. *Psychopharmacology*, **138** (1998), 305–310.

112. Hyman, S. E. and Nestler, E. J. Initiation and adaptation: a paradigm for understanding psychotropic drug action. *Am. J. Psychiatry*, **153** (1996), 151–162.

113. Duman, R. S., Heninger, G. R. and Nestler, E. J. A molecular and cellular theory of depression. *Arch. Gen. Psychiatry*, **54** (1997), 597–606.

114. Duman, R. S., Malberg, J. and Thome, J. Neural plasticity to stress and antidepressant treatment. *Biol. Psychiatry*, **46** (1999), 1181–1191.

115. Charney, D. S. and Redmond, D. E. Neurobiological mechanisms in human anxiety: evidence supporting central noradrenergic hyperactivity. *Neuropharmacology*, **22** (1983), 1531–1536.

116. Woodward, D. J., Moises, H. C., Waterhouse, B. D., Hoffer, B. J. and Freedman, R. Modulating actions of norepinephrine in the central nervous system. *Fed. Proc.*, **38** (1979), 2109–2116.

117. Korf, J., Aghajanian, G. K. and Roth, R. Increased turnover of norepinephrine in the rat cerebral cortex during stress: role of the locus coeruleus. *Neuropharmacology*, **12** (1973), 933–938.

118. Valentino, R. J., Foote, S. L. and Page, M. E. The locus coeruleus as a site for integrating corticotropin-releasing factor and noradrenergic mediation of stress responses. *Ann. N. Y. Acad. Sci.*, **697** (1993), 173–188.

119. Panksepp, J. Towards a general psychobiological theory of emotions. *Behav. Brain Sci.*, **5** (1982), 407–467.

120. Halgren, E. and Marinkovic, K. Neurophysiological networks integrating human emotions. In *The Cognitive Neurosciences*, ed. M. S. Gazzaniga. (Cambridge: MIT Press, 1995), pp. 1137–1151.

121. Van Bockstaele, E. J., Peoples, J. and Valentino, R. J. Anatomic basis for differential regulation of the rostrolateral peri-locus coeruleus region by limbic afferents. *Biol. Psychiatry*, **46** (1999), 1352–1363.

122. Stahl, S. M. Are two antidepressant mechanisms better than one? *J. Clin. Psychiatry*, **58** (1997), 339–340.

123. Santarelli, L., Saxe, M., Gross, C. *et al.* Requirements of hippocampal neurogenesis for the behavioral effects of antidepressants. *Science*, **301** (2003), 805–809.

13

Noradrenergic pathology and pain

Antti Pertovaara, M.D., Ph.D.
Institute of Biomedicine, University of Turku, Turku, Finland

Introduction

As early as a century ago, Weber[1] showed that adrenaline applied to the spinal cord of the cat attenuates a thermally evoked withdrawal response. A more extensive study of mechanisms underlying norepinephrine-induced pain suppression did not really start until the 1970s, following the development of several novel adrenergic compounds with a potential for clinical use. The role of various aspects of adrenergic mechanisms in pain control has been reviewed previously by a number of authors.[2–6] This chapter attempts to present what is currently known on pain modulatory actions of norepinephrine at various levels of the nervous system in physiological and pathophysiological conditions.

Peripheral mechanisms of noradrenergic pain modulation

Peripheral norepinephrine has only a minor role in modulation of pain in physiological conditions. Administration of norepinephrine in the skin of healthy subjects does not evoke pain, although it may induce a selective hyperalgesia to thermal stimulation.[7] However, in pathophysiological conditions, peripheral norepinephrine has a significant influence on nerve endings mediating pain. This is shown by the finding that administration of norepinephrine in inflamed or neuropathic skin in humans aggravates pain and hyperalgesia.[8–10]

Plasticity in peripheral noradrenergic systems following injury or inflammation

In line with psychophysical observations, neurophysiological studies in experimental animals indicate that under physiological conditions nociceptive primary afferent fibers are little, if at all, influenced by norepinephrine, sympathetic stimulation, or synthetic noradrenergic compounds.[11,12] However, following nerve injury

Brain Norepinephrine: Neurobiology and Therapeutics, ed. Gregory A. Ordway, Michael A. Schwartz and Alan Frazer. Published by Cambridge University Press.

particularly nociceptive C fibers, and to a lesser extent nociceptive A-delta fibers, are excited by norepinephrine and sympathetic stimulation.[12–14] Also inflammation leads to circumstances in which norepinephrine and sympathetic stimulation excite nociceptors.[15] These sensitivity changes are associated with injury-induced sympathetic sprouting in the dorsal root ganglion[16,17] and in the skin.[18] Additionally, novel adrenoceptors, particularly α_{2A}-adrenoceptors, develop in the peripheral terminals, perineurally at the injured site in the axon, and in the dorsal root ganglion.[19–21]

Evidence for pain facilitatory role of peripheral α_2-adrenoceptors

The novel α_2-adrenoceptors in the peripheral nerves may have a pronociceptive role as suggested by the electrophysiological findings that the excitatory effects induced by sympathetic stimulation and norepinephrine were attenuated by α_2-adrenoceptor antagonists, and that the excitation of nociceptors induced by norepinephrine was mimicked by α_2-adrenoceptor agonists.[12,15,22–24] The hypothesis that peripheral α_2-adrenoceptors have a pain-inducing role in hyperalgesic skin is also supported by the following behavioral findings. Both the norepinephrine-induced and a nerve injury-induced hypersensitivity were attenuated by an α_2-adrenoceptor antagonist but not by an α_1-adrenoceptor antagonist.[25–27] Rekindling of mechanical hypersensitivity in nerve-injured animals by cutaneous administration of an α_2-adrenoceptor agonist, but not by an α_1-adrenoceptor agonist, supports the interpretation that peripheral α_2-adrenoceptors mediate the norepinephrine-induced pain from the hyperalgesic skin.[28] The subtype A of the peripheral α_2-adrenoceptor may be involved in mediating the norepinephrine-induced enhancement of pain. This is suggested by the finding that a peripherally acting α_2-adrenoceptor antagonist selectively attenuated nerve injury-induced heat hyperalgesia in wild-type mice, and a knockout of α_{2A}-adrenoceptors selectively attenuated the development of heat hyperalgesia after nerve injury.[29]

Evidence for pain inhibitory role of peripheral α_2-adrenoceptors

Although there is a considerable amount of evidence suggesting that the norepinephrine-induced pain response in the periphery is due to activation of α_2-adrenoceptors, a number of studies report contrasting findings suggesting that it is the α_1-adrenoceptor that mediates the norepinephrine-induced pain response, and the peripheral α_2-adrenoceptor has a pain suppressive role. This is suggested by the neurophysiological finding that ectopic activity of injured nerve fibers was suppressed by an α_1-adrenoceptor antagonist but not by an α_2-adrenoceptor antagonist.[26] Also, an α_1-adrenoceptor agonist more effectively activated C-fiber nociceptors innervating a partially denervated skin than an α_2-adrenoceptor agonist.[30] The following behavioral results support the hypothesis that peripheral α_2-adrenoceptors have a pain inhibitory role. Peripheral administration of an

α_2-adrenoceptor agonist attenuated hypersensitivity in inflammatory and neuropathic conditions.[21,31–33] In line with this, topical administration of norepinephrine increased pain whereas an α_2-adrenoceptor agonist relieved pain in human patients with a reflex sympathetic dystrophy.[8] Moreover, intra-articular administration of a low dose of an α_2-adrenoceptor agonist significantly attenuated postoperative pain after surgery of the knee joint in humans.[34]

Noradrenergic pain modulation in the periphery: puzzling questions

Peripheral norepinephrine has minimal influence on pain in physiological conditions, but in inflamed and neuropathic conditions it aggravates pain. Sympathetic sprouting and development of novel adrenoceptors accompany this change in the function of peripheral norepinephrine. It has been postulated that norepinephrine released from sympathetic sprouts and acting on novel adrenoceptors may contribute to maintenance of chronic pain and hyperalgesia.[35] It is still a matter of debate which subtype of adrenoceptor in the periphery contributes to aggravation of pain and which one to suppression of pain. Also, there is conflicting evidence on whether norepinephrine in the periphery acts directly on peripheral nociceptors independent of the sympathetic nerve fibers,[36,37] indirectly on nociceptors by releasing prostaglandins from the postganglionic sympathetic nerve fibers,[38,39] or both.

Spinal mechanisms of noradrenergic pain modulation

Since the spinal dorsal horn is a critical link for all ascending pain pathways and since the spinal cord receives strong innervation from descending noradrenergic pathways, the pain modulatory action of norepinephrine has been more extensively studied in the spinal cord than elsewhere. There is strong evidence indicating that norepinephrine has an important role in suppressing pain at the spinal level, as indicated by the following findings.

Spinal norepinephrine suppresses nociception: behavioral evidence

There is abundant behavioral evidence from pharmacological studies in experimental animals indicating that administration of norepinephrine in the spinal cord produces a dose-dependent antinociceptive effect and that the norepinephrine-induced antinociception is due to action on α_2-adrenoceptors.[40–44] Epidural administration of an α_2-adrenoceptor agonist to the spinal cord has proved effective in alleviating pain in humans, also.[45,46] Synthetic α_2-adrenoceptor agonists, unlike norepinephrine, penetrate the central nervous system following systemic administration. Therefore, systemic administration of synthetic adrenergic compounds mimics central administration of norepinephrine. Like centrally administered norepinephrine, various α_2-adrenoceptor agonists administered systemically have produced significant dose-dependent antinociception.[47–51] α_2-adrenoceptor agonists with a preferential affinity for the α_{2A}-subtype were particularly effective in producing antinociception.[52] In line with this information, studies with knockout mice indicated that the α_{2A}-adrenoceptor has a major role in the pain suppressive effect induced by α_2-adrenergic compounds,[53–55] although the α_{2C}-adrenoceptor is also likely to contribute to pain regulation.[56] The attenuation of spinal nociceptive reflex and neuronal responses produced by systemically administered α_2-adrenoceptor agonists was independent of supraspinal structures, supporting the concept that their spinal antinociceptive effect is predominantly due to a direct action on the spinal cord.[57,58]

Noradrenergic pain inhibitory feedback loop

Intrinsic norepinephrine in the spinal cord is released from descending axons originating in noradrenergic nuclei of the brain stem such as the locus coeruleus, the A5 nucleus, and the A7 nucleus.[59] Stimulation of these noradrenergic structures produces spinal release of norepinephrine[60] and spinal antinociception that is attenuated by α_2-adrenoceptor antagonists.[2,5] The noradrenergic brain stem nuclei are involved also in release of norepinephrine induced by peripheral stimulation, as shown by the finding that noxious peripheral stimulation produced a spinal release of norepinephrine in animals with an intact spinal cord but not in spinalized animals.[61,62] Also a direct activation of some nonadrenergic structures in the brain stem, such as the periaqueductal gray, produced spinal antinociception that was associated with spinal release of norepinephrine.[63] This is explained by a recruitment of noradrenergic nuclei with descending projections.[64] The activation of descending noradrenergic pathways by peripheral noxious stimulation forms a negative feedback system that suppresses ascending pain signals at the spinal level.

The intrinsic noradrenergic pain regulatory system in the spinal cord has a low tonic activity, since α_2-adrenoceptor antagonists[65] or knockouts of various subtypes of α_2-adrenoceptors[66] did not consistently increase spinal pain responses to brief noxious stimulation in otherwise nonpainful conditions. A knockout of the *dopamine* β*-hydroxylase* gene led to absence of norepinephrine and to a mild hypersensitivity to acute thermal stimulation, but none to mechanical stimulation,[67] supporting the concept that tonic noradrenergic pain regulation has a weak and selective effect. During persistent stimulation, however, the role of noradrenergic pain regulation is more significant. This is shown by the finding that, following sustained noxious stimulation, an α_2-adrenoceptor antagonist or a knockout of the α_{2A}-adrenoceptor markedly increased spinal nociceptive responses. This finding indicates an involvement of noradrenergic feedback inhibition in the regulation of sustained pain.[68,69]

Spinal noradrenergic pain modulation in pathophysiological conditions

Inflammatory pain

In experimental arthritis, norepinephrine turnover in the spinal cord was increased indicating recruitment of descending noradrenergic pain modulatory pathways.[70] In line with this, a lesion of the noradrenergic locus coeruleus nucleus facilitated inflammatory hypersensitivity.[71] Painful inflammatory conditions may increase the number of spinal α_2-adrenoceptors.[72] These changes may underlie the finding that in inflammatory conditions the antinociceptive potency of systemically and spinally administered α_2-adrenoceptor agonists was enhanced.[73–77] The enhanced antinociceptive potency of an α_2-adrenoceptor agonist in animals with inflammation was due to a direct action on the spinal cord.[77,78] Unexpectedly, a knockout of α_{2A}-adrenoceptors did not influence development of inflammatory hyperalgesia induced by carrageenan treatment of the hindpaw,[79] although inflammation was expected to increase noradrenergic feedback inhibition. On the other hand, a knockout of α_{2A}-adrenoceptors did enhance hypersensitivity in animals with a neurogenic inflammation caused by capsaicin.[69] In the cited studies, carrageenan was probably a weaker stimulus than capsaicin as based on signs of spontaneous pain behavior. Thus, an appropriately strong inflammatory stimulus is probably needed to activate noradrenergic feedback inhibition. Moreover, the α_{2A}-adrenoceptor is significantly involved in noradrenergic inhibition of hypersensitivity.[69] Further studies are needed to determine a possible contribution of other adrenoceptor subtypes to intrinsic regulation of hypersensitivity in animals with inflammation.

Neuropathic pain

A peripheral nerve injury induces a neuropathic condition that involves hypersensitivity. Experimental animals with neuropathic hypersensitivity had a decrease of immunoreactivity for the α_{2A}-, but not α_{2C}-adrenoceptor in the spinal cord.[80] However, a peripheral nerve injury did not impair the antinociceptive efficacy of α_2-adrenergic compounds; if anything, it enhanced the central antinociceptive action of α_2-adrenergic compounds in some experimental models.[66,81–84] A knockout of the α_{2A}-, α_{2B}-, or α_{2C}-adrenoceptor had no effect on the development of neuropathic hypersensitivity following a peripheral nerve injury,[66] except for an α_{2A}-adrenoceptor-dependent component of thermal hypersensitivity of presumably peripheral origin.[29] In contrast, one model of peripheral neuropathy produced an increase in spinal norepinephrine, and this was associated with an α_2-adrenoceptor antagonist-induced increase of hypersensitivity.[85] It has also been reported that an α_2-adrenoceptor antagonist produced neuropathic hypersensitivity in those few animals that did not develop it following a peripheral nerve injury, whereas it had no effect in the majority of animals that developed neuropathic hypersensitivity.[86] These findings are partly contradictory, but in general they suggest that noradrenergic mechanisms do not have a major role in the development of neuropathic symptoms. The differences in results may be due to the use of different experimental models of neuropathy and also to possible genetic differences in noradrenergic innervation between strains.[87]

Neurophysiological studies on spinal mechanisms underlying noradrenergic analgesia

Early neurophysiological studies indicated that iontophoretic administration of norepinephrine in the spinal dorsal horn suppressed responses of spinal interneurons to afferent nerve stimulation.[88] Later on, it was shown that the norepinephrine-induced suppression was relatively selective for nociceptive responses of spinal dorsal horn neurons.[89,90] This norepinephrine-induced suppression was reversed by an α_2-adrenoceptor antagonist and mimicked by an α_2-adrenoceptor agonist.[91] In contrast to sensory neurons, microiontophoresis of norepinephrine decreased the threshold for glutamate-evoked motoneuron firing, indicating that norepinephrine has a facilitatory action on spinal motoneurons.[92] This is in line with the finding that, in the spinal ventral horn, norepinephrine facilitated nociceptive spinal reflex responses, whereas, in the spinal dorsal horn, norepinephrine suppressed nociceptive reflex responses.[93]

Presynaptic inhibition

The finding that α_2-adrenoceptor agonists, as well as norepinephrine, selectively attenuated nociceptive inputs to deep spinal dorsal horn neurons with convergent (nociceptive and mechanoreceptive) inputs supports the hypothesis that the pain inhibitory action was presynaptic to the deep spinal dorsal horn neurons.[57,89,90,94] Lack of a norepinephrine-induced effect on passive membrane properties of deep spinal dorsal horn neurons also fits the hypothesis that the norepinephrine-induced effect is rather pre- than postsynaptic to the deep spinal dorsal horn neurons.[95] The selective antinociceptive effect presynaptic to deep spinal dorsal horn neurons could be due to a suppressive action on central terminals of nociceptive primary afferent fibers. It has been shown that the release of neurotransmitters from terminals of nociceptive primary afferent fibers was decreased by norepinephrine and that the α_{2A}-adrenoceptor might have an important role in this presynaptic inhibitory effect.[96–98] Recent whole-cell patch-clamp studies performed in spinal cord slices give direct electrophysiological evidence for the concept that presynaptic mechanisms in the central terminals of primary afferent nociceptive fibers have an important role in antinociception induced by norepinephrine and α_2-adrenoceptor agonists. Namely, following administration of norepinephrine or an α_2-adrenoceptor agonist, postsynaptic responses evoked by dorsal root stimulation in lamina II neurons of the spinal dorsal horn were reduced, without influence on direct activation of neurons by excitatory amino acids.[99,100] In line with this, the responses evoked by direct application of an excitatory amino acid in a population of nociceptive spinal projection neurons were uninfluenced by norepinephrine.[91]

Immunohistochemical studies indicate that the primary localization of α_{2A}-adrenoceptors in the spinal cord is on the central terminals of capsaicin-sensitive, substance P-containing primary afferent fibers.[80] This finding supports the concept that a selective inhibitory effect on central terminals of nociceptive primary afferent fibers, through action on the α_{2A}-adrenoceptor, has a major role in noradrenergic pain suppression at the spinal cord level. Spinal distribution of the α_{2C}-adrenoceptor, which may also contribute to pain modulation,[56] was very different from that of the α_{2A}-adrenoceptor.[80,101] The α_{2C}-adrenoceptor was located on axon terminals of spinal interneurons that were likely to be excitatory ones. Additionally, the spinal interneurons with axon terminals containing the α_{2C}-adrenoceptor innervated nociceptive (neurokinin-1 receptor-containing) neurons with verified ascending projections to the medulla.[101] These anatomical findings support the hypothesis that the spinal α_{2C}-adrenoceptor also contributes to noradrenergic pain modulation by inhibiting release of excitatory amino acids from spinal interneurons that have a pronociceptive effect.

Postsynaptic inhibition

Postsynaptic mechanisms may also contribute to noradrenergic pain inhibition in the spinal cord, as suggested by anatomical and physiological evidence. Direct catecholaminergic innervation of the cell bodies of spinothalamic tract neurons fits the hypothesis that norepinephrine modulates ascending pain signals in the spinal dorsal horn due to a postsynaptic mechanism.[102] The role of postsynaptic mechanisms is also supported by the electrophysiological finding that a large majority of neurons in the substantia gelatinosa (lamina II) of the spinal dorsal horn were hyperpolarized due to an α_2-adrenoceptor-mediated action following administration of norepinephrine.[103] Also, a decrease of spontaneous activity[57,88,104] and neuronal discharge evoked by direct administration of an excitatory amino acid in a population of deep spinal dorsal horn neurons[88] by norepinephrine or an α_2-adrenoceptor agonist suggest that the noradrenergic pain inhibition was partly due to a postsynaptic action. It should be noted that since deep spinal dorsal horn neurons receive nociceptive inputs through the superficial layers, the selective norepinephrine-induced inhibition of nociceptive inputs to convergent deep spinal dorsal horn neurons might be explained by a postsynaptic suppression of nociceptive interneurons in the substantia gelatinosa, as well as by presynaptic inhibition of central terminals of primary afferent nociceptive fibers.

Activation of inhibitory interneurons

Superficial laminae of the spinal cord also contain some small, low-threshold units that are activated both by norepinephrine and by stimulation of the periaqueductal gray that is considered to be an important source of descending inhibitory controls.[105] These small units activated by norepinephrine are likely to be inhibitory interneurons of the substantia gelatinosa that act on pain-relay neurons of the spinal cord. Furthermore, norepinephrine release in the spinal cord may enhance γ-aminobutyric acid (GABA)ergic and glycinergic inhibitory synaptic transmission in the substantia gelatinosa through activation of presynaptic α_1-adrenoceptors.[106] This increase in efficacy of spinal inhibition may potentially contribute to norepinephrine-induced pain suppression.

Supraspinal mechanisms of noradrenergic pain modulation

Catecholaminergic brain stem nuclei innervate several supraspinal structures that are implicated in mediation and regulation of pain.[5] Various types of α-adrenoceptors are located in these supraspinal structures.[5,107] The pain regulatory role of supraspinal norepinephrine and adrenoceptors has been studied by microinjecting adrenergic compounds into the brain as described below.

The periaqueductal gray

Spinal antinociceptive effect induced by electrical or chemical stimulation of the periaqueductal gray of the midbrain was associated with spinal release of norepinephrine[63] and reduced by spinal administration of α_2-adrenoceptor antagonists.[108,109] However, direct administration of an α_2-adrenoceptor agonist into the periaqueductal gray had no significant pain modulatory effect.[110,111] These findings suggest that noradrenergic mechanisms within the periaqueductal gray are not involved in spinal antinociception, but the periaqueductal gray activates other noradrenergic nuclei of the brain stem that contribute to its descending antinociceptive action, and this involves α_2-adrenoceptors at the spinal cord level.

The rostroventromedial medulla

Stimulation of the medullary raphe magnus nucleus in the rostroventromedial medulla also produces spinal release of norepinephrine[112] and antinociception that is reduced by spinal administration of α_2-adrenoceptor antagonists.[6,113] Reciprocal connections between the raphe magnus and the noradrenergic A7 nucleus may be enough to explain the noradrenergic component of the spinal pain inhibitory action induced by stimulation of the raphe nuclei.[114,115] Microinjections of adrenergic compounds in the raphe nuclei have given somewhat conflicting results. An

α_2-adrenoceptor agonist[116,117] and an α_1-adrenoceptor antagonist in the raphe magnus nucleus had a spinal antinociceptive action, whereas an α_2-adrenoceptor antagonist and an α_1-adrenoceptor agonist had hyperalgesic actions.[117] Moreover, an α-adrenoceptor antagonist in the raphe magnus nucleus increased spinal release of norepinephrine.[118] These results suggest that noradrenergic receptors in the raphe magnus nucleus are involved in pain modulatory actions and that the norepinephrine-induced action in the raphe magnus nucleus varies in a complex way depending on the subtype of adrenoceptor. Another series of studies using a more selective drug reported that the spinal antinociceptive action induced by an α_2-adrenoceptor agonist in the raphe magnus nucleus could be explained by the spread of the drug to a spinal site of action.[110,119]

The lateral reticular nucleus of the medulla

The lateral reticular nucleus of the medulla overlaps with the catecholaminergic A1 cell group. Among brain stem sites where stimulation produces antinociception, production is reduced by spinal administration of α_2-adrenoceptor antagonists.[120] This spinal antinociceptive effect may be mediated by other brain stem nuclei.[5] Microadministration of α_2-adrenoceptor agonists in the lateral reticular nucleus of the medulla produced hyperalgesic actions in one study.[121] Another study reported no effect on spinal nociception by an α_2-adrenoceptor agonist or antagonist in the lateral reticular nucleus,[78,122] except that following inflammation, an α_2-adrenoceptor antagonist had a spinal antihyperalgesic effect.[78] These observations suggest that the descending noradrenergic pain inhibitory feedback loop may involve an α_2-adrenergic disinhibitory synapse at the medullary level; i.e., norepinephrine acting on α_2-adrenoceptors in the lateral reticular nucleus of the medulla may enhance ascending pain signals due to a spinal action, whereas a decrease of noradrenergic inputs on medullary α_2-adrenoceptors may have a spinal antinociceptive action.

The locus coeruleus

The locus coeruleus in the ponto-mesencephalic junction has an important role in noradrenergic pain modulation. This is shown by several studies in which its chemical or electrical activation produced antinociception that was reduced by spinal administration of α_2-adrenoceptor antagonists.[2,5] However, conflicting results have been reported concerning the pain modulatory role of adrenoceptors within the locus coeruleus. One study reported that a microinjection of an α_2-adrenoceptor agonist in the locus coeruleus produced spinal antinociception that was reduced by an α_2-adrenoceptor antagonist administered either spinally or in the locus coeruleus.[123] These authors concluded that activation of noradrenergic,

α_2-adrenergic mechanisms in the locus coeruleus produced analgesia that was mediated by norepinephrine in the spinal cord. In contrast, another group reported that an α_2-adrenoceptor agonist in the locus coeruleus did not modulate spinal reflex responses to phasic pain stimuli.[124,125] The spinal antinociceptive action in these studies was obtained only when the α_2-adrenoceptor agonist was expected to act directly on the spinal cord.[124,125] Since other studies have shown an α_2-adrenergic hyperpolarization of the cell bodies in the locus coeruleus,[126] it might be expected that activation of α_2-adrenoceptors in the locus coeruleus would suppress coerulo-spinal noradrenergic pain regulation, and that this suppression could be observed with a tonic pain stimulus that activates the noradrenergic feedback circuitry. Indeed, this was the case, since pain behavior induced by a tonic stimulus was enhanced by a presumably locally acting dose of an α_2-adrenoceptor agonist in the locus coeruleus.[124] Concerning variability in results between different laboratories, one particular cause for differences in coerulo-spinal pain modulation is the genetic differences in noradrenergic innervation between strains of animals.[87] Additionally, the well-known supraspinally mediated sedative effects of α_2-adrenergic drugs due to action on α_2-adrenoceptors in the locus coeruleus[127] provide a significant confounding factor in behavioral pain studies.

The thalamus

The results of a neurophysiological study indicated that nociceptive neurons in the medial thalamus, a presumably critical relay for the affective-motivational component of pain, were more sensitive to α_2-adrenergic antinociception than spinal neurons with verified projections to the medial thalamus.[57] An accompanying psychophysical study in humans reported that the affective-motivational component of pain was considerably more sensitive to suppression by a systemically administered α_2-adrenoceptor agonist than the sensory-discriminative component of pain.[128] Together these results suggest that a medial thalamic noradrenergic action contributes to attenuation of the affective component of pain. In line with this, another neurophysiological study showed that ascending projections of noradrenergic locus coeruleus neurons produce a dual modulation of nociceptive responses in medial thalamic neurons,[129] i.e., a facilitatory influence due to action on α_1-adrenoceptors and an inhibitory one due to action on α_2-adrenoceptors. In the ventroposterolateral thalamus that is considered an important relay for the sensory-discriminative component of pain and that also receives noradrenergic innervation,[130] α_2-adrenoceptors appeared not to be significantly modulating pain according to one study.[131] The role of noradrenergic pain modulatory mechanisms in other brain areas, beyond the thalamus, has been little studied and is not well understood.

Sympatho–adrenal system in pain modulation

Postganglionic sympathetic nerve fibers

In healthy subjects, the sympathetic nervous system has little effect on pain, but following a nerve-injury it may have a significant and complex role in regulation of pain.[132] In nerve-injured patients, the postganglionic sympathetic nerve fibers may interact with afferent neurons and induce activity in nociceptors. This pathologic interaction acts via norepinephrine released from sympathetic terminals and novel adrenoceptors on the nociceptive nerve fibers.[35] This pathological mechanism may contribute to development of chronic allodynia, hyperalgesia, and spontaneous pain. In some cases these nerve injury-induced symptoms are relieved by sympathectomy.[132] On the other hand, sympathectomy per se may sensitize peripheral nociceptors to circulating norepinephrine[133] and this sensitization may lead to post-sympathectomy neuralgia.[134] Sympathectomy per se also increased the peripheral antinociceptive action of a peripherally acting α_2-adrenoceptor agonist, which may provide a possibility to relieve sympathectomy-induced hyperalgesia without central side effects.[33]

The adrenal medulla

It has been reported that transplantation of adrenal medullary tissue into the subarachnoid space can reduce pain sensitivity in physiological and pathophysiological conditions.[135] This antinociceptive action by adrenal medullary grafts is at least partly due to release of cathecholamines, since the antinociceptive effect was attenuated by an α-adrenoceptor antagonist.[135] However, the antinociceptive effect by an adrenal medullary graft may be variable, since the effect has not been reproduced in all laboratories.[136] The adrenal medulla is also implicated in a pain facilitatory mechanism that involves the vagus nerve. This is shown by the finding that the hyperalgesic action induced by a vagotomy was reversed by denervation of the adrenal medulla.[137] It still has not been established whether epinephrine, norepinephrine, or some other hormone released by the adrenal medulla is involved in this hyperalgesic action.

Interactions of noradrenergic pain modulatory actions with other neurotransmitters

There is plenty of evidence indicating that the antinociceptive action by α_2-adrenergic compounds has a significant interaction with opiate-mediated antinociception. The antinociceptive effect induced by centrally administered morphine has been reduced by spinally administered α_2-adrenoceptor antagonists.[5,6] An

α_2-adrenoceptor agonist has potentiated the antinociceptive actions induced by morphine, and vice versa, probably due to an interaction at the spinal cord level.[73,138–140] In line with this, the inflammation-induced enhancement of the antinociceptive action by an α_2-adrenoceptor agonist was abolished in animals with a knockout of the opioid μ-receptor.[141] However, the α_{2A}-adrenoceptor was not necessary for the antihyperalgesic action induced by an opioid receptor agonist.[79] The finding that the antihyperalgesic action of an α_2-adrenoceptor agonist was attenuated by a small dose of an opioid receptor antagonist injected into the inflamed paw also suggests involvement of an opioidergic link in peripheral α_2-adrenergic actions.[31]

Serotonergic and noradrenergic pain modulatory systems were shown to interact at the spinal cord level to produce a more powerful antinociception.[142,143] A synergistic interaction at the spinal cord level was shown to take place between α_2-adrenoceptor agonists and serotonin $5HT_2$- and $5\text{-}HT_{1B}$- receptor actions.[144] Noradrenergic pain modulation interacted also with other neurotransmitters. For example, adenosine receptor agonists,[145,146] nitric oxide[147] and cholinergic compounds[148] enhanced α_2-adrenergic antinociception. Moreover, the antinociceptive action induced by a $GABA_B$ receptor agonist depended on activation of descending noradrenergic pathways.[149]

The role of noradrenergic mechanisms in anesthesia

The role of noradrenergic mechanisms in clinical anesthesia has been reviewed in detail elsewhere.[150] The following is only a brief outline of selected aspects on this topic. A decrease in activity of locus coeruleus neurons by activation of α_2-adrenoceptors leads to sedation.[127,151] This sedative effect in the locus coeruleus dissociates from the spinal antinociceptive effect induced by α_2-adrenoceptor agonists.[124–126] When an α_2-adrenoceptor agonist is administered systemically, it induces both sedation and antinociception, and provides a possibility to perform minor operations without general anesthetics. Therefore, α_2-adrenoceptor agonists have been widely used in veterinary anesthesia. Moreover, the effects induced by an α_2-adrenoceptor agonist can be quickly and effectively reversed by an α_2-adrenoceptor antagonist.[152] Thus, the animals are fully awake within a few minutes after the operation, an additional advantage in minor veterinary surgery.

An α_2-adrenoceptor agonist was shown to reduce the dose of general anesthetic required for surgical level of anesthesia.[153] Since then, α_2-adrenoceptor agonists have been successfully used as adjunctive anesthetic agents, both in veterinary and human practice.[150] When coadministered with general anesthetics, α_2-adrenoceptor agonists help to preserve hemodynamic stability, and reduce anxiety while increasing analgesia and sedation without respiratory depression. Also, α_2-adrenoceptor agonists may help to reduce narcotic-induced muscle rigidity. However, among their potential disadvantages is hypotension. Also, a bolus injection of an α_2-adrenoceptor agonist may cause acute hypertension and bradycardia.[150] Concerning the role of norepinephrine in actions induced by general anesthetics alone, it has been reported that isoflurane-induced antinociceptive effect was due to activation of descending noradrenergic neurons acting on spinal α_{2A}-adrenoceptors, while isoflurane also had a parallel pronociceptive effect due to action on supraspinal α_1-adrenoceptors.[154] Studies on noradrenergic mechanisms underlying the actions of another anesthetic agent used particularly by dentists, nitrous oxide, suggested that pontine α_{2B}-receptors are involved in its analgesic, but not its sedative, effects.[155] Additionally, activation of GABAergic interneurons through α_1-adrenoceptors appeared to contribute to the antinociceptive effect of nitrous oxide.[156]

Conclusions

Norepinephrine is involved in intrinsic regulation of pain, particularly in feedback inhibition of persistent pain. In pathophysiological conditions norepinephrine has a more complex role that varies from suppression to aggravation of pain. Under pathophysiological conditions, but not in healthy subjects, norepinephrine in the peripheral tissue aggravates pain. It is still a matter of debate which type of peripheral adrenoceptor mediates pain aggravation and which type of peripheral adrenoceptor suppresses the peripheral pain-induction mechanisms. The spinal dorsal horn is the most extensively studied site of norepinephrine-induced pain modulation. There is abundant evidence indicating that norepinephrine suppresses ascending pain signals at the spinal cord level. Norepinephrine released from terminals of descending pathways and acting on α_2-adrenoceptors underlies this spinal pain inhibitory effect. The evidence for the important pain inhibitory role of the α_{2A}-receptor subtype in the spinal cord is strong. However, α_{2C}-adrenoceptor is also likely to contribute, although further studies are needed to determine its more exact role. The results on the role of supraspinal noradrenergic actions in pain regulation are still controversial. However, it seems obvious that supraspinally, norepinephrine can either facilitate or inhibit pain depending on the supraspinal site of release, the type of adrenoceptor activated, and the intensity and duration of the pain stimulus. The existence of a pathophysiological condition may also be of importance in determining the action induced by supraspinal release of norepinephrine. Additionally, noradrenergic mechanisms are involved in analgesic actions of some anesthetics, and α_2-adrenoceptor agonists interact with general anesthetics by reducing anesthetic requirements.

REFERENCES

1. Weber, H. Über Anästhesie durch Adrenalin. *Verhandlungen der Deutschen Gesellschaft für innere Medizin*, **21** (1904), 616–619.
2. Jones, S. L. Descending noradrenergic influences on pain. *Prog. Brain Res.*, **88** (1991), 381–394.
3. Kingery, W. S., Davies, M. F. and Maze, M. Molecular mechanisms for the analgesic properties of alpha-2 adrenergic agonists. In *Molecular Neurobiology of Pain*, ed. D. Borsook. (Seattle: IASP Press, 1997), pp. 275–304.
4. Pertovaara, A., Bravo, R. and Herdegen, T. Induction and suppression of immediate-early genes in the rat brain by a selective alpha-2-adrenoceptor agonist and antagonist following noxious peripheral stimulation. *Neuroscience*, **54** (1993), 117–126.
5. Proudfit, H. K. Pharmacologic evidence for the modulation of nociception by noradrenergic neurons. *Prog. Brain Res.*, **77** (1988), 357–370.
6. Yaksh, T. L. Pharmacology of spinal adrenergic systems which modulate spinal nociceptive processing. *Pharmacol. Biochem. Behav.*, **22** (1985), 845–858.
7. Fuchs, P. N., Meyer, R. A. and Raja, S. N. Heat, but not mechanical hyperalgesia, following adrenergic injections in normal human skin. *Pain*, **90** (2001), 15–23.
8. Davis, K. D., Treede, R. D., Raja, S. N., Meyer, R. A. and Campbell, J. N. Topical application of clonidine relieves hyperalgesia in patients with sympathetically maintained pain. *Pain*, **47** (1991), 309–317.
9. Drummond, P. D. Noradrenaline increases hyperalgesia to heat in skin sensitized by capsaicin. *Pain*, **60** (1995), 311–315.
10. Torebjörk, H. E., Wahren, L., Wallin, G., Hallin, R. and Koltzenburg, M. Noradrenaline-evoked pain in neuralgia. *Pain*, **63** (1995), 11–20.
11. Hämäläinen, M. M., Jyväsjärvi, E. and Pertovaara, A. Can the alpha-2-adrenoceptor agonist-mediated suppression of nocitensive reflex responses be due to action on motoneurons or peripheral nociceptors? *Neurosci. Lett.*, **196** (1995), 29–32.
12. Sato, J. and Perl, E. R. Adrenergic excitation of cutaneous pain receptors induced by peripheral nerve injury. *Science*, **251** (1991),1608–1610.
13. Bossut, D. F. and Perl, E. R. Effects of nerve injury on sympathetic excitation of A-delta mechanical nociceptors. *J. Neurophysiol.*, **73** (1995), 1721–1723.
14. Petersen, M., Zhang, J., Zhang, J. M. and LaMotte, R. H. Abnormal spontaneous activity and response to norepinephrine in dissociated dorsal root ganglion cells after chronic nerve constriction. *Pain*, **67** (1996), 391–397.
15. Sato, J., Suzuki, S., Iseki, T. and Kumazawa, T. Adrenergic excitation of cutaneous nociceptors in chronically inflamed rats. *Neurosci. Lett.*, **164** (1993), 225–228.
16. Chung, K., Kim, H. J., Na, H. S., Park, M. J. and Chung, J. M. Abnormalities of sympathetic innervation in the area of an injured spinal nerve in a rat neuropathic model. *Neurosci. Lett.*, **162** (1993), 85–88.
17. McLachlan, E. M., Jänig, W., Devor, M. and Michaelis, M. Peripheral nerve injury triggers noradrenergic sprouting within dorsal root ganglia. *Nature*, **363** (1993), 543–546.

18. Ruocco, I., Cuello, A. C. and Ribeiro-da-Silva, A. Peripheral nerve injury leads to the establishment of a novel pattern of sympathetic fiber innervation in the rat skin. *J. Comp. Neurol.*, **422** (2000), 287–296.
19. Birder, L. A. and Perl, E. R. Expression of alpha$_2$-adrenergic receptors in rat primary afferent neurones after peripheral nerve injury or inflammation. *J. Physiol.*, **515**:2 (1999), 533–542.
20. Cho, H. J., Kim, D. S., Lee, N. H. *et al.* Changes in the α_2-adrenergic receptor subtypes gene expression in rat dorsal ganglion in an experimental model of neuropathic pain. *NeuroReport*, **8** (1999), 3119–3122.
21. Lavand'homme, P. M., Ma, W., De Kock, M. and Eisenach, J. C. Perineural alpha (2A)-adrenoceptor inhibits spinal cord neuroplasticity and tactile allodynia after nerve injury. *Anesthesiology*, **97** (2002), 972–980.
22. Chen, Y., Michaelis, M., Jänig, W. and Devor, M. Adrenoceptor subtype mediating sympathetic-sensory coupling in injured neurons. *J. Neurophysiol.*, **76** (1996), 3721–3730.
23. Leem, J. W., Gwak, Y. S., Nam, T. S. and Paik, K. S. Involvement of alpha$_2$-adrenoceptors in mediating sympathetic excitation of injured dorsal root ganglion neurons in rats with spinal nerve ligation. *Neurosci. Lett.*, **234** (1997), 39–42.
24. O'Halloran, K. D. and Perl, E. R. Effects of partial nerve injury on the responses of C-fiber polymodal nociceptors to adrenergic agonists. *Brain Res.*, **759** (1997), 233–240.
25. Banik, R. K., Sato, J., Yajima, H. and Mizumura, K. Differences between the Lewis and Sprague-Dawley rats in chronic inflammation induced norepinephrine sensitivity of cutaneous C-fiber nociceptors. *Neurosci. Lett.*, **299** (2001), 21–24.
26. Lee, D. H., Liu, X., Kim, H. T., Chung, K. and Chung, J. M. Receptor subtype mediating the adrenergic sensitivity of pain behavior and ectopic discharges in neuropathic Lewis rats. *J. Neurophysiol.*, **81** (1999), 2226–2233.
27. Tracey, D. J., Cunningham, J. E. and Romm, M. A. Peripheral hyperalgesia in experimental neuropathy: mediation by alpha-2-adrenoceptors on post-ganglionic sympathetic terminals. *Pain*, **60** (1995), 317–327.
28. Moon, D. E., Lee, D. H., Han, H. C. *et al.* Adrenergic sensitivity of the sensory receptors modulating mechanical allodynia in a rat neuropathic pain model. *Pain*, **80** (1999), 589–595.
29. Kingery, W. S., Guo, T. Z., Davies, M. F., Limbird, L. and Maze, M. The α_{2A} adrenoceptor and the sympathetic postganglionic neuron contribute to the development of neuropathic heat hyperalgesia in mice. *Pain*, **85** (2000), 345–358.
30. Ali, Z., Ringkamp, M., Hartke, T. V. *et al.* Uninjured C-fiber nociceptors develop spontaneous activity and alpha-adrenergic sensitivity following L6 spinal nerve ligation in monkey. *J. Neurophysiol.*, **81** (1999), 455–466.
31. Nakamura, M. and Ferreira, S. H. Peripheral analgesic action of clonidine: mediation by release of endogenous enkephalin-like substances. *Eur. J. Pharmacol.*, **146** (1988), 223–228.
32. Wei, H. and Pertovaara, A. Peripherally administered alpha$_2$-adrenoceptor agonist in the modulation of chronic allodynia induced by spinal nerve ligation in the rat. *Anesth. Analg.*, **85** (1997), 1122–1127.

33. Wei, H., Jyväsjärvi, E., Niissalo, S. *et al.* The influence of chemical sympathectomy on pain responsivity and α_2-adrenergic antinociception in neuropathic animals. *Neuroscience*, **114** (2002), 655–668.
34. Gentili, M., Juhel, A. and Bonnet, F. Peripheral analgesic effect of intra-articular clonidine. *Pain*, **64** (1996), 593–596.
35. Baron, R. Peripheral neuropathic pain: from mechanisms to symptoms. *Clin. J. Pain*, **16** (2000), S12–S20.
36. Koltzenburg, M., Kress, M. and Reeh, P. The nociceptor sensitization by bradykinin does not depend on sympathetic neurons. *Neuroscience*, **46** (1992), 465–473.
37. Sato, J., Suzuki, S., Tamura, R. and Kumazawa, T. Norepinephrine excitation of cutaneous nociceptors in adjuvant-induced inflamed rats does not depend on sympathetic neurons. *Neurosci. Lett.*, **177** (1994), 135–138.
38. Gold, M. S., White, D. M., Ahlgren, S. C., Guo, M. and Levine, J. D. Catecholamine-induced mechanical sensitization of cutaneous nociceptors in the rat. *Neurosci. Lett.*, **175** (1994), 166–170.
39. Levine, J. D., Taiwo, Y. O., Collins, S. D. and Tam, J. K. Noradrenaline hyperalgesia is mediated through interaction with sympathetic postganglionic neurone terminals rather than activation of primary afferent nociceptors. *Nature*, **323** (1986), 158–160.
40. Danzebrink, R. M. and Gebhart, G. F. Antinociceptive effects of intrathecal adrenoceptor agonists in a rat model of visceral nociception. *J. Pharmacol. Exp. Ther.*, **253** (1990), 698–705.
41. Kuraishi, Y., Harada, Y. and Takagi, H. Noradrenaline regulation of pain-transmission in the spinal cord mediated by alpha-adrenoceptors. *Brain Res.*, **174** (1979), 333–336.
42. Kuraishi, Y., Harada, Y., Satoh, M. and Takagi, H. Antagonism by phenoxybenzamine of the analgesic effect of morphine injected into the nucleus reticularis gigantocellularis of the rat. *Neuropharmacology*, **18** (1979), 107–110.
43. Kuraishi, Y., Hirota, N., Satoh, M. and Takagi, H. Antinociceptive effects of intrathecal opioids, noradrenaline and serotonin in rats: mechanical and thermal algesic tests. *Brain Res.*, **326** (1985), 168–171.
44. Reddy, S.V. and Yaksh, T. L. Spinal noradrenergic terminal system mediates antinociception. *Brain Res.*, **189** (1980), 391–401.
45. Eisenach, J. C., DuPen, S., Dubois, M., Miguel, R. and Allin, D. Epidural clonidine analgesia for intractable cancer pain. The epidural clonidine study group. *Pain*, **61** (1995), 391–399.
46. Gordh, T. Epidural clonidine for treatment of postoperative pain after thoracotomy. A double-blind placebo-controlled study. *Acta Anaesthesiol. Scand.*, **32** (1988), 702–709.
47. Dennis, S. G., Melzack, R., Gutman, S. and Boucher, F. Pain modulation by adrenergic agents and morphine as measured by three pain tests. *Life Sci.*, **26** (1980), 1247–1259.
48. Paalzow, L. Analgesia produced by clonidine in mice and rats. *J. Pharm. Pharmacol.*, **26** (1974), 361–363.
49. Pertovaara, A., Kauppila, T. and Tukeva, T. The effect of medetomidine, an α_2-adrenoceptor agonist, in various pain tests. *Eur. J. Pharmacol.*, **179** (1990), 323–328.
50. Pertovaara, A. Antinociception induced by alpha-2-adrenoceptor agonists, with special emphasis on medetomidine studies. *Prog. Neurobiol.*, **40** (1993), 691–709.

51. Skingle, M., Hayes, A. G. and Tyers, M. B. Antinociceptive activity of clonidine in the mouse, rat and dog. *Life Sci.*, **31** (1982), 1123–1132.
52. Millan, M. J., Bervoets, K., Rivet, J. M. *et al.* Multiple alpha-2 adrenergic receptor subtypes. II. Evidence for a role of rat R alpha-2A adrenergic receptors in the control of nociception, motor behavior and hippocampal synthesis of noradrenaline. *J. Pharmacol. Exp. Ther.*, **270** (1994), 958–972.
53. Hunter, J. C., Fontana, D. J., Hedley, L. R. *et al.* Assessment of the role of alpha-2-adrenoceptor subtypes in the antinociceptive, sedative and hypothermic action of dexmedetomidine in transgenic mice. *Br. J. Pharmacol.*, **122** (1997), 1339–1344.
54. Lakhlani, P. P., MacMillan, L. B., Guo, T. Z. *et al.* Substitution of a mutant alpha-2A-adrenergic receptor via 'hit and run' gene targeting reveals the role of this subtype in sedative, analgesic and anesthetic-sparing responses in vivo. *Proc. Natl. Acad. Sci. U. S. A.*, **94** (1997), 9950–9955.
55. Stone, L. S., MacMillan, L. B., Kitto, K. F., Limbird, L. E. and Wilcox, G. L. The alpha-2A-adrenergic receptor subtype mediates spinal analgesia evoked by alpha-2-agonists and is necessary for spinal adrenergic-opioid synergy. *J. Neurosci.*, **17** (1997), 7157–7165.
56. Fairbanks, C. A., Stone, L. S., Kitto, K. F. *et al.* Alpha-2C-adrenergic receptors mediate spinal analgesia and adrenergic-opioid synergy. *J. Pharmacol. Exp. Ther.*, **300** (2002), 282–290.
57. Pertovaara, A., Kauppila, T., Jyväsjärvi, E. and Kalso, E. Involvement of supraspinal and spinal segmental alpha-2-adrenergic mechanisms in the medetomidine-induced antinociception. *Neuroscience*, **44** (1991), 705–714.
58. Spaulding, T. C., Venafro, J. J., Ma, M. G. and Fielding, S. The dissociation of the antinociceptive effect of clonidine from supraspinal structures. *Neuropharmacology*, **18** (1979), 103–105.
59. Kwiat, G. C. and Basbaum, A. I. The origin of brainstem noradrenergic and serotoninergic projections to the spinal cord dorsal horn in the rat. *Somatosens. Mot. Res.*, **9** (1992), 157–173.
60. Hentall, I. D., Mesigil, R., Pinzon, A. and Noga, B. R. Temporal and spatial profiles of pontine-evoked monoamine release in the rat's spinal cord. *J. Neurophysiol.*, **89** (2003), 2943–2951.
61. Takagi, H., Shiomi, H., Kuraishi, Y., Fukui, K. and Ueda, H. Pain and the bulbospinal noradrenergic system: pain-induced increase in normetanephrine content in the spinal cord and its modification by morphine. *Eur. J. Pharmacol.*, **54** (1979), 99–107.
62. Tyce, G. M. and Yaksh, T. L. Monoamine release from cat spinal cord by somatic stimuli: an intrinsic modulatory system. *J. Physiol.*, **314** (1981), 513–529.
63. Cui, M., Feng, Y., McAdoo, D. J. and Willis, W. D. Periaqueductal gray stimulation-induced inhibition of nociceptive dorsal horn neurons in rats is associated with the release of norepinephrine, serotonin, and amino acids. *J. Pharmacol. Exp. Ther.*, **289** (1999), 868–876.
64. Proudfit, H. K. and Monsen, M. Ultrastructural evidence that substance P neurons form synapses with noradrenergic neurons in the A7 catecholamine cell group that modulate nociception. *Neuroscience*, **91** (1999), 1499–1513.

65. Sagen, J. and Proudfit, H. K. Effect of intrathecally administered noradrenergic antagonists on nociception in the rat. *Brain Res.*, **310** (1984), 295–301.
66. Malmberg, A. B., Hedley, L. R., Jasper, J. R., Hunter, J. C. and Basbaum, A. I. Contribution of α_2 receptor subtypes to nerve injury-induced pain and its regulation by dexmedetomidine. *Br. J. Pharmacol.*, **132** (2001), 1827–1836.
67. Jasmin, L., Tien, D., Weinshenker, D. *et al.* The NK1 receptor mediates both the hyperalgesia and the resistance to morphine in mice lacking noradrenaline. *Proc. Natl. Acad. Sci. U. S. A.*, **99** (2002), 1029–1034.
68. Green, G. M., Lyons, L. and Dickenson, A. H. Alpha$_2$-adrenoceptor antagonists enhance responses of dorsal horn neurones to formalin induced inflammation. *Eur. J. Pharmacol.*, **347** (1998), 201–204.
69. Mansikka, H., Lähdesmäki, J., Scheinin, M. and Pertovaara, A. Alpha(2A) adrenoceptors contribute to feedback inhibition of capsaicin-induced hyperalgesia. *Anesthesiology*, **101** (2004), 185–190.
70. Weil-Fugazza, J., Godefroy, F., Manceau, V. and Besson, J. M. Increased norepinephrine and uric acid levels in the spinal cord of arthritic rats. *Brain Res.*, **374** (1986), 190–194.
71. Tsuruoka, M. and Willis, W. D. Descending modulation from the region of the locus coeruleus on nociceptive sensitivity in rat model of inflammatory hyperalgesia. *Brain Res.*, **743** (1996), 86–92.
72. Brandt, S. A. and Livingston, A. Receptor changes in the spinal cord of sheep associated with exposure to chronic pain. *Pain*, **42** (1990), 323–329.
73. Hylden, J. L. K., Thomas, D. A., Iadarola, M. J., Nahin, R. L. and Dubner, R. Spinal opioid analgesic effects in a model of unilateral inflammation/hyperalgesia: possible involvement of noradrenergic mechanisms. *Eur. J. Pharmacol.*, **194** (1991), 135–143.
74. Idänpään-Heikkilä, J. J., Kalso, E. A. and Seppälä, T. Antinociceptive actions of dexmedetomidine and the kappa-opioid agonist U-50,488H against thermal, mechanical and inflammatory stimuli. *J. Pharmacol. Exp. Ther.*, **271** (1994), 1306–1313.
75. Kayser, V., Guilbaud, G. and Besson, J. M. Potent antinociceptive effects of clonidine systemically administered in an experimental model of clinical pain, the arthritic rat. *Brain Res.*, **593** (1992), 7–13.
76. Mansikka, H. and Pertovaara, A. Influence of selective alpha-2-adrenergic agents on mustard oil-induced central hyperalgesia in rats. *Eur. J. Pharmacol.*, **281** (1995a), 43–48.
77. Stanfa, L. C. and Dickenson, A. H. Enhanced α_2-adrenergic controls and spinal morphine potency in inflammation. *NeuroReport*, **5** (1994), 469–472.
78. Mansikka, H., Idänpään-Heikkilä, J. J. and Pertovaara, A. Different roles of alpha-2-adrenoceptors of the medulla versus the spinal cord in modulation of mustard oil-induced central hyperalgesia in rats. *Eur. J. Pharmacol.*, **297** (1996), 19–26.
79. Lähdesmäki, J., Scheinin, M., Pertovaara, A. and Mansikka, H. The α_{2A}–adrenoceptor subtype is not involved in inflammatory hyperalgesia or morphine-induced antinociception. *Eur. J. Pharmacol.*, **468** (2003), 183–189.
80. Stone, L. S., Broberger, C., Vulchanova, L. *et al.* Differential distribution of alpha2A and alpha2C adrenergic receptor immunoreactivity in the rat spinal cord. *J. Neurosci.*, **18** (1998), 5928–5937.

81. Pertovaara, A. and Wei, H. Attenuation of ascending nociceptive signals to the rostroventromedial medulla induced by alpha-2-adrenoceptor agonist, MPV-2426, following intrathecal application in neuropathic rats. *Anesthesiology*, **92** (2000), 1082–1092.
82. Xu, X. J., Puke, M. J. C. and Wiesenfeld-Hallin, Z. The depressive effect of intrathecal clonidine on the spinal flexor reflex is enhanced after sciatic nerve section in rats. *Pain*, **51** (1992), 145–151.
83. Xu, M., Kontinen, V. K. and Kalso, E. Effects of radolmidine, a novel alpha2-adrenergic agonist compared with dexmedetomidine in different pain models in the rat. *Anesthesiology*, **93** (2000), 473–481.
84. Yaksh, T. L., Pogrel, J. W., Lee, Y. W. and Chaplan, S. R. Reversal of nerve ligation-induced allodynia by spinal alpha-2 adrenoceptor agonists. *J. Pharmacol. Exp. Ther.*, **272** (1995), 207–214.
85. Satoh, O. and Omote, K. Roles of monoaminergic, glycinergic and GABAergic inhibitory systems in the spinal cord in rats with peripheral mononeuropathy. *Brain Res.*, **728** (1996), 27–36.
86. Xu, M., Kontinen, V. K. and Kalso, E. Endogenous noradrenergic tone controls symptoms of allodynia in the spinal nerve ligation model of neuropathic pain. *Eur. J. Pharmacol.*, **366** (1999), 41–45.
87. Clark, F. M. and Proudfit, H. K. Anatomical evidence for genetic differences in the innervation of the rat spinal cord by noradrenergic locus coeruleus neurons. *Brain Res.*, **591** (1992), 44–53.
88. Engberg, I. and Ryall, R. W. The inhibitory action of noradrenaline and other monoamines on spinal neurones. *J. Physiol.*, **185** (1966), 298–322.
89. Belcher, G., Ryall, R.W. and Schaffner, R. The differential effects of 5-hydroxytryptamine, noradrenaline and raphe stimulation on nociceptive and non-nociceptive dorsal horn interneurones. *Brain Res.*, **151** (1978), 307–321.
90. Headley, P. M., Duggan, A. W. and Griersmith, B. T. Selective reduction by noradrenaline and 5-hydroxytryptamine of nociceptive responses of cat dorsal horn neurones. *Brain Res.*, **145** (1978), 185–189.
91. Fleetwood-Walker, S. M., Mitchell, R., Hope, R. J., Molony, V. and Iggo, A. An alpha 2 receptor mediates the selective inhibition by noradrenaline of nociceptive responses of identified dorsal horn neurones. *Brain Res.*, **334** (1985), 243–254.
92. White, S. R. and Neuman, R. S. Facilitation of spinal motoneurone excitability by 5-hydroxytryptamine and noradrenaline. *Brain Res.*, **188** (1980), 119–127.
93. Bell, J. A. and Matsumiya, T. Inhibitory effects of dorsal horn and excitant effects of ventral horn intraspinal microinjections of norepinephrine and serotonin in the cat. *Life Sci.*, **29** (1981), 1507–1514.
94. Fleetwood-Walker, S. M., Hope, P. J. and Mitchell, R. Antinociceptive actions of descending dopaminergic tracts on cat and rat dorsal horn somatosensory neurones. *J. Physiol.*, **399** (1988), 335–348.
95. Garraway, S. M. and Hochman, S. Modulatory actions of serotonin, norepinephrine, dopamine, and acetylcholine in spinal cord deep dorsal horn neurons. *J. Neurophysiol.*, **86** (2001), 2183–2194.

96. Kuraishi, Y., Hirota, N., Sato, Y., Satoh, M. and Takagi, H. Noradrenergic inhibition of the release of substance P from primary afferents in the rabbit spinal cord dorsal horn. *Brain Res.*, **359** (1985), 177–182.
97. Li, X. and Eisenach, J. C. α_{2A}-Adrenoceptor stimulation reduces capsaicin-induced glutamate release from spinal cord synaptosomes. *J. Pharmacol. Exp. Ther.*, **299** (2001), 939–944.
98. Ueda, M., Oyama, T., Kuraishi, Y., Akaike, A. and Satoh, M. Alpha$_2$-adrenoceptor-mediated inhibition of capsaicin-evoked release of glutamate from rat spinal dorsal horn slices. *Neurosci. Lett.*, **188** (1995), 137–139.
99. Kawasaki, Y., Kumamoto, E., Furue, H. and Yoshimura, M. α_2-Adrenoceptor-mediated presynaptic inhibition of primary afferent glutamatergic transmission in rat substantia gelatinosa neurons. *Anesthesiology*, **98** (2003), 682–689.
100. Pan, Y. Z., Li, D. P. and Pan, H. L. Inhibition of glutamatergic synaptic input to spinal lamina II(o) neurons by presynaptic alpha$_2$-adrenergic receptors. *J. Neurophysiol.*, **87** (2002), 1938–1947.
101. Olave, M. J. and Maxwell, D. J. Neurokinin-1 projection cells in the rat dorsal horn receive synaptic contacts from axons that possess α_{2C}-adrenergic receptors. *J. Neurosci.*, **23** (2003), 6837–6846.
102. Westlund, K. N., Carlton, S. M., Zhang, D. and Willis, W. D. Direct catecholaminergic innervation of primate spinothalamic tract neurons. *J. Comp. Neurol.*, **299** (1990), 178–186.
103. North, R. A. and Yoshimura, M. The actions of noradrenaline on neurones of the rat substantia gelatinosa in vitro. *J. Physiol.*, **349** (1984), 43–55.
104. Pertovaara, A. and Kalmari, J. Comparison of the visceral antinociceptive effects of spinally administered MPV-2426 (fadolmidine) and clonidine in the rat. *Anesthesiology*, **98** (2003), 189–194.
105. Millar, J. and Williams, G. V. Effects of iontophoresis of noradrenaline and stimulation of the periaqueductal gray on single-unit activity in the rat superficial dorsal horn. *J. Comp. Neurol.*, **287** (1989), 119–133.
106. Baba, H., Shimoji, K. and Yoshimura, M. Norepinephrine facilitates inhibitory transmission in substantia gelatinosa of adult rat spinal cord (Part 1): effects on axon terminals of GABAergic and glycinergic neurons. *Anesthesiology*, **92** (2000), 473–484.
107. Scheinin, M., Lomasney, J. W., Hayden-Hixson, D. M. *et al.* Distribution of alpha 2-adrenergic receptor subtype gene expression in rat brain. *Mol. Brain Res.*, **21** (1994), 133–149.
108. Budai, D., Harasawa, I. and Fields, H. L. Midbrain periaqueductal gray (PAG) inhibits nociceptive inputs to sacral dorsal horn neurons through alpha2-adrenergic receptors. *J. Neurophysiol.*, **80** (1998), 2244–2254.
109. Peng, Y. B., Lin, Q. and Willis, W. D. Involvement of alpha-2 adrenoceptors in the periaqueductal gray-induced inhibition of dorsal horn cell activity in rats. *J. Pharmacol. Exp. Ther.*, **278** (1996), 125–135.
110. Hämäläinen, M. M. and Pertovaara, A. The rostroventromedial medulla is not involved in α2-adrenoceptor-mediated antinociception in the rat. *Neuropharmacology*, **32** (1993), 1411–1418.
111. Ossipov, M. H. and Gebhart, G. F. Absence of antinociceptive effect of alpha-2 agonists microinjected in the periaqueductal gray of the rat. *Brain Res.*, **289** (1983), 349–354.

112. Hammond, D. L., Tyce, G. M. and Yaksh, T. L. Efflux of 5-hydroxytryptamine and noradrenaline into spinal cord superfusates during stimulation of the rat medulla. *J. Physiol.*, **359** (1985), 151–162.
113. Barbaro, N. M., Hammond, D. L. and Fields, H. L. Effects of intrathecally administered methysergide and yohimbine on microstimulation-produced antinociception in the rat. *Brain Res.*, **343** (1985), 223–229.
114. Clark, F. M. and Proudfit, H. K. Projections of neurons in the ventromedial medulla to pontine catecholamine cell groups involved in the modulation of nociception. *Brain Res.*, **540** (1991), 105–115.
115. Yeomans, D. C., Clark, F. M., Paice, J. A. and Proudfit, H. K. Antinociception induced by electrical stimulation of spinally projecting noradrenergic neurons in the A7 catecholamine cell group of the rat. *Pain*, **48** (1992), 449–461.
116. Haws, C. M., Heinricher, M. M. and Fields, H. L. Alpha-adrenergic receptor agonists, but not antagonists, alter the tail-flick latency when microinjected into the rostral ventromedial medulla of the lightly anesthetized rat. *Brain Res.*, **533** (1990), 192–195.
117. Sagen, J. and Proudfit, H. K. Evidence for pain modulation by pre- and postsynaptic noradrenergic receptors in the medulla oblongata. *Brain Res.*, **331** (1985), 285–293.
118. Sagen, J. and Proudfit, H. K. Release of endogenous monoamines into spinal cord superfusates following the microinjection of phentolamine into the nucleus raphe magnus. *Brain Res.*, **406** (1987), 246–254.
119. Hämäläinen, M. M. and Pertovaara, A. The antinociceptive action of an α_2-adrenoceptor agonist in the spinal dorsal horn is due to a direct spinal action and not to activation of descending inhibition. *Brain Res. Bull.*, **37** (1995), 581–587.
120. Gebhart, G. F. and Ossipov, M. H. Characterization of inhibition of the spinal nociceptive tail-flick reflex in the rat from the medullary lateral reticular nucleus. *J. Neurosci.*, **6** (1986), 701–713.
121. Ossipov, M. H. and Gebhart, G. H. Opioid, cholinergic and alpha-adrenergic influences on the modulation of nociception from the lateral reticular nucleus of the rat. *Brain Res.*, **384** (1986), 282–293.
122. Mansikka, H. and Pertovaara, A. The role of α_2-adrenoceptors of the medullary lateral reticular nucleus in spinal antinociception in rats. *Brain Res. Bull.*, **37** (1995b), 633–638.
123. Guo, T. Z., Jiang, J. Y., Buttermann, A. E. and Maze, M. Dexmedetomidine injection into the locus coeruleus produces antinociception. *Anesthesiology*, **84** (1996), 873–881.
124. Pertovaara, A., Hämäläinen, M. M., Kauppila, T., Mecke, E. and Carlson, S. Dissociation of the α_2-adrenergic antinociception from sedation following microinjection of medetomidine into the locus coeruleus in rats. *Pain*, **57** (1994), 207–215.
125. Xu, M., Wei, H., Kontinen, V. K., Kalso, E. and Pertovaara, A. The dissociation of sedative from spinal antinociceptive effects following administration of a novel alpha-2-adrenoceptor agonist, MPV-2426, in the locus coeruleus in the rat. *Acta Anaesthesiol. Scand.*, **44** (2000), 648–655.
126. Aghajanian, G. K. and Van der Maelen, C. P. α_2-Adrenoceptor-mediated hyperpolarization of locus coeruleus neurons: intracellular studies *in vivo*. *Science*, **215** (1982), 1394–1396.

127. DeSarro, G. B., Ascioti, C., Froio, F., Libri, V. and Nistico, G. Evidence that locus coeruleus is the site where clonidine and drugs acting at alpha 1- and alpha 2-adrenoceptors affect sleep and arousal mechanisms. *Br. J. Pharmacol.*, **90** (1987), 675–685.
128. Kauppila, T., Kemppainen, P., Tanila, H. and Pertovaara, A. Effect of medetomidine, an alpha-2-adrenoceptor agonist, on experimental pain in humans. *Anesthesiology*, **74** (1991), 3–8.
129. Zhang, C., Guo, Y. Q., Qiao, J. T. and Dafny, N. Locus coeruleus modulates thalamic nociceptive responses via adrenoceptors. *Brain Res.*, **16** (1998), 116–122.
130. Westlund, K. N., Sorkin, L. S., Ferrington, D. G. *et al.* Serotoninergic and noradrenergic projections to the ventral posterolateral nucleus of the monkey thalamus. *J. Comp. Neurol.*, **295** (1990), 197–207.
131. Hirata, K., Koyama, N. and Minami, T. The effects of clonidine and tizanidine on responses of nociceptive neurons in nucleus ventralis posterolateralis of the cat thalamus. *Anesth. Analg.*, **81** (1995), 259–264.
132. Jänig, W. and Häbler, H. J. Sympathetic nervous system: contribution to neuropathic pain. *Prog. Brain Res.*, **129** (2000), 451–468.
133. Bossut, D. F., Shea, V. K. and Perl, E. R. Sympathectomy induces adrenergic excitability of cutaneous C-fiber nociceptors. *J. Neurophysiol.*, **75** (1996), 514–517.
134. Raskin, N. H., Levinson, S., Hoffman, P. M., Pickett, J. B. and Fields, H. L. Postsympathectomy neuralgia. Amelioration with diphenylhydantoin and carbamazepine. *Am. J. Surg.*, **128** (1974), 75–78.
135. Sagen, J., Wang, H. and Pappas, G. D. Adrenal medullary implants in the rat spinal cord reduce nociception in a chronic pain model. *Pain*, **42** (1990), 69–79.
136. Gulwadi, A. G., Hoane, M. R., Saydoff, J. A., Frydel, B. R. and Lindner, M. D. No detectable analgesic effects in the formalin test even with one million bovine adrenal chromaffin cells. *Pain*, **99** (2002), 263–271.
137. Khasar, S. G., Miao, F. J., Jänig, W. and Levine, J. D. Vagotomy-induced enhancement of mechanical hyperalgesia in the rat is sympathoadrenal-mediated. *J. Neurosci.*, **18** (1998), 3043–3049.
138. Drasner, K. and Fields, H. L. Synergy between the antinociceptive effects of intrathecal clonidine and systemic morphine in the rat. *Pain*, **32** (1988), 309–312.
139. Ossipov, M. H., Harris, S., Lloyd, P. *et al.* An isobolographic analysis of the antinociceptive effect of systemically and intrathecally administered combinations of clonidine and opiates. *J. Pharmacol. Exp. Ther.*, **255** (1990), 1107–1116.
140. Wilcox, G. L., Carlsson, K. H., Jochim, A. and Jurna, I. Mutual potentiation of antinociceptive effects of morphine and clonidine on motor and sensory responses in rat spinal cord. *Brain Res.*, **405** (1987), 84–93.
141. Mansikka, H., Zhou, L., Donovan, D. M., Pertovaara, A. and Raja, S. N. The role of μ-opioid receptors in inflammatory hyperalgesia and α_2-adrenoceptor-mediated antihyperalgesia. *Neuroscience*, **113** (2002), 339–349.
142. Archer, T., Jonsson, G., Minor, B. G. and Post, C. Noradrenergic—serotoninergic interactions and nociception in the rat. *Eur. J. Pharmacol.*, **120** (1986), 295–307.

143. Nakagawa, I., Omote, K., Kitahata, L. M., Collins, J. G. and Murata, K. Serotoninergic mediation of spinal analgesia and its interaction with noradrenergic systems. *Anesthesiology*, **73** (1990), 474–478.
144. Danzebrink, R. M. and Gebhart, G. F. Intrathecal coadministration of clonidine with serotonin receptor agonists produces supra-additive visceral antinociception in the rat. *Brain Res.*, **555** (1991), 35–42.
145. Chiari, A. I. and Eisenach, J. C. Intrathecal adenosine: interactions with spinal clonidine and neostigmine in rat models of acute nociception and postoperative hypersensitivity. *Anesthesiology*, **90** (1999), 1413–1421.
146. Sawynok, J., Reid, A. and Isbrucker, R. Adenosine mediates calcium-induced antinociception and potentiation of noradrenergic antinociception in the spinal cord. *Brain Res.*, **524** (1990), 187–195.
147. Chiari, A., Li, X. H., Xu, Z., Pan, H. L. and Eisenach, J. C. Formation of 6-nitro-norepinephrine from nitric oxide and norepinephrine in the spinal cord and its role in spinal analgesia. *Neuroscience*, **101** (2000), 189–196.
148. Gordh, T., Jr., Jansson, I., Hartvig, P., Gillberg, P. G. and Post, C. Interactions between noradrenergic and cholinergic mechanisms involved in spinal nociceptive processing. *Acta Anaesthesiol. Scand.*, **33** (1989), 39–47.
149. Sawynok, J. and Dickson, C. Evidence for the involvement of descending noradrenergic pathways in the antinociceptive effect of baclofen. *Brain Res.*, **335** (1985), 89–97.
150. Maze, M. and Tranquilli, W. Alpha-2 adrenoceptor agonists: defining the role in clinical anesthesia. *Anesthesiology*, **74** (1991), 581–605.
151. Aston-Jones, G. and Bloom, F. E. Activity of norepinephrine-containing locus coeruleus neurons in behaving rats anticipates fluctuations in the sleep–waking cycle. *J. Neurosci.*, **1** (1981), 876–886.
152. Jalanka, H. The use of medetomidine, medetomidine–ketamine combinations and atipamezole at Helsinki Zoo – a review of 240 cases. *Acta Vet. Scand.*, **85** (1989), 193–197.
153. Kaukinen, S. and Pyykkö, K. The potentiation of halothane anaesthesia by clonidine. *Acta Anaesthesiol. Scand.*, **23** (1979), 107–111.
154. Kingery, W. S., Agashe, G. S., Guo, T. Z. *et al.* Isoflurane and nociception: spinal alpha2A adrenoceptors mediate antinociception while supraspinal alpha1 adrenoceptors mediate pronociception. *Anesthesiology*, **96** (2002), 367–374.
155. Sawamura, S., Kingery, W. S., Davies, M. F. *et al.* Antinociceptive action of nitrous oxide is mediated by stimulation of noradrenergic neurons in the brainstem and activation of α_{2B}-adrenoceptors. *J. Neurosci.*, **20** (2000), 9242–9251.
156. Orii, R., Ohashi, Y., Guo, T. *et al.* Evidence for the involvement of spinal cord alpha1 adrenoceptors in nitrous oxide-induced antinociceptive effects in Fischer rats. *Anesthesiology*, **97** (2002), 1458–1465.

14

Norepinephrine and cognitive disorders

Amy F. T. Arnsten, Ph.D.
Yale University School of Medicine, New Haven, USA

Introduction

Traditionally, cognitive disorders have been associated with alterations in acetylcholine (e.g. Alzheimer's disease) or dopamine (e.g. attention-deficit/hyperactivity disorder). However, there is a long and consistent body of literature establishing that norepinephrine (NE) has profound effects on the cortical and subcortical cognitive functions that are disturbed in these syndromes. These findings from basic research, in concert with efficacy of clinical NE treatment modalities, indicate that NE plays a major role in many cognitive disorders.

The definition of "cognitive disorder" has expanded in recent years as neuroscience has identified impaired prefrontal cortical (PFC) cognitive function at the core of many neuropsychiatric illnesses. This chapter will review NE mechanisms influencing PFC, posterior cortical and subcortical functions, and review the evidence that NE is altered significantly in a range of disorders involved in higher cortical dysfunction.

NE innervation of cerebral cortex

The noradrenergic input to the cortex arises from the cells of the locus coeruleus (LC) at the pontine-midbrain junction in the brain stem. The noradrenergic fibers target layer I in the rodent cortex, but have an expanded, bilaminar distribution in the monkey cortex in both deep and superficial cortical layers.[1] In primates, the densest NE innervation targets the somatosensory cortex in the parietal lobe, while the most sparse innervation is of the primary visual cortex.[1] Higher cortical areas such as the PFC have a moderate NE input.[1] Interestingly, there is a reciprocal relationship between the PFC and the LC, as the PFC provides one of the few "intelligent" inputs back to LC neurons.[2–4]

Brain Norepinephrine: Neurobiology and Therapeutics, ed. Gregory A. Ordway, Michael A. Schwartz and Alan Frazer. Published by Cambridge University Press.

Norepinephrine interacts with three families of adrenergic receptors: the α_1-, the α_2-, and the β-adrenergic receptors. There are three subtypes of α_2-adrenergic receptors, the α_{2A}, the α_{2B}, and the α_{2C}. The α_{2A}, and to a lesser extent the α_{2C}, are found presynaptically on NE cells and terminals, while all three subtypes are found postsynaptically.[5] The α_{2A}-subtype is the most prominent in the cortex, although there are low levels of α_{2C}-receptors as well.[6,7] Binding studies suggest that α_2-receptors are most concentrated in superficial layers in primate PFC,[8] although electron microscopic (EM) analyses also reveal receptors in deep layers. These EM studies have found α_{2A}-receptors in a variety of cellular locations in the primate PFC, including over postsynaptic densities on dendritic spines.[6,7] α_2-receptors are generally coupled through G_i proteins to adenylate cyclase,[9] reducing intracellular cyclic adenosine monophosphate (cAMP) production and decreasing activation of protein kinase A (PKA). α_2-receptors can also enhance K^+ conductance, and thus can have an inhibitory effect on neuronal activity.[9]

There are also three subtypes of α_1-receptors, the α_{1A}, the α_{1B}, and the α_{1D}.[10] The α_{1A} and α_{1D} are most prominent in rodent PFC (these subtypes have not been localized in primate in PFC).[11,12] However, receptor binding studies of primate PFC have shown dense binding in superficial layers,[8] similar to α_2-receptors. The cellular localization of these receptors is not yet known. α_1-receptor stimulation has been found to enhance excitatory processes in many brain regions.[13,14] α_1-receptors are generally coupled through G_q proteins to phospholipase C and phosphatidylinositol intracellular signaling, resulting in activation of protein kinase C (PKC) and the release of intracellular Ca^{2+} via inositol 1, 4, 5-trisphosphate IP_3.[9]

Finally, there are three subtypes of β-receptors: β_1 (localized in heart); β_2 (localized in lung); and β-3, with the first two subtypes found in high concentration in the brain.[15] β-adrenergic receptors are densest in the intermediate layers of the PFC, where thalamic inputs are concentrated. Electron microscopic studies of monkey PFC have localized β_2-receptors on dendritic spines (presumably of PFC pyramidal neurons) and on γ-aminobutyric acid (GABA) ergic interneurons.[7] Electrophysiological studies have also shown that β-receptors can potentiate GABAergic processes.[16] Finally, many β-receptors are found on glia, where they have multiple actions including reducing the uptake of glutamate[17] and regulating glucose availability.[18] In general, β-receptors are coupled via G_s to adenylate cyclase, increasing cAMP and PKA activity.[9]

NE regulation of prefrontal cortex

The PFC is the cortical region most expanded in primates and one that subserves many of our highest order abilities (Figure 14.1). The PFC is critical for guiding our

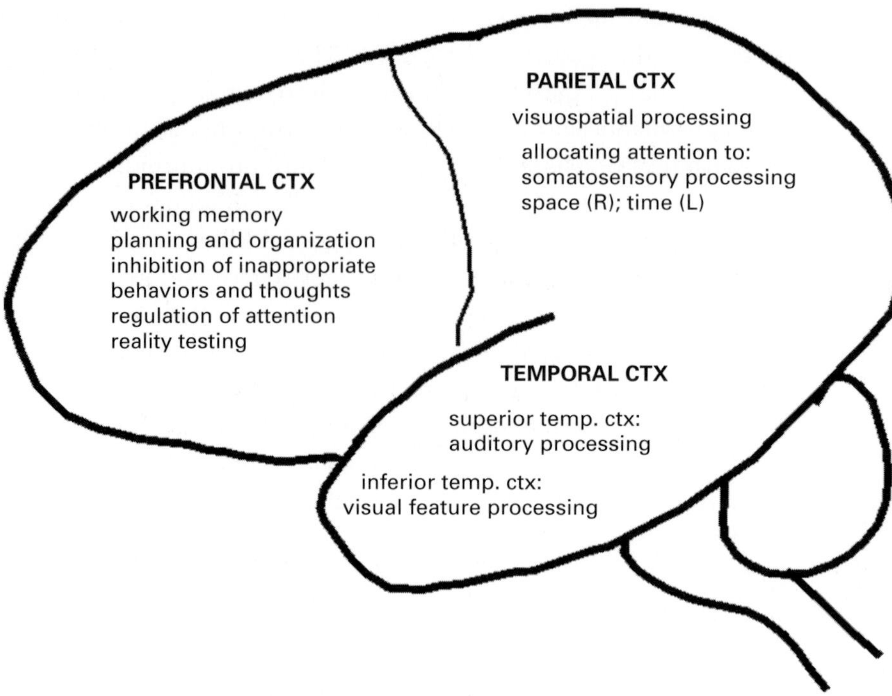

Figure 14.1 Schematic representation of some of the higher cognitive functions carried out by the prefrontal, parietal, and temporal association cortices in humans. Please note that in the interests of space, the occipital cortex and medial temporal cortices are not depicted. The occipital cortex is responsible for fundamental visual processing, which is then continued in the parietal and inferior temporal cortices. The medial temporal lobe includes the rhinal cortices, the hippocampus, and the amygdala, all of which are involved with memory consolidation. L, left; R, right; temp. ctx, temporal cortex.

behavior and thoughts using working memory, a process fundamental to cognitive functions such as language.[19] In humans, the PFC is critical for the production of language (syntax), for planning and organization, for the inhibition of inappropriate behaviors and thoughts, for empathy and social conscience, for the encoding and retrieval of memories (especially for nonsalient material), and for sustaining attention and inhibiting response to distractions.[20]

Over the last 20 years, research has shown that NE has powerful influence on the working memory and attention functions of the PFC. As illustrated in Figure 14.2, NE has widely varying affinities for adrenergic receptors – it has highest affinity for α_2-adrenoceptors, lower affinity for α_1-adrenoceptors, and lowest affinity for β-receptors – and it appears that NE has markedly different effects on working memory dependent on the receptor subtype engaged.

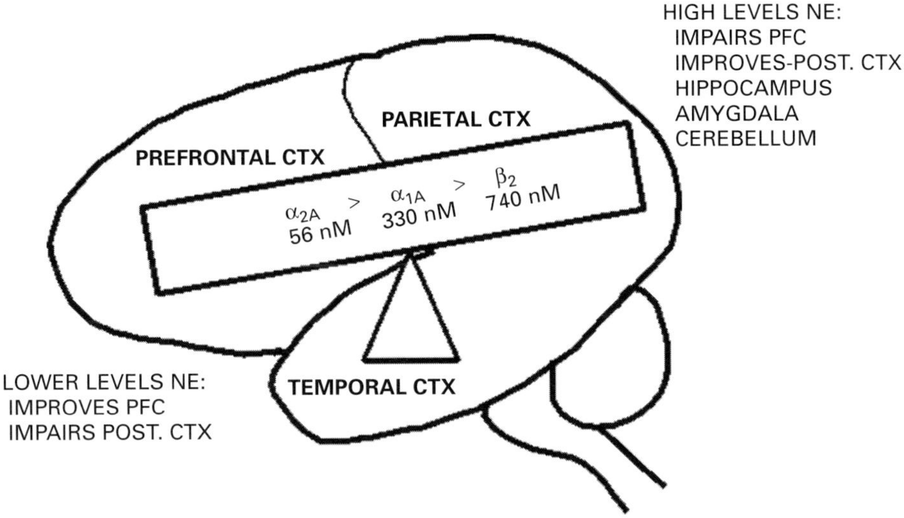

Figure 14.2 Depiction of NE's varying affinities for its different adrenergic receptors, and the opposing effects of NE on PFC vs. posterior cortical function via these receptors. Norepinephrine has higher affinity for α_{2A}-receptors than for α_1-receptors or β-receptors.[169–171] Thus, moderate levels of NE would engage α_{2A}-receptors and enhance PFC regulation of behavior and thought, while high levels of NE released during stress would predominately engage α_1- and β-receptors, impairing PFC and enhancing posterior cortical regulation. In this way NE can act as a switch for determining which brain regions control our behavior.

A highly consistent body of literature from studies of rodents, monkeys, and humans has established that NE has an essential, beneficial influence on PFC function through actions at postsynaptic, α_{2A}-adrenoceptors.[21] Alpha$_{2A}$-adrenergic agonists such as clonidine, guanfacine, and meditomidine improve the working memory and attention functions of the PFC,[22–25] and these agents become more potent and more efficacious in subjects with NE depletion.[26–29] They are especially effective in reducing distractibility,[30,31] a finding consistent with the long history of NE regulation of attention. Guanfacine can be particularly effective given its more benign side effect profile.[22] Studies in mutant mice have identified the α_{2A}-subtype as the receptor responsible for the cognitive enhancing effects of NE, and mice with a mutation of this receptor have weaker working memory abilities.[32] In monkeys, blockade of α_2-receptors in the PFC produces a striking, delay-related impairment in working memory,[33] while infusion of an α_2-agonist into the PFC of monkeys or rats enhances working memory.[34] A similar profile is observed at the cellular level, where an α_2-agonist increases and an α_2-antagonist decreases delay-related activity,[35] the cellular measure of working memory. These cellular changes are reflected in imaging studies, where systemic administration of guanfacine increases regional cerebral blood flow in the PFC but not in posterior association cortex.[36] Research in progress will determine whether these beneficial actions in the PFC involve coupling through G_i proteins, reducing the detrimental effects of cAMP/PKA intracellular signaling.[37]

In contrast to NE's beneficial actions at α_{2A}-adrenoceptors, high levels of NE release (such as occurs during stress) impair working memory via α_1-adrenoceptors. Thus, stress-induced working memory deficits can be prevented by blocking α_1-adrenoceptors in the PFC.[38] Conversely, one can mimic stress-induced PFC dysfunction by infusing an α_1-adrenergic agonist into the PFC. For example, infusion of phenylephrine into the dorsolateral PFC of the monkey induced a marked, delay-related impairment in performance.[34] Similar effects were observed in rat PFC, including perseverative responding that is characteristic of PFC deficits.[39] This mechanism may contribute to a variety of cognitive disorders that are caused or exacerbated by stress exposure (see below). Furthermore, recent studies indicate that the impairment arising from α_1-adrenoceptor stimulation involves activation of PKC intracellular signaling.[40] Thus, direct activation of PKC with the phorbol ester, phorbol-12-myristate-13-acetate (PMA), markedly impairs PFC function, while deficits induced by PMA, α_1-agonists, or stress can be reversed by the PKC inhibitor, chelerythrine. This work has immediate relevance to bipolar disorder.

In contrast to the α-adrenoceptors, work to date suggests that β-adrenergic receptor stimulation has little effect on PFC function. Systemic administration[26] or direct intra-PFC infusion[33] of a β-receptor antagonist in monkeys had no effect on working memory performance. Studies are needed to explore possible β_1- vs. β_2 receptor mechanisms, and to explain the role of chronic β-receptor stimulation. However, the negative results to date are in striking contrast to the robust effects of NE at β-receptors in posterior cortical and subcortical structures.

NE regulation of posterior and subcortical function

Norepinephrine has widespread effects on arousal state, on the processing of stimuli in the posterior sensory cortices (Figure 14.1), and on long-term memory consolidation and other forms of plasticity. Most of these NE actions require β- and/or α_1-adrenergic mechanisms (Figure 14.2). In contrast, α_2-receptor stimulation often erodes these functions, through both pre- and postsynaptic contributions. This chapter provides a very brief review of this research. More extensive discussion of these topics can be found in Chapter 8 of this text.

Arousal

Norepinephrine profoundly modulates the state of alertness through a variety of actions, many of which involve stimulation of β-adrenergic receptors.

β-adrenergic receptor stimulation in the cortex produces an awake state by potently blocking spike frequency adaptation through a reduction of a K^+ current, I_{AHP}.[41] In the thalamus, application of NE suppresses rhythmic burst activity and switches neurons to a single-spike firing mode capable of transmitting information. This effect is mediated by a β-receptor-mediated enhancement of the hyperpolarization-activated cation current, I_H, and an α_1-receptor mediated suppression of a resting leak K^+ current, I_{KL}.[41] Buzsaki and colleagues have shown that stimulation of α_1-receptors in the thalamus elicits an alert state, while stimulation of postsynaptic α_2-receptors produces sedation.[42] Activation of the electroencephalogram in cortex and hippocampus can also be produced by β- and α_1-adrenergic stimulation in the medial septal nucleus.[43,44] These thalamo-cortical mechanisms allow the cortex to be in a proper electrophysiological state for the processing of sensory stimuli.

Stimulus processing

It has long been appreciated that NE enhances "signal/noise" processing of sensory stimuli, for example, in the auditory cortex[45] and somatosensory cortex.[46] Waterhouse and colleagues have studied the receptor mechanisms underlying these effects and have found that β-adrenergic receptor stimulation can enhance inhibitory responses, while α_1-adrenergic receptor stimulation can enhance excitatory responses.[13,14,16] These actions are thought to contribute to enhanced attention and enriched processing during the alert state. In contrast, α_2-agonists have no effect that actually impairs the visuospatial attentional orienting tasks that depend upon the parietal cortex in either monkeys or in humans.[47–49]

Long-term memory consolidation

Classic studies demonstrated that NE increased signal/noise processing of memory formation in the hippocampus.[50] These mechanisms have focused on β-adrenergic receptors. For example, NE enhances long-term potentiation (LTP) in the mossy fiber-CA3 synapse via a β-adrenergic mechanism.[51] β-adrenergic receptor stimulation appears particularly important for the late phase of memory consolidation,[52] which is thought to involve a cAMP/PKA signaling pathway in hippocampus.[53] β-adrenergic mechanisms also appear to facilitate long-term memory consolidation in the entorhinal and parietal cortices.[54] α_1-adrenergic mechanisms may have a weaker but nonetheless beneficial effect on memory consolidation[55] and LTP in hippocampus.[56] In contrast to the beneficial effects of β- and α_1-adrenergic receptor stimulation, administration of α_2-adrenergic agonists has either no effect or impairs memory tasks (such as the Morris water-maze) that depend on hippocampal functioning.[57] Similar effects are observed at the cellular level, where α_2-receptor stimulation erodes place cell responses.[58]

The enhancement of memory by emotion is thought to occur through epinephrine release in the periphery and β-adrenergic receptor stimulation in the amygdala (reviewed in ref. 59). Thus, infusion of β-adrenergic antagonists into the amygdala impairs long-term memory consolidation, while infusion of a β-adrenergic agonist improves memory consolidation.[59] Recent findings suggest that α_1-adrenergic mechanisms may facilitate β-adrenergic effects.[60] In contrast, α_2-receptor stimulation impairs memory consolidation, and the α_2-agonist, clonidine, is used as an amnestic agent in basic research.[61] Thus, high levels of NE released during emotional events would engage β-receptors and enhance amygdala function, thus reinforcing the long-term memory of the emotional event.

More recently, several lines of research suggest that long-term memory consolidation is accomplished through activation of several intracellular signaling pathways, including PKC, PKA, mitogen-activated protein MAP kinase, and Ca^{2+}/calmodulin-dependent kinase II.[62–69] As β-receptors are generally positively coupled to the cAMP/PKA signaling pathway via G_s proteins, it is likely that NE enhances long-term memory processes via β-adrenergic stimulation of PKA.[53] In contrast, α_2-receptors are generally coupled to G_i proteins that reduce cAMP/PKA signaling. Thus, α_2-receptor stimulation may impair long-term memory consolidation through both presynaptic mechanisms (reducing catecholamine release), and postsynaptic mechanisms (reducing PKA activation).

Plasticity

In addition to β-adrenergic influences on LTP in the hippocampus, NE has been shown to modulate plasticity in the visual[70] and somatosensory[71] cortices. An interaction between noradrenergic and cholinergic mechanisms appears to be critical for plasticity of the visual cortex during development, permitting shifts in ocular dominance columns.[70] In the adult brain, NE α_1 mechanisms have been shown to be important for the modulation of long-term depression (LTD) in visual cortical neurons.[72] In contrast, α_2-receptor stimulation weakened LTD, and this was thought to occur through presynaptic reductions in NE release.[72] The authors speculated that NE modulation of LTD may have a critical modulatory influence on plasticity of receptive fields in the adult brain and during development. These mechanisms may contribute to associative learning throughout the life span.

Stress, NE, and brain systems regulating cognition

Norepinephrine is released in high levels during exposure to stressful events. Given the differential effects of NE at α_2- vs. α_1- or β-receptors, NE may have very different effects on cognitive function dependent upon the amount of NE being released. Thus, during modest levels of NE release (alert, nonstress), NE in cortex may primarily engage α_2-receptors and strengthen PFC cognitive functions, allowing us to inhibit responding to distractions and to plan thoughtfully for the future. In contrast, during stress when high levels of NE are released, α_1- and β-receptors would be substantially engaged, impairing PFC cognitive function and enhancing the functions of the amygdala, hippocampus, and sensory cortices. In this mode, one would be readily distracted by salient stimuli, sensory processing would be more vivid, and emotional events would be strongly etched in memory. Thus, NE may act as a chemical switch, determining which brain regions control our behavior and thoughts. Alterations in this "switch" may disrupt higher cortical functions.

Cognitive disorders with altered NE

The NE cells of the LC degenerate in a number of disease states, such as Alzheimer's and Parkinson's disease. Conversely, as described above, they become overactive under conditions of stress, and thus contribute to many stress-related cognitive disorders such as posttraumatic stress disorder (PTSD). The following is a brief review of NE alterations in these neuropsychiatric conditions, and how changes in NE may contribute to cognitive symptoms.

Disorders with too little NE

Normal aging

In contrast to pathological aging, the normal aging process does not appear to destroy the cells of the LC. Stereological counts suggest that NE cells do not degenerate significantly with normal aging (although subtle changes in subregions of the nucleus would not be detected using this technique). However, within aged animals, even the small degree of LC cell loss correlates strongly with the degree of memory impairment, suggesting that even small changes may have large ramifications for function.[73] Several studies indicate that the LC becomes less active with advanced age. Biochemical studies in rodents show reduced NE turnover in cortex even in late middle age. Recent studies of aged monkeys indicate the LC loses an important excitatory input from the hypothalamus; the orexin/hypocretin input reduces by 50%. Thus, the picture that emerges is an intact system that becomes sluggish with age. In rodents, electrical or pharmacological activation of the LC in middle age[74] or transplantation of NE neurons in old age[75] protected from age-related loss of memory. These findings again emphasize the importance of this system to memory in the aged brain.

In both human and nonhuman primates, the cognitive deficits of normal aging are best characterized by loss of PFC cognitive functions: e.g. impaired working memory, increased susceptibility to interference, and relatively preserved recognition memory. Many changes are likely to contribute to these deficits: e.g. depletion of dopamine, loss of spines and distal dendrites of PFC pyramidal cells, and disrupted myelin. However, reduced NE input to the PFC likely contributes as well, because NE α_2-agonists become increasingly effective in improving PFC functions with advancing age in both rats and monkeys. Reduced NE turnover in the hippocampus, entorhinal/perirhinal cortex, and amygdala likely weakens long-term memory consolidation and recognition memory as well; however these deficits are far more prominent in Alzheimer's disease.

Alzheimer's disease

The profound cognitive deficits of Alzheimer's disease primarily arise from the degeneration of the cortex. The degenerative process appears to originate in the medial temporal lobe (especially entorhinal cortex), and spreads to the nearby hippocampus, to surrounding association cortices, and to subcortical nuclei such as the cholinergic nucleus basalis and the LC. The rostral pole of the LC is particularly susceptible; interestingly, this is the region that receives a heavy input from the PFC. The progression of degenerative changes parallels the progression of cognitive symptoms: recent memory deficits (including impaired recognition memory), progressing to a dementia with global impairments in areas such as language, spatial abilities, logical and mathematical skills, and insight and judgement.

Although most treatments for Alzheimer's disease attempt to enhance cholinergic transmission, there has been some research suggesting that combined cholinergic and noradrenergic therapies may be more effective. However, no replacement therapy is truly helpful, as the major cause for dementia in Alzheimer's disease is the death of the cortex itself. Truly effective treatments will need to halt or reverse the degenerative process.

Interestingly, the aggressive symptoms that often arise late in the illness correlate with increased NE turnover in cerebrospinal fluid (CSF) and loss of NE cells in the rostral core of the LC, which has reciprocal connections with the cortex. Although the NE cells of the LC markedly degenerate in Alzheimer's disease, brain NE turnover actually appears to increase in patients with advanced dementia.[76–78] The appearance of agitated, inappropriate behaviors correlates with elevated plasma NE turnover.[77] An anterior subgroup of LC cells primarily project to the hypothalamus,[79] and it is possible that these remaining cells become disinhibited in Alzheimer's disease, overstimulating the hypothalamus and amplifying aggressive drives. Aggressive symptoms are among the most problematic in Alzheimer's patients, often requiring institutionalization and neuroleptic medication. In this regard, it is of interest that all effective medications have potent α_1-adrenoceptor blocking activity, perhaps strengthening PFC inhibition of inappropriate behaviors.

Parkinson's disease

Although most of the emphasis in Parkinson's disease is on the loss of dopamine cells in the substantia nigra, the NE cells of the LC also degenerate in this disease. Parkinson's patients have marked PFC cognitive deficits, which likely involve insufficient catecholamines in both the caudate nucleus and the PFC itself. Parkinson patients show a progressive profile of deficits resembling frontal lobe lesions. These include working memory deficits, impaired recency judgement, decreased planning, and impaired set-shifting.[80,81] Depletion of NE in the PFC may contribute to the array of PFC cognitive deficits, as neuropathological studies show that loss of the rostral LC cells correlates with cognitive decline.[82] Ironically, the medications given to Parkinson's patients often worsen their cognitive impairment. High doses of I-dopa (which increases both dopaminergic and NE transmission) or selective dopaminergic agonists are often required to restore movement, given the marked depletion of dopamine from the putamen in Parkinson's disease. However, these doses are often too high for the caudate and PFC. These sites have a more modest depletion, and in the case of the PFC, requires more delicate catecholaminergic manipulations.[83] Thus, patients are often left with substantial cognitive impairment.

Korsakoff's amnesia

Korsakoff's amnesia is the result of severe thiamine deficiency, most often found in alcoholics with improper diet. Memory problems in Korsakoff's patients arise at least in part from an increased sensitivity to proactive interference[84] and to shallow encoding, and are often accompanied by confabulation. There has been great debate as to the brain lesions responsible for these cognitive changes. Traditional views focused on the mammillary bodies and medial dorsal nucleus of the thalamus. More recent studies have examined the role of the intralaminar nuclei of the thalamus and the anterior nucleus of the thalamus.[85,86] Debate has been sparked by studies in rats emphasizing the critical role that the intralaminar nuclei play in memory.[87] All of these thalamic nuclei project to the PFC, and all receive a dense projection of NE from the LC.

Biochemical and pharmacological studies also suggest a role for NE in Korsakoff's amnesia. The memory deficits of Korsakoff's syndrome correlate with reduced NE turnover as measured in CSF,[88] even though there does not appear to be significant degeneration of the LC in this disorder.[89] Memory and performance of the Stroop task were improved by administration of the α_2-agonist, clonidine,[90] and clonidine's efficacy correlated with the degree of NE loss in CSF.[88] Imaging studies showed that clonidine enhanced blood flow in the anterior cingulate cortex. Increased blood flow in the left frontal lobe correlated with improved word fluency, a cognitive ability linked to the left PFC in humans.[91]

Attention-deficit/hyperactivity disorder (ADHD)

The cardinal symptoms of ADHD are poor attention regulation (with particular difficulty sustaining attention over a delay), poor impulse control, and locomotor hyperactivity. These symptoms are strongly associated with right PFC dysfunction.[92,93] Functional and structural imaging studies have documented alterations in the PFC of ADHD patients, as well as revealing changes in the cerebellum and striatum.[94] These three brain regions form an interconnecting circuit that plans and executes movements and thoughts. These regions are powerfully modulated by NE (PFC and cerebellum) and dopamine (striatum and PFC).

Many researchers have focused on the dopaminergic aspects of ADHD, and the dopaminergic effects of the stimulant drugs that are most commonly used to treat this disorder. However, more recent studies suggest that NE may be even more important than dopamine, particularly in regard to treatment. Indeed, some researchers have gone as far as calling ADHD an NE disorder.[95] The most common treatments for ADHD are the stimulants methylphenidate (Ritalin) and amphetamine (Adderall). Biochemical studies of methylphenidate and amphetamine actions in rats have shown that low doses (like those used in ADHD) preferentially block the reuptake of NE more than dopamine.[96] Furthermore, as there are few dopamine transporters in the PFC, methylphenidate should primarily alter NE in the PFC. Selective NE compounds are effective in treating ADHD. Norepinephrine antidepressants such as desipramine are very effective in treating this disorder,[95] but their potential cardiac side effects have made them less appropriate in a juvenile population. The more selective NE reuptake blockers currently available appear to have efficacy with a reduced side effect profile,[97] and are likely to become more prominent in the treatment of this disorder. The other selective NE agents commonly used to treat ADHD are the α_2-adrenergic agonists, clonidine and guanfacine. Clonidine was first found to be effective in patients with Tourette's Syndrome,[98] and was then used in ADHD as well, particularly in children who benefited from the strong sedative effects of this compound.[99] Indeed, many researchers assumed that clonidine was helpful simply because it sedated children who were otherwise hyperaroused. However, with the advent of guanfacine, it was found that the sedating and therapeutic effects of α_2-agonists could be dissociated, and it is now believed that these agents improve ADHD symptomology by strengthening PFC function.[92] Guanfacine is particularly useful in patients who cannot take stimulants, e.g. it is effective in ADHD patients with neurological tics,[100] and may also be useful in teenagers and adults with drug abuse problems. Although purely dopaminergic agents (Wellbutrin) have been used to treat ADHD, some experts feel these are less effective than NE agents.[95]

There is extensive literature demonstrating the importance of NE for attention regulation. These studies are consistent with a major role of NE in the etiology and treatment of ADHD. As described in Chapter 7, LC cells fire to stimuli that are of interest to the animal, and it is thought that this phasic NE release leads to enhanced signal/noise processing of stimuli in primary and secondary sensory cortices. Norepinephrine actions in PFC would further promote attention regulation, providing a more thoughtful orchestration of attentional focus. A classic study by Carli *et al.* demonstrated that depletion of forebrain NE leads to distractibility.[101] Distractibility is also seen under conditions of excessive NE release when LC cells fire rapidly in a tonic mode. The distractibility under these conditions is likely due to α_1-receptor-induced impaired PFC function, eroding the top-down regulation of attention and impulse control. Thus, there is a strong scientific basis for NE mechanisms contributing to attention disorders. Unfortunately, it is difficult to gather direct evidence from patients to support or refute this claim. The NE innervation of the PFC and remaining cortex is delicate and very difficult to image (in contrast to the dense dopamine innervation of striatum, which is readily imaged). However, there is one study of adults with ADHD that showed evidence of altered catecholamine innervation of the PFC.[102] This study examined the uptake of fluorodopa into catecholamine terminals using positron emission tomography (PET) imaging. They found reduced uptake in the PFC of ADHD patients. The authors interpreted this as a reduction in dopamine innervation of the PFC, as fluorodopa is a marker of dopamine terminals in the striatum where there is no NE input. However, in the PFC, there are very few dopamine transporters, and thus the fluorodopa would primarily be taken up by NE terminals. Thus, the reduction in fluorodopa uptake in the PFC of ADHD patients likely reflects altered NE transmission in the PFC. This alteration may be a reduction in NE transmission (e.g. reduced innervation, reduced numbers of transporters) or hyperactivity of endogenous NE competing for the fluorodopa. More research is needed in this very difficult arena.

Researchers have also tried to assess NE involvement in ADHD through measures of NE and its metabolites in plasma, urine, and CSF. These studies are obviously very indirect, and likely reflect NE changes in the sympathetic nervous system (plasma, urine), or more densely innervated brain regions that are close to the ventricles, such as the hypothalamus (CSF). However, there are a few studies which find reduced levels of NE turnover in ADHD, as reflected in reduced levels of the NE metabolite 3-methoxy-4-hydroxyphenylgycol (MHPG). Attention-deficit/hyperactivity disorder ADHD patients without reading disorder had lower plasma MHPG levels than controls, and this reduction related to success in school.[103] Similarly, boys with low plasma dopamine-β-hydroxylase activity (the synthetic enzyme for NE) had lower levels of urinary NE and MHPG, which correlated with ADHD symptoms.[104]

Recent studies suggest a possible genetic basis for these findings. Alterations in the *dopamine-β-hydroxylase* gene are associated with increased susceptibility to

ADHD,[105–107] suggesting that impaired ability to synthesize NE could contribute to ADHD symptoms in some individuals. There is controversy as to whether the gene for the α_{2A}-adrenergic receptor may be related as well.[107,108] Reduced signaling at this receptor would weaken PFC regulation of behavior and induce ADHD-like behaviors, a result observed in mice with a mutation of the α_2-adrenergic receptor.[32] Finally, polymorphisms in the dopamine D_4-receptor have been associated with ADHD.[109,110] Although the D_4-receptor is designated a dopamine receptor, NE has higher affinity for the D_4-receptor than any adrenergic receptor,[111] and D_4-receptors are found in high density in brain regions, such as the hippocampus, that are rich in NE and sparse in dopamine. Thus, alterations in this receptor in ADHD likely reflect disruptions in NE signaling. Subsequently, there are a number of NE mechanisms that are implicated in the etiology of ADHD, and a number of NE compounds that are used to treat this disorder.

Depression

Depression has a significant, cognitive component that likely involves dysregulation of NE and dysfunction of the PFC. Lesions of the PFC commonly produce poor concentration, impaired initiative, and perseverative thoughts, symptoms associated with depression. It has been noted for many years that strokes in the left PFC often induce depression.[112] In contrast, transmagnetic stimulation of the left PFC elevates mood.[113] Functional imaging studies have established that PFC activity is abnormal in depressed patients,[114–116] and these abnormalities correlate with reduced performance of PFC tasks.[117] Imaging studies have often focused on abnormalities of the ventromedial PFC,[115] a region tightly interconnected with limbic structures (such as the amygdala) that regulate emotional state. Neuropathological studies have confirmed PFC abnormalities in depression, including decreases in size and density of neurons and glia.[118]

Exposure to uncontrollable stress markedly impairs PFC function, and stress exposure is an important risk factor for depression, especially if the person feels no sense of control over the stressor.[119–121] The high levels of steroid release in both depression and stress have been of particular interest,[122] as has the role of NE. Stress activates the LC and elevates NE turnover throughout much of the neuroaxis. Weiss and Simson[123] showed that chronic, severe stress depleted NE from the LC in rats, reducing presynaptic regulatory control of LC firing and producing a disinhibited response to excitatory stimuli. This finding has been reinforced by recent studies showing that presynaptic α_2-receptors on LC neurons are rapidly and persistently down-regulated by chronic stress.[124] In this state, NE cells would likely release insufficient NE under basal conditions, but excessive NE when LC cells are excited. Thus, during conditions of chronic, uncontrollable stress, PFC neurons may intermittently receive either too little and too much NE, both of which are known to impair PFC function. Interestingly, antidepressant treatment (including selective NE reuptake inhibitors) appears to prevent stress-induced NE release,[125,126] and thus may be protective in this regard.

The role of NE in depression is discussed extensively in this volume in Chapter 11. Although there are consistent findings of abnormalities in the NE system of depressed patients, there remains debate as to whether the depressive state is associated with insufficient or excessive NE release.[127,128] As described above, both may be true under different conditions, and thus findings with animals exposed to chronic stress may apply to human patients with depression. Increased NE levels and α_1-receptor binding has been observed in the PFC and temporal cortex of suicide victims, but as MHPG was not measured, it is not known whether this reflected increased or decreased NE transmission.[129] A recent study reports increased NE in the CSF of patients with melancholic depression,[128] a disorder with prominent symptoms of stereotyped, negative thoughts. It is tempting to speculate that these injurious thoughts arise from lack of PFC inhibitory control, resulting from high levels of NE release engaging α_1-receptors, impairing PFC inhibitory functions, and inducing a perseverative cognitive profile. Conversely, depletion of brain monoamines with chronic reserpine treatment induces depressive symptomology in humans, and markedly impairs PFC functions in monkeys.[27] Reserpine's detrimental effects in monkeys were reversed by administration of the α_2-agonist, clonidine, indicating that insufficient postsynaptic α_2-receptor stimulation contributed significantly to the deficits in PFC function. Thus, it is possible that either excessive α_1- and/or insufficient postsynaptic α_2-receptor stimulation contributes to PFC dysfunction in depression.

A recent PET imaging study indicates that NE is altered in the PFC of patients with major depression.[130] Patients and controls performed a sustained attention task that challenges the right PFC while receiving an intravenous infusion of the α_2-agonist, clonidine, or saline vehicle. Clonidine increased regional cerebral blood flow in the right superior PFC of patients but not controls. The authors interpreted this finding as consistent with reduced presynaptic and supersensitive postsynaptic α_2 receptors in the PFC of depressed patients. Enhanced PFC response to clonidine has been observed in animals with both depleted NE (e.g. aged or reserpine-treated)[27] and in stressed animals with excessive NE release.[131] Thus, there appears to be dysregulated NE in the PFC of depressed patients, which may induce PFC dysfunction and contribute to depressive symptomology. Antidepressant treatment may prevent NE dysregulation and protect PFC regulation of affect and attention.

Disorders with too much NE

Mania

Traditionally, the manic state of bipolar disorder has been associated with excessive NE activity.[127] Recently, bipolar disorder has been associated with abnormally high levels of PKC activity, a critical enzyme in the phosphatidylinositol (PI) intracellular signaling pathway. Tissue from the brains of bipolar patients have revealed abnormalities throughout this signaling pathway, including increased levels of G_q proteins,[132] altered hydrolysis of PI,[133] and increased PKC activity.[134,135] A polymorphism that confers susceptibility to bipolar disorder has been identified in the gene encoding for phospholipase C, a critical enzyme in the PKC cascade, implicating a genetic contribution.[136] Importantly, all efficacious antimanic agents, such as lithium, reduce PKC activity.[137] Thus, there is a highly consistent literature implicating excessive PKC activity in bipolar disorder. High levels of NE activate PKC signaling via α_1-adrenergic receptors coupled to G_q proteins, thus relating the historical and recent research on this illness.

Many of the symptoms of mania (altered mood, impulsivity, poor judgement, flight of ideas, and distractibility) reflect dysfunction of the PFC, particularly of the right hemisphere. Converging evidence demonstrates that the PFC is dysfunctional in bipolar disorder. Functional imaging studies and postmortem analyses have shown abnormalities in the PFC of bipolar patients, with the right, medial PFC especially altered in the manic state.[138,139] Interestingly, manic episodes are often precipitated by exposure to stressors,[140] which would increase NE stimulation of α_1-receptors and thus PKC signaling. Our own data demonstrate that PKC activation markedly impairs PFC function,[40] indicating a direct mechanism by which excessive NE and PKC activity could produce many of the symptoms of mania.

Schizophrenia

Historically, research on neuromodulators and schizophrenia has focused on dopamine, with more recent emphasis on serotonin actions at 5-HT_{2A}-receptors. Much of this emphasis has arisen from the finding that effective antipsychotics block the D_2 family of dopamine receptors. The creation of more effective, atypical neuroleptics has brought focus to the 5-HT_{2A}-receptor, as all atypical neuroleptics block 5-HT_{2A}-receptors, while hallucinogens stimulate this receptor. However, all effective anti psychotics, typical or atypical, also block the NE α_1-receptor,[141] and there has been little discussion of the importance of this mechanism to therapeutic efficacy in schizophrenia. Baldessarini[142] has been one of the few to point out the potential importance of α_1-adrenergic receptor blockade to schizophrenia. Van Kammen has also emphasized the importance of NE hyperactivity in schizophrenic patients.[143] For example, increased NE metabolites in CSF are observed during acute psychosis, and these increases relate to both positive and negative symptoms.[144,145] Increased NE activity also relates to relapse.[146,147] We now know that excessive NE α_1-receptor stimulation impairs prefontal cortex function, suggesting that a hyper-NE state in schizophrenia could contribute to hypofrontality. Prefrontal cortex deficits in schizophrenic patients are well established,[148,149] including evidence linking PFC dysfunction to thought disorder, hallucinations and impaired cognitive control of eye movements.[150,151] Neuropathological studies of the schizophrenic brain have observed changes in a variety of PFC components, including pyramidal cells, GABAergic interneurons, and dopaminergic systems.[152] However, the role of NE in this disorder deserves further attention, given the link between α_1-receptor blockade and therapeutic efficacy of antipsychotic drugs. Preliminary evidence suggests that clonidine can improve thought disorder in haloperidol-treated patients, while guanfacine may be able to improve PFC functions in schizophrenic patients taking atypical antipsychotic medication.[153,154] However, given that many atypical neuroleptics block α_2-receptors, this treatment strategy may be counter productive. Breier[155] has found that indices of increased NE activity relate to clinical improvement in patients taking clozapine. Increased NE activity in the presence of α_1-receptor blockade may be helpful in this disorder.

Posttraumatic stress disorder (PTSD)

Previous research on PTSD has emphasized the important role of medial temporal lobe structures in PTSD symptomology. Heightened conditioned fear and enhanced memory of the traumatic event are likely mediated by the amygdala, while chronic exposure to high levels of cortisol may erode hippocampal morphology. The amygdala is capable of activating the hypothalamic–pituitary–adrenal axis, the sympathetic nervous system, as well as ascending arousal systems in the brain, and thus induces a hypervigilant and anxious state. More recent research has found evidence of marked PFC dysfunction as well. Posttraumatic stress disorder patients are impaired on tests of PFC function, and show reduced activity in the PFC during imaging studies.[156–159] Southwick has postulated that these PFC deficits may contribute to symptoms of intrusive memories and flashbacks, due to loss of PFC inhibitory properties and "reality testing" abilities, respectively.[160] Consistent with this idea is the recent report that injury to the right PFC induced flashbacks in a Vietnam veteran with PTSD.[161]

Norepinephrine plays a vital role in PTSD.[160] There is extensive evidence of hyperactivity of the NE system in PTSD, for example, increased NE in the CSF of PTSD patients.[162] Norepinephrine hyperactivity likely contributes to many symptoms of PTSD. It would heighten both the processing of sensory stimuli and the vividness of memories recalled from storage; it would strengthen conditioned fear responses mediated by the amygdala; and it would lead to weakening of PFC inhibitory functions. Thus, increased NE produces a "double whammy" in PTSD patients, strengthening traumatic memories and the fear response to current stimuli, while eroding the calming influences of the PFC. This idea was substantiated in a study where the α_2-antagonist yohimbine was given to PTSD patients to increase NE release (as well as blocking postsynaptic α_2-receptors in the PFC). Yohimbine induced flashbacks and reduced activity in the PFC of PTSD patients, while having little effect in controls.[163] It is likely that similar events occur due to endogenous NE activation when PTSD patients are reminded of a traumatic event.

Many of the agents used to treat PTSD impact the NE system. For example, antidepressant treatments that alter NE transmission are commonly used, as are agents that directly interact with adrenergic receptors. Examples include the α_2-agonists clonidine and guanfacine and the α_1-antagonist prazosin.[164–166] These agents should protect both the PFC and the posterior cortices from excessive NE stimulation of α_1-receptors, thus "flipping the switch" back to PFC regulation of thought and behavior (see Figure 14.2). Guanfacine and clonidine can both reduce NE release and directly strengthen PFC function through actions at pre- and postsynaptic receptors, respectively. Similar protective effects are seen in abused children given clonidine.[167,168] Thus, there are consistent findings that establish NE as an important factor in both etiology and treatment of PTSD.

Summary

Norepinephrine has powerful influences on higher cortical functions, orchestrating the cortical circuitry controlling our behavior and our thoughts. Recent research suggests that NE can act as a chemical "switch," strengthening PFC control of behavior under alert, nonstressful conditions, while impairing PFC function and strengthening posterior cortical and subcortical structures during stressful ones. Dysregulation of this NE "switch" appears to play a significant role in many cognitive disorders. Reduced NE transmission in degenerative or genetic disorders weakens PFC as well as posterior cognitive processes. Conversely, excessive NE release during stress may contribute to cognitive dysfunction in many neuropsychiatric disorders, impairing PFC functions while strengthening subcortical and posterior cortical regulation. Thus, NE plays a key role in the etiology and treatment of a variety of cognitive disorders.

ACKNOWLEDGEMENT

This work was supported by MERIT Award AG06036 to AFTA.

REFERENCES

1. Lewis, D. A. and Morrison, J. H. Noradrenergic innervation of monkey prefrontal cortex: a dopamine-beta-hydroxylase immunohistochemical study. *J. Comp. Neurol.*, **282** (1989), 317–330.

2. Arnsten, A. F. T. and Goldman-Rakic, P. S. Selective prefrontal cortical projections to the region of the locus coeruleus and raphe nuclei in the rhesus monkey. *Brain Res.*, **306** (1984), 9–18.
3. Sara, S. J. and Herve-Minvielle, A. Inhibitory influence of frontal cortex on locus coeruleus. *Proc. Natl. Acad. Sci. U. S. A.*, **92** (1995), 6032–6036.
4. Jodo, E., Chiang, C. and Aston-Jones, G. Potent excitatory influence of prefrontal cortex activity on noradrenergic locus coeruleus neurons. *Neuroscience*, **83** (1998), 63–79.
5. MacDonald, E., Kobilka, B. K. and Scheinin, M. Gene targeting – homing in on alpha-2-adrenoceptor subtype function. *Trends Pharmacol. Sci.*, **18** (1997), 211–219.
6. Aoki, C., Go, C.-G., Venkatesan, C. and Kurose, H. Perikaryal and synaptic localization of alpha-2A-adrenergic receptor-like immunoreactivity. *Brain Res.*, **650** (1994), 181–204.
7. Aoki, C., Venkatesan, C., Go, C.-G., Forman, R. and Kurose, H. Cellular and subcellular sites for noradrenergic action in the monkey dorsolateral prefrontal cortex as revealed by the immunocytochemical localization of noradrenergic receptors and axons. *Cereb. Cortex*, **8** (1998), 269–277.
8. Goldman-Rakic, P. S., Lidow, M. S. and Gallager, D. W. Overlap of dopaminergic, adrenergic, and serotonergic receptors and complementarity of their subtypes in primate prefrontal cortex. *J. Neurosci.*, **10** (1990), 2125–2138.
9. Duman, R. S. and Nestler, E. J. Signal transduction pathways for catecholamine receptors. In *Psychopharmacology: The Fourth Generation of Progress*, ed. F. E. Bloom and D. J. Kupfer. (New York: Raven Press, 1995), pp. 303–320.
10. Hieble, J. P., Bylund, D. B., Clarke, D. E. *et al.* International Union of Pharmacology. X. Recommendation for nomenclature of alpha 1-adrenoceptors: consensus update. *Pharmacol. Rev.*, **47** (1995), 267–270.
11. Pieribone, V. A., Nicholas, A. P., Dagerlind, A. and Hökfelt, T. Distribution of alpha1 adrenoceptors in rat brain revealed by in situ hybridization experiments utilizing subtype-specific probes. *J. Neurosci.*, **14** (1994), 4252–4268.
12. Day, H. E., Campeau, S., Watson, S. J. J. and Akil, H. Distribution of alpha 1a-, alpha 1b- and alpha 1d-adrenergic receptor mRNA in the rat brain and spinal cord. *J. Chem. Neuroanat.*, **13** (1997), 115–139.
13. Waterhouse, B. D., Moises, H. C. and Woodward, D. J. Alpha-receptor-mediated facilitation of somatosensory cortical neuronal responses to excitatory synaptic inputs and iontophoretically applied acetylcholine. *Neuropharmacology*, **20** (1981), 907–920.
14. Mouradian, R. D., Seller, F. M. and Waterhouse, B. D. Noradrenergic potentiation of excitatory transmitter action in cerebrocortical slices: evidence of mediation by an alpha1-receptor-linked second messenger pathway. *Brain Res.*, **546** (1991), 83–95.
15. Insel, P. A. Adrenergic receptors, G proteins, and cell regulation: implications for aging research. *Exp. Gerontol.*, **28** (1993), 341–348.
16. Waterhouse, B. D., Moises, H. C. and Woodward, D. J. Noradrenergic modulation of somatosensory cortical neuronal responses to iontophoretically applied putative transmitters. *Exp. Neurol.*, **69** (1980), 30–49.
17. Hansson, E. and Ronnback, L. Receptor regulation of the glutamate, GABA and taurine high-affinity uptake into astrocytes in primary culture. *Brain Res.*, **548** (1991), 215–221.

18. Fillenz, M., Lowry, J. P., Boutelle, M. G. and Fray, A. E. The role of astrocytes and noradrenaline in neuronal glucose metabolism. *Acta Physiol. Scand.*, **167** (1999), 275–284.
19. Goldman-Rakic, P. S. Circuitry of the primate prefrontal cortex and the regulation of behavior by representational memory. In *Handbook of Physiology, The Nervous System, Higher Functions of the Brain*, Vol. V, ed. F. Plum. (Bethesda: American Physiological Society, 1987), pp. 373–417.
20. Knight, R. T., Staines, W. R., Swick, D. and Chao, L. L. Prefrontal cortex regulates inhibition and excitation in distributed neural networks. *Acta Psychol. (Amst.)*, **101** (1999), 159–178.
21. Arnsten, A. F. T. Catecholamine modulation of prefrontal cortical cognitive function. *Trends Cognit. Sci.*, **2** (1998), 436–447.
22. Arnsten, A. F. T., Cai, J. X. and Goldman-Rakic, P. S. The alpha-2 adrenergic agonist guanfacine improves memory in aged monkeys without sedative or hypotensive side effects. *J. Neurosci.*, **8** (1988), 4287–4298.
23. Rama, P., Linnankoski, I., Tanila, H., Pertovaara, A. and Carlson, S. Medetomidine, atipamezole, and guanfacine in delayed response performance of aged monkeys. *Pharmacol. Biochem. Behav.*, **54** (1996), 1–7.
24. Tanila, H., Rama, P. and Carlson, S. The effects of prefrontal intracortical microinjections of an alpha-2 agonist, alpha-2 antagonist and lidocaine on the delayed alternation performance of aged rats. *Brain Res. Bull.*, **40** (1996), 117–119.
25. Jakala, P., Riekkinen, M., Sirvio, J. *et al.* Guanfacine, but not clonidine, improves planning and working memory performance in humans. *Neuropsychopharmacology*, **20** (1999), 460–470.
26. Arnsten, A. F. T. and Goldman-Rakic, P. S. Alpha-2 adrenergic mechanisms in prefrontal cortex associated with cognitive decline in aged nonhuman primates. *Science*, **230** (1985), 1273–1276.
27. Cai, J. X., Ma, Y., Xu, L. and Hu, X. Reserpine impairs spatial working memory performance in monkeys: reversal by the alpha-2 adrenergic agonist clonidine. *Brain Res.*, **614** (1993), 191–196.
28. Franowicz, J. S. and Arnsten, A. F. T. Treatment with the noradrenergic alpha-2 agonist clonidine, but not diazepam, improves spatial working memory in normal young rhesus monkeys. *Neuropsychopharmacology*, **21** (1999), 611–621.
29. Carlson, S., Tanila, H., Rama, P., Mecke, E. and Pertovaara, A. Effects of medetomidine, an alpha-2 adrenoceptor agonist, and atipamezole, an alpha-2 antagonist, on spatial memory performance in adult and aged rats. *Behav. Neural Biol.*, **58** (1992), 113–119.
30. Jackson, W. J. and Buccafusco, J. J. Clonidine enhances delayed matching-to-sample performance by young and aged monkeys. *Pharmacol. Biochem. Behav.*, **39** (1991), 79–84.
31. Arnsten, A. F. T. and Contant, T. A. Alpha-2 adrenergic agonists decrease distractability in aged monkeys performing a delayed response task. *Psychopharmacology*, **108** (1992), 159–169.
32. Franowicz, J. S., Kessler, L., Dailey-Borja, C. M. *et al.* Mutation of the alpha2A-adrenoceptor impairs working memory performance and annuls cognitive enhancement by guanfacine. *J. Neurosci.*, **22** (2002), 8771–8777.

33. Li, B.-M. and Mei, Z.-T. Delayed response deficit induced by local injection of the alpha-2 adrenergic antagonist yohimbine into the dorsolateral prefrontal cortex in young adult monkeys. *Behav. Neural Biol.*, **62** (1994), 134–139.
34. Mao, Z.-M., Arnsten, A. F. T. and Li, B.-M. Local infusion of alpha-1 adrenergic agonist into the prefrontal cortex impairs spatial working memory performance in monkeys. *Biol. Psychiatry*, **46** (1999), 1259–1265.
35. Li, B.-M., Mao, Z.-M., Wang, M. and Mei, Z.-T. Alpha-2 adrenergic modulation of prefrontal cortical neuronal activity related to spatial working memory in monkeys. *Neuropsychopharmacology*, **21** (1999), 601–610.
36. Avery, R. A., Franowicz, J. S., Studholme, C., van Dyck, C. H. and Arnsten, A. F. T. The alpha-2A-adrenoceptor agonist, guanfacine, increases regional cerebral blood flow in dorsolateral prefrontal cortex of monkeys performing a spatial working memory task. *Neuropsychopharmacology*, **23** (2000), 240–249.
37. Taylor, J. R., Birnbaum, S. G., Ubriani, R. and Arnsten, A. F. T. Activation of protein kinase A in prefrontal cortex impairs working memory performance. *J. Neurosci.*, **19** (1999), RC23.
38. Birnbaum, S. G., Gobeske, K. T., Auerbach, J., Taylor, J. R. and Arnsten, A. F. T. A role for norepinephrine in stress-induced cognitive deficits: alpha-1-adrenoceptor mediation in prefrontal cortex. *Biol Psychiatry*, **46** (1999), 1266–1274.
39. Arnsten, A. F. T., Mathew, R., Ubriani, R., Taylor, J. R. and Li, B.-M. Alpha-1 noradrenergic receptor stimulation impairs prefrontal cortical cognitive function. *Biol. Psychiatry*, **45** (1999), 26–31.
40. Birnbaum, S. G., Yuan, P. X., Wang, M. *et al.* Protein kinase C overactivity impairs prefrontal cortical regulation of working memory. *Science*, **306** (5697), 882–884.
41. McCormick, D. A., Pape, H. C. and Williamson, A. Actions of norepinephrine in the cerebral cortex and thalamus: implications for function of the central noradrenergic system. *Prog. Brain Res.*, **88** (1991), 293–305.
42. Buzsaki, G., Kennedy, B., Solt, V. B. and Ziegler, M. Noradrenergic control of thalamic oscillation: the role of alpha-2 receptors. *Eur. J. Neurol.*, **3** (1991), 222–229.
43. Berridge, C. W. and Foote, S. L. Enhancement of behavioral electroencephalographic (E. E. G), and electromyographic (E. M. G) indices of waking following stimulation of noradrenergic beta-receptors located within the medial septal region of the basal forebrain in the unanesthetized rat. *J. Neurosci.*, **16** (1996), 6999–7009.
44. Berridge, C. W. and Espana, R. A. Synergistic sedative effects of noradrenergic alpha(1)- and beta-receptor blockade on forebrain electroencephalographic and behavioral indices. *Neuroscience*, **99** (2000), 495–505.
45. Foote, S. L., Freedman, F. E. and Oliver, A. P. Effects of putative neurotransmitters on neuronal activity in monkey auditory cortex. *Brain Res.*, **86** (1975), 229–242.
46. Waterhouse, B. D., Moises, H. C. and Woodward, D. J. Phasic activation of the locus coeruleus enhances responses of primary sensory cortical neurons to peripheral receptive field stimulation. *Brain Res.*, **790** (1998), 33–44.
47. Witte, E. A. and Marrocco, R. T. Alterations of brain noradrenergic activity in rhesus monkeys affects the alerting component of covert orienting. *Psychopharmacology*, **132** (1997), 315–323.

48. Clark, C. R., Geffen, G. M. and Geffen, L. B. Catecholamines and attention II: pharmacological studies in normal humans. *Neurosci. Biobehav. Rev.*, **11** (1987), 353–364.
49. Coull, J. T., Nobre, A. C. and Frith, C. D. The noradrenergic alpha2 agonist clonidine modulates behavioural and neuroanatomical correlates of human attentional orienting and alerting. *Cereb. Cortex*, **11** (2001), 73–84.
50. Segal, M. and Bloom, F. E. The action of norepinephrine in the rat hippocampus. IV. The effects of locus coeruleus stimulation on evoked hippocampal unit activity. *Brain Res.*, **107** (1976), 513–525.
51. Hopkins, W. F. and Johnston, D. Noradrenergic enhancement of long-term potentiation at mossy fiber synapses in the hippocampus. *J. Neurophysiol.*, **59** (1988), 667–687.
52. Roullet, P. and Sara, S. Consolidation of memory after its reactivation: involvement of beta noradrenergic receptors in the late phase. *Neural Plast.*, **6** (1998), 63–68.
53. Bevilaqua, L., Ardenghi, P., Schroder, N. *et al.* Drugs acting upon the cyclic adenosine monophosphate/protein kinase A signalling pathway modulate memory consolidation when given late after training into rat hippocampus but not amygdala. *Behav. Pharmacol.*, **8** (1997), 331–338.
54. Ardenghi, P., Barros, D., Izquierdo, L. A. *et al.* Late and prolonged post-training memory modulation in entorhinal and parietal cortex by drugs acting on the cAMP/protein kinase A signaling pathway. *Behav. Pharmacol.*, **8** (1997), 745–751.
55. Puumala, T., Greijus, S., Narinen, K. *et al.* Stimulation of alpha-1 adrenergic receptors facilitates spatial learning in rats. *Eur. Neuropsychopharmacol.*, **8** (1998), 17–26.
56. Pussinen, R. and Sirvio, J. Minor role for alpha-1-adrenoceptors in the facilitation of induction and early maintenance of long-term potentiation in the CA1 field of hippocampus. *J. Neurosci. Res.*, **51** (1998), 309–315.
57. Sirviö, J., Riekkinen, P., Vajanto, I., Koivisto, E. and Riekkinen, P. J. The effects of guanfacine, alpha-2 agonist, on the performance of young and aged rats in spatial navigation task. *Behav. Neural Biol.*, **56** (1991), 101–107.
58. Tanila, H. Noradrenergic regulation of hippocampal place cells. *Hippocampus*, **11** (2001), 793–808.
59. Cahill, L. and McGaugh, J. L. Modulation of memory storage. *Curr. Opin. Neurobiol.*, **6** (1996), 237–242.
60. Ferry, B., Roozendaal, B. and McGaugh, J. L. Basolateral amygdala noradrenergic influences on memory storage are mediated by an interaction between beta- and alpha-1-adrenoceptors. *J. Neurosci.*, **19** (1999), 5119–5123.
61. Genkova-Papazova, M., Petkova, B. P., Lazarova-Bakarova, M., Boyanova, A. and Staneva-Stoytcheva, D. Effects of flunarizine and nitrendipine on electroconvulsive shock- and clonidine-induced amnesia. *Pharmacol. Biochem. Behav.*, **56** (1997), 583–587.
62. Paylor, R., Rudy, J. W. and Wehner, J. M. Acute phorbol ester treatment improves spatial learning performance in rats. *Behav. Brain Res.*, **45** (1991), 189–193.
63. Abeliovich, A., Chen, C., Goda, Y. *et al.* Modified hippocampal long-term potentiation in PKC gamma-mutant mice. *Cell*, **75** (1993), 1253–1262.
64. Bach, M. E., Hawkins, R. D., Osman, M., Kandel, E. and Mayford, M. Impairment of spatial but not contextual memory in CAMKII mutant mice with a selective loss of hippocampal L. T. P in the range of the theta frequency. *Cell*, **81** (1995), 905–915.

65. Bernabeu, R., Bevilaqua, L., Ardenghi, P. *et al.* Involvement of hippocampal cAMP/cAMP-dependent protein kinase signaling pathways in a late memory consolidation phase of aversively motivated learning in rats. *Proc. Natl. Acad. Sci. U. S. A.*, **94** (1997), 7041–7046.
66. Abel, T., Nguyen, P. V., Barad, M. *et al.* Genetic demonstration of a role for PKA in the late phase of LTP and in hippocampus-based long-term memory. *Cell*, **88** (1997), 615–626.
67. Bourtchouladze, R., Abel, T., Berman, N. *et al.* Different training procedures recruit either one or two critical periods for contextual memory consolidation, each of which requires protein synthesis and PKA. *Learn. Mem.*, **5** (1998), 365–374.
68. Barad, M., Bourtchouladze, R., Winder, D. G., Golan, H. and Kandel, E. Rolipram, a type IV-specific phosphodiesterase inhibitor, facilitates the establishment of long-lasting long-term potentiation and improves memory. *Proc. Natl. Acad. Sci. U. S. A.*, **95** (1998), 15 020–15 025.
69. Schafe, G. E., Nadel, N. V., Sullivan, G. M., Harris, A. and LeDoux, J. E. Memory consolidation for contextual and auditory fear conditioning is dependent on protein synthesis, PKA, and MAP kinase. *Learn. Mem.*, **6** (1999), 97–110.
70. Bear, M. F. and Singer, W. Modulation of visual cortical plasticity by acteylcholine and noradrenaline. *Nature*, **320** (1986), 172–176.
71. Levin, B. E., Craik, R. L. and Hand, P. J. The role of norepinephrine in adult somatosensory cortex (SmI) cortical metabolism and plasticity. *Brain Res.*, **443** (1988), 261–271.
72. Kirkwood, A., Rozas, C., Kirkwood, J., Perez, F. and Bear, M. F. Modulation of long-term synaptic depression in visual cortex by acetylcholine and norepinephrine. *J. Neurosci.*, **19** (1999), 1599–1609.
73. Leslie, F. M., Loughlin, S. E., Sternberg, D. B. *et al.* Noradrenergic changes in senescent memory loss. *Brain Res.*, **359** (1985), 292–299.
74. Zornetzer, S. F. Catecholamine system involvement in age-related memory dysfunction. *Ann. N. Y. Acad. Sci.*, **444** (1985), 242–254.
75. Collier, T. J., Gash, D. M. and Sladek, J. R. J. Transplantation of norepinephrine neurons into aged rats improves performance of a learned task. *Brain Res.*, **448** (1988), 77–87.
76. Gottfries, C.-G., Adolfsson, R., Aquilonius S.-M. *et al.* Biochemical changes in dementia disorders of Alzheimer's type (AD/SDAT). *Neurobiol. Aging*, **4** (1983), 261–271.
77. Lawlor, B. A., Bierer, L. M., Ryan, T. M. *et al.* Plasma MHPG and clinical symptoms in Alzheimer's disease. *Biol. Psychiatry*, **38** (1995), 185–188.
78. Elrod, R., Peskind, E. R., DiGiacomo, L. *et al.* Effects of Alzheimer's disease severity on cerebrospinal fluid norepinephrine concentration. *Am. J. Psychiatry*, **154** (1997), 25–30.
79. Loughlin, S. E., Foote, S. L. and Bloom, F. E. Efferent projections of nucleus locus coeruleus: topographic organization of cells of origin demonstrating three-dimensional reconstruction. *Neuroscience*, **18** (1986), 291–306.
80. Sagar, H. J., Sullivan, E. V., Gabrieli, J. D., Corkin, S. and Growdon, J. H. Temporal ordering and short-term memory deficits in Parkinson's disease. *Brain*, **111** (1988), 525–539.
81. Owen, A. M., James, M., Leigh, P. N. *et al.* Fronto-striatal cognitive deficits at different stages of Parkinson's disease. *Brain*, **115** (1992), 1727–1751.
82. Chan-Palay, V. and Asan, E. Alterations in catecholamine neurons of the locus coeruleus in senile dementia of the Alzheimer type and in Parkinson's disease with and without dementia and depression. *J. Comp. Neurol.*, **287** (1989), 373–392.

83. Arnsten, A. F. T. and Robbins, T. W. Neurochemical modulation of prefrontal cortical function in humans and animals. In *The Frontal Lobes*, ed. D. Stuss and R. T. Knight. (New York: Oxford University Press, 2002), pp. 51–84.
84. Butters, N., Tarlow, S., Cermak, L. S. and Sax, D. A comparison of the information processing deficits of patients with Huntington's chorea and Korsakoff's syndrome. *Cortex*, **12** (1976), 134–144.
85. Mennemeier, M., Fennell, E., Valenstein, E. and Heilman, K. M. Contributions of the left intralaminar and medial thalamic nuclei to memory. Comparisons and report of a case. *Arch. Neurol.*, **49** (1992), 1050–1058.
86. Harding, A., Halliday, G., Caine, D. and Kril, J. Degeneration of anterior thalamic nuclei differentiates alcoholics with amnesia. *Brain*, **123** (2000), 141–154.
87. Mair, R. G. On the role of thalamic pathology in diencephalic amnesia. *Rev. Neurosci.*, **5** (1994), 105–140.
88. McEntee, W. J. and Mair, R. G. The Korsakoff syndrome: a neurochemical perspective. *Trends Neurosci.*, **13** (1990), 340–344.
89. Halliday, G., Ellis, J. and Harper, C. The locus coeruleus and memory: a study of chronic alcoholics with and without the memory impairment of Korsakoff's psychosis. *Brain Res.*, **598** (1992), 33–37.
90. Mair, R. G. and McEntree, W. J. Cognitive enhancement in Korsakoff's psychosis by clonidine: a comparison with 1-dopa and ephedrine. *Psychopharmacology*, **88** (1986), 374–380.
91. Moffoot, A., O'Carroll, R. E., Murray, C. *et al.* Clonidine infusion increases uptake of Tc-exametazime in anterior cingulate cortex in Korsakoff's psychosis. *Psychol. Med.*, **24** (1994), 53–61.
92. Arnsten, A. F. T., Steere, J. C. and Hunt, R. D. The contribution of alpha2-noradrenergic mechanisms to prefrontal cortical cognitive function: potential significance to attention deficit hyperactivity disorder. *Arch. Gen. Psychiatry*, **53** (1996), 448–455.
93. Barkley, R. A. *ADHD and the Nature of Self-Control* (New York: Guilford Press, 1997).
94. Arnsten, A. F. T. and Castellanos, F. X. Neurobiology of attention regulation and its disorders. In *Pediatric Psychopharmacology: Principles and Practice*, ed. A. Martin, L. Scahill, D. Charney and J. Leckman. (New York: Oxford University Press, 2002), pp. 99–109.
95. Biederman, J. and Spencer, T. J. Genetics of childhood disorders: XIX. ADHD, Part 3: Is ADHD a noradrenergic disorder? *J. Am. Acad. Child Adolesc. Psychiatry*, **39** (2000), 1330–1333.
96. Kuczenski, R. and Segal, D. S. Locomotor effects of acute and repeated threshold doses of amphetamine and methylphenidate: relative roles of dopamine and norepinephrine. *J. Pharmacol. Exp. Ther.*, **296** (2001), 876–883.
97. Spencer, T., Biederman, J., Wilens, T. *et al.* Effectiveness and tolerability of tomoxetine in adults with attention deficit hyperactivity disorder. *Am. J. Psychiatry*, **155** (1998), 693–695.
98. Cohen, D. J., Young, J. G., Nathanson, J. A. and Shaywitz, B. A. Clonidine in Tourette's Syndrome. *Lancet*, **2** (1979), 551–553.
99. Hunt, R. D., Mindera, R. B. and Cohen, D. J. Clonidine benefits children with attention deficit disorder and hyperactivity: reports of a double-blind placebo-crossover therapeutic trial. *J. Am. Acad. Child Psychiatry*, **24** (1985), 617–629.

100. Scahill, L., Chappell, P. B., Kim, Y. S. *et al.* Guanfacine in the treatment of children with tic disorders and ADHD: a placebo-controlled study. *Am. J. Psychiatry*, **158** (2001), 1067–1074.
101. Carli, M., Robbins, T. W., Evenden, J. L. and Everitt, B. J. Effects of lesions to ascending noradrenergic neurons on performance of a 5-choice serial reaction task in rats: implications for theories of dorsal noradrenergic bundle function based on selective attention and arousal. *Behav. Brain Res.*, **9** (1983), 361–380.
102. Ernst, M., Zametkin, A. J., Matochik, J. A., Jons, P. H. and Cohen, R. M. DOPA decarboxylase activity in attention deficit disorder adults. A [fluorine-18]fluorodopa positron emission tomographic study. *J. Neurosci.*, **18** (1998), 5901–5907.
103. Halperin, J. M., Newcorn, J. H., Koda, V. H. *et al.* Noradrenergic mechanisms in ADHD children with and without reading disabilities: a replication and extension. *J. Am. Acad. Child Adolesc. Psychiatry*, **36** (1997), 1688–1697.
104. Rogeness, G. A., Maas, J. W., Javors, M. A. *et al.* Attention deficit disorder symptoms and urine catecholamines. *Psychiatry Res.*, **27** (1989), 241–251.
105. Roman, T., Schmitz, M., Polanczyk, G. V. *et al.* Further evidence for the association between attention-deficit/hyperactivity disorder and the dopamine-beta-hydroxylase gene. *Am. J. Med. Genet.*, **114** (2002), 154–158.
106. Daly, G., Hawi, Z., Fitzgerald, M. and Gill, M. Mapping susceptibility loci in attention deficit hyperactivity disorder: preferential transmission of parental alleles at DAT1, DBH and DRD5 to affected children. *Mol. Psychiatry*, **4** (1999), 192–196.
107. Comings, D. E., Gade-Andavolu, R., Gonzalez, N. and MacMurray, J. P. Additive effect of three noradrenergic genes (*ADRA2A, ADRA2C, DBH*) on attention-deficit hyperactivity disorder and learning disabilities in Tourette syndrome subjects. *Clin. Genet.*, **55** (1999), 160–172.
108. Xu, C., Schachar, R., Tannock, R. *et al.* Linkage study of the alpha2A adrenergic receptor in attention-deficit hyperactivity disorder families. *Am. J. Med. Genet.*, **105** (2001), 159–162.
109. Sunohara, G. A., Roberts, W., Malone, M. *et al.* Linkage of the dopamine D4 receptor gene and attention-deficit/hyperactivity disorder. *J. Am. Acad. Child Adolesc. Psychiatry*, **39** (2000), 1537–1542.
110. Tahir, E., Yazgan, Y., Cirakoglu, B. *et al.* Association and linkage of DRD4 and DRD5 with attention deficit hyperactivity disorder (ADHD) in a sample of Turkish children. *Mol. Psychiatry*, **5** (2000), 396–404.
111. Van Tol, H. H. M., Bunzow, J. R., Guan, H.-C. *et al.* Cloning of the gene for a human dopamine D4 receptor with high affinity for the antipsychotic clozapine. *Nature*, **350** (1991), 610–614.
112. Robinson, R. G. and Lipsey, J. R. Cerebral localization of emotion based on clinical-neuropathological correlations: methodological issues. *Psychiatr. Dev.*, **3** (1985), 335–347.
113. George, M. S., Speer, A. M., Molloy, M. *et al.* Low frequency daily left prefrontal rTMS improves mood in bipolar depression: a placebo-controlled case report. *Hum. Psychopharmacol.*, **13** (1998), 271–275.
114. Baxter, L. R., Schwartz, J. M., Phelps, M. E. *et al.* Reduction of prefrontal cortex glucose metabolism common to three types of depression. *Arch. Gen. Psychiatry*, **46** (1989), 243–250.

115. Drevets, W. C., Price, J. L., Simpson, J. R. J. *et al.* Subgenual prefrontal cortex abnormalities in mood disorders. *Nature*, **386** (1997), 824–827.
116. Mayberg, H. S., Liotti, M., Brannan, S. K. *et al.* Reciprocal limbic-cortical function and negative mood: converging PET findings in depression and normal sadness. *Am. J. Psychiatry*, **156** (1999), 675–682.
117. Elliott, R., Baker, S. C., Rogers, R. D. and O'Leary, D. A. Prefrontal dysfunction in depressed patients performing a complex planning task: a study using positron emission tomography. *Psychol. Med.*, **27** (1997), 931–942.
118. Rajkowska, G., Miguel-Hidalgo, J. J., Wei, J. *et al.* Morphometric evidence for neuronal and glial prefrontal cell pathology in major depression. *Biol. Psychiatry*, **45** (1999), 1085–1098.
119. Hammen, C., Ellicott, A. and Gitlin, M. Vulnerability to specific life events and prediction of course of disorder in unipolar depressed patients. *Can. J. Behav. Sci.*, **21** (1989), 377–388.
120. Weiss, E. L., Longhurst, J. G. and Mazure, C. M. Childhood sexual abuse as a risk factor for depression in women: psychosocial and neurobiological correlates. *Am. J. Psychiatry*, **156** (1999), 816–828.
121. Mazure, C. Life stressors as risk factors in depression. *Clin. Psychol.*, **5** (1998), 291–313.
122. Gold, P. W., Goodwin, F. K. and Chrousos, G. P. Clinical and biochemical manifestations of depression in relation to the neurobiology of stress. *New Engl. J. Med.*, **319** (1988), 413–420.
123. Weiss, J. M. and Simson, P. E. Neurochemical and electrophysiological events underlying stress-induced depression in an animal model. *Adv. Exp. Med. Biol.*, **245** (1988), 425–440.
124. Flugge, G. Alterations in the central nervous alpha 2-adrenoceptor system under chronic psychosocial stress. *Neuroscience*, **75** (1996), 187–196.
125. Dazzi, L., Ladu, S., Spiga, F. *et al.* Chronic treatment with imipramine or mirtazapine antagonizes stress- and F. G7142-induced increase in cortical norepinephrine output in freely moving rats. *Synapse*, **43** (2002), 70–77.
126. Dazzi, L., Vignone, V., Seu, E. *et al.* Inhibition by venlafaxine of the increase in norepinephrine output in rat prefrontal cortex elicited by acute stress or by the anxiogenic drug FG 7142. *J. Psychopharmacol.*, **16** (2002), 125–131.
127. Post, R. M., Lake, C. R., Jimerson, D. C. *et al.* Cerebralspinal fluid norepinephrine in affective illness. *Am. J. Psychiatry*, **135** (1978), 907–912.
128. Wong, M. L., Kling, M. A., Munson, P. J. *et al.* Pronounced and sustained central hypernoradrenergic function in major depression with melancholic features: relation to hypercortisolism and corticotropin-releasing hormone. *Proc. Natl. Acad. Sci. U. S. A.*, **97** (2000), 325–330.
129. Arango, V., Ernsberger, P., Sved, A. F. and Mann, J. J. Quantitative autoradiography of alpha 1- and alpha 2-adrenergic receptors in the cerebral cortex of controls and suicide victims. *Brain Res.*, **630** (1993), 271–282.
130. Fu, C. H., Reed, L. J., Meyer, J. H. *et al.* Noradrenergic dysfunction in the prefrontal cortex in depression: an [15O] H2O PET study of the neuromodulatory effects of clonidine. *Biol. Psychiatry*, **49** (2001), 317–325.

131. Birnbaum, S. G., Podell, D. M. and Arnsten, A. F. T. Noradrenergic alpha-2 receptor agonists reverse working memory deficits induced by the anxiogenic drug, FG7142, in rats. *Pharmacol. Biochem. Behav.*, **67** (2000), 397–403.
132. Mathews, R., Li, P. P., Young, L. T., Kish, S. J. and Warsh, J. J. Increased G alpha q/11 immunoreactivity in postmortem occipital cortex from patients with bipolar affective disorder. *Biol. Psychiatry*, **41** (1997), 649–656.
133. Jope, R. S., Song, L., Li, P. P. *et al.* The phosphoinositide signal transduction system is impaired in bipolar affective disorder brain. *J. Neurochem.*, **66** (1996), 2402–2409.
134. Wang, H. Y. and Friedman, E. Enhanced protein kinase C activity and translocation in bipolar affective disorder brains. *Biol. Psychiatry*, **40** (1996), 568–575.
135. Hahn, C. G. and Friedman, E. Abnormalities in protein kinase C signaling and the pathophysiology of bipolar disorder. *Bipolar Disord.*, **1** (1999), 81–86.
136. Turecki, G., Grof, P., Cavazzoni, P. *et al.* Evidence for a role of phospholipase C-gamma1 in the pathogenesis of bipolar disorder. *Mol. Psychiatry*, **3** (1998), 534–538.
137. Manji, H. K. and Lenox, R. H. Protein kinase C signaling in the brain: molecular transduction of mood stabilization in the treatment of manic-depressive illness. *Biol. Psychiatry*, **46** (1999), 1328–1351.
138. Blumberg, H. P., Stern, E., Ricketts, S. *et al.* Rostral and orbital prefrontal cortex dysfunction in the manic state of bipolar disorder. *Am. J. Psychiatry*, **156** (1999), 1986–1988.
139. Rajkowska, G., Halaris, A. and Selemon, L. D. Reductions in neuronal and glial density characterize the dorsolateral prefrontal cortex in bipolar disorder. *Biol. Psychiatry*, **49** (2001), 741–752.
140. Ellicott, A., Hammen, C., Gitlin, M., Brown, G. and Jamison, K. Life events and the course of bipolar disorder. *Am. J. Psychiatry*, **147** (1990), 1194–1198.
141. Cohen, B. M. and Lipinski, J. F. In vivo potencies of antipsychotic drugs in blocking alpha-1 noradrenergic and dopamine D2 receptors: implications for drug mechanisms of action. *Life Sci.*, **39** (1986), 2571–2580.
142. Baldessarini, R. J., Huston-Lyons, D., Campbell, A., Marsh, E. and Cohen, B. M. Do central antiadrenergic actions contribute to the atypical properties of clozapine? *Br. J. Psychiatry*, **160: S17** (1992), 12–16.
143. van Kammen, D. P. and Kelley, M. Dopamine and norepinephrine activity in schizophrenia. An integrative perspective. *Schizophr. Res.*, **4** (1991), 173–191.
144. Kemali, D., Maj, M., Galderisi, S., Grazia Ariano, M. and Starace, F. Factors associated with increased noradrenaline levels in schizophrenic patients. *Prog. Neuropsychopharmacol. Biol. Psychiatry*, **14** (1990), 49–59.
145. Maas, J. W., Contreras, S. A., Miller, A. L. *et al.* Studies of catecholamine metabolism in schizophrenia/psychosis – I. *Neuropsychopharmacology*, **8** (1993), 97–109.
146. van Kammen, D. P., Agren, H., Yao, J. K. *et al.* Noradrenergic activity and prediction of psychotic relapse following haloperidol withdrawal in schizophrenia. *Am. J. Psychiatry*, **151** (1994), 379–384.
147. van Kammen, D. P. The biochemical basis of relapse and drug response in schizophrenia: review and hypothesis. *Psychol. Med.*, **21** (1991), 881–895.

148. Weinberger, D. R., Berman, K. F. and Zec, R. F. Physiologic dysfunction of dorsolateral prefrontal cortex in schizophrenia. I. Regional cerebral blood flow evidence. *Arch. Gen. Psychiatry*, **43** (1986), 114–124.
149. Goldman-Rakic, P. S. Prefrontal cortical dysfunction in schizophrenia: the relevance of working memory. In *Psychopathology and the Brain*, ed. B. J. Carroll and J. E. Barrett. (New York: Raven Press, 1991), pp. 1–23.
150. Frith, C. The role of the prefrontal cortex in self-conciousness: the case of auditory hallucinations. *Phil. Trans. R. Soc. London ser. B*, **351** (1996), 1505–1512.
151. Holzman, P. S. Eye movements and the search for the essence of schizophrenia. *Brain Res. Brain Res. Rev.*, **31** (2000), 350–356.
152. Selemon, L. D., Rajkowska, G. and Goldman-Rakic, P. S. Abnormally high neuronal density in the schizophrenic cortex: a morphometric analysis of prefrontal area 9 and occipital area 17. *Arch. Gen. Psychiatry*, **52** (1995), 805–818.
153. Maas, J. W., Miller, A. L., Tekell, J. L. *et al.* Clonidine plus haloperidol in the treatment of schizophrenia/psychosis. *J. Clin. Psychopharmacol.*, **15** (1995), 361–364.
154. Friedman, J. I., Adler, D. N., Temporini, H. D. *et al.* Guanfacine treatment of cognitive impairment in schizophrenia: a pilot study. *Neuropsychopharmacology*, **25** (2001), 402–409.
155. Breier, A., Buchanan, R. W., Waltrip, R. W., 2nd. *et al.* The effect of clozapine on plasma norepinephrine: relationship to clinical efficacy. *Neuropsychopharmacology*, **10** (1994), 1–7.
156. Koenen, K. C., Driver, K. L., Oscar-Berman, M. *et al.* Measures of prefrontal system dysfunction in posttraumatic stress disorder. *Brain Cognit.*, **45** (2001), 64–78.
157. Southwick, S., Rasmusson, A., Barron, X. and Arnsten, A. F. T. Neurobiological and neurocognitive alterations in PTSD: a focus on norepinephrine, serotonin and the hypothalamic–pituitary–adrenal axis. In *Neuropsychology of PTSD*, ed. J. J. Vasterling and C. R. Brewin. (New York: Guilford Publications, 2005), pp. 27–58.
158. Zubieta, J. K., Chinitz, J. A., Lombardi, U. *et al.* Medial frontal cortex involvement in PTSD symptoms: a SPECT study. *J. Psychiatr. Res.*, **33** (1999), 259–264.
159. Bremner, J. D. Neuroimaging studies in post-traumatic stress disorder. *Curr. Psychiatry Rep.*, **4** (2002), 254–263.
160. Southwick, S. M., Bremner, J. D., Rasmusson, A. *et al.* Role of norepinephrine in the pathophysiology and treatment of posttraumatic stress disorder. *Biol. Psychiatry*, **46** (1999), 1192–1204.
161. Berthier, M. L., Posada, A. and Puentes, C. Dissociative flashbacks after right frontal injury in a Vietnam veteran with combat-related posttraumatic stress disorder. *J. Neuropsychiatry Clin. Neurosci.*, **13** (2001), 101–105.
162. Geracioti, T. D., Jr., Baker, D. G., Ekhator, N. N. *et al.* CSF norepinephrine concentrations in posttraumatic stress disorder. *Am. J. Psychiatry*, **158** (2001), 1227–1230.
163. Bremner, J. D., Innis, R. B., Ng, C. K. *et al.* Positron emission tomography measurement of cerebral metabolic correlates of yohimbine administration in combat-related posttraumatic stress disorder. *Arch. Gen. Psychiatry*, **54** (1997), 246–254.
164. Horrigan, J. P. Guanfacine for PTSD nightmares. *J. Am. Acad. Child Adolesc. Psychiatry*, **35** (1996), 975–976.

165. Taylor, F. and Raskind, M. A. The alpha1-adrenergic antagonist prazosin improves sleep and nightmares in civilian trauma posttraumatic stress disorder. *J. Clin. Psychopharmacol.*, **22** (2002), 82–85.
166. Raskind, M. A., Dobie, D. J., Kanter, E. D. *et al.* The alpha1-adrenergic antagonist prazosin ameliorates combat trauma nightmares in veterans with posttraumatic stress disorder: a report of 4 cases. *J. Clin. Psychiatry*, **61** (2000), 129–133.
167. De Bellis, M. D., Keshavan, M. S. and Harenski, K. A. Anterior cingulate N-acetylaspartate/creatine ratios during clonidine treatment in a maltreated child with posttraumatic stress disorder. *J. Child Adolesc. Psychopharmacol.*, **11** (2001), 311–316.
168. Harmon, R. J. and Riggs, P. D. Clonidine for posttraumatic stress disorder in preschool children. *J. Am. Acad. Child Adolesc. Psychiatry*, **35** (1996), 1247–1249.
169. O'Rourke, M. F., Blaxall, H. S., Iversen, L. J. and Bylund, D. B. Characterization of [3H]RX821002 binding to alpha-2 adrenergic receptor subtypes. *J. Pharmacol. Exp. Ther.*, **268** (1994), 1362–1367.
170. Mohell, N., Svartengren, J. and Cannon, B. Identification of [3H]prazosin binding sites in crude membranes and isolated cells of brown adipose tissue as alpha-1 adrenergic receptors. *Eur. J. Pharmacology*, **92** (1983), 15–25.
171. Pepperl, D. J. and Regan, J. W. Adrenergic receptors. In *G Protein-Coupled Receptors*, ed. S. Peroutka. (Boca Raton, Florida: CRC Press, 1994), pp. 45–78.

15

Norepinephrine in neurological disorders

Francesco Fornai, M.D., Ph.D.
University of Pisa, Italy and I. N. M. Neuromed., Pozzilli, Italy

Introduction: functional anatomy of central norepinephrine system in relation to neurological disorders

Catecholamine-containing neurons were first classified in the 1960s by Dahlstroem and Fuxe,[1] following the method of Falck. In this work they described altogether 12 cell groups (A1–A12) in the rat central nervous system (CNS). Although several neuronal nuclei produce norepinephrine (NE) in the brain stem, given the wide contribution of the locus coeruleus (LC) compared with other nuclei, this main NE nucleus of the pons is considered apart from other NE systems. Thus, we routinely distinguish the LC complex as the rostral NE system, keeping it distinct from other NE nuclei forming the so-called "caudal" NE nuclear formations. In fact, the latter consist of scattered cell groups located in the lower brain stem, deep within the medullary ventrolateral reticular formation (A1), or placed within the dorsal vagal complex and the nucleus of the solitary tract (A2). These caudal NE nuclei, which form an interconnected network, send their large varicose axons to restricted target areas, following a fairly specific pattern in which distinct nuclei innervate discrete regions of the brain.[2] This caudal component of the central NE system appears to be involved mainly in regulating vegetative functions and participates in neuroendocrine control.

The LC (A6) represents the most rostral NE complex, being localized in the pons (Figure 15.1a), in the upper part of the floor of the fourth ventricle. This rostral NE complex was originally called the nucleus pigmentosus pontis. It is present in all the mammalian species and represents the major source of NE for the CNS.[3] The LC is indeed a nuclear complex, being composed of various subnuclei located close to each other: the LC sensu stricto, A6; the nucleus subcoeruleus, A7; and scattered catecholamine neurons located close to the brachium conjunctivum (parabrachial nucleus, A5), respectively[3]. Catecholamine neurons belonging to this nuclear

Brain Norepinephrine: Neurobiology and Therapeutics, ed. Gregory A. Ordway, Michael A. Schwartz and Alan Frazer. Published by Cambridge University Press.

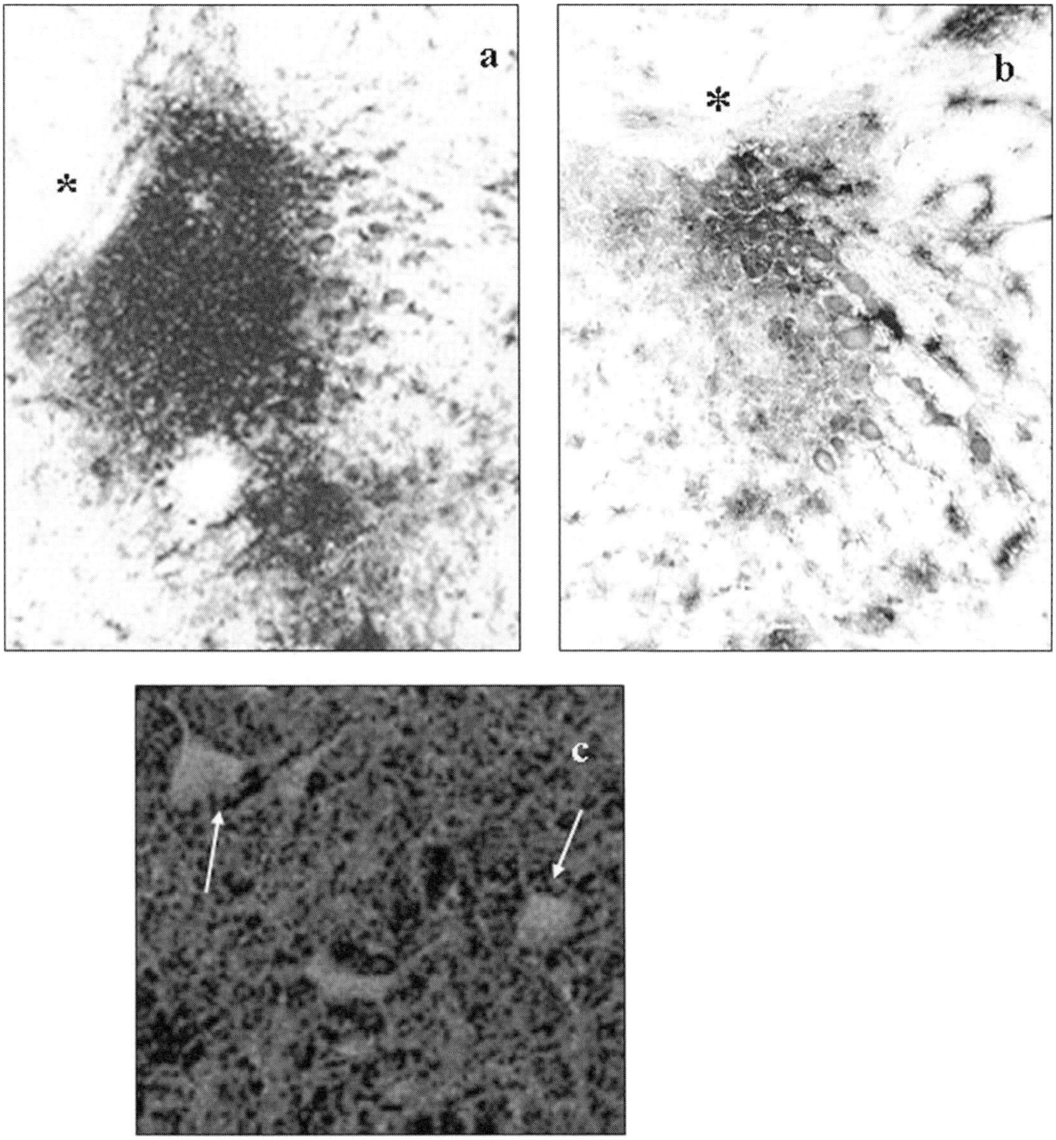

Figure 15.1 Selective staining of locus coeruleus neurons in normal conditions and during degenerative processes in rats. Dopamine-β-hydroxylase immunostaining shows NE neurons of LC, close to the floor of the fourth ventricle (*), in a control (a), and after systemic administration of *N*-(2-chloroethyl)-*N*-ethyl-2-bromobenzylamine (DSP-4), which is a neurotoxin selective for LC-arising axon terminals (b). The marked loss of staining within the LC a few days after DSP-4 is related to the loss of axon collaterals within the nucleus, while cell bodies survive. Nonetheless, when observed several months after DSP-4, apart from loss of NE axon terminals, a degeneration of cell bodies can be observed (c), as showed by the Fluoro-Jade B technique, which stains degenerating neurons (arrows).

complex widely branch their terminals up to a point in which a single NE axon can innervate all brain regions. Locus coeruleus efferents form two main ascending fiber systems: the dorsal bundle and the (much smaller) rostral limb of the dorsal periventricular pathway. Other LC efferents are distributed caudally to the cerebellum, the lower medulla, and the spinal cord.[3] The activity of the LC complex (from now on the complex will be considered simply as LC, unless specified as LC sensu stricto) modulates a variety of central functions through the release of NE into several brain areas. The general physiology of the LC has been extensively investigated in other chapters of this book and several review papers.[4] For example, the LC plays a central role in the sleep–waking cycle, anticipating fluctuations of electroencephalographic activity[5] and promoting a state of vigilance, in monitoring environmental stimuli with a special emphasis on alerting and orienting to novelty, as well as in the control of autonomic functions. Indeed, it is well known that stimulation of central NE receptors leads to changes in the state of vigilance, as well as in immediate early gene (IEG) transcription regulating the expression of *c-fos* and *nerve growth factor-induced A* (*NGF-I A*), *nur 77*, *tis-7*, *zif-268* and *tis-21* both in basal conditions, in seizures, and in general during stressful stimuli.[6] In line with the increased activity of the nucleus in the waking state, it has been shown that the LC plays a pivotal role in regulating the circadian rhythm of expression of a group of IEGs and the state of phosphorylation of cyclic adenosine monophosphate response element-binding (CREB) proteins. These findings suggest that this neuronal complex is critical for inducing cerebral plasticity related to the waking state.[7] In contrast, during rapid eye movement sleep, when the activity of the nucleus is suppressed, there is a marked reduction in the expression of these genes. Similarly, damage to the LC nucleus abolishes the waking-related early gene expression, and pharmacological agents, such as clonidine, that hyperpolarize NE cells decrease the expression of these genes below baseline values.

Accumulating data suggest a role for NE in the pathogenesis of various neurological disorders, including neurodegenerative diseases such as Parkinson's disease (PD); multiple system atrophy (MSA); and Alzheimer's disease (AD) and other dementia; and also nondegenerative neurological disorders such as epilepsy, neuroinflammatory diseases, sleep disorders and attention-deficit/hyperactivity disorder (ADHD). Further new evidence may lead to the inclusion of a variety of other neurological diseases.

For many years, it has been known that the LC degenerates in CNS disorders such as senile dementia, lobar atrophy, Down's syndrome, Pick's disease, and depression. In addition, various degenerative movement disorders feature a marked loss of LC cells: apart from PD, this is evident for specific forms of inherited ataxia, MSA, including Shy-Drager syndrome, and olivopontocerebellar atrophy. In contrast, a hyperactivity of LC neurons has been related to anxiety disorders and may produce panic attacks.

With regard to the involvement of NE in the onset and evolution of neurological diseases, it should be emphasized that NE, apart from its classic role as a neurotransmitter, may have additional functions. These potential roles will be clarified in the course of this chapter. Since NE terminals run in close contact to neurons, astrocytes, and microglia, it has been suggested that NE may also act as a neuromodulator or a paracrine agent, thus influencing a circumscribed microenvironment rather than distant projection areas. The LC produces widespread efferent projections all over the CNS. A striking feature of LC axons is the extreme divergence of their collaterals, which allows a single neuron to innervate the entire cerebral cortex.[8] This is consistent with the fact that LC axon terminals in the cerebral cortex make only a few direct synaptic contacts with postsynaptic neurons; rather, they exert a modulatory role on their targets. Indeed, NE activity leads to diffuse effects on target areas involving neurons, astrocytes, and cerebral blood vessels. For this reason, the physiology of the central NE system, apart from target neurons, also includes glia and brain vessels.[9] This finding is, again, consistent with the morphology of NE axons, which possess varicosities ("bouttons en passage") rather than classic "bouttons terminaux" typical of nonmonoaminergic axon terminals. Finer morphological studies of NE axons arising from the LC have shown that axon terminals are thin, with small varicosities. This contrasts with NE axons arising from the medullary A1 and A2 cell groups, which branch out with terminals featuring larger varicosities. It is worthwhile mentioning that monoaminergic axons with smaller beaded varicosities possess a lower threshold to various neurotoxic insults, and, at the same time, they are more affected in degenerative diseases. The involvement of LC neurons in neurological disorders is thus much more frequent than other NE nuclei.

Experimental tools to damage central NE pathways in modelling neurological disorders

Different experimental approaches could be used to obtain a selective lesioning of LC-arising fibers. The first consists in stereotactically guided microinjections of the catecholamine neurotoxin 6-hydroxydopamine (6-OHDA) within the LC. However, this experimental procedure is subject to unavoidable site variability, it is invasive, and it is not specific for NE neurons involving neighboring catecholamine cells (for review, see reference 10). Furthermore, the damage might involve ascending axons from lower NE-containing neurons (belonging to the A1–A2 cell groups), which run quite close to the LC and can easily be included in the lesion, therefore biasing the results. To overcome these limitations, a second approach consists in microinjecting 6-OHDA within the dorsal catecholamine bundle.[11] However, this procedure involves only the ascending NE projections. Therefore it does not result in the total loss of LC-arising axons, leaving intact the descending NE pathways. As emphasized in the clinical section of this chapter, these pathways are involved in a variety of PD symptoms. Furthermore, it is possible that microinjecting 6-OHDA within the dorsal catecholamine bundle leads to the diffusion of this neurotoxin to axons ascending from NE cells of the lower brain stem. Another method consists in administering 6-OHDA to newborn animals within the first week of life. During this time, the blood–brain barrier is immature and 6-OHDA can easily cross to the CNS. Using a selective schedule of administration (repeated injections of small doses), 6-OHDA becomes more selective in lesioning NE neurons compared with dopamine (DA) cells.[12] However, this procedure does not completely preserve DA neurons from toxicity and it does not discern between NE axons arising from the LC and extracoeruleus NE terminals. Moreover, this experimental approach needs to be carried out in newborn animals, and therefore is not feasible in experiments performed in adults. Additionally, similar invasive mechanisms do not happen in naturally occurring NE cell death, making this model useless in the understanding of the spontaneous mechanism underlying degeneration of NE neurons. Therefore, stereotactically guided destruction of LC neurons is useful only to produce a loss of NE neurons as an experimental model, leaving the pathogenesis of NE degeneration unsolved. In any case, the accuracy of this procedure requires a previous identification of NE LC neurons by electrophysiological recording. The model therefore seems a more appropriate tool when the experimental aim is to destroy the LC in some primate subjects.[13]

In rodent experiments, the above considerations call for a different experimental approach, one that will noninvasively reproduce the degeneration of NE LC axons in a large number of animals as might occur naturally. Such a noninvasive method has been used since the 1970s, and consists of systemic injections of the neurotoxin *N*-(-2-chloroethyl)-*N*-ethyl-2-bromobenzylamine (DSP-4). The intraperitoneal administration of DSP-4 (50 mg/kg) causes a long-lasting loss of NE terminals arising from the LC (Figures 15.1 and 15.2). Remarkably, this compound selectively destroys NE axons arising from LC neurons but spares extracoeruleus NE terminals (Figure 15.3). It is well established that after a single i.p. injection of this neurotoxin, NE axons degenerate in the LC target structures (Fig. 15.2 and Fig. 15.4). Furthermore, with an extended period after DSP-4 administration, a progressive decline in the number of LC cell bodies occurs (Fig. 15.1c), while the remaining NE terminals sprout to reinnervate the cerebral cortex. Following its uptake within NE neurons, the final steps responsible for DSP-4-induced toxicity remain unknown. As occurs with other selective catecholamine neurotoxins, knowledge about the mechanisms of DSP-4-induced neurotoxicity could provide us with new therapeutic strategies to slow down naturally occurring NE degeneration. In several studies, DSP-4 has been used merely as a

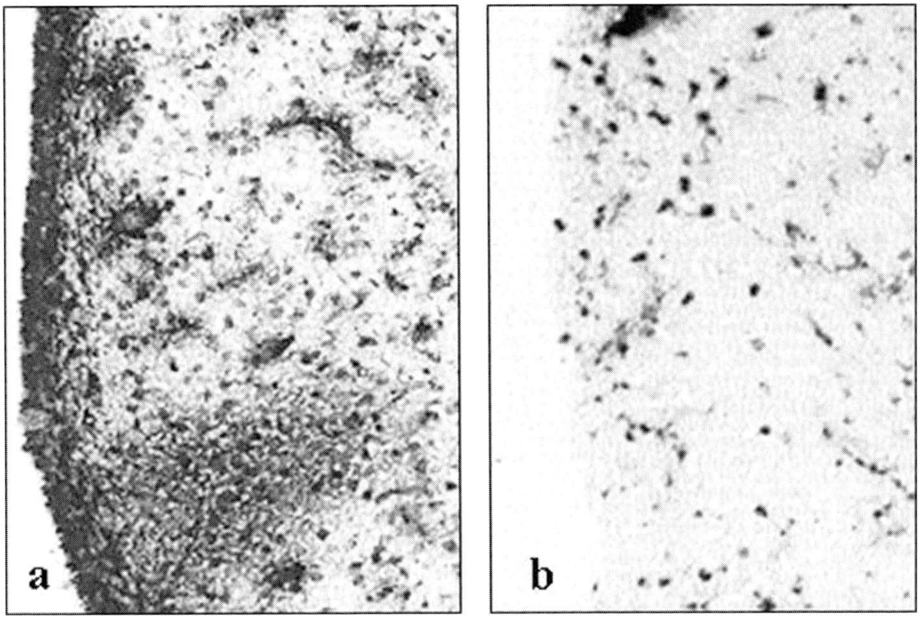

Figure 15.2 Norepinephrine nerve endings within the pyriform cortex of rats. A dense pattern of immunostaining is visible in the piriform cortex (a) by using the selective technique to stain NE axons using antibodies directed against the synthesizing enzyme dopamine-β-hydroxylase. Indeed, this brain area contains the highest levels of NE compared with other cortical regions. In particular, the difference is marked when comparing the pyriform cortex with neocortical areas. Incidentally, differing from this isocortex, the pyriform cortex is devoid of any thalamic input; therefore, it was suggested that in this area the NE pathways might substitute for the functional effects exerted by thalamic afferents on cortical regions. In (b) the piriform cortex is stained for the same antigens (dopamine-β-hydroxylase) after a treatment with the neurotoxin DSP-4. The dense trim of cortical NE fibers is massively lost with a special emphasis on the deep layers of the pyriform cortex, where immunostaining is no longer present.

denervation tool to reproduce the idiopathic NE depletion observed in humans (Table 15.1). This is also the aim of studies concerning the role of NE in experimental parkinsonism (described below). Nonetheless, our present knowledge calls for more in-depth studies addressing the mechanisms of naturally occurring NE cell death.

Markers to assess damage to NE pathways

There are several markers that allow documentation of damage to NE neurons at the level of either cell bodies, axon terminals, or both neuronal components

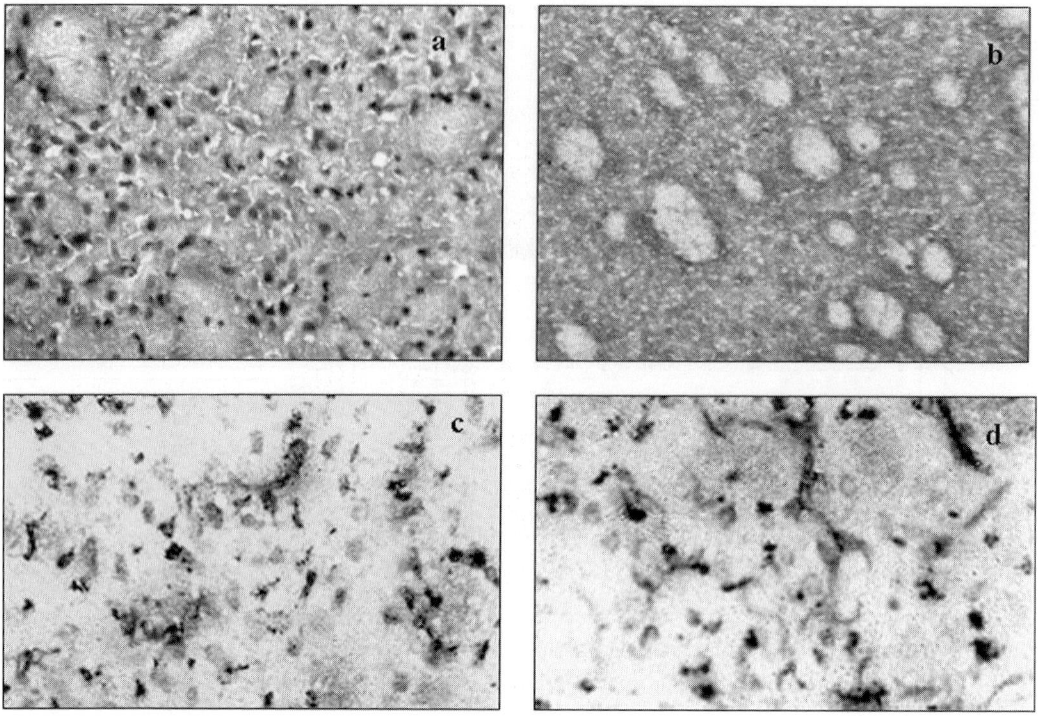

Figure 15.3 Analyzing catecholamine innervation within the striatum of rats. Catecholamine (DA and NE) innervation of striatum is showed by using antibodies against tyrosine hydroxylase (TH) (a), which is the rate-limiting step enzyme of catecholamine biosynthesis. However, in this way it not possible to distinguish between nerve endings containing DA and those containing NE. In fact TH is contained in both NE and DA nerve endings. Therefore, in order to selectively identify the DA nerve endings it is necessary to use antibodies directed against the DA transporter, which is located exclusively on DA axons. Similarly, in order to observe NE nerve endings antibodies against dopamine-β-hydroxylase can be used. Thus it is possible to carry out a comparison between the amount of DA and NE nerve endings within striatum (b and c, respectively). In this way it is evident that the dense pattern of DA nerve endings is far in excess compared with scattered NE terminals. This makes almost non relevant the amount of NE axons contributing to the TH immunostaining observed in the striatum and justifies the routine approach to measure the integrity of striatal DA axons by simply using TH immunohistochemistry, which almost overlaps with DA transporter immunohistochemistry. Conversely, the selective staining for NE axons, obtained by dopamine β-hydroxylase immunoperoxidase, shows only rare varicosities, which account for the scarce levels of striatal NE. Interestingly, striatal NE is not affected by the selective neurotoxin DSP-4 (d), which destroys LC-arising NE axons; thus suggesting that scattered striatal NE innervation derives from other NE nuclei placed more caudally in the brainstem (A1–A2 NE areas, as described in the text).

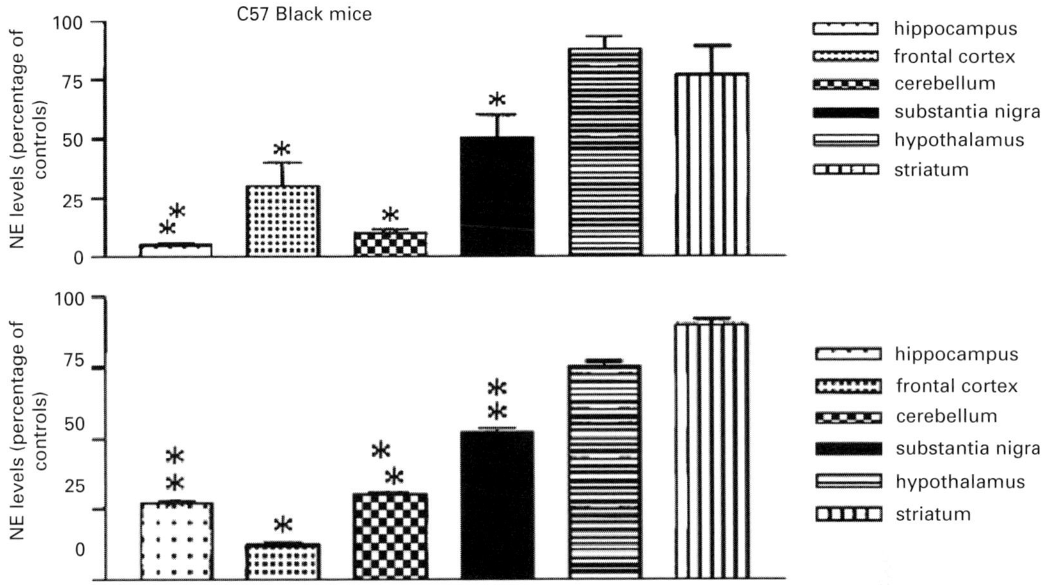

Figure 15.4 Regional pattern of DSP-4-induced NE loss in two strains of mice and depleting effect of NE neurotoxin DSP-4. Norepinephrine levels, expressed as percentage of controls, are showed in various brain areas of male C57Bl and Swiss-Webster mice, after DSP-4 neurotoxic treatment. The different amount of NE loss detectable in different brain areas depends on different density of NE terminals arising from the LC that are present within each region. In both strains of mice, DP-4 produces only a mild, non significant NE loss in the hypothalamus and striatum, while NE loss is significant in the substantia nigra. The most profound NE loss is detected in the hippocampus, frontal cortex and cerebellum. Furthermore, in these areas the amount of NE loss appears largely dependent on the animal strain. Error bars are expressed as SE of the percentage mean of 10 animals per group. *$p < 0.05$ compared with controls; **$p < 0.01$ compared with controls.

(Figure 15.5). The original studies used the histochemical method of Falck and Hillarp. This method was based on the fluorescence of catechol-containing molecules, but it did not obtain specific results and was not able to detect small amounts of catecholamines. Progress in neurochemistry and immunocytochemistry led to other methods that are more reliable and possess enough sensitivity to document the integrity of NE innervation, even in small brain areas. One method that can now be regarded as a classic tool consists of measuring the loss of endogenous NE within target areas. This procedure has improved in the last decade, leading to more precise experimental approaches.

At the present time, the routinely applied procedure consists in using high-performance liquid chromatography coupled with electrochemical detection. This method allows the measure of amounts of NE below a picomole. Generally the procedure is applied to specific homogenized brain regions and gives an estimation

1. TH, tyrosine hydroxylase (activity/immunostaining)
2. NE levels (homogenates)
3. α_2/β_2 pre synaptic NE receptors (binding/immunostaining)
4. NET, norepinephrine transporter (binding/immunostaining)
5. GFAP, glial fibrillary acidic protein (immunostaining)
6. VMAT-2, vesicular monoamine transporter (binding/immunostaining)
7. Extracellular NE (brain dialysis)

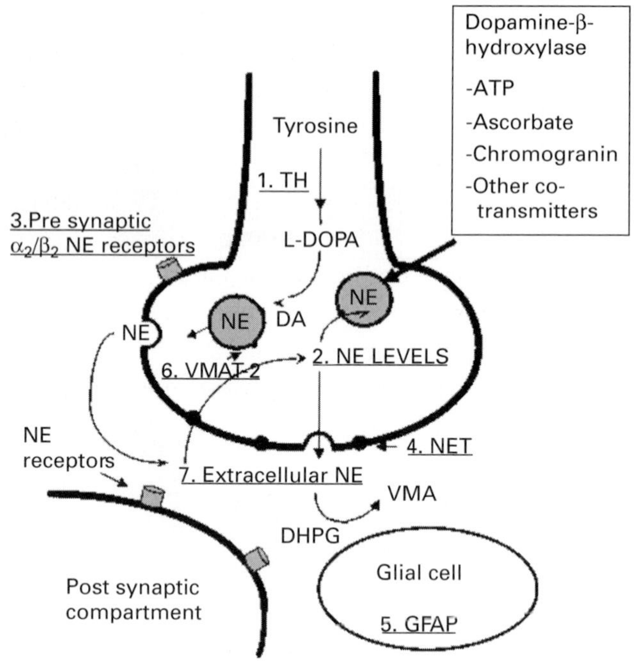

Figure 15.5 How to measure the integrity of the NE system: cartoon representing a NE terminal with its targets. The biosynthesis of **NE** starts with the aromatic amino acid tyrosine, which is then converted to dihydroxyphenylalanine (L-DOPA) by the enzyme tyrosine hydroxylase (**TH**), which represents the rate-limiting biochemical step in catecholamine biosynthesis. L-DOPA is then converted to dopamine (DA), which is taken up into the synaptic vesicle by the vesicular monoamine transporter type 2 (**VMAT-2**) where the intra vesicular enzyme dopamine-β-hydroxylase (**DBH**) converts DA to NE. This is the final product ready to be released together with ascorbic acid, adenosine triphosphate (ATP), adenosine and other various co transmitters in the extra cellular space where it is then able to interact either with post- or pre synaptic **NE receptors** or be taken up into the axon terminal by the specific NE transporter, **NET**. VMAT-2 thus represents a marker for all monoamine nerve terminals, TH is specific for catecholamine-containing neurons, while DBH and NET confer selectivity for NE-containing neurons. Damage to NE terminals can be also observed by increased levels of glial fibrillary acidic protein (**GFAP**) produced by surrounding glial cells. DHPG, dihydroxyphenylethylglycol; VMA, vanillylmandelic acid.

of the integrity of NE axon terminals, being that levels of NE stored within terminals far exceeds those measured in the extracellular space.

Measurement of the rate limiting enzyme – tyrosine hydroxylase (TH) – in NE biosynthesis is a useful tool to mark NE terminals in purely NE innervated regions. Tyrosine hydroxylase might, however, reveal a fallacy in translating results to the integrity of NE terminals, since dopaminergic nerve endings contain this enzyme as well (see an example in Figure 15.3). Therefore when aiming for an enzyme specific for NE nerve endings, it is necessary to target the enzyme dopamine-β-hydroxylase (DBH), which is responsible for the conversion of DA to NE. Targeting this enzyme allows the investigator to rule out DA nerve endings. This measure might involve enzyme activity or protein content (either visualized by immunocytochemistry or measured by semiquantitative or quantitative assays). Another molecular target that might be used to visualize or measure the amount of NE axons is the high-affinity selective NE transporter (NET). This can be done by using autoradiography to assay ^{3}H-NE uptake or labeled compounds that selectively bind the NET, such as desipramine or specific antibodies.

Despite the varying sensitivity of these methods, molecular targets, and procedures (biochemistry vs. morphology), results should always be considered cautiously. With this consideration in mind, a point that must be clarified is that visualizing or measuring NE terminal loss within target regions does not provide functional information about the occurrence of loss of NE activity. In fact, it is well established that even severe damage to NE neurons does not necessarily impair the NE activity in terms of postsynaptic efficacy. In other words, there are a number of compensatory mechanisms (ranging from postsynaptic NE receptor up-regulation, adaptation of second messengers, increased presynaptic turn over leading to increased NE synthesis, and release per surface unit) that are able to maintain the level of extracellular NE at a sufficient level to provide appropriate stimulation, at least in basal conditions. Therefore, apart from using the markers described above, one should never forget that in the presence of an NE lesion, there is a need to establish the occurrence of an authentic decrease of extracellular NE levels. This could be achieved by using brain microdialysis in selected NE target areas or by indirect imaging of displacement of labeled NE agonists.

Involvement of NE in specific neurological disorders

Parkinson's disease (PD)

General considerations

The LC appears to be consistently affected in PD, as documented by various neuropathological studies and biochemical findings. In view of these findings, it has been suggested that the pathological diagnosis of PD requires, together with the classic loss of DA neurons of the substantia nigra pars compacta (SNpc), the occurrence of a concomitant significant cell death in the LC. Despite the fact that the general physiology of the LC has been extensively investigated, the significance of lesioning of the LC in PD remains largely unknown. The following is a review of the latest data indicating that the loss of NE might play a double role in PD: (1) NE loss might sustain motor and nonmotor parkinsonian symptoms; and (2) NE loss might worsen the DA nigrostriatal damage. Within the context of the latter supposition, basic research has provided a new provocative hypothesis on the significance of the LC in shaping the course of PD. In fact, the loss of a trophic influence of these neurons is crucial in increasing the sensitivity of nigrostriatal DA axons to various neurotoxic insults.

Anatomical evidence for the involvement of NE neurons in PD

The LC nucleus is affected in PD, as demonstrated by the extensive cell loss occurring in patients.[14] In contrast, although controversial findings exist, the caudal NE nuclei seem to be only slightly affected in PD. In particular, whereas the A2 cell group of PD patients features only a few neurons containing Lewy bodies (a pathological hallmark of PD), the A1 nucleus appears to be completely spared.

The loss of LC cells in PD has been well documented by various neuropathological studies; however, only a few functional studies have focused on this alteration as a key element of PD. In fact, research on the disorder has focused on the pathophysiology of the nigrostriatal DA pathway and the basal ganglia, overlooking the constant loss of LC cells as contributing to the pathophysiology of the disorder. Despite this oversight, it should be noted that in their original paper describing the DA deficiency, Ehringer and Hornykiewicz[15] reported a concomitant marked NE depletion in PD patients (Table 15.1).

It has been suggested that the LC neuron loss is related to the early onset of PD; however, this correlation was not confirmed in further studies.[16] Consistently, pathological studies consider PD as a multifocal disorder involving the SNpc, and the LC. At present, it is widely accepted that in PD the loss of DA neurons in SNpc occurs concomitantly with significant cell loss in the LC.

Neuropathological findings on the involvement of LC neurons in PD have been confirmed as well-established data. These findings range from the observation of Lewy bodies within single cells, neurofibrillary tangles (NT) and reactive fibrosis in the nucleus to frank loss of neurons occurring at the same time in the SNpc and the LC of PD patients.[17] Remarkably, in non-Parkinsonian patients, NT occur in the LC much more frequently than in the SNpc. Again, NT appear in the LC at about the age of 30, progressively increasing with age until they are present in about 90% of individuals after the age of 90. The time of appearance of NT in the LC (early onset) and SNpc (later onset) differs strikingly from the appearance of Lewy bodies in these nuclei; indeed Lewy bodies appear later at approximately the same age in both nuclei, and they continue to progress at a similar rate.

In a study performed in the brains of 21 PD patients, it was estimated that the degree of LC cell loss compared with age-matched controls is about 70%. In this study, degeneration was present in the whole nucleus, although it was more

Table 15.1 Dopamine (DA), norepinephrine (NE), and serotonin (5-HT) levels in several brain areas of parkinsonian patients compared with age-mat ched controls (modified from Hornykiewicz and Pifl [1994][20].

Brain area	DA	NE	5-HT
Caudatus	**Nov**	Ov	Ov
Putamen	**Nov**	Ov	NS
Substantia nigra	**Nov**	Ov	Ov
Nucleus accumbens	**Nov**	NS	Ov
External globus pallidus	**Nov**	NS	NS
Internal globus pallidus	Ov	NS	NS
Subthalamic nucleus	Ov	NS	Ov
Ventral tegm. area	**Nov**	NS	NS
Hypothalamus	NS	Ov	NS
Motor cortex	NS	**Nov**	Ov
Premotor cortex	NS	**Nov**	NS
Supplem. motor area	NS	**Nov**	Ov
Hippocampus	Ov	Ov	Ov

Statistically significant reduction, with none (Nov) or some (Ov) of individual values overlapping with any of controls. NS = no statistically significant difference compared with controls. See text for detailed explanation.

pronounced in the middle and caudal segments of the LC. It has been suggested that in most cases of PD, the neuronal loss begins in the middle segment of the nucleus and then it spreads toward its caudal part.[18]

Locus coeruleus cells consist mainly of fusiform medium-sized neurons; a second group consists of smaller noncatecholamine neurons, and in PD only the former cell type appears to be affected. The cytopathology of catecholamine-containing LC neurons in PD ranges from loss of synaptic spines to a marked reduction in dendritic length. Consistent with the pattern of innervation of LC neurons, the loss of these cells in PD causes a degeneration of NE fibers. This process is most dramatic in the primary motor, premotor, supplementary motor, and prefrontal cortex. The loss of NE in the cortex involves all of the cortical layers.[19]

Neurochemical evidence concerning the involvement of NE neurons in PD

The biochemical basis of PD was originally defined by the loss of DA in the nigrostriatal pathway, and it has been stated that the massive DA deficiency that regularly occurs in the caudate-putamen of PD patients is sufficient to produce the classic symptoms of PD as a movement disorder. Nonetheless, as described above, it has been known for many years that in PD neurotransmitters other than DA are involved. In particular NE was shown to be reduced in PD. Understanding the significance of the loss of a given neurotransmitter in a specific disorder represents a difficult step in tracing the pathobiochemistry of neurological disorders. This concept was analyzed in depth by Hornykiewicz and his collaborators, who coined the concept of "overlapping" and "nonoverlapping" biochemical changes in PD (Table 15.1).[20]

In fact, it may happen that the brain levels of a given neurotransmitter, although significantly reduced in PD compared with normal age-matched populations, overlap in these two groups. Thus, despite the statistical significance of a decrease in such neurotransmitter, not all patients possess this alteration. On the other hand, some controls might have lower neurotransmitter levels compared with some PD patients. This latter case represents a so-called "overlapping" biochemical change. Therefore, an overlapping biochemical change, although significant, cannot be considered as necessary for the disease. Indeed, "overlapping decreases" occur in PD patients for various neurotransmitters, such as γ-aminobutyric acid (GABA), acetylcholine, serotonin, and a few neuropeptides. In contrast, striatal DA levels in PD brains never overlap with those of age-matched controls. Therefore, the loss of DA occurring in PD can be considered as a necessary neurochemical feature to induce specific clinical symptoms.

Similarly to DA, NE levels in several brain regions of PD patients are significantly reduced following a "nonoverlapping" pattern, thus suggesting that NE loss in PD could have a specific role in sustaining the disease itself. The nonoverlapping changes of NE levels in PD patients involve the motor, premotor, and supplementary motor cortex, the SNpc, and the hippocampus. Therefore, NE is a constant element of the pathobiochemistry of PD. The decrease in NE levels might vary in different brain areas depending on the specific patient. In conclusion, neuropathological findings defining PD as a multifocal disorder correlate with biochemical data showing the constant occurrence of a "nonoverlapping" NE loss, in addition to the commonly described "nonoverlapping" DA deficiency.

Clinical correlates for the involvement of NE neurons in PD

In the last two decades, the loss of NE neurons belonging to the LC has been related to some PD symptoms.[21] Table 15.2 summarizes the possible involvement of the LC in clinical symptoms of Parkinson's disease. Besides the classic triad of motor symptoms, akinesia, rigidity, and tremor, PD patients present with a variety of additional motor and nonmotor disturbances. Among them, dementia is the most frequent nonmotor symptom that complicates PD. The pathogenesis of dementia in PD has been related to the marked degeneration of the rostral part of the LC.[16,17,22] A number of studies indicate an involvement of NE neurons of the LC in the cognitive and memory impairment that characterizes Alzheimer's

Table 15.2 Possible involvement of locus coeruleus in selected clinical symptoms (motor and non-motor) of Parkinson's disease.

Symptom	Reference suggesting a role for LC lesion
Motor:	
Akinesia	Hornykiewicz, 1975;[69] Narabayashi *et al.*, 1987[70]
Freezing	Mizuno *et al.*, 1994;[71] Sandyk, 1996;[72] Rigendahl & Sierla, 1997[73]
"On-off" phenomenon	Sandyk, 1989a,b
Postural instability	Narabayashi *et al.*,1987[70]
Tremor	Yamazaki, 1979[76]
Autonomic:	Scatton *et al.*, 1986;[58] Goto & Hirano, 1991[77]
Cognitive:	
Dementia	Mann, 1983; Mann & Yates, 1983;[78] Chan-Palay and Asan, 1989;[22] German *et al.*, 1992;[16] Zweig *et al.*, 1993[79]
Depression	Chan-Palay & Asan, 1989;[22] Doonief *et al.*, 1992;[23] Chan-Palay, 1993[80]
Vigilance	Stein *et al.*, 1984[25]

disease (e.g. see reference 16). In recent years a number of studies focusing on the physiological role of LC in promoting learning and memory lend substance to the hypothesis of a cognitive impairment in LC-lesioned patients. In keeping with this, LC-lesioned animals have a marked suppression in the induction of *c-fos, NGF-I A* and in the phosphorylation of CREB proteins that are related to synaptic plasticity and memory formation. A close correlation has been found between the occurrence of dementia and the decrease in cortical NE in PD.[22] In addition to the loss of NE projection to cortical motor areas, there is also a marked loss of NE terminals projecting to the frontal cortex and to the parahippocampal regions. This leads to the suggestion that in demented PD patients, a loss of NE terminals projecting to associative areas of the frontal cortex might account for cognitive impairment.

Another major nonmotor feature in PD is depression.[23] The impairment of the NE system in both disorders could account for the frequent coincidence of parkinsonism and depression[22]. However, conclusive proof concerning this hypothesis is still lacking, and a role for brain serotonin seems at least to be equally important. Also, autonomic dysfunctions complicate PD,[24] although a severe autonomic failure is indicative of multiple system atrophy.

Finally, the marked decrease in NE observed in PD might also account for motor disturbances that often occur in PD patients. There are reports suggesting the involvement of the LC in sudden attacks of immobility ("freezing") occurring independently of the administration of L-dihydroxyphenylalanine (L-DOPA), which typically has its onset during the execution of motor activities such as walking, speaking, or hand movement. On this basis, administration of NE precursors (e.g., treo-dihydroxyphenylserine [treo-DOPS]) has been suggested; however, the ability of treo-DOPS to ameliorate freezing remains controversial. Lesions of LC neurons in PD and the subsequent NE reduction also has been implicated in the delayed reaction time and diminished vigilance of PD patients.[25]

Evidence for the involvement of NE in experimental parkinsonism

Locus coeruleus neuron lesions as a constant element of PD should also appear as a constant from an experimental point of view. It is currently accepted that experimental models of PD should solely reproduce the damage to the nigrostriatal DA pathway. This has been achieved either by microinfusing selectively into the SNpc and into the medial forebrain bundle the neurotoxin 6-OHDA, or by administering systemically specific DA neurotoxins such as 1-methyl-4-phenyl-1,2,3,6-tetrahydropyridine (MPTP) and methamphetamine. However, these procedures do not take into account the massive involvement of the NE system described above. Several reports demonstrate that, apart from lesioning DA neurons, MPTP damages the NE system in both primates[26] and mice.[27] Moreover, these studies showed that chronic treatment with small doses of MPTP produced an early lesion of the LC that possessed the same sensitivity to MPTP as SNpc DA cells.[27] Analogous results have been obtained in primates. However, even if the gross amount was comparable, the regional pattern of NE loss that is achieved through MPTP administration does not correspond to that occurring in PD patients. Similarly, experimental parkinsonism achieved through the injection of 6-OHDA into the median forebrain bundle produces a lesion of the NE fibers coming from the lower brain stem. Therefore, the pattern of NE loss is not specific for the different NE nuclei.

Moreover, the combined lesions of the two catecholamine systems mentioned above do not mimic the time-course of the spontaneous NE and DA degeneration occurring in humans. In addition to this temporal requirement, experimental models need to match the selective regional pathology concerning the two catecholamine pathways. This is well established for the DA system, which is massively involved as regards the mesostriatal pathway. By contrast, less attention has been given to reproducing the regional pattern of the NE lesion. Since the loss of NE innervation in PD does not occur randomly, but matches the anatomical course of LC innervation, the same pattern should be followed by experimental studies aiming to reproduce spontaneous NE cell death.

NE and the progression of parkinsonism

Apart from its role in sustaining specific PD symptoms, the loss of NE in PD represents a matter of interest also for the natural history of the disease.[28] In fact, recent experimental studies indicate that the impairment of LC neurons is crucial in worsening the nigrostriatal DA toxicity produced in various experimental models.[13,29] These findings suggest a key role for the NE system arising from the LC in the progression of PD.

Therefore, apart from being a neuropathological hallmark of PD, LC cell loss appears to have a causative role in the conditions for degeneration of the nigrostriatal DA pathway.

In a study performed in monkeys, Mavridis *et al.*[13] investigated the effects of an LC lesion on the recovery from experimental parkinsonism induced by MPTP. The authors suggested that the integrity of the LC was crucial for inducing the recovery of the DA nigrostriatal pathway after a neurotoxic lesion. In rodents, neurochemical studies concerning this phenomenon were performed for the first time by Marien *et al.*[30] and were histologically confirmed. These studies indicate that the LC lesion dramatically enhanced the toxic effects of MPTP on the nigrostriatal DA system. This indicates that a previous LC lesion increases the susceptibility of DA cells to neurotoxicity. Taken together, these results suggest that the LC has a marked protective activity on the SNpc DA neurons and consequently, the loss of LC cells may be involved in the etiology of PD.

To further study this hypothesis, investigators expanded this model from MPTP to other neurotoxins. It is well known that the administration of selected doses of methamphetamine to rodents induces a long-lasting depletion of striatal DA due to the loss of striatal DA axons. For this reason, methamphetamine is often used as a tool to induce experimental parkinsonism in rodents. Methamphetamine- and MPTP-induced experimental parkinsonism seem to differ with respect to several variables. For instance, these two neurotoxins possess a different mechanism of action that often leads to contrasting results in regard to their neurotoxic mechanisms. These findings are confirmed by the opposite effects of combined treatments, which, depending on the neurotoxin used, can produce either a protective or a deleterious effect.[31] Confirming the MPTP studies, damage to LC worsens the striatal DA toxicity induced by the injection of methamphetamine.[29] Moreover, a dose of methamphetamine not sufficient by itself to produce a nigrostriatal lesion can become toxic in animals with damage to LC.[32]

Sporadic PD tends to be a late-onset and variably progressive disorder, possibly due to repeated, mild, subtoxic insults that slowly accumulate and lead, with a variably progressive rate, to the massive loss of nigrostriatal neurons typical of the symptomatic stage of the disorder. Therefore, by using neurotoxins with different mechanisms of action, the demonstration that the LC is an endogenous modulator of the toxic threshold for DA neurons in different animal species suggests that the state of the endogenous NE system might condition the progression of PD.

There are several possibilities that could explain a protective role for NE against nigrostriatal toxicity. Endogenous NE could have an important role in the recovery process of lesioned nigrostriatal terminals based on the fact that the intensity of many parkinsonian signs decreased week by week in primates treated with MPTP alone, but not in pre-LC-lesioned animals. However, it seems that the major role of the LC consists in increasing the susceptibility to insults for DA neurons. Several reports concerning experimental parkinsonism also indicate an important role for excitatory amino acids in nigrostriatal neurotoxicity. As has been demonstrated, α_2-adrenergic receptor activation inhibits the release of glutamate in several brain regions,[33] and this may be a way in which endogenous NE protects nigrostriatal DA cells from toxic effects. In keeping with this hypothesis, the administration of clonidine (an α_2-agonist) in mice completely prevents DA neurotoxicity induced by MPTP, whereas the antagonist yohimbine enhances this toxicity and impairs the protective effects of clonidine. However, other mechanisms for α_2-adrenergic receptors may be involved as well.

It also seems likely that the mechanisms by which NE loss increases methamphetamine-induced DA depletion are due to an increase in the acute effects of methamphetamine, rather than to an impairment of the recovery. This conclusion was reached in a series of experiments performed using methamphetamine as a parkinsonism-inducing neurotoxin in mice.[34] In these experiments, a time-course was adopted to study the recovery of striatal DA levels after methamphetamine administration. The tendency of striatal DA to recover both in intact and in LC-lesioned animals administered methamphetamine was similar. This further indicates that the recovery of surviving DA axons is independent of the existence of an intact NE system. Thus, the mechanism by which the lesion of NE terminals enhances methamphetamine toxicity seems to depend on early biochemical events. Methamphetamine neurotoxicity has been related to many variables. For instance, it is well known that hyperthermia enhances, whereas hypothermia reduces, methamphetamine toxicity. However, it is unlikely that the enhancement of methamphetamine toxicity is related to an effect on body temperature, since reducing NE activity decreases body temperature despite enhancing methamphetamine-induced DA depletion.

It is more likely that NE depletion changes the physiology of DA terminals, making them more susceptible to the pharmacological effects produced by methamphetamine. This phenomenon might explain why, shortly (1 and 2 hours) after methamphetamine administration, the decrease in striatal DA measured in LC-damaged animals is more pronounced than that induced by methamphetamine alone.[34] Such a different pharmacological sensitivity to methamphetamine of DA terminals in NE-lesioned, compared with control animals, might be relevant for subsequent neurotoxicity. Furthermore, it is well established that DA released by methamphetamine from DA terminals plays a fundamental role in the subsequent damage of DA axons. In particular, it has been demonstrated using brain dialysis that the degree of damage to DA terminals is closely related to the total amount of DA released after methamphetamine administration.

It remains unknown how modifications in the biochemistry of DA neurons after NE denervation might take place. As reported in the introduction, several gene products that play a role in neuronal plasticity seem to be under the influence of the LC. Using mRNA differential display, modifications in the expression of several genes occur in DSP-4-pre treated rodents. It would be of interest to evaluate whether proteins (such as DA transporter) that are crucial for the early mechanism of action of MPTP or methamphetamine in DA terminals are modified in experimental models.

As stated above, the mechanisms by which the NE system protects from DA nigrostriatal lesions is far from being fully elucidated. One basic question still needs to be answered. It seems crucial to understand the brain regions where the effects of NE lesions on MPTP- and methamphetamine-induced neurotoxicity take place. As DSP-4 does not affect striatal NE levels, the loss of the NE system in the SN might be crucial in worsening the neurotoxicity induced by methamphetamine. Norepinephrine-containing varicosities arising from the LC are known to project through the SN pars compacta, and a significant loss of nigral NE after DSP-4 has been shown in mice. However, there is no electron microscopic evidence indicating that NE fibers actually synapse in this area. Similarly, functional studies on the role of nigral NE have never been carried out, and this topic should be investigated in detail.

Conclusions

The loss of central NE cells belonging to the LC in PD has been well documented by various neuropathological studies performed during the last few decades. Despite extensive anatomical investigations showing a massive loss of LC cells, only a few studies have focused on this alteration as a key element of the disorder.

After the discovery of DA depletion as the causal agent of PD, the vast majority of research of the disorder has focused on the nigrostriatal DA pathway and the basal ganglia, while overlooking the constant degeneration of LC cells. These observations, together with converging biochemical evidence, lead to the assumption that the NE loss in specific brain regions innervated by the LC is a constant element in the pathobiochemistry of PD. Remarkably, in recent years basic research provided innovative and thought-provoking hypotheses about the significance of the LC in PD. Experimental studies have investigated the role of NE depletion in the pathogenesis and progression of nigrostriatal DA damage. The data available seem to support the belief that the central NE system might significantly influence both the onset and the progression of damage to the DA nigrostriatal tract. Within this context, the lack of NE might play a double role in PD: (1) by itself, conditioning motor and nonmotor PD symptoms and (2) the worsening of the nigrostriatal DA damage.

In spite of the plethora of recent research data, the causes of PD and the mechanisms of neuronal degeneration are still unknown. Nonetheless, the role of LC cell loss, which occurs at an early stage in a high percentage of individuals, appears to be a critical mechanism in regulating the sensitivity of the nigrostriatal DA pathway. Therefore, LC cell loss could be an early conditioning factor for the natural history of PD. In conclusion, a lack of LC innervation enhances experimental DA neurotoxicity providing additional insight into the pathophysiology of PD. The use of therapeutic strategies leading to an increase in NE in the brain of PD patients may have a special place in the treatment of this degenerative disorder.

Epilepsy

General considerations

Preclinical research concerning the role of catecholamines in epilepsy demonstrated a prominent role for noradrenergic nuclei among other monoaminergic systems. Early studies were carried out on limbic kindling in which damage to the LC was achieved either mechanically or by neurotoxins and produced an acceleration of the rate of amygdala kindling. Soon after, the antiepileptic role of the LC was extended to other animal models, and it has been demonstrated on a wide spectrum of seizure types, including systemic pentylenetetrazole (PTZ), and maximal- and low-intensity electroshock (MES and LES, respectively). Apart from provoking seizures through the administration of systemic chemoconvulsants or electrical stimulations, a further approach consisted in analyzing the NE content and pattern of NE innervation present in animals with naturally occurring (noninduced) epilepsy.

During the last decades, improvement of experimental techniques allowed sophisticated approaches to disclose the specific role of NE in epilepsy. These studies ranged from in vitro electrophysiology and receptor binding to the latest genetic engineering of selective components of the NE system both in vitro and in vivo. This research led to a deeper insight into the mechanisms by which NE tunes seizure activity. In the remainder of this section, this material and the clinical relevance of these experiments will be reviewed.

NE and kindling

The original work of Arnold *et al.*[35] sparked substantial interest in the area of catecholamines and seizures, demonstrating for the first time that catecholamine depletion accelerates amygdala kindling in rats. Since both reserpine (a nonspecific catecholamine-depleting agent) and 6-OHDA (a nonspecific catecholamine neurotoxin) were used, it was not possible to discern the role of each specific monoamine in sustaining an anticonvulsant effect. This caveat was solved a few years later by McIntyre and Edson[36] who preadministered desipramine (DMI, an NE uptake inhibitor) before microinfusing 6-OHDA. In this way, 6-OHDA was no longer able to be taken up by the NET and, therefore, could not damage NE neurons. At the same time, the neurotoxicity to the DA system was left intact. In these experimental conditions (loss of DA and sparing of NE axons), the kindling process was no longer affected, contrasting the marked acceleration of amygdala kindling that occurred in rats not protected by DMI and featuring marked NE depletion. Further data confirmed the specific involvement of NE in kindling. This was obtained by restricted lesioning of NE pathways ascending in the dorsal forebrain bundle. These lesions produced a loss of the NE innervation to both limbic and neocortical areas and produced a significant increase of kindling rate, both in the amygdala and cortical areas.[37,38]

As expected, increasing NE activity produces the opposite effect (i.e. delayed kindling). This may be seen when extracellular NE levels are increased by blocking NE uptake (DMI administration),[39] or by electrical stimulation of the nucleus LC before each kindling session.[40]

In addition to the in vivo studies concerning kindling and the NE system, a different approach is possible in vitro, using slice preparations to examine the epileptic burst response in restricted neural circuits following repeated electrical stimulation with or without manipulation of the NE system. By using this approach, McIntyre and Wong[41] examined coronal slices including the amygdala-pyriform area. In these experimental conditions, neurons of the pyriform cortex produce a burst response following the stimulation of the adjacent amygdala. When the experiment is reproduced following previous kindling of the amygdala, a significant increase in the duration of the burst response occurs compared with nonkindled tissue. Finally, exposing the slice chamber to a low concentration of NE (2 μM) suppresses the burst response in the nonkindled preparation. When the dose of NE was increased, this effect was visible even in kindled slices. To examine the specific effect of NE receptors, further results were obtained by applying selective NE agonists and antagonists to the slice preparation. While the α_1-agonist, L-phenylephrine (10 μM) had no effect on the burst response, the α_2-agonist, clonidine (1 μM), reproduces the suppressive effect obtained with NE, and the β-agonist, isoproterenol (1 μM), increased the burst response. Taken together, these results suggest an inhibitory effect of NE on seizures, which might be mediated by α_2-adrenergic receptor subtypes.

NE in other seizure models

Apart from classic studies on the role of NE in kindling, the role of the NE system has been evaluated in various models of epilepsy. In line with data obtained in the kindling model, damage to NE pathways produces increased seizure susceptibility, while the stimulation of LC has a protective effect against epileptic seizures. Mason and Corcoran[42] demonstrated that a greater susceptibility to MES induced seizures in rats lesioned in the NE system by pretreatment with 6-OHDA. Pretreatment with 6-OHDA reduced the latency of myoclonic jerks, epileptic discharges, and epileptic convulsions in rats injected with penicillin.[43] The damage to the NE system using DSP-4 reduces the threshold for facial and forelimb clonus obtained by corneal electroshock stimulation in rats, while MES performed in NE-depleted rats confirms increased susceptibility to brain stem seizures. On the other hand stimulation of the LC suppresses penicillin-induced focal epileptiform activity in rats.

Evidence for the involvement of NE in experimental limbic status epilepticus

While the role of NE in modulating experimental seizures is well established, its effect on focally induced status epilepticus has been much less investigated. A recent paper[44] examined the effects produced by the loss of NE terminals originating from the LC on seizures induced focally by microinfusing chemoconvulsants in the anterior extent of the piriform cortex (APC). This cortical area was selected due to its high level of NE innervation (Figure 15.2). In fact, the pyriform cortex, among various cortical regions, possesses the highest NE content. The high sensitivity of APC to focal microinfusions of chemoconvulsants, as demonstrated in previous studies in rodents and primates,[45] makes it an optimal site to focally trigger limbic seizures. Thus, microinfusing bicuculline (a $GABA_A$ antagonist) in this brain region elicits sporadic limbic seizures resembling those originally described for amygdala kindling in intact rats.[45]

In rats deficient in NE innervation, we observed that the same dose of bicuculline prolonged seizure frequency and duration, leading to self-sustaining status epilepticus lasting up to several hours. The significance of these findings extends beyond a mere decrease in seizure threshold. This information allows the researcher to hypothesize a role for NE loss in modifying the responsiveness of neural epileptic circuits, thereby forming the anatomical substrates of self-sustaining status epilepticus in place of sporadic seizures. This modified response was evident when the trigger region was challenged by selective glutamate antagonists. While in intact rats limbic seizures were suppressed both by *N*-methyl-D-aspartate (NMDA) and non-NMDA antagonists with comparable efficacy, NE-lesioned rats had an observable loss of sensitivity to NMDA blockade. This was demonstrated by preinfusing the competitive NMDA antagonist AP-7 in NE-lesioned rats, which neither reduced limbic seizures nor prevented the onset of limbic status epilepticus. Conversely, the non-NMDA antagonist 2, 3-dihydroxy-b-nitro-7-sulfamoyl-benzo (F) quinoxaline

(NBQX) maintained its pharmacological efficacy, preventing the onset of status epilepticus and blocking sporadic seizure activity.

As stated above, the APC is probably the most densely NE-innervated cortical area. It is likely that loss of NE terminals innervating the APC is crucial for the conversion of sporadic seizures into limbic status epilepticus. In fact, the switch in sensitivity to NMDA antagonists takes place in the APC itself. However, the role of NE loss in other brain regions that are involved in the propagation and/or gathering of seizures evoked from APC cannot be excluded. It is interesting that the most significant effects of NE on seizure activity are obtained in the kindling model and in status epilepticus. These conditions are both characterized by long-lasting changes in the neural circuitry.

The implications for these findings extend beyond the role of NE in the field of experimental epileptology, to an understanding of the more generalized effects of NE on neuronal plasticity in the limbic system. For example, that loss of LC innervation does not modify baseline electroencephalogram or the sleep–waking cycle, but abolishes gene expression related to the sleep–waking cycle. Similarly, NE is important for producing long-term potentiation. Apart from epilepsy, NE input to the pyriform cortex, as to other parahippocampal limbic regions, plays a relevant role in learning and memory. Consistent with this information, an elegant study, carried out by Browning's group demonstrated that vagus nerve stimulation, which cures certain forms of drug-resistant epilepsy, does not produce any effect in the absence of the LC.[46] Vagus nerve stimulation, apart from producing antiepileptic effects, enhances recognition memory in humans. Altogether these data suggest that NE input to limbic regions is critical to maintenance of the ongoing physiological synaptic activity in this brain area.

Proconvulsant and anticonvulsant effects of adrenergic receptor agonists and antagonists

Numerous pharmacological studies have evaluated the role of specific NE receptors in experimental epilepsy. Weinshenker and Szot[47] comprehensively reviewed pharmacological experiments investigating the effects of NE agonists and antagonists in various models of epilepsy. The pharmacological approach gave conflicting results. They found that it was sometimes difficult to discern whether a given NE receptor possessed either pro- or anticonvulsant effects. There are several possible reasons for this difficulty. First, large differences exist in NE receptor distribution between animal species, strains, and brain regions, showing different effects due to the specific experimental model. Second, selectivity of NE agonists and antagonists for different subtypes of NE receptors vary with the dose of the drug. For example, clonidine might stimulate imidazoline receptors at doses even lower than those required to activate α_2-adrenergic receptors. Therefore, in some experimental conditions, a modulatory effect on seizures, which is attributed to a specific NE receptor, might result from a mixed effect on multiple receptors.

Despite discrepancies on the role of NE receptors in experimental epilepsy, there are some points which are well established. Most of the studies were carried out on α_2-adrenergic receptors, which produce an anticonvulsant effect in almost every experimental setting. For instance, systemic administration of clonidine produces a dose-dependent delay of the amygdala kindling. Conversely, kindling is accelerated by systemic administration of the α_2-antagonists idazoxan, yohimbine, or rauwolscine. The effects of clonidine on delaying amygdala kindling depend on the activation of postsynaptic α_2-receptors. In fact, this effect of clonidine is maintained following a loss of NE terminals.

Since the amygdala-pyriform region exhibits both high NE content and high density of α_2-receptors compared with other brain areas, the strong effects produced by clonidine in delaying amygdala kindling might be a physiological mechanism to delay kindling, while loss of these receptors in limbic regions might underly the development of the kindling process. Selective infusion of clonidine into the amygdala-pyriform region delays kindling, while the simultaneous infusion of idazoxan abolishes the clonidine-induced delay. Unexpectedly, selective infusions of the α_2-antagonists fail to accelerate kindling. It is likely that α_2-blockade in the stimulated amygdala is necessary, but not sufficient, to facilitate kindling; alternatively, the ongoing down-regulation of α_2-receptor reduces the efficacy of α_2-antagonists during amygdala kindling. In contrast to the effects on kindling development, neither α_2-antagonists nor clonidine modify seizures elicited from previously kindled animals. This evidence, joined with results mentioned above, indicates that endogenous NE (activating postsynaptic α_2-receptors) exerts a protective effect on the establishment of epileptic circuitry rather than on seizure susceptibility.

The role of α_1-adrenergic receptors has received less attention than that of α_2-receptors. However, considering even the few available pharmacological studies, the anticonvulsant effect of α_1-agonists appears consistent. For instance, the α_1-agonist methoxamine mimicks the anticonvulsant effect of high doses of clonidine against proconvulsants in rats. Studies are currently examining the role of α_1-receptor subtypes by using knockout and transgenic mice (see below), allowing selective studies of NE receptor subtypes despite the lack of selective pharmacological ligands.

The role of β-adrenergic receptors in epilepsy is unclear. The main confounding factor relates to the wide use of propranolol in various experimental models. This nonselective β-receptor antagonist also exhibits a weak intrinsic activity on β-receptors along with membrane stabilization properties and local anaesthetic activity. These effects alone are anticonvulsant and are thought to be unrelated to β-receptor blockade. Aside from the misuse of propranolol, a few general concepts can be drawn concerning the role of β-receptors in epilepsy. Most of the pharmacological studies indicate that β-agonists are anticonvulsant, while β-antagonists facilitate seizures. For instance, the β_1-agonist, dobutamine, as well as the β_2-agonist, terbutaline, decrease seizures in genetically epilepsy-prone rats (GEPR rats), with response being dose-dependent. Similarly, intrahippocampal injection of isoproterenol, a nonselective β-receptor agonist, causes a significant decrease of spikes in the hippocampus of rats. Further, there is general agreement that amygdala kindling produces a persistent down-regulation of neocortical β-receptors. This is likely due to an excess of NE, which is released during the kindling process. Since a reduction in β-receptor stimulation is associated with an increase in seizure susceptibility, anticonvulsant properties of β-receptors appear to be confirmed.

The role of NE in models based on selective breeding to epileptic phenotype

Additional evidence that NE has a strong inhibitory effect on development of a seizure circuitry by inhibiting seizure initiation and propagation was obtained by analyzing genetically epileptic strains. These consist of specific strains of mice, rats, primates, and dogs in which increased seizure susceptibility or spontaneous seizure activity occurs. Selective alterations in the NE system and the development of an epileptic phenotype was observed. Nonetheless, since these strains often bear various associated deficits it remains difficult to attribute a seizure-prone phenotype to a specific NE deficit. In the following paragraphs we will describe two classic genetic models of epilepsy: GEPR rats and tottering mice.

The GEPR colonies derive from progressive inbreeding of Sprague-Dawley rats susceptible to audiogenic seizures. To date, two inbred colonies of GEPRs, GEPR-3 and GEPR-9, have been developed. Seizure severity is higher in GEPR-9 compared with GEPR-3, as confirmed by the observation that after acoustic stimulation GEPR-3 show only mild clonic seizures, while GEPR-9 exhibit massive tonic and tonic–clonic convulsions. However, both strains possess a reduced threshold to various epileptic stimuli, such as electroshock, pentylenetetrazole, fluorothyl, and limbic kindling. There is a marked alteration of monoaminergic systems in the brains of GEPR, and the severity of the epileptic phenotype is associated with the degree of NE loss. In particular, in GEPR-9 there is a massive decrease of NE axons in various brain areas, as demonstrated by reduced NE levels, DBH activity, and immunohistochemistry, as well as NE uptake sites.[48] This massive loss is less pronounced for GEPR-3. Even though abnormalities in other neurotransmitters, such as serotonin and GABA, have been found in GEPR, restoration of NE activity by administering NE, NE reuptake inhibitors, or NE agonists reduce audiogenic seizures; thus implying that NE deficiency plays a major role in determining seizure susceptibility in these rats.

Tottering mice exhibit spontaneous absence seizures, with cortical spike-and-wave discharges late in the development. This epileptic phenotype is associated with ataxia and intermittent myoclonus. Histochemical analysis shows a significant increase in the number of NE axons in brain regions innervated by the LC. Despite this hypertrophy of axon branches from the LC to target brain regions, the number of cell bodies in the LC was similar in both wild-type and tottering mice. It has been suggested, then, that a higher self-inhibition due to increased branching of axon collaterals from the LC justifies the paradoxical association between epilepsy and higher NE levels. The abnormality in the NE system is not the unique alteration occurring in these mice. Mutations in the gene encoding the high-voltage-activated α1A Ca^{2+} channel subunit has been recently discovered in this model of epilepsy. The involvement of Ca^{2+} channels in numerous cellular functions, including cell proliferation and differentiation, membrane excitability, axon outgrowth, synaptogenesis, signal transduction, and gene expression might account for a possible molecular substrate predisposing to epilepsy in tottering mice.

NE in models of epilepsy based on selective gene manipulation

Recently, the role of specific NE receptor subtypes, the NET, and the specific NE biosynthetic enzyme were investigated in genetically engineered mice. This approach allowed the investigation of the role of these molecules in modulating seizures. These mice were obtained from selective gene manipulations, and may exhibit lack or overexpression of specific genes. They represent a new technique for study of the neurobiology of the NE system, and they are a valuable tool for the future of epilepsy research. In particular, these animals allow the researcher to overcome the uncertainties deriving from pharmacological approaches where, due to a lack of absolute selectivity, various concentrations of agonists/antagonists for specific NE receptors often lead to opposite results in the expression of seizure activity. Conversely, these models have the limitation of bearing mutations as inborn defects. Such conditions could lead to the development of epileptic circuitry independently by the NE alteration, as expressed in the phenotype, thereby producing an anatomical substrate that differed from controls but was not related to the aimed alteration in the NE system.

Dopamine-β-hydroxylase knockout mice are completely lacking endogenous NE; they were generated via a targeted disruption of the *DBH* gene, which codes for the enzyme converting DA to NE (Fig. 15.5). These mice show enhanced susceptibility to all the proconvulsant stimuli used to test seizure threshold (fluorothyl, PTZ, kainic acid, high decibel sounds).[49] These DBH null mutant mice indicate that NE by itself has anticonvulsant effects. In fact, demonstration of an increased seizure sensitivity obtained from studies lesioning the NE system was based on the loss of NE terminals. These terminals represent a complex anatomical structure containing NE, ATP, adenosine, and several neuropeptides that are coreleased with NE and might exert an effect on seizure activity. Therefore, one cannot rule out the possibility that anticonvulsant effects exerted by NE terminals are sustained by cotransmitters such as galanin, neuropeptide Y, or adenosine. Furthermore, various studies have demonstrated that these molecules possess an anticonvulsant effect. The same pitfall accompanies interpretation of data obtained by producing an anticonvulsant effect via the electrical stimulation of the LC. These latter considerations attenuate the hypothesis that NE, itself, is anticonvulsant. There was no pharmacological tool to discern this. The use of DBH knockout mice demonstrated for the first time that a substantial part of the anticonvulsant effect produced by NE terminals is really due to the presence of endogenous authentic NE.

By creating a transgenic mouse that overexpresses systemically either wild-type or constitutively active mutant forms of α_{1B}-adrenergic receptors, it was possible to investigate the role of α_{1B}-receptors on the onset of epilepsy. In transgenic mice, which exhibit spontaneous interictal epileptic spikes and seizures, seizure severity is related to the degree of activity of the over expressed receptor, being more severe in mutant than in wild-type mice. These seizures can be partially reversed in this model by terazosine, an α_1-antagonist, thus indicating that α_1-related signaling is responsible for seizure activity. In addition to epileptic phenotype, the mice show an age-progressive Parkinson's-like locomotor impairment with neurodegeneration of the cortex, hypothalamus, thalamus, cerebellum, striatum, and substantia nigra.[50,51] Even considering the extensive neurodegeneration and the complex phenotype described above, this model does not represent a "clean" model of epilepsy, since it is not clear whether seizures are due to the brain damage or are the consequence of reiterated epileptic events. Nevertheless, this model provides a tool which, in the absence of specific α_{1B}-adrenergic receptor ligands, indicates a role of α_{1B}-adrenergic receptors in epilepsy. In a recent paper confirming the role of α_{1B}-receptors, our lab demonstrated that knockout mice for these receptors possess a high resistance to methamphetamine-induced neurotoxicity[52] and we found these mice to be refractory to various forms of seizures with a special emphasis on chemical kindling.

The use of genetically modified animals has also helped to clarify the role of α_2-adrenergic receptors in the kindling phenomenon. Mice carrying a point mutation in the α_2-adrenergic receptor locus (the D79N mutant mice), markedly differ from wild-type mice concerning the development of amygdala kindling. In particular, compared with wild-type mice, the number of stimuli needed to achieve full kindling was dramatically lower, and seizure duration after kindling was longer, in D79N mutant mice.[53] Increased seizure severity was demonstrated in D79N mutant mice compared to wild-type mice. Administration of idazoxan (a pharmacological antagonist at α_2-adrenergic receptors) makes wild-type mice indistinguishable from D79N mice, and systemic administration of this agent to D79N mice does not modify the kindling process.

Clinical correlates for the involvement of NE in epilepsy

Data obtained from animal models of epilepsy since the 1980s suggests a causative role for NE in modulating seizures. This makes the study of the role of NE in epilepsy research extremely promising. However, there is still an enormous gap between data obtained in experimental neurology and clinical evidence for an antiepileptic role of NE. In particular, only a few papers examined this issue in epileptic patients, and the lack of specificity of several drugs does not yet allow extrapolation of a specific role of NE in human seizures. Nonetheless, the data from clinical studies seem to support basic science research. For instance, the anticonvulsant role of α_1-adrenergic receptors in animal models has been reviewed previously. Briere *et al.*[54] measured the amount of cortical α_1-adrenergic receptors from tissue obtained in patients undergoing temporal lobectomy due to intractable partial epilepsy. They found a decrease in α_1-receptor density in the epileptic focus, thus suggesting a reduced sensitivity to NE. This led to a decrease of inhibitory mechanisms focally within the epileptic tissue. Specific studies examined the concentration of NE within the cortex of patients with neocortical and mesial temporal lobe epilepsy. These studies demonstrated specific alterations of NE levels without involvement of DA content. These alterations were dependent on the type of seizures and the brain region under analysis. This data led the authors to hypothesize a global impairment of NE pathways in epileptic seizures in humans.[55]

A further issue that correlates the NE system to epilepsy is offered by observations that indicate that depressive disorders are frequent in patients with epilepsy, and it is thought that NE deficit might contribute to the onset of both epilepsy and depression because NE deficits occur in humans with depression, as well as in patients with epilepsy.[56] The relationship between the two disorders is also confirmed by antiepileptic drugs that are effective in the management of affective disorders, while antidepressants possess antiepileptic effects. Vagus nerve stimulation (VNS) represents a novel treatment for refractory seizures, and it has been recently approved for the treatment of refractory depression. Alternately, specific antidepressants produce or exacerbate seizures in epileptic patients. For instance, overdosing of tricyclic antidepressants causes seizures in humans and animals. However, this assumption is not justified in light of various studies showing that antidepressants at low doses have anticonvulsant activity both in humans and animal models. Further, the proconvulsant effect of overdosing is likely to be due to the blockade of α_1-receptors, which is a typical side effect of these drugs. In animal models, the anticonvulsant action of specific antidepressants has been related to their capacity to block NE and/or serotonin reuptake. As stated above, VNS represents a further clinical correlation of the involvement of NE in epilepsy. Vagal nerve stimulation exerts an antiepileptic effect both in experimental models of epilepsy and in humans. In clinical practice, VNS is indicated for treating medically refractory partial epilepsy, even though recent studies have shown its efficacy also in medically refractory generalized epilepsy, including Lennox-Gastaut syndrome.

Although the precise mechanism of the antiepileptic action of VNS is not fully understood, work focusing on the identification of the anatomic brain structures that mediate this antiepileptic effect demonstrates the importance of the NE system in epilepsy. Among several forebrain and brain stem regions activated by VNS in rats, the effects induced in the LC, which receives projections from the parabrachial nucleus, which, in turn, is innervated by the nucleus of the solitary tract, the visceral afferent nucleus of the vagus nerve, are significant. The most striking evidence for the crucial involvement of the noradrenergic system in the anticonvulsant effect of VNS is supported by the observation that bilateral ablation of the LC suppresses the anticonvulsant efficacy of VNS stimulation.[46]

Involvement of NE in other neurological diseases

Recent data suggest a contribution of the NE system in the pathogenesis of various neurological disorders. These include: neurodegenerative diseases such as AD and other forms of dementia including Lewy body disease, and neuroinflammatory disorders such as multiple sclerosis (MS).

German[16] pointed out a disease-specific pathological and regional pattern of neuronal loss in the LC. For instance, in Pick's disease, there is a frank LC cell loss, characterized by disease-specific inclusions (Pick bodies) within LC neurons. Some authors have suggested studying this differential pathology of the LC as a tool to distinguish between different subgroups of PD patients. In keeping with this, it was found that in PD not complicated with dementia, there is homogeneous degeneration involving both the caudal and the rostral portion of the nucleus. This pattern includes the loss of the projections from the LC to the spinal cord and to the cerebellum. In contrast, in PD complicated with dementia, as well as in AD, cell loss in the LC seems to be particularly pronounced in the most rostral portion of the nucleus that projects to the forebrain.

Multiple systems atrophy (MSA)

Multiple systems atrophy is a progressive neurological disease that typically includes parkinsonism or cerebellar atrophy in addition to autonomic failure. The cardinal symptom of autonomic failure is the orthostatic hypotension, although in atypical cases MSA may occur without a pathological decrease in blood pressure. This autonomic dysfunction is related to degeneration of the central NE system. In fact, the sympathetic postganglionic innervation remains relatively intact compared with loss of preganglionic cells in MSA patients,[57] while both postmortem and in vivo studies reveal an impairment in the central NE system. A marked decrease in NE content was observed in the spinal cord,[58] where it is known to modulate autonomic functions. Additionally, degeneration of LC neurons mainly in its caudal portion projecting to the spinal cord and the cerebellum has been shown to be markedly involved in MSA. Levels of NE were also measured in vivo within cerebrospinal fluid, long considered by neurologists as a "window on the brain." Patients with MSA with orthostatic hypotension show lower cerebrospinal fluid concentrations of NE than controls.[59] The α_{1B}-receptor subtype seems to play a relevant role in autonomic failure as well as in motor disturbances. The model of transgenic mice overexpressing α_{1B} receptors, discussed above in regard to epilepsy, was proposed as a model of MSA because transgenic mice develop hypotension, ataxia, and parkinsonism.[50]

The role of NE in dementia

Neuronal cell death in LC neurons and degeneration of NE projections are well-described features of AD. Cell loss was found in the rostral and central portion of the LC in patients with AD, as well as in PD complicated with dementia. These parts of the LC are considered to innervate mainly the hippocampal, frontal, and temporal cortex, i.e., the regions that are most affected by amyloid β plaques and neurofibrillary tangles. In addition, decreased LC neurons and degeneration of ascending noradrenergic axons lead to decreased NE levels in target brain projection areas, whereas an up-regulation of adrenergic receptors is observed subsequently in noradrenergic deafferentation.[60] Moreover, the LC cell loss in AD was reported to correlate significantly with the number of amyloid β plaques, neurofibrillary tangles, and the severity of dementia.

The involvement of an inflammatory contribution in the pathophysiology of AD received great attention in the last decade, leading to consideration of potential therapeutic strategies. Converging evidence exists describing a significant inflammatory component in AD.[61] The finding that activated microglia is interleukin (IL)-1 immunopositive and almost exclusively associated with amyloid deposits in the AD brain,[61] as well as in transgenic mouse models of AD, indicates an important role of microglia activation and cytokines in AD. Recent observations found a link between LC cell loss and neuroinflammatory events that may contribute to neuronal dysfunction in AD.[62] In particular, this work demonstrated that LC loss enhances the expression of amyloid β-dependent nitric oxide synthase, IL-1β, and IL-6. These findings suggest that LC loss contributes to increased inflammation in AD.

Evidence for NE involvement in multiple sclerosis

The first indication for a possible contribution of the NE system in demyelinating disorders comes from the experimental allergic encephalomyelitis (EAE), the animal model of MS. Norepinephrine concentrations were found depleted in the gray matter of spinal cord of rats with EAE.[63] Further information supporting the role of NE in EAE is represented by the observation that depletion of central NE levels by the use of 6-OHDA or electrolytic lesions of the LC modified the course of EAE.[64] More recent studies found a lack of β_2-adrenergic receptors in astrocytes of both normal white matter, and chronic active or inactive plaques in MS brains.[65] It has therefore been hypothesized that this lack of β_2-receptors might allow inflammatory gene expression.

Besides these studies showing a role for central NE in modulating the progress of experimental and human demyelinating diseases, there are a few data supporting a role, also, for peripheral NE. In fact, expression of β-adrenergic receptors is higher in peripheral blood monocytes of MS patients than in controls. Further, administration of β-receptor agonists including isoproterenol or terbutaline suppresses EAE symptoms. However, it is not clear whether this effect is mediated via peripheral (T cells) or central mechanisms (astrocytes or microglia). In vitro studies performed during the last decade clearly demonstrated an immunosuppressive role of NE on glial and endothelial cells. Although the mechanism remains to be determined, NE suppresses the transcription of inflammatory genes. The capacity of astro- and microglial cells to synthesize the proinflammatory cytokines such as IL-1β and tumor necrosis factor-α and other cytokines is strongly reduced by NE. Additionally, in astrocytes, NE has been found to mediate the inhibition in the expression of the enzyme nitric oxide synthase type II.

Sleep disorders

The possible involvement of the NE system has been evaluated in sleep disorders. Particular attention has been focused on narcolepsy, a debilitating neurologic disease consisting of cataplexy, hypnagogic hallucinations, daytime sleep attacks, and sleep paralysis. Although there is a consensus on the existence of neurochemical imbalance involving NE pathways in narcolepsy, the underlying mechanisms remain poorly understood. Most of the studies were performed on canine models of narcolepsy. This animal model exhibited an elevated number of α_2-receptors in the LC, thus suggesting enhanced self-inhibitory mechanisms leading to reduced NE release.[66] A prominent role for NE receptors is supported by the observation that treatment with the α_1-agonist methoxamine ameliorates cataplexy in dogs, whereas the α_1-antagonist prazosin exacerbates the attacks.[67] The preferential involvement of adrenergic mechanisms in the pathophysiology of the disorder is suggested by the evidence that all compounds affecting NE transmission (such as DMI, nisoxetine, nortriptyline, tomoxetine, and viloxazine) completely suppress cataplexy at low doses.[68]

Conclusions

Based on data coming from basic science and clinical research, it is evident that endogenous NE plays a protective role in a variety of neurological disorders. This is more pronounced when considering diseases characterized by selective cell death in which excitotoxicity is prominent. For this reason, PD and seizure disorders were emphasized in this chapter. The kind of influence played by NE in these neurological disorders recalls its physiological role in modulating long-term rather than short-term influence on neuronal activity. The antiepileptic effect of NE extends beyond a mere increase of seizure threshold and affects mostly the development of an epileptic circuitry. These plasticity-dependent antiepileptic mechanisms are enhanced by:

(1) the robust influence of NE in preventing the kindling of limbic structures where its antiepileptic role was first described;

(2) the ability of NE to prevent the conversion of sporadic into continuous (status epilepticus) limbic seizures.

In addition, the ability of an intact NE activity in the limbic system is important to reduce the damage produced by long-lasting status epilepticus. Similarly, when analyzing the basal ganglia, a long-term influence may underlie specific symptoms related to sensitization of motor circuits leading to abnormal involuntary movements.

Moving from plasticity-dependent mechanisms exerted by endogenous NE, this chapter emphasized the role of this neurotransmitter in determining and maintaining long-term potentiation (a long-lasting neuronal change produced by repetition of a stimulus that is supposed to underlie memory formation). These considerations open the field to a wider range of physiological and pathological conditions. For instance, it is well known that the same limbic circuits affected in the pathophysiology of epilepsy are also critical in the maintaining of physiological mechanisms involved in learning and memory. Similarly, an abnormal activation of the neostriatum might exacerbate motor learning, leading to dyskinesia. Among various brain regions within the limbic cortex, the role of NE is particularly crucial since these regions lack thalamic innervation and NE is supposed to substitute for thalamic inputs. The same brain areas exhibit pronounced excitability to serve learning and memory processes, thus lending substance to the crucial role of endogenous NE in sustaining plastic changes occurring in the limbic system beyond the mere regulation of seizure threshold. This matter forms the body of other specific chapters presented in this book (Chapters 12 to 14 and Chapters 19 to 21).

REFERENCES

1. Dahlstroem, A. and Fuxe, K. Evidence for the existence of monoamine neurons in the central nervous system. *Acta Physiol. Scand.*, **64** (1965), 1–36.
2. Woulfe, J. M., Fluferfelt, B. A. and Hrycyshyn, A. W. Efferent connections of the A1 noradrenergic cell group: a DBH immunohistochemical and PHA-L anterograde tracing study. *Exp. Neurol.*, **109** (1990), 308–322.
3. Brodal, A. (Ed.). The reticular formation and some related nuclei. In *Neurological Anatomy in Relation to Clinical Medicine.* (New York: Oxford University Press, 1981), pp. 394–447.
4. Foote, S. L., Bloom, F. E. and Aston-Jones, G. Nucleus locus coeruleus: new evidence of anatomical and physiological specificity. *Physiol. Rev.*, **63** (1983), 844–914.
5. Aston-Jones, G., Chiang, C. and Alexinski, T. Discharge of noradrenergic locus coeruleus neurons in behaving rats and monkeys suggests a role in vigilance. In *Progress in Brain Research*, Vol. 88, ed. C. D. Barnes and O. Pompeiano. (Amsterdam: Elsevier, 1991), pp. 501–520.
6. Bing, G., Filer, D., Miller, J. C. and Stone, E. A. Noradrenergic activation of immediate early genes in rat cerebral cortex. *Mol. Brain Res.*, **11** (1991), 43–46.
7. Cirelli, C., Pompeiano, M. and Tononi, G. Neuronal gene expression in the waking state: a role for the locus coeruleus. *Science*, **274** (1996), 1212–1215.
8. Nagai, T., Satoh, K., Immamoto, K. and Maeda, T. Divergent projections of catecholamine neurons of the locus coeruleus as revealed by fluorescent retrograde double labeling technique. *Neurosci. Lett.*, **23** (1981), 117–123.
9. Harik, S. I. and McGunigal, T., Jr. The protective influence of the locus ceruleus on the blood–brain barrier. *Ann. Neurol.*, **15**: 6 (1984), 568–574.
10. Reading, P. J. and Dunnett, S. B. 6-Hydroxydopamine lesions of nigrostriatal neurons as an animal model of parkinson's disease. In *Toxin-induced Models of Neurological Disorders*, ed. M. L. Woodruff and A. J. Nonneman. (New York: Plenum Press, 1994), pp. 89–119.
11. Lidbrink, P. and Jonsson, G. Noradrenaline nerve terminals in the cerebral cortex: effects on noradrenaline uptake and storage following axonal lesion with 6-hydroxydopamine. *J. Neurochem.*, **22** (1974), 617–626.
12. Luthman, J., Fredriksson, A., Sundstrom, E., Jonsson, G. and Archer, T. Selective lesion of central dopamine or noradrenaline neuron system in the neonatal rat: motor behaviour and monoamine alterations at adult stage. *Behav. Brain Res.*, **33** (1989), 267–277.
13. Mavridis, M., Degryse, A. D., Lategan, A. J., Marien, M. R. and Colpaert, F. C. Effects of locus coeruleus lesions on parkinsonian signs, striatal dopamine and substantia nigra cell loss after 1-methyl-4-phenyl-1,2,3,6-tetrahydropyridine in monkeys: a possible role for the locus coeruleus in the progression of Parkinson's disease. *Neuroscience*, **41** (1991), 507–523.
14. Forno, L. S. Neuropathology of Parkinson's disease. *J. Neuropathol. Exp. Neurol.*, **55** (1996), 259–272.
15. Ehringer, H. and Hornykiewicz, O. Verteilung von Noradrenalin und Dopamin (3-Hydroxytyramin) im Gehirn des Menschen und ihr verhalten bei Erkrankungen des Extrapyramidalen Systems. *Klin. Wschr.*, **38** (1960), 1236–1239.

16. German, D. C., Manaje, K. F., White, C. L. *et al.* Disease specific pattern of locus coeruleus cell loss. *Ann. Neurol.*, **32** (1992), 667–676.
17. Mann, D. M. The locus coeruleus and its possible role in ageing and degenerative disease of the human central nervous system. *Mech. Ageing Dev.*, **23** (1983), 73–94.
18. Bertrand, E., Lechowicz, W., Szpak, G. M. and Dymecki, J. Qualitative and quantitative analysis of locus coeruleus neurons in Parkinson's disease. *Folia Neuropathol.*, **35** (1997), 80–86.
19. Gaspar, P., Duickaerts, C., Alvarez, C., Javoy-Agid, F. and Berger, B. Alterations of dopaminergic and noradrenergic innervations in motor cortex of Parkinson's disease. *Ann. Neurol.*, **30** (1991), 365–374.
20. Hornykiewicz, O. and Pifl, C. The validity of the MPTP primate model for neurochemical pathology of idiopathic Parkinson's disease. In *Noradrenergic Mechanisms in Parkinson's Disease*, ed. M. Briley and M. Marien. (Boca Raton: CRC Press, 1994), pp. 59–71.
21. Gerlach, M. and Riederer, P. The pathophysiological basis of Parkinson's disease. In *Series of New Drugs*, Vol. 1. *Inhibitors of Monoamine Oxidase B*, ed. I. Szlenyi. (Basel: Birkhauser, 1993), pp. 25–50.
22. Chan-Palay, V. and Asan, E. Alterations in catecholamine neurons of the locus coeruleus in senile dementia of the Alzheimer's type and in Parkinson's disease with and without dementia and depression. *J. Comp. Neurol.*, **287** (1989), 373–392.
23. Dooneief, G., Mirabello, E., Bell, K. *et al.* An estimate of the incidence of depression in idiopathic Parkinson's disease. *Arch. Neurol.*, **49** (1992), 305–307.
24. Takahashi, A. Autonomic nervous system disorders in Parkinson's disease. *Eur. Neurol.*, **31** (Suppl.1): (1991) 141–147.
25. Stein, Y., Mayeux, L. and Cote, L. Reaction time and vigilance in Parkinson's disease: possible role of altered norepinephrine metabolism. *Arch. Neurol.*, **41** (1984), 1086–1089.
26. Forno, L. S., Langston, J. W., DeLanney, L. E., Irwin, I. and Ricaurte, G. A. Locus coeruleus lesions and eosinophilic inclusions in MPTP-treated monkeys. *Ann. Neurol.*, **20** (1986), 449–455.
27. Seniuk, N. A., Tatton, W. G. and Greenwood, C. E. Dose-dependent destruction of the coeruleus-cortical and nigral-striatal projections by MPTP. *Brain Res.*, **27** (1990), 7–20.
28. Gesi, M., Soldani, P., Giorgi, F. S. *et al.* The role of the locus coeruleus in the development of Parkinson's disease. *Neurosci. Biobehav. Rev.*, **24** (2000), 655–668.
29. Fornai, F., Bassi, L., Torracca, M. T., Scalori, V. and Corsini, G. U. Norepinephrine loss exacerbates methamphetamine-induced striatal dopamine depletion in mice. *Eur. J. Pharmacol.*, **283** (1995), 99–102.
30. Marien, M., Briley, M. and Colpaert, F. Noradrenaline depletion exacerbates MPTP-induced striatal dopamine loss in mice. *Eur. J. Pharmacol.*, **236** (1993), 487–489.
31. Sonsalla, P. K., Giovanni, A., Sieber, B. A., Delle Donne, K. and Manzino, L. Characteristics of dopaminergic neurotoxicity produced by MPTP and methamphetamine. *Ann. N. Y. Acad. Sci.*, **648** (1992), 229–238.
32. Fornai, F., Torracca, M. T., Bassi, L. *et al.* Norepinephrine loss selectively enhances chronic nigrostriatal dopamine depletion in mice and rats. *Brain Res.*, **735** (1996), 349–353.

33. Kamisaki, Y., Hamahashi, T., Hamada, T., Maeda, K. and Itoh, T. Presynaptic inhibition by clonidine of neurotransmitter amino acid release in several brain regions. *Eur. J. Pharmacol.*, **217** (1992), 57–63.
34. Fornai, F., Giorgi, F. S., Alessandrì, M. G., Giusiani, M. and Corsini, G. U. Effects of pretreatment with DSP-4 on methamphetamine-induced striatal dopamine losses and pharmacokinetics. *J. Neurochem.*, **72** (1999), 777–784.
35. Arnold, P. S., Racine, R. J. and Wise, P. A. Effects of atropine, reserpine, 6-hydroxydopamine and handling on seizure development in the rat. *Exp. Neurol.*, **40** (1973), 457–470.
36. McIntyre, D. C. and Edson, N. Facilitation of amygdala kindling after norepinephrine depletion with 6-hydroxydopamine. *Exp. Neurol.*, **74** (1981), 748–757.
37. Corcoran, M. E. and Mason, S. T. Role of forebrain catecholamines in amygdaloid kindling. *Brain Res.*, **190** (1980), 473–484.
38. Altman, I. M. and Corcoran, M. E. Facilitation of neocortical kindling by depletion of forebrain noradrenaline. *Brain Res.*, **270** (1983), 174–177.
39. McIntyre, D. C., Edson, N., Chao, G. and Knowles, V. Differential effect of acute vs chronic desmethylimipramine on the rate of amygdala kindling in rats. *Exp. Neurol.*, **78** (1982), 158–166.
40. Jimenez-Rivera, C., Voltura, A. and Weiss, G. K. Effect of locus coeruleus stimulation on the development of kindled seizures. *Exp. Neurol.*, **95** (1987), 13–20.
41. McIntyre, D. C. and Wong, R. K. Cellular and synaptic properties of amygdala-kindled pyriform cortex in vitro. *J. Neurophysiol.*, **55** (1986), 1295–1307.
42. Mason, S. T. and Corcoran, M. E. Catecholamines and convulsions. *Brain Res.*, **170** (1979), 497–507.
43. Sullivan, H. C. and Osorio, I. Aggravation of penicillin-induced epilepsy in rats with locus coeruleus lesions. *Epilepsia*, **32** (1991), 591–596.
44. Giorgi, F. S., Ferrucci, M., Lazzeri, G. *et al.* A damage to locus coeruleus neurons converts sporadic seizures into self-sustaining limbic status epilepticus. *Eur. J. Neurosci.*, **17** (2003), 2593–2601.
45. Piredda, S. and Gale, K. A crucial epileptogenic site in the deep prepyriform cortex. *Nature*, **317** (1985), 623–625.
46. Krahl, S. E., Clark, K. B., Smith, D. C. and Browning, R. A. Locus coeruleus lesions suppress the seizure-attenuating effects of vagus nerve stimulation. *Epilepsia*, **39** (1998), 709–714.
47. Weinshenker, D. and Szot, P. The role of catecholamines in seizure susceptibility: new results using genetically engineered mice. *Pharmacol. Ther.*, **94** (2002), 213–233.
48. Browning, R. A., Wade, D. R., Marcinczyk, M., Long, G. L. and Jobe, P. C. Regional brain abnormalities in noreprinephrine uptake and dopamine beta-hydroxylase activity in the genetically epilepsy-prone rat. *J. Pharmacol. Exp. Ther.*, **249** (1989), 229–235.
49. Szot, P., Weinshenker, D., White, S. S. *et al.* Norepinephrine-deficient mice have increased susceptibility to seizure-inducing stimuli. *J. Neurosci.*, **19** (1999), 10 985–10 992.
50. Zuscick, M. J., Sands, S., Ross, S. A. *et al.* Overexpression of the α1B adrenergic receptor causes apoptotic neurodegeneration: multiple system atrophy. *Nature Med.*, **6** (2000), 1388–1394.

51. Kunieda, T., Zuscik, M. J., Boongird, A. *et al.* Systemic overexpression of the alpha 1B-adrenergic receptor in mice: an animal model of epilepsy. *Epilepsia*, **43** (2002), 1324–1329.
52. Battaglia, G., Fornai, F., Buscati, C. L. *et al.* Alpha-1B adrenergic receptor knockout mice are protected against methamphetamine toxicity. *J. Neurochem.*, **86** (2003), 413–421.
53. Janumpalli, S., Butler, L. S., MacMillian, L. B., Limbird, L. E. and McNamara, J. O. A point mutation (D79N) of the alpha2A adrenergic receptor abolishes the antiepileptogenic action of endogenous norepinephrine. *J. Neurosci.*, **18** (1998), 2004–2008.
54. Briere, R., Sherwin, A. L., Robitaille, Y. *et al.* Alpha-1 adrenoceptors are decreased in human epileptic foci. *Ann. Neurol.*, **19** (1986), 26–30.
55. Pacia, S. V., Doyle, W. K. and Broderick, P. A. Biogenic amines in the human neocortex in patients with neocortical and mesial temporal lobe epilepsy: identification with in situ microvoltammetry. *Brain Res.*, **899**: 1–2 (2001), 106–111.
56. Jobe, P. C., Dailey, J. W. and Wernicke, J. F. A noradrenergic and serotonergic hypothesis of the linkage between epilepsy and affective disorders. *Crit. Rev. Neurobiol.*, **13** (1999), 317–356.
57. Goldstein, D. S., Holmes, C., Cannon, R. O., 3rd, Eisenhofer, G. and Kopin, I. J. Sympathetic cardioneuropathy in dysautonomias. *New Engl. J. Med.*, **336** (1997), 696–702.
58. Scatton, B., Dennis, T., L'Hereux, R. *et al.* Degeneration of noradrenergic and serotoninergic but not dopaminergic neurons in the lumbar spinal cord of parkinsonian patients. *Brain Res.*, **380** (1986), 181–185.
59. Goldstein, D. S., Holmes, C., Patronas, N. and Kopin, I. J. Cerebrospinal fluid levels of catechols in patients with neurogenic orthostatic hypotension. *Clin. Sci. (Lond.)*, **104** (2003), 649–654.
60. Kalaria, R. N., Andorn, A. C., Tabaton, M. *et al.* Adrenergic receptors in aging and Alzheimer's disease: increased beta 2-receptors in prefrontal cortex and hippocampus. *J. Neurochem.*, **53** (1989), 1772–1781.
61. Akiyama, H., Barger, S., Barnum, S. *et al.* Inflammation and Alzheimer's disease. *Neurobiol. Aging*, **21** (2000), 383–421.
62. Heneka, M. T., Galea, E., Gavriluyk, V. *et al.* Noradrenergic depletion potentiates beta-amyloid-induced cortical inflammation: implications for Alzheimer's disease. *J. Neurosci.*, **22** (2002), 2434–2442.
63. White, S. R., Bhatnagar, R. K. and Bardo, M. T. Norepinephrine depletion in the spinal cord gray matter of rats with experimental allergic encephalomyelitis. *J. Neurochem.*, **40** (1983), 1771–1773.
64. Jovanova-Nesic, K., Nikolic, V. and Jankovic, B. D. Locus coeruleus and immunity. II. Suppression of experimental allergic encephalomyelitis and hypersensitivity skin reactions in rats with lesioned locus coeruleus. *Int. J. Neurosci.*, **68**: 3–4 (1993), 289–294.
65. Zeinstra, E., Wilczak, N. and De Keyser, J. [3H]dihydroalprenolol binding to beta adrenergic receptors in multiple sclerosis brain. *Neurosci. Lett.*, **289** (2000), 75–77.
66. Fruhstorfer, B., Mignot, E., Bowersox, S. *et al.* Canine narcolepsy is associated with an elevated number of alpha 2-receptors in the locus coeruleus. *Brain Res.*, **500** (1989), 209–214.
67. Mignot, E., Guilleminault, C., Bowersox, S., Rappaport, A. and Dement, W. C. Role of central alpha-1 adrenoceptors in canine narcolepsy. *J. Clin. Invest.*, **82** (1988), 885–894.

68. Mignot, E., Renaud, A., Nishino, S. *et al.* Canine cataplexy is preferentially controlled by adrenergic mechanisms: evidence using monoamine selective uptake inhibitors and release enhancers. *Psychopharmacology (Berl.)*, **113** (1993), 76–82.
69. Hornykiewicz, O. Brain monoamines and parkinsonism. *Natl. Inst. Drug Abuse Res. Monogr. Ser.*, **3** (1975), 13–21.
70. Narabayashi, H., Kondo, T., Yokochi, F. and Nagatsu, T. Clinical effects of L-threo-3,4 dihydroxyphenylserine in cases of parkinsonism and pure akinesia. *Adv. Neurol.*, **45** (1987), 593–602.
71. Mizuno, Y., Kondo, T. and Mori, H. Various aspects of motor fluctuations and their management in Parkinson's disease. *Neurology*, **44** (1994), S29–S34.
72. Sandyk, R. Freezing of gait in Parkinson's disease is improved by treatment with weak electromagnetic fields. *Int. J. Neurosci.*, **85** (1996), 111–124.
73. Ringendahl, H. and Sierla, T. Freezing phenomenon in Parkinson disease. *Fortschr. Neurol. Psychiatr.*, **65** (1997), 435–445.
74. Sandyk, R. Locus coeruleus-pineal melatonin interactions and the pathogenesis of the "on-off" phenomenon associated with mood changes and sensory symptoms in Parkinson's disease. *Int. J. Neurosci.*, **49** (1989a), 95–101.
75. Sandyk, R. Hypothalamic-locus coeruleus mechanisms in the pathophysiology of "on-off" in L-dopa treated Parkinson's disease: a hypothesis. *Int. J. Neurosci.*, **47** (1989b), 303–308.
76. Yamazaki, M., Tanaka, C. and Takaori, S. Significance of central noradrenergic system on harmaline induced tremor. *Pharmacol. Biochem. Behav.*, **10** (1979), 421–427.
77. Goto, S. and Hirano, A. Catecholaminergic neurons in the parabrachial nucleus of normal individuals and patients with idiopathic Parkinson's disease. *Ann. Neurol.*, **30** (1991), 192–196.
78. Mann, D. M. and Yates, P. O. Pathological basis for neurotransmitter changes in Parkinson's disease. *Neuropathol. Appl. Neurobiol.*, **9** (1983), 3–19.
79. Zweig, R. M., Cardillo, J. E., Cohen, M., Giere, S. and Hedreen, J. C. The locus coeruleus and dementia in Parkinson's disease. *Neurology*, **43** (1993), 986–991.
80. Chan-Palay, V. Depression and dementia in Parkinson's disease. Catecholamine changes in the locus coeruleus, a basis for therapy. *Adv. Neurol.*, **60** (1993), 438–446.

16

Genetics of noradrenergic neurobiology

Inna Belfer, M.D., Ph.D.

National Institute of Dental and Craniofacial Research, Bethesda, USA

David Goldman, M.D.

National Institutes of Health, Bethesda, USA

Introduction

The genetic paradigm is one of inheritance and variation. A critical domain of application of the genetic paradigm is the common complex diseases, including several that involve noradrenergic function. These diseases, including hypertension and psychiatric disorders such as anxiety and depression, are substantially heritable, indicating a powerful role for inherited gene variation. The genetics tools, both informational and technical, have developed very rapidly during the past two decades and have created a scientific revolution. Whole-genome sequencing projects have been completed, beginning with microbial genomes and continuing with the eukaryotic species: human, chimpanzee, mouse, rat, dog, and fruit fly. The genome sequences provide an informational resource of linear sequence and sequence variation. A series of new technologies has been created to apply the knowledge, including bioinformatics, robotics, and genotyping chemistries.

Many of the principal aspects of the genome are still poorly understood, for example, the regulation of alternative splicing, control of transcription, role of intergenic material, and function of many noncoding RNAs. Gene prediction is uncertain. Although many common disorders (e.g., cardiovascular, autoimmune, psychiatric) have genetic etiologies, their inheritance is genetically complex, such that variations in more than one gene lead to a phenotype, prominent gene x environment interactions, and gene x gene interactions may also occur.

Despite these challenges, genes can be identified for complex traits by the use of either, or a combination of, two approaches: (1) positional cloning (reverse

This chapter was written by David Goldman, M.D. in his private capacity. No official support or endorsement by The National Institutes of Health, National Institute on Alcohol Abuse and Alcoholism is intended or should be inferred.

Brain Norepinephrine: Neurobiology and Therapeutics, ed. Gregory A. Ordway, Michael A. Schwartz and Alan Frazer. Published by Cambridge University Press.

genetics), using linkage to markers; and (2) forward genetics, using candidate genes involved in pathways implicated by biochemical or physiological differences. Susceptibility loci have been definitively identified in diverse complex diseases: hypertension; type I and type II diabetes; a variety of cancers; alcoholism; asthma; and malaria. Several of these discoveries have been made by the candidate gene approach. Identification of the functional locus in the candidate gene is also the endpoint of the positional cloning approach.

For diseases and phenotypes involving noradrenergic variation, the range of genes that are potentially involved is wide. The neurobiology of norepinephrine (NE) depends on multiple molecules involved in neuronal differentiation, connectivity, and plasticity, as well as more neurotransmitter-specific biosynthesis, storage, release, reuptake, metabolism, and signal transduction. In this chapter we describe NE-related genes, NE gene variation, and the influence of NE gene variation on expression and function.

Genetically variant traits influenced by noradrenergic function

Noradrenergic dysregulation is implicated in several complex disorders by a variety of studies in animal models and humans. Three main phenotypic domains influenced by noradrenergic function are anxiety/dysphoria, cardiovascular/respiratory, and pain. A common pathophysiological mechanism underlying all these disorders is stress. As discussed in chapters 9 to 11 of this volume, there is strong evidence for involvement of NE signaling in stress and stress-related behavior. Stress produces time-related regional increases in NE turnover in different brain regions. Uncontrollable (but not controllable) shock activates NE systems in the locus coeruleus (LC), hypothalamus, hippocampus, and frontal cortex. Chronic stress increases tyrosine hydroxylase (TH) activity in the LC. This contributes to enhanced response of the NE system to a subsequent stressor. Finally, uncontrollable (but not controllable) stress produces increased LC responsivity to excitatory stimulation via influences of α_2-adrenergic receptors on the LC.[1,2]

Anxiety/dysphoria domain

An increasing amount of data suggests that affective disorders are related to dysregulation of the hypothalamic–pituitary–adrenal (HPA) axis.[3] Disturbances of serotonin and NE neurotransmission along this axis are both implicated in anxiety and depression,[4] resulting in alteration of catecholaminergic function in many related disorders. For example, increased resting urinary and plasma levels of NE were reported in both posttraumatic stress disorder and panic disorder,[5] along with increased resting heart rate and blood pressure. Evidence for a role of NE in depression began with the discovery that several drugs, which either cause or alleviate depression alter NE metabolism. A role for NE in depression was predicted from clinical efficacy of older NE agents (desipramine, nortriptyline, and protriptyline) and newer noradrenergic-weighted or specific tricyclic antidepressants (venlafaxine, milnacipran). The efficacy of reboxetine, a highly selective NE reuptake inhibitor, in depression suggests that NE has an important role in this condition.[6]

The anxiety/dysphoria disorders and behaviors are familially transmitted, and the familiality largely results from genetic influences rather than shared family environment. Environmental influences specific to an individual are also etiologically significant. Depression and anxiety do not result from genetic or environmental influences acting alone, but rather from a combination of both.[7] Population-based twin studies as well as hospital registry-based twin studies find substantial heritability for affective disorders. Heritability of depressive symptoms or episodes in children and adolescents is estimated to be >50% on the basis of twin and adoption studies.[8] Although there is little doubt that life events play a precipitating role in depression, studies that have attempted to examine familial liability along with social adversity find that environmental measures tend to be contaminated by genetic effects. The tendency to experience (or report) life events appears to be influenced by shared family environment, and for certain types of events there is a genetic component.[9] Thus, the study of environmental influences in dysphoria would be greatly facilitated by the discovery of genes contributing to dysphoria and other behaviors.

The genes contributing to the heritability of depression have proven elusive.[10] However, several candidates originating from observed biochemical differences show promising results in the noradrenergic domain. Lower plasma activity of NE-related enzymes in patients with major depression and bipolar disorder[11,12] suggest that a dysfunction in catecholamine metabolism might be related to the etiology of depression. Japanese patients with depressive disorders showed a significantly higher rate of genotypes with the catechol-O-methyltransferase (COMT) low-activity Met158 allele compared to healthy controls.[13] The Met158 allele has also been linked to dimensionally measured anxiety in women[14] and to stress/pain response as detected in a neuroimaging study.[15] An excess of high-activity monoamine oxidase A (MAO-A) gene promoter alleles resulting in an elevated MAO-A activity may be a risk factor for major depressive disorder in females.[16] Both the COMT and MAO-A enzymes are essential for NE homeostasis. It can be hypothesized that the genes encoding these enzymes and additional genes involved in NE biosynthesis, metabolism, and noradrenergic signaling could interact to contribute to the etiology of anxiety/dysphoria disorders.

Cardiovascular and respiratory domain

A variety of environmental exposures are major contributors to cardiovascular diseases through effects on blood pressure variability and cardiac function. One important environmental factor in cardiovascular disease is stress. Along with other environmental exposures, stress is also a significant factor in asthma exacerbation, chronic obstructive pulmonary disease, and other respiratory disorders.[17]

Moderate heritability of onset and severity of both cardiovascular and respiratory diseases has been reported. A population-based study of a large cohort of Korean families estimated heritabilities of 26% for body mass index, 26% for high serum cholesterol, 19% for systolic blood pressure, and 9% for diastolic blood pressure.[18] These findings highlight the importance of considering genetic factors in studies of cardiovascular risk factors. A community-based study of Australian nuclear families confirmed the occurrence of familial aggregation of cardiovascular risk factors and the genetic basis for much of this aggregation. Estimated heritabilities were 27% for systolic and diastolic blood pressure, 37% for serum cholesterol, 52% for body mass index, and 23% for triceps fat fold.[19] A 36-year follow-up of Swedish twins found a moderate genetic effect on mortality related to coronary heart disease both in females and males, and that genetic factors were in operation throughout the entire life span.[20] Genetic factors also play an important role in the pathogenesis of essential hypertension.[21]

Noradrenergic signaling plays a significant role in development of congestive heart disease. Norepinephrine released from cardiac sympathetic nerves activates myocyte β_1-adrenergic receptors, causing an increase in cardiac inotropy, lusitropy, and chronotropy. Release of NE from cardiac sympathetic nerves is controlled by negative feedback from presynaptic α_2-adrenergic receptors.[22] Racial differences in the incidence of congestive heart failure, reduced penetrance in families with familial cardiomyopathy, and strain-specific effects in animal models all suggest that at least part of the variation in the general population with regard to susceptibility to heart failure is heritable.[23]

Several common polymorphisms within adrenergic receptor genes (α_{1B}, α_{2A}, β_1 and β_2) seem to contribute to interindividual variability in resting systolic and diastolic blood pressure, and in the increase in diastolic blood pressure in response to cognitive challenge.[24] Three psychosocial characteristics associated with cardiovascular disease – depression, hostility, and social support – correlate one with another. A study of young adult twins revealed that covariation among these characteristics was associated with a single common genetic factor.[25] An association between two polymorphisms in adrenergic receptor genes and the risk of congestive heart failure was found retrospectively in a black population.[26] Signaling through the β_1-adrenergic receptor is a key determinant of cardiac function. A common functional polymorphism in this gene was associated with resting heart rate in Chinese and Japanese individuals from nuclear families in whom the heritability of heart rate was 39.7%.[27]

High heritability of respiratory traits was shown in lambs[28] and in Holstein cattle.[29] In humans, the existence of important genetic determinants of adult lung function was shown in Caucasian nuclear families.[30] Polymorphisms in the β_2-adrenergic receptor gene has been associated with decreased airway responsiveness in a population-based study.[31] A prospective, case-control study of Chinese patients with chronic obstructive pulmonary disease found that genetic variation in the β_2-adrenergic receptor gene might increase disease susceptibility.[32] The same gene polymorphism was also associated with asthma.[33]

Pain

The genetics of pain and genetic–environmental contribution to pain production, sensitivity and maintenance are currently unknown.

The main components of the stress system, corticotropin-releasing factor and LC-NE-autonomic systems, stimulate the hypothalamic β-endorphin system, suppressing pain sensation.[34] However noradrenergic influences on pain processing and antinociception also directly implicate noradrenergic genes.[35] Release of NE, a mixed α_1-and α_2-adrenergic receptor agonist, from sympathetic terminals activates nociceptors, leading to the sensation of pain.[36] Studies in experimental animals suggest that nociceptors are not normally responsive to sympathetic stimulation, but develop catecholamine sensitivity after injury.[37] Sympathectomy by itself produced sensitivity to NE in 50% of rabbit C-fiber nociceptors.[38] Perineuromal administration of NE in patients with neuromas resulted in increased pain.[39] Norepinephrine injections produced pain in patients with sympathetically maintained pain (SMP) at doses that were at the threshold for producing vasoconstriction.[40] In patients with SMP who were rendered pain-free by a sympathetic block or surgical sympathectomy, NE injected into the skin rekindled the hyperalgesia-exaggerated pain in response to noxious stimuli.[41]

In contrast, in specific sites in the brain and spinal cord, NE is antinociceptive.[42–44] Such opposite actions can be explained by involvement of different types of adrenergic receptors in pain processing and regulation. Indeed, the pronociceptive effect of NE seems to be mediated via α_1-adrenoreceptors while α_2-adrenergic receptors are involved in its analgesic effect, primarily through a presynaptic inhibition of firing by nociceptive dorsal horn neurons in the spinal cord.[45,46] Several experiments support this hypothesis using selective α_1- or α_2-agonists and antagonists. Both α_1- and α_2-receptors are involved in nociception; however, α_2-adrenergic receptor agonists have a more important antinociceptive role. α_2-adrenergic receptor-mediated analgesia has been used clinically since 1984.[47] Clonidine, an α_2-adrenergic receptor agonist, produces analgesia by itself.[48] The antinociceptive action of the systematically administered clonidine was shown in a rat model of peripheral unilateral mononeuropathy.[49] Some patients with painful diabetic neuropathy benefit from systemically administered clonidine,[50] and intrathecal or epidural clonidine is used for peri- and postoperative pain management and in cancer pain[51] and is effective in most neuropathic pain patients.[52]

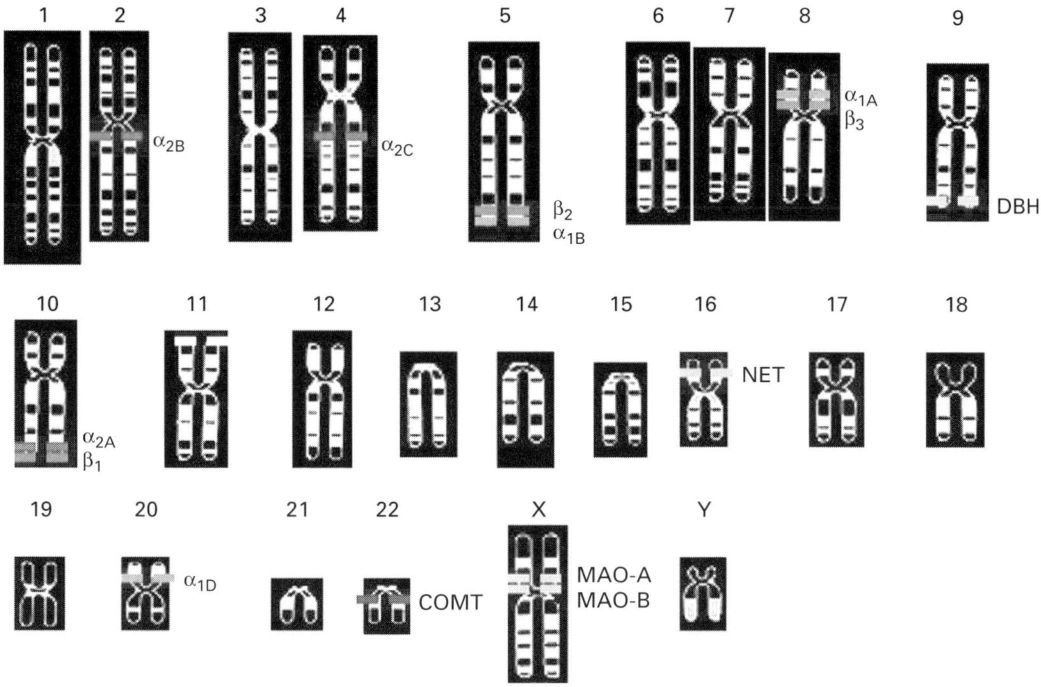

Figure 16.1 Genomic location of noradrenergic genes.

The involvement of adrenergic receptors in pain suggests that variability in pain sensitivity can be caused by variation in the encoding genes. Adrenergic receptor polymorphisms could also alter responses to antinociceptive drugs, such as ABT-594, a novel orally effective analgesic with affinity for α_{2B}- and α_{2C}-adrenergic receptors.[53]

Noradrenergic genes and their structural variation

All the genes encoding the proteins, which synthesize, transport, and metabolize NE, are represented in the available sequence of the human genome and all have apparently been identified. Gene targeting technology has been applied in the mouse to introduce gene knockouts for TH, dopamine β-hydroxylase; COMT, MAO-A, NE transporter; vesicular monoamine transporter (VMAT); and for α_{1A}-, α_{1B}-, α_{1D}-, α_{2A}-, α_{2B}-, α_{2C}-, β_1-, β_2-, and β_3-adrenergic receptors.[54] Furthermore, intracellular signal transduction systems of the catecholamines have been clarified using molecular techniques, including nine subtypes of adenylate cyclase, and pituitary adenylate cyclase-activating polypeptide.

Here we describe the main features of the genes known to be involved in the neurobiology of NE. These genes are dispersed widely across human chromosomes (Figure 16.1). The NE gene structures are known. Figure 16.2 shows intron–exon

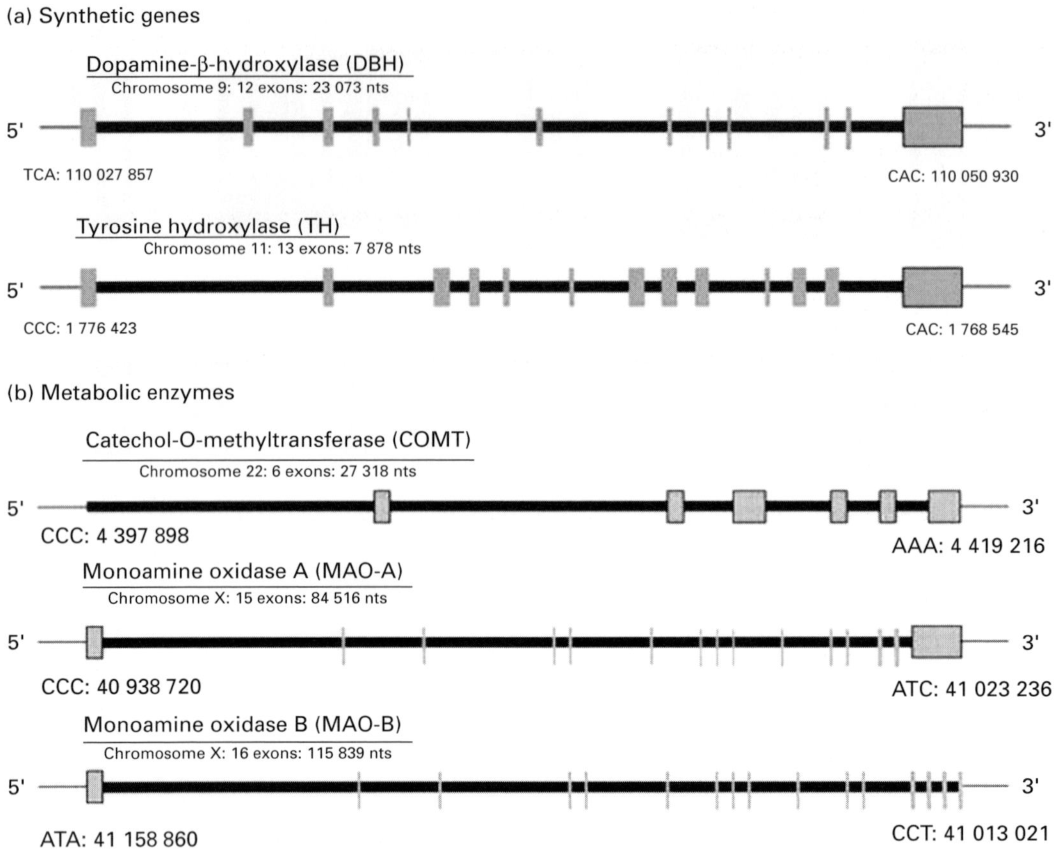

Figure 16.2 (a) Intron–exon organization of synthetic genes, (b) metabolic enzymes, and (c) *NET* and noradrenergic receptors.

organization of the biosynthetic genes (Figure 16.2a), metabolizing genes (Figure 16.2b) and receptors (Figure 16.2c). Each gene is a target for sequence variation. An extensive collection of sequence variation, primarily synonymous and noncoding in nature, is now available. Known missense substitutions are listed in Table 16.1.

Genes involved in NE synthesis: tyrosine hydroxylase and dopamine β-hydroxylase

Tyrosine hydroxylase converts tyrosine to dihydroxyphenylalanine (DOPA), and is rate-limiting in the synthesis of catecholamines. *TH* (hcG15485; GenBank) is located near the chromosome 11p telomere.[55] The length of *TH* is about 8 Kb, and it has 13 exons, 5 alternative transcripts, and encodes an enzyme that shares many structural features with other aromatic amino acid hydroxylases: phenylalanine hydroxylase, tryptophan hydroxylase 1, and tryptophan hydroxylase 2. Two functional polymorphisms in the *TH* gene were reported. The first is a $(TCAT)_n$ variable tandem repeat (VNTR) in intron 1.[56] The second is Met81Val, a missense

(c) ***NET* and Noradrenergic Receptors**

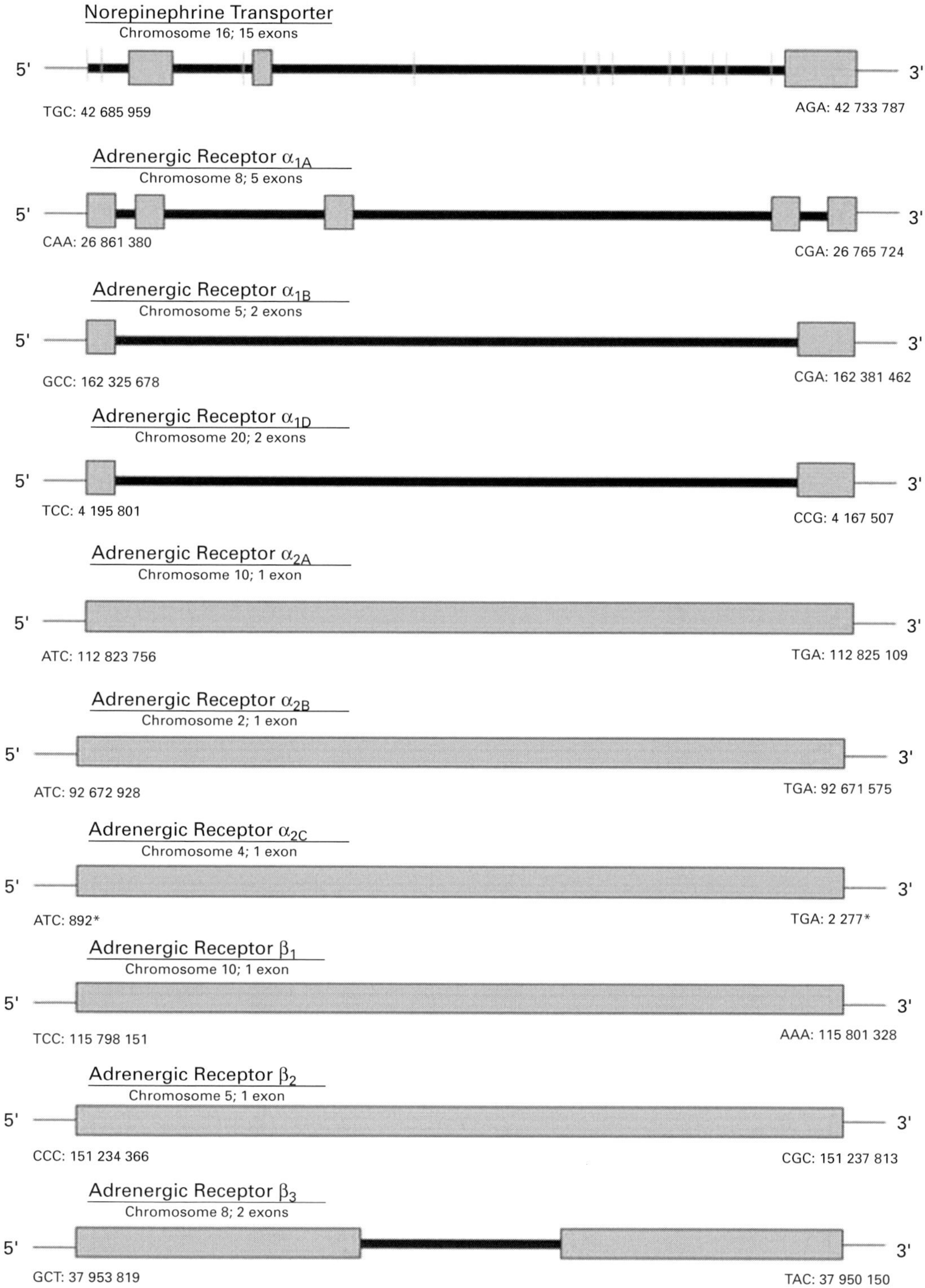

Figure 16.2 (*cont.*)

Table 16.1 List of missense mutations

		Adrenergic genes: missense mutations				
Gene	Snp ID (hCV)	Missense mut	Frequency	Amino acid change	Location	Function
NET	11361801	A/G	90/10	Hs503Tyr Hs548Tyr	43 011 052	
	11361802	A/C	90/10	Cys483Phe Cys528Phe	43 011 595	
	9610144	C/T	99/1	Ser433Gly Ser478Gly	43 012 731	
	11361806	T/C	90/10	Lys418Arg Lys463Arg	43 013 218	
	649159	C/G		Pro412Ala	43 013 237	
				Pro457Ala	exon 9	Orthostatic intolerance: plasma NE levels are disproportionately high in relation to sympathetic outflow, reduced reuptake of NE. Shannon *et al.* (2000).[165]
	11361848	C/T	99.93/0.07	Ile404Val Ile449Val	43 013 261	
	11361809	T/C	90/10	Asn330Ser Asn375Ser	43 015 863	
	11361811	C/G	95/5	Pro324Ala Pro369Ala	43 015 882	
	11361812	C/G	90/10	Leu311Val Leu356Val	43 015 921	
	11361814	G/T	10/90	Asn292Thr	43 019 233	
	9610112	C/T	99/1	Ile245Val	43 026 105	
	9610054	T/C	1/99	Ile69Val	43 054 459	

COMT	11731859	C/G		Ser34Cvs	4 412 828	
	11731855	A/C		Ala42Ser	4 412 941	
				Ala22Ser		
	2538751	C/T		Met158Val	4 413 949	Lachman *et al.* (1996).[157] 158met lowers activity of the enzyme
				Met33Val		
				Met108Val		
MAOA	8878826	T/G		Gly213Cys	41 004 723	
	11617986	G/A		Gln296Stop	41 008 201	Brunner *et al.* (1993).[70] MAO-A deficiency
	11617931	T/G		Val314Phe	41 008 255	
	8878814	G/T		Met445Ile	41 020 283	
	8878812	A/G	97/3	Lys520Arg	41 020 905	
MAOB	8878790	A/G		intron 13	41 045 170	Checkoway *et al.* (1998).[109]
α_{1A}	16172095	T/A		Glu465Asp	26 766 179	
	8795335	C/G		Gln431Glu	26 766 283	
	8795334	C/G		Arg376Gly	26 766 448	
	2696454	G/A		Arg347Cys	26 766 535	
	2696574	C/T		Arg247Gly	26 860 206	
	16172097	A/C		Ser200Ile	26 860 346	
α_{1B}	1738292	G/T	3/3	Lys13Thr	162 360 059	
α_{1D}			none			
α_{2A}	7611977	G/C		Ala25Gly	112 823 830	
	7611968	C/G		Asn251Lys	112 824 509	
	11741032	G/T		Glu369Stop	112 824 861	

(*cont.*)

Table 16.1 *(cont.)*

		Adrenergic genes: missense mutations				
Gene	Snp ID (hCV)	Missense mut	Frequency	Amino acid change	Location	Function
α_{2B}			none			
α_{2C}				322–325 deletion		Small *et al.* (2002)[26]
β_1	8898508	A/G		Ser49Gly	115 798 385	Ranade *et al.* (2002)[27]: Ser/Ser have the highest mean heart rate and Gly/Gly have the lowest
	2299974	C/A		Ser318Arg	115 799 192	
	8898499	A/G		Lys324Arg	115 799 211	
	2299973	A/G		Thr343Ala	115 799 267	
	8898495	T/G		Glu352Asp	115 799 296	
	8898494	C/G	75/25	Arg389Gly	115 799 405	Small *et al.* (2002).[26] Mason *et al.* (1999)[169]: Arg receptors had slightly higher basal levels of adenylate cyclase activities, agonist promoted binding was also increased
	2299971	G/T		Arg400Leu	115 799 439	
	2299969	C/A		Glu460Asp	115 799 620	
β_2	2084764	A/G	40/60	Arg16Gly	151 235 999	Meirhaeghe *et al.* (2001).[190] Large *et al.* (1997),[191] Turki *et al.*[192] (1995). Reihsaus *et al.* (1993)[94]: found that glycine at position 16 imparted enhanced agonist-promoted down-regulation
	2084765	C/G	60/40	Gln27Glu	151 236 032	Corbalan *et al.* (2002).[179] Hopes *et al.* (1998).[193] Dewar *et al.* (1997).[194] Large *et al.* (1997).[191] Martinez *et al.* (1997)[194]: form of the receptor Glu27 when coexpressed with Arg16 showed resistance to down-regulation when compared to Gln 27
	8950503	C/T	99/1	Thr164Ile	151 236 444	

β_3	12106152	G/A	90/10	Pro364Ser	37 952 495	
	12106151	G/A	95/5	Asp252Asp	37 952 829	
				Thr265Met		
	2215549	A/G	85/15	Pro51Leu	37 953 433	
				Arg64Trp		Walston *et al.* (1995)[102]: found Arg/Arg secrete significantly less insulin in response to a glucose influsion, have higher tasting glucose levels, and have lower glucose effectiveness. Festa *et al.* (1999).[196] Arner & Hoffstedt. (1999).[197] Mitchell *et al.* (1998).[198]
DBH	16199876	G/C		Ala74Gly	110 028 110	
	15810735	A/C		Asp92Ala	110 028 164	
	12020331	C/G	5/95	Gln167Glu	110 033 737	
	12020332	G/A	90/10	Ala197Thr	110 033 827	
	12020333	G/C	95/5	Lys225Asn	110 033 913	
	12020335	G/C		Glu236Gln	110 034 902	
	12020336	A/G	5/95	Asn276Asp	110 035 022	
	12020337	C/T	5/95	Pro303Leu	110 035 732	
	12020339	G/T	90/10	Ala304Ser	110 035 734	
	16200037	C/G		Ser530Trp	110 048 617	
	12020340	C/T		Arg535Cys	110048 631	
TH	579638	C/A		Val92Leu	1 764 443	
	579639	C/G		Hs89Arg	1 764 451	
		C/A		Pro89Arg		
		C/T				
	579640	G/C			1 765 360	
	579641	A/G			1 765 361	
	579642	C/G		His34Asp	1 765 404	
	8692685	C/T	99/1	Met468Val	1 768 945	
				Met409Val		
				Met495Val		

(*cont.*)

Table 16.1 *(cont.)*

Gene	Snp ID (hCV)	Adrenergic genes: missense mutations Missense mut	Frequency	Amino acid change	Location	Function
	12051780	G/A		Thr463Met	1 768 959	
				Thr404Met		
				Thr490Met		
	901878	G/T		Lys381Gln	1 770 347	Ludecke *et al.* (1995)[199]
				Lys322Gln		
				Lys408Gln		
	12051782	T/C		His306Arg	1 771 314	
				His247Arg		
				His333Arg		
	12051783	G/A		Thr283Met	1 771 383	
				Thr224Met		
				Thr310Met		
	12051790	T/G		Thr245Pro	1 771 616	
				Thr186Pro		
				Thr272Pro		
	12051791	A/G		Pro205Leu	1 772 517	Ludecke *et al.* (1996)[200]
				Pro148Leu		
				Pro232Leu		
	12051792	C/T		His202Arg	1 772 526	
				His143Arg		
				His229Arg		
	12051725	T/C	70/30	Met81Val	1 774 343	
				Met108Val		

SNP, single nucleotide polymorphisms.

polymorphism in exon 3. Met81Val shows significant linkage disequilibrium with the $(TCAT)_n$.[57] *TH* resequencing was performed in patients with infantile parkinsonism,[58] revealing four novel missense variants (exon 9, 14, 8, and 9), which did not occur in controls. Three are in the catalytic domain of TH and one may disturb tetramerization. In one patient, severe TH deficiency, resulting in axial hypotonia, hypokinesia, reduced facial mimicry, ptosis, and oculogyric crises from infancy, was due to homozygosity for *TH* missense mutation in exon 10 (cysteine to phenylalanine at codon 359).[59]

Dopamine β-hydroxylase (DBH) catalyzes the oxidation of dopamine to NE, and is present in synaptic vesicles of postganglionic sympathetic neurons. *DBH* (hCG27979) is located on chromosome 9q34 and is approximately 23 Kb in length. *DBH* has 12 exons and no known alternative splicing. Several *DBH* polymorphisms are known. A TaqI restriction fragment length polymorphism B1/B2 in intron 5 was found to be associated with serum DBH levels, suggesting linkage disequilibrium with a locus controlling expression of DBH. A $(CA)_n$ repeat polymorphism and a 19 base-pair insertion/deletion polymorphism located 5′ to the transcription start site are both common polymorphisms which are associated with changes in serum and plasma DBH levels.[60] An exon 2 polymorphism (G444A) was also associated with serum DBH, and also with cerebrospinal fluid DBH levels.[61] Clearly these results indicate that multiple loci are in linkage disequilibrium with each other and with a functional locus altering DBH expression.[62] C1021T in the 5′ flanking region of *DBH* accounts for 35% to 52% of variation in plasma DBH activity in Caucasian, African-American, and Japanese populations.[63] Within the DBH coding sequence, Ala304Ser results in lower DBH enzymatic activity (the frequency of the Ser304 DBH allele is 0.10).[64]

Genes for NE transport and metabolism: norepinephrine transporter, catechol-O-methyltransferase and monoamine oxidase

The NE transporter (NET) is involved in reuptake of NE via Na^+-dependent transport. *NET* [alternatively, *SLC6A2*] (hCG1742752) is located on chromosome 16q12.2. *NET* is 48 Kb in length, has 15 exons, and its structure substantially resembles other neurotransmitter transporters, genes, e.g. serotonin transporter (*HTT*) and *DAT*. *NET* has three alternative splice variants. Resequencing of *NET* identified 13 DNA sequence variants, among them five low-frequency missense substitutions.[65] The reported missense substitutions Val69Ile, Thr99Ile, Val245Ile, Val449Ile, and Gly478Ser are located at putative transmembrane domains 1, 2, 4, 9, and 10, respectively. The Thr99Ile substitution is at the fifth position of a putative leucine-zipper in transmembrane domain 2. The additional Ala457Pro missense substitution in exon 9 results in more than 98% loss of function compared to the wild type.[66]

A silent substitution in exon 9, G1287A, has been used in a series of association studies to NE-related phenotypes including hypertension and mood disorders.[65,67,68] However, these studies have in general been negative or equivocal. A haplotype approach combining abundant missense polymorphisms with a series of loci chosen for haplotype informativeness offers the potential for detection of linkage due to any allele of moderate abundance and effect size, regardless of whether the allele is presently known or unknown. Recently we developed a *NET* haplotype map for several populations, Finnish and American Whites, American Indians, and African-Americans, by genotyping a panel of 27 *NET* single nucleotide polymorphisms (SNPs) in these populations. Three conserved blocks showed little evidence for historical recombination in three populations, with some disruption of linkage disequilibrium in African Americans.[69]

Monoamine oxidase B and MAO-A are mitochondrial enzymes (of the outer membrane) involved in the degradation of biogenic amines. Monoamine oxidase A is the primary MAO found in fibroblasts, and it degrades serotonin and NE. Monoamine oxidase B is found in platelets and in the human brain. It degrades dopamine as well as other substrates such as phenylethylamine and benzylamine. *MAOA* and *MAOB* are located in close proximity to each other in the p11.3–11.4 region of the human X chromosome. *MAOA* (hCG16826) has 15 exons, spans 84.5 Kb, and is not alternatively spliced. Brunner *et al.* described an X-linked impulsivity syndrome in males of one Dutch family due to an *MAOA* stop codon variant.[70] Three *MAOA* polymorphisms – a 30-bp VNTR located in the promoter, a VNTR located in intron 1, and a SNP in exon 8 – define a particular haplotype associated with low MAO activity in human brain.[71] A $(CA)_n$ repeat in intron 2 is in linkage disequilibrium with a regulatory VNTR polymorphism in the *MAOA* promoter.[72] The *MAOB* (hCG16825) consists of 16 exons, spans > 115 Kb, and has two alternatively spliced forms. SNP in intron 13 [A to G substitution 36 bp upstream of the start position of exon 14] was associated with low MAO-B enzyme activity in human brain.[71]

Catechol-O-methyltransferase inactivates catecholamines by methylating the m-hydroxy moeity. *COMT* (hCG17889) is located on chromosome 22q11 and is 27 Kb in length, with 6 exons. It has four alternative transcripts. Two *COMT* forms – soluble and membrane-bound – result from alternative promoters, and two other forms result from alternative use of polyadenylation signal. *COMT* has a functional polymorphism: Val158Met in exon 6 (or Val108Met, depending on alternative splicing). Met158 is four fold less active and reduces the thermostability of the enzyme.[73] Systematic search for sequence variation within *COMT* identified five variants: C186T in exon 3, C408G in exon 4, G472A in exon 4, G597A in exon 5, and 821–827insC in the 3′ untranslated region,[74–76] and a common two-allele restriction fragment length polymorphism (RFLP).[77] Shu *et al.* have used transcribed COMT

SNPs to demonstrate differential expression of *COMT* gene copies in human lymphoblasts and brain.[78] These data indicate that at least another functional locus is likely to exist at *COMT*, other than Val158Met.

NE-binding receptors: α_{1A}, α_{1B}, α_{1D}, α_{2A}, α_{2B}, α_{2C} and β_1, β_2, and β_3

Three α_1-adrenergic receptors (α_{1A}-, α_{1B}- and α_{1D}-adrenergic receptors) are G-protein-coupled receptors that activate phosphoinositide hydrolysis, adenylate cyclase, or mitogen-activated protein kinase pathways after binding by NE. *ADRA1A* (hCG15476) is located on chromosome 8, and is the largest of the three adrenergic receptor genes, being more than 95.5 Kb in length. *ADRA1B* (hCG14750) is located on chromosome 5 and is 55.7 Kb and *ADRA1D* (hCG1646223) is located on chromosome 20 and is 27.5 Kb. The three α_1 genes all have two exons and a single known transcript, but despite having the same ligand they share higher sequence identities with G-protein-coupled receptors that are coupled to the same G proteins. *ADRA1A* Arg492Cys creates a putative palmitoylation site in the carboxy-terminal segment of the α_{1A}-adrenergic receptor.[79] However, the two alleles have similar pharmacological characteristics and binding capacities for NE. Two synonymous polymorphisms in exon 1, an amino acid insertion in exon 2 at position 368 (Arg) and a missense substitution (Arg371Gly) in the same exon, were found in *ADRA1B*.[80] A missense Cys492Arg substitution was described in *ADRA1D*.[81]

α_2-adrenergic receptors are G-protein-coupled receptors that bind to the G_i/G_0 family of G proteins. They operate as presynaptic inhibitory receptors by coupling to guanine nucleotide regulatory proteins and inhibiting adenylate cyclase. α_2-adrenergic receptors play a critical role in regulating and controlling NE release from sympathetic nerves and from adrenergic neurons in the central nervous system via presynaptic inhibition. These receptors are located on nerve terminals and on cell bodies and dendrites. Three genes (*ADRA2A*: hCG41806, *ADRA2B*: hCG37297, and *ADRA2C*: AF280399) are located on chromosomes 10, 2, and 4, respectively. They are all intronless and are approximately 1.3 Kb in length. Sequence variations within the coding region of each gene have been identified. These variants substitute or delete amino acids in the third intracellular loops of each receptor, affecting G-protein coupling, desensitization, and G-protein receptor kinase-mediated phosphorylation.[82] The prevalences of the polymorphisms differ across ethnic populations. The three polymorphisms are each relatively abundant, and two are functional in vitro.

The *ADRA2A* is most critical for adrenergic receptor autoregulation. Studies on mice with nonfunctional α_{2A}-adrenergic receptors reveal that this subtype is responsible for approximately 80% of the autoinhibition of NE release from sympathetic postganglionic neurons.[83] *ADRA2A* Asn251Lys is a functional polymorphism. Lys251 confers significantly increased agonist-promoted binding to G_i, leading to greater inhibition of adenylate cyclase, activation of mitogen-activated protein kinase signaling, and stimulation of inositol phosphate accumulation.[84] Lys251 has a frequency of 0.05 in African-Americans as compared to 0.004 in Whites, but is not associated with essential hypertension. Resequencing of the *ADRA2A* coding region also identified a rare 6-amino acid deletion involving codons 255–260,[85] and potentially altering function.

A polymorphism of *ADRA2B* consisting of a deletion of three glutamic acids (residues 301–303) from a glutamic acid repeat element in the third intracellular loop is more common in Whites (allele frequency 31%) relative to African-Americans (allele frequency 12%).[86] The presence of the Del 301–303 allele leads to a small decrease in coupling efficiency resulting in reduced inhibition of adenylate cyclase.[87]

Polymorphism in *ADRA2C* is an in-frame 12-nucleic acid deletion encoding a receptor lacking Gly-Ala-Gly-Pro in the third intracellular loop at codons 322–325.[88,89] This deletion allele has frequencies of approximately 44% in African-Americans and 3.5% in Whites. There is in vitro evidence that the deletion alters high-affinity agonist binding indicating impaired formation of the agonist–receptor–G protein complex.[89]

β-adrenergic receptors (β_1, β_2, and β_3) are also G-protein-coupled receptors. β_1- and β_2-adrenergic receptors are both normally coupled to G_s protein, and through this protein, they activate adenylate cyclase. β_2-adrenergic receptors couple to G_i, reducing adenylate cyclase. $Beta_1$ and β_2-adrenergic receptors are expressed in the heart, kidney, blood vessels, and many other tissues. β_3-adrenergic receptors are found mainly in adipose tissue, being involved in the regulation of lipolysis and thermogenesis. They seem to activate nitric oxide synthase. Genes encoding the β-adrenergic receptor subtypes (*ADRB1*: hCG39839, *ADRB2*: hCG36934 and *ADRB3*: hCG21141) are located on chromosomes 10, 5, and 8, respectively. *ADRB1* and *ADRB2* are non intronic and are 3.2 Kb and 3.5 Kb in length, respectively. *ADRB3* has two exons but is only 2.7 Kb. There are two common *ADRB1* polymorphisms: Ser49Gly and Gly389Arg.[90] Both alter agonist promoted down-regulation in vitro.[91] Resequencing of the 5′ untranslated region (UTR) of *ADRB1* identified variants at positions –93 C>T, −2146 T>C, and –210 C>T.[92]

ADRB2 is located on chromosome 5q31 and its 5′ untranslated region contains a number of genetic variants. Strong linkage disequilibrium exists across this gene, resulting in the occurrence of several common haplotypes. Thirteen SNPs were organized into 12 haplotypes out of a theoretically possible 8192 combinations.[93] Within the coding region of *ADRB2*, nine SNPs were identified,[94] five of which are synonymous. Missense polymorphisms were Arg16Gly, Gln27Glu, Val34Met, and Thr164Ile.[95,96] Gly16 was associated with desensitization to β_2-adrenergic receptor agonists in the Japanese population.[97] Gly16 leads to enhanced agonist-mediated down-regulation in vitro. Glu27 reduces agonist-mediated down-regulation.[98] The Ile164 allele in the fourth transmembrane spanning domain alters the proposed ligand-binding pocket and reduces efficiency of receptor coupling to downstream effector pathways.[99] Ile164 allele has an allele frequency of 2% in Whites. The 5′ UTR polymorphisms T-367C and C-47T are abundant in Whites[100] and are in strong linkage disequilibrium with each other and with the missense polymorphisms at residues 16 and 19.[101]

ADRB3 Trp64Arg is located in the first intracellular loop of the receptor. Arg64 has higher allele frequencies in Pima Indians (0.31) than in Mexican Americans (0.13), Blacks (0.12), and Whites (0.08), supporting the idea that this variant[102] could alter activation of thermogenesis in adipose cells, contributing to the high frequency of obesity and adult-onset diabetes in the Pima.[103] However, the linkage studies are contradictory, as discussed later.

Linkage and association studies with NE genes and loci

Norepinephrine is an important chemical messenger in the nervous system. It regulates affective states, learning and memory, endocrine, and autonomic functions. Noradrenergic neurons are implicated in the modulation of vigilance, improvement of visual attention, and initiation of adaptive response. Norepinephrine has been associated with depression, aggression, and addiction, as well as cardiac and thermal dysregulation. Several of the NE genes have variants that affect gene expression or function of the encoded protein. Therefore, it is not surprising that multiple linkage and association studies have been performed with these candidate genes and candidate functional loci and alleles using pedigree analysis and the population case/control approach. Table 16.2 summarizes positive linkage studies with NE genes. Although several of the findings are unreplicated or controversial, valid NE gene linkages have been established in several domains of normal function and disease.

Neurological disorders

Several NE-related genes lead to increased susceptibility for neurological diseases.

Movement disorders

A *TH* T24A IVS 11 substitution results in skipping of exon 12. This variant was reported in a child with a severe extrapyramidal movement disorder.[104] Four novel *TH* missense variants, located in exons 8, 9, and 14, were found in patients with infantile parkinsonism.[58] These appear to be examples of effects of relatively rare,

Table 16.2 Positive linkage studies with NE genes

	Phenotype	Gene/Marker	Sample	Reference
Neurological disorders	Severe extrapyramidal movement disorder	*TH*T24A1VS 11	One child	Janssen *et al.* (2000)[104]
	Infantile parkinsonism	Four *TH* missense variants in exons 8 9, and 14	Two families (Whites)	Swaans *et al.* (2000)[58]
	Parkinson's disease	*COMT* Val158/Val158	Japanese patients	Yoritaka *et al.* (1997)[105]
	Parkinson's disease	*MAOA* haplotype	Japanese patients	Nakatome *et al.* (1998)[106]
	Parkinson's disease	*MAOB* + *COMT* Met 158	Taiwanese patients	Wu *et al.* (2001)[107]
	Parkinson's disease	*MAOB*	Whites	Costa *et al.* (1997)[108]
	Parkinson's disease	*MAOB*1VS 2188$(GT)_n$	Australian cohort	Mellick *et al.* (1997)[110]
	Alzheimer's disease	*MAOA*1VS 113$(GT)_n$	Japanese patients	Takehashi *et al.* (2002)[111]
	Migraine	*DBH*repeat (intragenic)	Whites	Lea *et al.* (2000)[112]
	Migraine	*COMT* Met158	Turkish patients	Emin Erdal *et al.* (2001)[113]
	Pain	*COMT* Met158	Healthy volunteers	Zubieta *et al.* (2003)[15]
Psychiatric diseases	ADHD	*DBH* Taql	Irish family	Daly *et al.* (1999)[117]
	ADHD	*MAOA*(CA)n	Chinese patients	Jiang *et al.* (2001)[118]
	ADHD	*COMT* Val158 allele	Jewish families	Eisenberg *et al.* (1999)[119]
	Bipolar disorder	Three *MAOA* variants	Four families	Lim *et al.* (1995)[128]
	Bipolar disorder	*COMT* Met158 allele	US and British Whites and Chinese	Papolos *et al.* (1998)[137] Kinov *et al.* (1998)[138]; Li *et al.* (1997)[139]
	Panic disorder	Three *MAOA* variants	German and Italian female patients	Deckert *et al.* (1999)[129]
	Panic disorder	*COMT* Met158 allele	Korean patients	Woo *et al.* (2002)[140]
	Major depression	*MAOA* promoter repeat	German female patients	Schulze *et al.* (2000)[16]
	Major depression	*COMT* Met158	Japanese patients	Ohara *et al.* (1998)[13]

Obsessive-compulsive disorder	*COMT* Met158	Chinese patients	Karayiorgou *et al.* (1997;1999)[134, 135]; Karno *et al.* (1988)[136]
Anorexia nervosa	*COMT* Val158	Jewish families	Frisch *et al.* (2001)[141]
Anxiety	*COMT* Met158	Indian and White female patients	Enoch *et al.* (2003)[14]
Anxiety	*ADRB1* Ser49Gly	Whites	Stein *et al.* (2004)[144]
Schizophrenia	*COMT* haplotype	Ashkenazi Jews	Shifman *et al.* (2002)[151]
Schizophrenia	*TH*$(TCAT)_n$	French, Tunisians and Japanese females	Meloni *et al.* (1995);[153] Kurumaji *et al.* (2001)[154]
Addiction (Smoking)	*DBHG*1368/G1368, MAOA 1460 TT/TO	Whites	McKinney *et al.* (2000)[155]
Addiction (Cocaine)	*DBH* haplotype	Whites	Cubells *et al.* (2000)[62]
Addiction (Alcohol)	*TH*$(TCAT)_n$	German patients	Sander *et al.* (1998)[156]
Addiction (Alcohol and drugs)	*COMT* Met158	Whites	Lachman *et al.* (1996);[157] Karayiorgou *et al.* (1997);[134] Vandenbergh *et al.* (1997);[154] Papolos *et al.* (1998);[137] Tiihonen *et al.* (1999)[159]
Addiction (Heroin)	*COMT* Val158	Whites and Jewish	Vandenbergh *et al.* (1997);[158] Horowitz *et al.* (2000)[161]

(*cont.*)

Table 16.2 *(cont.)*

	Phenotype	Gene/Marker	Sample	Reference
Cardiovascular diseases	Heart failure	*ADRA2C* Del322–325 + *ADRB1* Arg389	Whites and African-Americans	Small *et al.* (2002)[26]
	Essential hypertension	*ADRB2* Arg16 and Glu27	Chinese patients	Ranade *et al.* (2001)[172]
	Essential hypertension	*ADRB2* RFLP	Whites and African-Americans	Svetkey *et al.* (1996)[173]
Metabolic disorders	Diabetes type 2	*ADRB2* Gly16Arg	Taiwanese population	Chang *et al.* (2002)[174]
	Diabetes type 2 and obesity	*ADRB3* Trp64Arg	Whites and Japanese	Corbalan *et al.* (2002a,b);[175,179] Oizumi *et al.* (2001)[176]
	Diabetes	*ADRB3* Met416Val	Chinese patients	Wang *et al.* (2002)[177]
	Hypertriglyceridemia	*ADR2* Gly16Arg	Japanese male patients	Iwamoto *et al.* (2001)[178]
	Abdominal obesity	*ADRB1*, *ADRB2* and *ADRB3* markers	Quebec population	Ukkola *et al.* (2000)[180]
Other disorders	Breast cancer	*COMT* Met158/Met158	Korean women	Yim *et al.* (2001)[184]
	Restless leg syndrome	*COMT* Val158	Quebec women	Desautels *et al.* (2002)[186]
	Narcolepsy	*COMT* Val158	White women	Dauvilliers *et al.* (2001)[187]
	Narcolepsy	*COMT* Met158	White men	
				Lavigne *et al.* (1997)[188]

ADHD, attention-deficit/hyperactivity disorder.

but strongly deterministic, alleles. In time, more subtle genotype effects on movement disorders may also be validated. *COMT* Val158/Val158 homozygotes were reported to be more abundant in Japanese patients with Parkinson's disease.[105] *MAOA* genotype and haplotype differed between controls and patients with Parkinson's disease.[106] An *MAOB* genotype (A to G substitution 36 bases upstream from the intron 13–exon 14 boundary) was implicated in Parkinson's disease in Taiwanese, and this association was augmented by *COMT* genotype (the Met158 allele, in this study), indicating a potential interaction of these two catecholamine metabolizing enzymes in the pathogenesis of sporadic Parkinson's disease.[107] The same *MAOB* association to Parkinson's disease was also seen in Whites.[108] For *MAOB* and Parkinson's disease, there is also a potential interaction with smoking. A reduced Parkinson's disease risk related to pack-years of smoking was detected for persons with the *MAOB* G allele, whereas an opposite effect was found among persons with the A allele.[109] Finally, in a large Australian cohort, a different *MAOB* locus (IVS 2 188 $(GT)_n$) was associated with Parkinson's disease.[110]

Neurodegenerative disease

An *MAOA* IVS 113 $(GT)_n$ allele (intron 2) was associated with Alzheimer's disease, and a Lewy body form of this disease.[111]

Migraine and pain

An intragenic dinucleotide repeat near *DBH* was associated with migraine in a case/control comparison. A transmission/disequilibrium test appeared to confirm transmission of the *DBH* allele to Caucasian migraine sufferers.[112] *COMT* Met158 was also overrepresented in migraine sufferers who had a family history of migraine.[113] The Met158 allele was recently shown to be associated with diminished activation of the brain's endogenous μ-opioid system response to sustained pain, and increased pain sensitivity.[15]

Behavior and psychiatric diseases

Attention-deficit/hyperactivity disorder and impulsivity

Pharmacological, biochemical, and neuropsychological findings have suggested that noradrenergic imbalances are likely to be important in attention-deficit/hyperactivity disorder (ADHD). Halperin *et al.* reported a significant increase in plasma NE in children having ADHD combined with reading and other cognitive disabilities.[114] There is preliminary evidence that enhanced central noradrenergic activity mediates behavioral anti-hyperactivity effects of reuptake inhibitors selective for NE in a rat model of ADHD.[115] Three NE-related genes, *ADRA2A*, *ADRA2C*, and *DBH*, have actually been reported to account for a total of

3.5% of the variance in ADHD score,[116] suggesting the possibility that more work remains to be done in this area. Daly *et al.* observed an association of the *DBH* TaqI polymorphism located in intron 5 in an Irish family.[117] A study in Chinese revealed linkage between ADHD and the *MAOA* (CA)n polymorphism.[118] The impulsive-hyperactive type of ADHD (excluding inattention) was associated with the *COMT* Val158 allele.[119] Several studies have also reported associations between *COMT* Val158Met allele and other directed aggression in schizophrenic and schizoaffective patients.[120–122] Significant associations were observed between *COMT* Met158 and history of violent suicide attempts.[123]

A *MAOA* stop codon variant is causal for impulsive behavior in males from one Dutch family,[124] but has not been reported in other families. A functional VNTR polymorphism in the *MAOA* promoter moderates the effect of maltreatment.[125] Maltreated children with a genotype conferring high levels of MAO-A expression were less likely to develop antisocial problems. Increased aggression and increased levels of brain NE were observed in transgenic mice in which *MAOA* was deleted,[126] providing a potential mouse genetic model for effects of MAO-A underexpression on impulsive and aggressive behaviors.

Anxiety, dysphoria, and affective disorders

Hyperactivity of central NE pathways was implicated in affective and anxiety disorders. *NET* disruption in knockout mice inhibited depression-like behavior in chronically stressed animals.[127] Lim *et al.* reported a genetic association between bipolar disorder and three *MAOA* variants including one associated with low MAO-A activity, but not with *MAOB* polymorphism.[128] The longer alleles of these *MAOA* markers were associated with panic disorder in German and Italian female patients.[129] Similar gender-related association between a functional *MAOA* promoter repeat polymorphism and major depressive disorder was observed in German female patients.[16] A synonymous substitution in *MAOA* exon 14 also has agender-dependent effect. The high enzyme activity-related allele at this locus was associated with depressed suicide in male victims[130] while the low enzyme activity allele was highly correlated with obsessive-compulsive disorder in female probands.[131–133] *COMT* Met158 was associated with obsessive-compulsive disorder in male patients.[134–136] Met158/Met158 homozygotes were associated with ultra-ultra rapid cycling bipolar disorder in Whites.[137] The results of this study were replicated in British Whites[138] and in Chinese.[139] Met158 was associated with depressive disorders in Japanese.[13] *COMT* Met158 allele and Met/Met genotype was associated with panic disorder and poorer treatment response of such patients compared to subjects with other genotypes.[140] *COMT* Val158/Val158 genotype was associated with a two fold increased susceptibility to anorexia nervosa.[141] Recently the *COMT* Met158 allele was found to be associated with higher anxiety scores in women.[14]

TH $(TCAT)_n$ in intron 1 was related to depressive symptoms in subjects with mood disorders.[142] A *DBH* G444A polymorphism in exon 2 was associated with paranoid ideation, interpersonal sensitivity, and psychoticism in patients with major depression.[61] Depressed patients with the G444/G444 genotype had lower scores for interpersonal sensitivity and paranoid ideation, suggesting that this genotype may be protective against the development of psychosis in the presence of a major depressive episode. Patients with posttraumatic stress disorder have elevated cerebrospinal fluid levels of NE as well as increased sensitivity to noradrenergic activation produced by the α_2-adrenergic receptor antagonist, yohimbine.[143] An association between *ADRB1* Ser49Gly and social anxiety has been identified. Social Interactional Anxiety scores in persons with Gly49 allele were significantly higher than in Ser49 homozygotes.[144]

Schizophrenia and prefrontal cortex function

Function of the prefrontal cortex (PFC) appears to be fundamental to the pathophysiology of many neuropsychiatric disorders. It is well recognized that dopamine binding to D_1-dopamine receptors located on pyramidal glutaminergic neurons is critical in the tuning and optimization of PFC functions in executive cognition and working memory.[145] However, other catecholamines may also be critical to PFC cognitive function. Increasing NE release may produce more optimal levels of NE receptor stimulation in the PFC.[146]

COMT Val158Met modulates frontal cognitive performance, metabolic activity in the frontal cortex and anterior cingulate cortex during working memory tasks, and it appears to be a vulnerability gene for schizophrenia.[147] The effect of Val158Met on frontal cognitive performance was replicated in several populations including normal controls,[148,149] schizophrenics,[149,150] and head injury patients (Lipsky and Ross, unpublished). A highly significant association between schizophrenia and a *COMT* haplotype (but not the Val158Met locus alone) was also found in Ashkenazi Jews.[151] Schizophrenia linkage has been reported to this chromosome 22 region, which is also involved in the velocardiofacial chromosome deletion syndrome. Neuromodulatory actions of amphetamine on prefrontal cortical function appear to be altered by *COMT* Val158Met: subjects with Val158/Val158 genotype showed the greatest PFC activation on placebo and least activation on amphetamine.[152] Individuals with the Met158/Met158 genotype had an adverse response to amphetamine. The Val158/Val158 genotype appears to be associated with higher activation of catecholaminergic neurons in midbrain, as indicated by a dramatic elevation of TH expression in postmortem brains with this genotype. All these data suggest enhanced cortical efficiency at baseline in individuals with Met158 genotypes, with consequences for frontal cortical activity, output, and associated behaviors. The role of NE in these effects remains to be determined. A *TH*

$(TCAT)_n$ was significantly associated with schizophrenia in French and Tunisians[153] and in Japanese females.[154]

Addictions

Multiple linkage and association studies have been conducted using markers for NE-related genes, although, as before, it should be added that several of these genes could act through other neurotransmitters. McKinney *et al.* showed that variations in *DBH* and *MAO* predict whether a person is a heavy smoker and how many cigarettes they consume.[155] Smokers with *DBH* G1368/G1368 genotype smoked fewer cigarettes than those with GA/AA, particularly in women. More heavy smokers (>20 a day) had the *DBH* A1368 allele when compared to light smokers. Smokers with *MAOA* 1460 TT/TO alleles smoked more cigarettes than those with CC/CT/CO genotype.[155]

Cubells *et al.* reported that the *DBH* haplotype associated with low enzyme activity (Del allele of *DBH* 5′-ins/del and A allele of *DBH* G444A), was also associated with cocaine-induced paranoia in Whites.[62] The frequency of the A10 allele of a polymorphic tetranucleotide motif in *TH* intron 1 was significantly increased in German alcoholics with withdrawal delirium compared with that in the controls.[156] A *COMT* Met158 allele was related to drug abuse and alcoholism[135,137,157–159] contributing to the development of late-[159] and early-onset alcoholism in men.[160] A previous case-control study by Vandenbergh *et al.* showed an association between *COMT* Val158 allele and poly-substance abuse in Whites.[158] This allele was also frequent in Israeli heroin addicts.[161] The low-activity three-repeat allele of *MAOA* functional regulatory promoter polymorphism was highly frequent in male alcoholics with antisocial and anxious-depressive personality disorder.[162] These results were replicated, confirming that this polymorphic allele confers increased susceptibility to antisocial behavior (rather than alcohol dependence per se) in alcohol-dependent males.[163] The longest alleles of *MAOA* VNTR polymorphism were associated with drug abuse.[164]

Cardiovascular disorders

An association was reported between the *TH* Val81Met polymorphism and increased left ventricular cavity dimension and mass in a young normotensive male population.[57] Study of *NET* genetic variation revealed that impairment of synaptic NE clearance might result in a syndrome characterized by excessive sympathetic activation in response to physiologic stimuli. Heterozygosity for a G237C substitution in *NET* exon 9 was reported to result in a more than 98% loss of NET function and was associated with the postural heart rate and abnormal plasma NE homeostasis leading to orthostatic intolerance.[165]

Subjects homozygous for the *ADRA2B* deletion allele[166] appeared to have an increased risk of impaired cardiac endothelial function, which may provide an explanation for the previously observed increased risk of myocardial infarction in male subjects with this genotype.[26] The Del/Del genotype might be a genetic risk factor for acute coronary events and flow-mediated dilatation of the brachial artery, but not for hypertension.[167] An *ADRB1* Arg389Gly substitution results in alterations of receptor–G_s interaction contributing to the conservation of cardiostimulant effects of NE.[168–170] Thus, this variant may be the basis of pathology or response to therapeutic β-adrenergic receptor agonists and antagonists in other cardiovascular diseases.

The combination of an *ADRA2C* Del322–325 genotype associated with decreased receptor function, and an *ADRB1* Arg389 associated with increased function of this receptor appears to predispose persons to heart failure,[26] potentially through increased synaptic NE release and enhanced receptor function at the myocyte. These two polymorphisms could thus act synergistically to increase risk of heart failure, particularly in African-Americans.

Adrenergic receptor polymorphisms may also have peripheral vascular effects. The *ADRB2* Arg16 allele was associated with enhanced agonist-mediated desensitization of the vasculature, while the Glu27 allele was associated with increased agonist-mediated responsiveness.[171] Therefore, polymorphisms in *ADRB2* are potentially important determinants of the vascular response to stress. These polymorphisms were shown to be associated with essential hypertension in a population of Chinese origin.[172] An *ADRB2* RFLP was associated with hypertension in Blacks and Whites.[173] In addition, there was association between *ADRA2A* genotype and hypertension in Whites. These findings suggest that *ADRB2* and *ADRA2A* may contribute, in a race-influenced manner, to the inheritance of essential hypertension.

Metabolic disorders

An *ADRB2* Gly16Arg substitution was associated with development and earlier onset of type 2 diabetes in a Taiwanese population.[174] *ADRB3* Trp64Arg polymorphism was associated in some, but not all, studies with an earlier onset of type 2 diabetes mellitus and features of insulin resistance syndrome. Individuals aged 20 to 35 years with the Arg64 allele had a substantially higher risk of obesity, independent of their gender or leisure-time physical activity.[175] Homozygosity for Arg64 was associated with both obesity and type 2 diabetes in a large Japanese sample.[176]

In the Chinese, an *ADRB3* Met416Val linkage was identified in a subgroup of diabetic subjects with high 2-hour post-glucose.[177] The *ADR2* Gly16Arg variant was associated with hypertriglyceridemia and with the development of fatty liver in Japanese males.[178] Another genetic marker, *ADRB2* Glu27 allele appeared to be a risk factor for abdominal obesity among male subjects, especially among those with lower high-density lipoprotein-cholesterol levels.[179] Gene–gene interactions between markers in *ADRB1, ADRB2* and ADRB*3* may contribute to abdominal obesity and plasma lipid and lipoprotein in a Quebec population.[180]

ADRA2A C1291G[181] was associated with a subnormal cortisol response to dexamethasone, elevated glucose levels and increased diastolic blood pressure.[182] The pathophysiology of this phenomenon could involve an altered density of the α_{2A}-adrenergic receptor that destabilizes the sympathetic–HPA systems in those with genetic vulnerability in gene promoter.

Other disorders

A study in vitro indicated that *ADRB2* Thr164Ile is a pharmacogenetic locus for differential response to the most commonly utilized agonists in the treatment of asthma.[183]

Yim *et al.* reported that obese women with *COMT* Met158/Met158 genotype had an increased risk for postmenopausal breast cancer compared to women with Val158/Val158 genotype,[184] whereas Thompson *et al.* observed that the obese women with at least one Met158 allele had an increased risk for premenopausal breast cancer.[185] Female (but not male) patients with *COMT* Val158 allele had a greater risk of being affected with restless leg syndrome than females carrying Met158 allele.[186]

A strong sexually dimorphic effect of *COMT* on narcolepsy severity was found. Women narcoleptics with the Val158 allele fell asleep twice as fast as those with the Met158 allele, while the opposite was true for men.[187,188] In another study, the Met158 allele was found to be associated with hypnotizability, in men.[189]

Summary and conclusions

The phenotypes selected for linkage studies with NE genes are diverse, reflecting the protean role of adrenergic function in behavior and physiology. The focus of this review has been on the positive linkage studies and on locus/phenotype combinations where positive findings have been accumulated. All of the findings need to be interpreted skeptically. However, several factors encourage the conclusion that several of the most replicated NE gene linkages are valid. A critical role is played by the availability of a relatively complete NE gene list and a panel of functional polymorphisms, many common, at these genes. In several cases, the availability of a functional polymorphism enables a directional hypothesis or a more specific hypothesis to be formulated (e.g., β_2-adrenergic receptor desensitization in asthma, e.g. enhanced NE release due to dysfunction of a presynaptic α_2-adrenergic receptors). A critical role is played by application of the linkage paradigm to the "correct" phenotype, the use of intermediate phenotypes whose causal heterogeneity is reduced, and by the ability to observe multiple findings converging on the same biology. For example, studies of COMT Val158 have now linked this allele to schizophrenia, to frontal lobe function, and to several convergent brain neurobiologic intermediate phenotypes. Similarly, the *COMT* Met158 allele has been linked to dimensionally measured anxiety, to pain/stress response, and to brain opioid response to pain measured using an imaging paradigm. Discovery of counterbalancing selective benefits of different alleles helps to explain the abundance of several of these polymorphisms across human populations. For example, the *COMT* Val158 allele has a frequency ranging from 0.4 to 0.6. Individuals who are *COMT* Met158/Met158 may be more efficient and effective in frontal cognition but more prone to stress and worry, as compared to Val158/Val158 homozygotes. The dichotomy, imperfect due to the probabilistic effect of such alleles on behavior, may be thought of as "Worrier vs. Warrior." For other common functional alleles, it is also likely that counterbalancing advantages and disadvantages also exist. The positive and negative effects of the alleles also appear to be substantially modified by many other factors: gene x gene interactions (e.g., adrenergic receptors in hypertension), gene x environment interactions (i.e. MAO-A in conduct disorder) and gender interactions (e.g., COMT and anxiety). Availability of genetic loci informative for intermediate processes in disease is informative for pathophysiology and may ultimately enable refined diagnoses and improved therapeutic targeting of common, poorly differentiated diseases (e.g. obesity, hypertension, anxiety) influenced by NE function. The identification of genetic polymorphisms involved in the synthesis, transport, storage, and metabolism of NE and other neurotransmitters also facilitates an understanding that an individual gene can simultaneously influence cardiovascular, endocrine, neurological, and psychiatric functions. The brain, the body, and the genome are not conveniently compartmentalized by disease.

REFERENCES

1. Southwick, S. M., Morgan, C. A., 3rd, Bremner, A. D. *et al.* Noradrenergic alterations in posttraumatic stress disorder. *Ann. N. Y. Acad. Sci.*, **821** (1997), 125–141.
2. Southwick, S. M., Bremner, J. D., Rasmusson, A. *et al.* Role of norepinephrine in the pathophysiology and treatment of posttraumatic stress disorder. *Biol. Psychiatry*, **46** (1999), 1192–1204.
3. Villafuerte, S. M., Del-Favero, J., Adolfsson, R. *et al.* Gene-based SNP genetic association study of the corticotropin-releasing hormone receptor-2 (CRHR2) in major depression. *Am. J. Med. Genet.*, **114** (2002), 222–226.
4. Baldwin, D. S., Evans, D. L., Hirschfeld, R. M. and Kasper, S. Can we distinguish anxiety from depression? *Psychopharmacol. Bull.*, **36** (2002), 158–165.

5. Southwick, S. M., Paige, S., Morgan, C. A. 3rd *et al.* Neurotransmitter alterations in PTSD: catecholamines and serotonin. *Semin. Clin. Neuropsychiatry*, **4** (1999), 242–248.
6. Brunello, N., Mendlewicz, J., Kasper, S. *et al.* The role of noradrenaline and selective noradrenaline reuptake inhibition in depression. *Eur. Neuropsychopharmacol.*, **12** (2002), 461–475.
7. Sullivan, P. F., Neale, M. C. and Kendler, K. S. Genetic epidemiology of major depression: review and meta-analysis. *Am. J. Psychiatry*, **157** (2000), 1552–1562.
8. Todd, R. D. and Botteron, K. N. Etiology and genetics of early-onset mood disorders. *Child Adolesc. Psychiatr. Clin. N. Am.*, **11** (2002), 499–518.
9. Malhi, G. S., Moore, J. and McGuffin, P. The genetics of major depressive disorder. *Curr. Psychiatry Rep.*, **2** (2000), 165–169.
10. Johansson, C., Jansson, M., Linner, L. *et al.* Genetics of affective disorders. *Eur. Neuropsychopharmacol.*, **11** (2001), 385–394.
11. Puzynski, S., Rode, A. and Zaluska, M. Studies on biogenic amine metabolizing enzymes (DBH, COMT, MAO) and pathogenesis of affective illness. I. Plasma dopamine-beta-hydroxylase activity in endogenous depression. *Acta Psychiatr. Scand.*, **67** (1983), 89–95.
12. Puzynski, S., Bidzinski, A., Mrozek, S. and Zaluska, M. Studies on biogenic amine metabolizing enzymes (DBH, COMT, MAO) and pathogenesis of affective illness. II. Erythrocyte catechol-O-methyltransferase activity in endogenous depression. *Acta Psychiatr. Scand.*, **67** (1983), 96–100.
13. Ohara, K., Nagai, M., Suzuki, Y. and Ohara, K. Low activity allele of catechol-o-methyltransferase gene and Japanese unipolar depression. *NeuroReport*, **9** (1998), 1305–1308.
14. Enoch, M. A., Xu, K., Ferro, E., Harris, C. R. and Goldman, D. Genetic origins of anxiety in women: a role for a functional catechol-O-methyltransferase polymorphism. *Psychiatr. Genet.*, **13** (2003), 33–41.
15. Zubieta, J. K., Heitzeg, M. M., Smith, Y. R. *et al.* COMT val158met genotype affects mu-opioid neurotransmitter responses to a pain stressor. *Science*, **299** (2003), 1240–1243.
16. Schulze, T. G., Muller, D. J., Krauss, H. *et al.* Association between a functional polymorphism in the monoamine oxidase A gene promoter and major depressive disorder. *Am. J. Med. Genet.*, **96** (2000), 801–803.
17. Bratel, T., Wennlund, A. and Carlstrom, K. Impact of hypoxaemia on neuroendocrine function and catecholamine secretion in chronic obstructive pulmonary disease (COPD). Effects of long-term oxygen treatment. *Respir. Med.*, **94** (2000), 1221–1228.
18. Jee, S. H., Suh, I., Won, S. Y. and Kim, M. Y. Familial correlation and heritability for cardiovascular risk factors. *Yonsei Med. J.*, **43** (2002), 160–164.
19. Knuiman, M. W., Divitini, M. L., Welborn, T. A. and Bartholomew, H. C. Familial correlations, cohabitation effects, and heritability for cardiovascular risk factors. *Ann. Epidemiol.*, **6** (1996), 188–194.
20. Zdravkovic, S., Wienke, A., Pedersen, N. L. *et al.* Heritability of death from coronary heart disease: a 36-year follow-up of 20 966 Swedish twins. *J. Intern. Med.*, **252** (2002), 247–254.
21. Zhuang, L., Cui, C., Chen, Y. *et al.* Genetic basis of essential hypertension in Tibetans. *Zhonghua Yi Xue Za Zhi*, **82** (2002), 1009–1012.

22. Aggarwal, A., Esler, M. D., Socratous, F. and Kaye, D. M. Evidence for functional presynaptic alpha-2 adrenoreceptors and their down-regulation in human heart failure. *J. Am. Coll. Cardiol.*, **37** (2001), 1246–1251.
23. Dries, D. L., Exner, D. V., Gersh, B. J. *et al.* Racial differences in the outcome of left ventricular dysfunction. *New Engl. J. Med.*, **340** (1999), 609–616.
24. McCaffery, J. M., Pogue-Geile, M. F., Ferrell, R. E., Petro, N. and Manuck, S. B. Variability within alpha- and beta-adrenoreceptor genes as a predictor of cardiovascular function at rest and in response to mental challenge. *J. Hypertens.*, **20** (2002), 1105–1114.
25. Raynor, D. A., Pogue-Geile, M. F., Kamarck, T. W., McCaffery, J. M. and Manuck, S. B. Covariation of psychosocial characteristics associated with cardiovascular disease: genetic and environmental influences. *Psychosom. Med.*, **64** (2002), 191–205.
26. Small, K. M., Wagoner, L. E., Levin, A. M., Kardia, S. L. and Liggett, S. B. Synergistic polymorphisms of beta1- and alpha 2C-adrenergic receptors and the risk of congestive heart failure. *New Engl. J. Med.*, **347** (2002), 1135–1142.
27. Ranade, K., Jorgenson, E., Sheu, W. H. *et al.* A polymorphism in the beta1-adrenergic receptor is associated with resting heart rate. *Am. J. Hum. Genet.*, **70** (2002), 935–942.
28. Gama, L. T., Dickerson, G. E., Young, L. D. and Leymaster, K. A. Genetic and phenotypic variation in sources of preweaning lamb mortality. *Anim. Sci.*, **69** (1991), 2744–2753.
29. Lyons, D. T., Freeman, A. E. and Kuck, A. L. Genetics of health traits in Holstein cattle. *J. Dairy Sci.*, **74** (1991), 1092–1100.
30. Palmer, L. J., Knuiman, M. W., Divitini, M. L. *et al.* Familial aggregation and heritability of adult lung function: results from the Busselton Health Study. *Eur. Respir. J.*, **17** (2001), 696–702.
31. Ramsay, C. E., Hayden, C. M., Tiller, K. J. *et al.* Polymorphisms in the beta2-adrenoreceptor gene are associated with decreased airway responsiveness. *Clin. Exp. Allergy*, **29** (1999), 1195–1203.
32. Ho, L. I., Harn, H. J., Chen, C. J. and Tsai, N. M. Polymorphism of the beta(2)-adrenoceptor in COPD in Chinese subjects. *Chest*, **120** (2001), 1493–1499.
33. Xu, X. and Weiss, S. T. Association of asthma with beta(2)-adrenergic receptor gene polymorphism and cigarette smoking. *Am. J. Respir. Crit. Care Med.*, **166** (2002), 775.
34. Tsigos, C. and Chrousos, G. P. Hypothalamic–pituitary–adrenal axis, neuroendocrine factors and stress. *J. Psychosom. Res.*, **53** (2002), 865–871.
35. Devor, M. Pain and 'state'-induced analgesia: an introduction. In *Stress-Induced Analgesia*, ed. M. D. Tricklebank and G. Curzon. (Chichester: Wiley, 1984), pp. 1–19.
36. Campbell, J. N. and Raja, S. N. Reflex sympathetic dystrophy. *Neurology*, **45** (1995), 1235–1237.
37. Fuchs, P. N., Meyer, R. A. and Raja, S. N. Heat, but not mechanical hyperalgesia, following adrenergic injections in normal human skin. *Pain*, **90** (2001), 15–23.
38. Bossut, D. F., Shea, V. K. and Perl, E. R. Sympathectomy induces adrenergic excitability of cutaneous C-fiber nociceptors. *J. Neurophysiol.*, **75** (1996), 514–517.
39. Chabal, C., Jacobson, L., Russell, L. C. and Burchiel, K. J. Pain response to perineuromal injection of normal saline, epinephrine, and lidocaine in humans. *Pain*, **49** (1992), 9–12.

40. Ali, Z., Ra, S. N., Wesselmann, U. *et al.* Intradermal injection of norepinephrine evokes pain in patients with sympathetically maintained pain. *Pain*, **88** (2000), 161–168.
41. Torebjork, E., Wahren, L., Wallin, G., Hallin, R. and Koltzenburg, M. Noradrenaline-evoked pain in neuralgia. *Pain*, **63** (1995), 11–20.
42. Eisenach, J. C., De Kock, M. and Klimscha, W. alpha(2)-Adrenergic agonists for regional anesthesia. A clinical review of clonidine (1984–1995). *Anesthesiology*, **85** (1996), 655–674.
43. Guo, T. Z., Poree, L., Golden, W. *et al.* Antinociceptive response to nitrous oxide is mediated by supraspinal opiate and spinal alpha 2 adrenergic receptors in the rat. *Anesthesiology*, **85** (1996), 846–852.
44. Swandulla, D., Hans, M., Zipser, K. and Augustine, G. J. Role of residual calcium in synaptic depression and posttetanic potentiation: fast and slow calcium signaling in nerve terminals. *Neuron*, **7** (1991), 915–926.
45. Pan, Y. Z., Li, D. P. and Pan, H. L. Inhibition of glutamatergic synaptic input to spinal lamina II(o) neurons by presynaptic alpha(2)-adrenergic receptors. *J. Neurophysiol.*, **87** (2002), 1938–1947.
46. Fukuhara, N., Ishikawa, T., Kinoshita, H., Xiong, L. and Nakanishi, O. Central noradrenergic mediation of nitrous oxide-induced analgesia in rats. *Can. J. Anaesth.*, **45** (1998), 1123–1129.
47. Tamsen, A. and Gordh, T. Epidural clonidine produces analgesia. *Lancet*, **28** (1984), 231–232.
48. Tejwani, G. A. and Rattan, A. K. Antagonism of antinociception produced by intrathecal clonidine by ketorolac in the rat: the role of the opioid system. *Anesth. Analg.*, **90** (2000), 1152–1156.
49. Kayser, V., Desmeules, J. and Guilbauda, G. Systemic clonidine differentially modulates the abnormal reactions to mechanical and thermal stimuli in rats with peripheral mononeuropathy. *Pain*, **60** (1995), 275–285.
50. Byas-Smith, M. G., Max, M., Muir, J. and Kingman, A. Transdermal clonidine compared to placebo in painful diabetic neuropathy using a two-stage 'enriched enrollment' design. *Pain*, **60** (1995), 267–274.
51. Eisenach, J. C., DuPen, S., Dubois, M., Miguel, R. and Allin, D. Epidural clonidine analgesia for intractable cancer pain. The Epidural Clonidine Study Group. *Pain*, **61** (1995), 391–399.
52. Eisenach, J. C., Hood, D. D., Tuttle, R. *et al.* Computer-controlled epidural infusion to targeted cerebrospinal fluid concentrations in humans. Clonidine. *Anesthesiology*, **83** (1995), 33–47.
53. Donnelly-Roberts, D. L., Puttfarcken, P. S., Kuntzweiler, T. A. *et al.* ABT-594 [(R)-5-(2-azetidinylmethoxy)-2-chloropyridine]: a novel, orally effective analgesic acting via neuronal nicotinic acetylcholine receptors: I. In vitro characterization. *J. Pharmacol. Exp. Ther.*, **285** (1998), 777–786.
54. Carson, R. P. and Robertson, D. Genetic manipulation of noradrenergic neurons. *J. Pharmacol. Exp. Ther.*, **301** (2002), 410–417.
55. Nielsen, D. A., Dean, M. and Goldman, D. Genetic mapping of the human tryptophan hydroxylase gene on chromosome 11, using an intronic conformational polymorphism. *Am. J. Hum. Genet.*, **51** (1992), 1366–1371.

56. Albanese, V., Biguet, N. F., Kiefer, H. *et al.* Quantitative effects on gene silencing by allelic variation at a tetranucleotide microsatellite. *Hum. Mol. Genet.*, **10** (2001), 1785–1792.
57. Linhart, A., Jindra, A., Golan, L. *et al.* Association between tyrosine hydroxylase polymorphisms and left ventricular structure in young normotensive men. *Br. J. Biomed. Sci.*, **59** (2002), 90–94.
58. Swaans, R. J., Rondot, P., Renier, W. O. *et al.* Four novel mutations in the tyrosine hydroxylase gene in patients with infantile parkinsonism. *Ann. Hum. Genet.*, **64** (2000), 25–31.
59. Brautigam, C., Steenbergen-Spanjers, G. C., Hoffmann, G. F. *et al.* Biochemical and molecular genetic characteristics of the severe form of tyrosine hydroxylase deficiency. *Clin. Chem.*, **45** (1999), 2073–2078.
60. Wei, J., Ramchand, C. N. and Hemmings, G. P. Possible control of dopamine beta-hydroxylase via a codominant mechanism associated with the polymorphic (GT)n repeat at its gene locus in healthy individuals. *Hum. Genet.*, **99** (1997), 52–55.
61. Wood, J. G., Joyce, P. R., Miller, A. L., Mulder, R. T. and Kennedy, M. A. A polymorphism in the dopamine beta-hydroxylase gene is associated with "paranoid ideation" in patients with major depression. *Biol. Psychiatry*, **51** (2002), 365–369.
62. Cubells, J. F., Kranzler, H. R., McCance-Katz, E. *et al.* A haplotype at the DBH locus, associated with low plasma dopamine beta-hydroxylase activity, also associates with cocaine-induced paranoia. *Mol. Psychiatry*, **5** (2000), 56–63.
63. Zabetian, C. P., Anderson, G. M., Buxbaum, S. G. *et al.* A quantitative-trait analysis of human plasma-dopamine beta-hydroxylase activity: evidence for a major functional polymorphism at the DBH locus. *Am. J. Hum. Genet.*, **68** (2001), 515–522.
64. Williams, H. J., Bray, N., Murphy, K. C. *et al.* No evidence for allelic association between schizophrenia and a functional variant of the human dopamine beta-hydroxylase gene (DBH). *Am. J. Med. Genet.*, **88** (1999), 557–559.
65. Stober, G., Nothen, M. M., Porzgen, P. *et al.* Systematic search for variation in the human norepinephrine transporter gene: identification of five naturally occurring missense mutations and study of association with major psychiatric disorders. *Am. J. Med. Genet.*, **67** (1996), 523–532.
66. Ivancsits, S., Heider, A., Rudiger, H. W. and Winker, R. Orthostatic intolerance is not necessarily related to a specific mutation (Ala457Pro) in the human norepinephrine transporter gene. *Am. J. Med. Sci.*, **325** (2003), 63–65.
67. Leszczynska-Rodziewicz, A., Czerski, P. M., Kapelski, P. *et al.* A polymorphism of the norepinephrine transporter gene in bipolar disorder and schizophrenia: lack of association. *Neuropsychobiology*, **45** (2002), 182–185.
68. Samochowiec, J., Kucharska-Mazur, J., Kaminski, R. *et al.* Norepinephrine transporter gene polymorphism is not associated with susceptibility to alcohol dependence. *Psychiatry Res.*, **111** (2002), 229–233.
69. Belfer, I., Phillips, G., Taubman, J. *et al.* Haplotype architecture of the norepinephrine transporter gene *SLC6A2* in four populations. *J. Hum. Genet.*, **49** (2004), 232–245.
70. Brunner, H. G., Nelen, M. R., van Zandvoort, P. *et al.* X-linked borderline mental retardation with prominent behavioral disturbance: phenotype, genetic localization, and evidence for disturbed monoamine metabolism. *Am. J. Hum. Genet.*, **52** (1993), 1032–1039.

71. Balciuniene, J., Emilsson, L., Oreland, L., Pettersson, U. and Jazin, E. Investigation of the functional effect of monoamine oxidase polymorphisms in human brain. *Hum. Genet.*, **110** (2002), 1–7.
72. Manuck, S. B., Flory, J. D., Ferrell, R. E., Mann, J. J. and Muldoon, M. F. A regulatory polymorphism of the monoamine oxidase-A gene may be associated with variability in aggression, impulsivity, and central nervous system serotonergic responsivity. *Psychiatry Res.*, **95** (2000), 9–23.
73. Lotta, T., Vidgren, J., Tilgmann, C. *et al.* Kinetics of human soluble and membrane-bound catechol O-methyltransferase: a revised mechanism and description of the thermolabile variant of the enzyme. *Biochemistry*, **34** (1995), 4202–4210.
74. Chen, C. H., Lee, Y. R., Liu, M. Y. *et al.* Identification of a BglI polymorphism of catechol-O-methyltransferase (COMT) gene, and association study with schizophrenia. *Am. J. Med. Genet.*, **67** (1996), 556–559.
75. Chen, C. H., Lee, Y. R., Wei, F. C. *et al.* Association study of NlaIII and MspI genetic polymorphisms of catechol-O-methyltransferase gene and susceptibility to schizophrenia. *Biol. Psychiatry*, **41** (1997), 985–987.
76. Chen, C. H., Lee, Y. R., Chung, M. Y. *et al.* Systematic mutation analysis of the catechol O-methyltransferase gene as a candidate gene for schizophrenia. *Am. J. Psychiatry*, **156** (1999), 1273–1275.
77. Winqvist, R., Lundstrom, K., Salminen, M., Laatikainen, M. and Ulmanen, I. The human catechol-O-methyltransferase (COMT) gene maps to band q11.2 of chromosome 22 and shows a frequent RFLP with BglI. *Cytogenet. Cell Genet.*, **59** (1992), 253–257.
78. Zhu, G., Lipsky, R. H., Xu, K. *et al.* Differential expression of human COMT alleles in brain and lymphoblasts detected by RT-coupled 5' nuclease assay. *Psychopharmacology (Berl.)*, **177** (2004), 178–184.
79. Shibata, K., Hirasawa, A., Moriyama, N. *et al.* Alpha 1a-adrenoceptor polymorphism: pharmacological characterization and association with benign prostatic hypertrophy. *Br. J. Pharmacol.*, **118** (1996), 1403–1408.
80. Buscher, R., Herrmann, V., Ring, K. M. *et al.* Variability in phenylephrine response and essential hypertension: a search for human alpha(1B)-adrenergic receptor polymorphisms. *J. Pharmacol. Exp. Ther.*, **291** (1999), 793–798.
81. Barr, C. L., Wigg, K., Zai, G. *et al.* Attention-deficit hyperactivity disorder and the adrenergic receptors alpha 1C and alpha 2C. *Mol. Psychiatry*, **6** (2001), 334–337.
82. Small, K. M. and Liggett, S. B. Identification and functional characterization of alpha(2)-adrenoceptor polymorphisms. *Trends Pharmacol. Sci.*, **22** (2001), 471–477.
83. Lakhlani, P. P., MacMillan, L. B., Guo, T. Z. *et al.* Substitution of a mutant alpha2a-adrenergic receptor via "hit and run" gene targeting reveals the role of this subtype in sedative, analgesic, and anesthetic-sparing responses in vivo. *Proc. Natl. Acad. Sci. U. S. A.*, **94** (1997), 9950–9955.
84. Small, K. M., Forbes, S. L., Brown, K. M. and Liggett, S. B. An asn to lys polymorphism in the third intracellular loop of the human alpha 2A-adrenergic receptor imparts enhanced agonist-promoted Gi coupling. *J. Biol. Chem.*, **275** (2000), 38 518–38 523.
85. Hamann, A., Brieske, C., Tafel, J. *et al.* Identification of a deletion variant in the gene encoding the human alpha(2A)-adrenergic receptor. *Eur. J. Endocrinol.*, **144** (2001), 291–295.

86. Small, K. M., Brown, K. M., Forbes, S. L. and Liggett, S. B. Polymorphic deletion of three intracellular acidic residues of the alpha 2B-adrenergic receptor decreases G protein-coupled receptor kinase mediated phosphorylation and desensitization. *J. Biol. Chem.*, **276** (2001), 4917–4922.
87. Makaritsis, K. P., Handy, D. E., Johns, C. *et al.* Role of the alpha2B-adrenergic receptor in the development of salt- induced hypertension. *Hypertension*, **33** (1999), 14–17.
88. Feng, J., Zheng, J., Gelernter, J. *et al.* An in-frame deletion in the alpha (2C) adrenergic receptor is common in African-Americans. *Mol. Psychiatry*, **6** (2001), 168–172.
89. Small, K. M., Forbes, S. L., Rahman, F. F., Bridges, K. M. and Liggett, S. B. A four amino acid deletion polymorphism in the third intracellular loop of the human alpha 2C-adrenergic receptor confers impaired coupling to multiple effectors. *J. Biol. Chem.*, **275** (2000), 23 059–23 064.
90. Maqbool, A. S., Hall, S. G., Ball, A. and Balmforth, J. Common polymorphisms of β1-adrenoceptor: identification and rapid screening assay. *Lancet*, **353** (1999), 897.
91. Wagoner, L. E., Craft, L. L., Zengel, P. *et al.* Polymorphisms of the beta1-adrenergic receptor predict exercise capacity in heart failure. *Am. Heart J.*, **144** (2002), 840–846.
92. Wenzel, K., Felix, S. B., Bauer, D. *et al.* Novel variants in 3 Kb of 5'U. T. R of the beta-1 adrenergic receptor gene (−93C>T, −210C>T, and −2146T>C): −2146C homozygotes present in patients with idiopathic dilated cardiomyopathy and coronary heart disease. *Hum. Mutat.*, **16** (2000), 534.
93. Drysdale, C. M., McGraw, D. W., Stack, C. B. *et al.* Complex promoter and coding region beta-2 adrenergic receptor haplotypes alter receptor expression and predict in vivo responsiveness. *Proc. Natl. Acad. Sci. U. S. A.*, **97** (2000), 10 483–10 488.
94. Reihsaus, E., Innis, M. and MacIntyre, N. Mutations in the gene encoding for ß2-adrenergic receptor in normal and asthmatic subjects. *Am. J. Respir. Cell. Mol. Biol.*, **8** (1993), 334–339.
95. Tomaszewski, M., Brain, N. J., Charchar, F. J. *et al.* Essential hypertension and beta2-adrenergic receptor gene: linkage and association analysis. *Hypertension*, **40** (2002), 286–291.
96. Yoshida, N., Sakane, N. and Umekawa, T. Mutation of beta-3-adrenergic-receptor gene and response to treatment of obesity. *Lancet*, **346** (1995), 1433–1434.
97. Makimoto, H., Sakaeda, T., Nishiguchi, K. *et al.* Beta2-adrenergic receptor genotype-related changes in cAMP levels in peripheral blood mononuclear cells after multiple-dose oral procaterol. *Pharm. Res.*, **18** (2001), 1651–1654.
98. Green, S. A., Turki, J., Bejarano, P., Hall, I. P. and Liggett, S. B. Influence of ß2-adrenergic receptor genotypes on signal transduction in human airway smooth muscle cells. *Am. J. Respir. Cell. Mol. Biol.*, **13** (1995), 25–33.
99. Green, S. A., Cole, G., Jacinto, M., Innis, M. and Liggett, S. B. A polymorphism of the human ß2-adrenergic receptor within the fourth transmembrane domain alters ligand binding and functional properties of the receptor. *J. Biol. Chem.*, **268** (1993), 23 116–23 121.
100. Lipworth, B., Koppelman, G. H., Wheatley, A. P. *et al.* Beta2 adrenoceptor promoter polymorphisms: extended haplotypes and functional effects in peripheral blood mononuclear cells. *Thorax*, **57** (2002), 61–66.
101. Wang, J., Mougey, E. B., David, C. J. *et al.* Determination of human beta(2)-adrenoceptor haplotypes by denaturation selective amplification and subtractive genotyping. *Am. J. Pharmacogenomics*, **1**:4 (2001), 315–322.

102. Walston, J., Silver, K., Bogardus, C. *et al.* Time of onset of non-insulin-dependent diabetes mellitus and genetic variation in the beta 3-adrenergic-receptor gene. *New Engl. J. Med.*, **333** (1995), 343–347.

103. Sakane, N., Yoshida, T., Umekawa, T. *et al.* Beta 3-adrenergic-receptor polymorphism: a genetic marker for visceral fat obesity and the insulin resistance syndrome. *Diabetologia*, **40** (1997), 200–204.

104. Janssen, R. J., Wevers, R. A., Haussler, M. *et al.* A branch site mutation leading to aberrant splicing of the human tyrosine hydroxylase gene in a child with a severe extrapyramidal movement disorder. *Ann. Hum. Genet.*, **64** (2000), 375–382.

105. Yoritaka, A., Hattori, N., Yoshino, H. and Mizuno, Y. Catechol-O-methyltransferase genotype and susceptibility to Parkinson's disease in Japan. Short communication. *J. Neural Transm.*, **104** (1997), 1313–1317.

106. Nakatome, M., Tun, Z., Shimada, S. and Honda, K. Detection and analysis of four polymorphic markers at the human monoamine oxidase (*MAO*) gene in Japanese controls and patients with Parkinson's disease. *Biochem. Biophys. Res. Commun.*, **247** (1998), 452–456.

107. Wu, R. M., Cheng, C. W., Chen, K. H. *et al.* The COMT L allele modifies the association between MAOB polymorphism and PD in Taiwanese. *Neurology*, **56** (2001), 375–382.

108. Costa, P., Checkoway, H., Levy, D. *et al.* Association of a polymorphism in intron 13 of the monoamine oxidase B gene with Parkinson disease. *Am. J. Med. Genet.*, **74** (1997), 154–156.

109. Checkoway, H., Franklin, G. M., Costa-Mallen, P. *et al.* A genetic polymorphism of MAO-B modifies the association of cigarette smoking and Parkinson's disease. *Neurology*, **50** (1998), 1458–1461.

110. Mellick, G. D., Buchanan, D. D., McCann, S. J. *et al.* Variations in the monoamine oxidase B (MAOB) gene are associated with Parkinson's disease. *Mov. Disord.*, **14** (1999), 219–224.

111. Takehashi, M., Tanaka, S., Masliah, E. and Ueda, K. Association of monoamine oxidase A gene polymorphism with Alzheimer's disease and Lewy body variant. *Neurosci. Lett.*, **327** (2002), 79–82.

112. Lea, R. A., Dohy, A., Jordan, K. *et al.* Evidence for allelic association of the dopamine beta-hydroxylase gene (DBH) with susceptibility to typical migraine. *Neurogenetics*, **3** (2000), 35–40.

113. Emin Erdal, M., Herken, H., Yilmaz, M. and Bayazit, Y. A. Significance of the catechol-O-methyltransferase gene polymorphism in migraine. *Brain Res. Mol. Brain Res.*, **94** (2001), 193–196.

114. Halperin, J. M., Newcorn, J. H., Koda, V. H. *et al.* Noradrenergic mechanisms in ADHD children with and without reading disabilities: a replication and extension. *J. Am. Acad. Child Adolesc. Psychiatry*, **36** (1997), 1688–1696.

115. Davids, E., Zhang, K., Kula, N. S., Tarazi, F. I. and Baldessarini, R. J. Effects of norepinephrine and serotonin transporter inhibitors on hyperactivity induced by neonatal 6-hydroxydopamine lesioning in rats. *J. Pharmacol. Exp. Ther.*, **301** (2002), 1097–1102.

116. Comings, D. E., Gade-Andavolu, R., Gonzalez, N. *et al.* Additive effect of three noradrenergic genes (*ADRA2A, ADRA2C, DBH*) on attention-deficit hyperactivity disorder and learning disabilities in Tourette syndrome subjects. *Clin. Genet.*, **55** (1999), 160–172.
117. Daly, G., Hawi, Z., Fitzgerald, M. and Gill, M. Mapping susceptibility loci in attention deficit hyperactivity disorder: preferential transmission of parental alleles at DAT1, DBH and DRD5 to affected children. *Mol. Psychiatry*, **4** (1999), 192–196.
118. Jiang, S., Xin, R., Lin, S. *et al.* Linkage studies between attention-deficit hyperactivity disorder and the monoamine oxidase genes. *Am. J. Med. Genet.*, **105** (2001), 783–788.
119. Eisenberg, J., Mei-Tal, G., Steinberg, A. *et al.* Haplotype relative risk study of catechol-O-methyltransferase (COMT) and attention deficit hyperactivity disorder (ADHD): association of the high-enzyme activity Val allele with ADHD impulsive-hyperactive phenotype. *Am. J. Med. Genet.*, **88** (1999), 497–502.
120. Strous, R. D., Bark, N., Parsia, S. S., Volavka, J. and Lachman, H. M. Analysis of a functional catechol-O-methyltransferase gene polymorphism in schizophrenia: evidence for association with aggressive and antisocial behavior. *Psychiatry Res.*, **69** (1997), 71–77.
121. Lachman, H. M., Nolan, K. A., Mohr, P., Saito, T. and Volavka, J. Association between catechol O-methyltransferase genotype and violence in schizophrenia and schizoaffective disorder. *Am. J. Psychiatry*, **155** (1998), 835–837.
122. Kotler, M., Barak, P., Cohen, H. *et al.* Homicidal behavior in schizophrenia associated with a genetic polymorphism determining low catechol O-methyltransferase (COMT) activity. *Am. J. Med. Genet.*, **88** (1999), 628–633.
123. Nolan, K. A., Volavka, J., Czobor, P. *et al.* Suicidal behavior in patients with schizophrenia is related to COMT polymorphism. *Psychiatr. Genet.*, **10** (2000), 117–124.
124. Brunner, H. G., Nelen, M., Breakefield, X. O., Ropers, H. H. and van Oost, B. A. Abnormal behavior associated with a point mutation in the structural gene for monoamine oxidase A. *Science*, **262** (1993), 578–580.
125. Caspi, A., McClay, J., Moffitt, T. E. *et al.* Role of genotype in the cycle of violence in maltreated children. *Science*, **297** (2002), 851–854.
126. Cases, O., Seif, I., Grimsby, T. *et al.* Aggressive behavior and altered amounts of brain serotonin and norepinephrine in mice lacking MAOA. *Science*, **268** (1995), 1763–1766.
127. Haller, J., Bakos, N., Rodriguiz, R. M. *et al.* Behavioral responses to social stress in noradrenaline transporter knockout mice: effects on social behavior and depression. *Brain Res. Bull.*, **58** (2002), 279–284.
128. Lim, L. C., Powell, J., Sham, P. *et al.* Evidence for a genetic association between alleles of monoamine oxidase A gene and bipolar affective disorder. *Am. J. Med. Genet.*, **60** (1995), 325–331.
129. Deckert, J., Catalano, M., Syagailo, Y. V. *et al.* Excess of high activity monoamine oxidase A gene promoter alleles in female patients with panic disorder. *Hum. Mol. Genet.*, **8** (1999), 621–624.
130. Du, L., Faludi, G., Palkovits, M. *et al.* High activity-related allele of MAO-A gene associated with depressed suicide in males. *NeuroReport*, **13** (2002), 1195–1198.

131. Alsobrook, J. P., 2nd, Zohar, A. H., Leboyer, M. *et al.* Association between the COMT locus and obsessive-compulsive disorder in females but not males. *Am. J. Med. Genet.*, **114** (2002), 116–120.
132. Camarena, B., Cruz, C., de la Fuente, J. R. and Nicolini, H. A higher frequency of a low activity-related allele of the MAO-A gene in females with obsessive-compulsive disorder. *Psychiatr. Genet.*, **8** (1999), 255–257.
133. Camarena, B., Rinetti, G., Cruz, C. *et al.* Additional evidence that genetic variation of MAO-A gene supports a gender subtype in obsessive-compulsive disorder. *Am. J. Med. Genet.*, **105** (2001), 279–282.
134. Karayiorgou, M., Altemus, M., Galke, B. L. *et al.* Genotype determining low catechol-O-methyltransferase activity as a risk factor for obsessive-compulsive disorder. *Proc. Natl. Acad. Sci. U. S. A.*, **94** (1997), 4572–4575.
135. Karayiorgou, M., Sobin, C., Blundell, M. L. *et al.* Family-based association studies support a sexually dimorphic effect of COMT and MAOA on genetic susceptibility to obsessive-compulsive disorder. *Biol. Psychiatry*, **45** (1999), 1178–1189.
136. Karno, M., Golding, J. M., Sorenson, S. B. and Burnam, M. A. The epidemiology of obsessive-compulsive disorder in five U.S. communities. *Arch. Gen. Psychiatry*, **45** (1988), 1084–1099.
137. Papolos, D. F., Veit, S., Faedda, G. L., Saito, T. and Lachman, H. M. Ultra-ultra rapid cycling bipolar disorder is associated with the low activity catecholamine-O-methyltransferase allele. *Mol. Psychiatry*, **3** (1998), 346–349.
138. Kirov, G., Murphy, K. C., Arranz, M. J. *et al.* Low activity allele of catechol-O-methyltransferase gene associated with rapid cycling bipolar disorder. *Mol. Psychiatry*, **3** (1998), 342–345.
139. Li, T., Vallada, H., Curtis, D. *et al.* Catechol-O-methyltransferase Val158Met polymorphism: frequency analysis in Han Chinese subjects and allelic association of the low activity allele with bipolar affective disorder. *Pharmacogenetics*, **7** (1997), 349–353.
140. Woo, J. M., Yoon, K. S. and Yu, B. H. Catechol O-methyltransferase genetic polymorphism in panic disorder. *Am. J. Psychiatry*, **159** (2002), 1785–1787.
141. Frisch, A., Laufer, N., Danziger, Y. *et al.* Association of anorexia nervosa with the high activity allele of the COMT gene: a family-based study in Israeli patients. *Mol. Psychiatry*, **6** (2001), 243–245.
142. Serretti, A., Macciardi, F., Verga, M. *et al.* Tyrosine hydroxylase gene associated with depressive symptomatology in mood disorder. *Am. J. Med. Genet.*, **81** (1998), 127–130.
143. Geracioti, T. D., Jr., Baker, D. G., Ekhator, N. N. *et al.* CSF norepinephrine concentrations in posttraumatic stress disorder. *Am. J. Psychiatry*, **158** (2001), 1227–1230.
144. Stein, M. B., Schork, N. J. and Gelernter, J. A polymorphism of the beta1-adrenergic receptor is associated with low extraversion. *Biol. Psychiatry*, **56** (2004), 217–224.
145. Arnsten, A. F. T., Cai, J. X., Murphy, B. L. and Goldman-Rakic, P. S. Dopamine D1 receptor mechanisms in the cognitive performance of young adult and aged monkeys. *Psychopharmacology (Berl.)*, **116** (1994), 143–151.
146. Pudovkina, O. L., Kawahara, Y., de Vries, J. and Westerink, B. H. The release of noradrenaline in the locus coeruleus and prefrontal cortex studied with dual-probe microdialysis. *Brain Res.*, **906** (2001), 38–45.

147. Egan, M. F., Goldberg, T. E., Kolachana, B. S. *et al.* Effect of COMT Val108/158 Met genotype on frontal lobe function and risk for schizophrenia. *Proc. Natl. Acad. Sci. U. S. A.*, **98** (2001), 6917–6922.
148. Malhotra, A. K., Kestler, L. J., Mazzanti, C. *et al.* A functional polymorphism in the COMT gene and performance on a test of prefrontal cognition. *Am. J. Psychiatry*, **159** (2002), 652–654.
149. Joober, R., Gauthier, J., Lal, S. *et al.* Catechol-O-methyltransferase Val-108/158-Met gene variants associated with performance on the Wisconsin Card Sorting Test. *Arch. Gen. Psychiatry*, **59** (2002), 662–663.
150. Bilder, R. M., Volavka, J., Czobor, P. *et al.* Neurocognitive correlates of the COMT Val(158)Met polymorphism in chronic schizophrenia. *Biol. Psychiatry*, **52** (2002), 701–707.
151. Shifman, S., Bronstein, M., Sternfeld, M. *et al.* A highly significant association between a COMT haplotype and schizophrenia. *Am. J. Hum. Genet.*, **71** (2002), 1296–1302.
152. Mattay, V. S., Goldberg, T. E., Fera, F. *et al.* Catechol O-methyltransferase val158-met genotype and individual variation in the brain response to amphetamine. *Proc. Natl. Acad. Sci. U. S. A.*, **100** (2003), 6186–6191.
153. Meloni, R., Laurent, C., Campion, D. *et al.* A rare allele of a microsatellite located in the tyrosine hydroxylase gene found in schizophrenic patients. *C. R. Acad. Sci. III*, **318** (1995), 803–809.
154. Kurumaji, A., Kuroda, T., Yamada, K., Yoshikawa, T. and Toru, M. An association of the polymorphic repeat of tetranucleotide (TCAT) in the first intron of the human tyrosine hydroxylase gene with schizophrenia in a Japanese sample. *J. Neural Transm.*, **108** (2001), 489–495.
155. McKinney, E. F., Walton, R. T., Yudkin, P. *et al.* Association between polymorphisms in dopamine metabolic enzymes and tobacco consumption in smokers. *Pharmacogenetics*, **10** (2000), 483–491.
156. Sander, T., Harms, H., Rommelspacher, H., Hoehe, M. and Schmidt, L. G. Possible allelic association of a tyrosine hydroxylase polymorphism with vulnerability to alcohol-withdrawal delirium. *Psychiatr. Genet.*, **8**:1 (1998), 13–17.
157. Lachman, H. M., Papolos, D. F., Saito, T. *et al.* Human catechol-O-methyltransferase pharmacogenetics: description of a functional polymorphism and its potential application to neuropsychiatric disorders. *Pharmacogenetics*, **6** (1996), 243–250.
158. Vandenbergh, D. J., Rodriguez, L. A., Miller, I. T., Uhl, G. R. and Lachman, H. M. High-activity catechol-O-methyltransferase allele is more prevalent in polysubstance abusers. *Am. J. Med. Genet.*, **74** (1997), 439–442.
159. Tiihonen, J., Hallikainen, T., Lachman, H. *et al.* Association between the functional variant of the catechol-O-methyltransferase (COMT) gene and type 1 alcoholism. *Mol. Psychiatry*, **4** (1999), 286–289.
160. Wang, T., Franke, P., Neidt, H. *et al.* Association study of the low-activity allele of catechol-O-methyltransferase and alcoholism using a family-based approach. *Mol. Psychiatry*, **6** (2001), 109–111.

161. Horowitz, R., Kotler, M., Shufman, E. *et al.* Confirmation of an excess of the high enzyme activity COMT val allele in heroin addicts in a family-based haplotype relative risk study. *Am. J. Med. Genet.*, **96** (2000), 599–603.

162. Schmidt, L. G., Sander, T., Kuhn, S. *et al.* Different allele distribution of a regulatory MAOA gene promoter polymorphism in antisocial and anxious-depressive alcoholics. *J. Neural Transm.*, **107** (2000), 681–689.

163. Samochowiec, J., Lesch, K. P., Rottmann, M. *et al.* Association of a regulatory polymorphism in the promoter region of the monoamine oxidase A gene with antisocial alcoholism. *Psychiatry Res.*, **86** (1999), 67–72.

164. Gade, R., Muhleman, D., Blake, H. *et al.* Correlation of length of VNTR alleles at the X-linked MAOA gene and phenotypic effect in Tourette syndrome and drug abuse. *Mol. Psychiatry*, **3** (1998), 50–60.

165. Shannon, J. R., Flattem, N. L., Jordan, J. *et al.* Orthostatic intolerance and tachycardia associated with norepinephrine-transporter deficiency. *New Engl. J. Med.*, **342** (2000), 541–549.

166. Baldwin, C. T., Schwartz, F., Baima, J. *et al.* Identification of a polymorphic glutamic acid stretch in the alpha2B-adrenergic receptor and lack of linkage with essential hypertension. *Am. J. Hypertens.*, **12** (1999), 853–857.

167. Snapir, A., Heinonen, P., Tuomainen, T. P. *et al.* An insertion/deletion polymorphism in the alpha2B-adrenergic receptor gene is a novel genetic risk factor for acute coronary events. *J. Am. Coll. Cardiol.*, **37** (2001), 1516–1522.

168. Tesson, F., Charron, P., Peuchmaurd, M. *et al.* Characterization of a unique genetic variant in the beta1-adrenoceptor gene and evaluation of its role in idiopathic dilated cardiomyopathy. CARDIGENE Group. *J. Mol. Cell. Cardiol.*, **31** (1999), 1025–1032.

169. Mason, D. A., Moore, J. D., Green, S. A. and Liggett, S. B. A gain-of-function polymorphism in a G-protein coupling domain of the human beta1-adrenergic receptor. *J. Biol. Chem.*, **274** (1999), 12 670–12 674.

170. Molenaar, P., Rabnott, G., Yang, I. *et al.* Conservation of the cardiostimulant effects of (-)-norepinephrine across Ser49Gly and Gly389Arg beta(1)-adrenergic receptor polymorphisms in human right atrium in vitro. *J. Am. Coll. Cardiol.*, **40** (2002), 1275–1282.

171. Dishy, V., Sofowora, G. G., Xie, H. G. *et al.* The effect of common polymorphisms of the beta2-adrenergic receptor on agonist-mediated vascular desensitization. *New Engl. J. Med.*, **345** (2001), 1030–1035.

172. Ranade, K., Shue, W. H., Hung, Y. J. *et al.* The glycine allele of a glycine/arginine polymorphism in the beta2-adrenergic receptor gene is associated with essential hypertension in a population of Chinese origin. *Am. J. Hypertens.*, **14** (2001), 1196–1200.

173. Svetkey, L. P., Timmons, P. Z., Emovon, O. *et al.* Association of hypertension with beta2- and alpha2c10-adrenergic receptor genotype. *Hypertension*, **27** (1996), 1210–1215.

174. Chang, T. J., Tsai, M. H., Jiang, Y. D. *et al.* The Arg16Gly polymorphism of human beta2-adrenoreceptor is associated with type 2 diabetes in Taiwanese people. *Clin. Endocrinol. (Oxf.)*, **57** (2002), 685–690.

175. Corbalan, M. S., Marti, A., Forga, L., Martinez-Gonzalez, M. A. and Martinez, J. A. The risk of obesity and the Trp64Arg polymorphism of the beta(3)-adrenergic receptor: effect modification by age. *Ann. Nutr. Metab.*, **46** (2002a), 152–158.

176. Oizumi, T., Daimon, M., Saitoh, T. *et al.* Genotype Arg/Arg, but not Trp/Arg, of the Trp64Arg polymorphism of the beta(3)-adrenergic receptor is associated with type 2 diabetes and obesity in a large Japanese sample. *Diabetes Care*, **24** (2001), 1579–1583.
177. Wang, G., Li, Q., Niu, T., Chen, C. and Xu, X. Association of G. Y. S1 and beta(3)-A. R gene with postprandial hyperglycemia and serum uric acid in type 2 diabetes mellitus. *Chin. Med. J.*, **115** (2002), 1308–1311.
178. Iwamoto, N., Ogawa, Y., Kajihara, S. *et al.* Gln27Glu beta2-adrenergic receptor variant is associated with hypertriglyceridemia and the development of fatty liver. *Clin. Chim. Acta*, **314** (2001), 85–91.
179. Corbalan, M. S., Marti, A., Forga, L., Martinez-Gonzalez, M. A. and Martinez, J. A. Beta(2)-adrenergic receptor mutation and abdominal obesity risk: effect modification by gender and H. D. L-cholesterol. *Eur. J. Nutr.*, **41** (2002b), 114–118.
180. Ukkola, O., Rankinen, T., Weisnagel, S. J. *et al.* Interactions among the alpha2-, beta2-, and beta3-adrenergic receptor genes and obesity-related phenotypes in the Quebec Family Study. *Metabolism*, **49** (2000), 1063–1070.
181. Lario, S., Calls, J., Cases, A. *et al.* Short repeat on D. N. A marker at candidate locus. MspI identifies a biallelic polymorphism in the promoter region of the alpha (2A)-adrenergic receptor gene. *Clin. Genet.*, **51** (1997), 129–130.
182. Rosmond, R., Bouchard, C. and Bjorntorp, P. A C-1291G polymorphism in the alpha2A-adrenergic receptor gene (*ADRA2A*) promoter is associated with cortisol escape from dexamethasone and elevated glucose levels. *J. Intern. Med.*, **251** (2002), 252–257.
183. Green, S. A., Rathz, D. A., Schuster, A. J. and Liggett, S. B. The Ile164 beta(2)-adrenoceptor polymorphism alters salmeterol exosite binding and conventional agonist coupling to G(s). *Eur. J. Pharmacol.*, **421** (2001), 141–147.
184. Yim, D. S., Parkb, S. K., Yoo, K. Y. *et al.* Relationship between the Val158Met polymorphism of catechol O-methyl transferase and breast cancer. *Pharmacogenetics*, **11** (2001), 279–286.
185. Thompson, P. A., Shields, P. G., Freudenheim, J. L. *et al.* Genetic polymorphisms in catechol O-methyltransferase, menopausal status and breast cancer risk. *Cancer Res.*, **58** (1998), 2107–2110.
186. Desautels, A., Turecki, G., Montplaisir, J. *et al.* Evidence for a genetic association between monoamine oxidase A and restless legs syndrome. *Neurology*, **59** (2002), 215–219.
187. Dauvilliers, Y., Neidhart, E., Lecendreux, M., Billiard, M. and Tafti, M. MAOA and COMT polymorphisms and gene effects in narcolepsy. *Mol. Psychiatry*, **6** (2001), 367–372.
188. Lavigne, J. A., Helzlsouer, K. J., Huang, H. Y. *et al.* An association between the allele coding for a low activity variant of catechol O-methyltransferase and the risk for breast cancer. *Cancer Res.*, **57** (1997), 5493–5497.
189. Lichtenberg, P., Bachner-Melman, R., Gritsenko, I. and Ebstein, R. P. Exploratory association study between catechol-O-methyltransferase (COMT) high/low enzyme activity polymorphism and hypnotizability. *Am. J. Med. Genet.*, **96** (2000), 771–774.
190. Meirhaeghe, A., Luan, J., Selberg-Franks, P. *et al.* The effect of the Gly16Arg polymorphism of the beta(2)-adrenergic receptor gene on plasma free fatty acid levels is modulated by physical activity. *J. Clin. Endocrinol. Metab.*, **86** (2000), 5881–5887.

191. Large, V., Hellstrom, L., Reynisdottir, S. *et al.* Human beta-2 adrenoceptor gene polymorphisms are highly frequent in obesity and associate with altered adipocyte beta-2 adrenoceptor function. *J. Clin. Invest.*, **100** (1997), 3005–3013.
192. Turki, J., Pak, J., Green, S. A., Martin, R. J. and Liggett, S. B. Genetic polymorphisms of the beta 2-adrenergic receptor in nocturnal and nonnocturnal asthma. Evidence that Gly16 correlates with the nocturnal phenotype. *J. Clin. Invest.*, **95** (1995), 1635–1641.
193. Hopes, E., McDougall, C., Christie, G. *et al.* Association of glutamine 27 polymorphism of beta 2 adrenoceptor with reported childhood asthma: population based study. *Br. Med. J.*, **316** (1998), 664.
194. Dewar, J. C., Wilkinson, J., Wheatley, A., *et al.* The glutamine 27 beta2-adrenoceptor polymorphism is associated with elevated IgE levels in asthmatic families. *J. Allergy Clin. Immunol.*, **100** (1997), 261–265.
195. Martinez, F. D., Graves, P. E., Baldini, M., Solomon, S. and Erickson, R. Association between genetic polymorphisms of the beta2-adrenoceptor and response to albuterol in children with and without a history of wheezing. *J. Clin. Invest.*, **100** (1997), 3184–3188.
196. Festa, A., Krugluger, W., Shnawa, N. *et al.* Trp64Arg polymorphism of the beta3-adrenergic receptor gene in pregnancy: association with mild gestational diabetes mellitus. *J. Clin. Endocrinol. Metab.*, **84** (1999), 1695–1699.
197. Arner, P. and Hoffstedt, J. Adrenoceptor genes in human obesity. *J. Intern. Med.*, **245** (1999), 667–672.
198. Mitchell, B. D., Blangero, J., Comuzzie, A. G. *et al.* A paired sibling analysis of the beta-3 adrenergic receptor and obesity in Mexican Americans. *J. Clin. Invest.*, **101** (1998), 584–587.
199. Ludecke, B., Dworniczak, B. and Bartholome, K. A point mutation in the tyrosine hydroxylase gene associated with Segawa's syndrome. *Hum. Genet.* **95** (1995), 123–125.
200. Ludecke, B., Knappskog, P. M., Clayton, P. T. *et al.* Recessively inherited L-DOPA-responsive parkinsonism in infancy caused by a point mutation (L205P) in the tyrosine hydroxylase gene. *Hum. Mol. Genet.* **5** (1996), 1023–1028.

Part IV

Psychopharmacology of norepinephrine

17

Action of antidepressant and other drugs on norepinephrine systems

Elliott Richelson, M.D.
Mayo Clinic of Jacksonville, Jacksonville, USA

Introduction

Researchers introduced the first antidepressants, tricyclic antidepressants and monoamine oxidase inhibitors, in the late 1950s to treat depression. It was the finding that chlorpromazine was an effective antipsychotic agent that led researchers to comb the shelves of drug companies for look-alike compounds to treat psychosis. They came upon imipramine, which was found to lack antipsychotic properties. However, astute clinicians, in particular, Doctor Roland Kuhn[1] in Switzerland, noted that patients who were both depressed and psychotic improved in their depression on imipramine although their psychosis was unaffected. The year after the report by Kuhn, Sigg[2] reported on the peripheral adrenergic effects of imipramine in laboratory animals. Among his findings was that there was an enhancement and prolongation of the vasopressor response to norepinephrine. This effect occurred with acute but not chronic treatment. Interestingly, without any direct evidence, he hypothesized that, based on the peripheral effects of imipramine on the adrenergic system, it exerted its antidepressant effect by sensitization of central adrenergic synapses.[2] Thus, we have known for over 40 years that antidepressants interact with noradrenergic systems. This chapter will review the evolution of our knowledge of the interactions of antidepressants and other drugs with the noradrenergic system. This overview will focus on antidepressants.

Temporal overview

The 1950s and 1960s

Aside from the 1950s and 1960s giving birth to rock and roll, this era gave birth to modern day psychopharmacology when researchers took advantage of the clinical insights made when patients were treated with psychotropic drugs. One of

Brain Norepinephrine: Neurobiology and Therapeutics, ed. Gregory A. Ordway, Michael A. Schwartz and Alan Frazer. Published by Cambridge University Press.

these insights mentioned above in the introduction was the use of imipramine. There was another clinical insight with the reports of the occasional euphoric tuberculosis patient who had been treated with the antituberculosis drug isoniazid, a monoamine oxidase inhibitor (MAOI). In this time frame, pharmaceutical firms were also researching antidepressant drugs in animal models of depression, which involved treatment of these animals with reserpine and testing for a drug's ability to reverse the sedation and ptosis caused by reserpine. Monoamine oxidase inhibitors reversed the effects of reserpine. Additionally, imipramine reversed reserpine-induced depression as Sigg and his colleagues found.[3] Reserpine depletes endogenous stores of norepinephrine, dopamine, and serotonin.

Reserpine itself is also a historically interesting compound. It has been known for centuries and was extracted from the plant *Rauwolfia serpentina* in the 1950s. In 1931, in the journal *Indian Medical World*, Sen and Bose[4] reported on a new Indian drug for the treatment of high blood pressure and insanity. This new Indian drug was not for the morose form of insanity, but for the excitable form, because reserpine will cause the development of this morose form of insanity, i.e. depression, in about 10% of the treated population.[5] This clinical observation and the known effects of reserpine on the depletion of biogenic amines is one of the cornerstones of the so-called biogenic amine hypothesis of affective illness, expounded upon by Schildkraut,[6] Maas,[7] and others in the late 1960s and 1970s.

Beginning in the early 1960s, research from the laboratories of Axelrod and others showed that imipramine-like drugs directly blocked uptake of norepinephrine[8,9] and serotonin[10] into neurons. These results thus implicated these two biogenic amines in the pathophysiology of depression and in the mode of action of antidepressants. Additionally, the work from Axelrod's group and others explained the results of Sigg,[2] reported a few years earlier: by blocking transport of norepinephrine into the nerve ending, imipramine acutely potentiated the effects of norepinephrine. By the late 1960s, the development of synaptosomal uptake methods[11] markedly facilitated the study of the transport-blocking effects of drugs.[12–14] Synaptosomes are pinched-off nerve terminals that contain the transporters for the neurotransmitters.

Also, consistent with the effects of imipramine-like drugs on transport of norepinephrine into nerve endings, is a now-classical pharmacodynamic drug interaction between the antihypertensive drug guanethidine and the tricyclic antidepressant desipramine.[15] Desipramine, a very potent antagonist of the norepinephrine transporter, blocks the antihypertensive effects of guanethidine by blocking its entry into the nerve endings, the site of action of this antihypertensive compound. Similar results have been found for some antipsychotic drugs, particularly, chlorpromazine.[16] Also, in the decade of the 1960s, it was found that tricyclic antidepressants blocked α_1-adrenoceptors,[17] which may explain their serious cardiovascular side effects of postural hypotension.[18]

The 1970s and 1980s

In the early 1970s, Prange and colleagues put forth the "permissive hypothesis of affective illness" to accommodate results in the literature on both norepinephrine and serotonin, and to explain their results of a carefully controlled trial that showed an antimanic effect of L-tryptophan.[19] Thus, low serotonin permits the expression of the affective state governed by the level of norepinephrine. In the presence of low serotonin, low norepinephrine results in depression, while high norepinephrine results in mania. Correcting the low serotonin level corrects the affective disease.

This idea relates to a "bi-dimensional" model of affective illness. This model was also developed to explain results of clinical studies showing lack of differential efficacy of drugs selective for uptake of norepinephrine vs. serotonin.[20] In this model there are two mood centers that influence one another. Each is also independently influenced pharmacologically.

Also, in the early to mid 1970s, Sulser and coworkers[21] resolved a question that researchers had long asked: if blockade of uptake of norepinephrine and serotonin at the synapse occurs very rapidly (within minutes after administration of drug), then why does the lifting of the depression take many days to weeks? What Sulser and colleagues found is that with long-term (4 to 8 weeks), but not short-term (1 to 2 weeks), treatment of laboratory animals with antidepressants, there is a loss of sensitivity (desensitization) of catecholamine receptors. More specifically, the assay involved the measurement of norepinephrine-stimulated cyclic adenosine monophosphate (cAMP) synthesis in tissue slices from the rat limbic forebrain. The sensitivity to norepinephrine was decreased only after several weeks of treatment with antidepressant compounds, as well as with electroconvulsive treatment of the laboratory animals. The time course for this desensitization to develop was more consistent with the time course for the onset of therapeutic activity of antidepressants in patients. Thus, these findings were a plausible explanation for the time lag for onset of activity of antidepressants in patients.

Although Sulser and colleagues could not clearly identify the noradrenergic receptor involved in the second messenger synthesis, their results were followed by the findings that β-adrenoceptor down-regulation, i.e. the loss of binding sites for this receptor, accompanies this desensitization.[22] The β-adrenoceptor that is down-regulated by antidepressants is the β_1-subtype.[23,24] Additionally, preclinical studies demonstrated that down-regulation of β-adrenoceptors requires intact serotonergic axons,[25] clearly linking the noradrenergic with the serotonergic systems. Either desensitization alone or in combination with down-regulation has been shown to occur with most antidepressants in preclinical studies.[26–28] Interestingly, the antidepressant venlafaxine, which blocks both the norepinephrine and the serotonin transporters, does not by itself cause desensitization of NE-stimulated cAMP formation.[29] Altogether the data suggest that elevating synaptic levels of norepinephrine by an antidepressant is not necessarily sufficient to cause either down-regulation of β-adrenoceptors or desensitization of norepinephrine-stimulated cAMP synthesis.[26,27,30]

Nonetheless, this research not only provided a possible explanation for the time lag to onset of antidepressant activity, it also changed the focus from the presynaptic (transport blockade) to the postsynaptic side of the synapse. However, desensitization of presynaptic α_2-adrenoceptors also occurs with chronic treatment of laboratory animals with antidepressants.[31,32] In the brain, presynaptic α_2-adrenoceptors can exist as both autoreceptors on noradrenergic neurons and heteroreceptors on serotonergic nerve endings. These autoreceptors inhibit the release of norepinephrine, while the heteroreceptors inhibit the release of serotonin. Thus, with desensitization of these receptors, there is an increase in synaptic levels of both norepinephrine and serotonin.

In other studies, chronic treatment with antidepressants has consistently decreased the number of binding sites for a subtype of serotonin receptors called 5-HT_{2A}. These changes were found in rat cerebral cortex as measured in binding studies with [^{3}H]spiperone.[33] However, neuroleptics also cause 5-HT_{2A}-receptor down-regulation.[34] Additionally, another consistent finding, predicted by Sigg[2], is the sensitization accompanied by up-regulation of α_1-adrenoceptors with chronic treatment with antidepressants.[35,36] Sensitization was demonstrated in behavioral studies, as well as in electrophysiological studies. Depression, then, is thought to be due to up-regulation and down-regulation of certain receptors for neurotransmitters, particularly those for norepinephrine.

Supporting the modern day permissive hypothesis of Prange, is the clinical study of fluvoxamine, a selective serotonin transport blocking antidepressant, and oxaprotiline, a selective norepinephrine transport blocking antidepressant.[20] In this study, depressed patients were treated with one or the other drug. Each group of patients had similar response rates. Those that did not respond to one drug were switched to the other – in each case only about 20% of the nonresponders to the first drug responded to the second drug. On the other hand, 90% of the responders to the first drug remained responders. Although the numbers of subjects was small ($n = 24$), these data do not support the idea that there are two biochemical subtypes of depression; however, they do suggest that either the noradrenergic or the serotonergic pathway can be involved in the alleviation of depression.

Consistent with these data[20] are the more recent studies. These reports involve depletion of L-tryptophan, the precursor of serotonin, or inhibition of tyrosine hydroxylase, the rate-limiting step in the synthesis of catecholamines, and were studied in patients treated for depression with either a drug that blocks mainly the serotonin transporter or one that blocks mainly the norepinephrine transporter.[37]

Depletion of L-tryptophan is achieved by having subjects drink a tryptophan-free cocktail of 15 other amino acids, which flood transport sites into the brain and block entry of any endogenous tryptophan into the brain. The data from these types of clinical studies show that the antidepressant effect of the drug can be reversed with depletion of L-tryptophan only if the patient was treated with a serotonin transport blocking drug. Alternately, reversal of the antidepressant effects of the norepinephrine transport blocking drug occurs only with the inhibitor of catecholamine biosynthesis. Thus, for example, α-methyl-p-tyrosine, an inhibitor of tyrosine hydroxylase, caused an increase in depressive symptoms, including depressed mood, decreased concentration, anhedonia, loss of interest, and feelings of worthlessness, helplessness, and hopelessness in the patient group treated with a norepinephrine transport blocker, but not in the patient group treated with a serotonin transport blocker.[37] These data again suggest that antidepressants do not work by a single monoaminergic mechanism. Additionally, these results suggest that a drug that elevates synaptic levels of both norepinephrine and serotonin (a dual-action antidepressant) would have superior efficacy in treating depression.

Is it true, in fact, that dual- or multi-action drugs, such as clomipramine, duloxetine, venlafaxine, and mirtazapine show improved efficacy over single-action drugs? There are data[38–40] to suggest that these compounds either have superior efficacy in treating depressed patients or have a more rapid onset of activity, compared to single-action compounds. Clomipramine, duloxetine, and high-dose venlafaxine will block transport of both serotonin and norepinephrine. Mirtazapine has no effects on transporters, but has actions on some presynaptic receptors (α_2-adrenoceptors) that can increase the activity of raphe serotonergic neurons and some postsynaptic receptors (5-HT_{2A}) that might be important for therapeutic effects.

From the locus coeruleus, neurons send fibers to cell bodies of raphe neurons, where they release norepinephrine to activate these cells through α_1-adrenoceptors. Drugs that block the transport of norepinephrine or the noradrenergic presynaptic α_2-adrenoceptors, which inhibit the release of norepinephrine, will increase synaptic levels of this neurotransmitter, increasing the activity of the raphe serotonergic neurons. Additionally, on raphe nerve endings in terminal fields, there are presynaptic α_2-adrenoceptors, which when activated by norepinephrine, inhibit the release of serotonin. Thus, mirtazapine, by enhancing release of norepinephrine, can enhance release of serotonin while drugs like clomipramine, duloxetine, and venlafaxine, by blocking transport of both norepinephrine and serotonin, can do so as well.

The work of Sulser and colleagues, as well as other groups, showed that chronic treatment of laboratory animals with antidepressants caused adaptation (desensitization) of brain neurotransmitter receptors. This work produced the hypothesis that supersensitivity of catecholamine receptors in the presence of low levels of serotonin is the biochemical basis of depression. Radioligand binding studies on postmortem brain tissue suggest that catecholamine receptors (both β- and α_2-adrenoceptors) are increased in depressed patients or in individuals who committed suicide.[41–45] However, with respect to receptor sensitivity changes, this thesis lacks evidence in support from studies with live patients, despite many years of experimentation. To the contrary, clinical studies suggest that there are no such changes in receptor sensitivity. For example, Heninger *et al.*[46] reported a study of plasma 3-methoxy-4-hydroxyphenylglycol (MHPG), behavioral, and cardiovascular responses to yohimbine, a potent and reasonably selective α_2-adrenoceptor antagonist, in depressed patients and controls. They found no support for the hypothesis for altered α_2-adrenoceptor sensitivity in depression. Although studies with an agonist of these receptors may be more relevant, such a study with the agonist clonidine was also negative.[47] Additionally, a study of cardiac β-adrenoceptors in depressed patients showed hyposensitivity.[48] It will likely continue to be difficult to validate the hypothesis of receptor sensitivity changes in depression. Evidence may come from imaging studies of brain receptors by positron emission tomography (PET) in depressed patients before and after treatment with medication or with electroconvulsive therapy. However, such PET studies focusing on serotonin receptors have thus far yielded conflicting data.[49–53]

Norepinephrine transport blocking effects of psychotropic drugs

Psychiatric drugs of various classes have long been known to block the norepinephrine transporter. However, only over the past ten years, through the molecular cloning of the human norepinephrine transporter, have we been able to obtain good data on the transport blocking effects of drugs at the human transporter level.[54] This has been due to the difficulty of using radioligand binding techniques with human brain tissue obtained at autopsy to study the norepinephrine transporter. There is absolutely no difficulty studying the rat norepinephrine transporter or the human serotonin and dopamine transporters ex vivo. The exact reasons for difficulty with the human norepinephrine transporter are not fully understood.

However, it may relate to the low density of these transporters in human brain. Thus, there are no published studies with homogenates of human brain tissue measuring binding to the norepinephrine transporter. Likewise, there are no published PET studies measuring binding to this transporter. However, there are a few published autoradiographic studies of the human norepinephrine transporter identified with [^{3}H]nisoxetine.[55,56] Interestingly, there was a statistically significant (32%) reduction in binding to the norepinephrine transporter in the midcaudal locus coeruleus in a postmortem study of the brains from depressed patients (n = 15) compared to those of controls (n = 12).[55] There was no difference between depressed patients and controls in the rostral locus coeruleus. All but two of the depressed patients had committed suicide by various means. The controls died of natural causes and only two had a past history of psychiatric illness. The authors speculate that this reduction in norepinephrine transporter binding is a compensation for low levels of norepinephrine in the synapse.[55]

Several years ago, our laboratory established a binding assay for the human norepinephrine transporter with the use of human brain tissue obtained at autopsy and the radioligand [^{3}H]desipramine, a tricyclic antidepressant and a potent, selective blocker of the norepinephrine transporter. However, specific binding was low and only detectable in the cerebellum. Our equilibrium dissociation constants for the several compounds tested were very similar to those obtained with the human molecularly cloned transporter. This data is yet unpublished.

Over many years we have published several studies on the effects of antidepressants at the norepinephrine transporter, as well as at the serotonin and dopamine transporters. Early work involved rat synaptosomal transport studies,[12,14,57] while more recent experimentation has involved binding studies with the molecularly cloned human transporters.[57,58] Additionally, we have also obtained binding data for antipsychotic drugs,[59] as well as other classes of compounds at these transporters.[58]

In a group of 19 compounds (17 antidepressants and 2 metabolites of antidepressants), it is interesting to compare their K_i (inhibitor constant) values at rat synaptosomal transport versus their K_d (equilibrium dissociation constant) values at the human norepinephrine transporter (Figure 17.1). Despite the differences in the methods and the differences in the species, there is an excellent correlation for these two variables for this group of antidepressant drugs and the two metabolites of antidepressants (desmethylsertraline and norfluoxetine). The equation for the regression line (Figure 17.1) is $Y = 1.04X + 0.290$, $R = 0.96$, $p < 0.001$. However, although this is an excellent correlation with a slope nearly equal to 1, the line of identity (Figure 17.1, dotted line) suggests that most of these compounds are weaker at the human norepinephrine transporter than at the rat norepinephrine

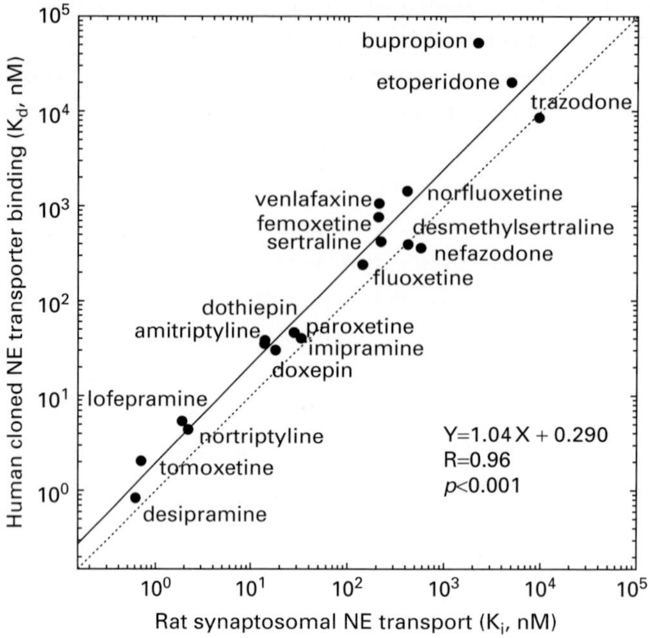

Figure 17.1 Antidepressant blockade of the human norepinephrine (NE) transporter: correlation with the blockade of rat synaptosomal uptake of [^{3}H]norepinephrine.

transporter. These results are consistent with our other studies[57] and those of others.[60] Thus, for example, venlafaxine, which is in the group of relatively potent compounds at the rat transporter, is fivefold weaker at the human norepinephrine transporter, making it a relatively weak norepinephrine transport blocker at this human transporter.

Consistent with these in vitro data for venlafaxine are clinical data showing that high-dose, but not low-dose, venlafaxine blocks the depressor effects of tyramine in humans.[61] Tyramine will elevate blood pressure by displacing stores of norepinephrine at noradrenergic nerve endings in the peripheral nervous system. In order for tyramine to cause this effect, it must be transported into the nerve endings by the norepinephrine transporter. Any drug that blocks this transporter will block the effects of tyramine.

Data for human norepinephrine transporter binding for antidepressants, antipsychotics, and some miscellaneous psychotropic compounds are presented in Tables 17.1, 17.2, and 17.3, respectively. The most potent compounds, of course, are antidepressants. However, combining the data for Tables 17.1 and 17.2 shows that, of the top 25 of the 75 antidepressants and antipsychotic drugs, 7 are antipsychotic drugs (listed in decreasing order of potency): chlorpromazine, zotepine, chlorprothixene, promazine, ziprasidone, triflupromazine, and fluperlapine. Therefore,

Table 17.1 Antidepressant binding at the molecularly cloned human norepinephrine transporter.[a]

Compound	K_d (Geometric mean * SEM, nM)	
desipramine	0.83	*0.05
protriptyline	1.41	*0.02
atomoxetine	2.03	*0.06
duloxetine	4.2	*0.4
nortriptyline	4.4	*0.1
oxaprotiline	4.9	*0.2
lofepramine	5.4	*0.4
reboxetine	7.2	*0.8
maprotiline	11.1	*0.3
nomifensine	15.7	*0.5
amoxapine	16.0	*0.3
doxepin	30	*1
amitriptyline	35	*2
imipramine	38	*2
clomipramine	38	*1
paroxetine	40	*2
dothiepin	46	*1
mianserin	71	*2
milnacipran	80	*10
viloxazine	155	*8
fluoxetine	240	*10
nefazodone	360	*40
desmethylsertraline[b]	393	*13
sertraline	420	*20
femoxetine	760	*40
venlafaxine	1060	*40
iprindole	1263	*7
fluvoxamine	1300	*30
norfluoxetine[b]	1427	*9
desmethylcitalopram[b]	1820	*50
trimipramine	2400	*30
citalopram	4070	*80
mirtazepine	4700	*300
butriptyline	5100	*400
tranylcypromine	5900	*200
trazodone	8500	*300
zimelidine	9400	*100
etoperidone	20 000	*2000
phenelzine	50 000	*5000
bupropion	52 000	*1000
iproniazid	>100 000	

[a] Data from ref. 58.

[b] Metabolite of an antidepressant.

Table 17.2 Antipsychotic binding at the molecularly cloned human norepinephrine transporter.[a]

Compound	K_d (Geometric mean * SEM, nM)	
chlorpromazine	19	*2
zotepine	20	*3
chlorprothixene	21	*1
promazine	25	*2
ziprasidone	59	*7
triflupromazine	110	*4
fluperlapine	128	*3
pimozide	370	*10
loxapine	380	*30
prochlorperazine	510	*10
sertindole	640	*40
perphenazine	740	*20
perazine	1540	*80
thioridazine	1620	*70
tiosperone	1800	*300
clozapine	2700	*300
trifluoperazine	3000	*100
clocapramine	3200	*200
oxypertine	4000	*200
olanzapine	4500	*100
fluphenazine	4600	*100
carpipramine	5100	*200
haloperidol	6500	*300
moperone	7690	*60
bromperidol	8600	*400
levomepromazine	13 500	*500
quetiapine	15 400	*800
timiperone	26 000	*2000
spiperone	27 000	*1000
cis-thiothixene	30 000	*2000
mesoridazine	38 400	*800
melperone	94 000	*2000
risperidone	98 000	*400
tiapride	≥100 000	
sultopride	≥100 000	
sulpiride	≥100 000	
molindone	≥100 000	

[a] Data from ref. 59.

Table 17.3 Binding of some miscellaneous compounds at the molecularly cloned human norepinephrine transporter.[a]

Compound	K_d (Geometric mean * SEM, nM)	
methylphenidate	234	*9
GBR 12935	310	*10
(+S)-amphetamine	530	*40
(+S)-methamphetamine	660	*20
diphenhydramine	960	*30
cocaine	1420	*50
chlorpheniramine	1440	*50
reserpine	15 300	*900
verapamil	48 000	*1000
valproic acid	>100 000	
nifedipine	>100 000	
liothyronine	>100 000	
esterone	>100 000	
lithium carbonate	>100 000	
alprazolam	>100 000	
clonazepam	>100 000	
carbamazepine	>100 000	
acetazolamide	>100 000	

[a] Data from ref. 58.

some antipsychotic drugs are more potent than many antidepressants at blocking the human norepinephrine transporter.

α-adrenoceptor blocking effects of antidepressants and antipsychotic drugs

Antidepressants and antipsychotic drugs also affect noradrenergic neurotransmission by blocking α-adrenoceptors, both α_1 (Tables 17.4 and 17.5) and α_2 (Tables 17.6 and 17.7). Relative to the reference compound, prazosin (an antihypertensive drug), the most potent antidepressants at α_1-adrenoceptors are some of the tricyclic antidepressants (Table 17.4). These agents are about one-hundredth as potent as the reference compound. However, for the antipsychotic drugs (Table 17.5), iloperidone is remarkably essentially equipotent to prazosin. Likewise, for the α_2-adrenoceptor, the reference compound, rauwolscine (Table 17.6) is more potent than any antidepressant by several fold. However, again, iloperidone (Table 17.7) is equipotent to the reference α_2-adrenoceptor compound.

Table 17.4 Antidepressant binding at the human brain α_1-adrenoceptor.[a]

Compound	K_d (Geometric mean * SEM, nM)	
doxepin	23	*1
amitriptyline	24	*2
trimipramine	24	*2
imipramine	32	*5
mianserin	34	*7
clomipramine	38	*6
etoperidone	39	*1
trazodone	42	*3
amoxapine	50	*3
nortriptyline	55	*2
maprotiline	90	*20
desipramine	100	*10
lofepramine	100	*3
protriptyline	130	*40
sertraline	380	*50
dothiepin	470	*20
butriptyline	600	*100
oxaprotiline	620	*80
mirtazapine	650	*30
femoxetine	650	*30
nomifensine	850	*80
desmethylsertraline[b]	1200	*100
norfluoxetine[b]	1300	*300
zirnelidine	1500	*50
clovoxamine	2200	*20
iprindole	2300	*400
fluoxetine	3001	*10
paroxetine	3500	*200
atomoxetine	3800	*200
bupropion	4300	*300
nisoxetine	5300	*500
fluvoxamine	8000	*1000
reboxetine	12 000	*800
viloxazine	14 000	*2000
milnacipran	15 000	*1000
venlafaxine	>35 000	
prazosin[c]	0.28	*0.01

[a] Data from refs. 63 and 64.

[b] Metabolite of an antidepressant.

[c] Reference compound.

Table 17.5 Antipsychotic binding at the human brain α_1-adrenoceptor.[a]

Compound	K_d (Geometric mean * SEM, nM)	
iloperidone	0.31	*0.02
ORG 5222	1.07	*0.06
spiperone	1.2	*0.2
mesoridazine	2.0	*0.5
chlorpromazine	2.6	*0.3
ziprasidone	2.6	*0.3
risperidone	2.7	*0.3
sertindole	3.9	*0.2
thioridazine	5	*1
promazine	6	*2
clozapine	7	*1
zotepine	7.3	*0.3
quetiapine	8	*1
fluphenazine	9	*2
perphenazine	10	*2
thiothixene	11	*0.9
haloperidol	17	*1
prochlorperazine	24	*7
trifluoperazine	24	*3
loxapine	28	*6
olanzapine	44	*4
d-butaclamol	56	*8
pimozide	76	*5
melperone	180	*20
molindone	2500	*600

[a] Data from refs. 65 and 66.

Clinical relevance of noradrenergic transporter and receptor blocking effects of antidepressants and antipsychotic drugs

Table 17.8 summarizes some possible clinical consequences of blocking transporters and receptors by antidepressants and antipsychotic drugs. While blocking the norepinephrine transporter may lead to a therapeutic effect in treating depression, it is not known what psychiatric benefit accrues from blocking either of the α-adrenoceptors. Nonetheless, there is a notion in the literature that a mechanism by which to explain the atypicality of some antipsychotic drugs is by the ability to block potently α_2-adrenoceptors.[62] In general, an atypical neuroleptic is a drug that has

Table 17.6 Antidepressant binding at the human brain α_2-adrenoceptor.[a]

Compound	K_d (Geometric mean * SEM, nM)	
mianserin	73	*6
mirtazapine	140	*20
trazodone	320	*10
etoperidone	570	*10
nefazodone	640	*30
amitriptyline	690	*20
trimipramine	700	*100
zimelidine	790	*40
doxepin	1270	*40
femoxetine	1970	*60
nortriptyline	2030	*30
dothiepin	2400	*500
amoxapine	2600	*200
lofepramine	2700	*100
imipramine	3100	*100
clomipramine	3200	*300
sertraline	4100	*200
butriptyline	4800	*900
desipramine	5500	*200
nomifensine	6500	*700
protriptyline	6600	*400
nisoxetine	6700	*600
desmethylsertraline[b]	7800	*200
iprindole	8600	*100
atomoxetine	8800	*100
maprotiline	9400	*100
clovoxamine	11 000	*1000
fluoxetine	13 900	*200
fluvoxamine	15 000	*4000
paroxetine	17 000	*400
norfluoxetine[b]	18 900	*700
oxaprotiline	42 000	*7000
viloxazine	44 000	*7000
bupropion	81 000	*8000
adinazolam	>35 000	
venlafaxine	>35 000	
rauwolscine[c]	2.5	*0.1

[a] Data from refs. 63 and 64.

[b] Metabolite of an antidepressant.

[c] Reference compound.

Table 17.7 Antipsychotic binding at the human brain α_2-adrenoceptor.[a]

Compound	K_d (Geometric mean * SEM, nM)	
iloperidone	3.0	*0.2
risperidone	8	*1
clozapine	15	*1
ORG 5222	16	*1
quetiapine	80	*10
melperone	150	*20
ziprasidone	154	*9
zotepine	180	*8
sertindole	190	*30
thiothixene	200	*20
olanzapine	280	*30
d-butaclamol	310	*40
perphenazine	510	*20
molindone	600	*100
haloperidol	600	*100
pimozide	650	*40
spiperone	660	*20
chlorpromazine	750	*50
thioridazine	800	*100
promazine	900	*100
fluphenazine	1550	*20
mesoridazine	1600	*120
prochlorperazine	1700	*100
loxapine	2400	*600
trifluoperazine	2600	*200

[a] Data from refs. 65 and 66.

antipsychotic efficacy and low propensity to cause extrapyramidal side effects (e.g., parkinsonism) with short-term treatment, and a low propensity to cause tardive dyskinesia (an involuntary movement disorder, largely involving muscles of the face) with long-term treatment. Alternately, blocking α-adrenoceptors can explain some adverse effects and certain drug interactions of these compounds (Table 17.8). Specifically, α_1-adrenoceptor blockade by antidepressants and antipsychotic drugs may be responsible for orthostatic hypotension, the most serious common cardiovascular effect of these drugs. This side effect can cause dizziness and a reflex tachycardia. In addition, this property of these drugs will result in the potentiation

Table 17.8 Possible therapeutic and adverse effects of receptor blockade by antipsychotic drugs.

	Norepinephrine transporter	α_1-Adrenoceptors	α_2-Adrenoceptors
Possible therapeutic effects	Antidepressant action	Unknown	Unknown
Possible adverse effects	Tremors Tachycardia Elevated blood pressure and heart rate Augmentation of pressor effects of sympathomimetic amines	Potentiation of the antihypertensive effect of prazosin, terazosin, doxazosin, and labetalol Postural hypotension, dizziness Reflex tachycardia	Blockade of the antihypertensive effects of clonidine hydrochloride, guanabenz acetate, and methyldopa

of the antihypertensive drugs (e.g., prazosin). The blockade of α_2-adrenoceptors can result in the blockade of the effects of other types of antihypertensive drugs (e.g., clonidine).

REFERENCES

1. Kuhn, R. The treatment of depressive states with G 22355 (imipramine hydrochloride). *Am. J. Psychiatry*, **115** (1958), 459–464.
2. Sigg, E. B. Pharmacological studies with tofranil. *Can. Psychiatr. Assoc. J.*, **4**: Suppl. (1959), 75–85.
3. Sigg, E. B., Gyermek, L. and Hill, R. T. Antagonism to reserpine induced depression by imipramine, related psychoactive drugs, and some autonomic drugs. *Psychopharmacologia*, **7** (1965), 144–149.
4. Sen, G. and Bose, K. C. Rawolfia serpentina – a new Indian drug for insanity and high blood-pressure. *Indian Med. World*, **11** (1931), 194–201.
5. Goodwin, F. K. and Bunney, W. E., Jr. Depressions following reserpine: a reevaluation. *Semin. Psychiatry*, **3** (1971), 435–448.
6. Schildkraut, J. J. Neuropsychopharmacology and the affective disorders. (First of three parts). *New Engl. J. Med.*, **281** (1969), 302–308.
7. Maas, J. W. Biogenic amines and depression. Biochemical and pharmacological separation of two types of depression. *Arch. Gen. Psychiatry*, **32** (1975), 1357–1361.

8. Axelrod, J., Whitby, L. F. and Hertting, G. Effect of psychotropic drugs on the uptake of H3-norepinephrine by tissues. *Science*, **133** (1961), 383–384.
9. Dengler, H. G., Spiegel, H. E. and Titus, Z. O. Effect of drugs on uptake of isotopic norepinephrine by cat tissues. *Nature*, **191** (1961), 816–817.
10. Alpers, H. S. and Himwich, H. E. An in vitro study of the effects of tricyclic antidepressant drugs on the accumulation of C14-serotonin by rabbit brain. *Biol. Psychiatry*, **1** (1969), 81–85.
11. Coyle, J. T. and Snyder, S. H. Catecholamine uptake by synaptosomes in homogenates of rat brain: stereospecificity in different areas. *J. Pharmacol. Exp. Ther.*, **170** (1969), 221–231.
12. Bolden-Watson, C. and Richelson, E. Blockade by newly-developed antidepressants of biogenic amine uptake into rat brain synaptosomes. *Life Sci.*, **52** (1993), 1023–1029.
13. Horn, A. S., Coyle, J. T. and Snyder, S. H. Catecholamine uptake by synaptosomes from rat brain. Structure-activity relationships of drugs with differential effects on dopamine and norepinephrine neurons. *Mol. Pharmacol.*, **7** (1971), 66–80.
14. Richelson, E. and Pfenning, M. Blockade by antidepressants and related compounds of biogenic amine uptake into rat brain synaptosomes: most antidepressants selectively block norepinephrine uptake. *Eur. J. Pharmacol.*, **104** (1984), 277–286.
15. Mitchell, J. R., Arias, L. and Oates, J. A. Antagonism of the antihypertensive action of guanethidine sulfate by desipramine hydrochloride. *JAMA*, **202** (1967), 973–976.
16. Janowsky, D. S., el-Yousef, M. K., Davis, J. M. and Fann, W. E. Antagonism of guanethidine by chlorpromazine. *Am. J. Psychiatry*, **130** (1973), 808–812.
17. Turker, R. K. and Khairallah, P. A. Demethylimipramine (desipramine), an alpha-adrenergic blocking agent. *Experientia*, **23** (1967), 252.
18. Glassman, A. H. and Bigger, J. T., Jr. Cardiovascular effects of therapeutic doses of tricyclic antidepressants. A review. *Arch. Gen. Psychiatry*, **38** (1981), 815–820.
19. Prange, A. J., Jr. , Wilson, I. C., Lynn, C. W., Alltop, L. B. and Stikeleather, R. A. L-tryptophan in mania. Contribution to a permissive hypothesis of affective disorders. *Arch. Gen. Psychiatry*, **30** (1974), 56–62.
20. Emrich, H. M., Berger, M., Riemann, D. and von Zerssen, D. Serotonin reuptake inhibition vs. norepinephrine reuptake inhibition: a double-blind differential-therapeutic study with fluvoxamine and oxaprotiline in endogenous and neurotic depressives. *Pharmacopsychiatry*, **20** (187), 60–63.
21. Vetulani, J., Stawarz, R. J., Dingell, J. V. and Sulser, F. A possible common mechanism of action of antidepressant treatments: reduction in the sensitivity of the noradrenergic cyclic AMP generating system in the rat limbic forebrain. *Naunyn-Schmiedeberg's Arch. Pharmacol.*, **293** (1976), 109–114.
22. Banerjee, S. P., Kung, L. S., Riggi, S. J. and Chanda, S. K. Development of beta-adrenergic receptor subsensitivity by antidepressants. *Nature*, **268** (1977), 455–456.
23. Ordway, G. A., Gambarana, C. and Frazer, A. Quantitative autoradiography of central beta adrenoceptor subtypes: comparison of the effects of chronic treatment with desipramine or centrally administered I-isoproterenol. *J. Pharmacol. Exp. Ther.*, **247** (1988), 379–389.

24. Ordway, G. A., Gambarana, C., Tejani-Butt, S. M. *et al.* Preferential reduction of binding of 125I-iodopindolol to beta-1 adrenoceptors in the amygdala of rat after antidepressant treatments. *J. Pharmacol. Exp. Ther.*, **257** (1991), 681–690.
25. Brunello, N., Barbaccia, M. L., Chuang, D. M. and Costa, E. Down-regulation of beta-adrenergic receptors following repeated injections of desmethylimipramine: permissive role of serotonergic axons. *Neuropharmacology*, **21** (1982), 1145–1149.
26. Garcha, G., Smokcum, R. W., Stephenson, J. D. and Weeramanthri, T. B. Effects of some atypical antidepressants on beta-adrenoceptor binding and adenylate cyclase activity in the rat forebrain. *Eur. J. Pharmacol.*, **108** (1985), 1–7.
27. Kopanski, C., Türck, M. and Schultz, J. Effects of long-term treatment of rats with antidepressants on adrenergic-receptor sensitivity in cerebral cortex: structure activity study. *Neurochem. Int.*, **5** (1983), 649–659.
28. Sulser, F. Mode of action of antidepressant drugs. *J. Clin. Psychiatry*, **44** (1983), 14–20.
29. Nalepa, I., Manier, D. H., Gillespie, D. D. *et al.* Lack of beta adrenoceptor desensitization in brain following the dual noradrenaline and serotonin reuptake inhibitor venlafaxine. *Eur. Neuropsychopharmacol.*, **8** (1998), 227–232.
30. Mishra, R., Janowsky, A. and Sulser, F. Action of mianserin and zimelidine on the norepinephrine receptor coupled adenylate cyclase system in brain: subsensitivity without reduction in beta-adrenergic receptor binding. *Neuropharmacology*, **19** (1980), 983–987.
31. Crews, F. T. and Smith, C. B. Presynaptic alpha-receptor subsensitivity after long-term antidepressant treatment. *Science*, **202** (1978), 322–324.
32. Yoshioka, M., Matsumoto, M., Numazawa, R. *et al.* Changes in the regulation of 5-hydroxytryptamine release by alpha(2)-adrenoceptors in the rat hippocampus after long-term desipramine treatment. *Eur. J. Pharmacol.*, **294** (1995), 565–570.
33. Peroutka, S. J. and Snyder, S. H. Long-term antidepressant treatment decreases spiroperidol-labeled serotonin receptor binding. *Science*, **210** (1980), 88–90.
34. Andree, T. H., Mikuni, M., Tong, C. Y., Koenig, J. I. and Meltzer, H. Y. Differential effect of subchronic treatment with various neuroleptic agents on serotonin 2 receptors in rat cerebral cortex. *J. Neurochem.*, **46** (1986), 191–197.
35. Maj, J., Klimek, V. and Nowak, G. Antidepressant drugs given repeatedly increase binding to α_1-adrenoceptors in the rat cortex. *Eur. J. Pharmacol.*, **119** (1985), 113–116.
36. Menkes, D. B. and Aghajanian, G. K. Alpha 1-adrenoceptor-mediated responses in the lateral geniculate nucleus are enhanced by chronic antidepressant treatment. *Eur. J. Pharmacol.*, **74** (1981), 27–35.
37. Miller, H. L., Delgado, P. L., Salomon, R. M. *et al.* Clinical and biochemical effects of catecholamine depletion on antidepressant-induced remission of depression. *Arch. Gen. Psychiatry*, **53** (1996), 117–128.
38. Anderson, I. M. Selective serotonin reuptake inhibitors versus tricyclic antidepressants: a meta-analysis of efficacy and tolerability. *J. Affect. Disord.*, **58** (2000), 19–36.
39. Thase, M. E., Entsuah, A. R. and Rudolph, R. L. Remission rates during treatment with venlafaxine or selective serotonin reuptake inhibitors. *Br. J. Psychiatry*, **178** (2001), 234–241.
40. Vestergaard, P., Gram, L. F., Kragh-Sorensen, P. *et al.* Therapeutic potentials of recently introduced antidepressants. Danish University Antidepressant Group. *Psychopharmacol. Ser.*, **10** (1993), 190–198.

41. Arango, V., Underwood, M. D. and Mann, J. J. Alterations in monoamine receptors in the brain of suicide victims. *J. Clin. Psychopharmacol.*, **12** (1992), 8S–12S.
42. De Paermentier, F., Mauger, J. M., Lowther, S. *et al.* Brain α-adrenoceptors in depressed suicides. *Brain Res.*, **757** (1997), 60–68.
43. Mann, J. J., Stanley, M., McBride, P. A. and McEwen, B. S. Increased serotonin 2 and beta-adrenergic receptor binding in the frontal cortices of suicide victims. *Arch. Gen. Psychiatry*, **43** (1986), 954–959.
44. Meana, J. J. and Garcia-Sevilla, J. A. Increased α_2-adrenoceptor density in the frontal cortex of depressed suicide victims. *J. Neural Transm.*, **70** (1987), 377–381.
45. Ordway, G. A., Widdowson, P. S., Smith, K. S. and Halaris, A. Agonist binding to alpha 2-adrenoceptors is elevated in the locus coeruleus from victims of suicide. *J. Neurochem.*, **63** (1994), 617–624.
46. Heninger, G. R., Charney, D. S. and Price, L. H. alpha 2-Adrenergic receptor sensitivity in depression. The plasma M. H. P. G, behavioral, and cardiovascular responses to yohimbine. *Arch. Gen. Psychiatry*, **45** (1988), 718–726.
47. Gann, H., Riemann, D., Stoll, S., Berger, M. and Muller, W. E. Growth hormone response to growth hormone-releasing hormone and clonidine in depression. *Biol. Psychiatry*, **38** (1995), 325–329.
48. Bertschy, G., Vandel, S., Puech, A. *et al.* Cardiac beta-adrenergic sensitivity in depression: relation with endogenous subtype and desipramine response. *Neuropsychobiology*, **21** (1989), 177–181.
49. Attar-Lévy, D., Martinot, J. L., Blin, J. *et al.* The cortical serotonin(2) receptors studied with positron-emission tomography and [F-18]-setoperone during depressive illness and antidepressant treatment with clomipramine. *Biol. Psychiatry*, **45** (1999), 180–186.
50. Biver, F., Wikler, D., Lotstra, F. *et al.* Serotonin 5-HT2 receptor imaging in major depression: focal changes in orbito-insular cortex. *Br. J. Psychiatry*, **171** (1997), 444–448.
51. Meltzer, C. C., Price, J. C., Mathis, C. A. *et al.* PET imaging of serotonin type 2A receptors in late-life neuropsychiatric disorders. *Am. J. Psychiatry*, **156** (1999), 1871–1878.
52. Meyer, J. H., Kapur, S., Houle, S. *et al.* Prefrontal cortex 5-HT2 receptors in depression: an [F-18]setoperone PET imaging study. *Am. J. Psychiatry*, **156** (1999), 1029–1034.
53. Yatham, L. N., Liddle, P. F., Shiah, I. S. *et al.* Brain $serotonin_2$ receptors in major depression: a positron emission tomography study. *Arch. Gen. Psychiatry*, **57** (2000), 850–858.
54. Pacholczyk, T., Blakely, R. D. and Amara, S. G. Expression cloning of a cocaine- and antidepressant-sensitive human noradrenaline transporter. *Nature*, **350** (1991), 350–354.
55. Klimek, V., Stockmeier, C., Overholser, J. *et al.* Reduced levels of norepinephrine transporters in the locus coeruleus in major depression. *J. Neurosci.*, **17** (1997), 8451–8458.
56. Ordway, G. A., Stockmeier, C. A., Cason, G. W. and Klimek, V. Pharmacology and distribution of norepinephrine transporters in the human locus coeruleus and raphe nuclei. *J. Neurosci.*, **17** (1997), 1710–1719.
57. Carlier, P. R., Lo, M. M. C., Lo, P. C. K. *et al.* Synthesis of a potent wide-spectrum serotonin-, norepinephrine-, dopamine-reuptake inhibitor (SNDRI) and a species-selective dopamine-reuptake inhibitor based on the gamma-amino alcohol functional group. *Bioorg. Med. Chem. Lett.*, **8** (1998), 487–492.

58. Tatsumi, M., Groshan, K., Blakely, R. D. and Richelson, E. Pharmacological profile of antidepressants and related compounds at human monoamine transporters. *Eur. J. Pharmacol.*, **340** (1997), 249–258.
59. Tatsumi, M., Jansen, K., Blakely, R. D. and Richelson, E. Pharmacological profile of neuroleptics at human monoamine transporters. *Eur. J. Pharmacol.*, **368** (1999), 277–283.
60. Paczkowski, F. A., Bryan-Lluka, L. J., Porzgen, P., Bruss, M. and Bonisch, H. Comparison of the pharmacological properties of cloned rat, human, and bovine norepinephrine transporters. *J. Pharmacol. Exp. Ther.*, **290** (1999), 761–767.
61. Harvey, A. T., Rudolph, R. L. and Preskorn, S. H. Evidence of the dual mechanisms of action of venlafaxine. *Arch. Gen. Psychiatry*, **57** (2000), 503–509.
62. Nutt, D. J. Putting the 'A' in atypical: does alpha(2)-adrenoceptor antagonism account for the therapeutic advantage of new antipsychotics? *J. Psychopharmacol.*, **8** (1994), 193–195.
63. Cusack, B., Nelson, A. and Richelson, E. Binding of antidepressants to human brain receptors: focus on newer generation compounds. *Psychopharmacology*, **114** (1994), 559–565.
64. Richelson, E. and Nelson, A. Antagonism by antidepressants of neurotransmitter receptors of normal human brain in vitro. *J. Pharmacol. Exp. Ther.*, **230** (1984), 94–102.
65. Richelson, E. and Nelson, A. Antagonism by neuroleptics of neurotransmitter receptors of normal human brain in vitro. *Eur. J. Pharmacol.*, **103** (1984), 197–204.
66. Richelson, E. and Souder, T. Binding of antipsychotic drugs to human brain receptors focus on newer generation compounds. *Life Sci.*, **68** (2000), 29–39.

18

The clinical role of norepinephrine antidepressants in depression and anxiety disorders

J. Craig Nelson, M.D.
University of California San Francisco, USA

Introduction

Since Roland Kuhn's observation of the antidepressant action of imipramine in 1958, noradrenergic antidepressants have played an important role in the treatment of depression.[1] Because imipramine has a central place in the discovery of treatments for depression and anxiety, understanding its pharmacology is crucial. Although imipramine itself has affinity for the serotonin (5-HT) transporter, its metabolites, desipramine and hydroxy-desipramine, have greater affinity for the norepinephrine (NE) transporter.[2] During chronic oral administration of imipramine in humans, median desipramine levels are about twice those of the parent compound, and hydroxy-desipramine, a NE reuptake inhibitor similar in potency to desipramine, is present at levels about 40% of those of desipramine.[3,4] The net effect is that, during chronic administration of imipramine, about three-quarters of the total concentration of the drug is present as metabolites that act primarily on the noradrenergic system.

Imipramine also played an important role in the development of pharmacologic treatments of anxiety. Klein observed that patients with episodes of intense anxiety received benefit from imipramine that was not seen with phenothiazines or sedatives.[5] He distinguished this syndrome, which later became known as "panic disorder," from persistent anticipatory anxiety. Two decades later, Hoehn-Saric *et al.* demonstrated that imipramine was in fact effective in generalized anxiety as well, even if the response was not as dramatic as in panic disorder.[6] These studies suggested the potential value of antidepressants for treatment of anxiety disorders. Ironically, imipramine was never approved for use in these anxiety syndromes.

Brain Norepinephrine: Neurobiology and Therapeutics, ed. Gregory A. Ordway, Michael A. Schwartz and Alan Frazer. Published by Cambridge University Press.

Norepinephrine agents for treatment of depression

Subsequent to the marketing of imipramine, other tricyclic compounds were introduced. The demethylated or secondary tricyclic compounds primarily inhibit reuptake of NE and have little effect on 5-HT. This group of agents included nortriptyline (Aventil, Pamelor), desipramine (Norpramin), and protriptyline (Vivactil). In Europe, lofepramine, another noradrenergic tricyclic was marketed. Maprotiline, a tetracyclic compound, is also a NE reuptake inhibitor.

During the past decade two second generation NE reuptake inhibitors – tomoxetine and reboxetine – were developed and tested in depression. Unlike their predecessors, these agents had no anticholinergic or antihistaminic properties and, perhaps most important, did not delay cardiac conduction. Initial studies of tomoxetine in depression were plagued by high placebo response rates and the drug failed to separate from placebo.[7] Subsequently the efficacy of this agent was demonstrated in attention-deficit/hyperactivity disorder[8,9] and it was marketed in the USA as atomoxetine. Reboxetine was found to be efficacious for depression in European studies, but failed to show superiority to placebo in the USA and was not marketed here.

Mechanism of action

Knowledge of the beneficial antidepressant effects of noradrenergic tricyclics, coupled with the observation that reserpine, which depletes NE from presynaptic vesicles, appeared to induce depression suggested that NE might be depleted in depression. These observations provided the basis for the catecholamine hypothesis of depression.[10] The observation that concentrations of the principal metabolite of NE, 3-methoxy-4-hydroxy phenylglycol (MHPG), were low in the urine of some depressed patients,[11] suggested that MHPG might be a marker for "noradrenergic" depression and that low MHGP levels might predict response to noradrenergic antidepressants. While several initial uncontrolled studies seemed to support this idea, a subsequent controlled comparison failed to show that the association was sufficiently predictive to be of clinical value.[12] This study, which employed amitriptyline and imipramine, might be criticized for the use of two drugs with overlapping effects. However, a later study with desipramine and fluoxetine also failed to show that there were NE and 5-HT depressions that were selectively responsive to NE and 5-HT antidepressants.[13]

The idea that NE reuptake inhibitors simply increased NE was complicated by the observation that MHPG fell rapidly after administration of a NE reuptake inhibitor – within 24 hours. The timing of this effect was not consistent with the observed antidepressant action of these drugs which occurred over several weeks.

Subsequently, Sulser and Vetulani[14,15] demonstrated that down-regulation of β-adrenergic receptors occurred with chronic administration of an antidepressant, and that this effect, which was more temporally related to the timing of antidepressant effects, might be the primary change during treatment. Later data suggested that this was likely a compensatory effect,[16] yet for the next decade this concept was one of the principal hypotheses of the mechanism of antidepressant action.

Beginning in the late 1980s Delgado and colleagues[17,18] explored the effects of blocking the synthesis of 5-HT and NE. In the first study they demonstrated that rapid depletion of tryptophan, the precursor of 5-HT, led to return of depressive symptoms in patients who had responded to treatment. The relapse symptoms included those initially present during the depression. Subsequently, they noted that relapse following tryptophan depletion tended to occur in those patients who had been treated with a 5-HT reuptake inhibitor. Those treated with noradrenergic antidepressants, like desipramine, were not as likely to relapse. In later studies, they demonstrated that AMPT (α-methyl-p-tyrosine), which interferes with the synthesis of NE, also led to relapse,[19] but this time relapse occurred in those patients who responded to an NE drug. Those treated with a 5-HT agent were not likely to be affected. These studies demonstrated that each class of drugs is dependent on the presence of the respective neurotransmitter. These investigators were careful to note that while these studies did inform us about the mechanism of action of antidepressant drugs, the findings did not necessarily address the pathophysiology of depression. In fact, many drugs used for treatment of medical disorders have mechanisms of action that have nothing to do with the etiology of the syndrome for which they are used. For example, there are a variety of effective antihypertensive drugs that lower blood pressure, but their mechanisms of action may have nothing to do with the etiology of the elevated blood pressure.

Ironically, the amine depletion studies might be construed as suggesting that rather than correcting deficits in NE and 5-HT systems, in fact the action of antidepressants may be dependent on the presence of a competent neurotransmitter system. Recent pharmacogenomic findings might be consistent with this view. It has been shown that the selective 5-HT reuptake inhibitors (SSRIs) are most effective in patients with the "LL" or "SL" genotype in the promoter region of the 5-HT transporter.[20–22] In these patients, the 5-HT transporter functions to rapidly take up serotonin into the presynaptic cell. Patients homozygous for the "S" allele have ineffective 5-HT reuptake. If the transporter is already functioning poorly, an inhibitor of the transporter might have limited effects. In fact, patients homozygous for the "S" allele show a blunted response to the SSRIs. In other words, genetic polymorphisms that interfere with a normally functioning amine system may interfere with the action of a selective antidepressant drug.

Evidence for the use of noradrenergic agents in depression

Imipramine is one of the best-studied drugs in depression. In part this is the result of its inclusion as the standard comparator in many studies. Its efficacy is well established and has been reviewed previously.[23,24] In 44 placebo-controlled studies, imipramine was more effective than placebo in 30. Pooled data from these studies indicated that 65% of 1334 patients *completing* treatment with imipramine were substantially improved while 30% of those on placebo improved. The Agency for Health Care Policy and Research review found that *intention-to-treat* response rates for imipramine were 48% in 53 studies and 18% higher than placebo.[24] The other secondary tricyclics have not been studied as frequently but have also been demonstrated to be effective in placebo-controlled studies.[24,25] The efficacy of reboxetine in depression is supported by several international placebo-controlled studies.[26–28]

In addition to placebo-controlled data, support for the efficacy of the noradrenergic tricyclic compounds was provided by studies demonstrating a relationship between drug plasma levels and response. Such relationships were demonstrated for imipramine, desipramine, and nortriptyline and were previously reviewed.[29] Nortriptyline was especially well studied, in part because of the popularity of this compound and because it had an interesting therapeutic window, with high concentrations being less effective.[30–32]

The tricyclic compounds were also effective for maintenance treatment. Early studies, previously reviewed,[33] demonstrated that maintenance treatment with a tricyclic reduced the relapse rate associated with placebo by 50%. Many of these early studies, however, employed lower doses during maintenance treatment. Subsequently, the Pittsburgh group (Frank *et al.*) demonstrated that imipramine, given at the full dose used for acute treatment, maintained nearly 80% of the depressed patients without relapse for a 3-year period compared with 21.8% for those on placebo.[34] In this study, maintenance psychotherapy had an intermediate effect. Although this landmark study demonstrated the importance of continuing a full dose, the magnitude of the findings may over estimate usual drug–placebo differences. Patients selected for this study were those with recurrent depression (who would be expected to have a higher relapse rate) and those who also had a history of symptom-free periods between episodes. In practice, clinicians may encounter patients with chronic depression or a history of residual symptoms for whom remission rates may be lower and relapse rates higher. More recent NE-selective agents, such as reboxetine, also appear to be effective for maintenance treatment.[35]

Comparative efficacy of the NE-selective antidepressants

The development of numerous antidepressants with different mechanisms of action led to questions of how their efficacy compared. Several studies were conducted that compared the efficacy of the tricyclic antidepressants (TCAs) and the SSRIs.

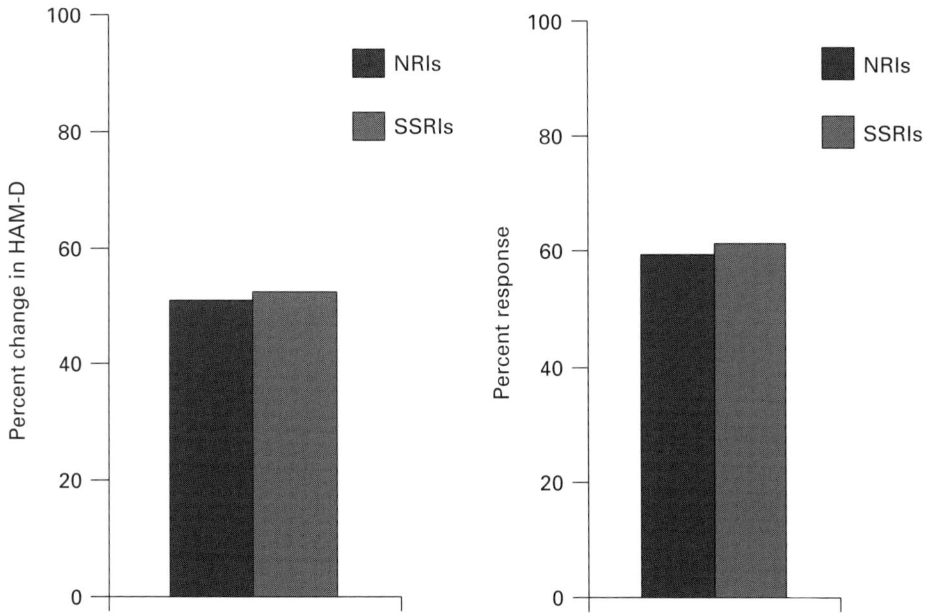

Figure 18.1 Response to NE uptake inhibitors and 5-HT uptake inhibitors in 15 comparison studies including 1563 patients. Adapted from Nelson (1999).[38]

For example, Anderson, in his reviews, found over 100 studies comparing TCAs and SSRIs.[36,37] In general these two drug classes appeared equally effective.

Interpretation of the TCA–SSRI comparison studies was hampered, however, because TCAs are a heterogeneous group of agents. To address this problem, I reviewed those studies that compared SSRIs with selective NE agents such as desipramine, nortriptyline, maprotiline, lofepramine, and reboxetine.[38] Fifteen comparison studies were found that included 1563 patients. Overall the efficacy of these compounds, using 50% improvement on the Hamilton (HAM-D) or Montgomery-Asberg Depression (MADRS) scales as the criterion for response, was very similar: 59.5% vs. 61.4% (NE and 5-HT agents, respectively – see Figure 18.1).

The clinical characteristics of responders to a NE-selective agent

The next question is whether efficacy differs among patient types and whether predictors of response could be identified. Several authors reviewed these questions.[39,40] Historically, patients considered ideal subjects for tricyclic treatment were those patients with endogenous or melancholic features. When Kuhn first tested imipramine, he thought it important to show that an antidepressant was effective in endogenous depression.[41] Unlike more recent clinical trials that are conducted almost entirely in outpatient subjects, many of the early TCA studies were conducted in inpatients, and in these samples, tricyclics were effective. Two studies, one with imipramine and one with desipramine, examined the relationship of drug plasma concentrations with response.[3,42] Both selected inpatients with endogenous features, reasoning that these patients should be TCA responsive. In these two samples, over 80% of the patients who achieved adequate blood levels responded. While these data were consistent with the concept that melancholic patients might be especially responsive to treatment with a noradrenergic TCA, they did not demonstrate that these agents are superior to antidepressants with different mechanisms of action.

Roose and colleagues provided some support for the concept that NE agents were superior to 5-HT agents in melancholia.[43] They compared the efficacy of nortriptyline in 42 depressed patients with heart disease with a subsequent sample of 22 patients treated with fluoxetine. They found nortriptyline more effective in general, but in the melancholic patients completing treatment, 83% responded to nortriptyline while only 10% responded to fluoxetine. While the patients in the two samples appeared to be similar, and were treated in the same setting using similar rating methods, this was not a randomized parallel comparison.

Recently Massana reviewed two randomized controlled 8-week trials comparing reboxetine and fluoxetine.[44] Both studies included inpatients and outpatients. In the pooled sample, reboxetine was more effective than fluoxetine in the more severely effected patients. This study and the Roose study cited above were both *post hoc* analyses. To my knowledge, there is no prospective study that demonstrates differential effectiveness of endogenous or severely effected patients comparing NE and 5-HT agents.

Another subtype thought to be predictive of response to a class of drugs was atypical depression. In a series of papers from Columbia University it was demonstrated that phenelzine was more effective than imipramine or placebo for treatment of atypical depression.[45,46] Although this might suggest that NE antidepressants are less effective in this disorder, another interpretation is that drugs with effects on all three major neurotransmitters (NE, 5-HT, and dopamine) are better than agents that effect primarily one neurotransmitter. The atypical distinction has been less predictive of response to the second-generation antidepressants. It does not appear to predict differential response to fluoxetine.[47] Bupropion, a drug with effects on both dopamine and NE, has been reported to be effective in atypical depression.[48,49]

Three studies with different tricyclics – imipramine, amitriptyline, and desipramine – reported that anxious depressives were less responsive than nonanxious depressives.[50–52] However, these studies did not indicate whether another agent would be better. At least one study found anxious depressives to have less treatment success with a serotonergic drug, fluoxetine, than nonanxious patients.[47]

Studies of the tricyclics in children have been disappointing. A recent review found 13 placebo-controlled studies that included over 500 children and adolescents.[53] The tricyclic agents were not more effective than placebo in this meta-analysis. Although this review did not specifically examine the value of NE antidepressants, the finding is consistent with the report that patients less than 35 years of age were less responsive to desipramine.[52]

Whether NE antidepressants have greater effects on certain individual symptoms is also relevant to the question of what predicts response. In 1997, Healy and McMonagle published a Venn diagram suggesting that NE, 5-HT, and dopamine act on overlapping but also separate dimensions of behaviors in depression.[54] The evidence for this hypothesis came primarily from animal models of behavior.

Data supporting this hypothesis in human subjects is sparse. In the previous review of 15 studies comparing NE and 5-HT agents, five studies did not examine how individual symptoms responded. One of these studies was available for review.[55] The data indicated that paroxetine and nortriptyline did not differ with respect to their effects on three symptoms – depressed mood, suicidality, and agitation – or four Hamilton factors (anxiety/somatization, cognition, retardation, and sleep disturbance). Eight of the studies did examine change in individual symptoms.[13,56–62] However, there was no consistent finding of greater response of an individual symptom to treatment with an NE agent vs. a 5-HT antidepressant.

The two remaining studies were examined in detail.[63] These two studies compared reboxetine and fluoxetine in patients with major depression.[27,64] The samples included 253 and 168 patients, respectively. The studies had similar 8-week, randomized, double-blind, parallel design. Reboxetine was adjusted to a dose of 8 to 10 mg/day and fluoxetine was adjusted between 20 and 40 mg/day. Effect sizes (mean change/s.d.) for symptom change on individual HAM-D scale items were determined. Figure 18.2 shows the effect sizes for the pooled samples. Both drugs showed a similar pattern of response, i.e., depressed mood showed the greatest change, followed by decreased interest in work and activity, and psychic anxiety. Symptom change was generally similar for the two drugs. No individual symptom differed significantly in either sample. Symptom change was also examined in those patients for whom the symptom was relatively more severe at baseline and again the degree of symptom change was similar for the two drug classes.

There are limitations to these comparisons. Both studies examined change on the HAM-D scale. Perhaps other dimensions not assessed on the HAM-D would show differences. In fact, the finding that social functioning showed greater change with reboxetine[65] would be one such example. It might be that much of the change was associated with nonspecific or placebo effects, thus true drug differences were obscured. While the limitations of these studies are noted, it is still the case that

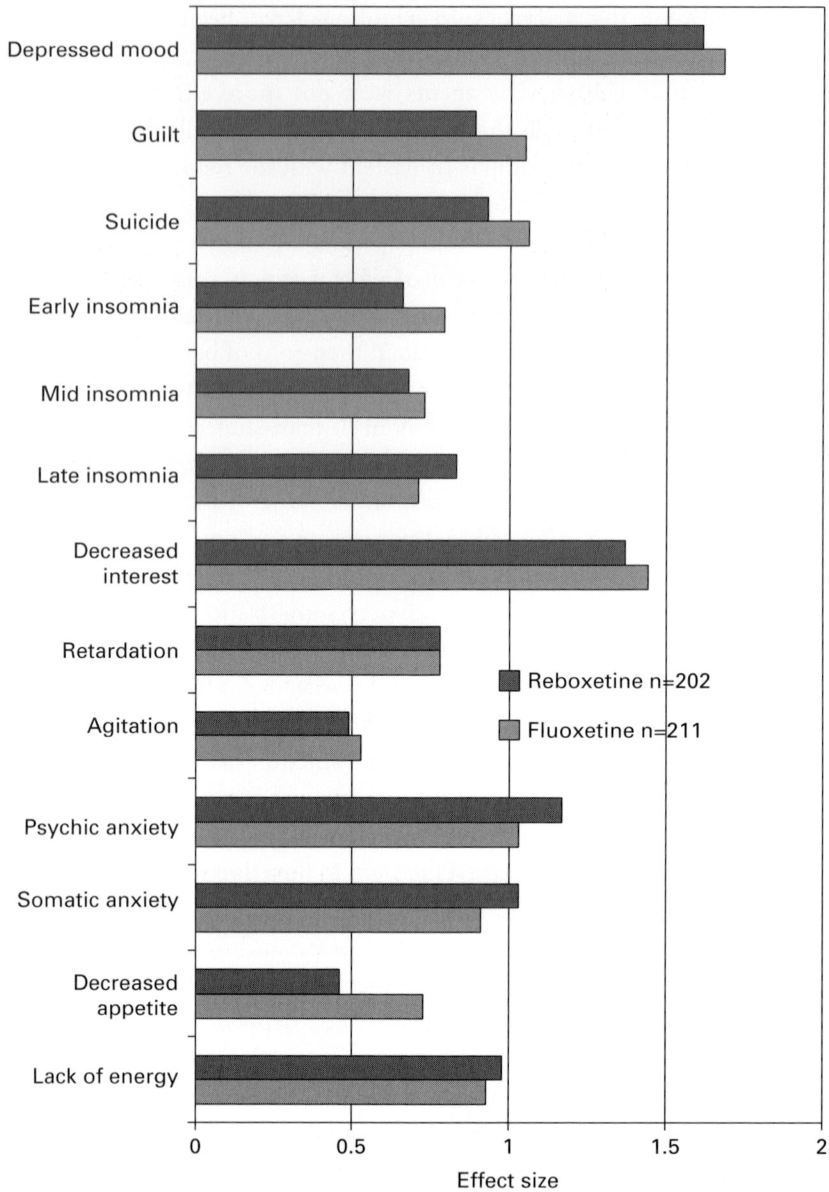

Figure 18.2 Mean effect sizes for change during treatment with fluoxetine and reboxetine in the pooled data from two similar 8-week, double-blind trials. Adapted from Nelson and Leon (2003).[63]

the 11 comparison studies in the literature that have examined symptom response during treatment with NE and 5-HT agents have not shown evidence of differences in the symptoms responding. Rather they suggest these agents appear to have a similar pattern of effects on the core symptoms of depression. This would support the idea that these different agents act through a similar final common pathway.

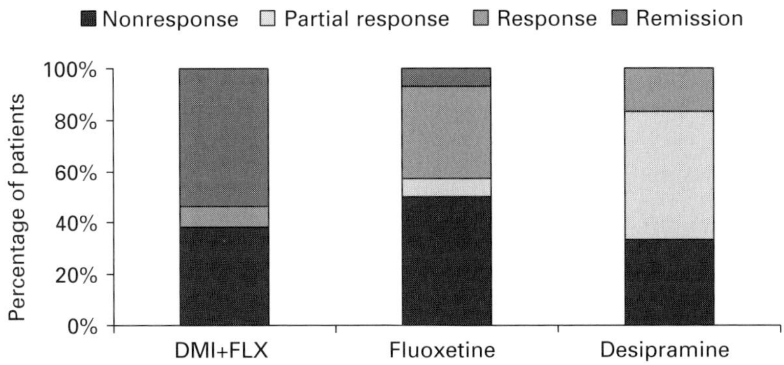

Figure 18.3 Level of response to a 6-week, double-blind, random assignment trial of desipramine, fluoxetine, and the combination (DMI + FLX) in inpatients with major depression. Remission = 75% improvement and MADRS $\leq$ 9; Response = 50% to 74% improvement; Partial response = 25% to 49% improvement; Nonresponse $<$ 25% improvement. X^2, treatment group by level of response, = 24.01, df = 6, p = 0.0005. Adapted from Nelson *et al.* (2004).[70]

Synergistic effects of NE and 5-HT antidepressants

In 1989 Weilberg *et al.* reported that the addition of fluoxetine to an ongoing antidepressant, usually a tricyclic, appeared to enhance efficacy.[66] Later other open studies supported this observation.[67,68] In 1991 we reported a systematic comparison of treatment of 14 inpatients with the combination of fluoxetine and desipramine with a historical control group of 52 inpatients with major depression.[69] This was not a double-blind randomized comparison, but the patients all met Diagnostic and Statistical Manual of Mental Disorders (DSM) III-R criteria for unipolar, nonpsychotic, major depression; were severely depressed inpatients; and were rated with the same modified HAM-D scale. The combination was more rapidly effective and more likely to result in remission.

We subsequently conducted a prospective randomized double-blind comparison of the combination of desipramine and fluoxetine, with either drug used alone in patients with nonpsychotic unipolar major depression (Figure 18.3). Again, subjects were severely depressed inpatients.[70] Patients were treated for 6 weeks, and the dose of desipramine was adjusted to a therapeutic level using a 24-hour desipramine blood level. In this study the combination was significantly more likely to result in remission. The combination was not significantly more effective at 2 weeks but the sample may have been too small to detect differences. It is possible that the greater number of remitters at 6 weeks reflected a shorter time to remission, and that with longer treatment, more patients with monotherapy would have achieved remission.

Given the similarity of symptom response with NE and 5-HT agents cited above, it seems unlikely that the value of combination treatment results from treating a broader array of symptoms. Rather, it appears that there is a pharmacologic synergism between these two drug classes. Consistent with this view, studies have demonstrated an advantage for the "dual-action" drugs – clomipramine, venlafaxine, and duloxetine. The Danish University Antidepressant Group reported that clomipramine was more effective than citalopram or paroxetine in two inpatient studies.[71,72] Thase *et al.* performed a meta-analysis of eight comparison studies of venlafaxine and SSRIs.[73] Recently a meta-analysis of six duloxetine studies was presented.[74] All of these reports suggested an advantage for the dual-action drug over placebo, and these differences appeared to be substantial. In the Danish University studies response rates for clomipramine were about twice those of the SSRIs. In the meta-analysis of venlafaxine vs. SSRIs vs. placebo, if comparisons are restricted to the four studies that included placebo in the study, remission rates were 41.7% (217/521), 30.3% (134/443), and 23.4% (104/446) respectively. The drug–placebo difference for venlafaxine (18.3%) was more than twice that for the SSRIs (6.9%). In the duloxetine studies, in all the randomized patients ($n = 1656$), remission rates for both duloxetine (43%) and the SSRIs (38%) were significantly better than that for placebo (28%)($p<0.05$). However, in the 963 patients with HAM-D scores > 18, the disparity between active agents was greater with remission rates of 38% for duloxetine, 29% for the SSRIs, and 18% for placebo. In this comparison the remission rate for duloxetine was significantly greater than that for the SSRIs ($p = 0.013$) and the drug–placebo difference for duloxetine (20%) was almost twice that for the SSRIs (11%).

Norepinephrine agents in anxiety disorders

The potential value of NE drugs in anxiety disorders was first observed for imipramine. In 1964, Klein described two drug-responsive anxiety syndromes.[5] In one, patients experienced the sudden onset of panic anxiety, often accompanied by severe symptoms of somatic anxiety. Phenothiazines and sedatives appeared to be of little use, but some patients had a dramatic response to imipramine. Panic anxiety was distinguished from expectant or anticipatory anxiety. Subsequently, other NE antidepressants were studied in panic disorder and in some other anxiety conditions. An important distinction is noted. In some cases the efficacy or lack of efficacy of the NE agents has been well established. In other cases, the questionable value of NE agents may be more related to limited study rather than lack of efficacy.

Panic disorder

Subsequent to Klein's observations, imipramine and other antidepressants were studied in panic disorder. Imipramine was shown to be more effective than placebo in several studies.[75–78]

Open studies suggested nortriptyline was of value in reducing panic attacks, as was desipramine.[79–82] Desipramine was also found effective in one placebo-controlled study.[83] A pharmacologic effect of desipramine was also suggested by the reported relationship of desipramine plasma concentrations with response: patients with plasma levels greater than 125 ng/ml were more likely to respond.[81] Alternatively, one study of imipramine administration found response more closely related to imipramine levels than desipramine levels.[84]

Perhaps the most definitive data supporting the value of a NE reuptake inhibitor for panic disorder come from studies of reboxetine. Reboxetine, a selective NE reuptake inhibitor, was reported to be of value in 24 patients who had been refractory to previous SSRI therapy.[85] In a placebo-controlled study of 82 patients with DSM III-R panic disorder, reboxetine 6 to 8 mg/day was significantly more effective than placebo for reducing the number of panic attacks and phobic symptoms.[86]

An important related question is this: what is the best treatment for panic? Two double-blind studies compared selective NE agents with SSRIs. In the first study, maprotiline and fluvoxamine were compared in 44 patients with panic disorder.[87] Fluvoxamine was found to have a potent effect on the number of panic attacks and the level of anxiety while maprotiline had "virtually no effect on the level of anxiety." In the second, reboxetine and citalopram were compared in a single-blind, crossover study of 19 patients.[88] Thirteen of 18 patients (54%) in the intent-to-treat sample responded to reboxetine while 9 of 11 responded to citalopram (82%). While the results generally favored citalopram, 1 citalopram nonresponder responded to reboxetine.

Two studies compared clomipramine with either desipramine or lofepramine. The first study was a 6-week double-blind parallel comparison of the two active drugs, with placebo, in 79 panic disorder subjects.[89] Both lofepramine and clomipramine were superior to placebo at 6 weeks. However, in the first 3 weeks, the dropout rate was significantly higher for clomipramine, 9 of 27 vs. 2 of 26 on lofepramine and 2 of 26 on placebo. The second study was a 16-week double-blind crossover study in 17 panic disorder patients.[90] Both drugs resulted in improvement in panic attack frequency and behavioral ratings, but clomipramine was superior to desipramine in reducing the frequency of panic attacks and on ratings of anxiety. Clomipramine is, of course, a dual-action drug: thus the findings of the latter study might suggest that dual-action agents have an advantage in panic disorder.

To summarize these data, NE agents do have efficacy in panic disorder, but there may be an advantage for serotonergic agents. The possible advantage for clomipramine may reflect its dual action. The advantage for the latter may be consistent with a study finding both 5-HT and NE abnormalities in panic patients. These abnormalities were not independent but occurred together.[91]

Obsessive-compulsive disorder

Noradrenergic agents are not effective in obsessive-compulsive disorder (OCD). Three studies found clomipramine more effective than a NE agent in OCD.[92–94] The first of these compared clomipramine and nortriptyline in a 5-week double-blind comparison.[92] Clomipramine, but not nortriptyline, was superior to placebo. In the second study, clomipramine and desipramine were compared in 48 children and adolescents with OCD in a 10-week double-blind crossover study.[93] Clomipramine was significantly more likely to reduce OCD symptoms than desipramine. Further, in patients showing improvement with clomipramine, 64% showed some evidence of relapse after crossover to desipramine. The third study examined relapse in 26 children who had been receiving clomipramine for long-term therapy of OCD.[94] Patients entering the trial were randomly assigned to continue clomipramine or receive desipramine for two months during double-blind treatment. Eight (89%) of 9 switched to desipramine relapsed, while only 2 of 11 (18%) relapsed when clomipramine was continued. While these studies demonstrate clear superiority of clomipramine over a NE reuptake inhibitor, they do not address the possible explanation that clomipramine's greater efficacy was the result of its dual effects on both NE and 5-HT.

Two studies have compared selective NE and 5-HT agents. Goodman *et al.* compared fluvoxamine and desipramine in 40 adult outpatients with OCD in an 8-week double-blind parallel design.[95] Fluvoxamine was significantly superior to desipramine in reducing OCD symptoms, and resulted in a higher response rate – 11 of 21 patients responded to fluvoxamine while only 2 of 19 responded to desipramine. The second study[96] compared sertraline and desipramine in 166 OCD patients with concurrent depression. This was a 12-week double-blind comparison. Sertraline was dosed to 200 mg/day and desipramine to 300 mg/day. In this study, symptoms of OCD and major depression responded significantly better to sertraline than desipramine. The SSRI was significantly more likely to result in response of OCD (> 40% improvement).

These data taken as a group indicate that 5-HT antidepressants are superior to NE reuptake inhibitors in OCD and indicate that it is not just a relative difference. In these studies NE uptake inhibitors were not effective in OCD.

The question of whether NE agents might have additive effects is more complicated. Noorbala *et al.* added nortriptyline to clomipramine in a double-blind placebo-controlled trial in 30 patients with OCD.[97] All patients received clomipramine 150 mg/day. Half also received desipramine at 50 mg and half received placebo. Patients in both groups of this study improved significantly, but the patients in whom nortriptyline was added did better. The study is somewhat puzzling because clomipramine is itself metabolized to desmethyl-clomipramine, which is a NE reuptake inhibitor. Another study[98] added desipramine or placebo to

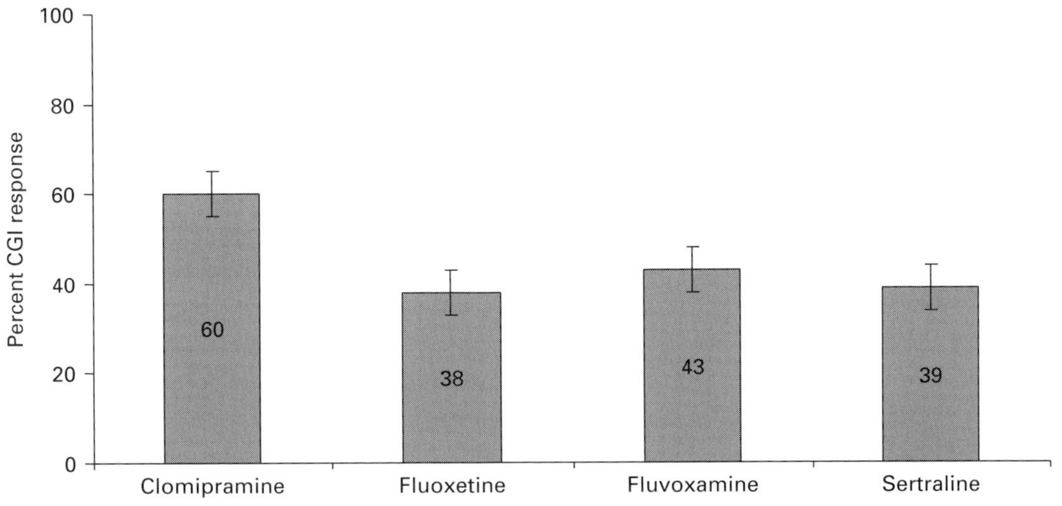

Figure 18.4 Percent responding on the Clinical Global Impression with "much" or "very much" improvement (last observation carried forward ratings). A meta-analysis of four drug trials in 1520 patients with OCD. Clomipramine was significantly superior to the SSRIs ($X^2 = 32.60$, $p < 0.001$). From Greist *et al.* (1995), with permission.[99]

a SSRI in the treatment of 30 patients with OCD. In this 6 to 10 week double-blind study, there were no significant differences between the two groups.

The final factor to consider is the possible difference between clomipramine and the SSRIs. In 1995 Greist reviewed the efficacy of clomipramine and the SSRIs in OCD.[99] He found that fluvoxamine, sertraline, and fluoxetine had beneficial effects and were similar in the degree of effectiveness. Clomipramine was also effective, and was substantially more effective than the SSRIs (Figure 18.4). This was not a parallel comparison study, but a meta-analysis of published studies. And because the SSRI studies were conducted after the clomipramine study, subjects in the SSRI studies were more likely to have failed prior treatment (19% of the fluvoxamine patients and 22% of the sertraline patients), and this may have reduced response in the SSRI samples. The data, in any case, might be interpreted as suggesting that the NE effects of clomipramine did add to the efficacy of clomipramine in OCD.

To summarize these data, NE agents do not appear to be useful as monotherapy for OCD. The question of whether NE adds something to efficacy of a 5-HT agent is unresolved. The review of clomipramine vs. SSRIs suggests this possibility. The fact that this review pools the results of a large number of patients might help to detect small differences. Yet this was not a parallel comparison. The negative results of the study adding desipramine to an SSRI argues against additive effects, but this was a small study.

Generalized anxiety disorder

The role of NE agents in generalized anxiety disorder (GAD) is not well established. Imipramine was one of the first antidepressants to be shown to be effective in GAD. In 1993, Rickels *et al.* reported the results of a double-blind 8-week parallel comparison of imipramine, trazodone, diazepam, and placebo.[100] They studied 230 patients with GAD in whom major depression and panic disorder had been excluded. Diazepam had earlier effects but the effects of imipramine were greater. Among completers, 73% of the patients on imipramine were responders (defined as moderate to marked improvement). Trazodone resulted in 66% response, diazepam 66% response, and placebo 47%. Another parallel comparison examined the efficacy of imipramine and alprazolam in 60 patients with GAD.[6] These two agents had equivalent effects, but alprazolam appeared more effective for somatic anxiety and imipramine for psychic anxiety items.

Imipramine also appears to be of value during benzodiazepine withdrawal. Rickels *et al.* reported the results of a double-blind parallel comparison of imipramine (mean dose 180 mg/day), buspirone (mean dose 38 mg/day), and placebo during benzodiazepine discontinuation in 107 patients with GAD.[101] Patients receiving imipramine were significantly more likely to successfully withdraw from benzodiazepines (82.6%) than placebo (37.5%). Buspirone was intermediate (67.9%). These two studies suggest that the NE effects of imipramine may be of value in GAD.

Again, the findings for imipramine are, to some extent, confounded by its multiple effects. The parent compound is relatively more serotonergic, while its metabolite desipramine is a selective inhibitor of NE uptake. In fact, McLeod *et al.* reported that during imipramine treatment, desipramine plasma levels were inversely related to response in 14 GAD patients.[102] This data suggested that the parent compound accounted for the beneficial effects of imipramine and that its metabolite desipramine may counteract these effects. These data would argue against a role for NE agents.

Recently, venlafaxine and paroxetine have been shown to be effective in GAD. These findings indicate that 5-HT reuptake inhibitors are effective in this disorder. Because venlafaxine has an increasing NE reuptake blockade at higher doses, data from a large venlafaxine study (541 patients) that compared fixed doses of 37.5 mg/day, 75 mg/day, and 150 mg/day, and placebo in a 24-week double-blind parallel trial were examined.[103] All doses of venlafaxine were superior to placebo. However, differences between 75 mg/day and 150 mg/day were slight. Even in the completer analysis, doses of 75 mg/day and 150 mg/day appeared to be equivalent. Because venlafaxine at 75 mg/day acts primarily as a SSRI, with little effect on NE uptake though it may inhibit NE at doses of 150 mg or more, a dose-response effect might suggest an additive effect of NE. The lack of such an effect for venlafaxine suggests no apparent advantage of "dual action" in GAD although this is clearly an indirect argument, and perhaps higher doses would have had more effect.

Posttraumatic stress disorder

The role of noradrenergic agents has not been well studied in posttraumatic stress disorder (PTSD). An early study of 17 combat veterans using concurrent or retrospective chart review found that 82% were rated as much improved after treatment with tricyclic antidepressants.[104] In an open study of response to antidepressants in 72 veterans with PTSD and major depression, substantial improvement was noted in 50% of the patients. However, antidepressants affecting 5-HT appeared to be more effective than those acting on NE, namely desipramine and nortriptyline.[105] One 4-week double-blind crossover study of desipramine in PTSD has been reported.[106] In that study of 18 male veterans, desipramine was found to have little advantage over placebo except for some symptoms of depression. Desipramine was not significantly better than placebo for symptoms of anxiety or PTSD. A Medline search found no studies of reboxetine, atomoxetine, lofepramine, or maprotiline in PTSD.

In summary, the literature on the use of NE agents in PTSD is very limited. Only one controlled study is reported and it found desipramine was not effective for PTSD symptoms.

Social anxiety disorder

The SSRIs have been shown to be effective in social anxiety disorder.[107] The NE reuptake inhibitors have not been well studied. In part, this may be the result of the report of Simpson *et al.* that found imipramine ineffective for treatment of social phobia.[108] This was an open study of 15 patients but the drug appeared to be of little value. A literature search of the selective NE reuptake inhibitors found no other studies of these agents in social anxiety disorder.

Summary of the use of NE agents in anxiety disorders

The literature on the use of NE agents in anxiety disorder is mixed (Table 18.1). In panic disorder NE agents appear superior to placebo; however, some evidence suggests that both SSRIs and dual action agents may be better. In OCD, NE agents have been fairly well studied and found to be ineffective. In fact desipramine appeared to result in minimal change. A meta-analysis of studies in OCD found evidence that the dual action agent clomipramine might be superior to the SSRIs suggesting NE effects may be of some value when added to 5-HT effects. In GAD, imipramine was one of the first antidepressants to be reported as efficacious. Yet, selective NE

Table 18.1 Efficacy of NE agents in anxiety disorders.

Disorders	Findings
Panic disorder	Moderately well studied. NE agents more effective than placebo but appear to be less effective than SSRIs or dual-action agents
OCD	Well studied. NE agents have minimal effects; less effective than SSRIs or dual-action agents
GAD	Imipramine superior to placebo; selective NE agents not well studied
PTSD	Not well studied; TCAs of some value; one controlled study found desipramine not effective
Social anxiety disorder	Not well studied; one open study reported imipramine to be of little value

antidepressants have not been well studied in this disorder. The NE antidepressants have not been well studied in social anxiety disorder and PTSD, but the data to date suggests they may be of limited value.

REFERENCES

1. Kuhn, R. The treatment of depressive states with G 22355 (imipramine hydrochloride). *Am. J. Psychiatry*, **115** (1958), 459–464.
2. Tatsumi, M., Groshan, K., Blakely, R. D. and Richelson, E. Pharmacological profile of antidepressants and related compounds at human monoamine transporters. *Eur. J. Pharmacol.*, **340** (1997), 249–258.
3. Glassman, A. H., Perel, J. M., Shostak, M., Kantor, S. J. and Fleiss, J. L. Clinical implications of imipramine plasma levels for depressive illness. *Arch. Gen. Psychiatry*, **3**: 4 (1977), 197–204.
4. Nelson, J. C., Bock, J. and Jatlow, P. The clinical implications of 2-hydroxy desipramine in plasma. *Clin. Pharmacol. Ther.*, **33** (1983), 183–189.
5. Klein, D. F. Delineation of two drug-responsive anxiety syndromes. *Psychopharmacologia*, **5** (1964), 397–408.
6. Hoehn-Saric, R., McLeod, D. R. and Zimmerli, W. D. Differential effects of alprazolam and imipramine in generalized anxiety disorder: somatic versus psychic symptoms. *J. Clin. Psychiatry*, **49** (1988), 293–301.
7. Preti, A. Tomoxetine. Eli Lilly & Co. *Curr. Opin. Invest. Drugs*, **3** (2002), 272–277. Review.
8. Michelson, D., Faries, D., Wernicke, J. *et al.* Atomoxetine in the treatment of children and adolescents with attention-deficit/hyperactivity disorder: a randomized, placebo-controlled, dose-response study. *Pediatrics*, **108** (2001), E83.

9. Spencer, T., Heiligenstein, J. H., Biederman, J. *et al.* Results from 2 proof-of-concept, placebo-controlled studies of atomoxetine in children with attention-deficit/hyperactivity disorder. *J. Clin. Psychiatry*, **63** (2002), 1140–1147.
10. Schildkraut, J. J. The catecholamine hypothesis of affective disorders: a review of supporting evidence. *Am. J. Psychiatry*, **122** (1965), 509–522.
11. Maas, J. W., Fawcett, J. and Dekirmanjian, H. 3-Methoxy-phenylglycol (MHPG) excretion in depressive states. A pilot study. *Arch. Gen. Psychiatry*, **19** (1968), 129–134.
12. Maas, J. W., Koslow, S. H., Katz, M. M. *et al.* Pretreatment neurotransmitter metabolite levels and response to tricyclic antidepressant drugs. *Am. J. Psychiatry*, **141** (1984), 1159–1171.
13. Bowden, C. L., Schatzberg, A. F., Rosenbaum, A. *et al.* Fluoxetine and desipramine in major depressive disorder. *J. Clin. Psychopharmacol.*, **13** (1993), 305–310.
14. Vetulani, J. and Sulser, F. Action of various antidepressant treatments reduces reactivity of noradrenergic cyclic AMP generating systems in limbic forebrain. *Nature*, **257** (1975), 495–496.
15. Sulser, F., Vetulani, J. and Mobley, P. L. Mode of action of antidepressant drugs. *Biochem. Pharmacol.*, **27** (1978), 257–261.
16. Duman, R. S., Heninger, G. R. and Nestler, E. J. A molecular and cellular theory of depression. *Arch. Gen. Psychiatry*, **54** (1997), 597–606.
17. Delgado, P. L., Charney, D. S., Price, L. H. *et al.* Serotonin function and the mechanism of antidepressant action: reversal of antidepressant-induced remission by rapid depletion of plasma tryptophan. *Arch. Gen. Psychiatry*, **47** (1990), 411–418.
18. Delgado, P. L., Miller, H. L., Salomon, R. M. *et al.* Monoamines and the mechanism of antidepressant action: effects of catecholamine depletion on mood of patients treated with antidepressants. *Psychopharmacol. Bull.*, **29** (1993), 389–396.
19. Miller, H. L., Delgado, P., Salomon, R. *et al.* Clinical and biochemical effects of catecholamine depletion on antidepressant-induced remission of depression. *Arch. Gen. Psychiatry*, **53** (1996), 117–128.
20. Smeraldi, E., Zanardi, R., Benedetti, F. *et al.* Polymorphism with the promoter of the serotonin transporter gene and antidepressant efficacy of fluvoxamine. *Mol. Psychiatry*, **3** (1998), 508–511.
21. Pollock, B. G., Ferrell, R. E., Mulsant, B. H. *et al.* Allelic variation in the serotonin transporter promoter affects onset of paroxetine treatment response in late-life depression. *Neuropsychopharmacology*, **23** (2000), 587–590.
22. Zanardi, R., Benedetti, F., Di Bella, D., Catalano, M. and Smeraldi, E. Efficacy of paroxetine in depression is influenced by a functional polymorphism with the promoter of the serotonin transporter gene. *J. Clin. Psychopharmacol.*, **20** (2000), 105–107.
23. Davis, J. M. and Glassman, A. H. Antidepressant drugs. In *Comprehensive Textbook of Psychiatry*, ed. H. I. Kaplan and B. J. Sadock. (Baltimore: Williams & Wilkins; 1989), pp. 1627–1655.
24. Depression Guideline Panel. *Depression in Primary Care*, Vol. 2, *Treatment of Major Depression.* Clinical Practice Guideline, Number 5. Rockville, MD. U.S. Department of Health and Human Services, Public Health Service, Agency for Health Care Policy and Research. AHCPR Publication No. 93–0551. April 1993.

25. Nelson, J. C. Tricyclic and tetracyclic antidepressants. In *Textbook of Psychopharmacology*, 3rd edn., ed. A. F. Schatzberg and C. B. Nemeroff. (Washington DC: American Psychiatric Press; 2004), pp. 207–230.
26. Versiani, M., Amin, M. and Chouinard, G. Double-blind, placebo-controlled study with reboxetine in inpatients with severe major depressive disorder. *J. Clin. Psychopharmacol.*, **20** (2000), 28–34.
27. Andreoli, V., Caillard, V., Deo, R. S. and Rybakowski, J. K. Reboxetine, a new noradrenaline selective antidepressant, is at least as effective as fluoxetine in the treatment of depression. *J. Clin. Psychopharmacol.*, **22** (2002), 393–399.
28. Ferguson, J. M., Wesnes, K. A. and Schwartz, G. E. Reboxetine versus paroxetine versus placebo: effects on cognitive functioning in depressed patients. *Int. Clin. Psychopharmacol.*, **18** (2003), 9–14
29. APA Task Force. Tricyclic antidepressants-blood level measurements and clinical outcome. *Am. J. Psychiatry*, **142** (1985), 155–162.
30. Asberg, M., Cronholm, B., Sjoqvist, F. and Tuck, D. Relationship between plasma level and therapeutic effect of nortriptyline. *Br. Med. J.*, **3** (1971), 331–334.
31. Kragh-Sorenson, P., Asberg, M. and Effert-Hansen, C. Plasma nortriptyline levels in endogenous depression. *Lancet*, **1** (1973), 113–115.
32. Kragh-Sorensen, P., Hansen, C. E., Baastrup, P. C. and Hvidberg, E. F. Self-inhibiting action of nortriptyline antidepressant effect at high plasma levels. *Psychopharmacologia*, **45** (1976), 305–312.
33. Davis, J. M. Maintenance therapy in psychiatry: II. affective disorders. *Am. J. Psychiatry*, **133** (1976), 1–13.
34. Frank, E., Kupfer, D. J., Perel, J. M. *et al.* Three-year outcomes for maintenance therapies in recurrent depression. *Arch. Gen. Psychiatry*, **47** (1990), 1093–1099.
35. Aguglia, E. Reboxetine in the maintenance therapy of depressive disorder in the elderly: a long-term open study. *Int. J. Geriatr. Psychiatry*, **15** (2000), 784–793.
36. Anderson, I. SSRIs versus tricyclic antidepressants in depressed inpatients: a meta-analysis of efficacy and tolerability. *Depress. Anxiety*, **7**:Suppl 1 (1998), 11–17.
37. Anderson, I. Selective serotonin reuptake inhibitors versus tricyclic antidepressants: a meta-analysis of efficacy and tolerability. *J. Affect. Disord.*, **58** (2000), 19–36.
38. Nelson, J. C. A review of the efficacy of serotonergic and noradrenergic reuptake inhibitors for treatment of major depression. *Biol. Psychiatry*, **46** (1999), 1301–1308.
39. Nelson, J. C. and Charney, D. S. The symptoms of major depressive illness. *Am. J. Psychiatry*, **138** (1981), 1–12.
40. Joyce, P. R. and Paykel, E. S. Predictors of drug response in depression. *Arch. Gen. Psychiatry*, **46** (1989), 89–99.
41. Kuhn, R. The imipramine story. In *Discoveries in Biological Psychiatry*, ed. F. J. Ayd and B. Blackwell. (Philadelphia: JB Lippincott Company; 1970), pp. 205–217.
42. Nelson, J. C., Jatlow, P., Quinlan, D. M. and Bowers, M. B. Desipramine plasma concentrations and antidepressant response. *Arch. Gen. Psychiatry*, **39** (1982), 1419–1422.
43. Roose, S. P., Glassman, A. H., Attia, E. and Woodring, S. Comparative efficacy of selective serotonin reuptake inhibitors and tricyclics in the treatment of melancholia. *Am. J. Psychiatry*, **151** (1994), 1735–1739.

44. Massana, J. Reboxetine versus fluoxetine: an overview of efficacy and tolerability. *J. Clin. Psychiatry*, **59** (1998), 8–10.
45. Liebowitz, M. R., Quitkin, F. M., Stewart, J. W. *et al.* Phenelzine vs. imipramine in atypical depression. *Arch. Gen. Psychiatry*, **41** (1984), 669–677.
46. Liebowitz, M. R., Quitkin, F. M., Stewart, J. W. *et al.* Antidepressant specificity in atypical depression. *Arch. Gen. Psychiatry*, **45** (1988), 129–137.
47. Fava, M., Uebelacker, L. A., Alpert, J. E. *et al.* Major depressive subtypes and treatment response. *Biol. Psychiatry*, **42** (1997), 568–576.
48. Goodnick, P. J., Dominguez, R. A., DeVane, C. L. and Bowden, C. L. Bupropion slow-release response in depression: diagnosis and biochemistry. *Biol. Psychiatry*, **44** (1998), 629–632.
49. Goodnick, P. J. and Extein, I. Bupropion and fluoxetine in depressive subtypes. *Ann. Clin. Psychiatry*, **1** (1989), 119–122.
50. Kupfer, D. J. and Spiker, D. G. Refractory depression: prediction of non-response by clinical indicators. *J. Clin. Psychiatry*, **42** (1981), 307–312.
51. Roose, S. P., Glassman, A. H., Walsh, B. T. and Woodring, S. Tricyclic nonresponders, phenomenology and treatment. *Am. J. Psychiatry*, **143** (1986), 345–348.
52. Nelson, J. C., Mazure, C. M. and Jatlow, P. I. Characteristics of desipramine refractory depression. *J. Clin. Psychiatry*, **55** (1994), 12–19.
53. Hazell, P., O'Connell, D., Heathcote, D. and Henry, D. Tricyclic drugs for depression in children and adolescents. *Cochrane Database Syst. Rev.*, **2** (2002), CD002317. Review.
54. Healy, D. and McMonagle, T. The enhancement of social functioning as a therapeutic principle in the management of depression. *J. Psychopharmacol.*, **11** (1997), S25–S31.
55. Nelson, J. C., Kennedy, J. S., Pollock, B. G. *et al.* Treatment of major depression with nortriptyline and paroxetine in patients with ischemic heart disease. *Am. J. Psychiatry*, **156** (1999), 1024–1028.
56. Aberg, A. Controlled cross-over study of a 5-HT uptake inhibiting and an NA uptake inhibiting antidepressant. *Acta Psychiatr. Scand.*, **63** (1981), 244–255.
57. Bouchard, J. M., Delaunay, J., Delisle, J.-P. *et al.* Citalopram versus maprotiline: a controlled, clinical multicentre trial in depressed patients. *Acta Psychiatr. Scand.*, **76** (1987), 583–592.
58. Poelinger, W. and Haber, H. Fluoxetine 40 mg vs. maprotiline 75 mg in the treatment of outpatients with depressive disorders. *Int. Clin. Psychopharmacol.*, **4** (1989), 47–50.
59. deJonghe, F., Ravelli, D. P. and Tuynmna-Qua, H. A randomized, double-blind study of fluoxetine and maprotiline in the treatment of major depression. *Pharmacopsychiatry*, **24** (1991), 62–67.
60. Bondareff, W., Alpert, M., Friedhoff, A. J. *et al.* Comparison of sertraline and nortriptyline in the treatment of major depressive disorder in late life. *Am. J. Psychiatry*, **157** (2000), 729–736.
61. Moon, C. Treatment of major depression in general practice: a double-blind comparison of paroxetine and lofepramine. *Br. J. Clin. Pract.*, **50** (1990), 240–244.
62. Robertson, M. M., Abou-Saleh, M. T., Harrison, D. A. *et al.* A double-blind controlled comparison of fluoxetine and lofepramine in major depressive illness. *J. Psychopharmacol.*, **8** (1994), 98–103.

63. Nelson, J. C. and Leon, A. SSRIs and SNRIs treat the same symptoms in depression. Presented at the 42nd Annual Meeting of the ACNP, December 10, 2003, San Juan, Puerto Rico.
64. Massana, J., Moller, H.-J., Burrows, G. D. and Montenegro, R. M. Reboxetine: a double-blind comparison with fluoxetine in major depressive disorder. *Int. Clin. Psychopharmacol.*, **14** (1999), 73–80.
65. Dubini, A., Dose, M. and Polin, V. Do noradrenaline and serotonin differentially affect social motivation and behavior? *Eur. Neuropsychopharmacol.*, **7** (1997), S49–S55.
66. Weilburg, J. B., Rosenbaum, J. F., Biederman, J. *et al.* Fluoxetine added to non-MAOI antidepressants converts nonresponders to responders: a preliminary report. *J. Clin. Psychiatry*, **50** (1989), 447–449.
67. Seth, R., Jennings, A. L., Bindman, J., Phillips, J. and Bergmann, K. Combination treatment with noradrenaline and serotonin reuptake inhibitors in resistant depression. *Br. J. Psychiatry*, **161** (1992), 562–565.
68. Zajecka, J. M., Jeffries, H. and Fawcett, J. The efficacy of fluoxetine combined with a heterocyclic antidepressant in treatment-resistant depression: a retrospective analysis. *J. Clin. Psychiatry*, **56** (1995), 338–343.
69. Nelson, J. C., Mazure, C. M., Bowers, M. B., Jr. and Jatlow, P. I. A preliminary, open study of the combination of fluoxetine and desipramine for rapid treatment of major depression. *Arch. Gen. Psychiatry*, **48** (1991), 303–307.
70. Nelson, J. C., Mazure, C. M., Jatlow, P. I., Bowers, M. B. Jr. and Price, L. H. Combining norepinephrine and serotonin reuptake inhibition mechanisms for treatment of depression: a double-blind, randomized study. *Biol. Psychiatry*, **55** (2004), 296–300.
71. Danish University Antidepressant Group. Citalopram: clinical effect profile in comparison with clomipramine. A controlled multicenter study. *Psychopharmacology (Berl.)*, **90** (1986), 131–138.
72. Danish University Antidepressant Group. Paroxetine: a selective serotonin reuptake inhibitor showing better tolerance, but weaker antidepressant effect than clomipramine in a controlled multicenter study. *J. Affect. Disord.*, **18** (1990), 289–299.
73. Thase, M. E., Entsuah, A. R. and Rudolph, R. L. Remission rates during treatment with venlafaxine or selective serotonin reuptake inhibitors. *Br. J. Psychiatry*, **178** (2001), 234–241.
74. Thase, M. E., Lu, Y., Joliat, M. *et al.* Remission in the placebo-controlled trials of duloxetine with an SSRI comparator. Presented at the 156th American Psychological Association Annual Meeting, May 17, 2003, San Francisco CA.
75. Sheehan, D. V., Ballenger, J. and Jacobsen, G. Treatment of endogenous anxiety with phobic, hysterical, and hypochondriacal symptoms. *Arch. Gen. Psychiatry*, **37** (1980), 51–59.
76. Pollack, M. H., Otto, M. W., Sachs, G. S. *et al.* Anxiety psychopathology predictive outcome in patients with panic disorder and depression treated with imipramine, alprazolam and placebo. *J. Affect. Disord.*, **30** (1994), 273–281.
77. Sheikh, J. I. and Swales, P. J. Treatment of panic disorder in older adults: a pilot study comparison of alpazolam, imipramine, and placebo. *Int. J. Psychiatry Med.*, **29** (1999), 107–117.

78. Rosenberg, R., Beck, P., Mellergard, M. and Ottosson, J. O. Alprazolam, imipramine and placebo treatment of panic disorder: predicting therapeutic response. *Acta Psychiatr. Scand. Suppl.*, **365** (1991), 46–52.
79. Munjack, D. J., Usigli, R., Zulueta, A. *et al.* Nortriptyline in the treatment of panic disorder and agoraphobia with panic attacks. *J. Clin. Psychopharmacol.*, **8** (1988), 204–207.
80. Nardi, A. D., Nascimento, I., Valenca, A. M. *et al.* Respiratory panic disorder subtype: acute and long-term response to nortriptyline, a noradrenergic tricyclic antidepressant. *Psychiatry Res.*, 120 (2003), 283–293.
81. Lydiard, R. B. Desipramine in agoraphobia with panic attacks: an open, fixed-dose study. *J. Clin. Psychopharmacol.*, **7** (1987), 258–260.
82. Kalus, O., Asnis, G. M., Rubinson, E. *et al.* Desipramine treatment in panic disorder. *J. Affect. Disord.*, **21** (1991), 239–244.
83. Lydiard, R. B., Morton, W. A., Emmanuel, N. P. *et al.* Preliminary report: placebo-controlled, double-blind study of the clinical and metabolic effects of desipramine in panic disorder. *Psychopharmacol. Bull.*, **29** (1993), 183–188.
84. Mavissakalian, M. R. and Perel, J. M. The relationship of plasma imipramine and N-desmethylimipramine to response in panic disorder. *Psychopharmacol. Bull.*, **32** (1996), 143–147.
85. Dannon, P. N., Iancu, I. and Grunhaus, L. The efficacy of reboxetine in the treatment-refractory patients with panic disorder: an open label study. *Hum. Psychopharmacol.*, **17** (2002), 329–333.
86. Versiani, M., Cassano, G., Perugi, G. *et al.* Reboxetine, a selective norepinephrine reuptake inhibitor, is an effective and well-tolerated treatment for panic disorder. *J. Clin. Psychiatry*, **63** (2002), 31–37.
87. Den Boer, J. A. and Westenberg, H. G. Effect of a serotonin and noradrenaline uptake inhibitor in panic disorder: a double-blind comparative study with fluvoxamine and maprotiline. *Int. Clin. Psychopharmacol.*, **3** (1988), 59–74.
88. Seedat, S., Van Rheede, Van Oudtshoorn, E., Muller, J. E., Mohr, N., Stein, D. J. Reboxetine and citalopram in panic disorder: a single-blind, cross-over, flexible-dose pilot study. *Int. Clin. Psychopharmacol.*, **18** (2003), 279–284.
89. Fahy, T. J., O'Rourke, D., Brophy, J., Schazmann, W. and Sciascia, S. The Galway study of panic disorder I: clomipramine and lofepramine in DSM III-R panic disorder: a placebo controlled trial. *J. Affect. Disord.*, **25** (1992), 63–75.
90. Sasson, Y., Iancu, I., Fux, M. *et al.* A double-blind crossover comparison of clomipramine and desipramine in the treatment of panic disorder. *Eur. Neuropsychopharmacol.*, **9** (1999), 191–196.
91. Asnis, G. M., Wetzler, S., Sanderson, W. C., Kahn, R. S. and Van Praag, H. M. Functional interrelationship of serotonin and norepinephrine: cortisol response to MCPP and DMI in patients with panic disorder, patients with depression, and normal control subjects. *Psychiatry Res.*, **43** (1992), 65–76.
92. Thoren, P., Asberg, M., Cronholm, B., Jornestedt, L. and Traskman, L. Clomipramine treatment of obsessive-compulsive disorder: a controlled clinical trial. *Arch. Gen. Psychiatry*, **37** (1980), 1281–1285.

93. Leonard, H. L., Swedo, S. E., Rapoport, J. L. *et al.* Treatment of obsessive-compulsive disorder with clomipramine and desipramine in children and adolescents. A double-blind crossover comparison. *Arch. Gen. Psychiatry*, **46** (1989), 1088–1092.
94. Leonard, H. L., Swedo, S. E., Lenane, M. C. *et al.* A double-blind desipramine substitution during long-term clomipramine treatment in children and adolescents with obsessive-compulsive disorder. *Arch. Gen. Psychiatry*, **48** (1991), 922–927.
95. Goodman, W. K., Price, L. H., Delgado, P. L. *et al.* Specificity of serotonin reuptake inhibitors in the treatment of obsessive-compulsive disorder: comparison of fluvoxamine and desipramine. *Arch. Gen. Psychiatry*, **47** (1990), 577–585.
96. Hoehn-Saric, R., Ninan, P., Black, D. W. *et al.* Multicenter double-blind comparison of sertraline and desipramine for concurrent obsessive-compulsive and major depressive disorders. *Arch. Gen. Psychiatry*, **57** (2000), 76–82.
97. Noorbala, A. A., Hosseini, S. H., Mohammadi, M. R. and Akhondzadeh, S. Combination of clomipramine and nortriptyline in the treatment of obsessive-compulsive disorder: a double-blind, placebo-controlled trial. *J. Clin. Pharm Ther.*, **23** (1998), 155–159.
98. Barr, L. C., Goodman, W. K., Anand, A., McDougle, C. J. and Price, L. H. Addition of desipramine to serotonin reuptake inhibitors in treatment-resistant obsessive-compulsive disorder. *Am. J. Psychiatry*, **154** (1997), 1293–1295.
99. Greist, J. H., Jefferson, J. W., Kobak, K. A., Katzelnick, D. J. and Serlin, R. C. Efficacy and tolerability of serotonin transport inhibitors in obsessive-compulsive disorder: a meta-analysis. *Arch. Gen. Psychiatry*, **52** (1995), 53–60.
100. Rickels, K., Downing, R., Schweizer, E. and Hassman, H. Antidepressants for the treatment of generalized anxiety disorder. A placebo-controlled comparison of imipramine, trazodone and diazepam. *Arch. Gen. Psychiatry*, **50** (1993), 884–895.
101. Rickels, K., De Martinis, N., Garcia-Espana, F. *et al.* Imipramine and buspirone in treatment of patients with generalized anxiety disorder who are discontinuing long-term benzodiazepine therapy. *Am. J. Psychiatry*, **157** (2000), 1973–1979.
102. McLeod, D. R., Hoehn-Saric, R., Porges, S. W., Kowalski, P. A. and Clark, C. M. Therapeutic effects of imipramine are counteracted by its metabolite, desipramine, in patients with generalized anxiety disorder. *J. Clin. Psychopharmacol.*, **20** (2000), 615–621.
103. Allgulander, C., Hackett, D. and Salinas, E. Venlafaxine extended release (ER) in the treatment of generalized anxiety disorder. *Br. J. Psychiatry*, **179** (2001), 15–22.
104. Falcon, S., Ryan, C., Chamberlain, K. and Curtis, G. Tricyclics: possible treatment for post-traumatic stress disorder. *J. Clin. Psychiatry*, **46** (1985), 385–388.
105. Dow, B. and Kline, N. Antidepressant treatment of posttraumatic stress disorder and major depression in veterans. *Ann. Clin. Psychiatry*, **9** (1997), 1–5.
106. Reist, C., Kauffman, C. D., Haier, R. J. *et al.* A controlled trial of desipramine in 18 men with posttraumatic stress disorder. *Am. J. Psychiatry*, **146** (1989), 513–516.
107. van der Linden, G. J., Stein, D. J. and Van Balkom, A. J. The efficacy of the selective serotonin reuptake inhibitors for social anxiety disorder (social phobia): a meta-analysis of randomized controlled trials. *Int. Clin. Psychopharmacol.*, **15** (2000), S15–S23.
108. Simpson, H. B., Schneier, F. R., Campeas, R. B. *et al.* Imipramine in the treatment of social phobia. *J. Clin. Psychopharmacol.*, **18** (1998), 132–135.

19

Norepinephrine in attention-deficit/hyperactivity disorder

Fred W. Reimherr, M.D.
University of Utah, USA

B.K. Marchant
University of Utah, USA

Michael A. Schwartz, M.D.
University of Hawai'i, USA

E. Higgins, M.D.
Medical University of South Carolina, USA

Introduction

Among the psychiatric disorders that involve norepinephrine (NE), attention-deficit/hyperactivity disorder (ADHD) is one of the most widespread and clinically significant. There is general agreement that neurological and genetic factors related to central nervous system NE and dopamine (DA) function are primary in its etiology. There have been multiple studies, some controlled and positive, of NE agents in the treatment of ADHD. However, the success of these studies has not diminished the wide popularity of the dual-acting stimulants, methylphenidate and dextroamphetamine, as first-line agents for the treatment of ADHD. Except for special populations or as second-line agents, many clinicians use NE agents infrequently. Many believe that NE's clinical effects are distinctly less robust than those of stimulants. The success of atomoxetine has produced questions about this belief and raised important theoretical concerns regarding the role for NE in ADHD. There is still controversy regarding its efficacy versus that of stimulants, and the relative efficacy of atomoxetine versus stimulants has not been carefully tested.

Historical overview

While there are earlier literary references to maladies similar to ADHD, the first description in the medical literature is attributed to Sir George Still in a 1902 publication[1]. Still described 20 children using adjectives such as aggressive, passionate, lawless, inattentive, impulsive, and overactive. Several of his most important observations regarding the disorder have been corroborated over the following

Brain Norepinephrine: Neurobiology and Therapeutics, ed. Gregory A. Ordway, Michael A. Schwartz and Alan Frazer. Published by Cambridge University Press.

decades. They include: (1) an overrepresentation in males; (2) alcoholism, criminal conduct, and depression among the relatives of these children; (3) a family predisposition to the disorder; and (4) a suggestion that the disorder may result from an injury to the nervous system.

The clinical description of the illness has not changed dramatically over the decades, but there has been clear progress in our understanding of its treatment, etiology, and the broad clinical significance of ADHD. Following the great influenza pandemic of 1917 and 1918 many children who contracted influenza and a secondary illness, Von Economo's or encephalitis lethargica, subsequently developed an ADHD-like syndrome. These patients would have met the criteria for Diagnostic and Statistical Manual of Mental Disorders (DSM-IV) ADHD, but they also had problems with anger, mood lability, and conduct. Postmortems in surviving adults who had developed Parkinson's disease showed damage in the substantia nigra with lesser changes in the lenticular nucleus and hypothalamus. These areas are rich in dopaminergic neurons.

In 1937, Bradley reported behavioral improvement in children receiving amphetamine. This paper refocused attention on Still's reports.[2] This study was conducted in the Emma Pendleton Bradley Home using children who were so severely impaired as to require hospitalization. Of the 30 children, the report described 14 as "spectacular" responders, particularly in school performance and "drive." Fifteen patients became distinctly "subdued" in their emotional responses, a change that was considered positive. Seven of the children showed improvement in both areas. Of interest is that the introductory notes for this paper refer to the use of medication in adults that led to its use in this setting, a theme that was generally ignored until Wender and his colleagues began studies of adult ADHD in the 1970s in Utah (see below). The observed responses were restated in a second report in 1950.[3] In this second paper, Bradley observed that most of the children demonstrated a preferential response to either Dexadrine or Benzedrine. The two drugs had similar effects on hyperactivity, but Benzedrine seemed more effective on attentional symptoms. This report would appear to be one of the first reports that discussed the use of a racemic medication (Benzedrine) versus a single isomer version of the same chemical entity in the treatment of a psychiatric disorder.

Bradley's cases and others with known causes (i.e. birth trauma, head injury, toxin exposure, and infections) led to the concept of the brain-injured child syndrome associated at times with mental retardation. Subsequently, this term was applied to children manifesting these same behavioral features but, importantly, without evidence of brain damage or retardation.[4–6] This concept of a brain-injured child syndrome without evidence for an actual brain injury would later evolve into the diagnostic term "minimal brain damage" and later "minimal brain dysfunction" (MBD). During the same period other researchers became more interested in the specific behaviors associated with the disorder, labeling the disorder "hyperkinetic impulse disorder" and later the "hyperactive child syndrome." They attributed the disorder to cortical over-stimulation resulting from inadequate thalamic filtering of stimuli entering the brain.[7] These writers shifted the focus away from the damaged brain to a description of the clinical manifestations of the disorder. Influential summaries of the period include Laufer and Denhoff,[8] and Connors.[9] As an example of the thinking of the time, Wender[10] included the primary symptoms of hyperactivity, attentional difficulties (including distractibility), "demandingness," impulsivity, problems in learning, coordination, emotion, and interpersonal impairment. He also strongly endorsed the role of monoamines (DA, serotonin, and NE) in the disorder and proposed that it should be viewed as a neurochemically based disorder.

During this period, the concepts of psychoanalysis held sway over much of child psychiatry. This school promoted the belief that children's mental disorders were a reaction to various environmental factors, especially the family environment. As a result, when DSM-II was written, all childhood disorders were described as "reactions" and the "hyperactive child syndrome" became the "hyperkinetic reaction of childhood." The disorder was described simply (and completely) as follows:

> This disorder is characterized by overactivity, restlessness, distractibility, and short attention span, especially in young children; the behavior usually diminishes in adolescence. If this behavior is caused by organic brain damage, it should be diagnosed under the appropriate nonpsychotic organic brain syndrome.[11]

By the 1970s, research was addressing not just the hyperactivity but also the difficulties in sustained attention and impulse control. Studies that emphasized attention, impulsiveness and the other cognitive sequelae of this disorder along with the educational benefits of the stimulant (methylphenidate) led in 1980 to renaming the disorder "attentional-deficit disorder" (ADD) in DSM-III. No longer was this viewed as a reaction to childhood trauma, but instead the cognitive, developmental, and genetic aspects of the illness were emphasized. As was done with most other disorders, reliable symptom lists and cutoff scores were developed to deal with the three major symptom areas. In 1987, in DSM-IIIR hyperactivity was again recognized in the name of the disorder, which was changed to "attentional-deficit/hyperactivity disorder."

In Utah, Wender and associates started initial investigations of "minimal brain dysfunction" in adults. They asked the parents of children with minimal brain dysfunction about similar problems when they were young. Many responded positively and added that they continued to experience these symptoms into adulthood. This line of inquiry led to clinical trials and the description of clinical symptoms appropriate for adults later called the "Utah Criteria."[12,13] Although initially controversial, Wender's research was subsequently replicated by several other research groups. Clinical trials, family studies, and longitudinal follow-up studies all support the notion that ADHD affects adults.

Also in Utah, a book by Camilla Anderson "*Society Pays: The High Costs of Minimal Brain Damage in America*"[14] was published for the general public in 1972. Anderson was a psychiatrist in practice in Utah and received little attention outside that state. Her book presented MBD as a condition with multiple causes, particularly genetic, that contributed to school problems, public assistance, other psychiatric illnesses, and legal problems. Like Wender, Anderson claimed that childhood ADHD was a precursor to adult impairment extending well beyond the initial symptoms specific to ADHD.

Starting in the 1980s, reports challenged the notion that ADHD was primarily a disturbance in attention per se. Instead, these reports focused on problems with motivation in general and, more specifically, insensitivity to response consequences.[15–17] It became clear that ADHD children responded to alterations in contingencies of reinforcement or punishment differently than normal children. Under conditions of continuous reward, the performances of ADHD children were seldom different from control groups. In contrast, under variable reward conditions or during extinction (no reward), ADHD children showed substantial declines in performance compared to controls.[18–20] Additionally, Barkley found that when rules were in conflict with prevailing immediate consequences, ADHD children responded to the immediate consequences more often than did control groups.[15,21,22] He hypothesized that the class of human behavior initiated and sustained by rules and language, called "rule-governed" by behavioral analysts, was impaired in children with ADHD.

As a result of these developments, numerous investigators have acknowledged similarities between ADHD and patients with prefrontal cortex (PFC) pathology. Barkley suggested that ADHD represented a developmental delay in the response inhibition processes related to PFC mechanisms.[23] According to Barkley, ADHD results from developmental delays in executive functioning. The ability to delay or inhibit one's response to environmental stimuli gives one an ability to delay and then modify emotional reactions to the stimuli. More importantly, this delay gives one an opportunity to shift behavior from control by the immediate environment to control by internally represented forms of information (internal self-regulation).

In the late 1990s, the National Institute of Mental Health Multimodal Treatment Study of ADHD (MTA), one of the most important studies ever conducted in childhood ADHD, helped better define the disorder and examined extended treatment with medication and/or behavior therapy. Almost 600 children were included in the study. At baseline, there were four major groups in the patient population: 30% with ADHD; 13% with ADHD plus an anxiety disorder; 32% with ADHD plus conduct disorder and/or oppositional defiant disorder; and 26% with all three disorders. Study investigators concluded that these groups were sufficiently distinct to warrant classification as ADHD subtypes.[24] This finding echoes the earlier conclusions of Still, Bradley, and Wender. They also carefully evaluated initial treatment with methylphenidate. Of these subjects, 77% of the children had "no room to improve" on methylphenidate.[25] Only 10% were nonresponders to methylphenidate and needed to be treated with dextroamphetamine. This result suggests that the more narrow neurochemical actions of methylphenidate are fully adequate for most patients with ADHD. Finally, the best long-term results were obtained with combination therapy at lower medication doses than in the medication only group.[26] The benefits of combination therapy were greatest in associated measures: oppositional/aggressive symptoms, internalizing symptoms, teacher-rated social skills, parent–child relations, and reading achievement.

Genetics

Multiple studies of dizygotic and monozygotic twins have been conducted to determine the heritability of ADHD.[27–34] Estimates of heritability range from a low of 60% to 68% by Hudziak[31] to a high of 75% to 91% by Levy *et al.*[32] These are quite high levels compared to other psychiatric illnesses. Twin studies also indicate that a shared environment has little impact on the expression of the illness, accounting for less than 5% of the variance, a level similar to other psychiatric illnesses.[32,33,35] Silberg *et al.*[35] reported that approximately 15% to 20% of the variance in hyperactivity-impulsive-inattentive behavior or ADHD symptoms can be attributed to nonshared environmental (nongenetic) factors.

The gene most frequently reported on in ADHD is the D_4-dopamine receptor gene (*DRD4*). This gene was first implicated by LaHosta *et al.* in 1996.[36] Faraone *et al.*[37] completed a meta-analysis of eight case-controlled studies and 14 family-based studies. While the individual studies had inconsistent findings, this meta-analysis supported the association between ADHD and the *DRD4* gene. The DA transporter gene (*DAT*) has been analyzed in several family studies of ADHD because a major action of stimulant medication is blockade of this transporter. The first study of the 480-bp allele of a DAT VNTR (variable number of tandem repeats) polymorphism, by Cook[38] in 1995, found a significant relationship between it and ADHD. Later studies did not reach significance. However, Kirley *et al.*[39] pooled all of the family-based studies and found a statistically significant relationship.

The most interesting reports dealing with NE are those of Comings. He and his colleagues compared the relative impact of NE, DA, and serotonin genes in ADHD, oppositional defiant disorder (ODD), and conduct disorder (CD). (A confounding variable in these studies is the fact that all of the subjects had Tourette's syndrome, as well as ADHD.) Using multivariate regression analysis, he examined the additive effects of multiple genes and concluded that NE plays a primary role in ADHD.[40,41] These studies were developed on the assumption that most complex behavioral disorders such as ADHD are inherited as polygenic traits. A major characteristic of polygenic inheritance is that each of the multiple genes involved contributes only a small effect. While the inheritance of a single gene is unlikely to have significant effect in the individual, as one inherits increasing numbers of these genes it puts them at increased risk of expressing a given trait. A major implication of such a model is that studies that examine one gene at a time will tend to produce a series of both successful and unsuccessful attempts at replication. In fact, only two of the NE genes used in this study have been associated with ADHD in more traditional genetic studies.

Neurobiology and the contribution of norepinephrine in the "prefrontal" abnormalities of ADHD

By the early 1970s, indicated earlier in this chapter, research implicated the frontal cortical system in the development of ADHD. In particular, research focused on PFC interactions with the caudate and the brain stem's DA and NE systems. Wender[10] hypothesized that ADHD children had "an abnormality in the metabolism of monoamines: serotonin, NE, or dopamine." While much of the subsequent research has focused on the DA pathways, the noradrenergic system has not been ignored.[42,43] Since that time, it has become clear that the DA and NE systems are interdependent, and changes in one usually impact the other. Detailed reviews of the role of catecholamines in ADHD and of animal studies of α_2-noradrenergic mechanisms on higher cortical functioning have been presented in recent literature.[44,45]

Although ADHD is a complex and varied disorder, many investigators have been impressed with the similarities between its symptoms and those of lesions to the PFC.[46] The symptoms of poor attention regulation, disorganized behavior, hyperactivity, and impulsivity found in ADHD can also be found in humans and animals with PFC lesions. Lesions involving the right dorsal PFC produce deficits in sustained attention regulation and inhibiting responses to distracting stimuli. Similarly, lesions to the right orbital PFC often produce affective deficits such as immature behavior and lack of restraint, as well as increased motor activity. Supporting a connection between PFC pathology and ADHD is the fact that ADHD patients often show impairment on psychological tests that purport to measure frontal lobe function (Continuous performance, Go/No-Go, and Stroop Interference Tasks). Imaging studies of patients with ADHD have shown a pattern of decreased activity in the PFC and striatum, and increased activity in the posterior sensory/motor cortices. Both are normalized by methylphenidate treatment.

A second historic element leading to interest in NE were reports regarding beneficial effects of α_2-adrenergic receptor agonists in young monkeys with experimentally induced NE depletion and aged monkeys with naturally occurring NE degeneration.[47–49] Three α_2-subtypes have now been cloned in humans: the α_{2A}, α_{2B} and α_{2C}, the genes for which reside on chromosomes 10, 2, and 4 respectively.[50–52] This discovery may contribute to an ability to separate the adverse effects of α_2-agonists from their beneficial drug effects. For example mRNA for the α_{2B}-receptor is found primarily in the thalamus, a major site for clonidine's sedative actions. In contrast the α_{2A}-receptor is densest in the PFC and the locus coeruleus (LC). The effects of α_2-agonists in monkeys further indicate these agents' cognitive-enhancing, sedative, and hypotensive actions can be dissociated according to their relative affinities for each receptor type. Thus, the ability of an α_2-agonist to improve PFC function without adverse effects was found to correspond to its selectivity for the α_{2A} site: guanfacine > UK-14304 > clonidine > BHT920.[48,49]

The cognitive abilities of the PFC are sensitive to α_2-receptor manipulation, and at least three α_2-agonists (guanfacine, clonidine, and medetomidine hydrochloride) have been shown to improve a variety of working memory tasks that use PFC abilities (delayed alternation, delayed response, and delayed match-to-sample with repeated stimuli) when administered to rats or monkeys. The α_2-agonists appear most effective when the PFC is challenged by conditions of interference or distraction. These findings buttress a belief in the importance of NE mechanisms in the regulation of attention. There is a general clinical belief that α_2-agonist treatments are not as effective as stimulants for ADHD, which is now supported by recent results in clinical trials (see below). This challenges the relevance of the α_2-agonist research in ADHD.

Norepinephrine and the pathophysiology of ADHD

Investigations of the pathophysiology of ADHD have focused mainly on the DA system. For example, the effectiveness of stimulants such as methylphenidate and amphetamine in ADHD is attributed to their capacity to enhance DA neurotransmission. However, pure DA agonists are ineffective as ADHD medications. Moreover, stimulants also enhance NE neurotransmission. Tricyclic antidepressants such as imipramine, desipramine and nortriptyline have fallen into disuse because of a perceived lack of efficacy despite studies supporting their usefulness (discussed later in this chapter, p. 566). The introduction of the selective NE reuptake inhibitor, atomoxetine, has reopened the issue of the relative efficacy of stimulants versus more selective NE medications in ADHD.

How might we understand the role of NE in ADHD and in ADHD treatment? We might begin with the observation that the NE system is intimately interconnected with the DA system and modulates DA release. Both systems of neurons arise as cell bodies in distinct nuclei in the midbrain (the LC in the case of the NE system) and project anteriorly and broadly to the frontal and PFC. In addition, another group of NE neurons from the LC project posteriorly to the parietal cortex. Norepinephrine cells in the LC have a phasic response to sensory stimuli and in turn phasically prime the response of cortical cells. A high level of basal activity of these LC cells will therefore establish a background of noise and dampen cortical responsiveness. Under these circumstances, medications can ameliorate symptoms of ADHD by reducing LC basal activity or by increasing LC tonic output at the cortical level. It has been hypothesized that posterior noradrenergic projections play a role in disengagement, changing focus, and reengagement, while anterior projections are involved in data analysis and response preparation.[44,53,54]

Stimulants presumably ameliorate symptoms of ADHD by increasing dopaminergic and noradrenergic levels and, consequently, catecholamine neurotransmission in the PFC and in other cortical regions. Nonstimulants such as imipramine and atomoxetine selectively inhibit NE reuptake and have no direct effect on DA. Nonetheless, the anatomy and physiology of the PFC enables these medications to robustly increase regional concentrations of extracellular DA.[55,56] It is important to understand that there is a paucity of DA transporters in this cortical region despite a plethora of DA nerve terminals. In contrast, the PFC has abundant NE transporters. Critically, NE transporters in this region are not selective to NE but also bind to and transport DA.[55] Consequently, in the PFC, DA is routinely transported into NE neurons, costored in synaptic vesicles with NE, and coreleased with NE when these neurons are stimulated.[55,56] Furthermore, the integrity of DA metabolism in the PFC is dependent upon NE. Following 6-hydroxydopamine-induced degeneration of the NE fibers projecting to the ventral tegmental area, DA utilization is decreased in the PFC but remains unaltered in the nucleus accumbens.[57]

Norepinephrine medications and ADHD

Review

Ultimately, evidence on the importance of NE in ADHD must come from treatment studies with agents with known NE effects. The tricyclic antidepressants, the monoamine oxidase inhibitors (MAOIs), the α_2-agonists, bupropion, venlafaxine, and atomoxetine, all have mechanisms of action that enhance the synaptic activity of NE. The included tables contain the published studies with these agents for patients with attentional problems and are described in more detail below. Since 1972 and Winsburg's study of imipramine versus dextroamphetamine and placebo,[58] numerous studies in children have been published comparing tricyclic antidepressants to placebo and/or amphetamines (Table 19.1).[58–86] The evaluation of MAOIs commenced in 1983 with the Utah group's report on pargyline[87] and later on selegiline.[88] These are summarized in Table 19.2.[87–94] The assessment of the α_2-adrenergic receptor agonists began in 1985 by Hunt.[95] These are summarized in Table 19.3.[84,95–106] Several negative studies of bupropion for childhood ADHD preceded a successful, open-label study of adult ADHD by Wender in 1990.[107] These are summarized in Table 19.4.[107–122] Venlafaxine[123–129] and atomoxetine[130–137] are the most recently assessed medications and they are summarized in Tables 19.5 and 19.6 respectively. Atomoxetine is the most successful and extensively studied of these NE treatments and has received approval for the treatment of ADHD.

In the last 40 years, the nomenclature for ADHD has gone from minimal brain damage, to the hyperactive child syndrome, to ADD with or without hyperactivity, and now to ADHD. We are assuming for the purposes of this analysis that these are different names for the same disorder, and will refer to them collectively as ADHD in our discussion of medications. The studies included in the tables were located in Medline searches, references in academic reviews, and from presentations in professional conferences.

Comparing these studies is difficult for many reasons. It is difficult to make a global assessment of a study when it entails a condition like ADHD that has features of motor activity and mental focus that can be unequally affected by a medication. There are significant population differences. For example Michelson's 2003 study eliminated almost all comorbid diagnoses,[137] while Jankovic 1993 used only patients with comorbid ADHD and Tourette's.[90] Outcome measures differ. There is no consistency in design. How can one compare an open-trial study with a double-blind, placebo-controlled, parallel groups design? Open-label studies are seldom published without robust results. The outcome of a robust open-label study is not equivelent to a robust double-blind placebo-controlled outcome. Most of these studies are short-term and involve few subjects. Given the chronicity of ADHD, larger, multiyear efficacy assessments for these medications are needed. Finally, while some of these studies are convincing, others seem extremely limited because of poor design, an observation reinforced by time. The particulars of each table are reviewed below.

Stimulants versus agents with NE effects

Stimulants remain the most common treatment for ADHD and the standard against which other treatments are judged. While research has focused on their actions within the DA system, it is clear that they also affect the NE system through blocking reuptake and stimulating neurotransmitter release. The tricyclic antidepressants and atomoxetine activate the DA system only as a secondary consequence of blocking the NE uptake pump, and consequently they primarily activate DA in limited areas (e.g., the PFC) while having minimal effects on DA in other parts of the brain (e.g., the striatum and the nucleus accumbens). As a result there are increases in DA levels in the PFC following stimulant use or NE reuptake inhibition. In contrast, side effects including tics in humans and other stereotypic motor activities in animal models are associated with increases in DA levels in the striatum. Given that DA levels increase in the striatum following stimulant use, but not administration of an NE reuptake inhibitor, there is a possible benefit of noradrenergic medications over stimulants. Dopamine levels in the nucleus accumbens – the "anticipated reward center" – are affected by stimulant use but not NE uptake inhibitors. In contrast to activation of the PFC and the striatum, the clinical consequences of activating the nucleus accumbens are less clear. Activation of this center is implicated in addictions, euphoria, and psychosis, suggesting that activation by stimulants may be problematic for vulnerable subgroups, such as the patient with comorbid or previously existing addiction, bipolar disorder, or schizophrenia.

Stimulants have proven to be a generally safe and effective treatment for ADHD in childhood. In adulthood, however, there is not a long history of stimulant use for ADHD that clearly demonstrates that they are safe. Exploration of more selective NE medications was undertaken in the hopes of lessening possible side effects of stimulants – the reality or (perhaps unwarranted) fear of tics, euphoria, abuse potential, and diurnal ups and downs.

Tricyclic antidepressants

The tricyclic antidepressants were the first medications with significant NE activity that were studied for ADHD. Some of the early reports in highly respected journals are extremely positive. The authors felt that they would soon surpass stimulants as medications for ADHD. All tricyclic medications studied have the ability to block the NE reuptake transporter. The early studies started in the middle 1960s using the tertiary amines such as imipramine and amitriptyline. Starting in 1980, more attention has been given to the secondary amines such as desipramine and nortriptyline. The secondary amines are more specific NE uptake inhibitors as opposed to imipramine and amitriptyline, which have more serotonin activity. While imipramine and amitriptyline are more selective for the serotonin transporter, they still have activity at the NE uptake pump, and their active metabolites are desipramine and nortriptyline, respectively.[138] The tricyclics (Table 19.1) have shown positive effects in the treatment of ADHD for children, adolescents and adults.

The two greatest criticisms of the studies (and most other studies in these tables, with the exception of atomoxetine) is the small number of participants.

Table 19.1 Published studies of tricyclic antidepressants for ADHD.

Study	Treatment	Control	Age	Diagnosis	N	Notes
OPEN-LABEL TRIALS						
Huessy & Wright (1970)[59]	IMP	None	Child	Hyper	52	This was an open-label, flexible-dose study in which 67% of the children showed "marked improvement." This was considered a very positive study and produced additional later studies and some clinical acceptance of imipramine.
Waizer *et al.* (1974)[60]	IMP for 8 weeks	None	Child	Hyper	19	This paper does not provide data on the percentage of responders. There were changes on parent, teacher, and psychiatrists ratings. They were treated for 8 weeks at an average dose of 173 mg/day and had changes on attention ratings by teachers and Stroop testing. They were then switched to placebo single-blind and deteriorated. The discussion presents this study as a confirmation of Huessy's study.[59]
Quinn & Rapoport (1975)[61]	IMP	MPH	Child	Hyper	76	This was a 1-year, open-label, follow-up evaluation of the Rapoport 1974 population. Those on IMP were doing as well as those on MPH. There was a higher discontinuation rate on IMP.
Riddle & Rapoport (1976)[62]	IMP	MPH	Child	Hyper	72	This was a 2-year, open-label, follow-up evaluation of the Rapoport 1974 population. Those on IMP were doing similar to those on MPH. Many patients on active treatment had continuing symptoms.
Gastfriend *et al.* (1984)[63]	DMI	None	Adol	ADD	12	Patients were treated in an open manner. Eleven of the 12 improved within one month. Nine of these patients sustained improvement for 6 to 12 months.
Gastfriend *et al.* (1985)[64]	DMI	None	Adol	ADD	12	Patients were treated with an average of 3.2 mg/kg of DMI for 6 to 12 months. Nine out of 12 responded based psychiatrist CGI.
Biederman *et al.* (1986)[65]	DMI	None	Child	ADD	18	Patients were treated in an open manner. Limited abstract. Positive results were based global judgements, not rating scale.

(*cont.*)

Table 19.1 *(cont.)*

Study	Treatment	Control	Age	Diagnosis	N	Notes
Ratey *et al.* (1992)[66]	DMI	None	Adult	ADHD	30	This positive open trial is not available on Medline.
Wilens *et al.* (1993)[67]	NT	None	Child & Adol	ADHD	58	This was a chart review suggesting that NT is effective in ADHD. A full 76% of patients were seen as having moderate to marked improvement.
Spencer *et al.* (1993a)[68]	NT	None	Child	ADHD & Tourette's	12	This was a chart review of children with a tic disorder and ADHD. In it, 92% of patients were seen as having moderate to marked improvement in ADHD symptoms.
Spencer *et al.* (1993b)[69]	DMI	None	Child	ADHD & Tics & Tourette's	33	This was a chart review of children with a tic disorder and ADHD. In it, 80% of patients were seen as having moderate to marked improvement in ADHD symptoms.
Wilens *et al.* (1995)[70]	DMI & NT	None	Adult	ADHD	37	This was a retrospective chart review claiming a 70% response rate with patients on DMI at 180 mg/day and 40% response rate to 90 mg of NT. Many patients were also on a stimulant.
Wilens *et al.* (1996)[71]	Protrip-tyline	None	Child & Adol	ADHD	13	This was a prospective open-label evaluation. All patients had failed to respond to at least one previous medication. Improvement was described as modest with substantial side effects.
Prince *et al.* (2000)[72]	NT	None	Child & Adol	ADHD	32	A 6-week open-label phase was followed (for responders) by a 3-week controlled discontinuation phase. Patients improved in both ADHD and oppositional symptoms in the open phase. Following discontinuation phase, symptoms returned.
POSITIVE CONTROL TRIALS – ALSO CALLED EQUIVALENCE TRIALS-						
Winsberg *et al.* (1972)[58]	IMP	DEX PLO	Child	Hyper & Aggress	41	Children were hyperactive and/or aggressive, some had a history of injury to the CNS. This was a cross over design with raters blind to IMP and DEX but not placebo. Response to IMP and DEX were similar.

Gross (1973)[73]	IMP	DEX or MPH	Child	Min brain damage	259	Single blind placebo preceeded MPH, then DEX, and then IMP for one week each. The doses were relatively low. Outcome was based on descriptions from patient and parents. The three active treatments were considered relatively equal.
Werry *et al.* (1980)[74]	IMP	MPH PLO	Child	Hyper	30	This study compares MPH with two levels of IMP in a parallel design. Both levels of IMP were comparable to MPH on learning, motor performance and social behavior
DOUBLE-BLIND PLACEBO-CONTROLLED TRIALS						
Rapoport *et al.* (1974)[75]	IMP	MPH PLO	Child	Hyper	76	This historic double-blind study compared IMP, MPH and placebo. Both active treatments were superior to placebo. MPH was somewhat superior to IMP in both efficacy and side effects. Dose levels of IMP were relatively low.
Kupietz & Balka (1976)[76]	AMT	MPH PLO	Child	Hyper & Aggress	20	This study found that both MPH and AMT improved vigilance performance equally but did not assess ADHD symptoms. It had both placebo and baseline comparisons.
Yepes *et al.* (1977)[77]	AMT	MPH PLO	Child	Hyper & Aggress	21	Two weeks of each were given in random order, MPH at 40mg/day, AMT at 90mg/day, and PLO. MPH did not separate from placebo, but AMT did. Both were positive on CPT testing. Ratings were done by blind raters; psychiatrists were not blinded.
Winsberg *et al.* (1980)[78]	IMP	PLO	Child	Hyper	10	Ten MPH nonresponders were placed in a double-blind IMP–PLO crossover design. Treatment differences were not significant.
Garfinkel *et al.* (1983)[79]	DMI & CMI	MPH PLO	Child	ADD	12	This was a 4-cell crossover design. MPH was superior to the other treatments. There were nonspecific positive emotional effects in the DMI and CMI groups. DMI (NE) and CMI (serotonin) were selected to contrast NE and serotonin effects and there were few differences.
Donnelly *et al.* (1986)[80]	DMI	PLO	Child	ADD-H	29	In this placebo-controlled, parallel, double-blind study reports of symptom levels improving were vague. Primary focus is on physical/medical changes.

(cont.)

Table 19.1 *(cont.)*

Study	Treatment	Control	Age	Diagnosis	N	Notes
Gualtieri & Evans (1988)[81]	IMP	PLO	Child	ADD-H	9	This study compared three levels of IMP with placebo. IMP exerted negative dose-response effects on motor performance (motor speed, motor pursuit), while it improved hyperactive behavior and attention.
Biederman *et al.* (1989)[82]	DMI	PLO	Child & Adol	ADD	62	This was a parallel, double-blind study of up to 6 weeks. Most patients had a history of non response for ADHD. 68% of DMI-treated patients were considered very much or much improved, compared with only 10% on placebo.
Rapport *et al.* (1993)[83]	DMI	MPH PLO	Child	ADHD & Mood	16	This study looked at the separate and combined effects of MPH and DMI in a double-blind, crossover design. Subjects were hospitalized children with both ADHD and mood disorders. Both treatments affected cognitive functioning. ADHD symptoms were not assessed.
Singer *et al.* (1995)[84]	DMI	CLON PLO	Child	ADHD & Tourette's	34	This was a double-blind comparison of DMI, CLON and PLO. Patients received all three treatments in a random order in 6-week periods. Children were comorbid for ADHD and Tourette's. DMI showed evidence of efficacy on ADHD symptoms.
Wilens *et al.* (1996)[85]	DMI	PLO	Adult	ADHD	41	At endpoint, 68% of DMI patients were improved compared to no placebo patients.
Spencer *et al.* (2002)[86]	DMI	PLO	Child & Adol	ADHD & Tics	41	Patients were comorbid for tics (including Tourette's) and ADHD in this 6-week, double-blind, placebo-controlled, parallel trial. Improvement in ADHD was 42% relative to placebo and improvement in tics was 30% relative to placebo.

Abbreviations: IMP, imipramine; AMT, amitriptyline; DMI, desipramine; CMI, clomipramine; NT, nortriptyline; DEX, dextroamphetamine; MPH, methylphenidate; CLON, clonidine; PLO, placebo; CCI, clinical global impression; CGI-S, clinical global impression-subject; CPT, continuous performance task; ADD-H, attention deficit disorder-hyperactivity; ADD-RT, attention deficit disorder – residual type.

However, several moderately sized, placebo-controlled studies have been conducted with the tricyclics and were positive. Perhaps the most informative of these studies are those by Judith Rapaport. Patients were assessed not only during the double-blind trial identified in the table, but follow-up assessments of these children were later published. Imipramine was compared with methylphenidate and placebo in the double-blind trial. Both treatments were superior to placebo; however, several measures favored methylphenidate over imipramine. At the one-year follow-up,[61] more patients had stopped imipramine due to side effects than methylphenidate. All three groups were improved over baseline, however both methylphenidate and imipramine were better than placebo in classroom evaluations with methylphenidate showing a slight advantage. At the two-year evaluation,[62] 65% of the sample were still in treatment although most had been switched to methylphenidate. Despite treatment, most of the sample continued to experience substantial social and academic problems at this second follow-up evaluation. Treatment did seem to affect ADHD symptoms. Since all patients had experienced treatment at some point in the previous two years, patients could not be compared to placebo. This set of studies has been very influential in leading to the belief in the superiority of stimulants over alternative medications. These studies were published at a time when the relative efficacy of stimulants versus antidepressants was being actively debated.

In all, ten studies compared a tricyclic to a stimulant. Unfortunately, two of these studies, including the largest,[73] did not use ADHD symptoms for their outcome measure. Many of the other studies reported that tricyclics and stimulants had similar impacts on the illness. Many also indicated that patients had higher levels of side effects on the tricyclics. Conversely, some included relatively sophisticated cognitive testing (including the Stroop Test) and continuous performance testing. A number of these studies reported positive results on such testing, not a finding one would predict would be caused by placebo response.

In summary, the later studies using the tricyclics that are more selective for NE seem generally more persuasive. But neither group of tricyclics has achieved any degree of acceptance among physicians treating ADHD. Many physicians acknowledge trying them with occasional patients, but being disappointed in their effects even when patients use them carefully for extended periods of time. Consequently, there is a startling conflict between these seemingly successful clinical trials and actual clinical practice. Until the introduction of atomoxetine, the generally accepted conclusion was that using an NE uptake inhibitor was simply not enough to effectively treat ADHD. It was felt that combined NE–DA activity was needed for an ADHD medication.

There are two potential explanations for this conflict. First, the problem could, in part, be due to inadequate dosing in many of these studies and by many clinicians.

The most successful studies were conducted by Biederman's group at Harvard.[82] They used much higher doses than employed in most other settings. A second explanation could be related to the postsynaptic effects of the tricyclics. Richelson[139] indicates that in addition to blocking reuptake, tricyclics have affinity for a number of postsynaptic NE receptors. This might blunt their activity in potentiating NE. The secondary amines (desipramine and nortriptyline) have a lower affinity for these receptors than the tertiary amines, imipramine and amitriptyline.[139] Additionally, it is believed that this affinity for postsynaptic receptors is associated with a negative side effect profile.

Monoamine oxidase inhibitors

The MAOIs increase monoamine neurotransmitters by blocking the metabolic enzyme MAO that degrades NE, serotonin, and DA. This mechanism of action suggests that MAOIs might be attractive candidates for the treatment of ADHD. Conversely, there is now an ongoing debate regarding the relative importance of this metabolic pathway versus others leading to the termination of action of the monoamines. Both family-based and associational genetic analyses[140] suggest a role for a MAO-A promoter-region polymorphism in conferring risk for ADHD. The MAOIs are available in several varieties – nonselective, MAO-A (which effects primarily serotonin and NE), MAO-B (which effects primarily DA and phenylalanine), and reversible inhibitors of MAO-A (which are not available in the United States).

Using adult populations, Wender[87] and Wood[88] were the first to look at MAOIs for the treatment of ADHD. Although these studies were positive, there are currently only nine published studies with MAOIs and ADHD (see Table 19.2). Unfortunately they are all small (29 patients in the largest study). They contain MAOIs from all the categories – one nonselective (tranylcypromine), one MAO-A (clorgyline), six MAO-B (selegiline and pargyline), and one reversible MAO-A (moclobemide). The uncontrolled studies found positive effects, but the two placebo-controlled studies found little or no difference between the active agent and placebo.

Of particular relevance to the topic of this chapter, the agents that increase NE (clorgyline, tranylcypromine and moclobemide) all showed positive effects in ADHD, although the total number for these three studies is just 26 subjects. Conversely, having performed several of these studies, and reviewed the results over the years, we were most impressed with pargyline, a relatively selective B inhibitor. This does have some "A" activity also, meaning that it has a combination of NE and DA effects. This compound was originally sold as an antihypertensive agent but has been removed from the market in the United States.

Table 19.2 Published studies of monoamine oxidase inhibitors (MAOIs) for ADHD.

Study	Treatment	Control	Age	Diagnosis	N	Notes
OPEN-LABEL TRIALS						
Wender *et al.* (1983)[87]	Pargyline	None	Adult	ADD-RT	22	This was an open-label assessment in adults with ADHD. In it, 59% showed moderate to marked response.
Wood *et al.* (1983)[88]	Selegiline	None	Adult	ADD-RT	12	This was an open-label study. Half the patients responded at 30 mg/day of selegiline.
Trott *et al.* (1992)[89]	Moclobemide	None	Child	ADHD	12	This open-label study used patients who had discontinued stimulants due to side effects. Outpatient children improved 41% while inpatients improved 31%.
Jankovic *et al.* (1993)[90]	Selegiline	None	Child	ADHD & Tourette's	29	This was an open-label study for children comorbid for Tourette's and ADHD. All patients were refractory to other ADHD treatments. Average duration of treatment was 6.7 months. 90% of patients were significantly improved.
POSITIVE CONTROL TRIALS						
Zametkin *et al.* (1985)[91]	Clorgyline	DEX	Child	ADD-H	6	This was a double-blind crossover study. Most children responded to both medications.
Zametkin *et al.* (1985)[91]	Tranylcypromine	DEX	Child	ADD-H	8	This was a double-blind crossover study. Most children responded to both medications.
DOUBLE-BLIND PLACEBO-CONTROLLED TRIALS						
Feigin *et al.* (1996)[92]	Selegiline	PLO	Child	ADHD & Tourette's	24	Patients were comorbid for Tourette's and ADHD. This was a double-blind placebo-controlled crossover study with 8-week treatment periods separated by a 6-week washout period. Fifteen of 24 subjects completed the trial. Treatment appeared effective in the first period but not overall.
Ernst *et al.* (1996)[93]	Selegiline	PLO	Adult	ADHD	24	This was a double-blind placebo-controlled trial of selegiline at two treatment levels. No treatment effects were noted.
Akhondzadeh *et al.* (2003)[94]	Selegiline	MPH	Child	ADHD	28	Two levels of selegiline (separated by age) were compared with MPH in a double-blind parallel condition. The three groups showed equivalent improvement. This study was done in Iran.

Abbreviations: DEX, dextroamphetamine; MPH, methylphenidate; PLO, placebo.

α_2-agonists

The α_2 agonists clonidine and guanfacine have been studied as alternative treatments for ADHD. The initial interest for this class of medication was generated in part by positive studies with clonidine for patients with Tourette's syndrome – a condition with significant overlap with ADHD. Animal studies suggest that the α_2-agonists improve cognition by enhancing the selective responsiveness of the LC to relevant stimuli and by indirect activation of the PFC.[45] Recent single photon emission computed tomography (SPECT) studies have demonstrated increased cerebral blood flow in the PFC in monkeys administered guanfacine.

The 13 studies are reviewed in Table 19.3 (nine clonidine and four guanfacine). The total number of subjects was 508 patients. Unfortunately, several of the studies are uncontrolled and two studies found no benefits. In general, it appears that with greater levels of control there were fewer benefits. This mixed pattern combined with very limited acceptance in the community argues against concluding that these agents are active in ADHD. The last study noted was an extremely well-designed trial and found no benefits in patients with ADHD plus a tic disorder.

Bupropion

Bupropion is a unique antidepressant with a mechanism of action that remains unclear. It has affinity for the DA and NE reuptake transporters although only mildly and not as strong as the other antidepressants (for NE) or methylphenidate (for DA). The effects of bupropion may be through metabolites that may concentrate in the central nervous system at higher doses. Overall, bupropion may be more noradrenergic than dopaminergic due to the effects of a metabolite.[141] Clearly dopaminergic and noradrenergic components play a role in the medication's mechanism of action, but the precise details remain unclear and the effects may not be robust. Most of the results of the 16 bupropion studies found mild/moderate improvement. Of particular interest is the Conner's multicenter study of 109 children.[118] He stated that while bupropion was effective compared to placebo, results were not as robust as stimulant studies. Nevertheless, as opposed to the other agents discussed so far, bupropion has achieved acceptance in the market either as a monotherapeutic agent for ADHD or used in combination with stimulants. Unfortunately, there are no studies assessing the usefulness of bupropion in combination with a stimulant. Following Wender's positive report in 1990,[107] the company that held ownership of the medication (at that time Burroughs Wellcome) decided against a testing program to obtain FDA approval for the medication for ADHD. It is likely that most psychiatrists use bupropion in conjunction with methylphenidate to treat ADHD with associated depressive symptoms. The 2003 Wilens study is particularly interesting.[112] Patients in this open-label study continued receiving adjunct

Table 19.3 Published studies using α_2 agonists for ADHD.

Study	Treatment	Control	Age	Diagnosis	N	Notes
OPEN-LABEL TRIALS						
Hunt (1985)[95]	Clonidine	PLO	Child	ADD-H	10	It is described as a double-blind crossover study, but it appears that patients were given a placebo lead in and 1 week of placebo was compared to 8 weeks of clonidine.
Hunt (1987)[96]	Clonidine	MPH PLO	Child	ADD-H	10	Open-label clonidine was compared to MPH and single-blind placebo. The effects of the active medications were equal.
Schvehla *et al.* (1994)[97]	Clonidine	None	Child	ADHD & CD	18	This was a retrospective study of boys comorbid for ADHD and CD who had not responded to stimulants. 61% showed marked improvement.
Chappell *et al.* (1995)[98]	Guanfacine	None	Child & Adol	ADHD & Tourette's	10	This was an open-label study of children comorbid for Tourette's and ADHD. Improvement was reported in tics and the continuous performance test but not in ADHD symptoms.
Hunt *et al.* (1995)[99]	Guanfacine	None	All	ADHD	13	This was an open-label assessment of guanfacine in the treatment of ADHD. Patients improved over baseline on all ADHD measures.
Steingard *et al.* (1993)[100]	Clonidine	None	Child & Adol	ADHD +/− Tics	54	This retrospective chart review compared the usefulness of clonidine in children with tics and/or ADHD. Children with comorbid tics/ADHD benefited more than those with single diagnoses.
POSITIVE CONTROL TRIALS						
Leckman *et al.* (1991)[101]	Clonidine	PLO	All	Tourette's	47	All subjects suffered from Tourette's in this 12-week double-blind trial. Patients on clonidine showed more improvement in tics, impulsivity and hyperactivity.
Connor *et al.* (2000)[102]	Clonidine	MPH	Child & Adol	ADHD & ODD or CD	24	This was a 3-month, randomized, blinded, group comparison of MPH monotherapy, clonidine monotherapy or combination therapy. Patients had ADHD with comorbid CD or ODD. The three groups showed similar improvement in attention deficits, impulsivity, oppositional, and conduct disorder symptoms.

(cont.)

Table 19.3 (*cont.*)

Study	Treatment	Control	Age	Diagnosis	N	Notes
DOUBLE-BLIND PLACEBO-CONTROLLED TRIALS						
Goetz *et al.* (1987)[103]	Clonidine	PLO	All	Tourette's	30	This was a 6-month placebo-controlled crossover study of clonidine in Tourette's patients. It was not effective in reducing the tics or symptoms.
Singer *et al.* (1995)[84]	Clonidine	DMI PLO	Child	ADHD & Tourette's	37	This was a double-blind comparison of DMI, clonidine and PLO. Patients received all three treatments in a random order in 6-week periods. Children were comorbid for ADHD and Tourette's. Clonidine showed little activity against either ADHD or tics.
Scahill *et al.* (2001)[104]	Guanfacine	PLO	Child	ADHD & Tics	34	Patients from a Tic Disorder Clinic were enrolled in an 8-week study. At endpoint guanfacine was associated with a 37% improvement compared to 8% for placebo. Tic severity showed similar improvement.
Taylor *et al.* (2001)[105]	Guanfacine	DEX PLO	Adult	ADHD	17	This is described as a double-blind placebo-controlled comparison of guanfacine and DEX. Both drugs demonstrated improvement over placebo on ADHD measures.
Tourette's Syndrome Study Group (2002)[106]	Clonidine	MPH PLO	Child	ADHD & Tics	136	This was a very well-done study comparing four groups, PLO, MPH, clonidine, and MPH+clonidine. MPH did not make tics worse; in fact the greatest reduction occurred in the combined treatment group. Clonidine was most helpful for hyperactivity/impulsivity and MPH for attention. In reviewing the article, it is unclear if clonidine improved attention measures.

Abbreviations: DEX, dextroamphetamine; MPH, methylphenidate; DMI, desipramine; PLO, placebo.

antimanic agents (mood stabilizers and antipsychotics). Treatment was associated with symptom reductions for ADHD, depression, and mania.

Glaxo has completed a large, multicenter evaluation of a once-a-day formulation (bupropion XL) in adults with ADHD. The study was clearly positive. There was no evidence of a loss of ADHD efficacy with the longer-acting formulation during the day. Further, in long-term evaluation in a small number of patients using bupropion SR, there was no loss of efficacy found over six months.[112]

Venlafaxine

Venlafaxine, indicated for the treatment of depression and generalized anxiety, is classified as a NE/serotonin reuptake inhibitor because it acts on both the NE and serotonin reuptake transporters.[139,142] Unlike the tricyclic antidepressants, venlafaxine does not bind to postsynaptic NE receptors. This predicts a different and more favorable side effect profile compared to the tricyclic antidepressants.[139]

All of the venlafaxine studies for ADHD are open-label using small numbers of patients (Table 19.5). The results suggest a positive effect but seem best with patients with comorbid disorders such as major depression or substance abuse. The Hedges report is different from the other studies due to its use of the Utah Criteria for ADHD.[124] The Utah Criteria include three emotional symptoms (temper, affective lability, and emotional over-reactivity) as part of ADHD. Venlafaxine seemed more active in reducing the emotional components compared to the attentional symptoms. Without more data, it is unclear how venlafaxine compares to other treatments and where its place should be as a treatment of ADHD.

Atomoxetine

Atomoxetine (originally called tomoxetine) is a selective NE reuptake inhibitor and is the first nonstimulant medication to receive an FDA indication for ADHD. Atomoxetine is chemically related to fluoxetine and has no other known binding affinities except for the NE uptake site. It was extensively tested for use in depression but was not effective in multiple, carefully conducted studies (Table 19.6). Following Spencer's initial exploratory, double-blind crossover study in adults,[133] it was tested first in large multicenter studies of children with ADHD.[134–136] The results (Table 19.6) were uniformly positive and seem to establish a significant role of atomoxetine, and thus NE, in ADHD. In fact, the compound came through registration studies with very few negative results, unusual in the field of psychiatry. The contrast between the uniformly positive studies in ADHD and the negative results in depression is very interesting.

We have previously discussed how NE reuptake inhibition increases extracellular DA as well as NE in the PFC (see p. 563), and it is likely that this invariant relationship at least partially explains the effectiveness of atomoxetine in ADHD. Since NE uptake

Table 19.4 Published studies using bupropion for ADHD.

Study	Treatment	Control	Age	Diagnosis	N	Notes
OPEN-LABEL TRIALS						
Simeon *et al.* (1986)[108]	Bupro	None	Child	ADD-H	17	This was an open-trial in ADHD children. Twelve of 17 were rated moderate or better improved. Two weeks following discontinuation 7 patients regressed.
Wender & Reimherr (1990)[107]	Bupro	None	Adult	ADHD	19	This was an open-trial of ADHD adults. Most patients had a history of prior ADHD treatment. Fourteen responded to bupropion and ten chose to remain on it after the study ended.
Riggs *et al.* (1998)[109]	Bupro	None	Adol	ADHD, CD & Substance abuse	13	Patients were boys in a residential program with comorbid ADHD, CD and substance abuse. This was a 5-week open-trial. Patients showed improvement in hyperactivity and CGI-S.
Reimherr *et al.* (2000a)[110]	Bupro	MPH	Adult	ADHD	18	This was an open-trial follow-up to Reimherr 2000b.[119] Patients who failed to improve on bupropion were tried on MPH. At the long-term assessment outcomes for the two treatments were similar.
Daviss *et al.* (2001)[111]	Bupro	None	Adol	ADHD & Depress	14	Patients were comorbid for ADHD and depression (MDD or dysthymia). 4% improved in ADHD only, 29% depression only and 58% both. However, teacher ratings of ADHD improvement were not significant.
Wilens *et al.* (2003)[112]	Bupro	None	Adult	ADHD & Bipolar	36	This was an open-label trial of bupropion for adults with comorbid ADHD and bipolar disorder. Patients remained on antimanic medications. Patients improved in ADHD symptoms as well as symptoms of mania and depression.
Levin *et al.* (2002)[113]	Bupro	None	Adult	ADHD & Cocaine abuse	12	Patients who were in treatment for cocaine abuse and comorbid ADHD entered this open-trial. Treatment compliance and retention for both conditions was good. Outcome was comparable to a previous MPH study with a similar population.
Upadhyaya *et al.* (2004)[114]	Bupro	None	Adol	ADHD & Nicotine Depend	11	This open-trial involved nicotine-dependent teens with ADHD. This was a 7-week study. There was counseling for the smoking. Smoking improved, ADHD did not.

POSITIVE CONTROL TRIALS						
Casat *et al.* (1989)[115]	Bupro	PLO	Child	ADD-H	30	In a 6-week trial, 60% responded at a dose of 6mg/kg.
Barrickman *et al.* (1995)[116]	Bupro	MPH	Child & Adol	ADHD	15	This was a double-blind, crossover, comparison of bupropion and MPH. Treatment periods were 6 weeks long. There were no significant differences between treatments for ADHD symptoms or other measures.
DOUBLE-BLIND PLACEBO-CONTROLLED TRIALS						
Casat *et al.* (1987)[117]	Bupro	PLO	Child	ADD-H	31	This was the first report from a four-center study of bupropion in ADHD. There were positive results on some measures but not on others.
Conners *et al.* (1996)[118]	Bupro	PLO	Child	ADD-H	109	This was a multisite double-blind placebo-controlled trial. Bupropion was superior to placebo in ADHD and other measures. Effect sizes were smaller than most stimulant studies.
Reimherr *et al.* (2000b)[119]	Bupro	PLO	Adult	ADHD	60	This was a double-blind placebo-controlled trial of bupropion SR. 41% of bupropion subjects were much or very much improved compared to 21% of placebo subjects. All measures favored bupropion but only some achieved significance.
Wilens *et al.* (2001)[120]	Bupro	PLO	Adult	ADHD	40	This was a double-blind placebo-controlled parallel 6-week trial using adults. Many patients were comorbid for depression. Bupropion was associated with a 42% reduction in symptoms compared to 24% in placebo.
Kuperman *et al.* (2001)[121]	Bupro	MPH PLO	Adult	ADHD	30	A randomized, double-blind, parallel design was used to compare bupropion SR with MPH and placebo in adults. There was a 7-week treatment period. Based on clinical global improvement scores: 64% on bupropion, 50% on MPH and 20% on placebo were considered responders.
Connor *et al.* (2004)[122]	Bupro	PLO	Adult	ADHD	162	This placebo-controlled, parallel-group design study compared the effectiveness of once-a-day bupropion XL with placebo. Effectiveness was measured morning, afternoon, and evening with a treatment effect at all three periods. There was an effect size = .61. HAM-D and HAM-A scores did not change.

Abbreviations: MPH, methylphenidate; PLO, placebo; MDD, major depressive disorder; HAM-A, Hamilton Rating Scale for Anxiety; HAM-D, Hamilton Rating Scale for Depression.

Table 19.5 Published studies using venlafaxine (VEN) for ADHD.

Study	Treatment	Control	Age	Diagnosis	N	Notes
OPEN-LABEL TRIALS						
Adler *et al.* (1995)[123]	VEN	None	Adult	ADD	16	This was an 8-week, open-trial of venlafaxine for ADHD adults. Four of 16 discontinued due to side effects, the rest demonstrated an average of 50% improvement in ADHD symptoms.
Hedges *et al.* (1995)[124]	VEN	None	Adult	ADHD	18	This was an open-trial of venlafaxine in ADHD adults. Of the 18 patients eight showed a good response and seven could not tolerate the medication.
Olvera *et al.* (1996)[125]	VEN	None	Child & Adol	ADHD	16	This was a 5-week, open-trial of venlafaxine in children and adolescents. Four patients could not tolerate the medication. Seven patients showed significant improvement.
Findling *et al.* (1996)[126]	VEN	None	Adult	ADHD	10	This was an 8-week, open-trial of venlafaxine in adults. Nine patients completed the trial, seven were considered responders.
Upadhyaya *et al.* (2001)[127]	VEN	None	Adult	ADHD & Sub abuse	10	Investigators report improvement in ADHD symptoms and reduction in alcohol use. Four out of ten completed the study.
Motavalli Mukaddes & Abali (2004)[128]	VEN	None	Child & Adol	ADHD	13	A 6-week open-trial of venlafaxine was conducted in 13 children and adolescents with ADHD and no depression. All subjects completed the trial with few complaints about side effects. Subjects showed significant improvement in ADHD symptoms.
RETROSPECTIVE EVALUATION						
Hornig-Rohan & Amsterdam (2002)[129]	VEN	Stimulant + Bupro	Adult	ADHD & Depress	17	This retrospective evaluation assessed the impact of venlafaxine and stimulants in adults with comorbid ADHD and major depressive disorder. Both VEN and VEN+MPH were associated with improvement in the combined symptoms of depression and ADHD.

MPH, methylphenidate.

Table 19.6 Published studies using atomoxetine (ATM) for ADHD.

Study	Treatment	Control	Age	Diagnosis	N	Notes
OPEN-LABEL TRIALS						
Spencer *et al.* (2001)[130]	ATM	None	Child	ADHD	22	This was a flexible dose, 11-week open-trial of atomoxetine. Subjects ranged from 7 to 14. Symptoms improved 39% over baseline.
Kratochvil *et al.* (2001)[131]	ATM	None	Child	ADHD	8	Eight of 10 subjects completed this 10-week, open-trial of ATM. Seven reached efficacy criteria at endpoint.
Kratochvil *et al.* (2002)[132]	ATM	MPH	Child	ADHD	228	184 children were randomized to ATM and 44 to MPH in this open-label comparison. Efficacy, safety and tolerability appear similar for the two treatments. However, the uneven numbers might decrease the sensitivity of the comparison.
DOUBLE-BLIND PLACEBO-CONTROLLED TRIALS						
Spencer *et al.* (1998)[133]	ATM	PLO	Adult	ADHD	22	This was a double-blind, placebo-controlled, crossover design. 52% of patients improved on ATM compared to 10% on placebo.
Michelson *et al.* (2001)[134]	ATM	PLO	Child & Adol	ADHD	297	This was a double-blind placebo-controlled comparison of three dose levels of ATM in children. 38% were comorbid for ODD. Higher doses were associated with more ADHD symptom improvement. Social and family functioning also improved. Discontinuations as a result of adverse events were <5%.
Spencer *et al.* (2002)[135]	ATM	MPH PLO	Child	ADHD	147	These are two identical double-blind placebo-controlled studies. Stimulant-naïve patients were randomized to ATM, placebo, or MPH.
Spencer *et al.* (2002)[135]	ATM	MPH PLO	Child	ADHD	144	Patients with prior stimulant exposure were randomized to atomoxetine or placebo. ATM was found to be superior to placebo for ADHD.

(cont.)

Table 19.6 *(cont.)*

Study	Treatment	Control	Age	Diagnosis	N	Notes
Michelson *et al.* (2002)[136]	ATM	PLO	Child & Adol	ADHD	171	This 6-week, double-blind placebo-controlled trial assessed the efficacy of once-daily ATM. Treatment was superior to placebo and similar to prior studies using BID dosing (effect size = .71). Discontinuations as a result of adverse events were <3%.
Michelson *et al.* (2003)[137]	ATM	PLO	Adult	ADHD	280	This pair of identical double-blind placebo-controlled studies assessed ATM in adult ADHD patients. All patients with other Axis I diagnoses were excluded. ATM was superior to placebo in all ADHD measures.
Michelson *et al.* (2003)[137]	ATM	PLO	Adult	ADHD	256	

Abbreviations: MPH, methylphenidate; PLO, placebo.

inhibitors enhance burst firing of ventral tegmental nuclei, there is no doubt that NE effects on DA occur in numerous brain regions. However, it is also clear that the role of NE on DA is focal and selective, involving the PFC but not the nucleus accumbens or the striatum.[55,56] This specificity may prove to be an advantage for atomoxetine – it affects the DA in this partial and selective way and may spare some patients problematic consequences such as tics, euphoria, and addiction. Alternatively, it may prove to be a disadvantage for atomoxetine, as this same partial and selective action may deprive some patients of benefits for ADHD-related issues such as aimlessness and lack of drive and motivation.

There have been no head-to-head studies of atomoxetine vs. stimulants designed with adequate numbers to provide accurate comparison. Lilly has conducted two studies that do provide some information on this topic. Kratochvil published results from a prospective, open-label, random assignment comparison of atomoxetine with methylphenidate.[132] The results showed no differences. A second set of Lilly studies were reported in the same year by Spencer.[135] These two studies were a core part of the FDA submission package produced by Lilly. The effects of atomoxetine were similar to methylphenidate, but slightly less positive. However, this numerical difference did not achieve significance.

The initial response to stimulants is clearly more rapid and robust than is the response to atomoxetine. In Spencer's study, significant effects from atomoxetine were apparent by 2 weeks and continued to increase thereafter. By week 6 these effects were comparable to the effects of methylphenidate. Furthermore, although the initial response to atomoxetine initially lags behind the response to stimulants, long-term studies show continuing improvement for extended time periods. In a reanalysis of an atomoxetine study by Adler,[143] adult ADHD patients improved for up to 16 weeks and sustained these gains for 34 weeks (Figure 19.1). Although atomoxetine has not proven effective in treating depression, it has proven effective in treating the emotional symptoms of ADHD. In a reanalysis of the Michelson 2003 adult studies,[137] Reimherr found that emotional dysregulation, as measured by the Wender–Reimherr Adult Attention Deficit Disorder Scale (WRAADDS) (anger, emotional over-reactivity and affective lability), improved as much as hyperactivity/impulsivity or attentional symptoms.[144]

Summary

In this chapter we reviewed the historical understanding of ADHD and how it has evolved since Sir George Stills' first monographs. Simplistic views of attention and activity levels have given way to increasingly specific measures often involving PFC executive functions. Current evidence suggests that ADHD results, in part,

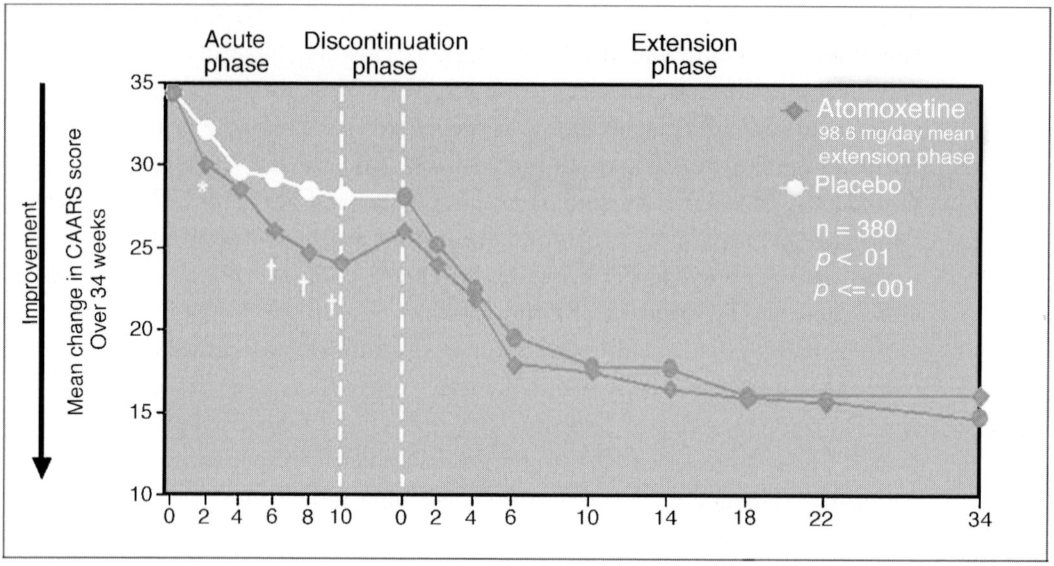

Figure 19.1 Open-label extension study of atomoxetine in adult ADHD (34 weeks).

from impairment of the NE system. This evidence includes: human genetic studies, neurobiological studies, animal studies, and medication response. This evidence does not negate the importance of the DA system in the etiology of ADHD since the two systems cannot be functionally separated within the PFC. Genetic studies have focused on the DA system particularly *DRD4* and *DAT* DA genes, but there is some evidence of NE system involvement. Given that ADHD is probably inherited via a polygenic mechanism, numerous genes will probably be identified.

Several studies implicate DA and noradrenergic impairment in the PFC. Further, the DA and NE systems are interdependent and changes in one usually impact the other. It is likely that the PFC regulates behavior in several ways. For example, connections between the inferior temporal lobe and ventrolateral PFC mediate working memory for visual feature information. Connections between the superior temporal cortex and anterior PFC may serve a similar role in the auditory domain. There may be parallel circuits extending to the orbital and medial PFC that interconnect with the amygdala and hypothalamus to serve working memory for affective behavior. It is likely that these pathways affect both "overt behaviors" (motor responses) and "covert behaviors" such as perceptions, thoughts and feelings.

Clinical research demonstrating the efficacy of medications with NE actions is perhaps the most important evidence implicating NE in ADHD. The evidence of this treatment effect is stronger for atomoxetine than tricyclics and bupropion. For other agents, the evidence is less clear. While methylphenidate often displays a treatment effect within a week and full response within a month, Wilens *et al.* found that the bulk of response to bupropion occurred in weeks 5 and 6[120] and Adler found improvement through the 14th week with atomoxetine.[143] Finally, the MTA study supports a general belief that medications are often insufficient.[26] Psychosocial intervention appears important for a variety of associated problems: oppositional/aggressive symptoms, internalizing symptoms, teacher-rated social skills, parent–child relations, and reading achievement.

REFERENCES

1. Still, G. F. Some abnormal psychical conditions in children. *Lancet*, **1** (1902), 1008–1012, 1077–1082, 1163–1168.
2. Bradley, C. The behavior of children receiving benzedrine. *Am. J. Psychiatry*, **94** (1937), 577–585.
3. Bradley, C. Benzedrine and Dexedrine in the treatment of children's behavior disorders. *Pediatrics*, **5** (1950), 24–37.
4. Strauss, A. A. and Lehtinen, L. E. *Psychopathology and Education of the Brain-injured Child* (New York: Grune & Stratton, 1947).
5. Dolphin, J. E. and Cruickshank, W. M. Pathology of concept formation in children with cerebral palsy. *Am. J. Ment. Defic.*, **56** (1951), 386–392.
6. Strauss, A. A. and Kephart, N. C. *Psychopathology and Education of the Brain-injured Child*, Vol. II. *Progress in Theory and Clinic* (New York: Grune and Stratton, 1955).
7. Knobel, M., Wolman, M. B. and Mason, E. Hyperkinesis and organicity in children. *Arch. Gen. Psychiatry*, **1** (1959), 310–321.
8. Laufer, M. W. and Denhoff, E. Hyperkinetic behavior syndrome in children. *J. Pediatr.*, **50** (1957), 463–474.
9. Connors, C. K. The syndrome of minimal brain dysfunction: psychological aspects. *Pediatr. Clin. North Am.*, **14** (1967), 749–766.
10. Wender, P. H. *Minimal Brain Dysfunction in Children* (New York: John Wiley and Sons, Inc., 1971).
11. Diagnostic and Statistical Manual of Mental Disorders DSM II (Washington, DC: American Psychiatric Press, Inc., 1968).
12. Wood, D. R., Reimherr, F. W., Wender, P. H. and Johnson, G. E. Diagnosis and treatment of minimal brain dysfunction in adults: a preliminary report. *Arch. Gen. Psychiatry*, **33** (1976), 1453–1460.
13. Wender, P. H., Reimherr, F. W. and Wood, D. R. Attention deficit disorder ('minimal brain dysfunction') in adults. *Arch. Gen. Psychiatry*, **38** (1981), 449–456.
14. Anderson, C. *Society Pays: The High Costs of Minimal Brain Damage in America* (New York: Walker Press, 1972).

15. Barkley, R. A. The problem of stimulus control and rule-governed behavior in children with attention deficit disorder with hyperactivity. In *Attention Deficit Disorders* ed. J. Swanson and L. Bloomingdale. (New York: Pergamon Press, 1989), pp. 203–234.
16. Glow, P. H. and Glow, R. A. Hyperkinetic impulse disorder: a developmental defect of motivation. *Genet. Psychol. Monogr.*, **100** (1979), 159–231.
17. Haenlein, M. and Caul, W. F. Attention deficit disorder with hyperactivity: a specific hypothesis of reward dysfunction. *J. Am. Acad. Child Adolesc. Psychiatry*, **26** (1987), 356–362.
18. Douglas, V. I. and Parry, P. A. Effects of reward on delayed reaction time task performance of hyperactive children. *J. Abnorm. Child Psychol.*, **11** (1983), 313–326.
19. Douglas, V. I. and Parry, P. A. Effects of reward and non-reward on attention and frustration in attention deficit disorder. *J. Abnorm. Child Psychol.*, **22** (1994), 281–302.
20. Parry, P. A. and Douglas, V. I. Effects of reinforcement on concept identification in hyperactive children. *J. Abnorm. Child Psychol.*, **11** (1983), 327–340.
21. Barkley, R. A. *Hyperactive Children: A Handbook for Diagnosis and Treatment* (New York: Guilford Press, 1981).
22. Barkley, R. A. *Attention-deficit Hyperactivity Disorder: A Handbook for Diagnosis and Treatment* (New York; Guilford Press, 1990).
23. Barkley, R. A. *ADHD and the Nature of Self-control* (New York; Guilford Press, 1997).
24. Jensen, P. S., Hinshaw, S. P., Kraemer, H. C. *et al.* ADHD comorbidity findings from the MTA study: comparing comorbid subgroups. *J. Am. Acad. Child Adolesc. Psychiatry*, **40** (2001), 147–158.
25. Vitiello, B., Severe, J. B., Greenhill, L. L. *et al.* Methylphenidate dosage for children with ADHD over time under controlled conditions: lessons from the MTA. *J. Am. Acad. Child Adolesc. Psychiatry*, **40** (2001), 188–196.
26. MTA Cooperative Group. MTA was a 14-month, randomized, clinical trial of treatment strategies for ADHD. *Arch. Gen. Psychiatry*, **56** (1999), 1073–1086.
27. Eaves, L. J., Silberg, J. L., Meyer, J. M. *et al.* Genetics and developmental psychopathology. 2. The main effects of genes and environment on behavioral problems in the Virginia Twin Study of Adolescent Behavioral Development. *J. Child Psychol. Psychiatry*, **38** (1997), 965–980.
28. Edelbrock, C., Rende, R., Plomin, R. and Thompson, L. A. A twin study of competence and problem behavior in childhood and early adolescence. *J. Child Psychol. Psychiatry*, **36** (1995), 775–785.
29. Gillis, J. J., Gilger, J. W., Pennington, B. F. and DeFries, J. C. Attention deficit disorder in reading-disabled twins: evidence for a genetic etiology. *J. Abnorm. Child Psychol.*, **20** (1992), 303–315.
30. Gjone, H., Stevenson, J. and Sundet, J. M. Genetic influence on parent-reported attention-related problems in a Norwegian general population twin sample. *J. Am. Acad. Child Adolesc. Psychiatry*, **35** (1996), 588–596.
31. Hudziak, J. J., Rudiger, L. P., Neale, M. C., Heath, A. C. and Todd, R. D. A twin study of inattentive, aggressive, and anxious/depressed behaviors. *J. Am. Acad. Child Adolesc. Psychiatry*, **39** (2000), 469–476.

32. Levy, F., Hay, D. A., McStephen, M. Wood, C. and Waldman, I. Attention-deficit hyperactivity disorder: a category or a continuum? Genetic analysis of a large-scale twin study. *J. Am. Acad. Child Adolesc. Psychiatry*, **36** (1997), 737–744.
33. Sherman, D. K., McGue, M. K. and Iacono, W. G. Twin concordance for attention deficit hyperactivity disorder: a comparison of teachers' and mothers' reports. *Am. J. Psychiatry*, **154** (1997), 532–555.
34. Thapar, A., Hervas, A. and McGuffin, P. Childhood hyperactivity scores are highly heritable and show sibling competition effects: twin study evidence. *Behav. Genet.*, **25** (1995), 537–544.
35. Silberg, J., Rutter, M., Meyer, J. *et al.* Genetic and environmental influences on the covariation between hyperactivity and conduct disturbance in juvenile twins. *J. Child Psychol. Psychiatry*, **37** (1996), 803–816.
36. LaHoste, G. J., Swanson, J. M., Wigal, S. B. *et al.* Dopamine D4 receptor gene polymorphism is associated with attention deficit hyperactivity disorder. *Mol. Psychiatry*, **1** (1996), 121–124.
37. Faraone, S. V., Doyle, A. E., Mick, E. and Biederman, J. Meta-analysis of the association between the 7-repeat allele of the dopamine D(4) receptor gene and attention deficit hyperactivity disorder. *Am. J. Psychiatry*, **158** (2001), 1052–1057.
38. Cook, E. H., Jr., Stein, M. A., Krasowski, M. D. *et al.* Association of attention deficit-disorder and the dopamine transporter gene. *Am. J. Hum. Genet.*, **56** (1995), 993–998.
39. Kirley, A., Hawi, Z., Daly, G. *et al.* Dopaminergic system genes in ADHD: toward a biological hypothesis. *Neuropsychopharmacology*, **27** (2002), 607–619.
40. Comings, D. E., Gade-Andavolu, R., Gonzalez, N. *et al.* Multivariate analysis of associations of 42 genes in ADHD, ODD and conduct disorder. *Clin. Genet.*, **58** (2000), 31–40.
41. Comings, D. E., Gade-Andavolu, R., Gonzalez, N. *et al.* Additive effect of three noradrenergic genes (*ADRA2a, ADRA2C, DBH*) on attention-deficit hyperactivity disorder and learning disabilities in Tourette syndrome subjects. *Clin. Genet.*, **55** (1999), 60–72.
42. Viggiano, D., Vallone, D., Ruocco, L. A. and Sadile, A. G. Behavioural, pharmacological, morpho-functional molecular studies reveal a hyperfunctioning mesocortical dopamine system in an animal model of attention deficit and hyperactivity disorder. *Neurosci. Biobehav. Rev.*, **27** (2003), 683–689.
43. Ohno, M. The dopaminergic system in attention deficit/hyperactivity disorder. *Congenit. Anom. (Kyoto)*, **43** (2003), 114–122.
44. Pliszka, S. R., McCracken, J. T. and Mass, J. W. Catecholamines in attention-deficit hyperactivity disorder: current perspectives. *J. Am. Acad. Child Adolesc. Psychiatry*, **35** (1996), 264–272.
45. Arnsten, A. F. T., Steere, J. C. and Hunt, R. D. The contribution of α2-noradrenergic mechanisms to prefrontal cortical cognitive function. *Arch. Gen. Psychiatry*, **53** (1996), 448–455.
46. Mattes, J. A. The role of frontal lobe dysfunction in childhood hyperkinesis. *Comp. Psychiatry*, **21** (1980), 358–369.
47. Arnsten, A. F. T. and Goldman-Rakic, P. S. Alpha-2 adrenergic mechanisms in prefrontal cortex associated with cognitive decline in aged nonhuman primates. *Science*, **230** (1985), 1273–1276.

48. Arnsten, A. F. T., Cai, J. X. and Goldman-Rakic, P. S. The alpha-2 adrenergic agonist guanfacine improves memory in aged monkeys without sedative or hypotensive side effects: evidence for alpha-2 receptor subtypes. *J. Neurosci.*, **8** (1988), 4287–4298.
49. Arnsten, A. F. T. and Leslie, F. M. Behavioral and receptor binding analysis of the alpha-2 adrenergic agonist, UK-14304 (5 bromo-6 [2-imidazoline-2-yl amino] quinoxaline): evidence for cognitive enhancement at an alpha-2 adrenoceptor subtype. *Neuropharmacology*, **30** (1991), 1279–1289.
50. Kobilka, B. K., Matsui, H., Kobilka, T. S. *et al.* Cloning, sequencing, and expression of the gene encoding for the human platelet alpha-2 adrenergic receptor. *Science*, **238** (1987), 650–656.
51. Regan, J. W., Kobilka, T. S., Yang-Feng, T. L. *et al.* Cloning and expression of a human kidney cDNA for an alpha-2 adrenergic receptor subtype. *Proc. Natl. Acad. Sci. U. S. A.*, **85** (1988), 6301–6305.
52. Lomasney, J. W., Lorenz, W., Allen, L. F. *et al.* Expansion of the alpha-2 adrenergic family: cloning and characterization of a human alpha 2 adrenergic receptor subtype, the gene for which is located on chromosome 2. *Proc. Natl. Acad. Sci. U. S. A.*, **86** (1990), 5094–5098.
53. Aston-Jones, G. Norepinephrine. In *Neuropsychopharmacology: The Fifth Generation of Progress*, ed. K. L. Davis, D. Charney, J. T. Coyle and C. B. Nemeroff. (Philadelphia: Lippincott Williams & Wilkins, 2002), pp. 47–57.
54. Mefford, I. N. and Potter, W. Z. A neuroanatomical and biochemical basis for attention deficit disorder with hyperactivity in children: a defect in tonic adrenaline mediated inhibition of locus coeruleus stimulation. *Med. Hypotheses*, **29** (1989), 33–42.
55. Yamamoto, B. K. and Novotney, S. Regulation of extracellular dopamine by the norepinephrine transporter. *J. Neurochem.*, **71** (1998), 274–280.
56. Bymaster, F. P., Katner, J. S., Nelson, D. L. *et al.* Atomoxetine increases extracellular levels of norepinephrine and dopamine in prefrontal cortex of rat: a potential mechanism for efficacy in attention deficit/hyperactivity disorder. *Neuropsychopharmacology*, **27** (2002), 699–711.
57. Roth, R. H. and Elsworth, J. D. Biochemical pharmacology of midbrain dopamine neurons. Anatomical and functional considerations. In *Neuropharmacology: The Fourth Generation of Progress*, ed. D. J. Kupfer and F. E. Bloom. (Philadelphia: Lippincott Williams & Wilkins, 1995), pp. 227–244.
58. Winsberg, B. G., Bialer, I., Kupietz, S. and Tobias, J. Effects of imipramine and dextroamphetamine on behavior of neuropsychiatrically impaired children. *Am. J. Psychiatry*, **128** (1972), 1425–1431.
59. Huessy, H. R. and Wright, A. L. The use of imipramine in children's behavior disorders. *Acta Paedopsychiatr.*, **37** (1970), 194–199.
60. Waizer, J., Hoffman, S. P., Polizos, P. and Engelhardt, D. M. Outpatient treatment of hyperactive school children with imipramine. *Am. J. Psychiatry*, **131** (1974), 587–591.
61. Quinn, P. O. and Rapoport, J. L. One-year follow-up of hyperactive boys treated with imipramine or methylphenidate. *Am. J. Psychiatry*, **132** (1975), 241–245.
62. Riddle, K. D. and Rapoport, J. L. A 2-year follow-up of 72 hyperactive boys. Classroom behavior and peer acceptance. *J. Nerv. Ment. Dis.*, **162** (1976), 126–134.

63. Gastfriend, D. R., Biederman, J. and Jellinek, M. S. Desipramine in the treatment of adolescents with attention deficit disorder. *Am. J. Psychiatry*, **141** (1984), 906–908.
64. Gastfriend, D. R., Biederman, J. and Jellinek, M. S. Desipramine in the treatment of attention deficit disorder in adolescents. *Psychopharmacol. Bull.*, **21** (1985), 144–145.
65. Biederman, J., Gastfriend, D. R. and Jellinek, M. S. Desipramine in the treatment of children with attention deficit disorder. *J. Clin. Psychopharmacol.*, **6** (1986), 359–363.
66. Ratey, J. J., Greenberg, M. S., Bemporad, J. R. and Lindem, K. J. Unrecognized attention-deficit hyperactivity disorder in adults presenting for outpatient psychotherapy. *J. Child Adolesc. Psychopharmacol.*, **2** (1992), 267–275.
67. Wilens, T. E., Biederman, J., Geist, D. E., Steingard, R. and Spencer, T. Nortriptyline in the treatment of ADHD: a chart review of 58 cases. *J. Am. Acad. Child Adolesc. Psychiatry*, **32** (1993), 343–349.
68. Spencer, T., Biederman, J., Wilens, T., Steingard, R. and Geist, D. Nortriptyline treatment of children with attention-deficit hyperactivity disorder and tic disorder or Tourette's syndrome. *J. Am. Acad. Child Adolesc. Psychiatry*, **32** (1993a), 205–210.
69. Spencer, T., Biederman, J., Kerman, K., Steingard, R. and Wilens, T. Desipramine treatment of children with attention-deficit hyperactivity disorder and tic disorder or Tourette's syndrome. *J. Am. Acad. Child Adolesc. Psychiatry*, **32** (1993b), 354–360.
70. Wilens, T. E., Biederman, J., Mick, E. and Spencer, T. A systematic assessment of tricyclic antidepressants in the treatment of adult attention-deficit hyperactivity disorder. *J. Nerv. Ment. Dis.*, **183** (1995), 48–50.
71. Wilens, T. E., Biederman, J., Abrantes, A. M. and Spencer, T. J. A naturalistic assessment of protriptyline for attention-deficit hyperactivity disorder. *J. Am. Acad. Child Adolesc. Psychiatry*, **35** (1996), 1485–1490.
72. Prince, J. B., Wilens, T. E., Biederman, J. *et al.* A controlled study of nortriptyline in children and adolescents with attention deficit hyperactivity disorder. *J. Child Adolesc. Psychopharmacol.*, **10** (2000), 193–204.
73. Gross, M. Imipramine in the treatment of minimal brain dysfunction in children. *Psychosomatics*, **14** (1973), 283–285.
74. Werry, J. S., Aman, M. G. and Diamond, E. Imipramine and methylphenidate in hyperactive children. *J. Child Psychol. Psychiatry*, **21** (1980), 27–35.
75. Rapoport, J. L., Quinn, P. O., Bradbard, G., Riddle, K. D. and Brooks, E. Imipramine and methylphenidate treatments of hyperactive boys. A double-blind comparison. *Arch. Gen. Psychiatry*, **30** (1974), 789–793.
76. Kupietz, S. S. and Balka, E. B. Alterations in the vigilance performance of children receiving amitriptyline and methylphenidate pharmacotherapy. *Psychopharmacology*, **50** (1976), 29–33.
77. Yepes, L. E., Balka, E. B., Winsberg, B. G. and Bialer, I. Amitriptyline and methylphenidate treatment of behaviorally disordered children. *J. Child Psychol. Psychiatry*, **18** (1977), 39–52.
78. Winsberg, B. G., Kupietz, S. S., Yepes, L. E. and Goldstein, S. Ineffectiveness of imipramine in children who fail to respond to methylphenidate. *J. Autism Dev. Disord.*, **10** (1980), 129–137.

79. Garfinkel, B. D., Wender, P. H., Sloman, L. and O'Neill, I. Tricyclic antidepressant and methylphenidate treatment of attention deficit disorder in children. *J. Am. Acad. Child Psychiatry*, **22** (1983), 343–348.
80. Donnelly, M., Zametkin, A. J., Rapoport, J. L. *et al.* Treatment of childhood hyperactivity with desipramine: plasma drug concentration, cardiovascular effects, plasma and urinary catecholamine levels, and clinical response. *Clin. Pharmacol. Ther.* **39** (1986), 72–81.
81. Gualtieri, C. T. and Evans, R. W. Motor performance in hyperactive children treated with imipramine. *Percept. Mot. Skills*, **66** (1988), 763–769.
82. Biederman, J., Baldessarini, R. J., Wright, V., Knee, D. and Harmatz, J. S. A double-blind placebo controlled study of desipramine in the treatment of ADD: I. Efficacy. *J. Am. Acad. Child Adolesc. Psychiatry*, **28** (1989), 777–784.
83. Rapport, M. D., Carlson, G. A., Kelly, K. L. and Pataki, C. Methylphenidate and desipramine in hospitalized children: I. Separate and combined effects on cognitive function. *J. Am. Acad. Child Adolesc. Psychiatry*, **32** (1993), 333–342.
84. Singer, H. S., Brown, J., Quaskey, S. *et al.* The treatment of attention-deficit hyperactivity disorder in Tourette's syndrome: a double-blind placebo-controlled study with clonidine and desipramine. *Pediatrics*, **95** (1995), 74–81.
85. Wilens, T. E., Biederman, J., Prince, J. *et al.* Six-week, double-blind, placebo-controlled study of desipramine for adult attention deficit hyperactivity disorder. *Am. J. Psychiatry*, **153** (1996), 1147–1153.
86. Spencer, T., Biederman, J., Coffey, B. *et al.* A double-blind comparison of desipramine and placebo in children and adolescents with chronic tic disorder and comorbid attention-deficit/hyperactivity disorder. *Arch. Gen. Psychiatry*, **59** (2002), 649–656.
87. Wender, P. H., Wood, D. R., Reimherr, F. W. and Ward, M. An open trial of pargyline in the treatment of attention deficit disorder, residual type. *Psychiatry Res.*, **9** (1983), 329–336.
88. Wood, D. R., Reimherr, F. W. and Wender, P. H. The use of L-deprenyl in the treatment of attention deficit disorder, residual type. *Psychopharmacol. Bull.*, **19** (1983), 627–629.
89. Trott, G. E., Friese, H. J., Menzel, M. and Nissen, G. Use of moclobemide in children with attention deficit hyperactivity disorder. *Psychopharmacology*, **106** (1992), S134–S136.
90. Jankovic, J. Deprenyl in attention deficit associated with Tourette's syndrome. *Arch. Neurol.*, **50** (1993), 286–288.
91. Zametkin, A., Rapoport, J. L., Murphy, D. L., Linnoila, M. and Ismond, D. Treatment of hyperactive children with monoamine oxidase inhibitors. I. Clinical efficacy. *Arch. Gen. Psychiatry*, **42** (1985), 962–966.
92. Feigin, A., Kurlan, R., McDermott, M. P. *et al.* A controlled trial of deprenyl in children with Tourette's syndrome and attention deficit hyperactivity disorder. *Neurology*, **46** (1996), 965–968.
93. Ernst, M., Liebenauer, L. L., Jons, P. H. *et al.* Selegiline in adults with attention deficit hyperactivity disorder: clinical efficacy and safety. *Psychopharmacol. Bull.*, **32** (1996), 327–334.
94. Akhondzadeh, S., Tavakolian, R., Davari-Ashtiani, R., Arabgol, F. and Amini, H. Selegiline in the treatment of attention deficit hyperactivity disorder in children: a double blind and randomized trial. *Prog. Neuropsychopharmacol. Biol. Psychiatry*, **27** (2003), 841–845.

95. Hunt, R. D., Minderaa, R. B. and Cohen, D. J. Clonidine benefits children with attention deficit disorder and hyperactivity: report of a double-blind placebo-crossover therapeutic trial. *J. Am. Acad. Child Psychiatry*, **24** (1985), 617–629.
96. Hunt, R. D. Treatment effects of oral and transdermal clonidine in relation to methylphenidate: an open pilot study in ADD-H. *Psychopharmacol. Bull.*, **23** (1987), 111–114.
97. Schvehla, T. J., Mandoki, M. W. and Sumner, G. S. Clonidine therapy for comorbid attention deficit hyperactivity disorder and conduct disorder: preliminary findings in a children's inpatient unit. *South. Med. J.*, **87** (1994), 692–695.
98. Chappell, P. B., Riddle, M. A., Scahill, L. *et al.* Guanfacine treatment of comorbid attention-deficit hyperactivity disorder and Tourette's syndrome: preliminary clinical experience. *J. Am. Acad. Child Adolesc. Psychiatry*, **34** (1995), 1140–1146.
99. Hunt, R. D., Arnsten, A. F. and Asbell, M. D. An open trial of guanfacine in the treatment of attention-deficit hyperactivity disorder. *J. Am. Acad. Child Adolesc. Psychiatry*, **34** (1995), 50–54.
100. Steingard, R., Biederman, J., Spencer, T., Wilens, T. and Gonzalez, A. Comparison of clonidine response in the treatment of attention-deficit hyperactivity disorder with and without comorbid tic disorders. *J. Am. Acad. Child Adolesc. Psychiatry*, **32** (1993), 350–353.
101. Leckman, J. F., Hardin, M. T., Riddle, M. A. *et al.* Clonidine treatment of Gilles de la Tourette's syndrome. *Arch. Gen. Psychiatry*, **48** (1991), 324–328.
102. Connor, D. F., Barkley, R. A. and Davis, H. T. A pilot study of methylphenidate, clonidine, or the combination in ADHD comorbid with aggressive oppositional defiant or conduct disorder. *Clin. Pediatr.*, **39** (2000), 15–25.
103. Goetz, C. G., Tanner, C. M., Wilson, R. S. *et al.* Clonidine and Gilles de la Tourette's syndrome: double-blind study using objective rating methods. *Ann. Neurol.*, **21** (1987), 307–310.
104. Scahill, L., Chappell, P. B., Kim, Y. S. *et al.* A placebo-controlled study of guanfacine in the treatment of children with tic disorders and attention deficit hyperactivity disorder. *Am. J. Psychiatry*, **158** (2001), 1067–1074.
105. Taylor, F. B. and Russo, J. Comparing guanfacine and dextroamphetamine for the treatment of adult attention-deficit/hyperactivity disorder. *J. Clin. Psychopharmacol.*, **21** (2001), 223–228.
106. Tourette's Syndrome Study Group. Treatment of ADHD in children with tics: a randomized controlled trial. *Neurology*, **58** (2002), 527–536.
107. Wender, P. H. and Reimherr, F. W. Bupropion treatment of attention-deficit hyperactivity disorder in adults. *Am. J. Psychiatry*, **147** (1990), 1018–1020.
108. Simeon, J. G., Ferguson, H. B. and Van Wyck Fleet, J. Bupropion effects in attention deficit and conduct disorders. *Can. J. Psychiatry*, **31** (1986), 581–585.
109. Riggs, P. D., Leon, S. L., Mikulich, S. K. and Pottle, L. C. An open trial of bupropion for ADHD in adolescents with substance use disorders and conduct disorder. *J. Am. Acad. Child Adolesc. Psychiatry*, **37** (1998), 1271–1278.
110. Reimherr, F. W., Strong, R. E., Hedges, D. W. *et al.* Six month, open-trial of bupropion SR in comparison to methylphenidate in the treatment of adults with ADHD. Poster

Presentation, National Institute of Mental Health, New Clinical Drug Evaluation Unit, 40th Annual Meeting, 2000a.

111. Daviss, W. B., Bentivoglio, P., Racusin, R. *et al.* Bupropion sustained release in adolescents with comorbid attention-deficit/hyperactivity disorder and depression. *J. Am. Acad. Child Adolesc. Psychiatry*, **40** (2001), 307–314.
112. Wilens, T. E., Prince, J. B., Spencer, T. *et al.* An open trial of bupropion for the treatment of adults with attention-deficit/hyperactivity disorder and bipolar disorder. *Biol. Psychiatry*, **54** (2003), 9–16.
113. Levin, F. R., Evans, S. M., McDowell, D. M., Brooks, D. J. and Nunes, E. Bupropion treatment for cocaine abuse and adult attention-deficit/hyperactivity disorder. *J. Addict. Dis.*, **21** (2002), 1–16.
114. Upadhyaya, H. P., Brady, K. T. and Wang, W. Bupropion SR in adolescents with comorbid ADHD and nicotine dependence: a pilot study. *J. Am. Acad. Child Adolesc. Psychiatry*, **43** (2004), 199–205.
115. Casat, C. D., Pleasants, D. Z., Schroeder, D. H. and Parler, D. W. Bupropion in children with attention deficit disorder. *Psychopharmacol. Bull.*, **25** (1989), 198–201.
116. Barrickman, L. L., Perry, P. J., Allen, A. J. *et al.* Bupropion versus methylphenidate in the treatment of attention-deficit hyperactivity disorder. *J. Am. Acad. Child Adolesc. Psychiatry*, **34** (1995), 649–657.
117. Casat, C. D., Pleasants, D. Z. and Van Wyck Fleet, J. A double-blind trial of bupropion in children with attention deficit disorder. *Psychopharmacol. Bull.*, **23** (1987), 120–122.
118. Conners, C. K., Casat, C. D., Gualtieri, C. T. *et al.* Bupropion hydrochloride in attention deficit disorder with hyperactivity. *J. Am. Acad. Child Adolesc. Psychiatry*, **35** (1996), 1314–1321.
119. Reimherr, F. W., Hedges, D. W., Strong, R. E. *et al.* 6-week, double-blind, placebo-controlled trial of bupropion SR for the treatment of adults with attention deficit hyperactivity disorder (ADHD). Poster Presentation, Annual Meeting of the American Psychiatric Association, Washington, DC: 2000b.
120. Wilens, T. E., Spencer, T. J., Biederman, J. *et al.* A controlled clinical trial of bupropion for attention deficit hyperactivity disorder in adults. *Am. J. Psychiatry*, **158** (2001), 282–288.
121. Kuperman, S., Perry, P. J., Gaffney, G. R. *et al.* Bupropion SR vs. methylphenidate vs. placebo for attention deficit hyperactivity disorder in adults. *Ann. Clin. Psychiatry*, **13** (2001), 129–134.
122. Connor, D. F., Hudziak, J. J., Wilens, T. E. *et al.* A controlled trial of extended-release bupropion in adult ADHD. NIMH: New Clinical Drug Evaluation Unit 44th Annual Meeting, 2004, Poster Abstract #003.
123. Adler, L. A., Resnick, S., Kunz, M. and Devinsky, O. Open-label trial of venlafaxine in adults with attention deficit disorder. *Psychopharmacol. Bull.*, **31** (1995), 785–788.
124. Hedges, D., Reimherr, F. W., Rogers, A., Strong, R. and Wender, P. H. An open trial of venlafaxine in adult patients with attention deficit hyperactivity disorder. *Psychopharmacol. Bull.*, **31** (1995), 779–783.
125. Olvera, R. L., Pliszka, S. R., Luh, J. and Tatum, R. An open trial of venlafaxine in the treatment of attention-deficit/hyperactivity disorder in children and adolescents. *J. Child Adolesc. Psychopharmacol.*, **6** (1996), 241–250.

126. Findling, R. L., Schwartz, M. A., Flannery, D. J. and Manos, M. J. Venlafaxine in adults with attention-deficit/hyperactivity disorder: an open clinical trial. *J. Clin. Psychiatry*, **57** (1996), 184–189.
127. Upadhyaya, H. P., Brady, K. T., Sethuraman, G., Sonne, S. C. and Malcolm, R. Venlafaxine treatment of patients with comorbid alcohol/cocaine abuse and attention-deficit/hyperactivity disorder: a pilot study. *J. Clin. Psychopharmacol.*, **21** (2001), 116–118.
128. Motavalli Mukaddes, N. and Abali, O. Venlafaxine in children and adolescents with attention deficit hyperactivity disorder. *Psychiatry Clin. Neurosci.*, **58** (2004), 92–95.
129. Hornig-Rohan, M. and Amsterdam, J. D. Venlafaxine versus stimulant therapy in patients with dual diagnosis ADD and depression. *Prog. Neuropsychopharmacol. Biol. Psychiatry*, **26** (2002), 585–589.
130. Spencer, T., Biederman, J., Heiligenstein, J. *et al.* An open-label, dose-ranging study of atomoxetine in children with attention deficit hyperactivity disorder. *J. Child Adolesc. Psychopharmacol.*, **11** (2001), 251–265.
131. Kratochvil, C. J., Bohac, D., Harrington, M. *et al.* An open-label trial of tomoxetine in pediatric attention deficit hyperactivity disorder. *J. Child Adolesc. Psychopharmacol.*, **11** (2001), 167–170.
132. Kratochvil, C. J., Heiligenstein, J. H., Dittmann, R. *et al.* Atomoxetine and methylphenidate treatment in children with ADHD: a prospective, randomized, open-label trial. *J. Am. Acad. Child Adolesc. Psychiatry*, **41** (2002), 776–784.
133. Spencer, T., Biederman, J., Wilens, T. *et al.* Effectiveness and tolerability of tomoxetine in adults with attention deficit hyperactivity disorder. *Am. J. Psychiatry*, **155** (1998), 693–695.
134. Michelson, D., Faries, D., Wernicke, J. *et al.* Atomoxetine in the treatment of children and adolescents with attention-deficit/hyperactivity disorder: a randomized, placebo-controlled, dose-response study. *Pediatrics*, **108** (2001), E83.
135. Spencer, T., Heiligenstein, J. H., Biederman, J. *et al.* Results from 2 proof-of-concept, placebo-controlled studies of atomoxetine in children with attention-deficit/hyperactivity disorder. *J. Clin. Psychiatry*, **63** (2002), 1140–1147.
136. Michelson, D., Allen, A. J., Busner, J. *et al.* Once-daily atomoxetine treatment for children and adolescents with attention deficit hyperactivity disorder: a randomized, placebo-controlled study. *Am. J. Psychiatry*, **159** (2002), 1896–1901.
137. Michelson, D., Adler, L., Spencer, T. *et al.* Atomoxetine in adults with ADHD: two randomized, placebo-controlled studies. *Biol. Psychiatry*, **53** (2003), 112–120.
138. Tatsumi, M., Groshan, K., Bakely, R. and Richelson, E. Pharmacological profile of antidepressants and related compounds at human monoamine transporters. *Eur. J. Pharmacol.*, **340** (1997), 249–258.
139. Richelson, E. Interactions of antidepressants with neurotransmitter transporters and receptors and their clinical relevance. *J. Clin. Psychiatry*, **64**: Suppl 14 (2003), 5–12.
140. Manor, I., Tyano, S., Mel, E. *et al.* Family-based and association studies of monoamine oxidase A and attention deficit hyperactivity disorder (ADHD): preferential transmission of the long promoter-region repeat and its association with impaired performance on a continuous performance test (TOVA). *Mol. Psychiatry*, **7** (2002), 626–632.
141. Ascher, J. A., Cole, J. O., Colin, J.-N. *et al.* Bupropion: a review of its mechanism of antidepressant activity. *J. Clin. Psychiatry*, **56** (1995), 395–401.

142. Béïque, J.-C., De Montigny, C., Blier, P. and Debonnel, G. Venlafaxine: discrepancy between in vivo 5-HT and NE reuptake blockade for reuptake sites. *Synapse*, **32** (1999), 198–211.
143. Adler, L. A., Spencer, T. J., Milton, D. R., Moore, R. J. and Michelson, D. Long-term, open-label study of the safety and efficacy of atomoxetine in adults with attention-deficit/hyperactivity disorder: an interim analysis. *J. Clin. Psychiatry*, **66** (2005), 294–299.
144. Reimherr, F. W., Marchant, B. K., Strong, R. E. *et al.* Emotional dysregulation in adult ADHD and response to atomoxetine. Poster Presentation, Annual Meeting of the American Psychiatric Association, Washington, DC: 2003.

20

Psychopharmacology of norepinephrine in eating disorders

Katherine A. Halmi, M.D. and Sun Young Yum

Weill Medical College of Cornell University, White Plains, USA

Basic research on norepinephrine and feeding

One of the first studies investigating the relationship of the noradrenergic system to feeding behavior was conducted by Ritter *et al.*[1] In this study the authors found that intraventricular injections of the α-adrenergic receptor agonist clonidine caused rats to eat voraciously; however, the injection of L-norepinephrine (L-NE) facilitated feeding but was not nearly as potent. The authors concluded that the relationship of noradrenergic mechanisms to feeding appeared to be a respondent one in which feeding was sensitized or disinhibited by activation of α-adrenergic receptors rather than an operant regulation that would be reinforced by the release of NE. Using a brain-cannula technique Leibowitz[2] found that injection of NE into the paraventricular nucleus (PVN) elicited feeding in satiated rats and enhanced feeding in already hungry rats. Using midbrain lesions, fluorescence histochemistry, and brain cannulation, Leibowitz and Brown[3] showed that the dorsal component of the central tegmental track contains the crucial noradrenergic axons that innervate the PVN and mediate noradrenergic stimulation of feeding behavior. They also determined that a strong reduction of loss of the feeding response was elicited by PVN injection of the presynaptic-acting drugs tranylcypromine and desipramine as well as showing a potentiation of the feeding response produced by injected NE. Russek *et al.*[4] found that intraperitoneal injections of NE produced a marked decrease in feeding in several different feeding conditions. The authors postulated that NE was stimulating hepatic adrenergic receptors.

In a series of microdialyses studies, Hoebel *et al.*[5] found that NE was released in the PVN during the active, feeding period of the circadian cycle in rats. Two other studies confirmed the hypothesis that within the PVN, stimulation of α_2-adrenergic receptors induces food intake whereas stimulation of α_1-adrenergic receptors suppresses food intake.[6,7]

Brain Norepinephrine: Neurobiology and Therapeutics, ed. Gregory A. Ordway, Michael A. Schwartz and Alan Frazer. Published by Cambridge University Press.

Changes in extracellular levels of endogenous NE were measured in the PVN of freely moving rats using microdialysis by Stanley *et al.*[8] The authors found a reliable peak at the beginning of the dark cycle and relatively stable levels at all other times. When the rats were food-deprived for 24 hours, a gradual rise in extracellular NE was observed and when these animals were re-fed, NE levels declined over the next 3 hours as food intake diminished. The authors concluded that endogenous NE in the PVN plays a role in the initiation and/or maintenance of normal eating behavior at the beginning of the nocturnal feeding period, as well as after food deprivation.

Norepinephrine also preferentially increases ingestion of carbohydrate over protein and fat. These studies were conducted by Leibowitz *et al.*[9] who found that peripheral and PVN injections of clonidine stimulated total food intake and preferentially increased ingestion of carbohydrate. In this same study they found that PVN injections of amitriptyline and peripheral injection of clomipramine also selectively enhanced carbohydrate intake. The authors concluded that noradrenergic neurons innervating the PVN in the rat play a role in regulating carbohydrate selection, and that the noradrenergic system mediates the stimulating action of clonidine and antidepressants on carbohydrate ingestion. In a review, Leibowitz[10] discusses the hypothesis that NE in the PVN, through activation of glucocorticoid and glucose-sensitive α_2-adrenergic receptors, is physiologically active in energy homeostasis, especially at the onset of the animal's active cycle. The author notes that α-noradrenergic stimulation of the PVN produces an increase in circulating levels of both corticosterone and glucose.

It appears that the NE neurotransmitter system affects eating behavior in areas of the brain other than the PVN as well. Gietzen and Jhanwar-Uniyal[11] showed the anorexia response to an amino acid deficiency diet was mediated through NE activity in the anterior pyriform cortex.

Norepinephrine is also affected by overfeeding. In a study by Scheurink *et al.*[12] overfeeding in rats was associated with an enhanced baseline plasma NE level in both ventromedial hypothalamus-lesioned rats and rats rendered obese by hyperalimentation. Plasma levels of NE during exercise were reduced in these two groups of animals. This attenuated outflow of NE in response to exercise was related to the size of the fat reserves, indicating that the feedback mechanism from the periphery to the central nervous system is altered in the overfed state. On the other end of the spectrum in the anorexia mutation mouse, Jahng *et al.*[13] found altered central NE levels and enervation associated with a decreased expression of NE transporter.

More recently the complex interactions between NE release in the PVN and leptin levels have been studied. For example, Francis *et al.*[14] found that systemic lipopolysaccharide stimulates NE release in the PVN and this in turn increases plasma concentrations of cortisone and leptin and the latter response is mediated in part by interleukin-1. Leptin levels in the human are directly related to the amount of adipose tissue. Flier[15] has proposed that decreasing leptin levels signal the brain to increase food intake.

Thus, in summary there is substantial evidence that the NE neurotransmitter system influences feeding behavior through a variety of complex interactive mechanisms.

Norepinephrine metabolism in anorexia nervosa

Anorexia nervosa is a psychiatric disorder which occurs predominantly in young women and is characterized by a markedly underweight state, behaviors to maintain the underweight state (fasting, overexercising, purging), a denial of the seriousness of the condition, and an overwhelming preoccupation in maintaining the underweight condition.[16] Anorexia nervosa also has the highest mortality rate of any psychiatric illness.[17]

Because animal studies demonstrated that NE influenced feeding behavior, an interest developed in the role of NE in the condition of anorexia nervosa. For example, NE, an α-adrenergic agonist, and propranolol, a β-adrenergic antagonist, canulated into the hypothalamus facilitate feeding whereas isoproterenol, a β-adrenergic agonist and phentolamine, an α-adrenergic antagonist, inhibit feeding.[18] Since acetylcholine produces a generalized behavior inhibition, a proposal was made that acetylcholine activation is dominant during satiety.[19] The authors then developed a concept of reciprocal inhibition between catecholamine and acetylcholine systems during feeding and satiety. Thus, during fasting noradrenergic activation occurs accompanied by cholinergic inhibition and in satiety, cholinergic activation is dominant and the noradrenergic system is inhibited. The authors hypothesized that anorexia nervosa resulted from a subcortical lesion involving a progressive depletion of brain NE. Studies had shown that a major metabolite of brain NE, 3-methoxy-4-hydroxyphenylglycol (MHPG), may reflect functional activity of the central noradrenergic neurons.[20] A significant fraction, 22% to 27% of urinary MHPG, was found to originate from metabolism of NE within the brain.[21] This urinary marker for brain NE metabolism was shown to be decreased in some patients with a primary depression.[22] These studies inspired the first measurements of catecholamine metabolism in anorexia nervosa by Halmi *et al.*[23] Urinary catecholamine levels were measured in 25 anorexia nervosa patients during acute illness and after treatment with weight gain. Urinary MHPG concentrations were significantly lower in the acutely ill anorectics than in a control group. An increase in urinary MHPG concentration after treatment was significantly correlated with a decrease in depressive symptomatology but was not correlated with change in the patient's activity level.

Soon after the above study, Gross *et al.*[24] showed that in the underweight anorectic, blood pressure, pulse rate, plasma levels of NE, urinary MHPG and homovanillic acid were lower than those of a group of normal volunteers. After weight gain, these parameters increased to near normal levels. Plasma dopamine-β-hydroxylase activity was within normal limits both during the underweight and normal-weight conditions. The authors suggested their results demonstrated the abnormalities in catecholamine metabolism in anorexia nervosa were caused by starvation and that neuronal functions dependent on aminergic neurotransmission may be altered as a result of starvation. Soon after these studies, Gerner and Gwirtsman[25] showed that underweight patients with anorexia nervosa had abnormally high cortisol levels and low MHPG levels regardless of their mood. The authors hypothesized that the low MHPG levels were not due to depression but that the decreased NE metabolism was related directly to the primary anorexia nervosa. These authors in another study[26] showed a consistent association between low urine MHPG and failure of adrenocorticotropic hormone (and ultimately cortisol) to suppress dexamethasone in anorexia nervosa patients. The authors also noticed that the families of these patients had a high percentage of relatives with primary affective disorder. They suggested there might be a genetic link between anorexia nervosa and primary affective disorder. That same year Abraham *et al.*[27] showed a positive relationship between urinary MHPG values and body weight in anorexia nervosa patients on a bed rest refeeding treatment program. Two other studies confirmed the relationship of decreased MHPG urinary levels in underweight anorexia nervosa patients and an increase of these levels with weight gain.[28,29]

The next methodology used to study NE function in anorexia nervosa was that of platelet aggregation. Luck *et al.*[30] found plasma NE concentrations in a group of malnourished anorexia nervosa patients to be significantly lower than those in an age- and sex-matched group of normal subjects. The platelet total α-adrenergic receptor densities of the patients significantly exceeded those of controls. The aggregation response of the patient's platelets was significantly enhanced after epinephrine challenge. The authors suggested the starvation-induced fall in plasma NE levels was associated with the up-regulation of platelet α-adrenergic receptors. This accounted for the exaggerated epinephrine-induced platelet aggregation in the anorectic patients. These findings were confirmed by Heufelder *et al.*[31] who found that the activity of the sympathetic nervous system was reduced in patients with anorexia nervosa and this was accompanied by an increased capacity and a decreased affinity of the platelet α_2-receptors and an increased prostaglandin E_1 stimulatory and epinephrine inhibitory effect on cyclic adenosine monophosphate (cAMP) production. These authors went on to suggest that the receptor changes were not limited to adrenergic receptors on platelets but may also occur in other peripheral and central tissues.

Additional studies of platelet hyperaggregability in response to epinephrine in anorexia nervosa patients showed a significantly greater aggregation and thromboxane A_2 release in patients with body weights below 75% of a normal weight range. These changes reverted to normal range with weight gain.[32] A study by Gill *et al.*[33] showed, in platelets from anorectic patients, a significant increase in NE–stimulated [Ca^{2+}] uptake, a significant diminution in epinephrine and isoproterenol–stimulated [Ca^{2+}] uptake, but no significant difference in dibutyryl cAMP-stimulated [Ca^{2+}] uptake, when compared with controls. The authors stated the relevance of these findings on Ca^{2+} mobilization in platelets to the symptomatology of anorexia was not clear.

Suppression of sympathetic activity was found to be not as pronounced at the time of admission to the hospital when body weight was lowest in anorectic patients, but rather some weeks later in a study by Pahl *et al.*[34] Only 8 of 22 patients had a decreased orthostatic response on admission, whereas 15 patients showed a pathologically low response 4 weeks later. The authors found the higher NE values in the anorectic patients on admission to the hospital were due to a general arousal that accompanied an adaptation to the new surrounding, which in many patients overcame the starvation–induced reduction of NE secretion.

The cerebrospinal fluid (CSF) level of NE in underweight anorectics and in these patients a few weeks after weight restoration was similar to that in normal subjects. However, long-term-weight-recovered anorectics had a 50% decrease in CSF NE levels compared with that of controls.[35] Plasma NE levels were not determined in the patients in this study. In view of the fact that plasma NE levels were low in underweight anorectics in other studies, the authors thought that underweight anorectics may have a dissociation between central and peripheral NE metabolism. The authors pointed out that only brain NE pathways have a role in the normal regulation of each of the neuroendocrine systems that become disturbed in underweight anorectics. Norepinephrine, for example stimulates the secretion of luteinizing hormone–releasing system. Release of NE in the hypothalamus decreases corticotropin–releasing factor secretion and appears to have an excitatory effect on the hypothalamic control of thyrotropin secretion. Central NE pathways regulate hypothalamic vasopressin release. The authors concluded that there is a possibility CSF NE metabolism has a vital role in anorexia nervosa. This study was replicated with 11 patients who had long-term weight maintenance (6 to 72 months). In this study, the long-term-weight-restored anorectics had lower plasma and CSF concentrations of NE and MHPG compared with healthy controls.[35] The authors suggested the reduced noradrenergic activity may be associated with long-term stability of weight recovery in patients with anorexia nervosa.

Norepinephrine function in anorexia nervosa patients has also been assessed with a series of perturbation tests. For example, Pirke *et al.*[36] found anorectic patients showed lower NE plasma concentrations before, during and after a bicycle ergometry test. The ratio of maximal NE / maximal lactate was significantly smaller in anorectic patients compared with controls and the ratio of maximal NE to maximal heart rate and the ratio of maximal NE to maximal systolic blood pressure were also significantly reduced in anorectics compared with controls. In another perturbation test Pirke *et al.*[37] measured plasma NE after a standardized test meal in acute and remitted patients with anorexia nervosa. Healthy controls had significantly higher NE baseline values than low-weight and weight-recovered patients. The increase in NE during the meal was significantly reduced in low-weight anorectics but there was no difference between weight-recovered patients and controls. Orthostatic NE response was investigated in 24 patients with anorexia nervosa during an inpatient treatment program.[38] Significantly worse mood was present in patients with pathologically low orthostatic NE response. The authors suggested the biological consequences of altered eating behavior may also effect psychological symptoms of depressed mood.

Assessment of pre and post-β-adrenergic receptor activity was conducted with an isoproterenol infusion test by Kaye *et al.*[39] They found increasing doses of isoproterenol was associated with a linear increase in plasma NE in healthy volunteers but anorectic patients, whether underweight or weight-recovered, had a significantly more erratic secretion of plasma NE in response to increasing doses of isoproterenol. The authors suggested that altered regulation of presynaptic adrenergic receptors may partly explain the large variance and lack of consensus between previous studies as to whether anorectic patients have reduced or normal plasma NE levels. In this study, weight-recovered anorectic patients needed a significantly higher dose of isoproterenol to increase basal heart rate by 25 beats/minute compared with underweight anorectic patients. The authors suggested that down-regulated postsynaptic cardiac β-adrenergic receptors during weight gain may protect against refeeding-induced exaggerated sympathetic activity.

Genetic research is the latest method of assessing NE function in anorexia nervosa. In a study by Urwin *et al.*[40] the authors hypothesized that the lower NE levels in long-term-weight-restored anorexia nervosa patients may reflect altered reuptake by the NE transporter (NET). The transmission disequilibrium test (TDT) with 87 Australian trios (patient plus parents) demonstrated significant preferential transmission of an L4 polymorphism from parent to child with restricting anorexia nervosa, suggesting that L4 or a DNA variant in the linkage disequilibrium doubles the risk for developing anorexia nervosa restricting type. The authors also found a preferential transmission of L4 to restricting anorexia nervosa patients and an S4 polymorphism to binge/purging anorexia nervosa patients. In the *NET* gene promoter region, a 4-bp deletion (S4) or insertion (L4) resulted in the net loss or gain respectively of a putative Elk-1 transcription factor site. The expression of *NET* is regulated by NE and due to the critical role played by NET in noradrenergic transmission it is reasonable to hypothesize that NET gene polymorphism contributes to the development of anorexia nervosa.

Norepinephrine metabolism in bulimia nervosa

Bulimia nervosa is a disorder in which predominantly women in late teenage years or early adulthood engage in binge eating, eating a large amount of food in a discrete period of time, followed by some compensatory behavior to counteract the caloric effect on weight gain. Most frequently this compensatory behavior is self-induced vomiting but may include laxative abuse, diuretic abuse, and vigorous exercising. In comparison with anorexia nervosa, bulimia nervosa occurs more frequently in older women and in males.

Since normal-weight bulimic patients have disturbed appetite, mood, and respond to antidepressants, it was reasonable to investigate brain monoaminergic abnormalities in bulimia nervosa. A study by Kaye *et al.*[41] showed that bulimic patients had a significantly lower mean CSF NE concentration compared to control subjects. Neither the bulimic patients nor control subjects showed a significant relationship between CSF 5-hydroxy-indole acetic acid (5-HIAA) and NE or between CSF homovanillic acid (HVA) and NE. The authors noted that levels of NE in CSF and plasma are normal or high in major depressive disorders. Thus, the authors concluded that normal-weight bulimia is not a subtype of depressive disorder and should be considered an illness with a unique and characteristic neurobiological profile.

In an effort to study the noradrenergic system in bulimia nervosa patients, Kaye *et al.*[42] measured plasma NE during bingeing and vomiting. They found this behavior activated the sympathetic nervous system. That is bulimia produced a significant increase in the duration and the peak increase of plasma NE when compared with normal controls eating a large meal. Shortly after admission to the hospital the bulimics were found to have normal basal plasma and CSF NE levels. After 30 days of hospitalization and abstinence from bingeing and vomiting the bulimics had a reduction of basal plasma and CSF NE levels compared with themselves on admission and compared with healthy controls. The authors hypothesize that dietary intake was related to noradrenergic activity. The noradrenergic disturbance in patients in this study was not related to weight, depression, physical activity or amino acid precursors. Those bulimic women who were amenorrheic were found to have lower CSF NE levels, suggesting that a noradrenergic disturbance may be associated with the frequent incidence of amenorrhea in bulimic women. It was unusual that reduced noradrenergic activity was found in normal-weight bulimic women after they had been abstinent from bingeing.

In order to assess if alterations in function of the peripheral sympathetic nervous system could contribute to the decreased metabolic rate in bulimia nervosa patients, George *et al.*[43] studied cardiovascular and hormonal responses to acute pharmacological challenge with intravenously administered isoproterenol. This was an assessment of β-adrenergic receptor sensitivity. In comparison to healthy controls, binge-abstinent bulimic patients had significantly reduced mean baseline plasma NE levels, pulse rate and systolic blood pressure. They also had significantly increased chronotropic responses to the isoproterenol infusion. The authors suggested that decreased sympathoneural activity may contribute to a tendency for bulimic patients to maintain body weight despite low calorie intake.

With a clonidine challenge test, Kaplan *et al.*[44] assessed α_2-adrenergic receptor activity in bulimia nervosa patients. There was no difference between bulimic patients and controls in this challenge response in the measures of blood growth hormone, cortisol, MHPG, and VMA. The authors noted these results were different than those found in depressed patients where there are blunted growth hormone responses to clonidine and differential responses in MHPG compared with controls, suggesting in depression a reduced responsiveness of α_2-adrenergic receptors. The authors concluded that data from their study do not support a shared biological abnormality in depression and bulimia nervosa.

Noradrenergic response to mental stress in bulimia nervosa patients was studied with the Trier Mental Challenge Test.[37] The degree of stress on self-ratings was not different between bulimics and controls during the mental challenge. The bulimic patients had significantly elevated mean cortisol levels and significantly reduced mean NE concentrations during the control session. During the stress session both NE and cortisol became significantly increased in the control group whereas the bulimic patients showed no significant response. Thus, although the bulimic patients felt as mentally stressed as the healthy controls, their plasma NE and cortisol levels did not increase significantly. This dissociation between subjective rating and endocrine response represented an abnormal stress response.

β-adrenergic receptor binding on circulating lymphocytes was evaluated in young female bulimic patients by Buckholtz *et al.*[45] There was a trend toward lower plasma NE levels in the bulimic patients compared with controls. The K_l/K_h ratio in bulimic patients was significantly greater than that for normal controls, indicating increased receptor coupling. The authors suggested that β-adrenergic receptors in bulimic patients may be more responsive than in normal subjects, without alteration of the traditional measures of receptor responses. They stated this difference cannot be explained on the basis of plasma NE.

Catecholamine stress reactivity was studied in another format in which women with bulimia nervosa were exposed to an interpersonally based speech stressor and a serial math task. The patients with bulimia nervosa had a blunted sympathetic activation in response to mental stress indicated by increased pre-ejection period responses and blunted systolic blood pressure, heart rate, and epinephrine responses. The women with bulimia nervosa perceived more stress, had worse coping skills, lower self-esteem and sense of mastery, reported less social support, had a worse mood, had greater anxiety, and were more depressed when compared with the control women. It is interesting to know that there was an equivalent performance during the stressors of women with bulimia nervosa and control women. The authors hypothesized that the combination of observed physiological and psychological response patterns reflected a dissociation of the sympathetic adrenal medullary and hypothalamic–pituitary–adrenal axes in the bulimia nervosa patients.[46]

Norepinephrine pharmacology in the treatment of anorexia nervosa

There are very few placebo-controlled randomly assigned trials of medications predominantly affecting NE metabolism in the treatment of anorexia nervosa. The use of tricyclic antidepressant (TCA) drugs, which block NE uptake and are regarded as NE agonists, seemed appropriate for the treatment of patients with anorexia nervosa because these patients had symptoms of depression. In addition, the side effects of TCAs, e.g. increasing appetite and weight gain, would be expected to be beneficial for anorectic patients. In the study by Halmi *et al.*[47] amitriptyline (a TCA) and cyproheptadine, a serotonin antagonist, were compared with placebo in the treatment of anorexia nervosa patients. The drugs were initiated after the patients had a pretreatment period of 7 days and had increased their weight to about 82% of their target weight. Patients on both drugs attained their target weights an average of 10.5 days earlier than patients taking placebo. The patients receiving amitriptyline had more side effects than those receiving cyproheptadine. In another study conducted by Biederman *et al.*[48] there was no difference between amitriptyline and placebo in the rate of weight gain of anorectic patients. This study however included both inpatients and outpatients in several different treatment centers; thus, the methodology was flawed. In this study the anorectic patients had a high refusal rate of entering the study and patients who did enter had a high prevalence of side effects with amitriptyline.

A study done much earlier by Lacey and Crisp[49] with clomipramine showed no difference in weight gain between those receiving drug and those receiving placebo. However, the dose of clomipramine was low, 50 mg. With no dramatic improvement in weight gain in the anorectic patients and the problem of side effects, especially hypotension, it no longer seems advisable to treat underweight patients with TCAs.

Norepinephrine pharmacology in the treatment of bulimia nervosa

Many more pharmacological treatment trials have been conducted with bulimic patients compared with anorectic patients. There are several reasons for this. First, bulimic patients are far more motivated for treatment and willing to enter studies. Second, anorectic patients frequently have medical complications, which then require withdrawal from the study. The initial rationale for the use of TCAs to treat bulimia nervosa patients was the empirical observation that bulimic patients were often depressed and TCAs were effective in treating depression. It soon became apparent, however, that TCAs were effective in reducing binge/purge behavior in bulimic patients who were not depressed. The specific pharmacological mode of action for the reduction of binge/purge behavior is not known. In a double-blind placebo-controlled study using amitriptyline, Mitchell *et al.*[50] found a 72% reduction in binge eating with amitriptyline and a 52% reduction in patients who received placebo with behavior therapy.

About the same time five randomly assigned placebo-controlled studies were conducted with imipramine. Pope *et al.*[51] found imipramine in a maximum dose of 200 mg to reduce binge eating by 70%, with a reduction of only 2% in the placebo group. Agras *et al.*[52] using a maximum dose of 300 mg imipramine found the drug reduced binge eating by 72% with a placebo reduction of 40% in bulimic patients. Mitchell *et al.*[53] found imipramine reduced binge eating by 50% with a dose of 300 mg with only a 2.5% reduction in the placebo group. Using a smaller dose of 200 mg imipramine, Alger *et al.*[54] obtained a 22% binge eating reduction with the drug and a 30% reduction with placebo.

There were six randomly assigned placebo-controlled studies with desipramine. Hughes *et al.*[55] obtained a 90% binge eating reduction on 200 mg of desipramine. Barlow *et al.*[56] in a crossover study with a dose of 150 mg obtained a 62% reduction in binge eating on desipramine. Blouin *et al.*[57] studied desipramine in a crossover study with fenfluramine and placebo. The end results of the study were given but the specific reduction of binge eating for desipramine was not measured. In a study involving a much larger sample of patients, Walsh *et al.*[58] obtained a 47% reduction in binge eating with desipramine and a 7% reduction with placebo. Finally, Agras *et al.*[52] compared desipramine with cognitive-behavior therapy and found a 12% reduction of binge eating after 16 weeks of drug therapy and a 44% reduction in binge eating after 24 weeks of drug therapy. There was no placebo control group in this study. All the treatment studies with TCAs mainly measured percent reduction of binge eating and only a few recorded total abstinence from binge eating. In the five studies that recorded total abstinence, the rate varied from 0% to 42%, the latter occurred with treatment of desipramine for 24 weeks.

A particular problem of treatment with TCAs is that of weight gain. This complicates the treatment goal of getting bulimia nervosa patients to overcome their fear that normal eating will result in significant weight gain. Other side effects of TCAs include dry mouth, constipation, and orthostatic hypotension. If the bulimia patient is fasting between binge eating episodes, constipation and hypotension will be further aggravated by TCAs. The most serious effect of TCAs are on cardiac rhythm since the TCAs may exacerbate arrhythmias due to hypokalemia or cardiomyopathy caused by severe ipecac abuse.

Tricyclic antidepressants are predominantly used today as a second trial after a selective serotonin reuptake inhibitor (SSRI) has failed. The reason for this is the SSRIs have a more benign side effect profile compared with the TCAs. It is generally recommended that treatment failures should be considered for TCAs only if adequate doses have been administered for 8 to 12 weeks. It is recommended that initial successful antidepressant treatment should continue at least 6 months in duration for bulimia nervosa. There is little evidence that continued treatment over an extended period of time such as a year with a single antidepressant is more effective than the use of placebo. Initial improvement and even remission is frequently followed by relapse while on the same medication. Open-label studies suggest that the use of different antidepressants after relapse may be a useful approach. Pope *et al.*[59] continued medication over a two-year period after the initial treatment with imipramine or placebo. Fifty percent of the patients achieved and maintained remission over that time and also showed a reduction in depressive symptoms. However, during that year many of the patients engaged in other therapies and took other medications.

Disturbances in NE function are present in both anorexia nervosa and bulimia nervosa. Whether these disturbances are a result of the disordered eating behavior or a cause of the disordered behavior is not known. Since eating behavior is highly influenced by other neurotransmitters and neuropeptides as well as NE, it is likely the genesis of the eating disorders is due to a complex interaction of the various neurotransmitters and neuropeptides with each other. Drugs that effect NE metabolism and neurotransmission have been modestly effective in treating bulimia nervosa and have had essentially no effect in treating anorexia nervosa. The most promising possibilities in developing more effective pharmacological agents for treating eating disorders are likely to come from genetic research.

REFERENCES

1. Ritter, S., Wise, C. and Stein, L. Neurochemical regulation of feeding in the rat: facilitation by alpha-noradrenergic, but not dopaminergic receptor stimulants. *J. Comp. Psychol. Psychology*, **88** (1975), 778–784.
2. Leibowitz, S. Paraventricular nucleus: a primary site mediating adrenergic stimulation of feeding and drinking. *Pharmacol. Biochem. Behav.*, **8** (1978), 163–175.

3. Leibowitz, S. and Brown, L. Histochemical and pharmacological analysis of noradrenergic projections to the paraventricular hypothalamus in relation to feeding stimulation. *Brain Res.*, **201** (1980), 289–314.
4. Russek, M., Vaga, C., Barrera, J. *et al.* Anorexia elicited by different catecolamines in rats. *Appetite*, **19** (1987), 119–126.
5. Hoebel, B., Hernendez, L., Schwartz, D., Mark, G. and Hunter, G. Microdialysis studies of brain norepinephrine, serotonin and dopamine release during ingestive behavior. Theoretical and clinical implications. *Ann. N. Y. Acad. Sci.*, **575** (1989), 171–191.
6. Davies, B., Wellman, P. and DiCarlo, B. Microinjection of the alpha 1-agonist methoxamine into the paraventricular hypothalamus induces anorexia in rats. *Brain Res. Bull.*, **28** (1992), 633–635.
7. Davies, B., Wellman, P. and Morien, A. An assessment of the involvement of paraventricular hypothalamic alpha 2-adrenergic adrenal captors in phenylpropanoloamine anorexia. *Physiol. Behav.*, **54** (1993), 121–128.
8. Stanley, B., Schwartz, D., Hernandez, L. Hoebel, B. and Leibowitz, S. Patterns of extracellular norepinephrine in the paraventricular hypothalamus: relationship to circadian rhythm and depravation-induced eating behavior. *Life Sci.*, **45** (1989), 275–282.
9. Leibowitz, S., Brown, O., Tretter, J. and Krischgessner, A. Norepinephrine, clonidine, and tricyclic antidepressants selectively stimulate carbohydrate ingestion through noradrenergic system of the paraventricular nucleus. *Pharmacol. Biochem. Behav.*, **23** (1985), 541–550.
10. Leibowitz, S. Hypothalamic paraventricular nucleus: interaction between alpha 2-noradrenergic system and circulating hormones and nutrients in relation to energy balance. *Neurosci. Behav. Rev.*, **12** (1988), 101–109.
11. Gietzen, D. W. and Jhanwar-Uniyal, M. Alpha 2 noradrenoceptors in the anterior pyriform cortex decline with acute amino acid deficiency. *Brain Res. Mol. Brain Res.*, **35** (1996), 41–46.
12. Scheurink, A., Balkan, B., Strubbe, J., Van Dijk, G. and Steffens, A. Overfeeding, autonomic regulation and metabolic consequences. *Cardiovasc. Drugs Ther.*, **10** (1996), 263–273.
13. Jahng, J., Houpt, T., Joh, T. and Son, J. Differential expression of monoamine oxidase A, serotonin transporter, tyrosine hydroxylase and norepinephrine transporter mRNA by anorexia mutation and food deprivation. *Brain Res. Dev. Brain Res.*, **107** (1998), 241–246.
14. Francis, J., MohanKumar, S. M. and MohanKumar, P. S. Correlations of norepinephrine release in the paraventricular nucleus with plasma corticosterone and leptin after systemic lipopolysaccharide: blockade by soluble IL-1 receptor. *Brain Res.*, **867** (2000), 180–187.
15. Flier, J. S. What's in a name? In search of leptin's physiologic role. *J. Clin. Endocrinol. Metab.*, **83** (1998), 1407–1413.
16. Halmi, K. A. Eating disorders: anorexia, bulimia nervosa and obesity. In *Textbook of Clinical Psychiatry*, 4th Ed, ed. R. E. Hales and S. C. Yudolfsky. (Washington, DC: American Psychiatric Publishing, Inc., 2003), pp. 1001–1021.
17. Keel, P., Dorer, D., Eddy, K. *et al.* Predictors of mortality and eating disorders. *Arch. Gen. Psychiatry*, **60** (2003), 179–183.
18. Leibowitz, S. Hypothalamic beta-adrenergic "satiety system" antagonizes an alpha-adrenergic "hunger" system in the rat. *Nature*, **226** (1970), 963–964.

19. Mawson, A. Anorexia nervosa and the regulation of intake: a review. *Psychol. Med.*, **4** (1974), 289–308.
20. Korf, J., Aghajanian, G. and Roth, R. Stimulation and destruction of the locus coeruleus: opposite effects on 3-methoxy-4 hydroxyphenylglycol sulfate levels in the rat's cerebral cortex. *Eur. J. Pharmacol.*, **21** (1973), 305–310.
21. Maas, J. W. and Landis, D. H. The metabolism of circulatory norepinephrine by human subjects. *J. Pharmacol. Exp. Ther.*, **177** (1971), 600–612.
22. Schildkraut, J., Keeler, B. and Papousek, M. MHPG excretion in depressive disorders: relation to clinical subtypes and desynchronized sleep. *Science*, **181** (1973), 762–764.
23. Halmi, K. A., Dekirmenjian, H., Davis, J., Casper, R. and Goldberg, S. Catecholamine metabolism in anorexia nervosa. *Arch. Gen. Psychiatry*, **35** (1978), 458–460.
24. Gross, H. A., Lake, C. R., Ebert, M. H., Ziegler, M. G. and Kopin, I. J. Catecholamine metabolism in primary anorexia nervosa. *J. Clin. Endocrinol. Metab.*, **49** (1979), 805–809.
25. Gerner, R. H. and Gwirtsman, H. E. Abnormalities of dexamethasone suppression test and urinary M. H. P. G in anorexia nervosa. *Am. J. Psychiatry*, **138** (1981), 650–653.
26. Gwirtsman, H. and Gerner, R. Neurochemical abnormalities in anorexia nervosa: similarities to affective disorders. *Biol. Psychiatry*, **10** (1981), 991–995.
27. Abraham, S., Beumont, P. and Cobbin, D. Catecholamine metabolism and body weight in anorexia nervosa. *Br. J. Psychiatry*, **138** (1981), 244–247.
28. Rieder, P., Toifl, K. and Kruzik, P. Excretion of biogenetic amine metabolites in anorexia nervosa. *Clin. Chim. Acta*, **123** (1982), 27–32.
29. Van Loon, G. Abnormal catecholamine mechanisms in hypothalamic disorders. *Metabolism*, **29** (1980), 1198–1202.
30. Luck, P., Mikhailidis, D., Dashwood, M. *et al.* Platelet hyperaggregability and increased alpha-adrenoceptor density in anorexia nervosa. *J. Clin. Endocrinol. Metab.*, **57** (1983), 911–914.
31. Heufelder, A., Warnhoff, M. and Pirke, K. M. Platelet alpha 2-adrenoceptor and adenylate cyclase in patients with anorexia nervosa and bulimia. *J. Clin. Endocrinol. Metab.*, **61** (1985), 1053–1060.
32. Mikhailidis, D., Barradas, M., DeSouza, V. *et al.* Adrenaline-induced hyperaggregability of platelets and enhanced thromboxane release in anorexia nervosa. *Prostaglandins Leukot. Med.*, **24** (1986), 27–34.
33. Gill, J., DeSouza, V., Wakeling, A., Dandona, P. and Jeremy, J. Differential changes in alpha and beta adrenoceptor linked calcium uptake in platelets from patients with anorexia nervosa. *J. Clin. Endocrinol. Metab.*, **74** (1992), 441–446.
34. Pahl, J., Pirke, K., Schweiger, U. *et al.* Anorectic behavior, mood, and metabolic and endocrine adaptation to starvation in anorexia nervosa during inpatient treatment. *Biol. Psychiatry*, **20** (1985), 874–887.
35. Kaye, W., Jimerson, D., Lake, C. and Ebert, M. Altered norepinephrine metabolism following long-term weight recovery in patients with anorexia nervosa. *Psychiatry Res.*, **14** (1985), 333–342.
36. Pirke, K., Eckert, M., Ofers, B. *et al.* Plasma norepinephrine response to exercise in bulimia, anorexia nervosa and controls. *Biol. Psychiatry*, **15** (1989), 799–802.

37. Pirke, K., Platt, E. P., Laessle, R., Seidl, M. and Fichter, M. The effect of a mental challenge test of plasma norepinephrine and cortisol in bulimia nervosa and in controls. *Biol. Psychiatry*, **32** (1992), 202–206.
38. Laessle, R., Schweiger, U. and Pirke, K. Mood and orthostatic norepinephrine response in anorexia nervosa. *Psychiatry Res.*, **24** (1988), 87–94.
39. Kaye, W., George, D. T., Gwirtzman, H. *et al.* Isoproterenol infusion test in anorexia nervosa: assessment of pre and post beta noradrenergic receptor activity. *Psychopharmacol. Bull.*, **26** (1990), 355–359.
40. Urwin, R., Bennetts, B., Wilcken, B. *et al.* Anorexia nervosa (restrictive subtype) is associated with a polymorphism in the novel norepinephrine transporter gene promoter polymorphic region. *Mol. Psychiatry*, **7** (2002), 652–657.
41. Kaye, W., Ballenger, J., Lydiard, R. *et al.* CSF monoamine levels in normal-weight bulimia: evidence for abnormal noradrenergic activity. *Am. J. Psychiatry*, **147** (1990), 225–229.
42. Kaye, W., Gwirtzman, H., George, D. *et al.* Disturbances of noradrenergic systems in normal weight bulimia: relationship to diet and menses. *Biol. Psychiatry*, **1** (1990), 4–21.
43. George, D., Kaye, W., Goldstein, D., Brewerton, T. and Jimerson, D. Altered norepinephrine regulation in bulimia: effects of pharmacological challenge with isoproterenol. *Psychiatry Res.*, **33** (1990), 1–10.
44. Kaplan, A., Garfinkel, P., Warsh, J. and Brown, G. Clonidine challenge test in bulimia nervosa. *Int. J. Eat. Disord.*, **8** (1989), 425–435.
45. Buckholtz, N. S., George, D. T., Davis, A., Jimerson, D. and Potter, W. Lymphocyte beta-adrenergic receptor modification in bulimia. *Arch. Gen. Psychiatry*, **45** (1988), 479–482.
46. Koo-Loeb, J., Pedersen, C. and Girdler, S. Blunted cardiovascular and catecholamine stress reactivity in women with bulimia nervosa. *Psychiatry Res.*, **27** (1998), 13–27.
47. Halmi, K. A., Eckert, E., LaDu, T. and Cohen, J. Anorexia nervosa: treatment of efficacy of cyproheptadine and amitriptyline. *Arch. Gen. Psychiatry*, **43** (1986), 177–181.
48. Biederman, J., Herzog, D., Rivinus, T. *et al.* Amitriptyline in the treatment of anorexia nervosa: a double-blind, placebo-controlled study. *J. Clin. Psychopharmacol.*, **5** (1985), 10–17.
49. Lacey, J. H. and Crisp, A. H. Hunger, food intake and weight: the impact of clomipramine on a refeeding anorexia nervosa population. *Postgrad. Med. J.*, **56** (1980), 79–85.
50. Mitchell, J. and Groat, R. A placebo-controlled, double-blind trial of amitriptyline in bulimia. *J. Clin. Psychopharmacol.*, **4** (1984), 186–193.
51. Pope, H., Hudson, J., Jonas, J. and Yurgelun-Todd, D. Bulimia treated with imipramine: a placebo-controlled, double-blind study. *Am. J. Psychiatry*, **140** (1983), 554–558.
52. Agras, W., Rossiter, E., Arnow, B. *et al.* Pharmacologic and cognitive-behavioral treatment for bulimia nervosa: a control comparison. *Am. J. Psychiatry*, **149** (1992), 82–87.
53. Mitchell, J., Pyle, R., Eckert, E. *et al.* A comparison study of antidepressants and structured group therapy in the treatment of bulimia nervosa. *Arch. Gen. Psychiatry,* **47** (1990), 149–157.
54. Alger, S. A., Schwalberg, M. D., Bigaouette, J. M., Michalek, A. V. and Howard, L. J. Effect of a tricyclic antidepressant and opiate antagonist on binge-eating behavior in normoweight bulimic and obese, binge-eating subjects. *Am. J. Clin. Nutr.*, **53**: 4 (1991), 865–871.
55. Hughes, P., Wells, L., Cunningham, C. and Ilstrup, D. Treating bulimia with desipramine. *Arch. Gen. Psychiatry*, **43** (1986), 182–186.

56. Barlow, J., Blouin, J., Blouin, A. and Perez, E. Treatment of bulimia with desipramine: a double-blind crossover study. *Can. J. Psychiatry*, **33** (1988), 129–133.
57. Blouin, A., Blouin, J., Perez, E. *et al.* Treatment of bulimia with fenfluramine and desipramine. *J. Clin. Psychopharmacol.*, **8** (1988), 261–269.
58. Walsh, T., Hadigan, C., Devlin, M., Gladis, M. and Roose, S. Long-term outcome of antidepressant treatment for bulimia nervosa. *Am. J. Psychiatry*, **148** (1991), 1206–1212.
59. Pope, H., Hudson, J., Jonas, J. and Yurgelun-Todd, D. Antidepressant treatment of bulimia: a two-year follow-up study. *J. Clin. Psychopharmacol.*, **5** (1985), 320–327.

21

Role of norepinephrine in substance abuse

Susan L. Broom, Ph.D. and Bryan K. Yamamoto, Ph.D.
Boston University School of Medicine, USA

Introduction: defining substance abuse

Drug abuse is a serious medical and social problem affecting millions of people each year. Substance abuse is usually defined as the misuse of drugs, including taking a drug for purposes other than those prescribed or intended. Dependent or addicted individuals engage in compulsive drug seeking and drug taking despite the consequences to self and others. A recent national survey (in the US) on drug abuse estimated 16.6 million individuals classified as substance abuser and/or dependent, with 1.1 million seeking treatment for substance abuse disorders.[1] Both conditions are recognized for their deleterious consequences to individuals as well as society. Therefore, the primary goal of clinicians is to optimize therapies for maladaptive drug-taking behaviors.

This chapter provides an overview of the role of noradrenergic mechanisms involved in the development and maintenance of substance abuse, as well as the integration of the mechanisms mediating the potential adverse consequences of maladaptive drug-taking behavior (i.e., tolerance, dependence, and withdrawal). A wide array of commonly abused substances is discussed. This chapter focuses on the stimulants (i.e., amphetamine and cocaine), but also includes opiates, alcohol, and nicotine. Although some studies to date do not support the clinical efficacy of noradrenergic-based therapies for the treatment of stimulant abuse,[2] others demonstrate a therapeutic potential of noradrenergic compounds for their ability to alleviate the adverse consequences of other abused drugs such as nicotine[3] and opiates.[4]

A substantial literature implicates the role of dopamine in the subjective "rewarding" properties of abused drugs (as defined below).[5,6] However, this system alone does not explain substance abuse behaviors. For example, stimulant-induced norepinephrine (NE) release, as opposed to stimulant-induced dopamine release,

Brain Norepinephrine: Neurobiology and Therapeutics, ed. Gregory A. Ordway, Michael A. Schwartz and Alan Frazer. Published by Cambridge University Press.

correlates significantly with reported subjective effects of these drugs, suggesting that noradrenergic activity serves as an essential component of the rewarding aspects of psychostimulants.[7] In fact, recent experiments indicate that dopamine transporter knockout mice exhibit stimulant-induced dopamine release in reward regions[8] and stimulant-induced conditioned place preference.[9] These findings support the conclusion that other systems (i.e., noradrenergic), in addition to dopamine, mediate the rewarding aspects and/or maladaptive behavior patterns associated with abused substances. Therefore, attempts to define the role of NE in mediating the balance between the rewarding and aversive properties of abused drugs may lead to the development of novel pharmacotherapies that target noradrenergic systems for the treatment of substance abuse disorders.

Role of norepinephrine in reward

An important element of substance abuse is the ability of a drug to produce feelings of euphoria or positive reinforcement. The rewarding aspects of the drug during acute exposure contribute to repeated use, whereas these subjective effects become attenuated with sustained periods of exposure due to the development of tolerance. The rewarding aspects have been defined historically as being mediated by dopamine. However, recent efforts have been directed toward defining the role of NE in the modulation of the mesolimbic dopamine system. This interaction is suggested by the significant anatomical overlap of the NE and dopamine systems in the brain. Figure 21.1 illustrates the primary reward pathway, otherwise termed the mesolimbic dopamine pathway, with additional structures associated with modulating activity of this system. The key elements in the mesolimbic path include the ventral tegmental area, which provides the predominant source of dopamine projection neurons to the nucleus accumbens. The medial forebrain bundle represents the collection of projection fibers exiting both the substantia nigra and ventral tegmental pathway and projecting to the striatum and nucleus accumbens, respectively. The increase in dopamine activity in the mesolimbic dopamine pathway to nucleus accumbens is associated with the pleasurable and locomotor stimulatory effects of psychostimulants, as well as the pleasurable effects of opiates, alcohol, and nicotine.[5,6,10]

Both dopamine and NE innervate the nucleus accumbens shell region, an area implicated in the rewarding and aversive aspects of drug use.[11] The noradrenergic innervation of the nucleus accumbens shell arises primarily from the A2 region of the nucleus solitary tract (NTS), with minor inputs from the A1 region of the solitary tract and the locus coeruleus (LC).[12] Moreover, α_1-adrenergic receptors have been shown to modulate dopamine activity in the prefrontal cortex[13] and ventral tegmental area,[14] the former brain regions known to be involved in the

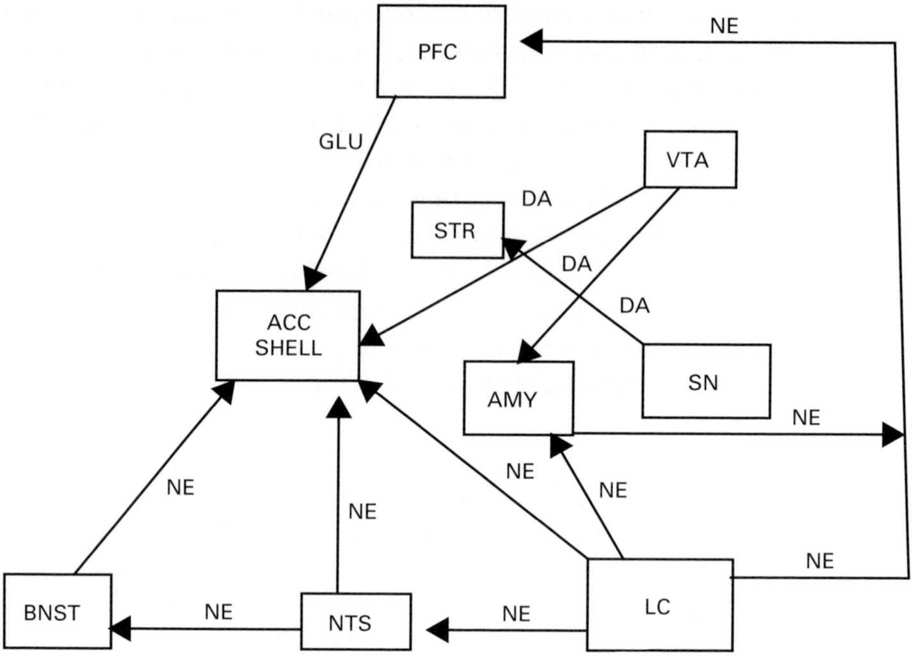

Figure 21.1 Mesolimbic dopamine pathway with accompanying noradrenergic inputs known to modulate activity on this system. PFC, prefrontal cortex; ACC SHELL, nucleus accumbens shell; VTA, ventral tegmental area; SN, substantia nigra; STR, striatum; NTS, nucleus solitary tract; AMY, amygdala; LC, locus coeruleus; BNST, bed nucleus of the stria terminalis; DA, dopamine; Glu, glutamate; NE, norepinephrine.

cognitive aspects of drug abuse.[15] In addition, both β_1- and β_2-adrenergic receptors are present in other dopamine-rich regions such as the substantia nigra, cortex, and medulla.

Two additional noradrenergic regions associated with the mesolimbic dopamine system and brain regions involved in drug reward include the bed nucleus of the stria terminalis (BNST) and the amygdala. The BNST has a high density of NE transporters (NET),[16] and recent evidence from a primate self-administration model indicates that the transporter binding sites in this region are up-regulated following chronic cocaine exposure.[17] Furthermore, electrical and chemical (glutamate) stimulation of the BNST produced significant activation of dopamine neurons in the ventral tegmental area,[18] while reduced noradrenergic activity in the BNST attenuated stress-induced reactivation of morphine place preference in rats.[19] The amygdala contains a high density of α_2-adrenergic receptors,[20] and stimulation of this receptor in the amygdala appears to attenuate stress-induced cocaine-seeking behavior in rats.[21] Furthermore, reduced noradrenergic activity in the BNST attenuates stress-induced reactivation of morphine place preference in rats,[22] suggesting an important role of the BNST and amygdala in mediating aspects of stress-induced drug-seeking behavior. Collectively, given the noradrenergic innervation of these structures and their association with brain reward regions such as the accumbens shell, these data suggest that NE regions may be involved in modulating aspects of drug reward and stress-induced relapse.

In addition to the significant anatomical overlap between the NE and dopamine pathways, there are direct neurochemical interactions between these systems. Norepinephrine has been shown to mediate dopamine release in brain reward regions. The α_1-adrenergic receptors located in the prefrontal cortex alter stimulant-induced dopaminergic activity in the nucleus accumbens.[13] α_1-receptors located on dopamine neurons of the ventral tegmental area enhance firing rates of these neurons through the attenuation of inhibitory postsynaptic potentials (IPSPs) on dopaminergic neurons.[14] Furthermore, the α_{1B}-receptor subtype knockout mice show attenuated cocaine and morphine-induced increases in locomotor activity, as well as an attenuated development of stimulant-induced conditioned place preference.[22] In addition, substantial clinical evidence suggests noradrenergic metabolic activity is strongly correlated with the propensity for substance abuse (i.e., reward dependence)[23,24] and actual drug-taking behavior.[25]

Recent studies suggest a link between the activation of noradrenergic receptors and stimulant-induced changes in brain reward regions. Local administration of amphetamine into the ventral tegmental area of rats produced increases in extracellular dopamine in both the nucleus accumbens and medial prefrontal cortex, an effect that can be attenuated by the local infusion of either the α-receptor antagonist phentolamine or the selective β-adrenergic receptor antagonist alprenolol.[26] Moreover, α_1-receptor activation in dopamine neurons of the ventral tegmental area is associated with enhanced burst firing of dopaminergic neurons in the ventral tegmental area following amphetamine administration.[14] This activating effect of the α_1-receptor on dopamine burst firing appears to be mediated through the attenuation of IPSPs produced by the metabotropic glutamate subtype 1 receptor.

Cortical α-adrenergic receptors also appear to be involved in mediating subcortically mediated stimulant-induced locomotor activity.[13] For example, pretreatment with either the α_1-receptor antagonist prazosin or WB-4101 into the prefrontal cortex attenuates amphetamine-induced increases in nucleus accumbens dopamine transmission and locomotor hyperactivity in a rat model of stimulant abuse.[13]

The α_{1B}-adrenergic receptor subtype also mediates stimulant and opiate-induced locomotor behavior, as well as behavioral assays of reward behaviors such as conditioned place preference and drug intake.[22] Alpha$_{1B}$-knockout mice exhibit alterations in both locomotor and rewarding properties of stimulants and morphine. More specifically, α_{1B} knockout mice exhibit less sensitization to the locomotor stimulant effects of amphetamine, cocaine, and morphine following repeated injections.[22] In addition, these knockout mice showed less oral intake of a cocaine solution compared to wild-type controls, as well as a marked absence of conditioned place preference to morphine injections.[22] Thus, there are significant roles for α_1- and β-adrenergic receptors in mediating the rewarding aspects of amphetamine perhaps through the enhancement of dopaminergic activity in brain reward regions.[14,26]

The noradrenergic modulation of dopamine transmission can also be mediated by heterotransporter-mediated mechanisms. The NET has an equal affinity for dopamine and NE.[27,28] Therefore, dopamine and NE transmission in brain regions that are coinnervated by dopamine and NE terminals can be explained by the balance of activities of NET and the dopamine transporter (DAT). There is in vitro evidence that blockade of NET increases the extracellular concentrations of dopamine in the substantia nigra that could be mediated by heterotransporter mechanisms.[29] Norepinephrine transporter has been implicated in mediating the extracellular dopamine concentrations in the prefrontal cortex and nucleus accumbens shell,[30] with a greater influence of NET on the control of extracellular dopamine in the prefrontal cortex. These results were interpreted in light of the finding that there is a relatively lower concentration of DAT relative to NET in the cortex, and thus a greater contribution of NET to the control of extracellular concentrations and volume transmission of dopamine in the cortex. It is interesting to note that in the nucleus accumbens shell, in which there is a greater concentration of DAT compared to the cortex, there is a smaller but significant contribution of NET to the modulation of extracellular dopamine. In contrast, there is no influence of the NE uptake blocker, desipramine, on dopamine in the nucleus accumbens core, a region with very little, if any, noradrenergic innervation. Similar evidence of heterotransporter-mediated mechanisms was observed using measures of extracellular dopamine and NE within the ventral tegmental area following the administration of DAT and NET blockers.[31]

Research with genetic knockout mice supports a role for NET in psychostimulant reward. In fact, recent studies demonstrate that DAT blockade alone is insufficient for explaining cocaine reward; however this remains arguable.[32] Both heterozygous and homozygous DAT knockouts retained a significant conditioned place preference to cocaine and methylphenidate.[9] Furthermore, blockade of NET increases extracellular dopamine in the nucleus accumbens of DAT knockout mice,[8] suggesting that NET can participate in modulation of dopamine transmission within the accumbens in the absence of DAT. Overall, these collective findings support a role for NET in mediating dopamine activity in the mesolimbic reward regions, particularly with regard to psychostimulants.[8,30,31]

Norepinephrine may be the predominant mediator of the cognitive aspects of a variety of phenylethylamine stimulants, as well as nicotine. Rothman and colleagues showed that the reported subjective effects of stimulants correlated with their potency ranges to enhance NE – and not necessarily dopamine – release (amphetamine [AMPH], methylenedioxymethamphetamine [MDMA], methamphetamine [METH], ephedrine, phentermine, and aminorex).[7] In addition, although nicotine is known to enhance dopamine release, NE release in the paraventricular nucleus of the hypothalamus is positively correlated with nicotine self-administration.[33]

Norepinephrine markers have also been associated with the personality trait related to reward dependence. Although little relationship appears to exist between reward dependence scores and measures of NET, α_{1A}-, α_{2A}-, or α_{2C}-adrenergic receptor variants in nonalcoholic populations,[34,35] both Garvey *et al.*[23] and Curtin *et al.*[24] reported a significant positive correlation between 3-methoxy-4-hydroxyphenolglycol (MHPG) levels (present in urine) and reward dependence scores as measured on the Tridimensional Personality Questionnaire. In addition, Valverius *et al.*[25] demonstrated a positive correlation between ethanol consumption and MHPG levels. These investigators suggested that MHPG levels are predictive of actual ethanol consumption and a predisposition to alcoholism. More recent evidence by Samochowiec and colleagues[36] reported a significantly higher occurrence of NET allele carriers in alcoholic subjects scoring high on reward dependence. Collectively, these data suggest an association may exist between alcohol dependence, MHPG levels, and NET alleles with regard to the personality trait of reward dependence. Future research is needed to fully characterize these changes in males and females, as well as in alcoholic versus nonalcoholic populations.

Nicotine consumption is closely related to changes in enzymes involved in the biosynthesis and metabolism of NE. Recent genetic studies indicate that specific allele markers for dopamine-β-hydroxylase positively correlate, while monoamine oxidase (MAO) negatively correlate, with nicotine intake. Specifically, these genetic alleles were significantly correlated with the type of smoker (heavy versus light) and the number of cigarettes smoked per day.[37]

In conclusion, a growing body of literature suggests that NE contributes to the dopaminergic system in producing changes in the activities of the brain reward pathways induced by psychostimulants. In addition, strong positive correlations between NE metabolite levels (MHPG) and self-administration of alcohol have emerged. Several studies implicate NE projections to reward areas, β-adrenergic receptor activity, and the NET in mediating the subjective aspects of psychostimulant and nicotine use, whereas other genetic studies show that changes in synthesis and enzymatic degradation of NE correlate with nicotine intake. The ability of these mechanisms to contribute to the rewarding effects of other drug classes, such as opiates, remains to be determined, although results from the α_{1B} genetic knockout studies suggest a common denominator related to stimulant- and opiate-induced reward. Regardless, recent studies suggest a more substantial role of NE in the development of the adverse consequences such as tolerance and withdrawal symptoms associated with opiate, as well as cocaine, alcohol, and nicotine, abuse.

Role of norepinephrine in dependence

Dependence on an abused drug is characterized by the appearance of symptoms generally opposite to those typically apparent during acute drug exposure, and those symptoms generally occurring after the abrupt withdrawal of the drug. The noradrenergic system appears to contribute significantly to the onset and expression of these withdrawal symptoms. Most of the published evidence in this regard has focused on the role of noradrenergic systems in mediating both physical and affective signs of opiate withdrawal.[38]

The affective (i.e., place aversion) and somatic/physical symptoms of naloxone-precipitated opiate withdrawal appear to be brain region-dependent. Frenois *et al.*[39] showed that low-dose (30 and 120 *mg*) naloxone-precipitated withdrawal is manifested as a naloxone-induced conditioned place aversion in morphine-dependent rats and is associated with increased *c-fos* mRNA in limbic regions. These regions include the central nucleus of the amygdala, BNST, and nucleus accumbens. In addition, local administration of the α_2-agonist clonidine into the central amygdala nucleus attenuated the affective symptom of irritability (i.e., vocalizations) accompanying opiate withdrawal.[40] In contrast, high-dose naloxone-precipitated withdrawal is characterized by weight loss, jumping, diarrhea, and salivation, and is associated with increased *c-fos* mRNA expression in the LC and NTS and in dopaminergic regions such as the ventral tegmental area and caudate. Moreover, decreasing noradrenergic transmission within the LC by the local administration of clonidine alleviates some of the somatic signs of opiate withdrawal, such as wet dog shakes, weight loss, and diarrhea.[41] Thus, the somatic or physical symptoms of morphine withdrawal are mediated in both noradrenergic and dopaminergic brain areas, whereas the affective symptoms of opiate withdrawal appear to be mediated primarily by NE-containing brain regions.[37,39]

Although some studies have shown that blockade of NE transmission in the LC using 6-hydroxydopamine lesions[42,43] in morphine-dependent rats failed to reverse the expression of spontaneous or naloxone-precipitated withdrawal from morphine, the balance of the evidence appears to suggest that the LC is intimately involved in mediating the somatic symptoms of opiate withdrawal. For example, electrolytic lesions of the LC reduced methylnaloxonium-induced jumping, wet dog shakes, and teeth chattering in morphine-dependent rats.[44] Moreover, several studies have shown a positive correlation between the time course of the physiological and biochemical activity of the LC and expression of the somatic signs of opiate withdrawal. The electrophysiological activity of the LC appears to mirror the expression of the behavioral signs (i.e., jumping, teeth chattering, salivation) of naltrexone-induced withdrawal.[45] Along biochemical lines, the hyperactivity of LC neurons observed during opiate withdrawal can be attenuated by cyclic adenosine monophosphate or protein kinase A inhibitors.[46]

Brain regions other than the LC have also been implicated. Most studies have focused on regions that innervate the LC. In particular, the paragigantocellularis nucleus (PGi) sends excitatory amino acid projections to the LC.[47] Lesions of the PGi or administration of the nonselective excitatory amino acid antagonist, kynurenic acid, attenuates the hyperactivity of LC neurons in morphine-dependent rats during withdrawal.[48] Conversely, electrical stimulation of the PGi in naïve rats produces behavioral responses similar to those observed following naloxone-precipitated withdrawal in opiate-dependent rats.[49] The excitatory inputs to the LC from the PGi are counterbalanced by parallel inhibitory enkephalin-opioid inputs from the PGi.[50] Changes in these inhibitory inputs to the LC have also been implicated in opiate withdrawal. In fact, results from Western blot analysis revealed reduced enkephalin levels in both the LC and NTS following chronic morphine treatment. In addition, decreased pre-proenkephalin mRNA expression was seen in the PGi.[50] Therefore, increases in the activity of excitatory inputs and decreases in the inhibitory inputs from the PGi and perhaps the NTS to the LC contribute to the hyperactivity of LC neurons seen during opiate withdrawal and may account for the somatic symptoms observed during this period.[50]

The BNST has been implicated in mediating the affective aspects of opiate withdrawal. The noradrenergic afferent projections from the A1 and A2 regions to the BNST showed significant increases in Fos-related antigen expression following opiate withdrawal-induced place aversion in rats.[51] Conversely, lesions of the A1 path, or the administration of an α_2-autoreceptor agonist or a combination of β-receptor antagonists into the BNST attenuated this place aversion suggesting that noradrenergic input to the BNST and stimulation of α_2-receptors and/or β-receptor antagonists mediates the affective aspects of opiate withdrawal.[51,52]

Collectively, NE appears to mediate the precipitating events associated with the behavioral expression and affective aspects of opiate withdrawal. The LC and its innervation by glutamatergic and enkephalinergic projections from the PGi play important roles in the onset of the somatic signs of opiate withdrawal, whereas noradrenergic transmission within the central nucleus of the amygdala and the noradrenergic inputs to the BNST from the NTS appear to be involved primarily in the motivational/affective aspects of opiate withdrawal.

The role for NE in withdrawal from chronic exposure to psychostimulants and alcohol is less significant than its role in withdrawal from opiates. However, chronic self-administration of cocaine in nonhuman primates produced up to 52% increases in NET-binding site densities relative to controls. Furthermore, cerebral metabolism in the BNST, as measured through glucose utilization, was significantly decreased following 100 days of self-administration.[17] These data collectively suggest that functional metabolic and noradrenergic changes occur in the BNST following chronic cocaine exposure in nonhuman primates in a brain region associated with stress-induced cocaine relapse.

Chronic in vitro amphetamine exposure of HEK-293 cells transfected with human NET cDNA significantly reduced NE uptake as well as NET-immunoreactivity and nisoxetine binding.[53] In contrast, chronic cocaine exposure modestly decreased NE uptake only. Therefore, these data suggest that although both of these psychostimulants are known to acutely affect noradrenergic transmission, there is a differential effect of these drugs on NE uptake processes following chronic exposure. Chronic use of MDMA by humans also has been shown to affect NE such that peripheral plasma NE concentrations are positively correlated with the frequency of Ecstasy use when examined in individuals that were occasional to moderate abusers of MDMA.[54] Studies in rats show that chronic administration of MDMA to rats results not only in a significant reduction in the concentration of serotonin in the brain (i.e., prefrontal cortex, nucleus accumbens, and striatum), but also decreases the concentration of NE in the nucleus accumbens that persists for at least 4 weeks after MDMA exposure.[55] However, it remains to be determined if the decreases in NE in the nucleus accumbens represent a long-term neurotoxic effect of MDMA.

The role for NE in alcohol abuse also is not well established. Exposure of rats to chronic ethanol results in decreased β-adrenergic receptor Bmax (density)[56] and decreased Kd (increased affinity)[57] in the hippocampus. It has been shown that bilateral lesions of the LC or clonidine administration to intact rats attenuate the signs of ethanol withdrawal,[58] similar to the reported effects of these manipulations on opiate withdrawal.[44] In studies on human tissue, postmortem analysis of the brains of alcoholics revealed a significant decrease in the number and density of LC neurons compared to controls.[59] Overall, although the role of NE in withdrawal from opiates is well established, there is a paucity of research and a lack of significant findings related to the role of NE in withdrawal from either chronic psychostimulant or alcohol use.

Interaction between norepinephrine and stress in relapse

As more evidence emerges supporting the role of noradrenergic systems in precipitating withdrawal, particularly opiate withdrawal, greater attention is being focused on the relationship between noradrenergic systems and relapse into drug addiction. It is known that both stress and drug preexposure can induce drug-seeking and drug-taking behaviors.[60] Given that physiological (adrenal changes) and psychological (relapse) responses to stress involve noradrenergic activation, NE has been suggested to modulate stress-induced changes in abuse behaviors.[61] For example, stress-induced (conditioned footshock) cocaine-seeking behaviors in rats can be attenuated by infusion of the combination of the β_1-receptor antagonist betaxolol and the β_2-receptor antagonist ICI-118,551 into the BNST, with a complete blockade produced following infusions into the central nucleus of the amygdala during a withdrawal period from cocaine.[61] In contrast, however, the same receptor antagonists failed to attenuate reinstatement behaviors resulting from preexposure to cocaine. This suggests a dissociation of the mechanisms mediating stress-induced and drug-induced reinstatement behaviors, and that NE contributes primarily to stress-induced, and not drug-induced, drug-seeking behavior.[60,61]

α_2-receptors have been implicated in stress-induced drug reinstatement. Treatment with clonidine or lofexidine (α_2-receptor agonists) blocked stress-induced cocaine reinstatement behavior in rats.[21] This study also showed that guanabenz, an α_2-receptor agonist with affinity for imidazoline type-1 receptors, produced similar, though less robust results, in response to footshock-induced reinstatement of cocaine seeking. In addition, stress can also induce the maintenance of morphine place preference that, in turn, can be attenuated by pretreatment with clonidine into the BNST. It is interesting to note that these effects were observed when clonidine was administered directly into the BNST and not the LC.[19] Moreover, recent evidence suggests that activation of corticotropin-releasing factor (CRF) receptors specifically in the BNST is a necessary component of footshock-induced reinstatement of cocaine seeking.[62] Collectively, these results suggest stress-induced drug-seeking behavior is mediated via α_2-receptor activation or β-receptor antagonism. In addition, BNST mediation of stress-induced reinstatement behavior requires the presence of CRF activation,[62] suggesting an additional system involved in mediating stress-induced reinstatement drug behaviors. Furthermore, stress-induced maintenance of morphine-conditioned place preference is likely mediated by activation of α_2-receptors in the BNST and NE projections from the ventral noradrenergic bundle.[19]

Despite the growing amount of literature to support a role of NE in psychostimulant reward, the enhancement of NE transmission via the blockade of NET may be more specifically related to the aversive rather than rewarding properties of cocaine. This conclusion is based on the finding that chronic cocaine exposure in primates increases NET binding density in the BNST,[17] a region identified for its involvement with stress-induced drug relapse.[21,61,63]

In summary, a substantial literature documents a role of the LC and accompanying structures, including the PGi and NTS, as the critical system mediating the physical and affective signs of opiate withdrawal. Although some evidence supports a role of the LC in mediating affective aspects of opiate withdrawal,[64] the extended amygdala, which includes the BNST, is also critically involved in the expression of affective components of opiate withdrawal.[51,52] Although less characterized, changes in noradrenergic functioning (i.e., receptor subtype regulation and metabolite activity) may also mediate withdrawal from stimulants and alcohol. The information gained from research on the mechanisms mediating opiate withdrawal may lead to the identification of specific pharmacological targets for the development of noradrenergic-based therapies related to the treatment of opiate withdrawal. With increasing knowledge accumulating to support common systems associated with withdrawal from stimulants and alcohol, the development of similar therapeutic strategies in the treatment of cocaine and alcohol abuse may also be a promising avenue for the development of noradrenergic compounds for the treatment of substance abuse disorders.

Noradrenergic treatment of substance abuse

Based on the above literature regarding the role of NE in substance abuse behaviors, research efforts have been directed toward the use of noradrenergic compounds as a possible treatment of substance abuse disorders.

Nicotine dependence has been treated with the tricyclic antidepressant nortriptyline, the α_2-agonist clonidine, and the combined NET and DAT antagonist, buproprion.[65] Unlike other noradrenergic compounds used for treating nicotine addiction, clonidine has been used frequently and deemed effective for reducing the craving associated with nicotine withdrawal. However, the effectiveness of clonidine in smoking cessation appears limited to special populations and its use sensitive to gender, history of depression, and type of addiction.[66]

A recent clinical trial[3] showed potential for nortriptyline in treating nicotine dependence in nondepressed patients. Nortriptyline significantly reduced some withdrawal symptoms such as irritability, anxiousness, impatience, and difficulty concentrating at eight days into withdrawal. These data suggest that nortriptyline may be effective in treating acute nicotine withdrawal in nondepressed populations, although more research is needed to assess the long-term potential of nortriptyline therapy for nicotine dependence. Buproprion also has demonstrated therapeutic benefit for nicotine addiction, particularly in producing high abstinence rates from 12 to 26 weeks,[67,68] in addition to reducing aversive affective signs such as tobacco craving.[68] It has been suggested that buproprion reportedly shows promise in combating nicotine addiction through decreasing LC-mediated noradrenergic activation and promoting dopamine increase.[65]

The treatment of psychostimulant abuse with noradrenergic uptake blockers has been tried but with very limited or no success. Although some tricyclic antidepressants such as desipramine[69] and imipramine[70] have shown potential for treating some symptoms of cocaine abuse, the overall consensus seems to be that blockade of dopamine release, as opposed to NE release, produced by the amphetamine-like psychostimulants remains the critical factor in attenuating the rewarding effects of this class of drugs. The major benefit of blocking central and peripheral NE release is primarily related to the reduction of the adverse cardiovascular effects of stimulants.

One interesting line of research has attempted to identify the utility of selective DAT inhibitors for the treatment of substance abuse disorders.[71] Recent studies assessed the therapeutic potential of the benzotropine class of selective DAT inhibitors for the treatment of cocaine abuse. To date, the literature supporting the utility of selective DAT inhibitors in substance abuse treatment based on animal models of cocaine abuse are inconclusive, since they do not appear to consistently reduce stimulant-induced locomotor and reinforcing behaviors.[71]

Although binding to the DAT is accepted as a primary mechanism of cocaine reward, the inability to eliminate cocaine-induced place preference in DAT knockouts suggests that additional mechanisms are involved. Moreover, since NET could also serve the same function as DAT, and blockade of NET can effectively elevate extracellular dopamine, the selective blockade of DAT may be insufficient to alleviate drug craving. Blockade of NET and serotonin transporter may be necessary to reduce or eliminate the adverse consequences such as cravings and depression associated with stimulant withdrawal. Rothman and colleagues[72] assessed the ability of indatraline, a compound with high affinity for DAT, serotonin transporter, and NET, and slow dissociation rates, to inhibit both methamphetamine and MDMA-induced dopamine release in vitro. Although indatraline alone had no effect on release, it did decrease methamphetamine and MDMA-induced dopamine, serotonin, and NE release. These data suggest that indatraline may prove useful for the treatment of stimulant abuse with regard to both the rewarding aspects and aversive consequences following withdrawal through its ability to prevent stimulant-induced NE, dopamine, and serotonin release.

In summary, little evidence exists at the present supporting an exclusive role of NE in treatment of psychostimulant abuse. However, more structure-activity-based research is needed to determine whether high-affinity DAT inhibitors could serve as a sole treatment for stimulant abuse disorders.[71] In fact, a more nonselective approach may prove more beneficial as evidenced by the studies of indatraline.[72]

Although insufficient evidence exists concerning the treatment of stimulant abuse with noradrenergic compounds, other research demonstrates potential for the treatment of adverse consequences of opiate abuse. A review by Maldonado[64] showed that inhibition of NE transmission effectively alleviates signs of opiate withdrawal. Administration of either α_2-agonists clonidine in mice[73] or guanfacine in rats[74] effectively decreased some signs of opiate withdrawal syndrome, such as jumping and wet dog shakes, while administration of the β_1-receptor antagonist atenolol in rats attenuated wet dog shakes, teeth chattering, and diarrhea.[75] More recent studies indicate that the noradrenergic and serotonergic uptake inhibitor, venlafaxine can effectively attenuate aspects of opiate withdrawal and relapse to opiate dependence in morphine-dependent rats.[76] Collectively, these data suggest that noradrenergic compounds are effective in treating the adverse consequences associated with opiate withdrawal.

Conclusions

There is a strong relationship between brain noradrenergic systems and the development of drug reward, particularly with regard to psychostimulant drugs of abuse. In addition, a substantial body of literature is supportive of the conclusion that NE in select brain regions such as the LC and BNST is a major precipitating factor for events associated with opiate withdrawal and stress-induced drug relapse, respectively. A positive association also exists between NE activity and both alcohol and nicotine abuse. Moreover, the finding that a personality trait associated with reward dependence is positively correlated with alcohol consumption and NE activity further supports the role of NE in substance abuse disorders. The elucidation of the noradrenergic mechanisms mediating the development of substance abuse, and the aversive consequences related to withdrawal symptoms and stress-induced relapse may highlight new alternative targets other than the traditional components involving the dopamine system for the treatment of substance abuse disorders.

REFERENCES

1. Substance Abuse and Mental Health Services Administration. *2001 National Household Survey on Drug Abuse (NHSDA)* (United States Department of Health and Human Services, 2001).
2. Nunes, E. V., McGrath, P. J., Quitkin, F. M. *et al.* Imipramine treatment of cocaine abuse: possible boundaries of efficacy. *Drug Alcohol Depend.*, **39** (1995), 185–195.
3. Prochazka, A. V., Weaver, M. J., Keller, R. T. *et al.* A randomized trial of nortriptyline for smoking cessation. *Arch. Int. Med.*, **158** (1998), 2035–2039.
4. Gold, M. S., Pottash, A. C., Sweeney, D. R. and Kleber, H. D. Opiate withdrawal using clonidine. A safe, effective, and rapid nonopiate treatment. *JAMA*, **243**: 4 (1980), 343–346.
5. Tomkins, D. M. and Sellers, E. M. Addiction and the brain: the role of neurotransmitters in the cause and treatment of drug dependence. *Can. Med. Assoc. J.*, **164** (2001), 817–821.
6. Leshner, A. I. and Koob, G. F. Drugs of abuse and the brain. *Proc. Assoc. Am. Physic.*, **111** (1999), 99–108.

7. Rothman, R. B., Baumann, M. H., Dersch, C. M. *et al.* Amphetamine-type central nervous system stimulants release norepinephrine more potently than they release dopamine and serotonin. *Synapse*, **39** (2001), 32–41.
8. Carboni, E., Spielewoy, C., Vacca, C. *et al.* Cocaine and amphetamine increase extracellular dopamine in the nucleus accumbens of mice lacking the dopamine transporter gene. *J. Neurosci.*, **21** (2001), RC141: 1–4.
9. Sora, I., Wichems, C., Takahashi, N. *et al.* Cocaine reward models: conditioned place preference can be established in dopamine- and in serotonin-transporter knock-out mice. *Proc. Nat. Acad. Sci. U. S. A.*, **95** (1998), 7699–7704.
10. Wise, R. A. and Bozarth, M. A. Brain mechanisms of drug reward and euphoria. *Psychiatr. Med.*, **8** (1985), 445–460.
11. Di Chiara, G. Nucleus accumbens shell and core dopamine: differential role in behavior and addiction. *Behav. Brain Res.*, **137** (2002), 75–114.
12. Delfs, J. M., Zhu, Y., Druhan, J. P. and Aston-Jones, G. S. Origin of noradrenergic afferents to the shell subregion of the nucleus accumbens: anterograde and retrograde tract-tracing studies in the rat. *Brain Res.*, **806** (1998), 127–140.
13. Blanc, G., Trovero, F., Vezina, P. *et al.* Blockade of prefronto-cortical alpha 1-adrenergic receptors prevents locomotor hyperactivity induced by subcortical D-AMPH injection. *Eur. J. Neurosci.*, **6** (1994), 293–298.
14. Paladini, C. A., Fiorillo, C. D., Morikawa, H. and Williams, J. T. Amphetamine selectively blocks inhibitory glutamate transmission in dopamine neurons. *Nat. Neurosci.*, **4** (2001), 275–281.
15. Goldstein, R. Z. and Volkow, N. D. Drug addiction and its underlying neurobiological basis: neuroimaging evidence for the involvement of the frontal cortex. *Am. J. Psychiatry*, **159** (2002), 1642–1652.
16. Tejani-Butt, S. M. [3H]Nisoxetine: a radioligand for quantitation of norepinephrine uptake sites by autoradiography or by homogenate binding. *J. Pharmacol Exp. Ther.*, **260** (1992), 427–436.
17. Macey, D. J., Smith, H. R., Nader, M. A. and Porrino, L. J. Chronic cocaine self-administration upregulates the norepinephrine transporter and alters functional activity in the bed nucleus of the stria terminalis of the rhesus monkey. *J. Neurosci.*, **23** (2003), 12–16.
18. Georges, F. and Aston-Jones, G. Potent regulation of midbrain dopamine neurons by the bed nucleus of the stria terminalis. *J. Neurosci.*, **21**: RC160 (2001), 1–6.
19. Wang, X., Cen, X. and Lu, L. Noradrenaline in the bed nucleus of the stria terminalis is critical for stress-induced reactivation of morphine-conditioned place preference in rats. *Eur. J. Pharmacol.*, **432** (2001), 153–161.
20. Unnerstall, J. R., Fernandez, I. and Orensanz, L. M. The alpha-adrenergic receptor: radiohistochemical analysis of functional characteristics and biochemical differences. *Pharmacol. Biochem. Behav.*, **22**: 5 (1985), 859–874.
21. Erb, S., Hitchcott, P. K., Rajabi, H. *et al.* Alpha-2 adrenergic receptor agonists block stress-induced reinstatement of cocaine seeking. *Neuropsychopharmacology*, **23** (2000), 138–150.
22. Drouin, C., Darracq, L., Trovero, F. *et al.* Alpha1b-adrenergic receptors control locomotor and rewarding effects of psychostimulants and opiates. *J. Neurosci.*, **22** (2002), 2873–2884.

23. Garvey, M. J., Noyes, R., Jr., Cook, B. and Blum, N. Preliminary confirmation of the proposed link between reward-dependence traits and norepinephrine. *Psychiatry Res.*, **65** (1996), 61–64.
24. Curtin, F., Walker, J. P., Peyrin, L. *et al.* Reward dependence is positively related to urinary monoamines in normal men. *Biol. Psychiatry*, **42** (1997), 275–281.
25. Valverius, P., Hogstrom-Brandt, A. M. and Borg, S. Norepinephrine metabolite in CSF correlates with ethanol consumption and heredity in humans. *Alcohol*, **10** (1993), 499–503.
26. Pan, W. H., Sung, J. C. and Fuh, S. M. Local application of amphetamine into the ventral tegmental area enhances dopamine release in the nucleus accumbens and the medial prefrontal cortex through noradrenergic neurotransmission. *J. Pharmacol Exp. Ther.*, **278**: 2 (1996), 725–731.
27. Horn, A. S. Structure-activity relations for the inhibition of catecholamine uptake into synaptosomes from noradrenergic and dopaminergic neurones in rat brain homogenates. *Br. J. Pharmacol.*, **47** (1973), 332–339.
28. Raiteri, M., del Carmine, R., Bertollini, A. and Levi, G. Effect of sympathomimetic amines on the synaptosomal transport of noradrenaline, dopamine, and 5-hydroxytryptamine. *Eur. J. Pharmacol.*, **41** (1977), 133–143.
29. Kelly, E., Jenneer, P. and Marsden, C. D. Evidence that [^{3}H]-dopamine is taken up and released from the nondopaminergic nerve terminals in the rat substantia nigra in vitro. *J. Neurochem.*, **45** (1985), 137–144.
30. Yamamoto, B. K. and Novotney, S. Regulation of extracellular dopamine by norepinephrine transporter. *J. Neurochem.*, **71** (1998), 274–280.
31. Chen, N. H. and Reith, M. E. Effects of locally applied cocaine, lidocaine, and various uptake blockers on monoamine transmission in the ventral tegmental area of freely moving rats: a microdialysis study on monoamine interrelationships. *J. Neurochem.*, **63**: 5 (1994), 1701–1713.
32. Uhl, G. R., Hall, F. S. and Sora, I. Cocaine, reward, movement and monoamine transporters. *Mol. Psychiatry*, **7** (2002), 21–26.
33. Fu, Y., Matta, S. G., Brower, V. G. and Sharp, B. M. NE secretion in the hypothalamic paraventricular nucleus of rats during unlimited access to self-administered nicotine: an in vitro microdialysis study. *J. Neurosci.*, **21** (2001), 8979–8989.
34. Tsai, S.-J., Wang, Y.-C. and Hong, C.-J. Allelic variants of the alpha$_{1a}$ adrenoreceptor and the promoter region of the α_{2a} adrenoreceptor and temperament factors. *Am. J. Med. Genet.* (Neuropsychiatr. Genet.), **105** (2001), 96–98.
35. Tsai, S.-J., Wang, Y.-C. and Hong, C.-J. Norepinephrine transporter and α_{2c} adrenoreceptor allelic variants and personality factors. *Am. J. Med. Genet.* (Neuropsychiatr. Genet.), **114** (2002), 649–651.
36. Samochowiec, J., Kucharska-Mazur, J., Rybakowski, F. *et al.* Norepinephrine transporter polymorphism and personality trait of reward dependence in male alcoholics. *Pharmacopsychiatry*, **35** (2002), 195–196.
37. McKinney, E. F., Walton, R. T., Yudkin, P. *et al.* Association between polymorphisms in dopamine enzymes and tobacco consumption in smokers. *Pharmacogenetics*, **10** (2000), 483–491.

38. Koob, G. F., Maldonado, R. and Stinus, L. Neural substrates of opiate withdrawal. *Trends Neurosci.*, **15** (1992), 186–191.
39. Frenois, F., Cador, M., Caillé, S., Stinus, L. and Le Moine, C. Neural correlates of the motivational and somatic components of naloxone-precipitated morphine withdrawal. *Eur. J. Neurosci.*, **16** (2002), 1377–1389.
40. Taylor, J. R., Punch, L. J. and Elsworth, J. D. A comparison of the effects of clonidine and CNQX infusion into the locus coeruleus and the amygdala on naloxone-precipitated opiate withdrawal in the rat. *Psychopharmacology*, **138** (1998), 133–142.
41. Taylor, J. R., Elsworth, J. D., Garcia, E. J. *et al.* Clonidine infusions into the locus coeruleus attenuate behavioral and neurochemical changes associated with naloxone-precipitated withdrawal. *Psychopharmacology*, **96** (1988), 121–134.
42. Britton, K. T., Svensson, T., Schwartz, J., Bloom, F. E. and Koob, G. F. Dorsal noradrenergic bundle lesions fail to alter opiate withdrawal or suppression of opiate withdrawal by clonidine. *Life Sci.*, **34** (1984), 133–139.
43. Caillé, S., Espejo, E. F., Reneric, J.-P. *et al.* Total neurochemical lesion of noradrenergic neurons of the locus coeruleus does not alter either naloxone-precipitated or spontaneous influence ability of clonidine to reverse opiate withdrawal. *J. Pharmacol. Exp. Ther.*, **290** (1999), 881–892.
44. Maldonado, R. and Koob, G. F. Destruction of the locus coeruleus decreases physical signs of opiate withdrawal. *Brain Res.*, **605** (1993), 128–138.
45. Rasmussen, K., Beitner-Johnson, D. B., Krystal, J. H., Aghajanian, G. K. and Nestler, E. J. Opiate withdrawal and the rat locus coeruleus: behavioral electrophysiological, and biochemical correlates. *J. Neurosci.*, **10**: 7 (1990), 2308–2317.
46. Ivanov, A. and Aston-Jones, G. Local opiate withdrawal in locus coeruleus neurons in vitro. *J. Neurophysiol.*, **85** (2001), 2388–2397.
47. Rasmussen, K. Afferent effects on locus coeruleus in opiate withdrawal. *Prog. Brain Res.*, **88** (1991), 207–216.
48. Rasmussen, K. and Aghajanian, G. K. Withdrawal-induced activation of locus coeruleus neurons in opiate-dependent rats: attenuation by lesions of the nucleus paragigantocellularis. *Brain Res.*, **505** (1989), 346–350.
49. Rockhold, R. W., Liu, N., Coleman, D. *et al.* The nucleus paragigantocellularis and opioid withdrawal-like behavior. *J. Biomed. Sci.*, **7**: 3 (2000), 270–276.
50. Van Bockstaele, E. J., Menko, A. S. and Drolet, G. Neuroadaptive responses in brainstem noradrenergic nuclei following chronic morphine exposure. *Mol. Neurobiol.*, **23** (2001), 155–171.
51. Delfs, J. M., Zhu, Y., Druhan, J. P. and Aston-Jones, G. Noradrenaline in the ventral forebrain is critical for opiate withdrawal-induced aversion. *Nature*, **403**: 6768 (2000), 430–434.
52. Aston-Jones, G., Delfs, J. M., Druhan, J. and Zhu, Y. The bed nucleus of the stria terminalis: a target site for noradrenergic actions in opiate withdrawal. *Ann. N. Y. Acad. Sci.*, **877** (1999), 486–498.
53. Zhu, M. Y., Shamburger, S., Li, J. and Ordway, G. A. Regulation of the human norepinephrine transporter by cocaine and amphetamine. *J. Pharmacol. Exp. Ther.*, **295** (2000), 951–959.

54. Stuerenburg, H. J., Petersen, K., Bäumer, T. *et al.* Plasma concentrations of 5-HT, 5HIAA, norepinephrine, epinephrine and dopamine in ecstasy users. *Neuroendocrinol. Lett.*, **23** (2002), 259–261.
55. Mayerhofer, A., Kovar, K. A. and Schmidt, W. J. Changes in serotonin, dopamine and noradrenaline levels in striatum and nucleus accumbens after repeated administration of the abused drug MDMA in rats. *Neurosci. Lett.*, **308** (2001), 99–102.
56. Wigal, S. B., Amsel, A. and Wilcox, R. E. Fetal ethanol exposure diminishes hippocampal beta-adrenergic density while sparing muscarinic receptors during development. *Brain Res. Dev. Brain Res.*, **55** (1990), 161–169.
57. Hadjiivanova, K., Petkov, V. D., Alova, L., Petkov, V. V. and Vuglenova, Y. Changes in brain muscarinic and beta-adrenoreceptors of the offspring of ethanol-consuming mothers (experiments on rats). *Acta Physiol. Pharmacol. Bulg.*, **17** (1991), 91–97.
58. Kostowski, W. and Trzaskowska, E. Effect of lesion of the locus coeruleus and clonidine treatment on ethanol withdrawal syndrome in rats. *Pol. J. Pharmacol. Pharm.*, **32** (1980), 617–623.
59. Arango, V., Underwood, M. D. and Mann, J. J. Fewer pigmented neurons in the locus coeruleus of uncomplicated alcoholics. *Brain Res.*, **650** (1994), 1–8.
60. Stewart, J. Pathways to relapse: the neurobiology of drug- and stress-induced relapse to drug-taking. *J. Psychiatry Neurosci.*, **25** (2000), 125–136.
61. Leri, F., Flores, J., Rodaros, D. and Stewart, J. Blockade of stress-induced but not cocaine-induced reinstatement by infusion of noradrenergic antagonists into the bed nucleus of the stria terminalis or the central nucleus of the amygdala. *J. Neurosci.*, **22** (2002), 5713–5718.
62. Erb, S. and Stewart, J. A role for the bed nucleus of the stria terminalis, but not the amygdala, in the effects of corticotropin-releasing factor on stress-induced reinstatement of cocaine seeking. *J. Neurosci.*, **19** (1999), RC35: 1–6.
63. Shaham, Y., Erb, S. and Stewart, J. Stress-induced relapse to heroin and cocaine seeking in rats: a review. *Brain Res. Brain Res. Rev.*, **33** (2000), 13–33.
64. Maldonado, R. Participation of noradrenergic pathways in the expression of opiate withdrawal: biochemical and pharmacological evidence. *Neurosci. Biobehav. Rev.*, **21** (1997), 91–104.
65. Sullivan, M. A. and Covey, L. S. Nicotine dependence: the role for antidepressants and anxiolytics. *Curr. Opin. Invest. Drugs*, **3** (2002), 262–271.
66. Glassman, A. H. Cigarette smoking: implications for psychiatric illness. *Am. J. Psychiatry*, **150** (1993), 546–553.
67. Jorenby, D. E., Leischow, S. J., Nides, M. A. *et al.* A controlled trial of sustained-release bupropion, a nicotine patch, or both for smoking cessation. *New Eng. J. Med.*, **340** (1999), 685–691.
68. Tashkin, D., Kanner, R., Bailey, W. *et al.* Smoking cessation in patients with chronic obstructive pulmonary disease: a double-blind, placebo-controlled, randomised trial. *Lancet*, **57** (2001), 1571–1575.
69. Carroll, K. M., Rounsaville, B. J., Gordon, L. T. *et al.* Psychotherapy and pharmacotherapy for ambulatory cocaine abusers. *Arch. Gen. Psychiatry*, **51** (1994), 177–187.
70. Nunes, E. V., McGrath, P. J., Quitkin, F. M. *et al.* Imipramine treatment of cocaine abuse: possible boundaries of efficacy. *Drug Alcohol Depend.*, **39** (1995), 185–195.

71. Newman, A. H. and Kulkarni, S. Probes for the dopamine transporter: new leads toward a cocaine abuse therapeutic - A focus on analogues of benztropine and rimcazole. *Med. Res. Rev.*, **22** (2002), 429–464.
72. Rothman, R. B., Partilla, J. S., Baumann, M. H. *et al.* Neurochemical neutralization of METH with high-affinity nonselective inhibitors of biogenic amine transporters: a pharmacological strategy for treating stimulant abuse. *Synapse*, **35** (2000), 222–227.
73. Valeri, P., Martinelli, B., Pimpinella, G. and Severini, C. Effects of dapiprazole, clonidine and yohimbine on the development of dependence and withdrawal behaviour in mice. *Drug Alcohol Depend.*, **23** (1989), 73–77.
74. Coupar, I. M. Effect of alpha 2-adrenoceptor agonists on the expression of morphine-withdrawal in rats. *Naunyn-Schmiedeberg's Arch. Pharmacol.*, **345** (1992), 553–557.
75. Harris, G. and Aston-Jones, G. Beta-adrenergic antagonists attenuate somatic and aversive signs of opiate withdrawal. *Neuropsychopharmacology*, **9** (1993), 303–311.
76. Lu, L., Su, W. J., Yue, W. *et al.* Attenuation of morphine dependence and withdrawal in rats by venlafaxine, a serotonin and noradrenaline reuptake inhibitor. *Life Sci.*, **69** (2001), 37–46.

Index

LC = locus coeruleus; NE = norepinephrine